IMPACT ASSESSMENT

IMPACT ASSESSMENT
Practical Solutions to Recurrent Problems and Contemporary Challenges

SECOND EDITION

David P. Lawrence
Lawrence Environmental

WILEY

Published by John Wiley & Sons, Inc., Hoboken, New Jersey

Published simultaneously in Canada

Library of Congress Cataloging-in-Publication Data:

Lawrence, David P. (David Phillip), 1947–
[Environmental impact assessment]
Impact assessment / David P. Lawrence. – Second edition.
pages cm
Previously published: Hoboken, New Jersey : Wiley-Interscience, 2003, under title Environmental impact assessment.
"Published simultaneously in Canada"–Title page verso.
Includes bibliographical references and index.
ISBN 978-1-118-09737-3 (cloth)
1. Environmental impact analysis. I. Title.
TD194.6.L39 2013
333.7′14–dc23

2013004575

ISBN: 9781118097373

10 9 8 7 6 5 4 3 2 1

To Barbara

Contents

Preface

Impact assessment (IA), as a field of theory and practice, has changed substantially since 2003 when the first edition of this book was published. Numerous changes have been made to regulatory requirements and guidelines, both within and across jurisdictions. The conceptual foundation of IA, as represented by IA literature and by conference and symposium proceedings, has greatly evolved, especially in terms of good practice guidance and insights. IA also has undergone something of a reorientation during this period, with a greater emphasis, for example, on strategic environmental assessment, sustainability assessment, human health impact assessment, and impact assessment from an inter-jurisdictional, international and global perspective. This reorientation has resulted in a title change (impact assessment rather than environmental impact assessment) for this book and the integration of the perspectives and practices of various IA types (sustainability assessment, strategic environmental assessment, project-level environmental impact assessment, ecological impact assessment, social impact assessment, health impact assessment) into each chapter.

Many issues and challenges have arisen during this period—issues and challenges that have been addressed, to varying degrees and often from conflicting perspectives. Notwithstanding these advances, IA remains a field whose relevance and legitimacy continues to be challenged from a variety of quarters. Within this context, the recurrent problems identified in the first edition of the book are still present and the solutions offered still remain largely valid. The challenge for this new edition then has been to retain and update those still relevant elements of the original book while, at the same time, reflecting the new reality of IA as it has become and is becoming. Changes to pertinent IA requirements and guidelines have been incorporated into the regulatory level analysis. Insights from relevant IA literature have been integrated into the applied level analysis. A new set of practice-based stories, with a greater orientation toward strategic level decision making (i.e., policies, plans, and programs), has been included. A series of contemporary challenges are addressed. They include good practice SEA (Chapter 2), good practices for influencing decision making (Chapter 3), follow-up good practices (Chapter 4), siting "locally unwanted land uses" (Chapter 5), horizontal integration (Chapter 6), cumulative effects assessment good practices (Chapter 7), multijurisdictional IA (Chapter 8), capacity building good practices (Chapter 9), significance determination good practices (Chapter 10), climate change IA good practices (Chapter 11), and IA context adjustments good practices (Chapter 12).

The continuing challenge of the influence of IA on decision making has been addressed in a separate chapter. The remaining part of the book is divided into the same chapters as the first edition. However, significant changes have been made to each chapter. Substantially more good practice guidance has been added. A range of major, often problematic, issues and contemporary challenges are explored. A host of additional concepts and frameworks are presented. Collectively, it is hoped that these refinements, additions, and modifications, together with the retained portions of the book, will represent a worthwhile contribution to the resources available to IA practitioners and to other IA process stakeholders.

The analyses and solutions offered in this book are far from definitive. Hopefully, they are practical. I believe that sufficient knowledge and experience now exist regarding the recurrent problems and contemporary challenges that major pitfalls can be identified and possible improvements suggested. I am not sufficiently naive to suggest that we are on the brink of delineating that elusive core body of common knowledge that is supposed to characterize "mature" fields. I have serious doubts as to whether such a quest is even desirable. I also appreciate that there are immense impediments to significant improvements to IA practice, many of which lie beyond the control or influence of IA practitioners. But I still believe that sufficient operating room remains within which IA practice enhancements are possible. I also maintain that the IA process is at the core of many such improvements. Hopefully, this book will contribute to such efforts.

I wish to thank the following people for contributing their thoughtful and insightful stories for inclusion in this book: Charles Alton, Jos Arts, Alan Bond, Larry Canter, Peter Croal, Michael Doelle, Peter Duck, Richard Fuggle, Robert Gibson, Davide Geneletti, Thomas Fischer, John Glasson, Jill Gunn, Peter Homenuck, Tony Jackson, Arend Kolhoff, Angus Morrison-Saunders, Bruce Mitchell, Bruce Muir, Peter Mulvihill, Hens Runhaar, Luis Sánchez, Luke Sandham, Laura Sims, Marielle Rowan, Jaap Rozema, Gary Schneider, John Sinclair, Harry Spaling, Gernot Stoeglehner, Tom Swor, and Zhenghong Tang. I especially wish to thank Richard Fuggle, Davide Geneletti, Bob Gibson, and John Sinclair for their additional thoughtful comments, inputs, and advice.

I also wish to thank the anonymous reviewers and the staff of John Wiley & Sons (most notably Michael Leventhal and Bob Esposito) for their constructive suggestions and guidance.

A great many colleagues have provided encouraging comments and/or have influenced my thinking and writing regarding IA process management over the past several years. A very partial list includes Rabel Burdge, Dave Cressman, Peter Croal, Bob Dorney, George Francis, Bob Gibson, Eric Hunter, Peter Homenuck, Larry Martin, Jim Micak, Greg Michelenco, John Page, Donna Pawlowski, Barry Sadler, Paul Scott, John Sinclair, Graham Smith, Margaret Smith, Roger Suffling, Richard Szudy, and Tom Wlodarczyk.

My thanks to the following for permission to reproduce, without charge, the following copyright material in this edition of the book (full details are provided in the reference list at the back of the book):

- Elsevier Science, Inc. (Figures 4.1, 10.3, and 10.4),
- Imperial College Press (Figures 2.3–2.6 and Table 2.4),
- International Association for Impact Assessment (IAIA) (Table 10.3),
- Springer-Verlag (Figures 6.2 and 6.3),
- Taylor & Francis (Figures 5.5 and 7.5 and Table 5.5), and
- Wiley (Figures 4.2, 5.4, 6.6, 7.6, 8.5, 9.6, 10.2, 11.6, and 12.2).

Finally and most importantly, I am especially grateful to my wonderful wife, Barbara, for her enduring and endearing patience, encouragement, and support throughout this lengthy process.

David P. Lawrence
Lawrence Environmental
(lawenv@telus.net)

Chapter 1

Introduction

1.1 HIGHLIGHTS

This book is intended to enhance impact assessment (IA) practice. It utilizes IA process design and management choices to provide practical solutions to IA practitioners for major, recurrent problems and contemporary challenges encountered in daily IA practice.

- In Section 1.2 we present a scenario that highlights the problems and challenges. The scenario illustrates how a failure to adequately anticipate and respond to varying perspectives and challenges can contribute to the collapse of a seemingly well-designed and managed IA process.
- In Section 1.3 we "go back to the fundamentals." We use an IA definition and definitions of various IA types—environmental impact assessment (EIA), ecological impact assessment (EcIA), social impact assessment (SIA), health impact assessment (HIA), strategic environmental assessment (SEA), and sustainability assessment (SA)—to identify implications for overall IA process management. We also describe the current "state-of-the-art" of IA process management themes and issues. These analyses illustrate the widespread nature of the recurrent problems and contemporary challenges in IA theory and practice.
- In Section 1.4 we frame the IA process and identify IA regulatory and applied design choices as the bases for building a strategy. We explain why the IA process in general, and alternative IA processes and process types and variations, in particular, are essential to the effort.
- In Section 1.5 we present a strategy for facilitating more effective IA process management.
- In Section 1.6 we suggest how IA stakeholders could use this book.
- In Section 1.7 we highlight the major themes and conclusions.

1.2 A "NOT SO HYPOTHETICAL" SCENARIO

1.2.1 Brave Beginnings

A private proponent decides to establish a new hazardous waste treatment facility. It realizes that there will be numerous licensing requirements including the preparation and approval of an EIA (a type of IA applied to projects). Accordingly, a consulting team is hired to prepare the EIA documentation and to ensure that all approval requirements are satisfied. A preliminary design is prepared for a "state-of-the-art" facility. An overview of available properties is conducted. A site is selected in a general industrial park a couple of miles outside a medium-sized community. An option is taken out on the property. Local community officials express a willingness to accept the facility because of the tax revenue to be generated and a promise to share a portion of the facility revenues with the local community. Two municipal councilors express reservations because of a fear that the facility might stigmatize the community. They also question whether the proposal might be premature on the grounds that need and alternatives have not been addressed at higher decision-making levels.

The EIA process has a promising beginning. A core study team is assembled with ample EIA and regulatory approval experience. The project manager, a civil engineer, is experienced in the design, approval, and construction of similar projects. The EIA team's experience and expertise derive from a working familiarity with pertinent EIA requirements and guidelines, and the experience acquired from several similar projects.

A variety of engineering and environmental specialists, together with an expert in public participation, are added to the team. A preliminary study design is prepared. Initial scoping sessions are conducted with government officials to identify regulatory requirements, concerns, and priorities. An initial set of public meetings and open houses is convened to identify public concerns and preferences. The study program is modified to accommodate public and agency concerns. The EIA is divided into a clearly defined sequence of steps. Provision is made for public and agency input into each step.

The focus, in the early months of the process, is on establishing a sound environmental baseline and on refining facility characteristics. Several mitigation options are screened and compared in the ongoing effort to prevent and ameliorate adverse impacts. Initial background papers are prepared documenting baseline conditions, study

Impact Assessment: Practical Solutions to Recurrent Problems and Contemporary Challenges, Second Edition. David P. Lawrence.

methodology, the analysis of alternatives, and preliminary impact predictions. Impact predictions are then refined, and impact significance ratings are determined, for both individual and cumulative impacts. A concerted effort is made to mitigate potentially significant, adverse impacts. In a few cases, this necessitates comparing mitigation options. These various analyses are consolidated first in working and background papers and then in a draft impact statement. Summary reports are prepared for each document. Documents are circulated for initial agency comment and are used as the basis for discussions and presentations at public meetings and open houses. All comments and suggestions are recorded. Responses are provided to each comment received including a detailing of how and where the comments are addressed in the EIA documentation.

1.2.2 Cracks in the Foundation

Public opposition begins to mount during this period. Initially, this opposition comes from individuals. It is not long before a local opposition group is formed. Local and then regional environmental organizations quickly join the fray. The local community groups are concerned about accidents and long-term potential human health effects, possible declining property values, and community stigma. They strongly criticize the limited, closed, and informal procedure adopted for selecting the preferred site. The environmental groups question the need for the facility, arguing that it is "old technology" that should be superseded by waste reduction, reuse, and recycling initiatives. They challenge the "growth ethic" inherent in the predicted usage of the facility, express concerns about possible climate change and cumulative effects from the facility and other industrial activities in the area, and argue that the proposed facility undermines the cause of environmental sustainability.

Several faculty members from the local university also voice their opposition. They focus their comments on the scientific validity of the impact predictions. They especially point to the failure to use control communities, the lack of peer review, the excessively descriptive analysis, the questionable statistical analyses, the crude models employed, and the short duration of the baseline studies. They stress that the studies fail to adequately address uncertainties, low probability–high consequence risks, and perceived risks. They argue that effects are defined too narrowly, noting that the socioeconomic effects analysis focuses on and overestimates short-term benefits, while addressing only superficially adverse direct and indirect ecological and socioeconomic impacts. They question the absence of a policy framework, the lack of waste management strategy for the region, the absence of a coherent approach to climate change impacts, and the failure to evaluate need and alternatives to the project. They are especially critical because of the lack of a comprehensive impact management strategy. They also wonder whether the net contribution of the facility to environmental sustainability will be positive or negative. The opposition to the facility culminates in a raucous public meeting.

Many members of the public attending the meeting stress that public involvement in the process has been at best "tokenism" and at worst "manipulation." Considerable frustration is expressed about what is seen as a loss of community control. Many participants argue that the process is neither open nor fair. They complain that the major decisions (i.e., need, alternatives to the project, alternative locations) were already determined before the EIA process was initiated. They suggest that it is unfair to locate such a facility in an area, which generates such a small proportion of the waste, has several similar facilities, and which is social and economically disadvantaged. Frequent reference is made to the mixed, "track record" of the proponent in other communities. The EIA process is described as unfair, especially giving the overwhelming resource advantage of the proponent and the lack of capacity of local groups and communities to effectively participate in the IA process. Both the municipality and the state are criticized for failing to have in place waste management and sustainability strategies. Such strategies, it is argued, would have provided a policy and program context for evaluating the proposed project. The EIA documents are criticized for lacking a broader planning perspective. The process is criticized for failing to adequately engage senior governments. Several municipal councilors soon reconsider their initial support for the facility.

Several potential participants decide that the EIA process is pointless. Some environmental groups argue that the EIA procedures are a wasteful distraction, with little, if any, environmental benefit. They decide either to focus on direct action against the project or to focus their limited resources on other environmental causes, where they believe they can make a difference. Many local individuals fail to become involved because of the resource disparity favoring the proponent and the perception that the major decisions have already been made. The local community groups are of mixed minds concerning whether it is better to participate in the process and risk being "co-opted" or oppose the project outside the process and risk being even further "marginalized" from decision making. Several academics, with potentially valuable knowledge, decide to bypass the process on the grounds that the data generated by the process is of no scientific value and of negligible utility in predicting or managing environmental changes. They also assert that the process is biased, subjective, and a waste of resources.

It becomes increasingly evident that there is a major disconnect between the proponent/EIA team and the public. The study team emphasizes the limited likelihood of an accident or spill. The public focuses on the severity of consequences if an accident or spill occurs. The team asserts that the scale and duration of impacts will be very limited. The public argues that the significance of ecological and community-related impacts will be high. The team talks in terms of project-related impacts. The public concentrates on the cumulative ecosystem and community impacts from the project and from other projects and activities. The team points to significant waste management benefits and limited project-related impacts. The public questions whether the net environmental contribution will be positive or negative. The team stresses the many opportunities for public involvement. The public maintains that its decision-making role is negligible. The team demonstrates its compliance with environmental requirements. The public asserts that the project is unacceptable to the community, regardless of regulatory compliance. The team seeks public input into facility design and management options. The public focuses on whether the facility is necessary, whether there are better ways to reduce and manage the wastes, and whether other facility locations would be more suitable.

This pattern of "talking past each other" leads to considerable frustration on both sides. Both sides redouble their efforts to get their points across while expressing consternation that they are not being heard. An attitude of mutual contempt becomes increasingly evident. The proponent and the study team conclude that they have "done their bit" by informing and involving the public. They decide that offering up further opportunities will only "ramp up" the intensity of opposition. Many members of the public decide that continuing to attend "token" public participation events only legitimizes a "tone deaf process" impervious to genuine public involvement. Initial agency reactions to the EIA documents are mixed at best.

Some reviewers have difficulty in determining whether specific regulatory requirements and policies have been explicitly addressed, as they work their way through the lengthy documents. Other reviewers question the clarity of the methodology, challenge the methods or data sources used, and assert that the criteria and indicators selected for predicting impacts are inappropriate for the setting. Some also argue that the methods have been misapplied or suggest that conclusions are insufficiently substantiated. Several reviewers are troubled by unexplained format and methodological inconsistencies among disciplinary analyses, and by the failure to systematically address interconnections among disciplines, climate change, and cumulative effects. The EIA, they point out, is essentially a compilation of independently prepared analyses.

The alternatives analysis becomes a focal point of criticism. Several reviewers argue that a wider range of alternatives should have been considered, criteria are not explicitly defined or consistently applied, criteria are not ranked, and sensitivity analyses have not been undertaken to explore the implications of alternative criteria rankings and varying interpretations of mitigation potential and the implications of uncertainty. The superficial and arbitrary approach to interpreting impact significance also is widely condemned. Some reviewers, who could make a worthwhile contribution to the process, have little to offer. EIA review, they note, is a largely secondary function of their agency. And yet, they assert, it absorbs far too much valuable time and resources. They too see EIA requirements and procedures as of little substantive value to either agency decision making or to environmental sustainability.

Substantial document modifications are made to address public and agency concerns and preferences. However, it is apparent that document modifications alone will not be sufficient to quell the tide of opposition that is building against the facility.

1.2.3 Hasty Repairs

The proponent decides, in the face of this mounting opposition, to retrench and reconsider how best to proceed. A community advisory committee is established to ensure the ongoing involvement of all affected interests. A community conciliator, acceptable to all parties, is hired to chair the committee. Funding is provided to the committee to hire specialists to peer review all the major technical analyses. A separate subcommittee is established to formulate an impact management and local benefits strategy. The strategy is to ensure a greater level of local participation and control in facility operations, management, monitoring, and contingency planning. It also is to formulate local benefits and compensation policies and procedures for both local residents and for the overall community. A parallel government advisory committee is established to better coordinate regulatory interactions.

1.2.4 Too Little Too Late

The costs and the duration of the process have greatly increased—to the considerable exasperation of the proponent. The reformulated approach has some success in addressing many of the technical, scientific, and community control concerns. Broader environmental sustainability and social equity concerns are largely beyond the committee's mandate. Several options advanced by facility opponents are not addressed on the grounds that they are impractical or beyond the control of the proponent. The negative perceptions of the proponent, the facility, and the EIA process are only slightly ameliorated by these efforts. Some environmental and community groups either refuse to participate in the modified process or opt out when it becomes evident that the committee agenda will be largely confined to refinements to technical analyses and to impact management. Several municipal councilors come to the conclusion that the likelihood of a satisfactory middle ground is remote and decide to add their voices to those of the facility opponents. More parties withdraw from the community advisory committee under a barrage of criticism from the groups they ostensibly represent.

Demonstrations, media interviews, and e-mail and letter campaigns by community and environmental groups further contribute to the momentum shift in political and public opinion against the project. The proponent attempts, behind the scene, to exert direct political pressure on local and senior government officials and on elected representatives.

These contacts are made public, further undermining the credibility of both the proponent and the EIA process.

It is increasingly evident that it is virtually impossible to reverse the momentum that has built up against the facility. Faced with the prospect of continued intense local opposition and protracted legal battles, the proponent decides that the costs of proceeding are simply too great and the likelihood of project approval is too low. The application is withdrawn and the proponent decides that it will concentrate instead on upgrading and expanding existing facilities in other communities.

1.2.5 Recurrent Problems and Contemporary Challenges

The scenario illustrates a range of recurrent problems and contemporary challenges that are broadly evident in IA practice.

With regard to the recurrent problem of decision-making influence:

- The proponent and the government (at all levels) failed to see the relevance of strategic level decision making.
- Some environmental groups chose not to participate because they saw the IA process as lacking in environmental substance and, therefore, irrelevant to their objectives.
- Some environmental and community groups saw the IA process as so peripheral to decision making that direct action was considered a more effective strategy.
- The proponent attempted to influence decision making outside the IA process by exerting political influence.
- Some community groups and individuals decided that the process was so biased and inequitable that any involvement would be ineffective.
- Some government officials stayed out of the IA process on the grounds that both the process and its outcomes were of negligible relevance to agency objectives.
- Some community groups and individuals withdrew from the process because their views and proposals were not being considered.
- Some community groups and individuals failed to participate in or withdrew from the process because the major strategic decisions had already been made.
- Some academics chose not to be involved because they viewed IA as unconnected to theoretical or applied science.
- Some academics chose not to be involved because they challenged whether IA methods could predict or manage, with any degree of certainty, environmental changes, with or without a proposed action.

With regard to other recurrent problems encountered in the IA process:

- The local university faculty members and some peer reviewers argued that the IA process, documents, and methods should have been more scientifically rigorous.
- The environmental groups and some government reviewers made the case that alternatives were too narrowly defined and were not systematically and consistently evaluated.
- The environmental groups concluded that the IA process and documents failed to adequately advance long-term environmental quality and sustainability principles and goals.
- The proponent and some reviewers felt that the process and documents were too lengthy and costly. They also were concerned that the IA documents were insufficiently linked to specific regulatory approval requirements, policies, and guidelines.
- Local community groups and some politicians expressed the view that they were losing control over their lives and their community. They did not trust the proponent and had little faith in the government.
- Local community groups and individuals took the position that the IA process was largely closed and that their views and positions were not seriously considered.
- Local community groups and politicians argued that the IA process was unfair and that the benefits and costs from the proposed facility were unfairly distributed.
- The environmental groups, some local residents, and some government reviewers felt that the risks and uncertainties associated with the proposed facility were not adequately anticipated or managed.

The IA process also failed to effectively grapple with a range of contemporary challenges:

- The study team failed to learn from other experiences in siting "locally unwanted land uses."
- No effort was made, by drawing upon other experiences, to enhance the decision-making influence of the IA process and documents.
- The IA process neither drew upon nor, by using effective impact management, contributed to other IA theory-building efforts.
- The choices made regarding the scope of environment and effects were not effectively derived or substantiated.
- The IA process was undermined by limited, flawed, and overly simplistic methods for determining significance, for addressing cumulative effects, and for addressing climate change impacts.
- No effort was made to enhance the capacity of local communities, area residents, and other key stakeholders to effectively participate in the IA process.
- The IA process was not effectively embedded within broader level, strategic planning, and decision making.

This limitation greatly inhibited the ability of the process to effectively address alternatives, cumulative effects, and transboundary effects.
- The IA process and related methods were poorly designed for and adapted to the local context.

These flaws are not inherent to IA practice. Better results can be achieved if appropriate changes are made to IA process design and execution at the regulatory and applied levels.

1.3 THE BASICS

This section uses IA definitions and an overview of historical and emerging IA themes and issues to identify implications for overall IA process management and to provide a foundation for addressing the recurrent problems and contemporary challenges described in the scenario.

1.3.1 IA Definitions

IA, in its most basic form, is the process of identifying the future consequences of a current or future action. The "impact" is the difference between what would happen with the action and what would happen without it (IAIA, undated b). This definition is general enough to encompass (or not) a range of current or proposed, planned or unplanned initiating actions (e.g., projects, plans, programs, policies, legislation, activities, operational procedures). It also may or may not include various "reasonable" alternatives to the action and/or alternative means of carrying out the action.

The "impact" part of the definition, with the impacts taking place in the future, implies a cause–effect relationship, and an ability to predict future condition changes, with and without the proposed action. Future conditions, with and without an action, cannot be predicted with certainty. Therefore, IA involves identifying and managing risks and uncertainties. It must necessarily adapt to unanticipated changes, which, in turn, entails a postapproval component. The definition of "impact" is broad enough to encompass (or not) a diversity of potential future, current or actual, direct and indirect, individual and cumulative, positive and negative, and likelihood of occurrence impacts upon a broadly or narrowly defined range of environmental conditions (e.g., physical, chemical, biological, ecological, human health, cultural, heritage, social, economic, financial, built, interrelationships, sustainability). The "assessment" part of the definition can be very general (e.g., limited to identifying future consequences) or more specific activities can be identified (e.g., identifying, describing, measuring, predicting, interpreting, mitigating, enhancing, integrating, and controlling impacts).

IA is often, but not always, explicitly guided and bounded by values, goals, and objectives, and ethical imperatives and standards. There can be both direct and indirect, and both procedural (e.g., decision making informed, decision making influenced, decision making made more open, systematic, transparent, inclusive, substantiated, coordinated and unbiased, power shared or redistributed) and substantive (e.g., adverse impact reduced, positive impacts enhanced, environmental or social justice served or undermined, sustainability facilitated) outcomes from the IA.

IA precedes decision making, prior to irrevocable commitments. It can be directed and shaped by action-forcing and strictly defined institutional arrangements and requirements. Or the connections between IA and its institutional setting can be more passive, informal, general, voluntary, and discretionary. IA institutional arrangements can entail interconnections across decision-making authorities (e.g., among departments, among governments for addressing transboundary impacts), and among decision-making levels (e.g., tiering).

The assessment process involves interested and affected parties who participate in and influence, to varying degrees, the conduct of the process and its outcomes. The process can be designed and managed to facilitate or inhibit stakeholder participation, influence, and control. The links between IA and decision making mean that the exercise and allocation of political power is inherent to IA practice. IA can adopt a largely passive political role (e.g., documents and process inform decision making). Or it can proactively seek to influence, shape, bound, or direct decision making by either reinforcing the existing distribution of power or by facilitating the redistribution of power among parties. Impact assessment is adapted to (sometimes well and sometimes poorly) context. Inasmuch as IA is intended to positively influence decision making and environmental outcomes, IA practice affects context, either negatively (e.g., a wasteful diversion of resources and attention, sustainability undermined) or positively (e.g., reforms decision making, influences environmental perceptions, values, attitudes and behaviors, net environmental gains). The boundaries between IA theory and practice, among IA types, and between IA and related fields of theory and practice are fluid and fuzzy. Interconnections flow in multiple directions, not always explicitly. Figure 1.1 highlights the major elements encompassed, to varying degrees, within or implied by IA definitions. The treatment of these elements varies among IA types.

EIA, the longest standing and most institutionalized form of IA, tends to be largely limited to physical projects and activities. Action-forcing institutional requirements tend to dictate the scope of EIA requirements, although these requirements vary by jurisdiction, and have evolved over time. The definition of environment generally includes a partial (as when limited to areas of jurisdiction) or more comprehensive list of physical and ecological impacts. The extent to which other direct and indirect social, health, and economic aspects of the environment are included varies by jurisdiction. Direct, indirect, negative, and individual impacts are generally addressed. Whether positive, cumulative, and sustainability impacts are addressed again varies by

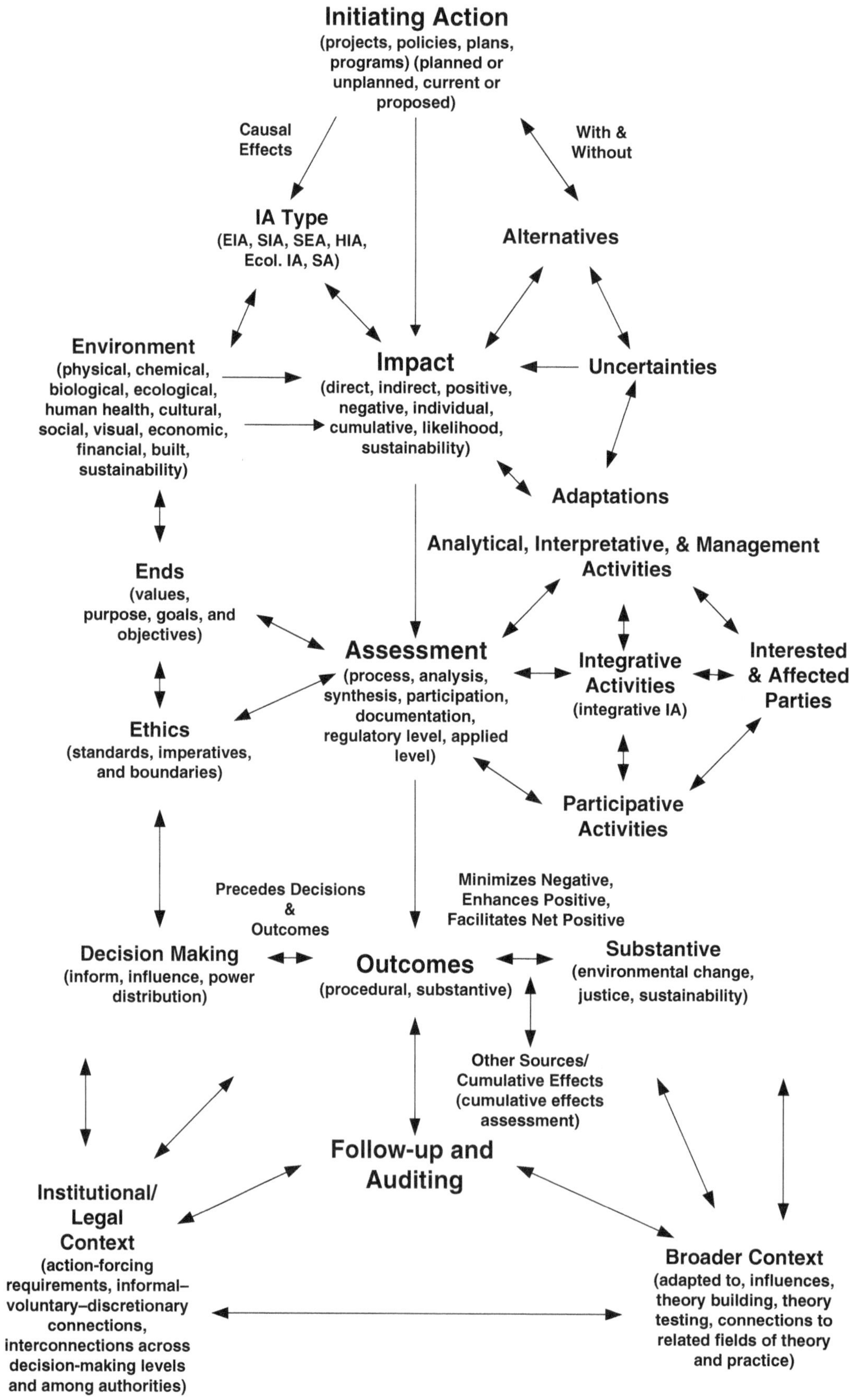

Figure 1.1 IA definition elements.

setting. There is considerable variation in if and in what manner alternatives are addressed, in the treatment of uncertainties, and in the extent to which the process is guided and directed by explicit ends and ethical boundaries.

EIA definitions and characterizations tend to include a few basic activities or steps (e.g., screening, scoping, baseline analysis, impact identification, impact prediction, impact interpretation, alternatives analysis, mitigation, public involvement, decision making, follow-up), although, as detailed in Chapter 2, there are an immense array of choices available in process design and management. EIA operates at a regulatory (e.g., requirements and guidelines, document review, review meetings) and at an applied (e.g., process design and management, interim and final document preparation, and review) level. It tends to center around the preparation and review of assessment documents. It informs and involves, to varying degrees, interested and affected parties at key decision points in the assessment process. Inasmuch as the assessment process is generally "triggered" by a proposed action, the involvement of parties, other than the proponent, in defining the essential elements of the proposed action is often limited.

The EIA process occurs, or should occur, prior to decision making. It, however, tends to assume a fairly passive approach to decision making (e.g., documents provided to decision makers as information) and an indirect route to environmental change. Follow-up to effectively manage anticipated and unanticipated impacts, to ensure the effective implementation of commitments, to involve interested and affected parties in postapproval activities, and to ascertain the accuracy of predictions is increasingly becoming a vital EIA activity. EIA generally occurs under the auspices of legal, action-forcing, institutional arrangements. Increasing attention is being devoted to the interconnections among decision-making levels and among authorities. There is now a considerable body of EIA regulatory and applied "good practices." There is substantial variation in the extent to which these "good practices" are being applied. There also are misgivings regarding if and the extent to which these "good practices" are always appropriate in different contexts. Although there is an active interchange between EIA theory and practice, and between EIA and related fields of theory and practice (including other IA types), there remains a substantial gulf between EIA theory and practice, and a considerable variation in the quality and effectiveness of EIA institutional arrangements, and of EIA documents and processes. Although the underlying purpose of EIA is to facilitate the integration of environmental concerns and values into decision making, a recurrent criticism of EIA institutional arrangements and practices has been the predominance of procedural concerns at the expense of and sometimes exclusion of substantive concerns. This explains, in part, the emergence of other IA types with a more substantive focus. Notwithstanding differences in emphasis, definitions of other IA types tend to largely follow the lead of EIA, with a few notable exceptions and variations.

EcIA is "concerned with the process of identifying, quantifying, and evaluating the potential impacts of defined actions on ecosystems and their components" (IEEM, 2006). If properly implemented, it provides a scientifically defensible approach to ecosystem management (Treweek, 1999). Increasingly, EcIA focuses on the maintenance of biodiversity. EcIA can take the form of a standalone IA (if, for example, the "trigger" is an environmentally sensitive area), a component of EIA, or a support document for another IA type. EcIA emphasizes the effective application of scientific knowledge, the importance of interactions and cumulative impacts, the desirability of net ecological gains and ecological sustainability, the need to operate within ecological limits, the necessity of systematically identifying and adaptively managing risks and uncertainties, and the critical role of follow-up. It broadly defines ecological environment and impacts, encompasses a body of research, knowledge, and techniques practiced by ecological professionals, entails close natural science and community knowledge links, adheres to ecological ethical imperatives, and explicitly recognizes links to the human environment and sustainability.

Although more commonly assuming a supportive role to EIA, EcIA can assume an especially valuable role in support of SEA (e.g., in addressing regional cumulative effects), HIA (e.g., in addressing health implications of ecological change), and SIA (e.g., in addressing social implications of ecological change). EcIA definitions and characterizations appear to devote less attention to the systematic generation and evaluation of alternatives, to community knowledge and participation, to decision-making influence, and to process adaptations to contextual variations. More attention could be devoted to interconnections with other components of the environment and other types of impact.

SIA includes "the processes of analyzing, monitoring, and managing the intended and unintended consequences, both positive and negative, of planned interventions (policies, programs, plans, projects) and any social change processes invoked by those interventions" (Vanclay, 2003). SIA is "best understood as an umbrella or overarching framework that embodies the evaluation of all impacts on humans and all ways in which people interact with their sociocultural, economic, and biophysical surroundings" (Vanclay, 2003). The primary purpose of SIA is "to bring about a more sustainable and equitable biophysical and human environment" (IAIA, 2003). It seeks to "help individuals and communities, as well as governments and private sector organizations, understand and better anticipate possible social consequences on human populations and communities of planned and unplanned social changes" (Burdge, 2003a).

SIA definitions generally broadly define the initiating action, although there is more SIA experience at the project level. They recognize that assessing impacts on people can rarely be limited to simple cause–effect relationships. It is necessary to consider both the consequences of both planned interventions and social change processes, to evaluate how people interact with their surroundings, and to facilitate individual and community understanding and management of social change processes. Individual and community perceptions can have a considerable effect on the incidence and severity of social impacts. The recognition of the close interconnections between SIA and public participation underscores the ethical responsibility to involve interested and affected parties, and the contributions that those parties can make in preventing and ameliorating potential negative social impacts, and in facilitating the realization of net positive social impacts.

The SIA definitions broadly define the human environment and impacts, include a body of research, knowledge, and techniques practiced by SIA professionals, provide for close links to the social sciences and the community knowledge base, appreciate the value-full nature of IA, explicitly recognize links to the biophysical environment and sustainability, and acknowledge the importance of contextual adaptations. SIA is often, but not always, encompassed within EIA and SEA requirements and procedures. Depending on the situation, it can assume a secondary status to biophysical impacts (e.g., only indirect social impacts considered, receives less decision-making weight) and/or selective socioeconomic effects (e.g., employment, investment, community benefits) can assume a more prominent decision-making role than biophysical impacts. SIA definitions vary in their treatment of decision-making influence and the distribution of power. Technical/scientific and participative SIA characterizations tend to give it limited attention. More politically oriented SIA characterizations consider the exercise and redistribution of political power more fully. The generation and evaluation of alternatives, the management of uncertainties, and interconnections with the institutional setting tend to be featured less prominently in SIA definitions and characterizations.

HIA is defined as "a combination of procedures, methods, and tools that systematically judge the potential, and sometimes unintended, effects of a policy, plan, program, or project on the health of a population and the distribution of those effects within the population. HIA identifies appropriate actions to manage those effects" (IAIA, 2006a). It addresses induced changes in health determinants and resulting health outcomes. It seeks to proactively address health hazard risks and health improvement opportunities (IAIA, 2006a).

HIA is a multidisciplinary activity. It cuts across the traditional boundaries of public health, medical services, and environmental and social sciences. HIA can operate as a standalone form of IA or as a component of EIA or SEA. It defines health broadly (e.g., health determinants, health consequences, positive, negative, direct, indirect, cumulative). It explicitly considers the distribution of health effects within and among populations and communities. It includes interconnections with biophysical effects, socioeconomic–cultural effects, and sustainability. It acknowledges the need to adaptively manage risks and uncertainties. HIA definitions and characterizations do not always devote as much explicit attention to alternatives, to the role of interested and affected parties, to the shaping and bounding roles of ends and ethics, to proactive measures to influence decision making, and to contextual adaptations.

SEA "is generally understood as an impact assessment process that aims to mainstream environmental, social, economic, and health issues and ensure the sustainability of strategic decisions" (IAIA, undated b). Early SEA definitions tended to be more strongly rooted in project EIA. SEA was commonly depicted as a "formalized, systematic, and comprehensive process of evaluating the environmental effects of a policy, plan, or program, and its alternatives, including the preparation of a written report on the findings of the evaluation, and using the findings in publicly accountable decision making" (Thérivel and Partidário, 1996). These early SEA characterizations adopted a largely positivistic–consequentialist perspective to impact assessment (Jiliberto, 2011). They emphasized the action-forcing nature of SEA institutional arrangements. In common with EIA characterizations, they depicted a planning process exemplified by formal goals, and objectives, the systematic generation and evaluation of alternatives, and the formal identification, scoping, analysis, interpretation, and management of environmental effects. They stressed the need for comprehensive documentation, the explicit incorporation of the views and preferences of interested and affected parties, and informed, transparent, and fully substantiated decision making.

Over time, there has been a widespread acknowledgement of the SEA process implications of higher level decision-making characteristics. SEA is more strategic, visionary, and conceptual than EIA (Fischer, 2007b). Outputs are generally less detailed. SEA entails larger scales and longer time horizons. It is more qualitative, more uncertain, vaguer, and more distant from the public than EIA. Planning and policy-making processes vary dramatically among SEA types (e.g., policies, programs, and plans). SEA is highly context dependent. The boundaries between planning/policy making and SEA are blurred and fluid. These differences have led to an acknowledgement that SEA is best described as a family of approaches using a variety of tools—approaches and tools that must be adapted and combined to suit the situation (Ahmed and Sánchez-Triana, 2008a). There also has been a tendency to replace terms such as consequences and impacts with vaguer and broader concepts such as environmental aspects, effects, and issues (Jiliberto, 2011). The conceptualization of SEA as a tool for "mainstreaming" environment and sustainability reflects a less formalized-rationalistic and more collaborative–integrative–adaptive impact assessment approach. Typically, the SEA proponent is the decision maker (Thérivel, 2010).

This orientation shift is far from uniform. By and large, there is a propensity to employ a broad definition of the environment (i.e., biophysical, social, economic, health, sustainability) and to seek to infuse higher level planning and decision making with environmental values, perspectives, and ethical imperatives. Beyond this basic, shared objective, there is considerable variability in the approaches adopted. Consequently, SEA characterizations may or may not, and to varying degrees, systematically address alternatives, manage uncertainties, and identify, predict, interpret, and manage direct, indirect, and cumulative positive and negative impacts. The desirability of involving interested and affected parties is widely accepted. But the extent of actual involvement varies from virtually none or public information only to highly collaborative even joint decision-making approaches. The considerable variability evident in SEA characterizations necessarily reflects the considerable differences among SEA types and among contextual characteristics. But it also illustrates genuine perspectives differences in actual and preferred process attributes.

SA is "a process that seeks to identify the future consequences of a proposed action in a manner that directs future decision making toward sustainability" (Pope and Dalal-Clayton, 2011). SA is a range of processes, rather than a single process, all of which share the same aim of integrating sustainability concepts into decision making (De Ridder et al., 2010; Pope, 2006). SA, consistent with the stress placed on integration into decision making and among environmental components and impacts, is a form of integrated assessment (De Ridder et al., 2010). SA can be applied to any initiating action and can, and arguably should, subsume other IA types. SA is directed and bounded by an underlying set of values and ethical precepts. The assessment of sustainability necessitates the consideration of direct–indirect, positive–negative, and individual and cumulative effects. Alternative courses of action for achieving sustainability must be explored. Other essential attributes of SA include the adaptive and precautionary management of uncertainties, the systematic consideration of procedural and substantive equity, a proactive approach to influencing decision making, a demonstrable contribution to the cause of sustainability, the full and ongoing involvement of interested and affected parties, a vital role for follow-up and auditing, and contextual adaptations.

Not immediately evident from SA definitions and characterizations are answers to a host of thorny conceptual, methodological, and applied issues. How, for example, should SA best subsume or blend with other IA types? How can such a broad range of environmental components and impacts be effectively integrated without "watering down" the often small number of critical components essential to sustainability? How is SA to move beyond brave, but vague, intentions to dealing effectively with the hard trade-offs inherent in public and private decision making? How are the concerns and preferences of existing interested and affected parties to be traded off with those of future generations? How can SA effectively influence decision making and redistribute power when sustainability imperatives so often run counter to the interests of those in positions of power? How can flexibility and adaptability be maintained without compromising away critical limits and thresholds? How can overriding sustainability principles be adhered to at the same time as making the necessary regional and local contextual adaptations? The field of sustainability has devoted considerable attention to these issues and many others. However, the necessarily broad and admittedly ambiguous definitions of SA and sustainability leave open the potential for many forms of SA that ultimately do little to facilitate sustainability, and may, in some circumstances, undermine its basic purpose.

Several synergies are evident when the IA definitions are considered collectively. EIA offers a well-established set of action-forcing institutional arrangements, a host of process design and management choices, and a well-tested array of frameworks, concepts, methods, and "good practices." The more focused forms of IA (e.g., EcIA, SIA, and HIA) provide an opportunity to more effectively integrate substantive environmental concerns into EIA and SEA. They also offer more direct and effective links to the natural and social science knowledge base, a greater sensitivity to and experience with addressing interactions among changing patterns of baseline conditions, cumulative effects, and management actions, well-defined ethical principles and imperatives, and an enhanced ability to appreciate, adapt to and manage risks and uncertainties. EcIA provides a foundation for integrating ecological sustainability concerns. SIA provides an effective, direct and ongoing connection, both in an analytical and in a collaborative sense, to public values, attitudes, preferences, concerns, and knowledge. HIA and SIA provide a foundation for integrating human sustainability concerns. SEA can frame and direct project-level EIA. EIA, especially when it fully integrates HIA, SIA, and ecological IA, can inform and ground SEA. SA can facilitate the infusion of sustainability frameworks, values, principles, objectives, thresholds, and trade-off rules into all forms of IA. SA provides a mechanism for integrating IA types into all decision-making levels. Each IA type can benefit from a greater understanding of the perspectives, concepts, frameworks, methods, and experiences associated with other IA types.

All IA types struggle, in different ways and utilizing different solutions, with recurrent problems. Difficult decisions must be made regarding the selection, integration, and adaptation of appropriate regulatory and process design and management choices. Each attempts, with varying degrees of success, to influence decision making. Each, again to varying degrees, draws upon and contributes to a theoretical and applied knowledge base. Each identifies, blends, and achieves ends, in part through the systematic generation and evaluation of alternatives. Each seeks to protect and enhance the environment. Each attempts to use resources effectively and efficiently. Each aims to make decision making more democratic, transparent, and collaborative. Each introduces and applies, some more explicitly than others, ethical procedural and substantive ethical principles and standards.

Each must identify, adapt to, and manage risks and uncertainties. All must address interconnections and how best to adapt to and influence context.

Contemporary challenges cut across the IA types. All IA processes encompass significance determination and follow-up activities. Each can contribute to more effective approaches for siting "locally unwanted land uses," and for managing effects on and from climate change. Cumulative effects assessment (CEA) and multijurisdictional IA are pertinent to and potentially advantageous to all IA types. All struggle with the potential and limits of "good practices" at the strategic level. All are concerned with how best to influence decision making, with enhancing capacity building, and with ensuring that the approaches and methods applied are appropriate to the context. Each encompasses varying, sometimes conflicting, perspectives regarding if and the extent to which the definitions of the environment and effects should be broadly or more narrowly defined.

The definitions and characterizations of IA types presented in this section collectively demonstrate the need to systematically identify and explore interconnections among IA types. The varying characteristics, strengths, and limitations of the IA types suggest the potential for both complementary and conflicting interactions. The whole can be either more or less than the "sum of the parts." Inasmuch as the recurrent problems and contemporary challenges are shared by, but in different ways, by each IA type, when addressing each problem or challenge it is essential that the potential contribution of each IA type be fully addressed.

1.3.2 Themes and Issues

The roots of IA, as a formalized (i.e., action-forcing) procedure for assessing the potential environmental effects of proposed actions, are not deep. First introduced in the United States in 1969 (the National Environmental Policy Act—NEPA), the field has expanded rapidly, especially over the past three decades. IA is now applied in more than 100 countries, often at multiple decision-making levels. The definition of the environment has broadened from an early emphasis on physical and biological effects to an increased concern with biodiversity, social, cultural, heritage, human health, and ecological effects. At the same time there has been intense debate regarding whether environmental values are better served when IA requirements and processes are broadly or narrowly defined.

In the early years direct impacts, usually from large capital projects, tended to be the focus. Much greater emphasis is now placed on indirect, positive, and cumulative effects, on substantive gain and compensation (e.g., mitigation banking, compensation pools) rather than simply on procedural compliance and minimizing negative effects, on integrating IA into earlier decision making (e.g., need, reasonable alternatives), on different IA types (e.g., SEA, SA, SIA, EcIA, and HIA), and on interconnections with other existing and proposed actions. The effectiveness of more holistic regulatory and applied approaches, especially with reference to facilitating sustainability, has received considerable attention. Progress in the effort to shift the orientation of IA practice from process to substance, however, has been halting, uncertain, and uneven. In some jurisdictions, the range of proposed actions subject to IA requirements has been narrowed, and the scope of substantive environmental concerns subject to IA requirements has been reduced.

IA requirements have been applied to policies, plans, programs, legislative proposals, technologies, development assistance, products, and trade agreements. The application of IA to higher decision-making levels, through SEA and SA, has increased dramatically in recent years, albeit unevenly among jurisdictions. The standard of proposal acceptability has increasingly shifted, again with exceptions, from positive economic/limited negative ecological and indirect social effects to substantial, net positive direct and indirect ecological, social and economic gains, especially for global commons concerns. The role of trade-offs in IA-related decision making has been a particular focal point. IA also has been adapted to different jurisdictional types and settings (e.g., third world and transitional economies, indigenous decision making regimes, the global commons—oceans, Arctic, Antarctic), to inter- and multijurisdictional situations (e.g., harmonization, duplication), and to private sector decision making. Increasing attention has been devoted to the political nature, relevance, and effectiveness of IA instruments, both in general, and relative to other means of political influence and control.

Boundary spanning and integration, both within IA and to related fields, has received considerable attention. Frameworks and strategies have been formulated for better addressing and transcending interrelationships among impacts, disciplines, and IA types. The identification and management of systemic and cumulative effects has been a priority. Links between IA and decision making, in varying institutional settings, has been a particular concern. IA has been linked to resource management, urban and regional planning, landscape planning, risk assessment, corporate planning, environmental management systems (EMS), and efforts to protect threatened and endangered species, communities, and ecosystems. The reconciliation of broadly based effectiveness standards and the need for contextual adaptations has received considerable attention. Such efforts have been aided by comparisons across jurisdictions and setting types.

From an initial focus on procedural requirements, much greater stress is now often placed on procedural and substantive fairness, equity, justice, and vulnerability, on the integration of substantive environmental concerns such as biodiversity and sustainability, on the adaptation of precautionary and pollution prevention principles, and on the application of traditional knowledge. The roles of complexity, precaution, diversity, reversibility, resilience, uncertainty management, anticipation and rapid appraisal and response, and harm reduction and catastrophe (human and natural, deliberate and accidental) avoidance and minimization are increasingly stressed in the management of risks and uncertainties. This orientation shift reflects a greater humility in acknowledging ignorance and its implications, and a greater openness to accept and correct the biases and limitations of conventional risk and uncertainty management procedures.

The political dimension of IA has received more emphasis in recent years. Particular attention has been paid to whether, and in what ways, IA can more effectively and more proactively influence decision making and reshape institutional arrangements, procedures, and values. An especial effort has been made to draw upon lessons and insights from IA decision making "successes and failures." Courts and review panels have assumed a greater role in defining and interpreting IA requirements. Prominent, politically related, contemporary issues include the effectiveness of tiering arrangements, the management of transboundary impacts, the potential role of IA in community development and democratization, and the desire to resolve disputes and achieve sociopolitical acceptance, without negotiating away environmental interests.

The conceptual foundation of IA was largely neglected in the early years when the priority was on establishing and refining institutional arrangements and "good practices." In more recent times, connections to the natural and social sciences and to philosophy, IA theory building and testing, explorations of the role of causality, the strengthening of follow-up procedures, the potential role of community-based and indigenous research and knowledge, and the facilitation of knowledge and information sharing, training and learning have received more attention. The role of rationalism, in all its variations, has been the subject of ongoing debate, and a fruitful source of conceptual frameworks and procedures.

The grounding of IA in practice has been another priority. IA documents and review procedures have been streamlined, and made more efficient and effective. Good practice screening, scoping, and mitigation procedures and tools, together with insights from problematic efforts to focus IA requirements and procedures, have received particular attention. An especial effort has been made to enhance effectiveness and efficiency at strategic decision-making levels. Greater use is made of IA document standards (e.g., document format and style, electronic publishing, use of geographic information systems). An increased effort has been made to integrate practice-based knowledge through the greater use of IA quality and effectiveness analyses, practitioner and other state-of-practice surveys, case studies, quality control and assurance initiatives, monitoring and external auditing procedures, and applied research, especially with reference to the tracking of sustainability progress.

Effective and meaningful public access, understanding, communications, learning, and participation remain priorities. Greater efforts are being made to facilitate IA capacity building and training, the pooling of information, knowledge and experience, collaboration and conflict management, and community-based participatory and sustainability research and knowledge development. Methods and procedures for facilitating earlier and more collaborative and empowering public participation have been refined. Especial attention has been devoted to the meaningful involvement of indigenous peoples and to effective participation in developing and transitional countries. Oversight of IA practices and procedures by courts and civil bodies is increasingly evident. Greater use is being made of social networking and of online discussion groups to extend the reach and the role of the public in IA knowledge building and testing.

More effective significance determination has emerged as a major priority. The role of subjectivity, perceptions, values, and ethics in IA practice has received increased attention. A greater effort has been made to more effectively integrate environmental and social distributional, vulnerability, equity, and justice concerns. Distributional analysis requirements and tools have been refined. Impacts on and the perspectives of women and indigenous peoples are more prominently featured. As IA has matured as a field of practice, more thought has been given to professional and corporate roles, responsibilities and ethical standards, including the possible need for and accreditation of IA professionals.

The spotlight in IA literature and practice is now being turned on new challenges. Transboundary impacts, the "tiering" of IA requirements, and the integration of macro-environmental issues, such as climate change and biodiversity, are particular concerns. IA is being adapted to protect global environmental resources (e.g., the oceans, the Arctic, and the Antarctic), to address the implications of global economic activity, to integrate international agreements, conventions, principles, standards, directives and treaties, and to explore the consequences of economic trends such as privatization.

IA is now a large, complex, and rapidly changing field. The themes and issues, described above, are far from clearly defined, uniform among jurisdictions, or unambiguous. Countervailing trends and patterns are commonplace. The gulf between theory and practice remains immense. Notwithstanding numerous examples of "good practice," "poor practice" examples are even more evident. There is far from a consensus regarding many "appropriate" good practice standards, even regarding such basics as IA aims and definitions. Perspectives vary greatly regarding how the field is evolving, and whether the changes that are occurring are desirable or undesirable. It could equally well be argued that, depending on one's perspective and the setting, little of substance has changed, there is no discernible general pattern or trends, the field swings back and forth in pendulum-like fashion in concert with political changes, multiple, often conflicting and counterbalancing trends and patterns are occurring, or a gradual, albeit halting, uneven and "fuzzy" trend can be discerned. Arguably, the greatest danger is complacency. Simply assuming that "things are getting better" and positive reforms will remain in place and not be "watered down" or reversed is, at best, naïve. A more prudent approach would be to identify and systematically explore a select series of recurrent problems and contemporary challenges, with ample provision for multiple approaches and perspectives, and for contextual adaptation. Such an exploration should be extremely cautious in reaching conclusions or in making recommendations. It also should progressively move back and forth between theory and practice, and should encompass multiple regulatory and applied experiences and perspectives.

1.4 A STRUCTURE

The preceding sections indicate that there is a genuine need to come to grips with recurrent problems and contemporary challenges that plague IA practice. They illustrate that there is a considerable knowledge base and many promising initiatives to draw upon. They also make it clear that these resources, although necessary, are not likely to resolve recurrent problems and contemporary challenges in IA practice. The shortest path to improvement does not lie with simply "tossing" an array of potentially relevant methods and procedures at each problem, either individually or collectively. A coherent structure is required. The IA process, embedded with a broader framework, provides that structure. This section briefly frames and describes the IA process. It then explains why the IA process is important and why multiple IA processes are necessary to any strategy to address the problems and challenges.

1.4.1 Framing the IA Process

IA process and design, as illustrated in Figure 1.2, operates at both the regulatory (e.g., requirements, guidelines) and the applied (e.g., individual proposals) levels. It entails both strategic (e.g., legislative proposals, policies, plans, and programs) and project (e.g., proposed projects and activities) planning. It encompasses an array of partially overlapping IA types. Interconnections between the regulatory and applied levels, and among IA types, are crucial. IA process design and management draws upon (e.g., recurrent problems and contemporary challenges, effectiveness reviews, stories, case studies, and surveys) and contributes to (e.g., good practices, methods and procedures, process design choices and variations) IA practice. It selects from and adapts regulatory and applied process design choices and variations, and draws upon (e.g., concepts, frameworks and principles, theory testing, definitions and distinctions) and contributes to (e.g., theory building, theory testing) IA theory. IA theory and practice draw upon and contribute to related fields of practice and knowledge. IA is adapted to and contributes to the reshaping of context. These elements are all linked and integrated through continuous analysis and synthesis. IA process design and management, to function effectively, must be systematically linked to and integrated within a broader framework.

An especially important aspect of framing the IA process, as illustrated in Figure 1.3, pertains to interconnections between strategic and project-level IA, and between SEA/EIA regulatory and applied practice and decision making. EIA can be defined sufficiently broadly, albeit with some difficulty, to encompass various strategic level choices and decisions. SEA can scope and frame EIA level planning and decision making. EIA level planning and decision making can inform and refine SEA planning and decision making. EIA and SEA can be linked or even merged at one or even at all decision-making levels. The linking and integration of SEA and EIA level planning and decision making can facilitate the setting of goals and ethical boundaries, the assessment of need and strategic options, the analysis and management of cumulative effects, the tiering of decision making, more comprehensive follow-up, and the more systematic treatment of issues such as proportionality.

1.4.2 The IA Process

At a most basic level a process is a series of actions directed toward an end. The end, in this case, is fourfold: (1) the more effective management of recurrent problems, (2) the more effective management of contemporary challenges, (3) the more effective integration of good practices, practical solutions to issues and challenges, and relevant concepts, principles, frameworks, definitions, distinctions, methods, procedures, and insights, and (4) more effective regulatory and applied practice for various IA types, both individually and collectively. The primary instrument, within which these ends can be achieved, is the IA process, as expressed through regulatory and applied IA design choices and variations.

Some important distinctions, as illustrated in Figure 1.4, must be made before the choice and arrangement of IA process actions can be considered. IA operates within a framework established by institutional arrangements. The institutional arrangements provide the basis for determining whether IA requirements are applied (i.e., screening). They also offer guidance regarding administrative procedures, documentation, and various planning activities. The planning process activities interact with proposal and alternative characteristics, and with environmental conditions. The IA administrative, planning, and documentation processes are all reviewed and contribute to decision making.

IA requirements usually provide only general guidance regarding the conduct of each process type and each IA type. This is especially the case for IA planning process activities. Major decisions may be identified. Early public involvement is usually encouraged. Specific guidance is commonly offered for conducting individual planning process activities. More detailed requirements and guidelines are often provided for administrative (such as document circulation and agency review) procedures and the content of IA documents. IA practitioners then design and manage the IA process within the framework established by IA requirements.

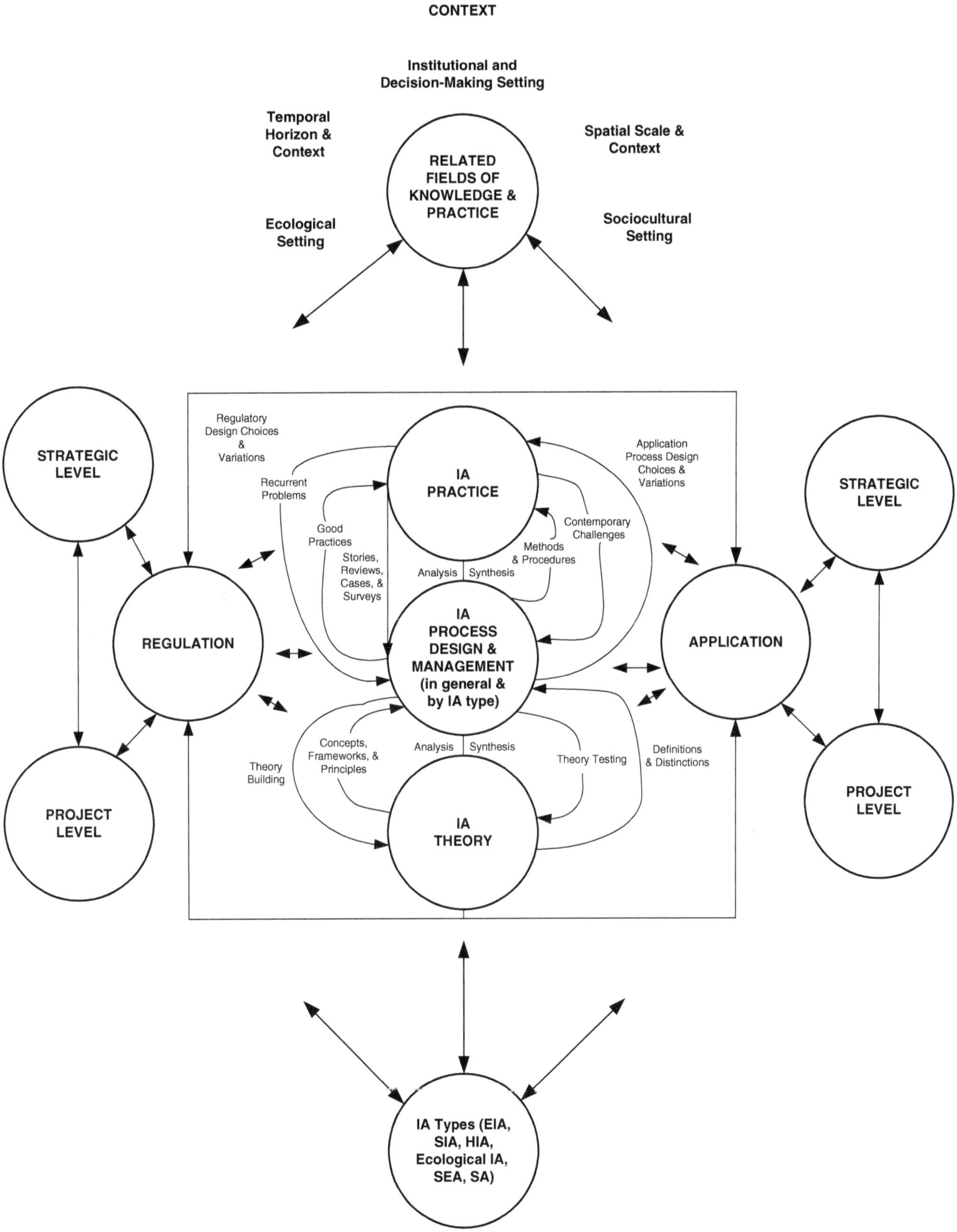

Figure 1.2 Framing the IA process.

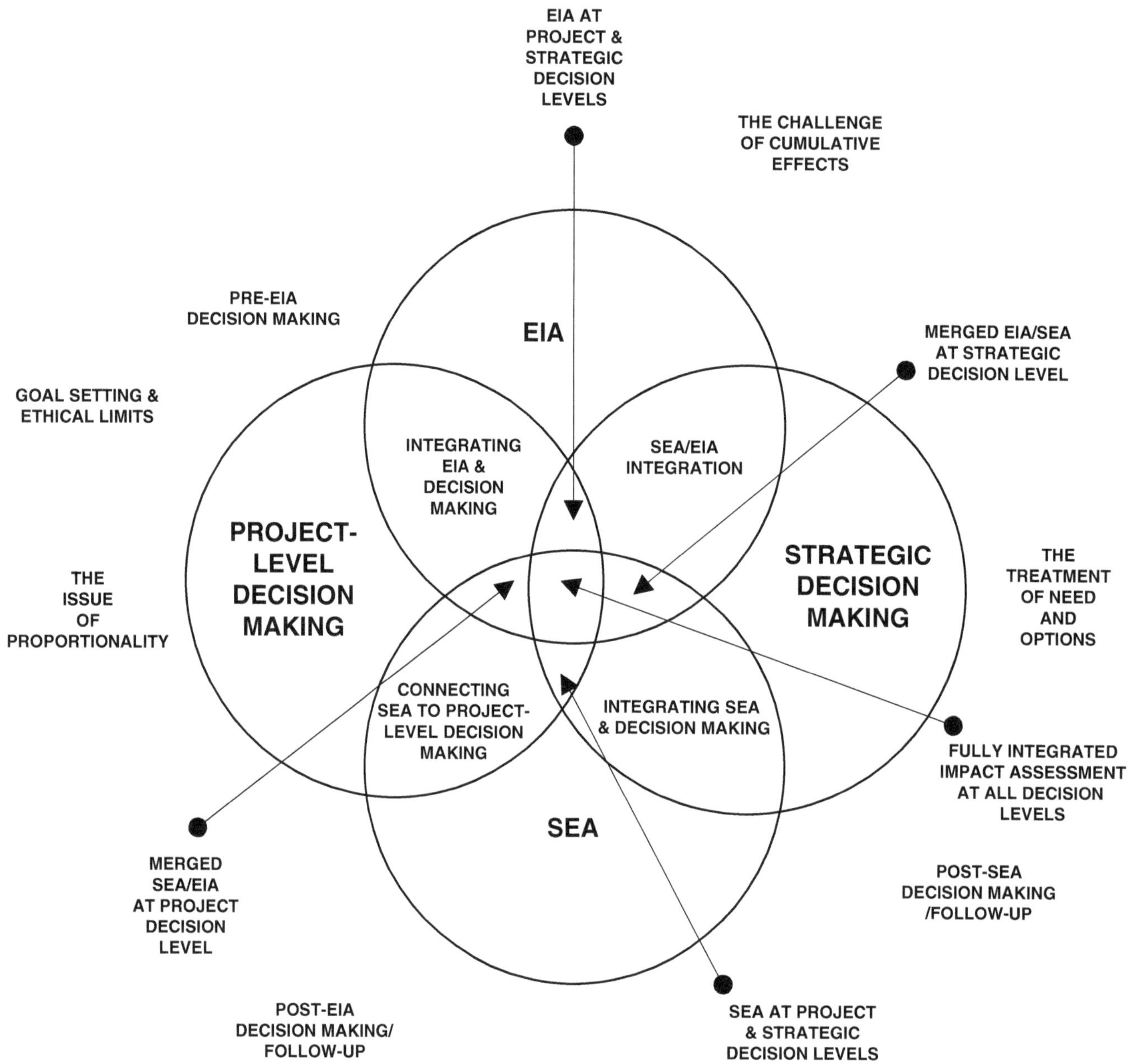

Figure 1.3 EIA, SEA, and decision-making linking and integrating.

As explored more fully in Chapter 2, there are many ways in which the IA process can be managed at the regulatory and applied levels for both strategic and project planning. The IA process also varies among IA types. Some management procedures are better able to address the problems and challenges that arise from diverging perspectives than others.

1.4.3 Why Is the IA Process Important?

IA practitioners have a great deal of discretion, in the sequence in which IA activities are arranged, in the extent to which IA activities are subdivided, in the choice of activity inputs and outputs, in the choice and nature of interconnections among IA activities, and in how IA types are linked and integrated. IA practitioners, in concert with other stakeholders, determine if and to what extent potential problems and challenges that result from varying perspectives will be resolved. They also determine if and how the IA knowledge base is drawn upon and adapted (theory testing), and if and how IA practice contributes to the IA knowledge base (theory building). The varying perspectives generally focus on whether or not IA requirements and processes are relevant, and, if so, how IA processes are or are not conducted. Considerable potential for preventing and ameliorating the problems and challenges stemming from varying perspectives lies with better IA process management.

A great many methods, an extensive knowledge base, a vast pool of experience, and a diverse array of values and

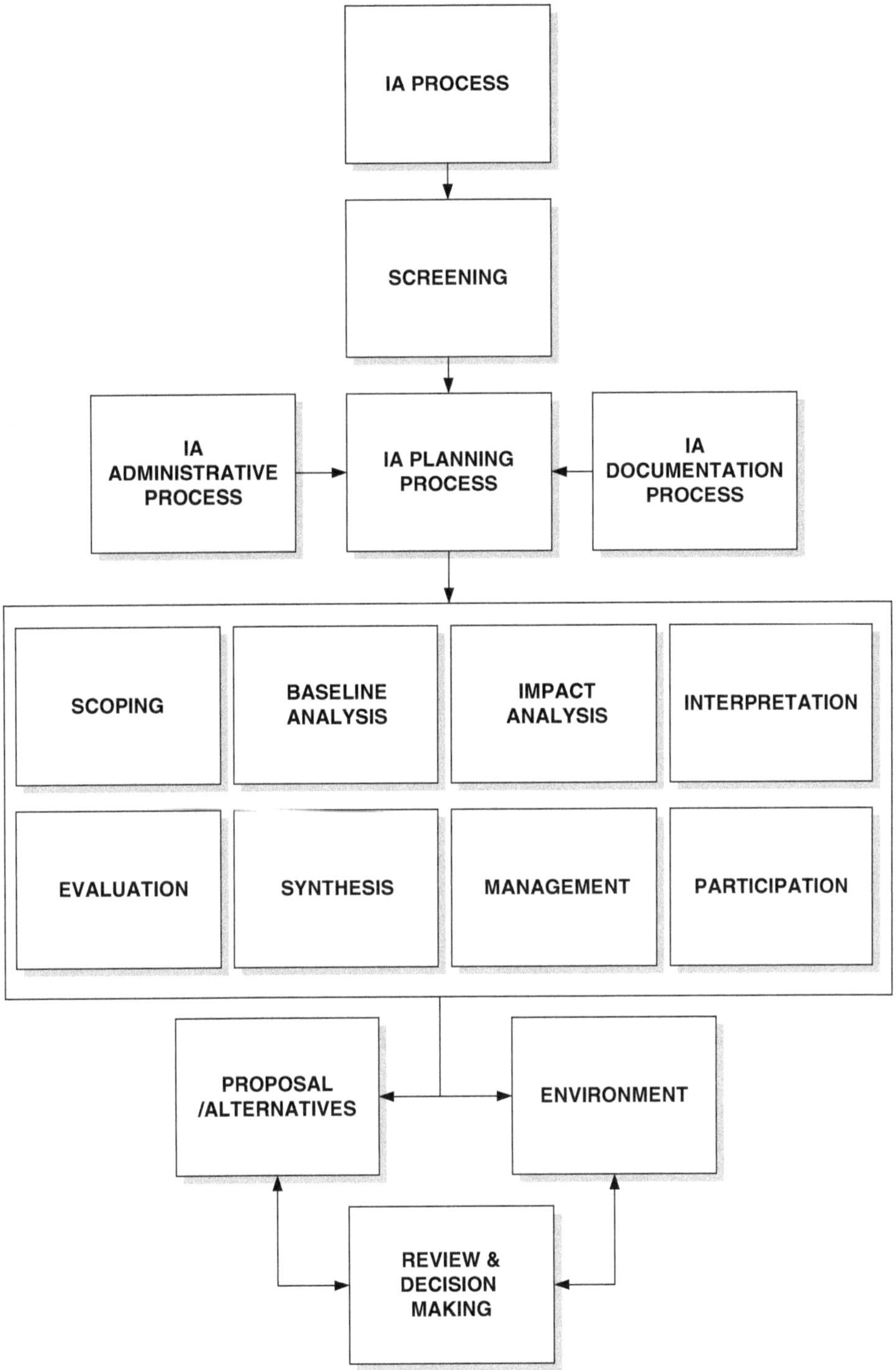

Figure 1.4 IA process activities.

perspectives, are available to IA practitioners. The IA process, to varying degrees, integrates these inputs. Poorly managed processes will more than the offset any benefits resulting from, for example, the rigorous application of the "latest" methods. The IA process bridges IA regulatory requirements and IA practices, and cuts across decision-making levels. It also helps link and integrate IA theory and practice. IA requirements, because of the breadth of their application, must necessarily be general. At best IA requirements will reduce the incidence and severity of "bad" IA practice. Highly effective IA practice and high-quality IA documents will only occur when effectively framed by IA processes.

The IA process varies greatly depending on the type of proposal, the type of IA, the purposes that drive and structure the IA, the decision-making level, the local and regional environmental conditions, and the types of anticipated impacts. The effectiveness of an IA is often highly dependent on how well the IA process fits the context. The roles assumed by stakeholders, both planned and unplanned, are often crucial. The IA process determines which stakeholders are involved and when and how they are involved. It determines how stakeholder concerns and suggestions are solicited and addressed. The IA process is the bridge between IA practice and decision making. It determines the decision points and establishes how IA technical and procedural activities and decision making are linked and integrated. Decision makers often reject IA analyses and documentation when those interconnections are inappropriately established and managed.

IA texts, manuals, and guides usually provide only very general IA process characterizations. These characterizations are rarely tested and refined in practice. When the IA process is discussed, it tends to be in terms of document preparation and the satisfaction of government requirements. IA resource materials generally concentrate on IA institutional arrangements, IA types, IA methods, and the adaptations required for predicting different types of impacts and for assessing various types of proposals. This deficiency in the literature (notwithstanding the central role of the IA process) underscores the need to formulate and apply more effective and adaptable IA processes.

1.4.4 Why Are Multiple IA Processes Important?

IA processes should be designed to encompass stakeholder perspectives. A multipronged approach can narrow the gap between process characteristics and stakeholder values. As illustrated in Figure 1.5, the first priority of IA process managers is to ensure that the IA process influences decision making, and is legitimate and credible to all potentially interested and affected parties, to pertinent objectives, to all decision-making levels, and to the accumulation of IA-related knowledge. The effectiveness of an IA process can be severely, sometimes fatally, undermined when major stakeholders choose to operate outside the IA process. The integration of both strategic and project-level planning with decision making also is critical.

Different processes will be appropriate in different situations. IA practitioners could benefit from being able to "pick and choose" from a suite of IA processes as they seek to meet multiple demands, and to match process and context. The procedures for addressing any one demand are many and varied. A coherent understanding of how each demand can be addressed is essential before strategies for counterbalancing overlapping, emerging, and conflicting demands can be explored and applied.

The IA process has been scrutinized before. Numerous variations have been suggested. There are many references to alternative approaches scattered across IA literature—some specific to various IA types. Process differences between project-based EIA and SEA processes have received particular attention. Debates and discussions regarding the relevance and legitimacy of IA, as an environmental management instrument, have occurred since the inception of the field, and still persist, in some ways with even greater vigor. The effectiveness of various IA types in influencing decision making has been a critical and recurrent theme. There is an ongoing debate between advocates of a more scientific as compared with a more practical IA process. A similar debate has occurred, especially within SIA, between proponents of technical systems, consensus-based processes, and political, conflict-based processes. Much has been written regarding study team roles and responsibilities within the IA process. Alternative processes, such as adaptive environmental assessment and management, have been advanced. The application of alternative models (e.g., advocacy planning) from related fields, such as planning theory, has been suggested. What has not occurred is the consolidated presentation, analysis, comparison, and integration of the available concepts and insights into IA process forms readily applicable to IA practice.

Other related fields of practice (such as urban and regional planning and environmental management) and other related disciplines (such as sociology and philosophy) have demonstrated the many insights that can be acquired when multiple models or frameworks are applied to a complex problem or situation. IA, as a field of theory and practice, is very much the exception in its traditional assumption that a single process model (with minor adaptations) can be applied to any and all situations. This assumption has been challenged in IA literature in recent years, especially at the SEA level, and as reflected in alternative SA, SIA, and HIA models and approaches. A rich array of applied research, methods, and conceptual frameworks (both within IA and in related fields) could be readily integrated into multiple IA processes. Once multiple IA processes are consistently formulated, presented, and analyzed, it becomes much easier to formulate composite processes that balance and integrate a range of perspectives. A foundation also is established for effectively integrating good practice principles and procedures into IA process design and management.

1.5 A STRATEGY

This section presents a strategy for facilitating more effective IA process management. As indicated in Figure 1.6, the IA process provides the organizing structure for both the regulatory and applied level analyses. Both the conventional IA process (in a diversity of manifestations) (Chapter 2) and multiple IA processes (one for each perspective) are presented and assessed (Chapters 3–11). Practice-based stories set the stage for each of Chapters 3–12 analyses. Variations among IA types (EcIA, SIA, HIA, SEA, SA), in the expression and management of the recurrent problems are addressed in each of Chapters 3–11. The contemporary challenges addressed in Chapters 2–12, respectively, include SEA good practice, strategies for influencing decision making, follow-up, the siting of "locally unwanted land uses," the issue of how broadly the environment and effects should be defined, the

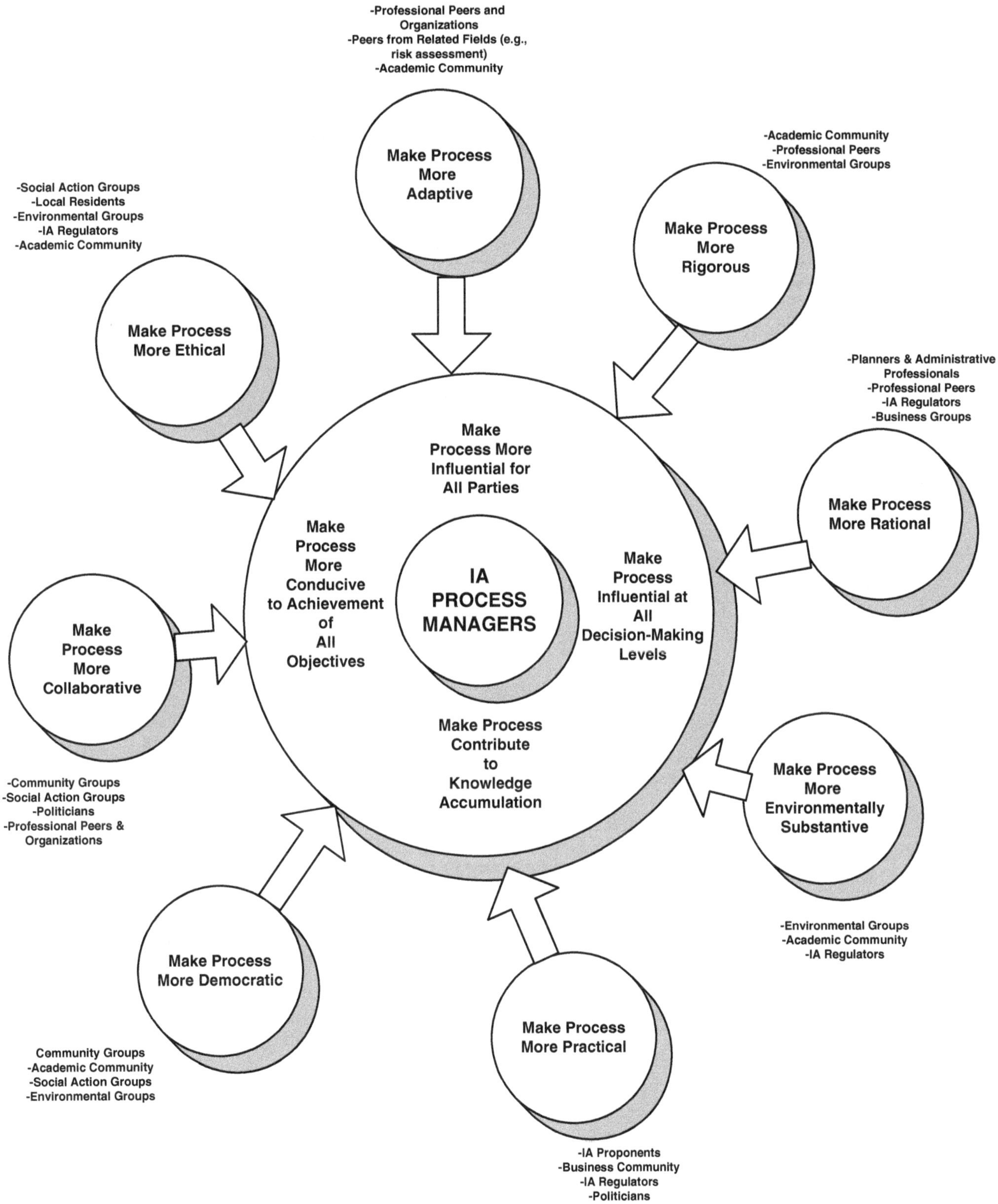

Figure 1.5 Demands on IA process managers.

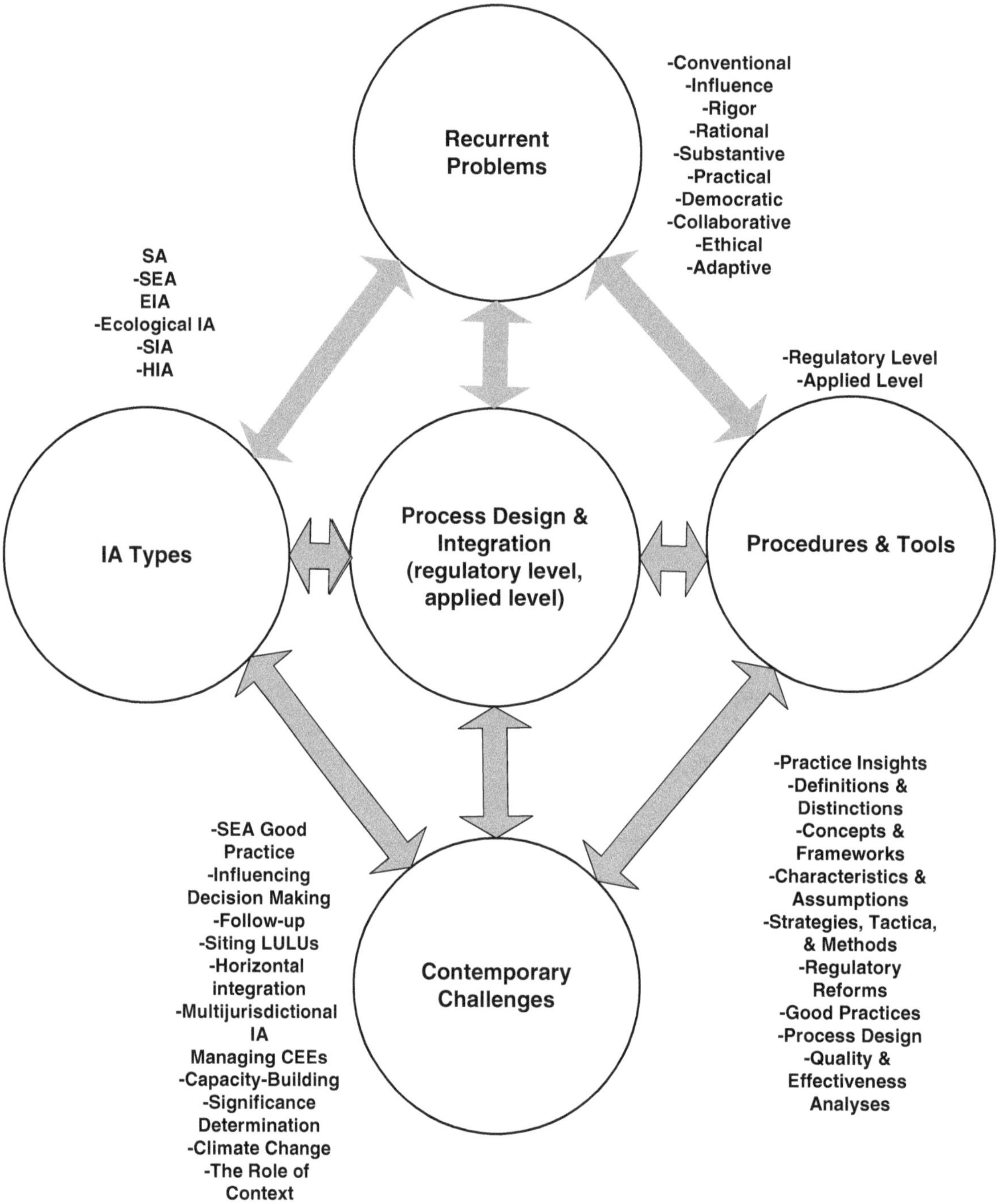

Figure 1.6 Reforming the IA process.

management of cumulative effects, multijurisdictional impact assessment, capacity building, significance determination, climate change, and the role of context. Composite IA processes are also described (Chapter 12).

Applied examples, IA literature, and selective literature from related fields contribute to the major definitions, concepts, frameworks, characteristics, assumptions, strategies, tactics, and methods presented. The regulatory analysis is based on an overview of readily available (i.e., largely on the Internet) Federal or senior level IA legislation, regulations, guidelines, reforms, and proposed reforms from four jurisdictions—the United States, Canada, Australia, and the European Union. The regulatory analysis also includes the approach taken, in each jurisdiction, to EcIA, SIA, HIA, SEA, and SA. Positive and negative examples, derived from the four jurisdictions, for addressing each recurrent problem are briefly highlighted as the basis for a broader discussion of potential regulatory choices.

An iterative, analysis–synthesis relationship is inherent to IA practice. It also is intrinsic to this book. Interactions between IA requirements and IA practice are explored in each of Chapters 3–11. Anecdotal examples from IA practice are integrated into each of Chapters 3–11. Major conclusions and lessons are integrated for conventional IA (for different IA types) (at the end of Chapter 2) and for individual IA processes and contemporary challenges (at the end of each of Chapters 3–11). Chapter 12 presents multiple approaches for synthesizing individual IA processes, for integrating IA types, for establishing and maintaining decision-making links, and for better bridging theory and practice. Chapter 12 also addresses contextual adaptations and identifies residual challenges and priorities.

Each of Chapters 2–11 uses relevant concepts, frameworks, definitions, and distinctions to establish a foundation for the regulatory and applied analyses. Good practice guidance is provided for addressing recurrent problems and contemporary challenges. Particular attention is devoted to process adaptations and adjustments required when undertaking IA, in all its forms. Each chapter seeks to build and test theory.

A comprehensive treatment of the subject matter addressed by this book, at a uniform level of detail, is neither practical nor appropriate. The analysis is selective, sometimes arbitrarily so. It was necessary to focus on some subjects, treat other subjects more selectively and at a broader level of detail, and not consider other potentially relevant topics.

This analysis focuses on managing the IA process at both the regulatory and applied levels. It seeks to ameliorate the negative consequences that sometimes stem from perspective differences. The regulatory analysis is based on senior level IA requirements and guidelines in the United States, Australia, Canada, and the European Union that bear directly on IA process management choices. The applied analysis focuses on the roles and responsibilities of IA process managers. It assesses and reformulates conventional IA processes, formulates, analyzes, integrates, and refines multiple alternative IA processes, explores process variations and commonalities among IA types, presents practical approaches for managing a range of contemporary challenges, and formulates composite IA processes. Both the regulatory and applied analyses are largely based on pertinent IA literature and practice. A broader and more selective analysis is undertaken of other IA aspects connected to IA process management and literature and practice in related fields.

IA requirements outside the four jurisdictions are not considered. It is possible that insights and lessons from this analysis may be of value to other jurisdictions. Adaptations to address, for example, IA practice in developing and transitional countries have only been briefly considered. No effort has been made to trace the long and complex intellectual traditions that underlie the various IA processes. These traditions span hundreds of years and encompass tens of thousands of references. Although intriguing (albeit a major undertaking), it would be impossible to provide a just and accurate treatment of those traditions without detracting from the central thrust of the book, which is to provide a succinct and practical reference to IA practitioners.

Finally, any treatment of the IA process, no matter how practical, will never fully match the subtle complexities encountered in IA practice. Each proposal and each setting are, in important, often not readily apparent respects, unique. Hopefully, much can be learned from IA references such as this book. Nevertheless, the IA process and methods must always be designed and adapted, jointly with stakeholders, to fit unique setting related circumstances. Both the approach and the circumstances will change and evolve—preferably in tandem.

1.6 A ROAD MAP

Table 1.1 identifies examples of how IA process managers, IA study team specialists, IA regulators, nongovernment organizations (NGOs), and IA instructors might wish to use this book.

Table 1.1 Suggested Approaches for Using This Book

IA process managers	• Read the book for an in-depth review of alternative IA process management approaches and approach combinations • Use Chapters 3–11, as appropriate, to avoid and minimize the recurrent problems • Use the good practice guidance, provided in Chapters 3–11, to enhance the application and adaptation of IA regulatory approaches and methods • Use the contemporary challenges guidance, provided in Chapters 2–11, for more effective responses to contemporary challenges • Use the guidance for various IA types, provided in Chapters 2–11, to more effective address, for example, strategic, social, health, and ecological choices in IA process design and management • Use sustainability assessment guidance, in Chapters 2–11, for addressing various ways in which sustainability concerns can be integrated into process design and management • Read the highlights and summing up subsections for an initial impression of each chapter • Use the figures and tables, with text support, for process design and management • Draw upon individual chapters as issues are identified • Take a particular look at Chapter 3 to more effectively influence decision making

(*continued*)

Table 1.1 (*Continued*)

	• Take a particular look at Chapter 7 for study team management strategies • Use Chapter 12 for composite planning process management and for more effective integration within and external to the IA process • Use references for follow-up
IA study team specialists	• Review highlights and summing up subsections to obtain an initial impression of available processes • Review descriptions of process activities, inputs and outputs in Chapter 2, to understand specialist integration procedures • Review IA types guidance, provided in Chapters 2–11, to facilitate process adaptations and IA type integration for addressing ecological, social, health, and sustainability concerns • Review good practice guidance, provided in Chapters 2–11, to enhance discipline-specific application of IA methods • Review contemporary challenges guidance, provided in Chapters 2–11, to better understand potential role in anticipating and responding to contemporary challenges • Review Chapters 3 and 7 to understand decision-making links and roles within the study team • Natural and social scientists may be especially interested in Chapters 2, 4, 6, and 11 • Urban and regional planners and resource managers may be especially interested in Chapters 5 and 6 • Social scientists may wish to take a close look at Chapter 10 • Public participation specialists may find Chapters 8 and 9 especially relevant • Any specialist, concerned with risk and uncertainty management, may wish to take a closer look at Chapter 11
IA regulators	• Review Chapter 2 and regulatory sections of each of Chapters 3–11 • Review SEA sections in Chapters 2–11 for guidance in assessing policies, programs, plans, and legislative proposals • Use SA guidance in Chapters 2–11 for addressing how best to integrate sustainability concerns into IA requirements and guidelines, and for evaluating the adequacy of the treatment of sustainability concerns in individual IA processes and documents • Review IA type guidance (e.g., SIA, HIA, EcIA) for addressing discipline-specific concerns • Peruse highlights, tables, summing up subsections, and figures in determining possible regulatory reform approaches • Use flow charts for evaluating IA documents • Use good practice guidance, provided in Chapters 2–11, for evaluating IA methods and processes • Use contemporary challenges guidance, provided in Chapters 2–11, for working with other stakeholders in anticipating and responding to contemporary challenges • Review definition of the problem sections to alert proponents and IA managers to potential and emerging problems and challenges • Take a closer look at Chapter 7 for examples of procedures for expeditiously satisfying regulatory requirements • Review Chapter 12 for an overview of composite process approaches and for integration procedures within and external to the IA process
Nongovernment organizations	• Take an especially close look at Chapters 8 and 9 • Use the good practice guidance, provided in Chapters 2–11, for evaluating the choice and application of regulatory approaches and applied methods • Use the IA type guidance in Chapters 2–11 for assessing the treatment of ecological, social, health, and sustainability concerns • Use the contemporary challenges guidance, provided in Chapters 2–11, for anticipating and responding, jointly with other stakeholders, to contemporary challenges • Use SA guidance for evaluating and challenging the treatment of sustainability concerns in IA guidelines, documents, and processes • NGOs are likely to find Chapters 6 and 8–11 especially pertinent to their concerns • Peruse highlights, definition of problem, direction, and summing up sections and Chapter 12

1.7 SUMMING UP

This book is intended to enhance IA practice. It is especially directed toward the needs of IA practitioners who manage IA processes. It provides practical solutions to major, recurrent problems encountered in daily IA practice. It also provides good practice guidance for addressing a range of contemporary challenges.

A scenario is used to identify several, recurrent, IA process-related problems and contemporary challenges.

The IA definitions provide an initial sense of the perspectives and of the requisite characteristics of an IA process. They also provide the basis for identifying the recurrent shortcomings and contemporary challenges. An overview of IA issues and themes demonstrates that the shortcomings are unlikely to resolve themselves through the normal evolution of IA practice.

The framing, design, and management of the IA process, at both the regulatory and applied levels, are pivotal to effectively managing many recurrent shortcomings of IA practice, and to effectively addressing a host of contemporary challenges. Formulating, analyzing, and synthesizing multiple IA processes can further enhance IA process management.

A strategy is presented for more effectively managing regulatory and applied IA process design choices. The strategy initially assesses how well various versions of the conventional IA process (for each IA type), with appropriate modifications, respond to major recurrent problems. It then formulates, assesses, and combines IA processes that respond to each problem. Practical, "good practice" advice is provided in Chapters 2–12 for addressing a range of contemporary challenges. Pertinent literature and reviews of IA requirements in four jurisdictions support the analyses. A focused and selective approach is required because of the breadth of the subject matter.

Chapter 2

Conventional IA Processes

2.1 HIGHLIGHTS

In this chapter, we describe and assess generic project EIA choices for controlling and shaping the EIA process through EIA legislation, regulations, and guidelines, and conventional EIA choices for designing and managing the IA process. We also describe regulatory and applied choices encompass choices at the SEA level, and for various IA types: EcIA, SIA, HIA, and SA. We give particular consideration to the challenging task of providing good practice SEA guidance.

- The analysis begins in Section 2.2 with the problem, which is the need for better IA process management. The desire is to reduce the incidence and severity of recurrent process-related problems, through both regulatory guidance and applied practice. The approach taken to address the problem is described and substantiated.
- In Section 2.3 we provide an overview of generic EIA choices. In Section 2.3.1 we address regulatory level choices. The analysis encompasses screening guidance, guidance for individual activities, and integration and coordination guidance. In Section 2.3.2 we describe, integrate, and enhance the EIA process as conventionally portrayed in IA literature. This analysis encompasses general process management, IA process inputs, outputs, and links, and IA process adaptations.
- In Section 2.4 we present regulatory and IA process design choices and characteristics for other IA types (EcIA, SIA, HIA, SEA, SA).
- In Section 2.5 we address the contemporary challenge of SEA good practice guidance. An overview of suggested good practices is presented. Some of the issues and dilemmas associated with providing good practice SEA guidance for an IA subfield that encompasses such a diverse array of proposal types, context types, and conceptualizations of SEA, as it is and as it should be, are then discussed.
- In Section 2.6 we describe the major insights and lessons derived from the analysis, including the potential for conventional IA regulatory and applied processes to adequately address the recurrent problems and contemporary challenges.

2.2 DEFINING THE PROBLEM AND DECIDING ON A DIRECTION

The problem in this chapter is twofold. First, there is the question of whether conventional characterizations of the IA process, as typically portrayed in IA literature, adequately reflect the choices available. Second, there is the question of whether conventional IA process guidance and practice, even if substantially reformed, can adequately respond to the recurrent problems and contemporary challenges identified in Chapter 1. The direction is, first, to present a range of available choices at the regulatory level; second, to present a range of available choices at the applied level; and, third, to assess whether those choices can provide a comprehensive response to the recurrent problems and contemporary challenges.

2.2.1 Regulatory Level

The point of departure for the regulatory analysis is IA process control and guidance. The focus is on regulatory guidance of the IA process as it is (in terms of broad patterns), as it could be (in terms of good practices), and as it could be (if reconstructed and reformed). The focus is on helping IA regulators more effectively guide IA process management.

There is a tendency in IA literature to focus on the details of individual IA regulatory systems. IA regulators must do so if they are to effectively administer the system. IA practitioners must do so if they are to effectively and efficiently operate within the system. There are numerous texts, which seek to aid practitioners in understanding and operating within individual IA systems, all of which is well and good. However, sometimes a broader perspective is required.

IA regulatory systems are not static. Reforms, refinements, and modifications are commonplace. Sometimes change takes the form of a fundamental restructuring of or replacement of IA and related legislation. IA regulators, while administering the system as it is, must determine how the system can be improved either by modifications and refinements or by more fundamental changes. If the benchmark is simply the system as it is, the basic questions that lead to more fundamental changes may never or rarely be asked. Individual modifications and refinements may,

Impact Assessment: Practical Solutions to Recurrent Problems and Contemporary Challenges, Second Edition. David P. Lawrence.

moreover, when aggregated, result in an inefficient and ineffective patchwork of reforms and existing practices. Individual elements may operate at cross-purposes. Significant flaws, gaps, and inconsistencies may (and based on Chapter 1 do) remain. Even if the existing system is not replaced, the repairs should be based on basic principles and objectives, an appreciation of practical alternative approaches and experiences elsewhere, a systematic evaluation of the existing system, and a coherent strategy that builds on strengths and ameliorates weaknesses.

Insights can be acquired from reviewing the requirements and experiences of other jurisdictions, appreciating the need for contextual adjustments. However, whether the intent is to replace, reform, or just "fine-tune" IA requirements and procedures, the ultimate benchmark should be "good regulatory practice."

IA legislation and regulations should spell out both aspirations (e.g., goals, objectives, principles, policies, priorities) and minimum requirements (e.g., thresholds, standards, criteria, areas of application, roles and responsibilities). IA legislation and regulations should result in a consistent and acceptable level of IA practice. Ideally, the gap between IA aspirations and requirements should be narrow. IA guidelines can further diminish that gap by facilitating compliance and by contributing to the quality of IA documents and to the effectiveness of the IA process. IA guidelines have the additional benefit of flexibility. They can be adapted as the state of the art and practice of IA evolves. They also can be adjusted for different setting and proposal types, and for individual applications. A delicate balancing act is required. Too general IA legislation and regulations will contribute to a low, or at least highly inconsistent, level of IA practice. IA legislation and requirements, which "micromanage" every aspect of IA practice, are likely to stifle innovation and inhibit necessary adaptations.

2.2.2 Applied Level

Identifying IA regulatory choices is not enough. There is considerable discretion, within the framework established by IA requirements and guidelines, to effectively and efficiently (or not) manage the IA process. Moreover, IA requirements and guidelines do not and cannot adequately convey the many process management choices potentially available to IA practitioners.

It also is not enough, once IA requirements are satisfied, to simply apply "the IA process" as presented in any one of several IA texts. Although these process depictions can be very helpful, numerous versions of the IA process are presented. Most IA process descriptions are not identified as one among many contributions to an ongoing debate and discussion. Alternative processes and process variations are rarely described, compared, or critically evaluated. Instead, it is commonly assumed that there is only one IA process or that the process presented is superior to other available processes. Occasionally, allowance is made for individual process variations. Usually, little importance is attached to these variations. The IA practitioner is left in the difficult position of designing and managing the IA process, without the benefit of an array of readily accessible process guidance choices that can be integrated, combined, adapted, and applied to suit the circumstances.

The applied analysis describes the broad patterns in how the IA process is conventionally portrayed in IA literature. It then integrates and enhances the IA process characterizations, with a particular emphasis on identifying choices available to IA process managers at the SEA and project EIA levels and for various IA types. Finally, it identifies residual gaps and priorities (with special reference to the recurrent problems and contemporary challenges) to be pursued in subsequent chapters.

Many perspectives on how the IA process should be designed and managed are integrated into the analysis. Various ways of framing, identifying, and structuring IA process activities and activity components are considered. Particular consideration is given to blending and integrating procedural and substantive elements in IA process design and management for various IA types. Procedures for integrating key inputs into the IA process, such as IA requirements, public and agency concerns, substantive environmental priorities, methods and pertinent values, knowledge and experiences, are addressed. Key IA process outputs are identified. Consideration is given to how to establish links to proposal planning, to decision making, and to related actions and fields. Process management choices are consolidated in tabular form. Several tables and figures are presented that extend beyond conventional IA process portrayals. A separate analysis is presented of approaches for confronting the particularly difficult challenge of SEA good practice guidance.

The applied analysis is based on an overview of major IA texts and mainstream IA journals. It focuses on prescriptive portrayals of the IA process. A broader range of IA literature and literature in related fields is considered in the Chapters 3–12 analyses. The emphasis is less on comparing the IA process depictions than on identifying and illustrating the range of conventional process management choices available to IA practitioners at both the regulatory and applied levels. The integration and refinement of conventional IA process portrayals provide the baseline for assessing whether the recurrent problems and contemporary challenges cited in Chapter 1 require further consideration. The specific nature of the recurrent problems and contemporary challenges, and how they might best be addressed, is explored in greater detail in Chapters 3–12.

2.3 CONVENTIONAL EIA CHOICES

2.3.1 Regulatory Level Choices

The first step in designing or adapting any project-level EIA regulatory system is determining (1) what should trigger the application of EIA requirements and (2) which particular set

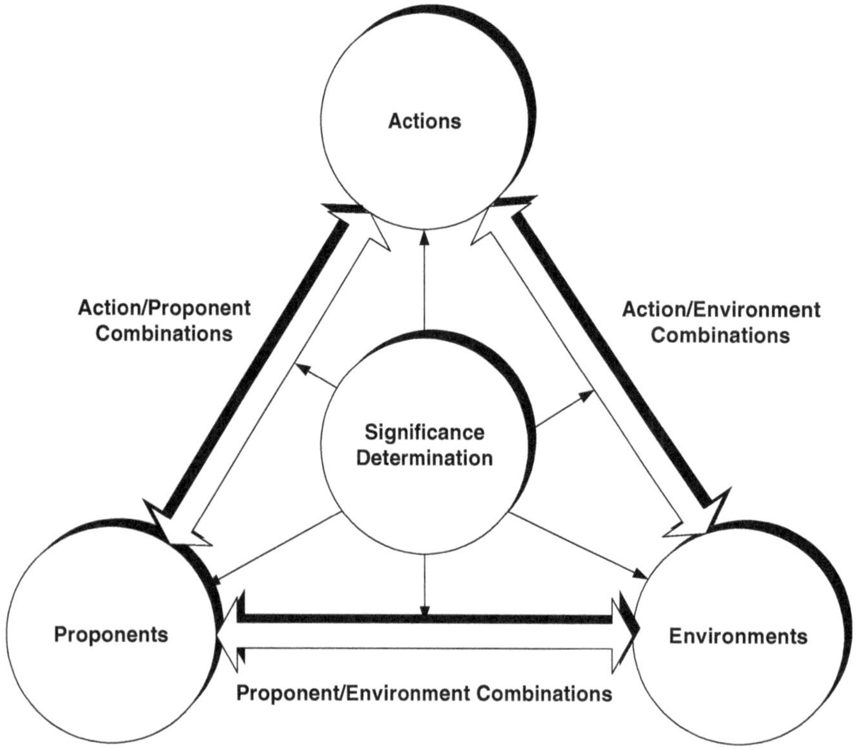

Figure 2.1 EIA screening combinations.

of EIA requirements should be applied. These two screening steps, as highlighted in Figure 2.1, focus on various actions (what), proponents (who), and environments (where). Often screening decisions are based on action, proponent, and environmental combinations. Each decision involves a significance determination (i.e., has a level of importance been reached where EIA requirements should be instituted?). Added to the mix is an effort to achieve two not always complementary purposes (1) building environmental considerations into proponent and action-related decision making and (2) protecting and enhancing the environment. Table 2.1 lists examples of good practice screening characteristics for each of the three elements (individually and collectively) and for significance determination.

Regulating and guiding the EIA process does not end with screening and significance determination. It also is necessary to control and guide individual IA activities. Table 2.2 lists examples of good regulatory practice, general and specific (for each EIA activity), characteristics. Overall guidance, across EIA process activities, necessitates a delicate balancing act. Ideally, EIA requirements (legislation and regulations) should identify objectives, spell out minimum requirements, and include general performance standards or criteria. Guidelines can then provide more specific guidance. Successively more specific guidance can be provided, first, through environment and proposal type guidelines and, second, through proposal and environment-specific requirements and guidelines. Guidance for individual activities can be offered through both general guides (that address all EIA activities but at a broad level of detail) and specific guides (that provide more detailed guidance for individual activities). Guidelines that are too superficial are likely to be of little value in facilitating good EIA practice. Overly specific requirements and guidelines can inhibit good practice innovations and adaptations, especially when there are multiple and changing perspectives regarding good practice standards and methods. General EIA process guides are helpful because they provide an overview of all EIA activities at a consistent level of detail. They also can give the reader a sense of how the individual activities fit together into an overall process.

It is not sufficient to implement EIA requirements and prepare IA guidelines in a manner consistent with the criteria listed in Table 2.2. While EIA continues to evolve rapidly, as a field of theory and practice, sufficient knowledge and experience have been acquired to be able to distinguish between good and inadequate practice. EIA guidelines that systematically draw upon applied research and case studies can identify reasonable minimum standards coupled with good practice performance standards. Most EIA jurisdictions, for example, have prepared public involvement guidelines. Such guidelines tend to summarize regulatory requirements, identify a few general principles, and provide an overview of the characteristics, strengths, and limitations of a standard set of consultation procedures. This often dated and largely descriptive approach fails to convey a state of practice that is much further along in providing for earlier and more extended public involvement, for sharing decision making, for consensus building and conflict resolution, and for facilitating the involvement of traditionally underrepresented groups and organizations. A more concerted effort could be made to formulate and refine EIA requirements and

Table 2.1 Examples of Good Practice Screening Characteristics

Proponents	Actions	Environments	Interactions	Significance Determination
Applies to public and private proponents	Automatic triggers for major actions that almost always have significant environmental effects	Automatic rejection if unacceptable environments	(A-E) action thresholds adjusted to significance/sensitivity of environment	Significance defined
Exclusion of proponents that do not perform environmentally significant actions	Triggers for proposal characteristics likely to induce significant effects (e.g., emissions, waste disposal)	Requirements adjusted depending on environmental and resource carrying and regenerative capability	(A-E) action thresholds adjusted where cumulative effects a concern	Provision for early exclusion of unacceptable effects
Funding and approval triggers early in process and consistent	Exclusion of actions with minimal adverse effects	Broad definition of environment and effects	(A-E) provisions to group proposals geographically (e.g., mineral exploration)	Well-defined significance thresholds
Scope of funding and approval triggers consistent with scope of likely effects	Adjustment of requirements for different action types and characteristics	Provisions for extraterritorial application	(A-E) focuses requirements on action characteristics likely to significantly affect sensitive/significant environmental components, processes, interactions, species, and habitats	Provisions for degrees of significance (scaling systems)
Triggers adapted to proponent characteristics (e.g., agency characteristics, indigenous EIA regimes)	SEA-project EIA links (e.g., tiering)	Provisions to apply to global commons (e.g., Antarctic, oceans)	(P-E) requirements and guidelines focus on proponent characteristics most likely to significantly affect environment	Explicit significance criteria
Provisions to group agencies (e.g., coproponents)	Action type guidelines (e.g., best practice)	Effect type requirements and guidelines (e.g., biodiversity, social, health, environmental justice)	(P-E) requirements and guidelines focus on locations and location types favored by proponents	Explicit and consistent significance determination process
Provisions to coordinate levels of government (e.g., federal, state/provincial/territorial, municipal, indigenous)	Individual proposal requirements/guidelines (e.g., terms of reference)	Transboundary effect requirements	(P-A) requirements and guidelines focus on actions undertaken by proponents that are most likely to result in significant effects	Significance determinations adjusted based on consideration of; magnitude/intensity, mitigation and enhancement potential, availability of reasonable alternatives, risks and uncertainty, public policies, requirements and priorities, degree of controversy, reversibility, carrying capacity, sensitivity, and significance of potentially affected environment and context
Provisions to prevent and minimize duplication and overlap (e.g., delegation, harmonization)	Provisions to group actions by type (e.g., site remediation, weapons decommissioning)	Provisions for areawide assessments	(P-E) requirements and guidelines seek to integrate environmental considerations into all proponents' actions	Recognition that significance can be positive or negative
Voluntary participation provisions	Provisions to combine related actions (e.g., precedent setting)	Setting type requirements/guidelines		Recognition that individually insignificant actions can be cumulatively significant
Consideration of proponent's environmental record	Selective application for expansions to existing projects where significant effects likely	Adjustments in requirements to match regional characteristics		Provision for stakeholder involvement in significance determination
Individual agency requirements/guidelines	Provisions to apply to existing action (with consent of proponent)	Regional guidelines		Significance determination integrated into each IA process step
Assessment of privatization implications	Small project guidance	Special requirements for specific sensitive/significant/hazardous environments		Provision for integration of significance determination methods
Guidance for business (e.g., links to environmental management systems)		Endangered/threatened/rare species and habitat triggers		
		Contaminated site guidance		
		Special provisions for culturally sensitive lands (e.g., archaeologically significance, indigenous peoples' reserves, land claims)		

Table 2.2 Examples of Good EIA Regulatory Practice: Individual Process Activities

General Characteristics
Objectives, minimum requirements, and performance standards/criteria included in legislation or regulations
Identifies activity objectives and principles
Defines key terms
Addresses methodological issues (e.g., level of detail)
Describes the process for undertaking the activity, including possibility of alternative approaches
Provides examples of the role of methods within the activity
Describes potential stakeholder roles and responsibilities within the activity
Identifies links to EIA regulatory requirements and to related activities
Requires consideration of uncertainties and associated implications
Provides good practice examples and case studies
Identifies potential pitfalls and obstacles
Identifies follow-up references and sources
Provides more specific guidance (e.g., environment types, proposal types, effects type, area specific, proposal specific)
Maintains a balance between good practice control/guidance and ensuring sufficient flexibility to apply alternative approaches, innovate, and make necessary adaptations to suit local and proposal specific conditions
Scoping
Provides for scoping as a formal decision-making step in the EIA process (e.g., approval of terms of reference, potential for proposal rejection)
Identifies the role of scoping (e.g., focusing) in each EIA process activity
Provides for the scoping of significant environmental components and processes, data sources, effects, issues, alternatives, proposal characteristics, stakeholders, uncertainties, and proponent characteristics
Ensures sufficient flexibility to adjust process after scoping
Proposal Characteristics
Focuses on proposal characteristics most likely to induce significant environmental effects
Identifies minimum information requirements for proposal characteristics (e.g., status, location, scale, stages, service, land and resource requirements, components, processes, design, emissions, effluents, residuals, and interactions among proposal characteristics)
Provides for links to alternatives, mitigation, land use planning, and other related proposals
Provides for early and ongoing links between proposal planning and EIA process
Recognizes that proposal characteristics will evolve and change
Baseline Analysis
Broad definition of environment
Requires justified boundaries for analysis (e.g., temporal, spatial, ecological, administrative)
Provides for the consideration of patterns over space and time (e.g., existing environmental degradation and hazards, environmental carrying capacity)
Facilitates focusing on sensitive and significant environmental components and processes most likely to be affected
Identifies potentially significant environmental components and processes
Facilitates consideration of links among environmental elements (e.g., physical, biological, ecological, social, economic)
Provides for links to impact prediction, monitoring, and state-of-the-environment reporting
Impact Analysis and Synthesis
Broad definition of effects
Provisions for characterizing impact dimensions (e.g., intensity, duration, frequency, reversibility, direct, indirect, and cumulative)
Impact identification and prediction guidance (including examples of methods)
Refined guidance for effect (e.g., biodiversity, social, cultural, noise, environmental quality, health) types
Explicit consideration of transboundary effects
Provisions for considering interactions among activities and effects
Linked to alternatives and mitigation analyses
Alternatives Analysis
Requires identification of purpose and need
Provides guidance for alternatives identification
Provides overview of alternatives generation and evaluation process (including possible stakeholder roles)
Identifies and defines alternatives that must be considered (e.g., no action, environmentally preferred, alternatives to proposal, alternative means, alternatives outside jurisdiction)
Indicates when alternatives must be considered (e.g., when potentially significant effects)
Identifies types of alternatives that could be considered depending on circumstances (e.g., siting)

Table 2.2 *(Continued)*

Identifies possible approaches to screening alternatives
Provides criteria examples
Points out need to consider differences in criteria importance
Identifies possible approaches and methods for comparing alternatives
Links alternatives analysis to scoping, significance interpretation, and mitigation
Mitigation and Enhancement
Broad definition (e.g., prevention, amelioration, rehabilitation, restoration, compensation, enhancement, local benefits)
Provides examples of typical methods
Requires consideration and documentation of mitigation measures when potentially significant effects
Requires consideration of feasibility, effectiveness, and consequences of methods
Linked to proposal characteristics, significance determination, monitoring (e.g., mitigation effectiveness), and legal requirements (e.g., compliance and enforcement)
Provisions to integrate individual measures into action plan
Methods
Guidance and examples for each process activity and for major environmental components and types of effects
Identifies characteristics, strengths, and limits of methods
Provisions for integration of traditional knowledge
Sponsoring of methods research, methods symposiums, and research institutions
Links to technical guides in related areas of jurisdiction and related fields
Documentation
Identifies documentation requirements for each IA decision-making step
Provides for interim documentation; encourages documentation that traces EIA process
Guidance: style, format, level of detail, length, cover sheet, contents, list of preparers, rationale for interpretations, conclusions and recommendations, treatment of uncertainties, summaries, use of graphics and mapping, cross-references to other documents, source and reference list, use of appendices, indexes and keywords, and electronic publishing standards
Guidance: document circulation procedures
Guidance: documentation of agency and public involvement (including treatment of comments and suggestions)
Guidance: appendices, draft, and final reports; supplemental studies
Contents guidance: notifications, project registry/referral forms, decision-making record, approval requirements, hearings record, and postapproval documents

guidelines that are conducive to EIA practice as it could and should be (and sometimes is) rather than as it too often is or was 10–20 years ago.

Occasionally, EIA requirements and guidelines are too precise. Such an approach is problematic. There are many procedures available, each with a different mix of advantages and disadvantages. Proponents need the flexibility to be able to select, integrate, and adapt methods, jointly with stakeholders, which are appropriate to local circumstances. Maintaining such flexibility does not preclude EIA requirements and guidelines, spelling out objectives, principles, and performance standards. Examples of methods, including their strengths and limitations, criteria and decision rules for method selection in varying settings, and good and bad practice examples also can be provided.

A similar danger exists with EIA document requirements. Document requirements can be extremely helpful in facilitating consistency and in ensuring that minimum content requirements are satisfied. At some point, however, very detailed document format and content requirements can reduce EIA to a "fill-in-the-blank cookbook" exercise. The focus should not be exclusively on preparing IA documents in accordance with requirements. The primary emphasis should remain on protecting and enhancing the environment and on facilitating more environmentally sound decision making and undertakings. There also is the question of whether the EIA process suffers when too much stress is placed on document preparation requirements. EIA documents should be outputs from and should reflect the EIA process structure. EIA document requirements are worthwhile, but only if they reinforce EIA objectives and do not inhibit innovative process design and management approaches that seek to better achieve EIA objectives.

Integration and coordination are central attributes of regulatory EIA process management (Lawrence, 1994). Figure 2.2 illustrates interconnections among various integration and coordination categories. Table 2.3 lists examples of good regulatory practices pertaining to integration and coordination with international EIA activities, vertical integration and coordination, horizontal integration and coordination, and knowledge base links.

2.3.2 Applied Level Choices

Figure 2.3 categorizes EIA process management choices. Decisions can be made regarding the appropriate activities and activity components (including choices regarding activity sequence, form, frequency, duration, and interactions),

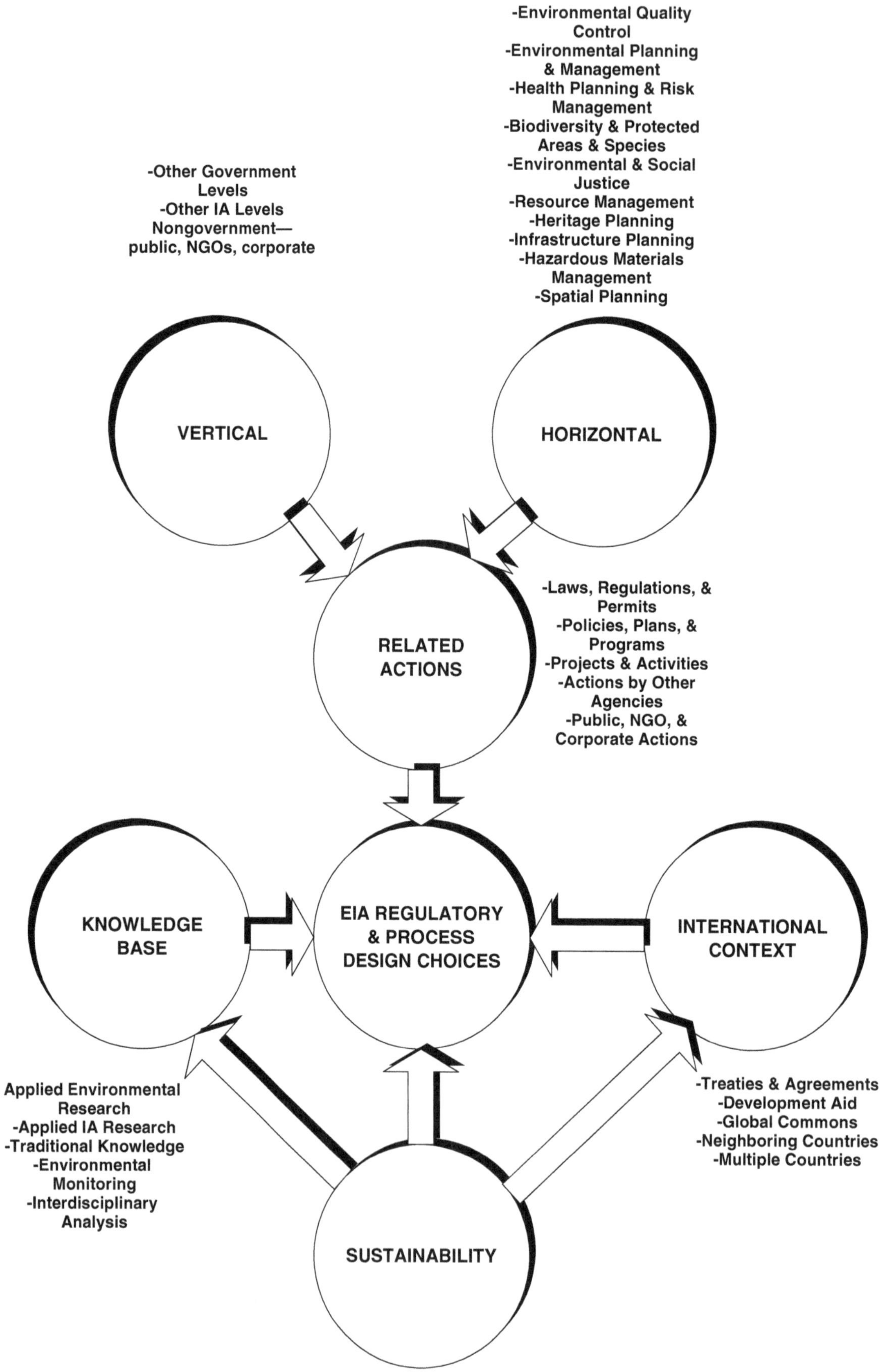

Figure 2.2 EIA integration and coordination.

Table 2.3 Examples of Good Regulatory Practice: EIA Integration and Coordination

International—Global Commons	Vertical—Other Government Levels	Horizontal—Related Policies, Programs, and Plans	Knowledge Base—Applied Research
Extends EIA requirements to actions by nations that might affect global commons Extends EIA requirements to boundary areas of commons Coordinates EIA requirements and strategies for global commons Integrates global commons concerns (e.g., climate change, rare and endangered species) Integrates international conventions for protection of commons into EIA requirements	Minimizes EIA duplication and overlap among government levels (e.g., accreditation) Integrates approval requirements of all government levels in EIAs Harmonizes EIA requirements and procedures (to ensure a consistently high level of EIA practice and to address potential conflicts) Provides for joint or cooperative EIAs Coordinates EIA procedures and activities Provides for Indigenous EIA regimes (establishment and coordination) EIA guidelines and assistance for lower government levels	Provides for and encourages areawide assessments Undertakes SEAs of regional and sectoral plans Establishes EIA systems based on natural boundaries (e.g., watersheds) Integration of EIAs within multilevel sectoral and spatial planning Integrates EIA requirements into regional and land use planning Systematically and formally integrates into EIA requirements sustainability and other substantive environmental concerns	Establishes and supports EIA research centers Sponsors EIA research by universities, research institutions, and consultants Coordinates EIA research among jurisdictions Provides for case studies of good EIA practice
International—Multiple Countries	**Vertical—Other IA Types**	**Horizontal—Related Projects and Activities**	**Knowledge Base—Traditional Knowledge**
Harmonizes EIA requirements among countries Harmonizes environmental legislation, strategies, and plans that provide context for and substantive content to EIAs Facilitates multijurisdictional EIA coordination, consultation, and planning Extends applications to actions by nationals in other countries Integrates international EIA and other environmental conventions and agreements Assesses environmental impacts of trade agreements	SEAs (policies, program) frame project-level EIA Tiering provisions (e.g., project-level EIAs summarize and incorporate by reference higher level SEA documents) Legislative requirements for both SEAs and project-level EIAs (recognizing differences) Provides for IAs of legislative proposals	Includes cumulative effects requirements Provides cumulative effects guidance Provides for regional environmental databases to support cumulative effects assessment	Formally recognizes in EIA legislation potential value of traditional knowledge to EIA practice Integrates traditional knowledge into EIA practice

(continued)

Table 2.3 (*Continued*)

International—Neighboring Countries	Government/Nongovernment	Horizontal—Actions by Other Government Departments and Agencies	Knowledge Base—Environmental Monitoring
Transboundary EIA agreements and coordination Adherence to transboundary EIA conventions Conducts EIAs for projects in neighboring jurisdictions Assesses environmental impacts of trade agreements	Links and integrates EIA and EMS (environmental management systems) EIA application to public and private projects (recognizing differences) Provides forums and mechanisms (e.g., round tables, committees) for multistakeholder EIA dialogue Provides for voluntary EIA preparation Assessment of EIA implications of privatization Provides for research, coordination, and auditing roles by independent nongovernment organizations Considers environmental record of proponent	Provides for interagency coordination agreements (to address timetables, disputes, etc.) Provides for forums (e.g., task forces, committees) to facilitate coordination Integrates (where appropriate) and coordinates IA and permitting Includes well-defined interagency EIA circulation, review and consultation requirements, procedures, and mechanisms (e.g., review teams)	Integrates EIA monitoring into areawide monitoring Integrates results from areawide monitoring into EIA practice Provides for cumulative effects monitoring Links EIA outputs to state-of-the-environment and environmental pressure indicators, where practical
International—Development Aid	**Horizontal—Related Laws, Regulations, and Permits**	**Knowledge Base—Applied Environmental Research**	**Knowledge Base—Interdisciplinary Analysis**
Applies, with adjustments, EIA requirements to development assistance undertakings Assists in establishing and reforming EIA systems including capacity building	Merges EIA and other environmental regulatory requirements (where appropriate); formally linked and coordinated; guidance provided Formally incorporates substantive environmental requirements into baseline and significance determination IA activities Analyzes related requirements to ensure complementary Brings EIA requirements under the umbrella of sustainability requirements	Provides mechanisms (e.g., networks, committees) to integrate relevant applied research into EIA requirements and guidelines Initiatives to sponsor applied environmental research of relevance to IA practice	Emphasizes EIA requirements and guidelines on importance of transcending disciplinary boundaries Sponsors interdisciplinary EIA research

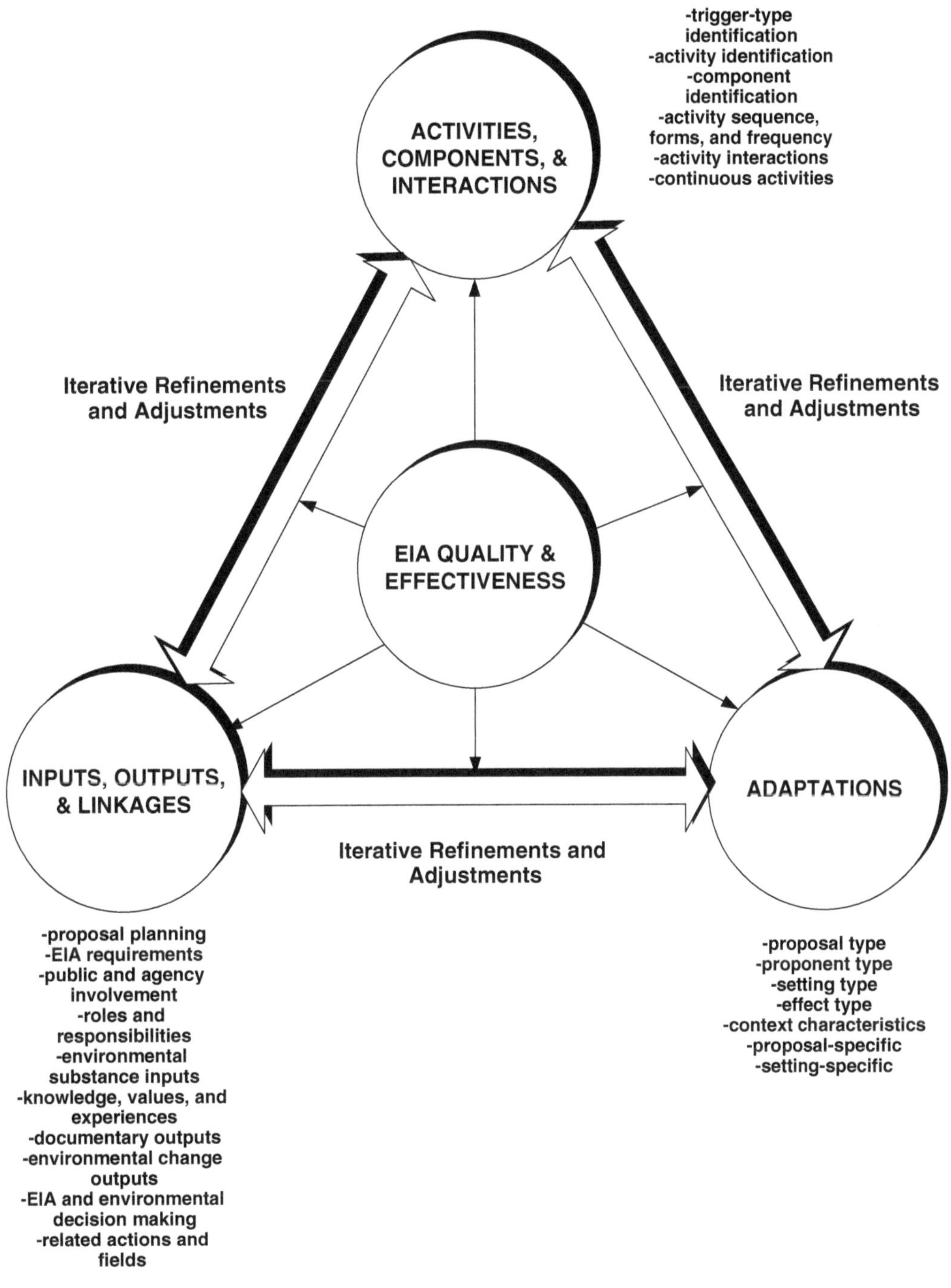

Figure 2.3 EIA process management elements. Adapted from Lawrence (2001).

inputs to the process, outputs from the process, and links between the EIA process and decision making and related activities and fields, adapting the process for different proposal, proponent, effect and setting types and to match proposal and effect specific characteristics, and how to build into process management insights and lessons from EIA quality and effectiveness analyses. The choices presented are far from definitive. However, they do offer a cross section of possibilities.

Table 2.4 identifies examples of EIA process design choices. Figure 2.4 illustrates a basic EIA process. Figure 2.5 delineates a more complex EIA process. Figure 2.6 displays a more iterative EIA process. There are many choices available regarding interconnection roles in the EIA process. Table 2.5 provides examples of activity frequency and sequence choices. EIA assumes many forms and includes numerous subfields. One size does not fit all for EIA process management. Adaptations are likely to be necessary.

Table 2.4 Examples of EIA Process Design Choices

General EIA Process Design Choices					
Choice Types	Examples of Choices				
Trigger type	Proposal (Figure 2.4)	Ends (Figure 2.5)	Setting (Figure 2.6)	Contingency model	Class or categorical assessment
Range of activities and components	Narrow (Figure 2.4)	Front-end elaborations (e.g., screening, scoping, ends) (Figure 2.5)	Mid-process elaborations (e.g., impact identification, impact prediction) (Figure 2.6)	Process extension (e.g., monitoring, management, auditing) (Figures 2.5 and 2.6)	Combinations (Figures 2.5 and 2.6)
Treatment of alternatives	No alternatives	Alternatives in general (Figures 2.4 and 2.5)	Alternatives by type (e.g., alternative means, alternatives to, alternative locations) (Figure 2.6)	Alternative combinations	–
Treatment of proposal characteristics	Determined from outset (Figure 2.4)	Concept and refinement (Figure 2.5)	Progressive refinement (Figure 2.6)	Integrated with proposal planning	–
Sequence and frequency	Linear (Figure 2.4)	Linear and activities in different forms (Figure 2.6)	Nonlinear and activities in single form (Figure 2.5)	Nonlinear and activities in different forms	Continuous activities (Figures 2.6)
Range of interactions	None	Limited interactions (Figure 2.4)	Extensive interactions (Figures 2.5 and 2.6)	Iterative (Figure 2.6)	Restructured (Figure 2.6)

Examples of Input, Output, and Linkage EIA Process Design Choices					
Choice Types	Examples of Choices				
Proposal planning links	No links to proposal planning (Figure 2.4)	Periodic inputs from proposal planning (Figures 2.5 and 2.6)	General recognition of ongoing two-way links (Figure 2.6)	Specific periodic two-way links	EIA and proposal planning fully integrated
EIA requirements inputs	No connections specified	Identified as a general input (Figure 2.6)	Linked at major decision points	All connections specified	Fully integrated
Public and agency involvement provisions	None	Limited and late (Figure 2.4)	Major decision points (Figure 2.5)	Continuous involvement provisions (Figure 2.6)	Continuous involvement with consensus building and conflict resolution provisions (Figure 2.6)
Roles and responsibility inputs	Roles and responsibilities unspecified	Study team roles specified (Figure 2.6)	Study team and reviewers roles specified	Roles for all participants specified (Figure 2.6)	Peer review provisions (Figure 2.6)
Environmental substance inputs	No connection acknowledged (Figures 2.4 and 2.5)	Identified as a general input	Specific categories of substantive environmental concerns identified (Figure 2.6)	Environmental concerns fully integrated (Figure 2.6)	–
Knowledge, values, and experience inputs	Not acknowledged (Figures 2.4 and 2.5)	Need to integrate values acknowledged (Figure 2.6)	Values integrated as ends	Need for applied research recognized (Figure 2.6)	Need to integrate knowledge, values, and experiences into each IA activity recognized
Methods	Not acknowledged (Figures 2.4 and 2.5)	Identified as a general input	General roles of methods in process specified	Selective role of methods identified (e.g., applied research, peer review) (Figure 2.6)	Role of methods in each process activity specified

Documentary inputs and outputs	Final report only (Figure 2.4)	Draft and final reports (Figure 2.5)	Draft, final, and multiple interim reports (Figure 2.6)	Documentation inputs and outputs as a continuous activity	–
Environmental outputs	Documents assumed to be only outputs (Figure 2.4)	Recognized through impact management (Figure 2.5)	Recognized through impact management and auditing (Figure 2.6)	Explicit links to environmental quality (Figure 2.5 and 2.6)	–
Decision-making links	No connections	Approvals at end (Figure 2.4)	Approvals at end with conditions and follow-up (Figure 2.5)	Staged approval	Multiple decision points specified
Related environmental decisions	No connections (Figures 2.4 and 2.5)	General recognition of need to link to other decision-making levels (Figure 2.6)	General recognition of need to link to related decisions (Figure 2.6)	Specific links to other decision-making levels	EIA integrated into related decisions
Related fields	No connections (Figures 2.4 and 2.5)	General recognition of need to link to related fields (Figure 2.6)	Specific links between process and related fields	IA and related fields integrated within broader systems framework	Sustainability assessment
Related activities	No connections (Figure 2.5)	Addressed through cumulative effects assessment (Figure 2.5)	Addressed through comparable activity review (Figure 2.6)	Addressed through comparable setting review	Combinations

Examples of Adaptation EIA Process Design Choices

Choice Types	Examples of Choices				
Effects type	No recognition of need to adapt process (Figure 2.4)	General recognition of need to adapt process (Figure 2.5)	Recognition that effects types could shape EIA process	Recognition of specific effects types	Specific EIA process modifications for effect types
Proposal type	No recognition of need to adapt EIA process for different proposal types (Figures 2.4–2.6)	General recognition of need to adapt EIA process for different proposal types	Recognition that proposal types could shape EIA process	Recognition of specific proposal types	Specific EIA process modifications for different proposal types
Setting and context type	No recognition of need to adapt EIA process for different setting types (Figures 2.4–2.6)	General recognition of need to adapt EIA process for different settings and contexts	Recognition that settings and contexts could shape EIA process	Recognition of specific settings and contexts	Specific EIA process modifications for different settings and contexts
Proponent type	No recognition of need to adapt EIA process for different proponent types (Figures 2.4–2.6)	General recognition of need to adapt EIA process for different proponent types	Recognition that proponent type differences could shape EIA process	Recognition of specific proponent types (e.g., private sector, indigenous communities)	Specific EIA process modifications for different proponent types
Quality and effectiveness provisions	No provisions	General auditing provisions (Figure 2.6)	Specific document quality provisions	Specific process effectiveness provisions	Specific quality and effectiveness provisions (Figure 2.6)
Overlaps and interconnections	No provisions	Need to consider overlaps and interconnections acknowledged (Figure 2.6)	Specific interconnections and areas of overlap identified	Integrative frameworks within EIA (including process implications)	Integrative frameworks include EIA and related forms of environmental management

Source: Adapted from Lawrence (2001).

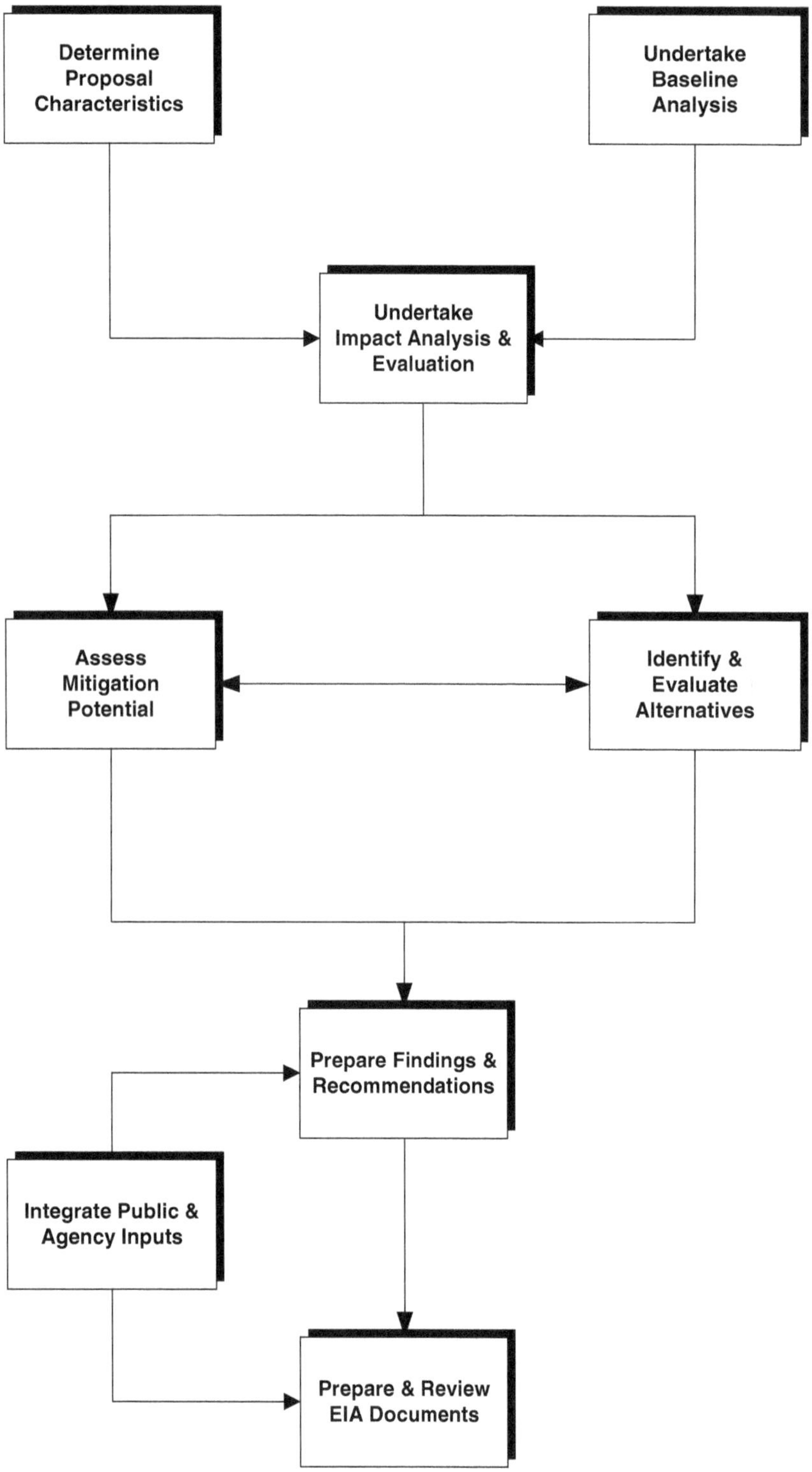

Figure 2.4 Basic EIA process. Adapted from Lawrence (2001).

Examples of EIA process design adaptation choices are listed in Table 2.4.

2.4 CHOICES FOR OTHER IA TYPES

The IA process varies by IA type. Many of the regulatory and applied level choices described in Section 2.3 can be adapted and applied for IA types. Much can be learned of value when designing and applying processes for other IA levels and types, from experiences at the project EIA level. There is, however, a danger in proceeding from the assumption that EIA requirements and processes represent the baseline for the formulation of requirements and processes for other IA types. Instead, there is likely to be value in considering separately requirements and guidelines for other IA types, both because of the differences among IA types

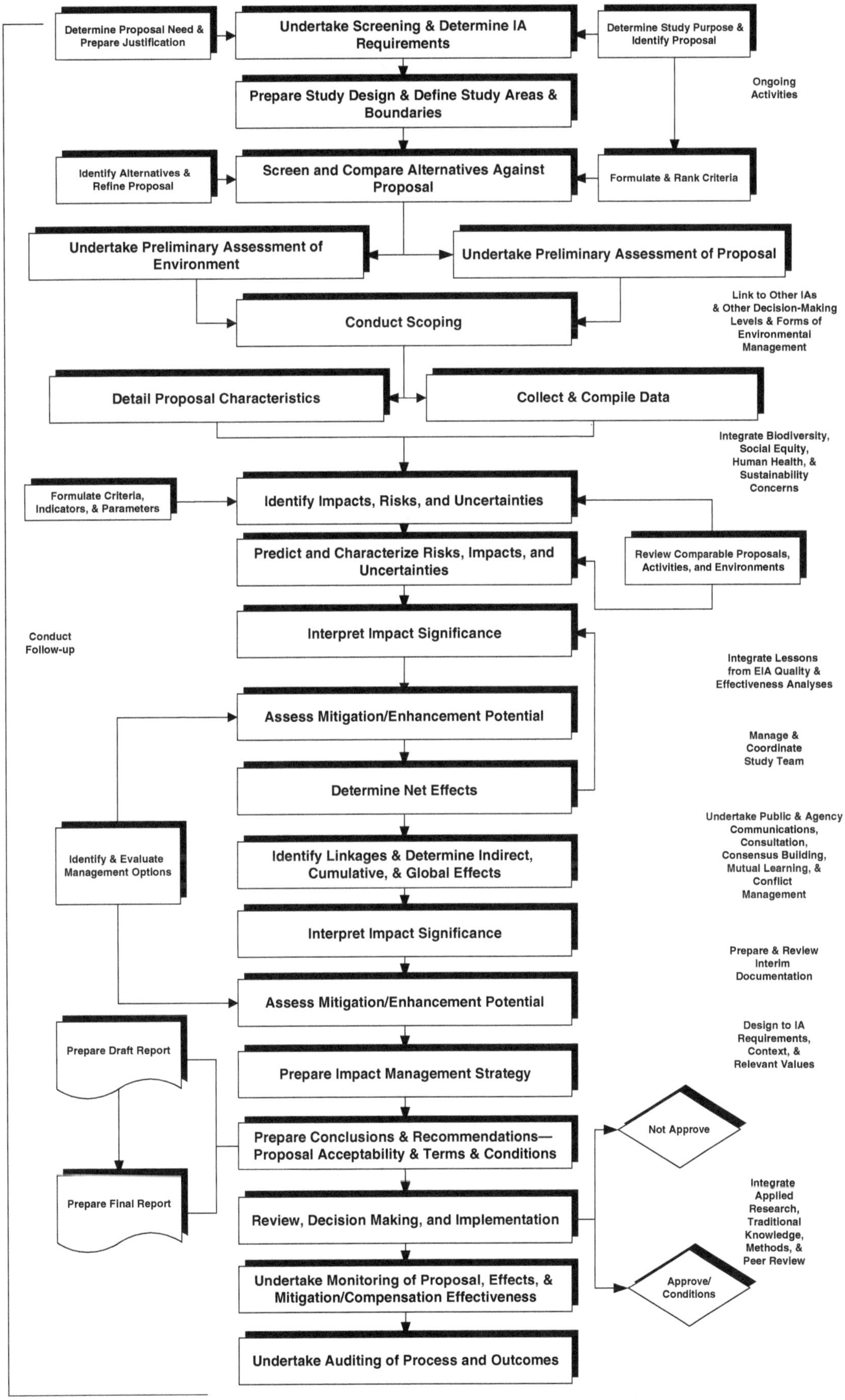

Figure 2.5 Example of a complex EIA process. Adapted from Lawrence (2001).

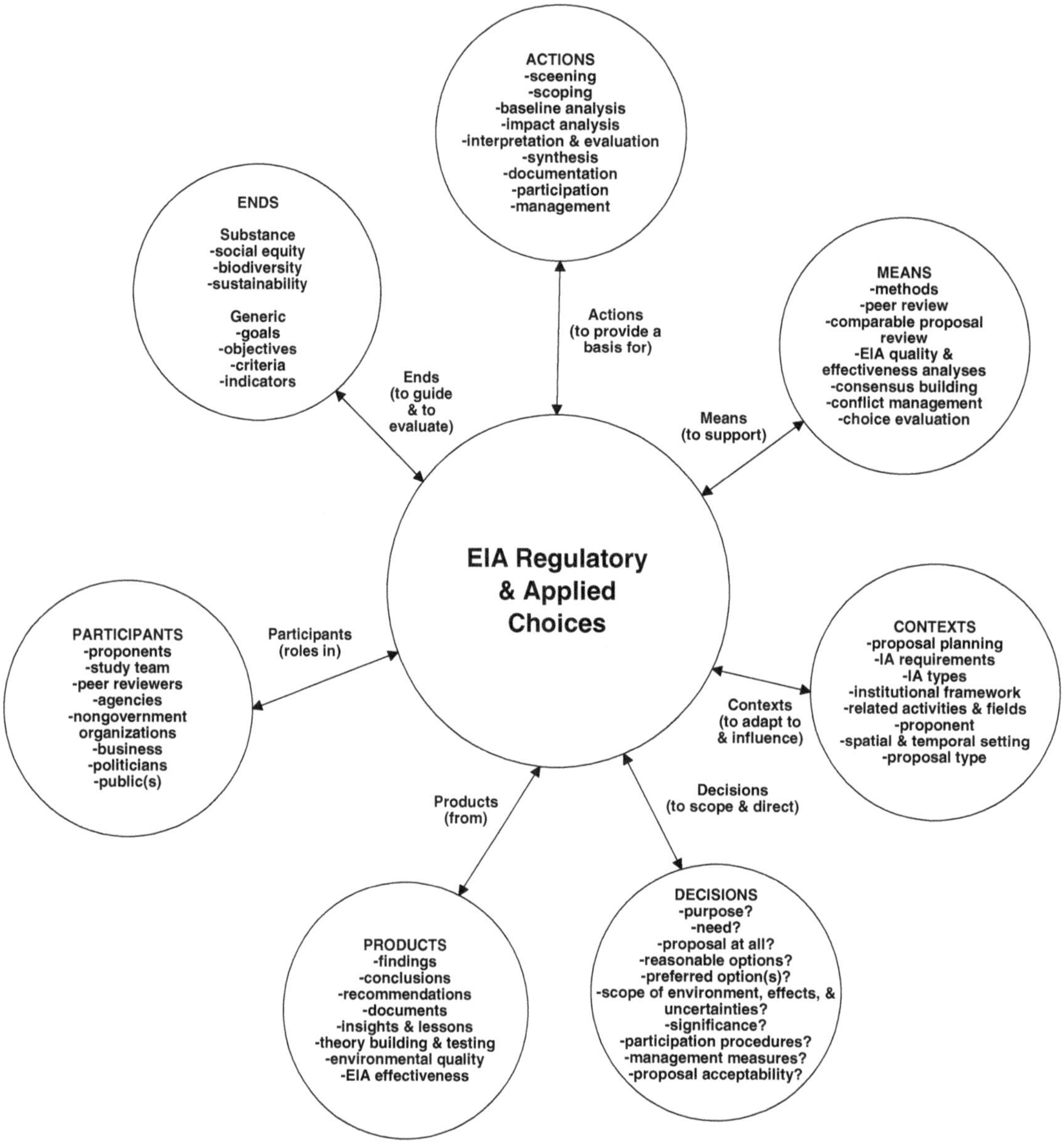

Figure 2.6 Reconstructed EIA process. Adapted from Lawrence (2001).

and because the historical evolution of EIA requirements and processes may have unduly narrowed and distorted the range of choices. At the same time, insights can be gained from EIA experiences (while appreciating the need for adaptation), and linking and integration possibilities among IA types at the regulatory and applied levels should be explored.

As illustrated in Figure 2.7, IA process design and management, for various IA types, is framed in both a conceptual and institutional sense. Process design for each IA type incorporates procedural and substantive elements. Each IA type may be fully integrated, partially integrated (at the SEA or EIA levels, by means of a tiered IA system, at one or more decision points), or not integrated with other IA types. Table 2.6 summarizes the conceptual framing sources, institutional framing choices, IA process types, procedural elements, and substantive elements for each of SEA, EcIA, SIA, HIA, and SA.

2.4.1 Strategic Environmental Assessment (SEA)

SEA, especially the early formulations, drew heavily upon project-level EIA. But it also shares the same conceptual roots as other related forms of public and private strategic policy analysis, planning, and management. More recent SEA characterizations have tended to favor more strategic, adaptive, and decision-centered procedural formulations. At the same time, in common with EIA, SEA is motivated by a desire to more effectively integrate substantive environmental concerns and priorities into decision making.

Table 2.5 Examples of EIA Activity: Frequency and Sequence

Activity	Sequence	Frequency
Screening	Commonly at the outset of process (requirements applied or not) At the end of process—proposal acceptability	Throughout process—application of exclusionary criteria to screen alternatives, environmental effects, effects and mitigation/management measures
Scoping	At the beginning after screening Sometimes after need and alternatives identification Selects issues, interests, major alternatives, environmental components, effects, and uncertainties to focus on	Throughout when treated as focusing or bounding At the outset of every activity Addresses what to be considered and to what level of detail
Proposal characteristics	Initially after screening but progressively refined Analysis can lead to reconsideration—sensitivity analyses	Successively greater levels for scoping, for various levels of alternatives analysis, for impact identification, prediction, and interpretation, for determining mitigation potential, for determining proposal acceptability, and for impact management
Baseline analysis	Initial environmental overview for screening and scoping Detailed environmental evaluation for impact prediction and interpretation	Occurs whenever supplementary environmental data incorporated into analysis—alternatives evaluation, mitigation measure determination, and in conjunction with monitoring and auditing
Impact analysis	Primarily, detailed impact identification and prediction	Also, part of screening, scoping, alternatives analyses, mitigation analysis, and in determining and applying monitoring measures
Interpretation and evaluation	Commonly refers to interpreting environmental components (e.g., valued environmental components) and impacts	Significance determinations also made during screening (need for EIA), scoping (major issues), proposal characteristics (most likely to induce impacts), alternatives analysis (to screen and compare alternatives), and impact management (preferred measures)
Synthesis	Rarely an explicit activity Implied in determining environmental system characteristics, in addressing impact interconnections and cumulative effects, in overall management strategy, and in consolidation of findings and recommendations	Synthesis, in common with analysis, a recursive activity that repeats through the EIA process
Alternatives	Characterization varies Sometimes excluded, at outset (near scoping—reasonable alternatives) or at end (as part of mitigation/management)	Recurrent activity—wherever there are choices prior to decision points (e.g., alternatives to, alternative means, baseline scoping, proposal characteristics, impacts, mitigation, management, synthesis, documentation)
Management	Mitigation in parallel with impact analysis Impact management strategy preapproval Monitoring and auditing after approval	Mitigation sometimes part of alternatives analysis Project management, an ongoing function Management aspects in conduct of all activities
Participation	Often limited to review of EIA documents and, sometimes, scoping	Sometimes integrated into alternatives assessment, significance interpretations, impact management, and proposal acceptability Arguably, should occur prior to all decisions
Documentation	Early characterizations limited to single EIA document More recently, also, pertains to scoping, to draft reports, and to postapproval documents	Sometimes additional documentation—technical support reports, documentation of public consultation activities, and summary reports Arguably, should parallel, in stages, EIA process
Decision making	Conventionally, screening (if and decision stream), scoping (what should IA encompass), and EIA review (acceptable, conditions)	

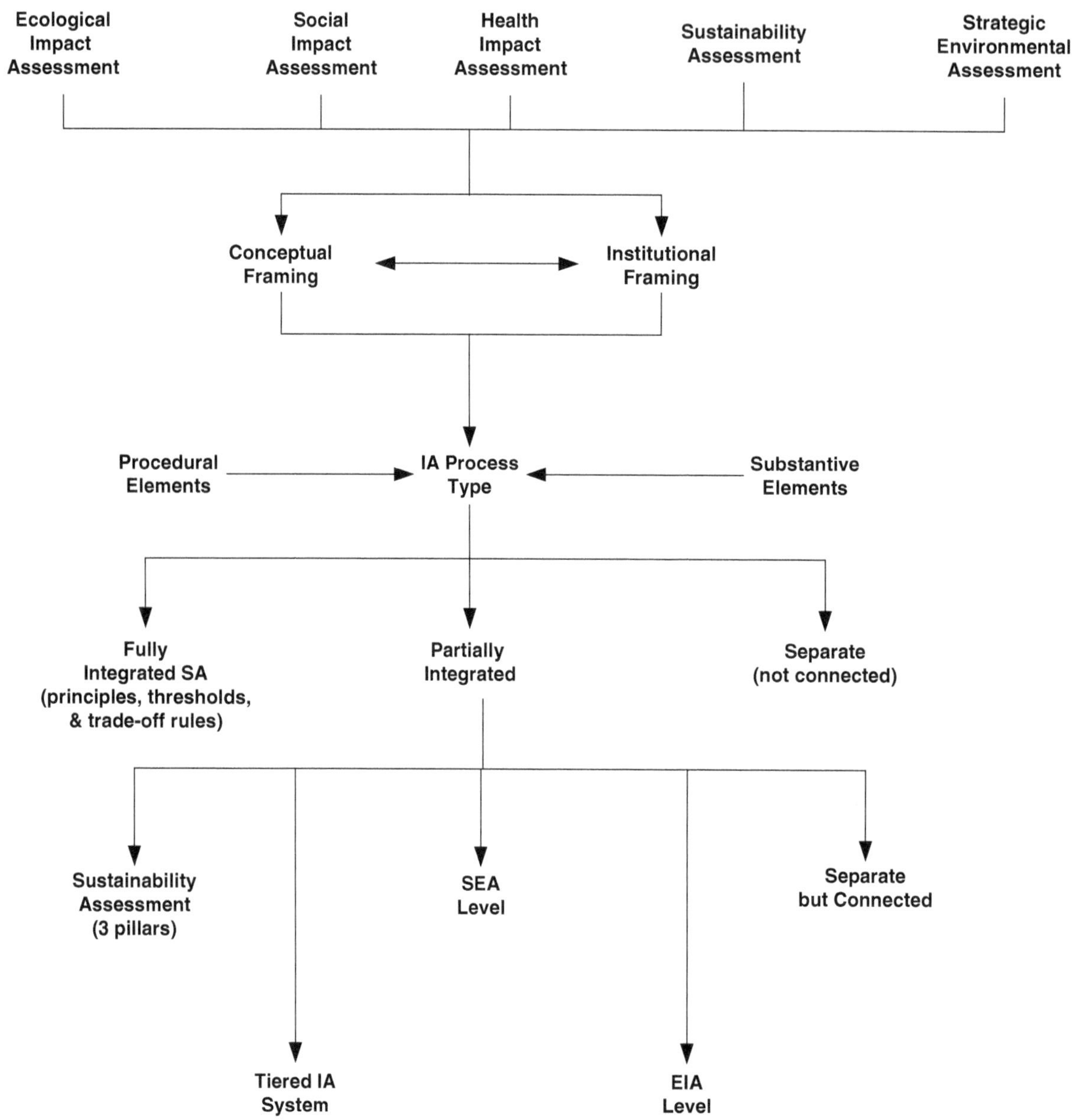

Figure 2.7 IA type: process choices and characteristics.

Characterizations of the relationship between SEA and decision making have ranged from fully integrated, through separate but with multiple connections, to separate and only connected prior to decision making. SEA institutional arrangements take a variety of forms. Examples include being largely subsumed under EIA requirements (as occurs in Australia and the United States), being separate but parallel to the EIA system with some similarities but also distinct differences (as occurs in Europe), and a fully separate and decidedly different system (as occurs in Canada). Varying SEA systems have been developed for public policy, legislation, regulations, trade agreements, plans, and programs, between the public and private sectors, and for varying scales of analysis (e.g., international, national, and regional). SEA institutional arrangements vary depending on the extent to which they are "action-forcing." They can involve, for example, mandatory, quasimandatory (administrative), or strictly advisory procedures. Interconnections within and among SEA and EIA institutional systems are often depicted, in theory, as a formal tiered structure with well-defined roles, and horizontal and vertical interconnections. In practice, the structure is usually more informal, incomplete, and with poorly or only partially defined roles and interconnections. SEA, because of its close connection to decision making, is inherently political.

A host of procedural choices have been advanced for SEA. Early formulations tended to favor effects-driven EIA models, with a strong emphasis on formality, rigor, rationality, and technical analysis. Lower levels of spatial and program planning often utilize elements of a rational–

Table 2.6 Other IA Types: Examples of Process Choices and Characteristics

	Conceptual Framing	Institutional Framing	IA Process Types	Procedural Elements	Substantive Elements
SEA	Policy analysis Strategic planning Strategic decision making Political science Philosophy Environmental management Feedback of knowledge and good practice Feedback from effectiveness assessments Environmental impact assessment	Parallel (linked at each step or only at end) or fully integrated into policy, plan, or program-making process Under EIA legislation, under separate administrative process, duel track SEA/EIA, and integrated policy/planning Sustainability appraisal Legislative or mandatory provision, quasimandatory (administration or directive), and advisory or policy provision Tiered SEA/EIA system Overlap SEA and EIA especially for large projects SEA with all elements integrated or non-SEA elements separate Public, semipublic, and private agencies International, national, and regional Inherently political	EIA based, EIA modified (separate elements), and SEA supersedes (broader appraisal) EIA based and decision-centered Top-down and bottom-up Policy, plan (sector, spatial, regional), and program SEAs; different levels of each Regulatory impact assessment (RIA), cabinet level SEA, strategic EIA, strategic environmental appraisal Technical–rational, strategic integrative, collaborative–communicative, and political–social struggle Rational–positivistic, procedural and transformative Regional (CEA oriented) (SEA or EIA driven) Equity oriented, community based, participative, value driven, and equity–justice Formal or informal Objectives led or baseline led Private sector versus public sector, administration led, analytical, by proponent, or by independent entity Combinations; a family of approaches	EIA based versus strategic action based Highly iterative process Importance of links to other strategic actions Sometimes includes alternative scenarios Emphasis on ameliorating the current limitations of the policy, plan-making process; need to influence not just decision making but also institutions and environmental governance mechanisms Process varies by context, also modifies context Collaboration, negotiation, and persuasion central to process Importance of organizational learning and capacity; often requires strengthening Generally characterized by high uncertainty levels; often not recognized	Based on holistic understanding of natural and social environment Accommodates physical, ecological, social, health, economic, and cultural issues Emphasis on sustainability framework and progress toward sustainability Importance of encompassing a range of temporal and spatial scales Cumulative effects cutting across SEA types

(*continued*)

Table 2.6 (*Continued*)

	Conceptual Framing	Institutional Framing	IA Process Types	Procedural Elements	Substantive Elements
EcIA	Natural sciences Ecological science Natural resource management Conservation planning Research methods, analytical techniques, ecological models, and concepts Sometimes linked to ecological sustainability Linked to good practices	Part of EIA or SEA or in support of Identification of laws and regulations related to issues Loop back from IA to legal, policy, and development control implications Informs decision making Link to capacity building and institutional arrangement reform	Biological IA Ecological IA Biodiversity IA Ecosystem approach Ecosystem services in IA Threatened and endangered species IA Ecological risk assessment Protected area IA Ecological evaluation for EIA Distinction among policy level, program/plan level, and project-level ecological IA Combinations	Largely EIA- and SEA-based steps Sometimes includes mitigation/enhancement options, goals, objectives, criteria, and indicators, alternative steps Analysis of implications of uncertainties and integration of adaptive management; occasional references to public and agency consultation and collaboration Collaboration of ecologists and other specialists Understanding and support by nonspecialists Iterative process	Impact on conservation status of habitat or species, valued ecological resources or features, priority ecosystem services, natural areas, and biological diversity Concern with ecological vulnerability, capacity, and adaptability Impacts on ecological processes, composition, functions and structure, and on unique, endemic, threatened, and declining species, habitats and ecosystems Biodiversity levels—habitat, species, and genetic; spatial levels—international, national, regional, and local Links from social and economic drivers of change (direct and indirect) and to human well-being
SIA	Social sciences Social services Values and ethical standards Political science (e.g., empowerment) Loop back—best practice and applied research Direct and strong links to public participation knowledge and methods Importance of local knowledge and experience	Part of EIA or SEA or in support of or separate from Role in guiding legislation, policies, plans, programs, and projects Recognizes political link (e.g., capacity building, empowerment, decentralization) Varying perspectives—society—consensus based or conflict based Acknowledgement that poorly integrated into decision making	Technical/scientific (product, action oriented—centralized management, research oriented), participatory (from perspective of those affected), process oriented, action oriented, social development and advocacy, and issue oriented Socioeconomic IA Poverty IA Gender IA Community-based SIA Social/psychological IA Cultural/heritage IA	EIA- and SEA based Modified steps—profiling, social context, and determining probable responses of affected parties Heavy emphasis on community involvement and participation Acknowledges uncertainties and need to manage Scientific/technical—stresses research methods, assumptions, models, and controls Participatory/political approaches—stresses facilitation, inclusiveness,	Guided by values and bounded by ethical limits Awareness of differential impacts among population groups Priority concern—assessing and redressing impacts on most vulnerable groups Changes to people's ways of life, their culture, their political system, their personal and property rights, and their fears and aspirations Social, cultural, heritage, and economic effects Social services and facility impacts-emphasis on positive

Table 2.6 (*Continued*)

	Conceptual Framing	Institutional Framing	IA Process Types	Procedural Elements	Substantive Elements
			Political (focus on capacity building and empowerment) Equity oriented Vulnerability oriented Combinations	conflict resolution, coping strategies, skills development, accountability, mutual learning, advocacy, vulnerability and inequities, community buy-in, influence and control, community values and priorities, and capacity building	outcomes (realization of human and social potential and capital) and community acceptance (people as beneficiaries) Distinction—micro- and macrosocial and societal change, with and without intervention; distinction between social change processes and social impacts Distinctions—individuals/sites, families, population, community, and social structure; political, social, and community resources Sometimes includes health and biophysical links and sustainability
HIA	Public health Medical services Medical, social, and environmental sciences Environmental health Risk assessment Social sciences Stakeholder knowledge Links to health research system	Integrated into or separate from EIA and SEA (connected to or not connected to) Mandated, decision support (voluntary), advocacy, and community lead Links to health regulations, policies, services, and promotion Government commitment to public health promotion critical Importance of organizational partnerships and capacity building	Biomedical model (epidemiological/toxicology, emphasis on diseases and ill-health and quantification) Socioeconomic model (qualitative, social sciences, stakeholder knowledge, broader determinants) Technical versus procedural orientation Health equity IA Policy, plan/program, and project-level HIA Risk assessment Mini, standard, and maxi—levels of intensity (e.g., desktop, rapid-participatory and nonparticipatory, comprehensive) Combinations	EIA and SEA steps Emphasis on population based, systems approach Sometimes includes alternatives analysis, especially more participatory approaches Sometimes risk assessment is a separate step Participatory approach—strong links to SIA and to public participation and democratization Recognizes policy context Generally assesses limitations and uncertainties Systems based, structured but flexible	Physical, mental, and social well-being; not just absence of disease or infirmity Behavioral, physical, community, economic, cultural, and social health determinants Individual factors, social and community networks, living and working conditions, socioeconomic conditions, cultural and environmental conditions Distribution of health effects critical, especially with reference to vulnerable groups Health service implications Importance of intersectoral action and linkages Occurrence and importance of health determinants and outcomes

(*continued*)

Table 2.6 (*Continued*)

Conceptual Framing		Institutional Framing	IA Process Types	Procedural Elements	Substantive Elements
SA	Interdisciplinary perspectives, concepts, frameworks, and models Values and ethics Frameworks and scales (e.g., technocentric to ecocentric) Environmental management instruments Necessitates additional scientific research Linked to state-of-the-environment reporting	Integration into EIA and SEA legislation and guidelines Separate SA system Can be applied to all decision-making levels (integrated into or applied to final proposal) Tiering necessary Transboundary Formal and informal External (for approval) and internal (for improved internal decision making) Need for SA training and capacity building	Sustainability appraisals or assessments of spatial plans, trade agreements, and development strategies Private sector integrated assessment Goal or effects oriented Pillars, principles based, thresholds, and trade-off rules (e.g., impact minimization for each pillar, "win-win-win," net gains, threshold tests for each criterion, full integration) Broad to narrow (e.g., biophysical only to all sustainability themes) As type of SEA or SEA driven As type of EIA or EIA driven Many approaches, levels, sectors, and combinations	Feeds into each SEA stage; same for EIA Systems perspective Includes rules (e.g., net conservation benefits) Includes some additional steps (e.g., refining to maximize benefits and minimize negativity, within context of range of future scenarios, global effects analysis, intergenerational effects analysis, application of sustainability thresholds, targets and trade-off rules) Iterative, flexible, and integrative Integrates precautionary principle; favors diversity, reversibility, and adaptability Adapted to context Ample stakeholder involvement	Guided by holistic concept of sustainability Encompasses and transcends physical, ecological, social, economic, and political Varying frameworks for displaying and linking Sustainability reference points and perspectives Includes cumulative local, regional, and global effects Future oriented Seeks substantive integration Use of environmental offsets Directs decision making toward sustainability Explicit key sustainability requirements Includes intra- and intergenerational equity Challenges status quo and prevailing perspectives

Sources: Ahmed and Sánchez-Triana (2008b), Azcarate and Balfors (2009), Becker et al. (2004), Bina et al. (2011), Bond et al. (2011a), Buchan (2003), Burdge (2003b, 2004), Canter and Ross (2010), Chaker et al. (2006), Clark et al. (2011), Cole and Fielding (2007), Connelly (2011), Cooper and Sheate (2004), Dalal-Clayton and Sadler (2004), Dalkmann et al. (2004), Dora (2004), Eccleston (2008), Elkin and Voiturez (2009), Elling (2011), Esteves and Vanclay (2009), Fischer (2003, 2007b, 2010), Fischer et al. (2010), Geneletti (2002), Genter et al., (2008), Gibson et al. (2005), Gibson and Hanna (2009), Gunn and Noble (2009a, b), IEEM (2006), ICPGSIA (2003), Gontier et al. (2006), Gunn and Noble (2009a), Hacking and Guthrie (2008), Harriman and Noble (2008), Harris et al. (2003), Harris-Roxas and Harris (2011), IAIA (2006a), Ison (2004), Jackson and Illsley (2007), Jay (2005), Jones and Slinn (2008), Kemm and Parry (2004b), Khera and Kumar (2010), Kvaener et al. (2006), Kwiatkowski (2011), Landsberg et al. (2011), Lobos and Partidário (2010), Mackenbach et al. (2004), McCarthy and Utley (2004), McCluskey and João (2011), Morrison-Saunders and Thérivel (2006), Noble and Gunn (2009), OECD (2006), Palmer (2004), Partidário (2007, undated), Pope (2006), Pope and Dalal-Clayton (2011), Pope et al. (2004, 2005), Posas (2011), Puschchak and Farrugia-Uhalde (2009), Putters (2005), Rauschmayer and Risse (2005), Retief (2007a), Sadler (2005b, 2011b), Schirmer (2011), Sheate (2011), Shepherd (2008), Sinclair et al. (2009), Slootweg et al. (2010), Slootweg and Kolhoff (2003), Söderman and Saarela (2010), Taylor et al. (2004), Thérivel et al. (2010), Thérivel and Ross (2007), Treweek et al. (2011), US NRC (2011a,b), Vanclay (2003, 2010), Wale and Yalew (2010), Walker (2003), Wegner et al. (2005), Wiek and Binder (2005).

technical approach, but also tend to be more goal driven, employ longer time horizons and broader spatial boundaries, be more strategic and adaptive, and stress the importance of interconnections (cumulative effects, for example). Policy and higher program/plan levels tend to be even broader in scope, less rooted in analysis and effects (e.g., environmental issues and concerns), even more adaptive, and more closely integrated with policy making, planning, and decision making. Even these broad patterns are far from uniform among jurisdictions and among SEAs. SEA process types vary between sectors/programs and regional/spatial planning, and within individual sectors, programs, and regional/spatial planning levels and settings. Also, SEA encompasses numerous procedural (e.g., analytic, administrative, technical, participative, communicative, political, community based), substantive (e.g., sustainability oriented, ecologically oriented, equity–justice oriented), proponent (e.g., preparation by proponent, preparation by independent entity), and decision type and direction (e.g., formal, informal, top-down, bottom-up, cabinet level, regulatory, policy based) choices. These choices are far from clearly defined. Often they overlap or combine elements or approaches. Moreover, any approach or approach combination necessitates contextual adaptations. Given the range of SEA permutations and combinations available, and the diversity of context types, it is not surprising that it is difficult to reach a consensus regarding good SEA practices (see Section 2.5).

Procedural SEA elements vary with the selection, adaptation, and integration of SEA process types. Still there are some shared characteristics. Given the nature of strategic planning and decision making, for example, it is broadly acknowledged that SEA is necessarily iterative, integrative, adaptive, uncertain, and context dependent. The close connection between SEA and decision making, moreover, necessarily means that SEA must adopt a proactive approach to ameliorating the prevailing limitations of policy and plan making. This means, for example, seeking to make decision making more transparent, open, inclusive, collaborative, informed, and substantiated. In some cases, this effort can entail contributing to organizational learning and capacity through effective follow-up, capacity building, institutional reform, and altering the institutional culture and value system.

SEA, in common with EIA, has a substantive purpose. The substantive purpose of SEA can be defined narrowly (e.g., focus on ecological issues), more broadly (e.g., social, economic, ecological issues and effects including cumulative effects, long time horizons, large spatial scales), or more broadly still (e.g., holistic environmental perspective, emphasis on sustainability, and progress toward sustainability). As detailed in Chapter 6, there is far from a consensus within the SEA community regarding how broadly or narrowly SEA should be defined. Although SEA was originally intended to remedy some of EIA's substantive limitations (e.g., preoccupation with procedure over substance, weak treatment of cumulative effects), in practice, SEA also has struggled with a tendency to become a "proforma" procedural requirement that adds little in terms of substantive, demonstrable environmental enhancements.

2.4.2 Ecological Impact Assessment (EcIA)

EcIA provides a mechanism for more effectively drawing upon the natural sciences, environmental planning and management, and ecological science and sustainability. EcIA can be subsumed within EIA or SEA, where EIA/SEA is either broadened to include EcIA or where the scope of SEA/EIA is limited to ecological concerns. EcIA can perform a support role to or be fully independent from EIA/-SEA. Outputs from EcIA can result in legal, policy, and development control reforms. To ensure effective implementation, they also can necessitate capacity building and changes in institutional arrangements.

Although broadly concerned with maintaining and enhancing ecological systems, EcIA can be focused, to a lesser or greater degree, upon such concerns as specific physical or biological effects or risks, threatened and endangered species, protected areas, biological diversity, and ecological services. The ecosystem approach may, to varying degrees, assume a central role in the process. Process characteristics vary, as with SEA, depending on whether the assessment is undertaken at the policy, plan, program, or project level. Characterizations of the EcIA process closely parallel more technical and scientifically oriented depictions of SEA and EIA processes. Particular emphasis tends to be placed on systematic scoping, adaptability, ecological vulnerability and significance, uncertainty management, cumulative effects assessment, effective mitigation, enhancement and compensation, and rigorous monitoring and follow-up procedures. The assessment of alternatives and participation, beyond the involvement of those with specialist ecological knowledge, tends to be stressed to a lesser extent.

Consistent with its substantive orientation, EcIA focuses on the integration and adaptation of frameworks and models for systematically assessing and managing impacts on ecosystem composition, processes, functions, and structure. Particular attention tends to be devoted to systematically characterizing ecological value and services, and vulnerability, capacity, and adaptability, at various biodiversity levels (e.g., habitat, species, genetic) and spatial scales (e.g., international, national, regional, local).

2.4.3 Social Impact Assessment (SIA)

SIA provides a means of more effectively introducing and integrating social science and social service knowledge into IA practice. SIA also draws heavily upon (and has the potential to contribute to) philosophy, political science, and public participation. This is because of the central role in SIA of values and ethics, the close connection between human impacts and the exercise of political power, and the vital role of public participation in informing IA, in interpreting impact significance and context, in avoiding and reducing potential adverse impacts, and in facilitating the realization of social and community objectives. As with EcIA, SIA can be integrated with, in support of, or separate from EIA or SEA. Perspectives regarding the institutional and societal context for SIA range from consensus to conflict based. SIA can potentially advance political objectives such as capacity building, community empowerment, and decentralization. It also can contribute to institutional reform. In practice, when SIA is integrated within, supportive of, or even separate from EIA or SEA, it tends to have limited decision-making influence, except in terms of the driving role of economic and financial benefits.

Several SIA process models have been advanced (e.g., technical–scientific, issue oriented, participatory, community based, and social action). SIA also can be oriented to favor particular social and ethical concerns and priorities (e.g., poverty IA, sociopsychological IA, cultural heritage IA, equity-oriented IA, socioeconomic IA, social development IA). Characterizations of the SIA process largely model good practice SEA and EIA models, with some variations in emphasis. Particular stress, for example, is placed on understanding the community (through community profiling and scoping), on anticipating the probable responses of affected parties, on assessing the distribution of human impacts (individuals, families, groups, communities, society, for example), on avoiding and minimizing impacts on the most vulnerable, on effective and inclusive community involvement and participation (including mutual learning and aided by capacity building, skills development, and conflict resolution and coping strategies), on avoiding and redressing inequities, on facilitating community support and influence, and on consistency with and the furthering of community values and aspirations.

SIA provides a mechanism for incorporating a diversity of substantive issues and concerns into IA theory and practice. It addresses the magnitude and distribution of effects on people's way of life, their culture, their political system, their personal and property rights, and their fears and aspirations. It includes social, economic, heritage, cultural, and social service and facility impacts. It differentiates impacts at, for example, the individual, family, group, population, community, and societal levels. It distinguishes between social change processes and impacts. It devotes particular attention to impacts on the most vulnerable, emphasizes the need to identify impacts on and reinforce the supportive role of political, social, and community resources, and stresses the need for long-term positive social outcomes (i.e., social sustainability) and community support. SIA systematically explores interrelationships among social impacts and connections to health effects, ecological effects, and sustainability.

2.4.4 Health Impact Assessment (HIA)

HIA represents a means of integrating knowledge and practice from such fields and disciplines as the medical, social and environmental sciences, medical services, environmental health, risk assessment, and health-related social sciences. HIA, much like EcIA and SIA, can be undertaken as part of, as separate but connected to, or as fully distinct from SEA or EIA. It can be mandatory (as a regulatory or statutory requirement), voluntary (in a decision support sense), an advocacy initiative, or community led. HIA only tends to be effective if there is clear government commitment to public health promotion and adequate institutional and organizational support. Such support could, for example, include partnerships with medical service providers, capacity building, applied research, and regulatory reform.

HIA procedures can be structured around a biomedical model (quantification, disease and ill health, and epidemiological–toxicological knowledge) or a socioeconomic model (qualitative analysis, social science and stakeholder knowledge, broader health determinants). The HIA process tends to vary depending on whether it is integrated with or in support of policies, plans, programs, or projects. The HIA process can be procedural or technical (e.g., structured around formal risk assessment) in orientation, applied at varying levels of intensity (e.g., desktop, rapid-participatory or nonparticipatory, comprehensive), and with or without an emphasis on health equity. Process combinations also are possible. Characterizations of the HIA process largely conform to one or more of the SEA or EIA process models. The more technical–scientific HIA process types tend to be similar to rational–technical IA processes. More participatory socioeconomic HIA approaches tend to be similar to comparable SIA processes. Most HIA process characterizations recognize the need for a systems and integrative perspective, for the effective management of risks and uncertainty, for adaptability, and for effective management and follow-up.

HIA represents a vehicle for integrating substantive, health-related knowledge into IA practice. Health tends to be broadly defined (i.e., physical, mental, and social well-being not just the absence of disease and infirmity). Pathways and effects linking health determinants (e.g., behavioral, physical, community, economic, cultural, and social) and health outcomes (occurrence, importance) are systematically considered. Health service implications are fully addressed. The distribution of health effects, especially for vulnerable groups, is considered critical.

2.4.5 Sustainability Assessment (SA)

SA is an inherently interdisciplinary, arguably transdisciplinary, form of applied knowledge. It relies heavily on concepts, frameworks, and models derived from a range of disciplines, applied fields, and research initiatives that transcend conventional academic disciplines and environmental planning and management instruments. Also, because of its strong normative–ethical core, it is highly reliant on value-based and applied ethical research and application initiatives. The sustainability knowledge base is expanding rapidly. Thus, "feedback" from state-of-the-environment reporting and from sustainability initiatives at both the regulatory and applied levels is critical to the advancement and application of the field. The institutional framing of SA is still in flux. It has been grafted, to a limited extent, onto existing SEA and EIA institutional arrangements, but largely in terms of broad intentions. The means for achieving those intentions tend to be much more narrowly defined. There also are a few scattered examples of SA as a separate set of institutional arrangements. SA, in common with other substantive IA types, can be applied at any or preferably all decision-making levels, and can be separate, aligned with, or connected to policy making, planning, and decision making. SA, given its substantive nature, frames and encompasses EcIA, HIA, and SIA. SA reforms and redefines other IA types when fully and systematically applied and integrated. The transformative nature of SA inevitably necessitates training and capacity building.

SA process types, in broad terms, parallel those of other IA types, especially SEA. SA can be effects or objectives driven. It can be derived from first principles, modeled after or grafted onto SEA or EIA process types, or SEA or EIA process types can be progressively reformed to eventually conform to ideal SA process types. SA can be applied to different proposed actions (e.g., trade agreements, spatial plans), undertaken by the public or private (e.g., integrated IA) sectors, and defined narrowly (e.g., ecological sustainability only) or broadly (e.g., full integration of ecological, social, and economic concerns). SA is generally structured by and around conceptual frameworks (e.g., three or five pillars, principles based, objectives driven, thresholds, and trade-off rules). SA processes are adapted to context types and to individual contexts. SA frequently integrates and applies explicit principles and trade-off rules (e.g., impact minimization for each pillar, "win-win-win" net gains, thresholds tests for each criterion, fully integrated trade-off rules). The choice, adaptation, and manner of application of such frameworks and tools is critical in terms of whether the SA is little more than "tokenism" or genuinely facilitates sustainability.

SA incorporates many of the same procedural elements found with other IA process types, albeit with important modifications and additions. The SA process is applied against a framework of objectives, criteria, thresholds, principles, targets, and rules for achieving sustainability. This contrasts with EIA, which tends to focus, from the outset on means of avoiding and minimizing potential adverse effects. It contrasts with SEA, which may apply objectives and principles, but which tends not to apply clearly defined thresholds, targets, and trade-off rules with a tangible "bottom line" (i.e., sustainability).

SA processes are inherently holistic, integrative, iterative, and innovative. Both SEA and EIA processes tend to be more analytic, discipline-bound, inflexible, and prone to incremental adjustments to the status quo and prevailing perspectives. SA processes broaden temporal (e.g., future generations) and spatial (e.g., global perspective) boundaries, comprehensively define effects (e.g., positive and negative, direct, indirect and cumulative, local, regional, and global, ecological, social, and economic), systematically identify and evaluate a wide array of choices (e.g., multiple future scenarios, evaluation against sustainability requirements, choices that challenge the status quo and prevailing perspectives), make ample provision for stakeholder involvement and collaboration, and adopt a cautious approach to risk and uncertainty management (e.g., precautionary principle, adaptive management). SA fully integrates the procedural and the substantive. The procedural elements of sustainability are framed within a holistic sustainability concept. Individual disciplines are linked and transcended. Choices are derived from and tested against sustainability reference points and principles. Sustainability provides both a direction for change and boundaries within which potential changes are considered acceptable.

2.4.6 Patterns and Connections

The IA types (SEA, EcIA, SIA, HIA, and SA), described above, all have much to contribute to IA theory and practice. Each draws upon a substantial knowledge base—a knowledge base not likely to be as familiar to other IA type practitioners. Knowledge sharing across IA types is, therefore, essential.

Each IA type is connected to, informs, and (should) enhances institutional arrangements and decision making. Institutional arrangements for different IA types overlap to a considerable degree. The resources available for impact assessment are invariably constrained. It is, therefore, crucial that institutional arrangements for various IA types are linked and blended in a manner that is mutually beneficial, efficient, and effective.

Processes, for the various IA types, share many common elements but also exhibit distinct differences. There is considerable potential for mutual learning. Efforts to link and integrate procedural choices within IA types and the processes of different IA types need to carefully consider which procedural and substantive elements are complementary and which could operate cross-purposes. In some cases it may be wise to limit or bound integration when conflicts are evident and cannot be reconciled in an acceptable manner. The "force-fitting" of procedural or substantive elements from one IA type into the frameworks and procedures of another, in the name of consistency, may distort or undermine the effectiveness of both, while providing minimal efficiency or effectiveness benefits. At the same time, there is enough overlap among IA types (especially if each is broadly defined), and a sufficient range of shared procedural and substantive elements, that further efforts to link and partially integrate the institutional arrangements and practices of different IA types, are worth pursuing, albeit cautiously.

2.5 CONTEMPORARY CHALLENGE—SEA GOOD PRACTICE GUIDANCE

SEA could be characterized as a loose collection of approaches for integrating environmental concerns into higher level decision making (e.g., policies, plans, programs). The range of approaches, the variety of perspectives, and the diversity of contexts within which SEA is applied suggest that it may be difficult to agree upon a common set of good practices. Table 2.7 lists examples of suggested SEA good practices structured by recurrent IA problems.

2.5.1 Crosscutting Criteria

Several major themes addressed by the criteria appear sufficiently broad to include all SEA types and settings (e.g., appropriate to context, adequate for and well in advance of decision making, sufficient, reliable, and useable information, efficient, cost and time effective, accountable, impartial, tiered as appropriate, independent review, public and government informed and involved, clarity and availability of documentation, feedback on actual impacts). They also offer guidance regarding such transcending matters as throughout initiative life cycle, specific roles and responsibilities, effective follow-up, methods justified, legal compliance, focuses on environmental sensitivity and significance, transparent process, acknowledges uncertainties and complexities, justifies preferred options, identifies environmental opportunities and constraints, identifies mitigation and enhancement actions, interprets significance, concentrates on key issues, assets, sensitive areas and threats, addresses inequities including future generations, analyzes risks, seeks to enhance communications and collaboration among stakeholders, seeks just and equitable outcomes, enhances context awareness without undermining global aims, customizes to tier, objectives, constraints, nature, proposal type, mandate, legal, and policy framework, and customizes to spatial, temporal, biophysical, social, cultural, economic, institutional, and political context.

The criteria leave considerable latitude for interpretation to encompass a wide array of SEA types and contexts. Within these discretionary areas (e.g., varying interpretations of efficiency, effectiveness, and accountability), it is possible that different parties could reach different conclusions concerning "good practice."

2.5.2 Particular Values

Those who argue that SEA should be more narrowly focused (e.g., ecological effects only) might dispute the requirement that SEA should address interrelationships among biophysical, social, and economic effects, assess alternatives and strategic decisions against a sustainability standard, focus on key sustainability issues, and assess how sustainability issues were addressed in documentation and decision making. The same argument could be made regarding such criteria as ensure relevancy for achieving sustainability, assess effects, options, and risks within a sustainability framework, promote conservation and the sustainable use of biodiversity, ensure biophysical effects at least on an equal basis with social and economic concerns, foster democratization, ensure a democratic decision-making process, apply specific ethical principles, and explicitly address climate change. The counter to this argument would be that SEA (in common with all forms of IA) is inherently value driven—specifically environmental values and, more broadly, sustainability values. That being the case, SEA practitioners should explicitly identify and apply, as a matter of good practice, specific environmental/sustainability values.

2.5.3 Particular SEA Tier, Type, Approach, or Setting

Some criteria might be appropriate for a particular SEA tier or type (e.g., policy level, program level, regional or spatial plan, sector plan), a particular approach to undertaking SEA (e.g., technical–rational, participatory, community driven, effects or EIA based, objectives led, integrated with or parallel to policies, plans, or programs), or a particular setting type (e.g., developed nation, developing nation). Examples of such criteria include integrate with plans, policies, or programs, employ strong enforcement measures, establish clear environmental and sustainability goals and objectives, propose and apply environmental and sustainability, criteria, targets, thresholds, indicators, and trade-off rules, facilitate the search for the best alternative or scenario, consider the no-change alternative, justify the selection of preferred options, explicitly address indirect and cumulative effects, promote discourse reflection by stakeholders, use explicit criteria and procedures to assess significance, apply the precautionary principle, incorporate adaptive management, emphasize visions and visioning, stress mutual learning, establish independent oversight of implementation, compliance, and performance, include effective, preferably independent, quality assurance systems, and apply experimental design, where practical. Arguably, rather than excluding such criteria, on the grounds that they are not "universal," it would be more appropriate to identify and group SEA "good practices" by SEA tier, SEA type, SEA approach, and setting type. It also may be argued that some criteria (e.g., consider no-change alternative, search for the best alternative, justify selection of preferred options, address indirect and cumulative effects, apply precautionary principle, incorporate adaptive management) should be universal, and should not be confined to particular classes of situations.

2.5.4 Enhancing the Political Influence and Effectiveness of SEA

Several suggested criteria pertain to actions that extend beyond simply informing decision making and decision makers to include measures to proactively analyze, bound,

Table 2.7 Examples of Suggested SEA Good Practices

Influential
Integrate within plans, policies, and programs, wherever practical, and initiate early in process
Undertake throughout initiative life cycle
Apply to all strategic decisions (tier in manner appropriate to level)
Is the responsibility of the leading agencies for the strategic decision to be taken
Provide sufficient information on the actual impacts of implementing a strategic decision, to judge whether this decision should be amended, and to provide a basis for future decisions
Link to project EIA and to decision making
Ensure availability of the assessment results early enough to influence the decision-making process and inspire future planning
Seek to enhance understanding of "real" decision-making processes
Assign specific roles, responsibilities, and accountability, keyed to decision points
Promote use of SEA
Explore strategies for overcoming institutional resistance to instilling environmental values
Employ strong enforcement mechanisms
Rigorous
Carry out systematically and rigorously
Emphasize effective follow-up; include monitoring and adaptation strategies
Establish independent oversight of implementation, compliance, and performance
Include effective, preferably independent, quality assurance systems
Learn from comparative studies
Apply experimental design, where practical
Apply appropriate spatial and temporal scales
Share SEA knowledge and experiences
Ensure appropriate level of detail
Delineate methods by which findings are obtained, including uncertainties and associated implications
Subject to independent checks and verification
Seek to strengthen the science–policy link
Rational
Establish clear environmental and sustainability goals and objectives
Propose and apply environmental and sustainability criteria, targets, thresholds, indicators, and trade-off rules for evaluating effects of PPP and alternatives
Facilitate the search for the best alternative or scenario (more sustainable)
Consider no-change alternative
Systematically assess effects and options
Justify selection of preferred options
Substantive
Ensure relevant to achieving sustainability
Explicitly and systematically address indirect, cumulative, and transboundary effects, and life cycle issues
Seek to integrate SEA and CEA (cumulative effects assessment)
Promote conservation of and sustainable use of biodiversity
Explicitly address climate change
Focus on environmental sensitivity, significance, and sustainability
Address interrelationships of biophysical, social, and economic aspects
Ensure biophysical effects at least on an equal basis with social and economic
Identify environmental opportunities and constraints
Apply environmental objectives, targets, and indicators
Assess effects, options, and risks against a sustainability framework
Document and justify how sustainability issues were taken into account in decision making
Identify mitigation and enhancement actions
Practical
Concentrate on key issues, assets, sensitive areas, and threats
Provide sufficient, reliable, and usable information for planning and decision making
Ensure sufficient resources to implement
Apply as proactive environmental management tool
Concentrate on performance effectiveness
Foster ownership of outcomes

(*continued*)

Table 2.7 (*Continued*)

Focus on key issues
Provide specific SEA methodological guidance (e.g. decision rules, guidance for methods selection)
Ensure cost and time effective
Undertake with professionalism
Democratic
Ensure a democratic decision-making process
Integrate institutional and political factors
Comply with legal requirements and address legal gaps
Recognize SEA as sociopolitical struggle among diverging interests
Promote discourse reflection by stakeholders
Facilitate open and accountable political and organizational system
Provide opportunity for appeal of process or decision output
Consider direct and indirect democratic effectiveness
Foster democratization
Collaborative
Facilitate coordination within government and between governmental and nongovernmental organizations
Ensure transparent process
Inform and involve interested and affected public and government bodies throughout the decision-making process
Explicitly address public and agency inputs and concerns in documentation and decision making
Seek to enhance communications and collaboration among stakeholders
Have clear, easily understood information requirements
Ensure sufficient access to all relevant information
Stress mutual learning
Foster organizational/institutional learning
Build capacity for undertaking and using SEA
Ethical
Interpret significance of effects and trade-offs
Use explicit criteria and procedures to assess significance
Emphasize visions and visioning
Address inequities including future generations
Identify and apply ethical principles (e.g., no net loss, priority—most vulnerable)
Seek just and equitable outcomes
Undertake with fairness, impartiality, and balance
Adaptive
Use an iterative, creative, and flexible process
Analyze risks of policy, plan, or program
Apply precautionary approach
Incorporate adaptive management
Acknowledge and address uncertainties and complexities, and associated implications
Integrative/Contextual
Enhance context awareness and sensitivity without undermining global aims
Customize to tier, objectives, constraints, nature, proposal type and characteristics, legal and policy framework, and mandate
Customize to values and policies of a country
Customize to spatial, temporal, biophysical, social, cultural, economic, institutional, decision-making, and political context
Link to state-of-the-environment reporting
Link to national and institutional environmental and sustainability policies and strategies
Link to other assessments, and to other planning and environmental management tools
Strive to institutionalize SEA values
Undertake institutional analysis
Provide an appropriate and necessary regulatory framework
Ensure organizations and infrastructure can support SEA implementation

Sources: Bonifazi et al. (2011), Buuren and Nooteboom (2010), Clark et al. (2011), Connelly (2011), Cooper (2011), Croal et al. (2010), Donnelly et al. (2007), Dusik and Sadler (2004), Eccleston (2008), Elling (2000), Fischer (2005, 2007a,b), Fischer and Gazzola (2006), Franz and Kirkpatrick (2007), Gunn and Noble (2009b), Hildén et al. (2004), Hindling-Rydevik and Bjarnadóttar (2007), IAIA (2002a), Jackson and Illsley (2007), Jha-Thakur et al. (2009), Jiliberto (2011), Jiricka and Pröbstl (2008), Kørnøv and Thissen (2000), Landry et al. (2009), McCluskey and João (2011), Nilsson et al. (2009), Noble (2000a, 2003, 2008, 2009a), Noble and Gunn (2009), Noble et al. (2012), OECD (2006), Partidário (2007, undated), Partidário and Arts (2005), Partidário and Coutinho (2011), Retief (2007b), Retief et al. (2008), Runhaar (2009), Sadler (2005b), Sadler and Jurkeviciute (2011), Sheate and Partidário (2010), Slootweg et al. (2010), Stoeglehner et al. (2009), Treweek et al. (2005), Vicente and Partidário (2006), Weiland (2010), Wirutskulshai et al. (2011), Zhou and Sheate (2009), Zhu et al. (2010).

shape, influence, and direct decision making. Examples include the following: enhance understanding of real decision making, recognize SEA as a sociopolitical struggle among diverging interests, seek to overcome institutional resistance to change, provide sufficient resources to implement, integrate institutional and political factors, facilitate an open and accountable political and organizational system, foster organizational learning, link to state-of-the-environment reporting, require an institutional analysis, ensure an appropriate and necessary regulatory framework, ensure organizations and infrastructure can support SEA implementation, link to national and institutional environmental and sustainability policies and strategies, concentrate on performance effectiveness, provide opportunities for appeal of process or decision outputs, consider direct and indirect democratic effectiveness, enhance democratization, facilitate coordination between government and nongovernment, and foster ownership of outcomes. It could be suggested that such proactive measures are beyond the mandate of SEA practitioners. Alternatively, it might be argued that if SEA practitioners are to avoid being relegated to the "sidelines" of decision making, they must be more active in seeking to better understand and to bring about the legal, institutional, and decision-making conditions conducive to effective SEA practice.

2.5.5 Advancing SEA as an Effective Field of Practice

Several possible criteria pertain to actions by SEA practitioners that extend beyond managing and participating in the SEA process. Such actions relate more to the professional responsibilities of SEA practitioners to elevate the quality and effectiveness of SEA as an applied form of environmental management. Examples of such criteria include the following: promote SEA use, learn from comparative studies, share SEA knowledge and experiences, build capacity for undertaking and using SEA, strive to institutionalize SEA values, apply as a proactive environmental management tool, strengthen the science–policy link, seek to integrate SEA and CEA, and link to related tools. It could be argued that such actions, while laudable, are not "good practices." Or, it could be maintained that only through such actions will the level of SEA "good practice" be raised, or indeed, remain relevant.

2.5.6 Arguably Too Vague

Certain criteria might be disputed on the grounds that they are too open-ended in terms of alternative interpretations (e.g., professionalism, fairness, rigor, balance, appropriate level of detail, carry out systematically, appropriate temporal and spatial scales, systematically assess effects and options, apply an iterative, creative, and flexible process). Rather than rejecting such criteria, it may be a case of elaborating on the criteria such that minimal standards of acceptable and of good practice criteria can be determined. Alternatively, if this is not possible, either the criteria could be eliminated or only defined for particular SEA or context types.

The problem of vague SEA criteria is symptomatic of the broader issue of vague SEA guidance—guidance, which tends to remain at the level of broad principles. Good practice SEA knowledge needs to be integrated into more specific methodological guidance regarding the selection, refinement, adaptation, and application of SEA methods (Noble et al., 2012). The preparation, circulation, and adaptation of such guidance material could contribute to the refinement of context-specific SEA effectiveness criteria and, over time, help enhance the quality and effectiveness of SEA practice.

2.5.7 Structuring the Criteria

If it is accepted that SEA good practice criteria, to a considerable extent, run the risk of "falling between two stools" (i.e., either too broad to be of much practical value or too narrow to apply across SEA and context types), one way around this conundrum is to classify SEA dimensions. Table 2.8 represents an initial effort along those lines. Such a classification system might engender a dialogue among SEA theorists and practitioners regarding both "crosscutting" criteria and criteria that might suit particular categories. Such a dialogue might contribute to healthy debates regarding the purpose, scope, underlying values, aspirations, roles, boundaries, perspectives, focus, limits, appropriate approaches, and external connections of SEA as it is and as it could or should be.

2.6 SUMMING UP

This chapter addresses the question of whether conventional IA regulatory and process characterizations adequately convey the available choices. It also considers whether conventional IA process guidance and practice, even if substantially reformed, can adequately respond to the recurrent problems and contemporary challenges. These questions are addressed by characterizing a range of IA regulatory and IA process design and management choices both for EIAs and for other IA types (SEAs, SA, EcIA, HIA, SIA).

The EIA regulatory analysis identifies generic regulatory choices and good practices for screening, individual IA process and integration, and coordination activities. The applied EIA analysis presents a variety of process characterizations portrayed in IA literature. Tables and summary text identify examples of possible choices pertaining to general process design and management, IA process inputs, outputs, and linkages, IA process adaptations, IA frequency and sequence, and IA types. Support tables and figures indicate relevant distinctions and illustrate process management approaches.

Alternative SEA, EcIA, SIA, HIA, and SA approaches to conceptual framing, institutional framing, IA process design,

Table 2.8 SEA Dimensions

SEA Level	Policy (Upper Tier)	Policy (Lower Tier)	Plan/Program (Upper Tier)	Plan/Program (Lower Tier)
Decision-making integration	Fully integrated	Parallel (connected at each decision point)	Parallel (connected at end)	Not connected
Decision-making influence	Controls—shapes decision making	Influences decision making	Informs decision making	Minimal decision-making influence
Formality	Mandatory (legislation and regulations)	Quasimandatory (policy advice with auditing)	Advisory only (no auditing)	Informal
Direction of control	Top-down	Top-down and bottom-up	Bottom-up	Control–diffuse
Tiering	Fully tiered (policy, plan, program, project)	SEA tiered (policy, plan, program)	SEA and EIA coordinated or integrated	Minimal connections
Policy/planning connections	Fully integrated	Parallel—multiple connections	Parallel—one connection	Not connected
Institutional connections	Controls	Influences	Informs	Ignored or resisted
SEA/EIA dominates	SEA supersedes	SEA modified	EIA modified	EIA based
Scope of environment	Sustainability (fully integrated)	Ecological/social/economic connected	Ecological/social/economic separate (three pillars)	Only one or two of economic, social, and ecologic
Participation form	Community based	Collaborative	Two-way communications	One-way communications or closed
Generality	Broad issues	Specific	General region or sector analysis	Detailed region or sector analysis
Role of values	Value driven (normative)	Values explicit and diverse	Values explicit but narrowly defined	Values implicit or ignored (value-free–objective)
Dominant values	Technocratic—growth oriented	Environmental management	Conservation	Deep ecology
Sociopolitical perspective	SEA as political struggle or social advocacy	SEA as negotiations and collaborations	SEA as communicative planning or social advice	SEA as technical advice
Relationship to context	Transcends—universal good practice	Policy-, plan-, or program-type specific	Policy, plan, or program specific	Unique to context
Role of context	Shapes context	Influences context	Adapts to context	Ignores context
Development status	Developed nation	Transitional nation	Developing nation	
Role of rationality	Technical rationality	Strategic rationality	Communicative rationality	Political or ecological or social rationality
Degree of change	Transformative	Major reform	Incremental adjustment	Negligible change—status quo
Role of ethics	Ethics driven	Explicit ethical principles	Implied ethical principles	Ethics implicit or ignored
Responsibility	Proponent entirely	Proponent with external advice	Proponent jointly with external assistance	Prepared by external, independent entity
Qualitative–quantitative	Analytic quantitative	Largely quantitative	Largely qualitative	Completely qualitative
Nonmandated elements	SEA integrates	Separate but connected	Separate not connected	
Public–private	Public	Semipublic	Semiprivate	Private
Spatial perspective	International	National	Subnational or regional	Local
Temporal perspective	Future generations	Long term	Medium term	Short term
Cumulative effects	CEA focused—detailed	CEA focused—general	Direct and indirect effects	Direct effects only

procedural elements, and substantive elements are summarized. Patterns and interconnections among these IA types are explored at both the conceptual and applied levels.

Particular consideration is devoted to the contemporary challenge—SEA good practice. Suggested SEA good practices for each recurrent problem are presented. The analysis of issues surrounding SEA good practice addresses such matters as crosscutting criteria, inherent values, adaptations to SEA tier, type, approach, and setting, measures to enhance SEA political influence and effectiveness, advancing SEA as an effective field of practice, level of detail, and criteria structuring approaches.

The IA choices described in this chapter could be supplemented by interjurisdictional comparisons of IA requirements, guidelines, and practices. Information, knowledge, and experience sharing would be highly beneficial. Workshops, joint studies, and collaborative efforts (such as the joint preparation of IA proposal and setting type guidelines) are likely to lead to regulatory and applied enhancements well beyond what is practical within individual jurisdictions.

More frequent and comprehensive effectiveness analyses (from multiple stakeholder perspectives) of IA requirements, guidelines, and practices also are conducive to enhanced IA process management. Such reviews need to ask basic questions regarding what is and is not working and why and to assess the options available for enhancing the levels of regulatory and applied practice. Oftentimes, it is far from clear whether and to what extent the control and guidance provided and the level of practice is adequate, appropriate, or has unintended secondary consequences. Frequently, only a narrow range of choices for elevating regulatory and applied practice is systematically considered. The search for potential approaches can be advanced by case study analyses and by applied research. The scope of potential improvements should not be limited to refinements. Basic regulatory restructuring and a redefinition of what is considered adequate and good IA practice should always be a possibility.

The generic regulatory and applied IA choices presented only partially address the recurrent problems and contemporary challenges identified in Chapter 1. Although broadening the range of available choices, it is unlikely that the recurrent problems and contemporary challenges can be fully addressed by further adaptations and refinements to conventional IA requirements and guidelines and to conventional IA processes. Equally important, having a wide array of choices, while helpful, provides little guidance regarding which choices or combination of choices would be most appropriate, and would be most likely to effectively grapple with the recurrent problems and contemporary challenges. A more in-depth analysis, potentially including fundamental reorientations, will likely be needed. This is the role of Chapters 3–12.

Chapter 3

How to Make IAs More Influential

3.1 HIGHLIGHTS

In this chapter, we address how to avoid and ameliorate the tendency of some parties to not participate in IA processes, to circumvent IA requirements and procedures, to not adequately draw upon the IA knowledge base, or to participate in "good faith" in IA requirements or procedures but to have little or no discernible impact on decision making.

- The analysis begins in Section 3.2 with four applied anecdotes. These stories describe applied experiences in which (1) the major parties decide to circumvent the IA process on the grounds that the proposal is too "important," (2) SEA is used as an instrument for enhancing public policy-making influence, (3) various approaches are explored in one jurisdiction (the Netherlands) for making IA more relevant and influential, and (4) the EIA did not address the key environmental issue of whether the proposal should proceed.
- The analysis in Section 3.3 then defines the problem—three negative perspectives that undermine IA's effectiveness.
- In Section 3.4 we explore the legitimacy of the three negative perspectives and present measures to prevent and offset these negative perspectives. This analysis is then extended by establishing a foundation (using concepts, frameworks, and research priorities) for making IA requirements and processes more relevant and influential.
- In Section 3.5 we present an overview of selected characteristics and reforms from the four jurisdictions potentially conducive to and inhibiting of greater IA decision-making influence. We then illustrate how an influential IA process could be expressed at the applied level. We also address process and good practice variations among IA types (SEA, project-level EIA, SA, EcIA, SIA, and HIA).
- In Section 3.6 we address the contemporary challenge of good practice approaches for making IA more influential. Good practices are grouped by criteria at both the regulatory and the applied levels.
- In Section 3.7 we highlight the major insights and lessons derived from the analysis.

3.2 INSIGHTS FROM PRACTICE

3.2.1 What Happens When a Proposal Is Too Important for IA?

A challenge in IA arises related to large projects with national and international economic and environmental implications because many stakeholders often do not believe in or trust that an IA process will truly inform and influence decision making. Why? Senior governments and opposition parties publicly declare their views and preferences at the outset. As a result, proponents lobby governments and sign agreements with affected Aboriginal and other local communities to bring them "on side." Opponents also lobby, often extensively using direct political action—rallies, demonstrations—to put pressure on decision makers and to influence the general public. All sides actively promote their views through the media. In brief, almost all parties conclude that relying only on IA reports, submissions, and review procedures is unlikely to result in the decision they want because such projects are viewed to be bigger or more important than "IA." Thus, how can IA reports and processes become serious components in decision-making environments that are highly charged and polarized?

Two high-profile examples highlight this conundrum. Both are usually characterized as "energy initiatives," but their scope for IA is much broader. The first example is the Keystone XL 2736 km pipeline from Alberta to Oklahoma and then to the Texas Gulf Coast. At full capacity, it will transport up to 830,000 barrels of oil daily. Canadian Foreign Affairs Minister John Baird has stated that the pipeline is "all about jobs and economic growth for Canada," and that the federal government would "continue to be an active supporter of the project." Various leaders and groups in the United States support the proposed pipeline for the same reasons related to jobs and economic development in the United States. In contrast, President Barack Obama stopped the $7.6 billion project in mid-January 2012 because of concerns about negative environmental impacts on the environmentally sensitive Sandhills Ogallala aquifer area in Nebraska. Yet, he also indicated that an alternative route could be proposed by TransCanada, the proponent of the pipeline project, and, by the end of the winter in 2012,

Impact Assessment: Practical Solutions to Recurrent Problems and Contemporary Challenges, Second Edition. David P. Lawrence.

TransCanada had developed a proposal for a 176 km rerouting of the pipeline that it was claimed would "jog around" the Sandhills. Thus, the Keystone XL pipeline brings together interests and concerns about energy, economic development, management of aquatic systems and environmentally sensitive areas, and political positioning in the run up to US Presidential election later in 2012. The behavior and conundrum highlighted in the introduction above both are clearly visible related to this project.

The second example is the Enbridge Northern Gateway proposed pipeline ($5.5 billion, 1177 km twin pipeline) from Edmonton to a super tanker port in Kitimat, BC. The federal government has been a vocal supporter of this pipeline, arguing that it will generate thousands of jobs and bring economic benefits to numerous communities along the pipeline route. The federal government went even further, referring, within a March 2011 report, to environmental groups and aboriginals as "opponents." Indeed, some commentators claimed the federal government had categorized stakeholders into two categories: allies, and adversaries. This interpretation was reinforced by remarks by Natural Resources Minister Joe Oliver, in an open letter to Canadians in early January 2012, when he called opponents "ideological" and opposed to all major projects. In contrast, those expressing opposition to the pipeline pointed to potential for significant negative environmental impacts if there were a pipeline break or if there were a major oil spill in the enclosed sound through which the super tankers would travel to reach Kitimat.

Another dimension is reflected in January 2012 comments by Prime Minister Harper, who stated that environmentalists were funded by foreign money and were trying to hijack the National Energy Board public hearings that had started in mid-January 2012. Others observed that those promoting the Northern Gateway pipeline also had significant financial support from foreign petroleum corporations, and wondered why such foreign support was not also a problem. The federal government never directly responded to the apparent contradiction that some foreign financial support was desirable and acceptable, while other foreign support was not.

A further aspect in the discussions was a view by some that the federal government was using the Northern Gateway project, which will export crude oil to China and other Asian markets, to pressure the American government to approve the Keystone XL pipeline. In such a context, questions emerge about the role and value of impact assessments when the ultimate approving authority has openly declared support for a project and has dismissed groups expressing concerns because they are "ideological."

The Keystone XL and Northern Gateway pipelines highlight ongoing challenges for IA processes. Such projects usually involve the same government actively supporting the project while also being the regulator and approver of it. In such contexts, IA processes are not adequately designed and developed to address such challenges in a credible way. Solutions are not obvious, but resolution will require attention to governance, decision-making arrangements, and regulatory processes. Where is the necessary research and work being done on these matters to allow IA reports and processes to be credible components in decision-making environments that are highly charged and polarized?

BRUCE MITCHELL

Department of Geography and Environmental Management, University of Waterloo, Waterloo, Ontario, Canada

3.2.2 Using SEA to Enhance Public Influence over Policy Making

The Scottish government has extended the scope of the 2001 European Union SEA Directive to cover virtually all new Scottish public sector programs, plans, and strategies (PPSs). Describing the Environmental Assessment (Scotland) Act 2005 as "world-leading legislation," Scottish ministers saw this as meeting three goals:

- contributing to their aims of improving the Scottish environment and making Scotland more sustainable;
- improving policy making by ensuring that environmental effects were fully considered at an early stage in policy formulation and that the environmental effects of different options were assessed; and
- promoting more open government by allowing the public and interested organizations to comment on environmental reports, and obliging public bodies to explain how they have taken such comments into account.

The new facilities created to support the additional SEAs required included:

- an electronic SEA gateway linking the public bodies (responsible authorities) undertaking an SEA with the three consulting authorities statutorily obliged to offer comments and advice (Scottish Natural Heritage, Scottish Environment Protection Agency, and Scottish Heritage), which provides a repository for reports and comments on all stages of the process that is fully open to public scrutiny (www.scotland.gov.uk/Topics/Environment/environmental-assessment/sea);
- an SEA forum that meets several times a year, bringing together Scottish SEA practitioners to exchange ideas on good practice;
- an annual statistical report on the operation of the legislation presented to the Scottish Parliament; and
- an SEA Toolkit, also in electronic format and regularly updated.

The Scottish government also commissions research and offers formal guidance on various aspects of SEA practice, such as its use in assessing the climate change implications of PPSs.

A good example of how SEA is making Scottish plan making more transparent is demonstrated by its role in producing the second Scottish National Planning Framework (NPF). The SEA Scoping Report was made available for consultation early in 2007, accompanied by a briefing note that explained that SEA enabled the public to become involved at the outset in choosing an optimal development strategy for constructing the NPF:

> SEA is not just a test of how 'environmentally friendly' the NPF is, applied after its content has already been decided. Importantly, environmental impacts are being identified (and where possible avoided) as the NPF is being written, so that the SEA really influences its content. SEA is required to assess the environmental impacts of the proposed NPF, and to compare this with a range of 'reasonable alternatives'. This allows us to explore a wide range of ideas and opportunities before deciding on the best solution and, if possible, including it in the NPF.

The team responsible for drafting the NPF included an SEA specialist who championed the role of public participation at this initial stage, arguing that "it's all about getting involvement as early on as possible. There's no point in doing separate consultation on SEA. If you're doing a proper consultation on the plan the SEA should be integral to that." While drafting the Scoping Report, the team went out to meet the public, holding a series of open meetings across Scotland:

> allowing us just to listen to what people were expecting from the exercise and to draw initial ideas. It's all about front-loading the exercise and getting those views in before you start drafting. There should be an awful lot of work done before you get to the Discussion Draft stage. You need early and effective engagement. In the early part our role was quite passive, hearing views rather than going out and leading too much on the SEA. It was still so open it was impossible to pre-judge that. Alongside that we also did the scoping work and spoke to the consulting authorities—quite low key really but it was to get their views on how we were going to do the SEA.

A key innovation stemming from this process of front-loading was the production of an interim environmental assessment of the strategic alternatives which emerged from initial discussions:

> We came back from all those listening events and it was clear that there were lots of different ways that the NPF could go. There needed to be some structure for thinking through what the different environmental performance of different options and choices identified would be. I'd used scenarios in other work that I'd done before and felt it was quite a good way of setting out some of the key choices. I worked with the rest of the NPF team to crystallize scenarios based on the consultees' views that we'd got from the listening events. We then ran an assessment on the environmental effects of these before choosing the preferred scenario for the Consultation Draft of the NPF and its full Environmental Report, both of which were then put out for discussion.

These interim reports were published simultaneously at the start of 2008, allowing the completed NPF to come into effect late in 2008 accompanied by a Finalized Environmental Report indicating how the views of consultees on potential environmental effects had been accommodated.

This case study epitomizes the Scottish approach to SEA. Because its all-embracing legislation includes voluntary as well as statutory PPSs, SEA is central to Scottish public sector policy making. As a consequence, Scottish SEA practitioners are establishing a "community of practice" through regular exchanges of views on techniques, guidance, and policy initiatives via the SEA gateway and meetings of the SEA forum.

Scottish official guidance stresses that SEA should focus on the environmental aspects of PPSs, reflecting the Scottish government's views that including social and economic considerations "risks obscuring the environmental considerations that we are setting out to identify." As our case study demonstrates, its emphasis on public participation in SEA is central to discharging the Scottish government's obligations under the Aarhus Convention to deliver "environmental justice."

A.A. JACKSON

Town and Regional Planning, School of the Environment, University of Dundee, Dundee, Scotland, UK

3.2.3 Making Impact Assessment More Influential: Lessons from the Netherlands

When discussing how impact assessment can be made more influential, often the term "effectiveness" is employed. *Procedural effectiveness* usually addresses the extent to which an IA is conducted in line with established procedures (e.g., the extent to which opportunities for stakeholder participation are provided). *Substantive effectiveness* relates to the extent to which IA contributes to environmental awareness of decision makers (i.e., competent authorities and developers), the explicit consideration of environmental values in decision making, and eventually the environmental performance of projects, plans, and policies (PPPs).

An often advanced claim is that the *context* in which IA operates matters for the level of effectiveness achieved. In addition, IA regulations vary between countries and even within countries. Hence, in this story, we briefly reflect on the Dutch situation.

In the Netherlands, IA formally got a legal status in 1987, but experiments had already started a decade earlier. Internationally, the Netherlands has often been perceived as a frontrunner in IA because the Dutch IA legislation covered many of the elements that are considered essential for good

IA systems. It should be noted that the Dutch IA system was heavily inspired by the Canadian model. Elements that were considered to make the Dutch IA system unique included the requirement to develop an "alternative most friendly to the environment" (AMFE), and the Netherlands Commission for EA charged with quality review. In the present situation, the Dutch IA system no longer seems to be front-running, among other things, due to the abolishment of the AMFE requirement and the fact that other countries have expanded their IA systems.

In the Netherlands, similar to many other countries, *conducting* an IA is required for particular PPPs, but decision makers are not challenged to actually incorporate the outcomes of the assessment—for example, choose the most environmentally friendly alternatives or mitigation measures identified during the assessment. "Follow-up" requirements are in place but often are not complied with in practice, due to, among other things, difficulties in attributing changes in the natural environment to the decision at issue. Therefore, in particular, the *substantive* effectiveness of IA largely depends on the willingness of decision makers to take environmental concerns into account, which of course is influenced by the interplay between them and other actors involved.

Our 2010 survey of over 440 Dutch IA practitioners (representing all roles in the IA process) and 20 in-depth supplementary interviews revealed that IA substantive effectiveness is reasonably high. A majority of respondents perceive that IA contributes to environmental awareness among decision makers. In addition, IA often influences PPPs subject to IA at an early stage of decision making: anticipating an IA, environmental concerns are more consciously taken into account (the "prevention effect"). In addition, in about 60% of the cases IA results in adjustments of PPPs. It should be noted that these adjustments are seldom fundamental (e.g., in terms of abolishing initiatives or choosing alternatives developed in IA). The legal requirement to conduct IA appears to be the main explanation for IA substantive effectiveness: IA has an impact because it has to be conducted, not because actors choose to do so voluntarily. This also becomes visible in the prevention effect and in how our respondents perceive IA is applied: a mandatory exercise rather than a tool to design and optimize the PPP at stake. The result is that they seldom move beyond what is minimally required by IA or environmental legislation, which puts a serious constraint on opportunities to make IA more influential. Earlier evaluations conducted in 1990, 1996, and 2003 have similar outcomes, suggesting IA substantive effectiveness is rather stable over time.

Despite its mandatory character, the predominant attitude toward IA is quite positive. For most respondents, IA namely is instrumental in providing transparency of decision making and in minimizing legal risks of not complying with environmental laws. This also explains why procedural effectiveness is quite high: formal procedures are usually complied with and there is not much evidence of "salami tactics" (i.e., avoiding IA by subdividing PPPs into smaller ones). Perceptions of and expectations regarding IA largely converge. This "common ground" provides a good basis for working with IA but, at the same time, does not stimulate creativity in decision making and optimization of environmental values. Our survey and in-depth interviews with IA practitioners with experience from various policy fields provide a few suggestions for improving IA effectiveness, such as more transparency in IA legislation in the screening stage, promoting a more concise *scoping*, and providing an environment in which creative use of IA as a design tool is promoted.

The lesson that we derive from our study is that expectations of procedural as well as substantive effectiveness of IA should be realistic. Although conducting IA is mandatory for specific initiatives, it is up to decision makers to weigh the insights from an IA against other concerns. IA practitioners may facilitate decision making by clarifying IA regulations, ensuring high quality of IA and pointing to discriminating results. They, however, can never replace decision makers.

DR. HENS RUNHAAR

Copernicus Institute of Sustainable Development, Utrecht University, Utrecht, The Netherlands

PROF. DR. JOS ARTS

Centre for Transport and Navigation, Rijkswaterstaat, Ministry of Infrastructure and Environment, Delft, The Netherlands

3.2.4 When an Impact Assessment Misses the Point

This story describes an example of IA process and documents, which had no to negligible effects on decision making. The process and documents were treated as a procedural "annoyance." The story describes what went wrong and why. It addresses the ineffectiveness of EIA when compared to local authority political will.

A new shopping mall was constructed in Potchefstroom in 2007. Potchefstroom is a medium-sized town of approximately 163,000 people situated about 130 km south-west of Johannesburg in South Africa, on the banks of the Mooi River.

The Mooi River is a relatively small, perennial stream flowing through the center of town. Such perennial streams are uncommon in the semiarid conditions of large parts of South Africa. It is, therefore, an important part of the unique character and heritage of Potchefstroom. However, the area surrounding the river in the center of town was largely derelict land, mostly used by truckers for overnight parking, with associated waste and litter problems, in the absence of proper ablution facilities. The rest of the particular area was overgrown, rather unsightly, and not safe.

A developer applied for permission to build a modern and upmarket shopping mall on the most central portion of the derelict land, which straddles the stream. The proposed mall entailed a building that connected both banks and channeled the perennial river through a tunnel below the ground floor.

A comprehensive EIA was conducted with a number of specialist studies including an ecological assessment of the riparian vegetation and associated fauna, a traffic impact study, and a geo-hydrological study. It was accompanied by an extensive public participation exercise. Apart from the weak consideration of location alternatives, it can be regarded as a very well conducted EIA. The EIA was also aligned with a Water Use application by virtue of the location of the site in a floodplain and within the 1-in-50 years flood line.

The EIA was approved, and an Environmental Authorization and a Water Use License were issued, subject to a number of environmental management conditions being carried out in accordance with the environmental management plan.

On the basis of the Impact Assessment, various elaborate mitigation measures were implemented, including semi- or partly translucent floor panels in the mall to allow a limited amount of natural light into the tunnel by which the stream flowed beneath the mall, as well as allowing some of the natural banks of the stream to remain undisturbed. The construction phase also required careful maintenance of the river banks during construction of the mall and associated parking lots, roads, and bridges.

The EIA can therefore be seen to have been conducted correctly and to a good best practice standard. However, despite the standard of the EIA, it didn't address the bigger and more important issue of whether it is in the best interest of the town and the environment for a commercial enterprise to be constructed in a unique urban green belt situated around a perennial stream. Ideally, this should have been dealt with at the level of Strategic Environmental Assessment, but there was neither SEA nor an Environmental Management Framework nor a Green Belt Policy.

It was clearly the intent of the City Council to develop this mall with the associated income from the sale of municipal services and land rental in an area which the City Council was unable or unwilling to develop as an environmental resource, for example, by way of a an attractive and safe urban park and water area. In terms of the Constitution, as an organ of state, the City Council is the guardian of the environment for the citizens of the City, but by encouraging the development of this mall, the City Council delegated certain environmental maintenance responsibilities to the developer, and abdicated environmental governance responsibility to the Environmental Authority who had to approve or reject the EIA application.

The abdication of environmental responsibility by the City Council was facilitated by the administrative mismatch between municipal planning, which operates at the municipal level, whereas Environmental Authorisation operates at the provincial level. One consequence is that the Environmental Authority (situated 200 km away) does not have sufficient understanding of the full spectrum of environmental governance issues in the municipality.

Therefore, we see an example of a situation where the City Council had decided that the mall had to be built, and the EIA was therefore reduced to a procedural exercise, and had negligible effect on the decision.

LUKE SANDHAM

Geography and Environmental Management, School of Geo and Spatial Sciences, North West University, Potchefstroom, South Africa

3.3 DEFINING THE PROBLEM AND DECIDING ON A DIRECTION

3.3.1 Some Insights from Practice

The four stories, presented in the previous section, address the issue of IA influence at both the regulatory and the applied levels. The first story demonstrates the tendency of major stakeholders, including the government, to adopt positions regarding major proposals from the outset, and then to actively seek to "win the contest," largely operating outside the IA process. Even though the IA process proceeds, the expectation by all parties tends to be that, regardless of the outcome of the process, major decisions concerning if and how the proposal might proceed will be made outside the IA process. The dilemma then is how to make the IA process relevant to the needs and interests of all parties. The second story highlights a creative approach to using SEA to enhance public influence over policy making. It describes the creative and proactive application of a range of procedures for making Scottish plan-making more transparent, open, collaborative, and inclusive. The story demonstrates that, provided there is an early and ongoing commitment to public participation, practical methods are available for opening up policy making to public scrutiny, involvement, and influence. The third story illustrates the types of "real world" issues related to IA influence that tend to arise within organizations and institutions because of the perceptions, attitudes and behaviors of administrators and decision makers. The story uses the prevailing practice, an IA practitioner survey, and in-depth interviews to paint a realistic picture of both constraints to and opportunities for enhancing the procedural and substantive effectiveness of IA in informing and affecting decision making. The fourth story describes a situation in which the key environmental issue (i.e., whether the mall should be built) was not addressed by the EIA, and the EIA process and documents had a negligible influence on decision making.

3.3.2 IA Relevance and Influence Still Being Questioned

The fact that the relevance and influence of IA is still being questioned after close to 40 years of practice and application in nearly 100 countries is something of a surprise. Arguably, Impact Assessment, in all its forms, should be broadly viewed as a well-established form of environmental management—distinct from and complementary to other forms of environmental management. It should be generally acknowledged as necessary and worthwhile. IA practitioners should be widely seen as accomplished and highly trained professionals essential to IA design and application. The IA methodological foundation, at both the regulatory and the applied levels, should, for the most part, be regarded as conceptually sound, substantive, and proven. Questions regarding the utility and net benefits of IA largely should have subsided. Instead, the focus should be on broadening and deepening the IA knowledge base, and on refining and polishing IA practice.

An overview of IA-related literature tends to echo this generally positive impression. However, perceptions in IA practice are more mixed. One perspective that tends to undermine IA's effectiveness is the view that there is very little to IA methodology. As such, other professionals and specialists (e.g., engineers, specialists in the natural and social scientists, lawyers), it is maintained, can readily acquire (on the job) the requisite knowledge and skills. This perspective is sometimes found among project managers responsible for capital projects subject to IA requirements, and among legal professionals responsible for advising clients on how best to satisfy the government requirements.

A second perspective questions whether IA is necessary, worthwhile, or even feasible. This perspective is often found within the academic community (that questions the validity of IA as a body of knowledge), among environmental advocates (who question whether IA, on balance, advances or undermines environmental objectives), and among politicians, proponents, and sometimes administrators (who see IA as, at best, a necessary procedural "nuisance" that is largely unrelated to decision making).

A third perspective argues that the purpose of IA (i.e., making planning and decision making more environmentally substantive) is already being achieved through a combination of actions (policies, plans, programs, and projects) directed toward environmental ends and actions where environmental considerations are systematically integrated into the planning and decision-making process. This perspective maintains that IA requirements are either superfluous or a negative influence (e.g., wasted resources, distorts planning and decision making). This view tends to be held by policy makers and planners responsible for formulating and managing the preparation and review of policies, plans, programs, and projects.

The identification of these negative perspectives immediately raises a series of questions. What is the nature of these perspectives? How widely are they held? To what degree are they valid? To what degree are they undermining IA's effectiveness? What steps can be taken, at the conceptual, regulatory, and applied levels, to avoid and counter these negative perspectives? In what ways can these individual initiatives be integrated into a coherent set of reforms to IA requirements and processes? This section begins by characterizing the perspectives (i.e., defining the problem). It then establishes a general direction for resolving the problem.

3.3.3 Perspective 1—IA—What Could Be Simpler?

According to Perspective 1, as illustrated by Figure 3.1, and as described below, IA is a relatively simple and straightforward procedure. It is a procedure capable of effective management with only a cursory knowledge of IA methodology. The IA process, for example, is simple and logical. It closely mirrors rational planning formulations familiar to anyone with even a superficial exposure to public policy, planning or administration. A proposal is brought forward. Regulatory requirements are determined through screening. The overall IA process is designed and focused through scoping. Proposal characteristics are determined. Environmental data are collected, analyzed, and interpreted. Future environmental conditions, with and without the proposed action, and relative to other actions affecting the same environmental conditions (i.e., cumulative effects), are identified, predicted, and interpreted. Impacts are avoided and minimized, to the extent practical, by considering alternatives and mitigation measures. Measures to monitor future environmental conditions and to manage potential impacts are formulated. Agency requirements and public concerns are integrated into the process prior to key decisions. All of the above is documented in technical and summary reports that provide the basis for agency review, and for political decision making. The proposed action is then rejected or approved. Approval, when it takes place, is generally subject to an array of conditions.

Impact Assessments are generally prepared by teams of proposal-related and environmental specialists. These teams are generally headed by experienced proposal or environmental experts (e.g., engineers, planners), supplemented by procedural experts (e.g., lawyers, public participation). The management team is expected to work closely with proponent representatives, and with a team of proposal and environmental specialists. Most, often all, of the IA methodological knowledge employed in the process is derived from reviews of IA regulatory requirements and guidelines, from discussions with government agencies, from experiences with previous IA processes, and from reviews of IA documents associated with other projects. The IA knowledge base is assumed to be limited, simple, and largely static. IA team managers and members believe that they can acquire the essentials of IA methodology "on the job" in the course of preparing IA documents, conducting IA processes, and interacting with colleagues. It is further assumed that there is no need for additional methodological guidance beyond what can be accessed from a quick perusal of existing IA guidelines and introductory texts. All that good practice requires is for IA practitioners to "polish" their knowledge through experience and, where practical, through discussions with peers in the same and related fields. Anyone with good general project manager

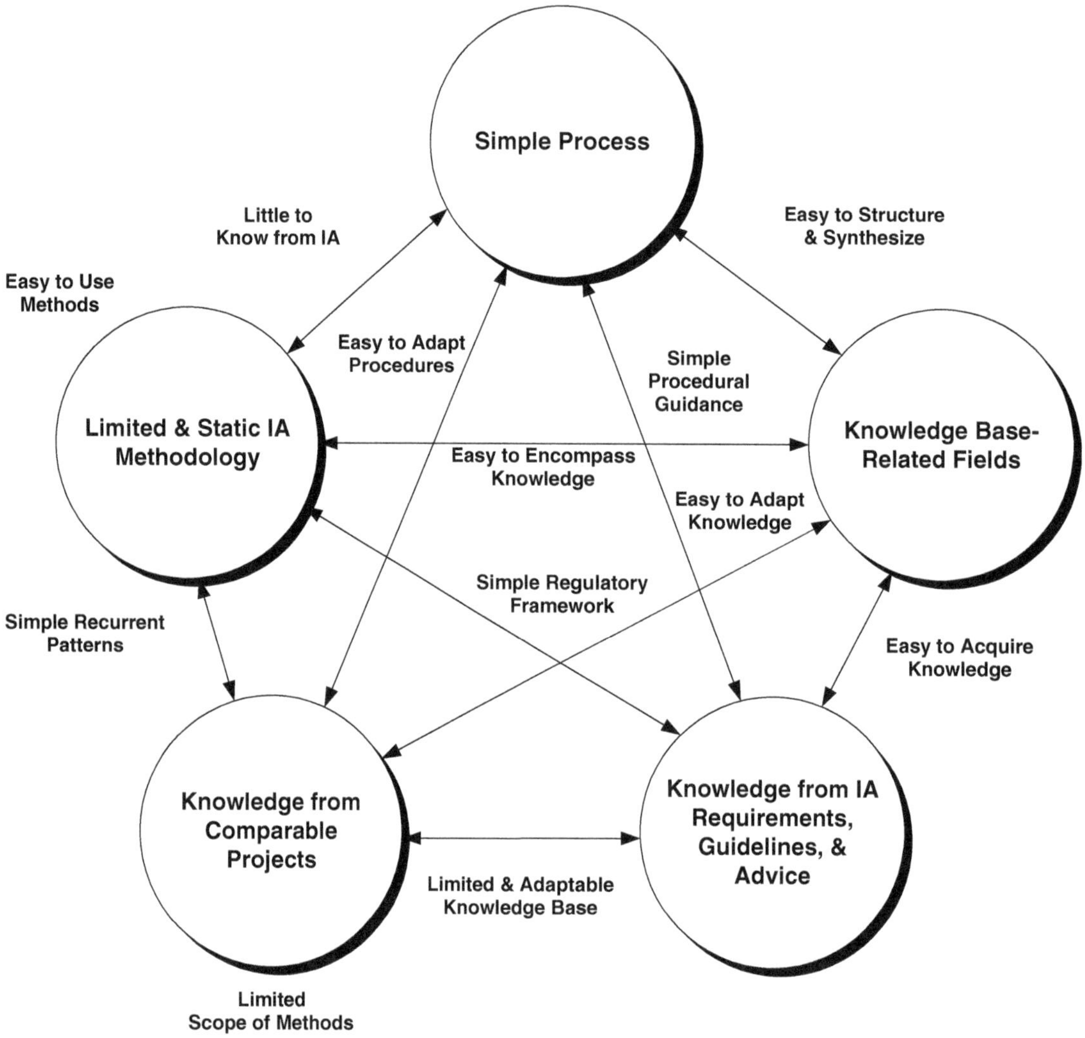

Figure 3.1 Perspective 1: the "what could be simpler" perspective.

skills is capable of combining the inputs of the specialists into a coherent set of documents, and within a sound planning process.

Perspective 1 is not widely embraced in IA literature as an appropriate model for practice, notwithstanding calls for IA practice to exhibit more "common sense" (Ross et al., 2006). This does not, however, mean that it is not a widely held view among those who actually prepare IA documents and manage IA processes. This could partially explain the inadequate theoretical basis of IA (Cashmore, 2004), and the frequently cited gulfs between theory and practice (Cashmore, 2004; Cashmore et al., 2010; Lee, 2006), practice and decision making (Bond and Morrison-Saunders, 2011; Cashmore et al., 2010; Deelstra et al., 2003; Galbraith et al., 2007), and process and substance (Benson, 2003; Cashmore, 2004). It also might provide reasons for the extremely slow and limited IA learning curve (some would maintain the absence of a learning curve) (Tzoumis, 2007).

3.3.4 Perspective 2—IA—It Can't, Won't, or Shouldn't Be Done

Perspective 2 (see Figure 3.2) asserts that an IA cannot, will not, and should not continue to be institutionalized and applied as a decision-making aid. A variety of overlapping and mutually reinforcing reasons have been offered up to support this conclusion.

IA boundaries are inherently artificial. It can always be argued that more alternatives, further aspects of the environment, more proposed action information, additional direct, indirect and cumulative effects, and more impact management measures should have been considered, or considered at a greater level of detail. Opinions also tend to vary, often dramatically, regarding if and which thresholds should be used in screening requirements, concerning the appropriate procedures for defining study areas and time horizons, and

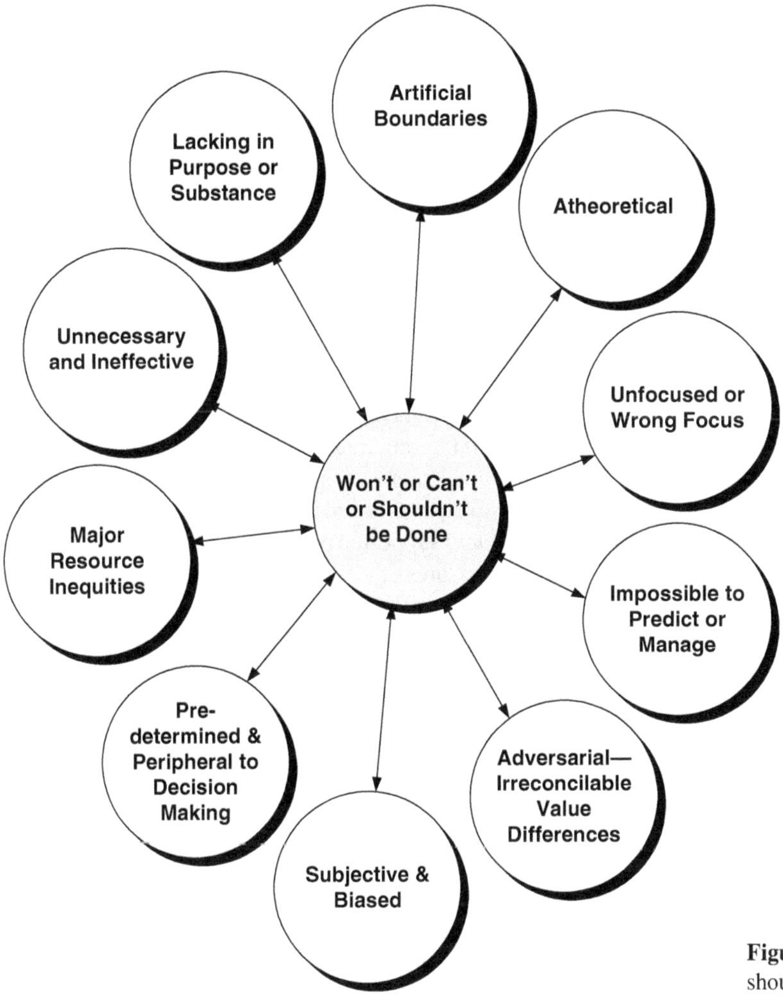

Figure 3.2 Perspective 2: the "it can't, won't, or shouldn't be done" perspective.

pertaining to how best to scope environmental attributes, impacts, and the extent of involvement of various participants in the process. Any attempts to circumscribe such analyses to manageable proportions are subjective and arbitrary.

IA is atheoretical. The theoretical state of the art of IA is not adequately developed. Also, there remains a persistently wide gulf between the conceptual foundation of IA and IA practice and between potential and outcome. To the extent that the conceptual basis for IA is applied through, for example, the application of "good practices," there is a tendency to apply frameworks, concepts, and methods without the necessary contextual adjustments. With major theoretical gaps and little interaction between theory and practice, it is little wonder that the IA "learning curve" is so gradual and erratic that flaws and pitfalls continue to be repeated in practice.

IA lacks focus or tends to focus on the wrong things. Time and resources are never unlimited. A misplaced attempt to be comprehensive can mean far too much attention and resources are devoted to aspects of the environment and potential impacts of limited significance, and far too little resources and time are devoted to those aspects of the environment most susceptible to proposal-induced changes and most likely to experience severe, proposal-induced changes. At the same time, focusing too early in the process, and then remaining inflexible, can result in critical concerns being missed or receiving insufficient attention. This combination of lack of focus and inappropriate focus has contributed to an IA performance that all too often is well below potential, and has failed to show improvement consistently over time and among agencies.

Environmental condition changes and effects are impossible to predict or manage over the long term. Forecasting changes in proposal characteristics, in environmental conditions, and, most importantly, in the interactions between the two, for many decades into the future is extremely difficult and commonly plagued by major, difficult to bound, uncertainties. It verges on the impossible as the orientation shifts from the "is" and "is likely to be" to the even more subjective realm of "what's important" and "what should be done." Cumulative Effects Assessment (CEA) can even more problematic. The state-of-IA practice, as evidenced by the major gap between theory and practice, reinforces this perspective. The net result of all these limitations and questionable practices could be seen as an amorphous mass of glib generalizations, baseless speculations, wishful thinking, and untestable assertions.

IA practice is confounded by irreconcilable value differences and is inherently adversarial. Value systems invariably in IA practice are almost always extremely diverse. Conflicting perspectives, values, and belief systems abound. Often diametrically opposed and irreconcilable positions are adopted from the outset, and maintained throughout the IA process. Adversarial IA processes, if anything, compound these divisions. Public consultation is commonly viewed as a mechanism from informing and educating the public regarding the merits of the proposed action. Without meaningful public participation, the IA process lacks legitimacy. When the IA process is not considered legitimate or relevant by interested and affected parties, it is little wonder that those parties so often choose to advance their interests outside the IA process.

IA processes and practices are inherently biased and subjective. IA tends to depict itself as an "impartial" decision-making tool. If anything, it is ostensibly inclined toward more environmentally sound and more open and inclusive decision making. These aspirations are rarely achieved in practice. Notwithstanding its environmental aspirations, IA proposal review processes continue to be dominated by entrenched interests. Although cloaked in an aura of objectivity and impartiality, subjective judgments are inherent to every step of the IA process. When those judgments are not substantiated, but rather are simply presented as expert pronouncements or conclusions, bias can rapidly creep into and dominate IA processes and documents.

IA-related decision making tends to occur prior to and/or outside the IA process. The IA process is "triggered," in most cases, by a proposed action on behalf of a proponent. Usually the proponent's planning is already well advanced. External political and economic decision-making factors generally hold sway when it comes to deciding whether a proposed action should or should not proceed. The best that can be hoped for is slightly less environmentally intrusive actions where the most severe or, at least the easiest to measure and/or manage, adverse effects are ameliorated by mitigation measures, grafted on to a largely predetermined action. The worst that can occur is the illusion of environmentally responsible decision making while the real, often environmentally irresponsible, decision making is taking place outside the IA process.

Inequities in resource distribution result in a largely meaningless IA process. IA resources tend to be concentrated with the proponent (who arguably has a decided bias), and, to a lesser degree, with the government regulators (who are prone to political influence and whose perspectives are confined by narrowly defined, regulatory compliance boundaries). Other interested and affected parties must "make do" with a tiny fraction of the resources available to the proponents and regulators. Resources for IA practice also are invariably "front-end loaded." Once a proposed action is approved, the resources remaining for monitoring, enforcement, contingency measures, and auditing tend to be minimal.

IA requirements and processes are neither necessary nor effective. IA has been around long enough that it is reasonable to ask whether it has been worth the effort. The list of practice-related flaws is both lengthy and, arguably, all encompassing (e.g., undisclosed methods, scoping too narrow and inflexible, insufficient information, the weak treatment of alternatives, uncertainty and cumulative and transboundary effects, over-simplified analyses, poor integration of health and sustainability concerns, vague, arbitrary, and inconsistent significance interpretations, preoccupied with document preparation, inadequate adaptations to context, documents inaccessible to the public and decision-makers, token, belated and inadequately supported public involvement, non-specific mitigation measures, the tendency to ignore postapproval monitoring and management). These deficiencies have occurred, notwithstanding a range of IA texts, journals and applied research studies, and forums seeking to provide methodological advice to IA practitioners and other stakeholders. If IA practitioners are only using the available applied resources to a very limited extent and those with real decision-making power hardly at all, are continuing efforts to enhance the IA state-of-practice really worth the effort?

IA has either not made a substantive environmental contribution or it is unclear what the nature and extent of that contribution might be. IA has long been criticized for an almost complete preoccupation with procedure and documents over either substantive goals or outcomes. Even the modest objective of taking into account and ameliorating the likely environmental consequences of proposed developments has only been achieved to a limited degree. When measured against sustainability objectives, the outcomes are even more dubious and limited. These very modest, at best, achievements have occurred notwithstanding the vast resources consumed by a largely procedural (arguably vacuous) set of institutional arrangements, largely divorced from actual decision making.

3.3.5 Perspective 3—We Already Do That

Perspective 3, as illustrated by Figure 3.3, starts from the premise that the major purpose of IA (i.e., more environmentally sound decision making) is or can be achieved through environmentally driven and shaped policies, plans, programs, and projects, without the imposition of "action-forcing" IA requirements. Proponents of this perspective point to the many public policies, plans, and programs designed and implemented to achieve environmental aspirations—initiatives launched and undertaken without the necessity of grafting on IA requirements. They maintain that there already is a long tradition of infusing environmental perspectives and knowledge into public and private

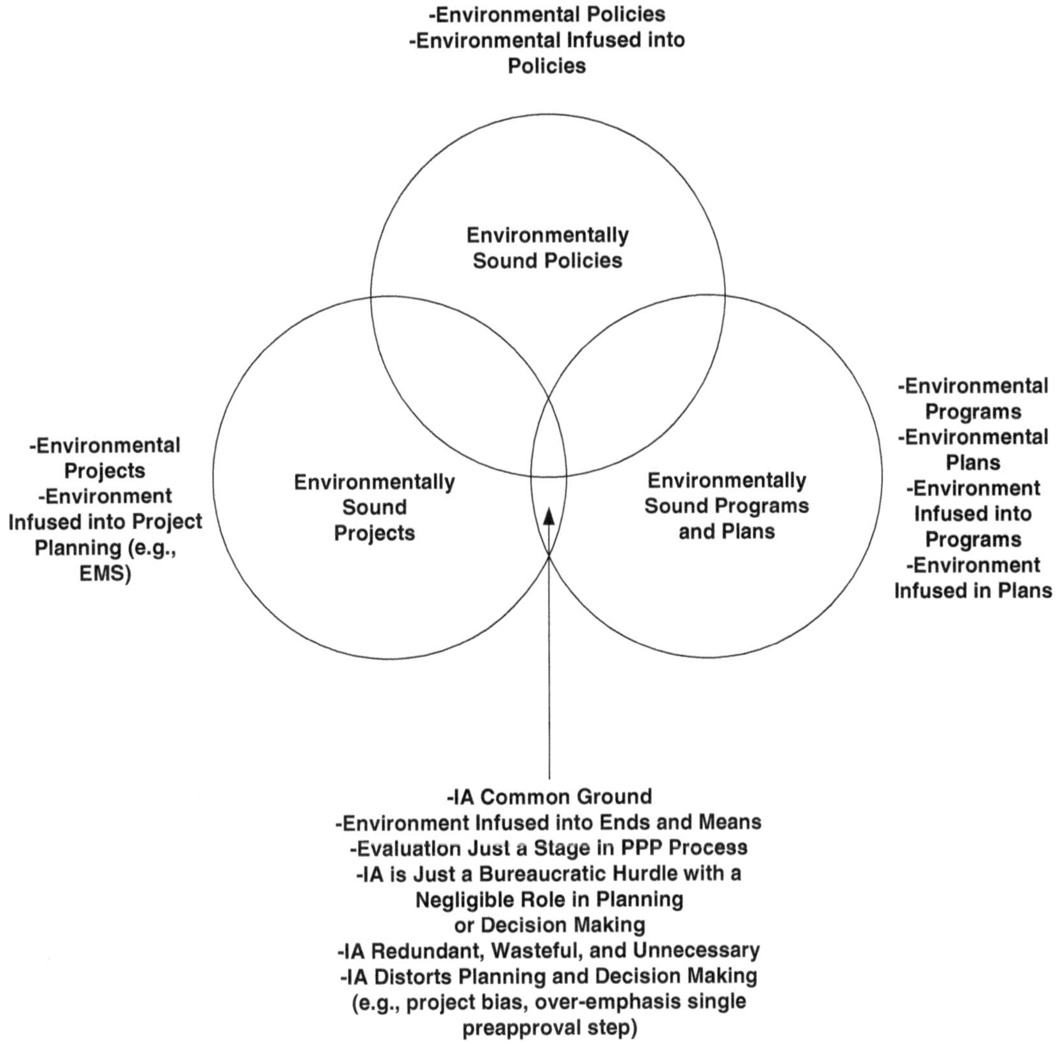

Figure 3.3 Perspective 3: we already do that (mostly or even better).

decision making. Reference often is made, for example, to land use and resource management plans that avoid and minimize cumulative effects on environmentally sensitive areas and features; to infrastructure and urban development plans and programs that seek to minimize the environmental "footprint" of human activities; to the integration of climate change concerns into public decision making at all levels; and to numerous initiatives (e.g., EMS) that effectively integrate environmental concerns into private-sector planning and decision making.

Perspective 3 is often common to public and private policy, plan, program, and project specialists. These individuals tend to assert that they have applied environmental values, principles, and good practices in their professional activities for many years. Therefore, they argue, they neither need nor want external advice or direction regarding how best to reorient their actions toward environmental ends. They tend to resent the presence of IA requirements, arguing that such requirements are an additional and unnecessary burden. They stress that their time and other resources are already tightly constrained. The addition of "paper pushing" IA exercises only hinders, they suggest, their ability to perform their already environmentally oriented functions in an efficient and effective manner.

Advocates of Perspective 3 tend to see IA as the artificial separation or rather duplication of the planning process evaluation step—a step well established in public and private planning and decision-making processes and fully supported by an extensive plan, program, and project evaluation literature. Not only does IA artificially segment planning processes, they argue, it distorts planning processes by introducing biases that have been carefully avoided in public policy and planning research and practice. IA, they maintain, notwithstanding efforts to the contrary, still reflects a heavy project bias. This bias infuses, they argue, all aspects of IA theory and practice. As a consequence, when applied to the evaluation of policies, programs, and plans, it impedes the effective execution of the evaluation step. IA further distorts and impedes planning and decision making because it over-emphasizes a single preapproval step. As a result it does not reflect the continuity and multiplicity of planning and decision-making steps characteristic of both public and private decision making. These biases and distortions, they argue, tend to mean that not only is IA wasteful, unnecessary, and redundant, it is often a negative influence on environmentally sound planning and decision making.

3.3.6 Where Do We Go from Here? (The Direction)

Sections 3.3.3–3.3.5 detail the problem as represented by the three negative perspectives. The next step (presented in Section 3.4) is to explore the validity of these perspectives together with exploring ways in which these perspectives can be refuted (to the extent that they are not valid) and offset (to the extent that they are valid). Included in this analysis are key concepts and integrative frameworks, together with knowledge base and research priorities. Section 3.5 summarizes how IA can be made more influential at the regulatory level. Examples of pertinent reforms from the four jurisdictions are included. An overview of an influential IA process is then presented. Adaptations for different IA types also are highlighted. Section 3.6 addresses the contemporary challenge of good practice guidance for making IA more influential over decision making. Section 3.7 presents an overall summary of the analysis.

3.4 SELECTING THE MOST APPROPRIATE ROUTE

3.4.1 The Legitimacy of Perspective 1

Perspective 1 is correct in maintaining that the basic IA process and general IA methodology (in a "broad brush" sense) is not and should not be complex. All parties in the IA process must be able to understand the major steps in the process and make their contributions prior to major decisions in the process. It also must be possible to prepare summary materials and documents in a form that can be readily understood, and promptly reviewed by government agencies, the public, and elected representatives. But it does not follow that how best to design and apply the IA process is self-evident. Nor does it follow that IA methodology constitutes no more than compiling (in consistently formatted technical documents), and summarizing (in user-friendly summary reports) specialist analyses, and the inputs of government reviewers and the public.

Over the close to 40 years that IA requirements have been in place, a considerable IA methodological knowledge base has been constructed. That knowledge base pertains to IA institutional arrangements, to IA activities (e.g., impact prediction, impact management, the evaluation of alternatives), to the many ways in which individual IA activities can be subdivided, linked and aggregated into overall IA processes, and to a variety of forms of IA (e.g., SIA, SEA, technology assessment). The range and scope of IA theory and practice has broadened considerably, especially over the past two decades. Although a consensus is gradually emerging regarding many aspects of good and poor practice, overlapping and competing perspectives still abound regarding how IA should and should not be conducted. To some extent these differences can be attributed to contextual differences (e.g., social, cultural, political, and ecological variations). However, they also reflect deeply ceded value, perspective, and ideological differences—differences that often are expressed in varying interpretations of IA's effectiveness. The IA knowledge base is, in addition, highly permeable. Concepts, theories, frameworks, and distinctions are frequently borrowed and adapted, not always successfully, from various social and natural science disciplines, and from related fields of practice.

Admittedly then, the IA knowledge base is complex. It necessitates careful interpretation, and a considerable sensitivity to subtle distinctions and the potential implications of uncertainties. It does not, however, follow that the IA knowledge base is, therefore, irrelevant, impractical, or irreparably compromised. To gloss over all this considerable body of knowledge and experience, as if it were irrelevant to IA practice, as proponents of Perspective 1 would have, is dangerously naïve. Surveys of IA practitioners and analyses of IA practice demonstrate that there is considerable room for improvement. The deficiencies, cited in practice in recent years, often mirror those identified decades ago. Arguably, after more than three decades of experience, this should not be happening with the same frequency, and to the same degree. At the same time there are numerous examples of sound and effective IA processes, built to a considerable degree upon a foundation of well-designed, applied, and adapted methodology.

Although each IA process has its unique aspects, it should be possible to steadily improve IA quality and effectiveness if IA practitioners are systematically learning from the positive and negative experiences of other practitioners, and from insights available through IA and related literature. The knowledge base of related fields can provide many valuable insights and relevant methods. Substantial experience and working knowledge, however, is required when drawing upon and adapting procedures and methods from other forms of environmental management and from other fields of study and practice. Adapting discipline-specific knowledge and methods to applied, prescriptive, and interdisciplinary situations can be especially difficult. IA, therefore, is far from simple, static, mature or easily acquired "on the run." Any additional advice that systematically draws upon and supplements the conceptual foundation of IA, and is grounded in IA practice, should be welcomed rather than dismissed as unnecessary.

3.4.2 The Legitimacy of Perspective 2

Perspective 2 argues that IA contributes next to nothing to our understanding of the environment, and is a fatally flawed environmental management tool. It offers up numerous shortcomings, some related to the conceptual foundation of IA, some that represent external constraints, and some that constitute practice-based deficiencies. Each assertion needs to be carefully explored to determine its legitimacy and, where potentially valid, offsetting measures need to be instituted.

Boundaries Judgments must, indeed, be made in bounding and focusing IA requirements and processes. Subjective interpretations are inherent to IA as they are in all applied fields of practice. The issue then is not objective versus subjective judgments but rather how well the subjective judgments are substantiated by clear and explicit methods and by consistent, thorough, thoughtfully reasoned, and broadly supported arguments. Subjective judgments, in particular, tend to have greater validity if they are strengthened by the direct participation of and support by all interested and affected parties.

Atheoretical The data collected through IA practice is primarily for decision-making purposes. As such, in IA data collection, analysis, and interpretation, more emphasis is placed on relevance, prescription, and application than on rigor, explanation, and scientific knowledge accumulation. IA also has many subjective elements, and is highly context dependent (i.e., more relative that absolute). These decision-centered tendencies constrain but do not preclude an IA knowledge contribution role for environmental management purposes (e.g., best practical science) and for an enhanced understanding of environmental systems. Many proposals have been made and constructive examples documented for more effectively grounding IA theory in practice, for enhancing the connections from practice back to theory, for narrowing the gap between decision-making and scientific requirements, and for establishing a coherent and comprehensive IA research strategy.

Unfocused or Wrong Focus IA requirements can facilitate greater focus through well-defined roles and responsibilities, mandatory scoping, and ongoing efforts to streamline and simplify procedures. In a substantive sense, IA can become more focused if substantive goals, especially sustainability, become the touchstone against which outcomes from IA institutional arrangements and IA processes are assessed. Procedurally, IA practitioners, organizations, and professional bodies can "raise the bar" concerning IA practice and can demand better quality assurance. Closer connections between IA research and decision-making demands are needed. An enhanced understanding of power and politics is required as a means of bridging IA research and decision making. Sufficient resources also need to be devoted to postapproval monitoring and management in order to reduce impacts, facilitate better impact prediction, and contribute to the IA knowledge base.

Impossible to Predict and Manage Uncertainty is inherent to all forms of decision making. This is especially the case for actions undertaken over lengthy periods of time, and whose effects may extend even further into the future. Even greater uncertainties are associated with cumulative effects. Multiple procedures are available for addressing, or, at least containing to manageable levels, these uncertainties and risks. IA mechanisms are likely to be even more effective if they encompass multiple decision-making levels, if they broadly define the environment, if they are guided by substantive environmental objectives, if uncertainties are acknowledged and their associated implications acknowledged, if they devote sufficient time and resources to impact management and IA auditing, if they effectively integrate the perspectives of all interested and affected parties, if they allow for the possibility of proposal rejection when there is a combination of major uncertainties and potentially severe consequences, and if proposed policies, plans, and projects are designed and undertaken with uncertainty management and adaptation as fundamental guiding principles.

Adversarial—Irreconcilable Value Differences Diverse, often conflicting, value systems are, indeed, an inherent feature of IA practice. This fact does not need to unduly constraint or confound IA practice. It can be viewed as a positive and necessary feature. For, given the complexity of IA practice and decision-making environments, it is necessary and desirable to integrate a wealth of knowledge and diverse perspectives. It can be easier to contain and manage conflicts within the IA process if, from the outset, the process is guided by a broadly discussed and supported vision and complementary set of goals regarding what the IA is intended to achieve. Early and ongoing discussion among interested and affected parties, supported by an inclusive and interactive public participation program, is essential if shared interests are to be identified and if conflicts are to be explored in a manner that does not derail the IA process.

Biased and Subjective IA decision making can, indeed, be problematic. Numerous potential biases can constrain, sometimes severely, sometimes even fatally, IA practice at both the regulatory and applied levels. But these biases can, in many cases, be prevented or, at least, partially offset. Numerous subjective judgments also must be made throughout the IA process. Subjective bounding and interpretative judgments must be justified, preferably by means of transparent procedures that integrate a diversity of perspectives. Such procedures are available to IA practitioners. At the regulatory level, bias and subjectivity can be ameliorated with clear legal standards and applicability rules, and a broad definition of the environment and effects. At the applied level, bias and weakly supported subjective judgments are less likely if the public assumes an early, active, and collaborative role in the IA process.

Predetermined and Peripheral to Decision Making IA requirements and processes are intended to inform decision making. This does not mean that the advice offered will always be followed nor does it mean that other factors, outside the IA process, will not influence decision making. Concomitantly, it does not follow that IA processes will have no or negligible influence on decision making. All too often the lack of IA decision-making influence can be traced to poor IA practice. There are many reforms, conducive to narrowing the gap between IA and decision making, which can be made to IA regulatory requirements and processes. Sincere and capable proponents, environmental and IA specialists, regulators and other stakeholders can and have collectively contributed to better IA practice. Given the contentious and political nature of IA, approval and implementation is far more likely to occur if supported by a coalition of interested and affected parties.

Major Resource Inequities Resource distribution inequities are commonplace in IA practice. However, it does not follow that such inequities cannot be offset nor does it follow that steps cannot be taken to ensure that the perspectives, values, and interests of those with more limited resources cannot have a vital, in some cases decisive, role in IA related decision making.

The role of public participation in the IA process can be crucial in offsetting resource inequities. Regulatory reforms (e.g., broadly define the public, provide effective notice, provide participant funding, provide processes to address policy issues, assess public participation effectiveness) can help set the stage for an open, balanced, and collaborative IA process. The public needs to be involved early in the IA process, their concerns need to be taken seriously, and funding and assistance need to be provided for independent peer reviews, to allow stakeholders to conduct or fund their own IA analyses, to support process enablers such as facilitators and mediators, and for independent IA decision making (e.g., adjudicators, arbitrators). The provision of sufficient resources for IA auditing and during the postapproval stage also is crucial to ensuring that commitments are adhered to, and to facilitate ongoing public involvement in impact management.

Unnecessary and Ineffective On the question of whether IA has anything to contribute to environmental decision making, IA provides an environment-driven, "action-forcing" mechanism for evaluating proposed actions. Other environmental management instruments, in conjunction with a multilevel IA system, can contribute to more environmentally sound decisions. What they lack, without IA, though, is a centralized, in-depth, and "action-forcing" decision-making framework for evaluating proposed plans, programs, projects, and activities with potentially significant adverse environmental impacts. It is, moreover, hyperbole to suggest that IA has been totally or overwhelmingly ineffective. There is an ample record of IA leading to the enhanced consideration of environmental factors in decision making and to environmental enhancements. Improvements, albeit incremental, have been made to the government, institutional, and corporate perspectives, procedures, and proposed actions. Public and private decision making, for many proposed actions with potentially significant environmental effects, has been made more transparent and more open to public involvement. Institutional and corporate learning has occurred. There are numerous examples of environmentally unsound proposals being withdrawn, being improved to the point of becoming environmentally acceptable, and being rejected. Cost savings have been realized, the role of legal proceedings has been ameliorated, and public acceptance has been increased. Notwithstanding such positive examples, it is not an understatement to suggest that IA practice has fallen well short of its potential, and remains much less relevant, to decision making and in a substantive environmental sense, than it should be.

Lacking in Purpose or Substance The theme that IA is obsessed with process, almost to the complete exclusion of substance, is almost endemic in IA literature, and among IA critics and stakeholders. Although valid to a point, the dichotomy is not as sharp as is commonly portrayed. Process and substance can be complementary—one can facilitate the other. A well-designed and managed IA process can contribute to the achievement of environmental objectives. Environmental objectives can, in turn, guide and bound IA requirements and processes. It, however, is fair to argue that more attention needs to be devoted, in IA theory and practice, to environmental, especially sustainability, outcomes. Such a reorientation would entail, at the regulatory level, giving IA a statutory purpose and a clearly defined set of environmental and sustainability objectives, acceptability thresholds, and trade-off rules. At the level of the individual SEA and EIA process, greater attention could be devoted to collaboratively formulating environmentally sustainability visions, objectives, and criteria, and to more fully and effectively integrating social, economic, cultural, and health concerns.

3.4.3 The Legitimacy of Perspective 3

Perspective 3 is partially valid. Many policies, plans, programs, and projects are directed toward environmental ends. An environmental perspective is reflected in many policies, plans, programs, and projects. The evaluation step is intrinsic to policy making, program formulation, plan making, and project planning. Oftentimes, it encompasses environmental considerations. IA, when it simply is a belated "add-on" to existing planning procedures, can be of minimal planning or decision-making value while consuming valuable and limited resources. IA is indeed prone to a bias toward project-related methods and to a preoccupation toward a single preapproval stage. However, Perspective 3 overstates the extent to which the planning of policies, plans, programs, and projects, without the integration of IA requirements, fully and adequately addresses environmental concerns. It also overstates the deficiencies and limitations of IA as a field of theory and practice.

In an ideal world, IA would not be necessary. Planning and decision making at all levels would be directed towards environmental ends and would fully integrate environmental considerations into every planning and decision-making step. Such, however, is rarely the case. The reason for this shortfall can be partially attributed to external (e.g., political, institutional) factors that strongly influence planning at every level. However, it also is a reflection of deficiencies in contemporary planning and decision making—deficiencies that can be partially remedied through IA requirements and good practices. These deficiencies, as described below, stem in part from fallacies regarding environmental practice, both with and without IA requirements.

Environmental Proposals Have Negligible Adverse Environmental Impacts It is often assumed, erroneously, that a proposed action, in the form of a policy, plan, program or project, which is intended to achieve an environmental purpose (e.g., "green" projects, policies, or programs to reduce dependence on fossil fuels), are inherently positive in an environmental sense. For this reason, such proposals tend to be exempted from or "fast-tracked" with respect to environmental approvals. However, such proposed actions often generate adverse environmental consequences—sometimes to the point that the net environmental effects are negative. Also, there are the possibilities that alternative actions could represent a more environmentally effective allocation of resources or potential adverse environmental effects can be prevented or ameliorated. IA requirements and practices, when applied to such proposed actions, can effectively address whether, on balance, such proposed actions are environmentally desirable, whether they are superior to other potential proposed actions, and whether associated adverse effects can be prevented or reduced.

Environmental Considerations and Aspirations Are Enough There tends to be the expectation, at the policy-/planning/program level, that broad environmental goals and the consideration of environmental concerns is sufficient to ensure the adequate integration of environmental perspectives into planning and decision making at all levels. Intentions, however, are not the same as accomplishments. Nor is the consideration of environment equivalent to the "good practice" systematic integration of environmental concerns into every process activity in a manner that is transparent, systematic, comprehensive, consistent, accountable, and inclusive. Laudable aims can be undermined by ineffective means. Intended outcomes can be more than offset by unintended or inadequately considered consequences. IA requirements and practices enhance the potential for environmental aspirations to be realized, reduce the likelihood and severity of unintended environmental consequences, and facilitate the realization of a level of practice that is unlikely to be achieved when environmental concerns are addressed inconsistently and in a manner that lacks transparency and rigor, fails to fully include the perspectives of all interested and affected parties, falls short of good practice standards, and does not provide for independent and transparent external scrutiny.

Action-Forcing IA Requirements Are Unnecessary Advocates of this perspective tend to start by citing examples of environmentally conscious policies, plans, programs, and projects. Examples, however, are not necessarily the norm. Being environmentally conscious, moreover, does not mean that such examples are consistent with either good practice standards or even rise to the level of the minimally acceptable. Ample, arguably many if not most, proposed actions, with or without the environmental moniker, tend to fall well short of the possible when addressing such matters as cumulative effects, the systematic generation and evaluation of alternatives, a traceable approach to significance determination, and the management of impacts. IA requirements provide minimal performance standards. IA guidelines provide good practice guidance. Without such requirements and guidance the assessment and management of environmental effects is likely to be of highly variable quality, lacking in transparency, less open and inclusive than it should be, and less than systematic.

Impact Assessment Is Inherently Biased and Has Nothing to Contribute Methodologically The fact that IA originated as a field of theory and practice largely at the project level and has a tendency to focus on preapproval requirements does not mean that these are inherent, fixed qualities, arguably biases, of the field. IA has expanded rapidly in recent years to encompass all decision-making levels (e.g., policies, plans, programs) and other forms of impact assessment (e.g., SEA, SA, HIA, EcIA, SIA). Over the course of this expansion, considerable thought and effort has been given to procedural and methodological implications of operating at different decision-making levels and of addressing varying substantive environmental concerns and aspirations. These deliberations have resulted in a major reorientation of the field at both conceptual and applied levels. Major differences in approach and orientation are now evident among IA types and in the application of IA at varying decision-making levels. Particular care has been given to avoiding and minimizing any biases instilled in the early years of the field and to the implications of linking, merging, and transferring approaches among various IA types and levels.

The potential contribution of IA to policy making, to plan and program making, and to project planning is considerable. Substantial IA methodological development has occurred in addressing such matters as the prediction and interpretation of individual and cumulative effects, the generation and evaluation of alternatives, stakeholder involvement, and impact management. These advances complement rather than duplicate methodological advances in such related fields as policy making, plan and program making, and project planning. Methodological synthesis across these and other related fields offers major potential benefits to each field individually and jointly.

3.4.4 Perspectives 1–3 Conclusions

Perspectives 1, 2, and 3 are valid to a point. They exemplify clear, often recurrent, deficiencies in IA practice. These deficiencies are not inherent or inevitable to IA practice. They also are overstated. Many positive examples, which run counter to Perspectives 1–3, are evident in IA practice. Moreover, there remains considerable operating room for IA practitioners and other parties to further facilitate, within the boundaries established by IA requirements and external restraints, good IA practice. Good IA practice also can contribute to regulatory improvements. It is possible, to some degree, for both regulators and practitioners, especially if they work cooperatively with other stakeholders, to push back and even redefine the "limits" that tend to inhibit IA quality and effectiveness. This does not mean that no constraints will remain regarding what is possible through good IA practice. But it does suggest that there is considerable middle ground between what is currently being achieved and what is possible to achieve. The question then becomes how best to narrow the gap between the possible and actual and, in the process, to make IA practice more relevant and influential. This entails shifting the orientation from the reactive (countering negative perspectives) to the proactive (constructive, systematic efforts to make IA more relevant and influential). The first step in this bridging process is the identification of critical concepts, frameworks, and research priorities.

3.4.5 Establishing a Foundation

This section bridges the reactive (responding to Perspectives 1–3) and the proactive (presenting an influential IA regulatory structure and an influential IA process). Concepts, integrative frameworks and knowledge base, and research priorities, pertinent to designing and managing relevant and influential IA regulatory and applied systems and processes, are described.

Concepts The term *relevance* encompasses a demonstrable and logical connection (implying a traceable link), a threshold for distinguishing between the relevant from the irrelevant (implying a subjective, value-full interpretation), a process (for establishing and applying relevance thresholds), one or more perspectives (as in who makes and is affected by the connection), and an outcome (as in the implications for decision making and the environment). Accordingly, then, IA requirements and practices are more likely to be "relevant" if they are logical, understandable, and credible; if the value-basis and procedures for applying those values are clear, consistent, impartial, and inclusive; if the values that drive IA requirements and processes match up with the perspectives and values of interested and potentially affected parties; and if the outcomes, both procedural and substantive, from IA accord with environmental objectives and the needs and preferences of stakeholders (IAIA, 1999).

Influence, in the context of IA, is about the power to affect decisions and actions that have potentially significant impacts on the environment. The traditional and conventional assumption in IA practice is that the only appropriate means of influencing such decisions is to inform decision makers through IA documents that objectively present relevant facts and analyses. IA documents are, in turn, the product of a systematic, rational planning process. This model for influencing decision making is tempered to the extent that it is recognized that summary documents are needed to facilitate understanding by decision makers and the public, and the process needs to be opened up to incorporate (usually only in a responsive sense) the concerns, knowledge, and preferences of interested and affected parties, including politicians and the public.

Not surprisingly, given this model, IA has, at best, a mixed and often limited record of influencing decision making. IA literature and, to a lesser extent, IA practice have increasingly recognized that the nexus between IA and decision making requires closer scrutiny. The route between IA and decision makers, for example, is generally filtered through a complex array of legal and administrative, formal and informal, institutional arrangements. Administrative actors assume a key role in the interpretation and treatment of IA-related outputs. Rather than a single decision maker, decisions are made by ill-defined and often rapidly changing coalitions of interests and individuals. These multiple decision-making parties (some elected and some not) bring to decisions their own values, preconceptions, interests, perspectives, beliefs, knowledge, and priorities. They may or may not be open to new information, knowledge and values. Decision makers also are influenced by many other circumstances, advisors, and (internal and external) stakeholders. The pattern of influences that shape decision-making processes and persuade decision makers often occur outside the boundaries of IA requirements and procedures. Even within IA processes "direct action" and advocacy by potentially affected parties, predefined positions by key actors, and informal political persuasion by, for example, proponents and industry representatives may be far more instrumental in IA-related decisions than IA documents.

The decision-making influence of IA documents and processes tends to be diminished even further if there is a major discord between the characteristics of IA processes and "real" decision making; if IA practitioners have minimal understanding of and make no effort to adapt to the organizational, institutional, or political environments in which they operate; if no acknowledgement is made of the subjective and value-full nature of IA interpretations and judgments; if IA "influence" is limited to preapproval phases and excludes postapproval decisions and implementation; if the professional status and credibility of key IA professionals are not widely accepted and recognized; and if IA practitioners adopt an exclusively passive role in the advancement of environmental values and imperatives. IA practice can be especially ineffective in seeking to influence decision making if the IA process is closed, rational, and technocratic rather than open, collaborative, and democratic. Interested and affected parties are much more likely to "buy-in" to the conclusions and recommendations stemming from an IA process if all stakeholders have participated and shared in, from the outset, the generation and refinement of IA process outputs.

Even when IA professionals recognize the limits of the conventional approach to connecting IA practice to decision making, they are likely to have limited training in or knowledge about how best to go about narrowing the gap. They also are likely to have serious questions and concerns regarding if and the extent to which they should be responsible for analyzing, influencing, and redefining institutional and political institutions and procedures. As an "environmental" professional, the IA practitioner seeks to advance environmental values and perspectives. But questions commonly arise regarding how far environmental advocacy can be taken without transgressing professional ethical boundaries. Enhancing the decision-making effectiveness of IA, while clearly desirable, therefore, is a complex task with many potential dilemmas and uncertainties.

Purpose Driven A common critique of IA practice is that it is a largely vacuous paper-generating process—without direction and largely without a measurable positive outcome. Although ostensibly created as an instrument for environmental change, as illustrated in Figure 3.4, the route to such change is often so indirect (e.g., better environmental information and understanding eventually leading to an enhanced environment) that the environmental influence of the outcomes is cast into doubt. IA practice is more likely to be influential if, as illustrated in Figure 3.4, IA requirements and processes are directed and bounded by clearly defined environmental and sustainability objectives, criteria, thresholds, and trade-off rules (Gibson et al., 2005). Purpose-driven IA requirements and processes could be more directly linked to stakeholder values and beliefs (influential over whom?), to broader environmental purposes (facilitates the achievement of what?), and to related environmental management instruments (influential in conjunction with what?). Such tangible benchmarks could help in assessing the effectiveness of IA instruments in contributing to positive environmental outcomes. Interpretations of the influence of IA practice also would be served by clearly defined procedural (e.g., inclusiveness, collaboration) and substantive (e.g., empowerment) ethical standards.

Political Power and Discourse Political power is concerned with influencing the behavior of others. If IA requirements are largely circumvented and/or if the IA decision-making influence is minimal or untraceable, then it is not surprising if many consider IA requirements and processes to be irrelevant to the advancement of their interests. The political influence of IA requirements and processes is likely to be limited if they are designed and managed as if they were apolitical; if key stakeholders are able to shape outcomes by operating outside the political system (e.g., by lobbying, through direct political action); if IA outputs are not in a form conducive to stakeholder understanding and application (e.g., dense, obtuse documents); if key interested and affected parties are largely excluded (e.g., public involvement limited to public information and education); and if IA processes are closed, biased, discretionary, unsubstantiated, and imbalanced (Cashmore et al., 2010). IA requirements and processes are more likely to be influential if they proactively integrate political knowledge and concepts; if they encompass all potentially significant actions and effects; if the potential for circumvention is minimized; if they are designed and managed to facilitate constructive political dialogue, reflection, consensus seeking, and coalition-building; and if the exercise of political power is open, explicit, nondiscretionary, and substantiated (Runhaar et al., 2010; Devlin and Yap, 2008).

Frameworks *The framing of IA knowledge* is essential for IA practice to achieve greater influence. As illustrated in Figure 3.5, IA practice is likely to have limited influence if it is not effectively adapted to context, and if it makes no use of and no contribution to the IA knowledge base and to the knowledge derived from related fields of knowledge and practice.

Influential IA practice should suit the decision-making, institutional, political, spatial, temporal, societal, and environmental context (Fischer and Gazzola, 2006). It necessarily takes into account community values, attitudes, and beliefs; is designed and applied to fit within institutional arrangements; seeks to integrate community and environmental visions, inspirations, and knowledge; and incorporates the concerns, priorities, and challenges identified by interested and affected parties (Lee, 2006; Sinclair et al., 2008). It should provide decision makers and other parties with valuable information and knowledge. It should lead to greater understanding, involvement, collaboration, and empowerment by interested and affected parties, especially by parties traditionally excluded or under-represented in public and private decision making. To be influential in realizing the basic purpose of IA, it should be possible to demonstrate a direct path from IA practice to tangible environmental enhancements, and to positive sustainability contributions.

The ties between IA practice and the IA knowledge are critical when addressing IA influence. Influential IA practice makes systematic and appropriate use of objectives, good practices, principles, frameworks, concepts, and models derived from the IA knowledge base (i.e., theory testing). It contributes to the IA knowledge base by means of applied

Figure 3.4 The ordering of IA objectives.

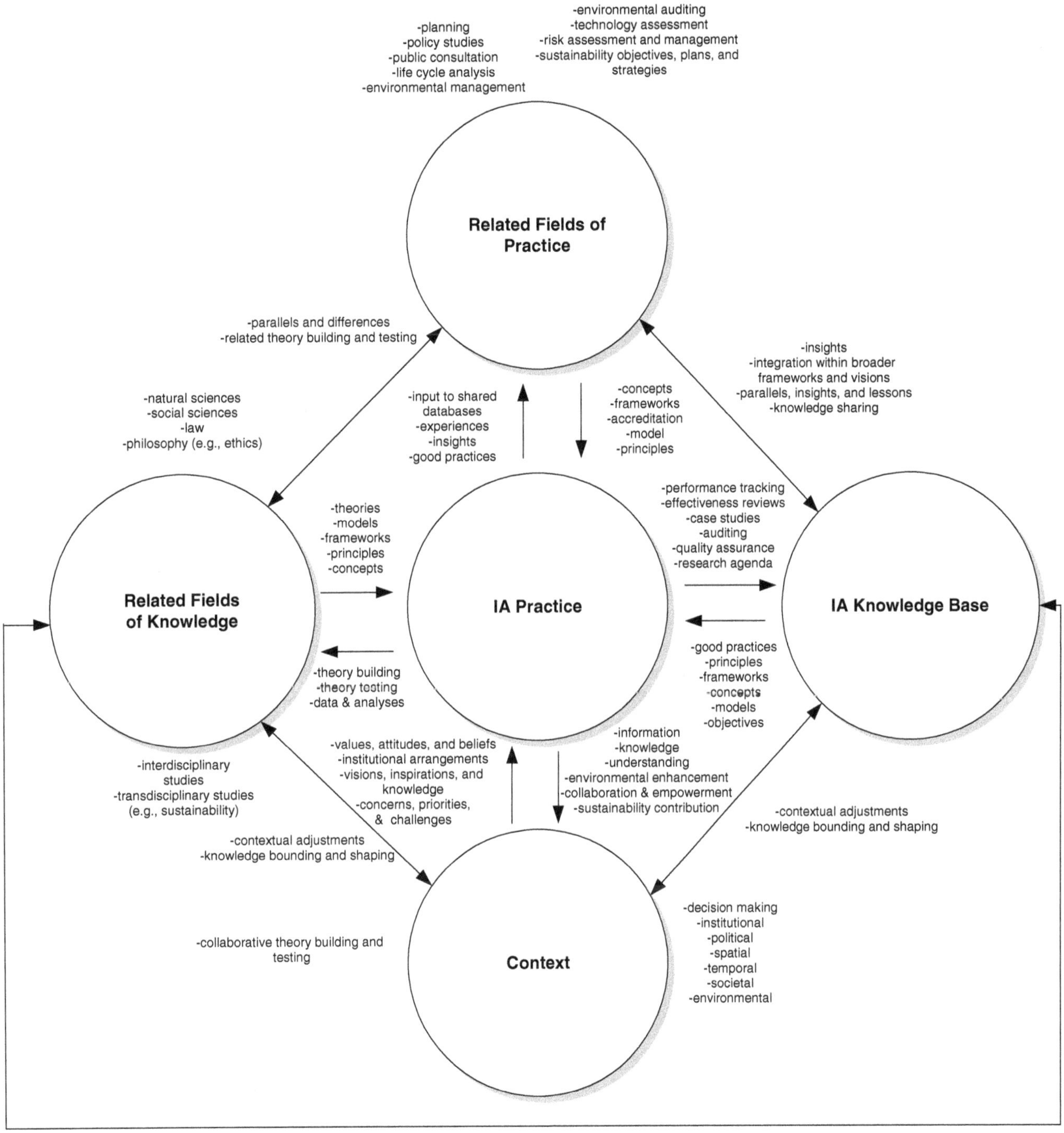

Figure 3.5 Greater influence by framing IA knowledge.

research, effectiveness reviews, case studies, the auditing of outcomes, performance tracking, and quality assurance initiatives (Cashmore et al., 2004; Doelle and Sinclair, 2006; Noble and Storey, 2005; Ross et al., 2006; Tzoumis, 2007). Influential IA practice also draws upon and contributes to related fields of practice (e.g., planning, policy studies, risk assessment, sustainability strategies, life cycle assessment, environmental management), and related fields of knowledge (e.g., natural sciences, social sciences, philosophy, interdisciplinary, and transdisciplinary studies) (Benson, 2003; Cashmore et al., 2010; Hanna, 2005; Devlin and Yap, 2008).

With related fields of practice, influential IA practice provides inputs to shared databases; shares experiences, good practices, and insights; and refines and adapts concepts, frameworks, models, principles, and accreditation procedures. With related fields of knowledge, influential IA practice contributes data and analyses and assists in theory building and testing. It also applies, refines, and adapts pertinent theories, models, frameworks, principles, and concepts. In addition, influential IA practice is cognizant of and takes into account linkages among context, the IA knowledge base, related fields of practice, and related fields of knowledge.

The decision making–IA connection is critical if IA practice is to achieve greater influence. As illustrated in Figure 3.6, it is much more than simply IA practice informing decision makers, who, in turn, openly make more environmentally informed and justified decisions, which then leads to an enhanced environment. The possible interactions among stakeholders, decision making, IA institutional arrangements, and IA practice are many, varied, complex, indirect, and often subtle. If IA practice and institutional arrangements are to be more influential over decision making, there needs to be an enhanced understanding of the actual and potential nature of these interactions.

The connection between IA practice and decision making is largely mediated through a diverse array of IA institutional arrangements (e.g., legislation, regulations, policies, guidelines, objectives, standards). IA institutional arrangements can further, structure, focus, and legitimize the role of IA practice in influencing decision making. However, they also can constrain, bound, and inhibit the ability of IA practice to inform and guide decision making. Decision making can, in a positive way, implement, apply, interpret, adjust, fund, and direct IA institutional arrangements. But it also, in a negative sense, can bypass, exempt, ignore, politicize, and limit IA institutional arrangements. IA practice, to the extent that it is systematically audited, can help test, refine, reform, educate, facilitate, legitimize, and enhance the design and application of IA institutional arrangements. To the extent that there is a considerable gulf between the potential and achievements of IA practice, IA institutional arrangements can be undermined and ultimately jeopardized.

IA practice takes many forms (e.g., SEA, EIA, SIA, CEA, SA). IA institutional arrangements, together with decision-making links, vary, often dramatically, depending on the type of IA. Interconnections between IA practice and decision making (both direct and indirect) can be strongly affected by the nature and effectiveness of interconnections among IA types. IA decision-making links also vary depending on the decision-making level, whether the decisions are taking place before, during, or after the IA, and whether the IA role is to inform policy, program, planning, or project decisions. Oftentimes informal and indirect decision-making links are as or more important than direct, formal decisions. Ideally, IA practice, directly and indirectly, informs, inspires, focuses, structures, supplements, constrains, transforms, and substantiates decision making. Unfortunately, at times, it also can depoliticize (in the sense of masking the allocation of power) and rationalize (in the sense of propping up preexisting decisions) decision making (Cashmore et al., 2010). Again, ideally decision making helps to support, fund, ground, test, and transform IA practice. Again, unfortunately, decision making sometimes, reverses, bypasses, limits, and trivializes IA practice.

Stakeholders (e.g., politicians, regulators, industry, professionals, members of the public) assume a crucial role in determining if and to what extent IA practice and institutional arrangements influence decision making (Lee, 2006). Decision-making roles naturally vary, often dramatically, among stakeholder groups. Stakeholders can help focus, legitimize, facilitate, interpret, challenge, test, and refine IA practice (Alton and Underwood, 2003). IA practice can, in turn, help inform, educate, energize, and empower stakeholders. Or, IA practice can confuse and distract stakeholders from the "real" decision making. Stakeholders can help to inform, test, challenge, reform, and legitimize IA institutional arrangements. Stakeholders also will ignore or bypass IA institutional arrangements if they consider those arrangements to be irrelevant or contrary to their interests. IA institutional arrangements inform, educate, facilitate the involvement of, legitimize, and empower stakeholders. Or, they can distract and marginalize the decision-making role and effectiveness of stakeholders. The connections between stakeholders and decision-making connections are not always mediated through IA practice and institutional arrangements. Stakeholders often will seek to influence decision making through, for example, direct action, pressure, and lobbying (Devlin and Yap, 2008). They also sometimes seek to directly inform and educate decision making and decision makers (Cashmore et al., 2010; Devlin and Yap, 2008). Decision makers, in turn, can seek to inform, educate, legitimize, and empower various stakeholders. Or they can ignore or marginalize various stakeholders.

Choices for the design of an influential IA system, as illustrated by Figure 3.7, are many and varied. There are a host of public actions that can be subject to IA requirements (as with policies, plans, programs, and public works) or which can form part of IA institutional arrangements (as with legislation, regulations, guidelines, principles). Also, there are many private actions and joint public/private actions (e.g., major resource and energy projects) to which IA requirements can be applied. Many public actions and instruments frame IA requirements, indirectly influence IA (e.g., as with priority setting and the allocation of resources and funding), or have no connection to IA. Private plans and actions (e.g., environmental management) also can influence how IA requirements are applied or can serve as alternative or complementary instruments for infusing environmental values into decision making. Critical questions bearing on the influence of IA practice concern such matters as which public and private actions should be subject to IA requirements; what form IA institutional arrangements should assume; how IA requirements should be adapted for private actions; how IA should be connected to other public and private actions (e.g., environmentally quality standards, sustainability initiatives); the nature of international,

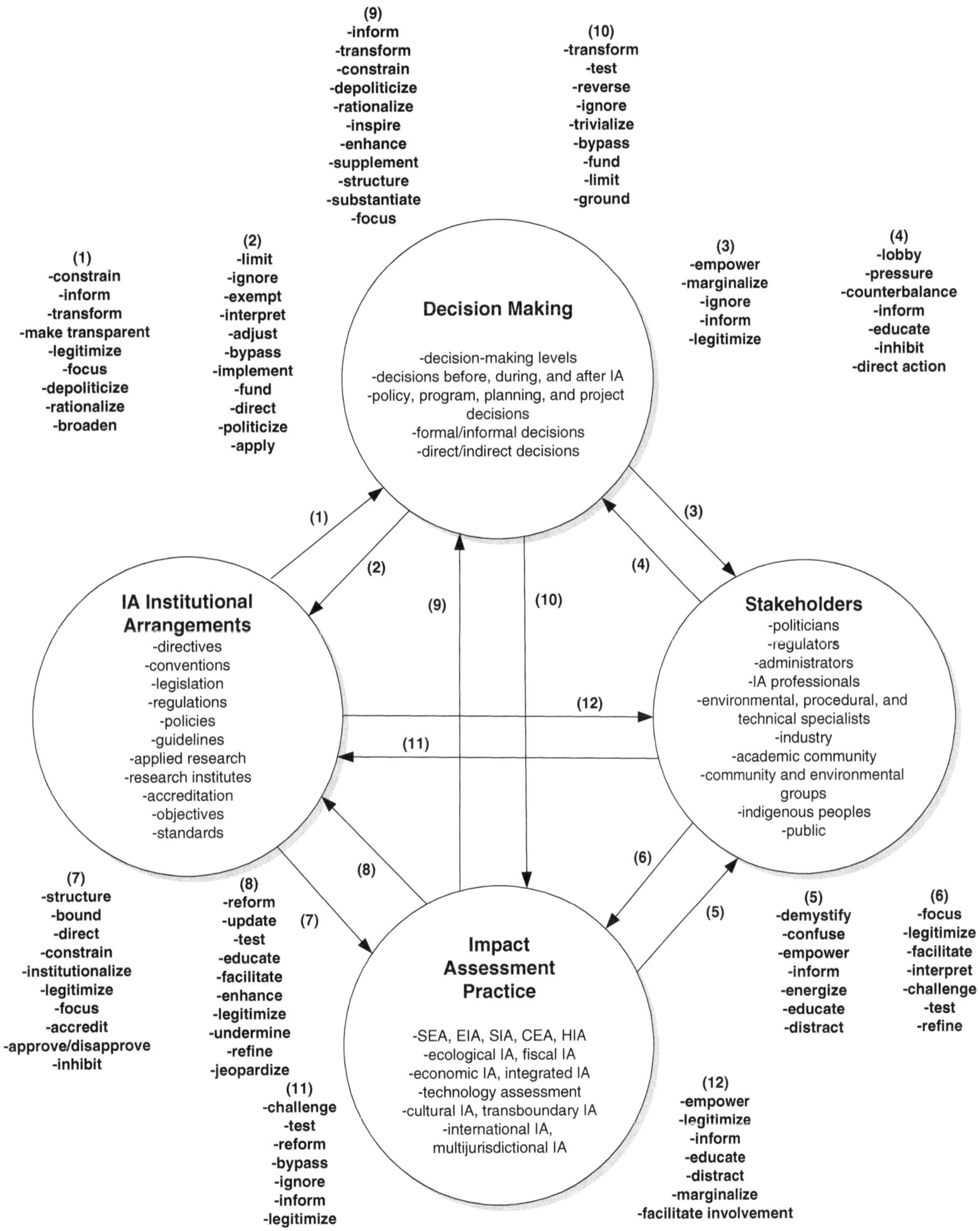

Figure 3.6 Connecting influence, IA, and decision making.

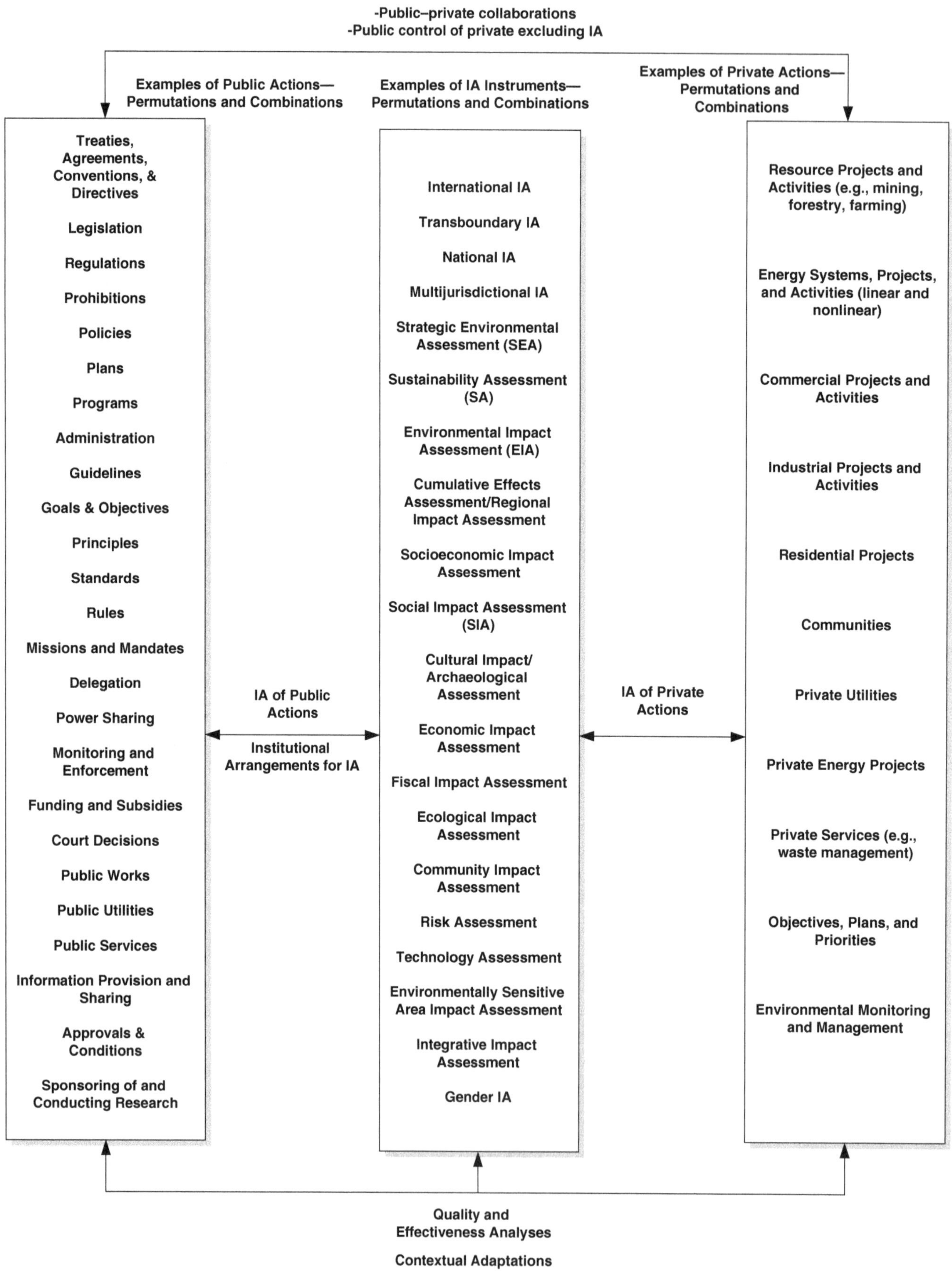

Figure 3.7 Choices for designing an influential IA system.

intergovernmental, and public/private partnership connections; the establishment of links among decision-making levels; the nature of interconnections among public/private actions (e.g., complementary, conflicting); and if and how IA purposes (e.g., a more environmentally sound and sustainable environment) can be achieved through IA and/or through alternative institutional arrangements (Doelle and Sinclair, 2006).

There are numerous IA types. The institutional arrangements (if any) for each type can vary considerably. No IA system can realistically encompass all IA types as separate entities. The question then becomes how best to combine IA types within integrated requirements, procedures, and documents. The influence of IA is open to challenge to the extent that major (or cumulatively significant) public and private actions, with potentially significant environmental consequences, are not subject or not adequately subject to either IA or comparable requirements. The suitability of IA requirements and procedures, both individually and collectively, can vary significantly depending on contextual characteristics. The question of which mix of IA and non-IA institutional arrangements is the most appropriate and influential from the perspectives of all interested and affected parties, under which contextual conditions, can only be determined based on the systematic application of IA quality and effectiveness analyses addressed from multiple perspectives (Cashmore et al., 2010; Ross et al., 2006; Tzoumis, 2007). The design of an influential IA system, accordingly, is a complex matter as each element must be designed separately, connected to complementary elements in different ways, adapted to context, tested through effectiveness analyses, viewed from multiple perspectives, and refined and adjusted over time as conditions change (Morrison-Saunders and Sadler, 2010).

Procedures for reorienting, linking, and integrating IA types can assume a vital role in making IA more relevant and influential. As illustrated in Figure 3.8, there are a diversity of IA instruments and actions encompassed within the overlap between SEA and EIA. EIA has frequently been criticized for not adequately addressing purpose and alternatives to the proposed action and/or for making decisions about such matters before EIA requirements are triggered and then rationalizing those decisions within the EIA in a perfunctory manner. Either way, potentially affected parties are effectively excluded from and are irrelevant to decision making. Conventional EIA practice, at the project level, also has been criticized for the weak treatment of cumulative, regional, and transboundary effects, and for a failure to adequately address interjurisdictional and international concerns and perspectives. SEA in its many forms (e.g., plan, program, policy) can frame the treatment of such issues at the EIA project level, and ensure that all stakeholders are directly involved in such front-end decisions when they matter (Benson, 2003). EIA, in turn, can ground and extend SEA at the project and activity levels, and provide a more regional and local perspective. SEA and EIA, effectively integrated with each other and with decision making, are, accordingly, likely to become more credible and relevant.

Greater IA influence also is more likely to occur if effective interconnections are drawn among SEA types and among EIA types. The linking of SEA types and EIA types tends to be more effective if structured by systematic, transparent, and collaboratively formulated conceptual and regulatory frameworks. Sustainability assessment (SA) has the potential to transform both SEA and EIA. SA, either as separate strategies and plans or as an array of visions, values, principles, thresholds, and trade-off rules, can encompass, structure, reform, complete, and make more influential SEA and EIA practice (Bond and Morrison-Saunders, 2011; Hanna, 2005; Noble and Storey, 2005; Sinclair et al., 2008). SEA and EIA, in turn, can test and refine SA concepts, principles, and methods (Noble and Storey, 2005). The same type of mutually beneficial relationship is possible between SA, SEA, and EIA and related fields such as risk assessment, technology assessment, environmental auditing, and life cycle assessment (Benson, 2003). Greater influence also is more likely to be achieved if IA types, both individually and collectively, are reoriented with an increased emphasis upon purpose and substantive outcomes over process, maximizing the positive over minimizing the negative, synthesis, and interdisciplinary integration over disciplinary analysis, adaptation over the rigid and deterministic, postapproval management over just preapproval analysis, multiple over unitary perspectives, political mobilization and empowerment over apolitical expert-centered analyses, learning/and collaboration over rational and scientific analyses, precaution over risk reduction, and value-full/subjective over value-free/objective approaches (Cashmore, 2004; Cashmore et al., 2004; Doelle and Sinclair, 2006; Jones and Slinn, 2008; Morrison-Saunders and Sadler, 2010; Noble and Bronson, 2006; O'Faircheallaigh, 2009).

Knowledge Base and Research Priorities Making IA more influential necessitates a better understanding of the extent to which IA practitioners hold and embrace Perspective 1, as well as the reasons for and legitimacy of those views, and the implications of Perspective 1 for IA practice and for decision making. It also is necessary to understand how widely and profoundly Perspective 2 is held by various IA stakeholders. Again, understanding the reasons for and legitimacy of those views, together with IA practice and decision-making implications, is crucial to making IA theory and practice more influential. To the extent that Perspectives 1 and 2 reflect the reality within which IA operates, it is important to draw upon the constructive insights that the two perspectives offer, and to build upon available initiatives for countering the perspectives when they tend to inhibit and undermine IA's effectiveness. Particular emphasis should be placed on an enhanced understanding of how and why various stakeholders choose to influence decision making outside IA requirements and processes, and which measures

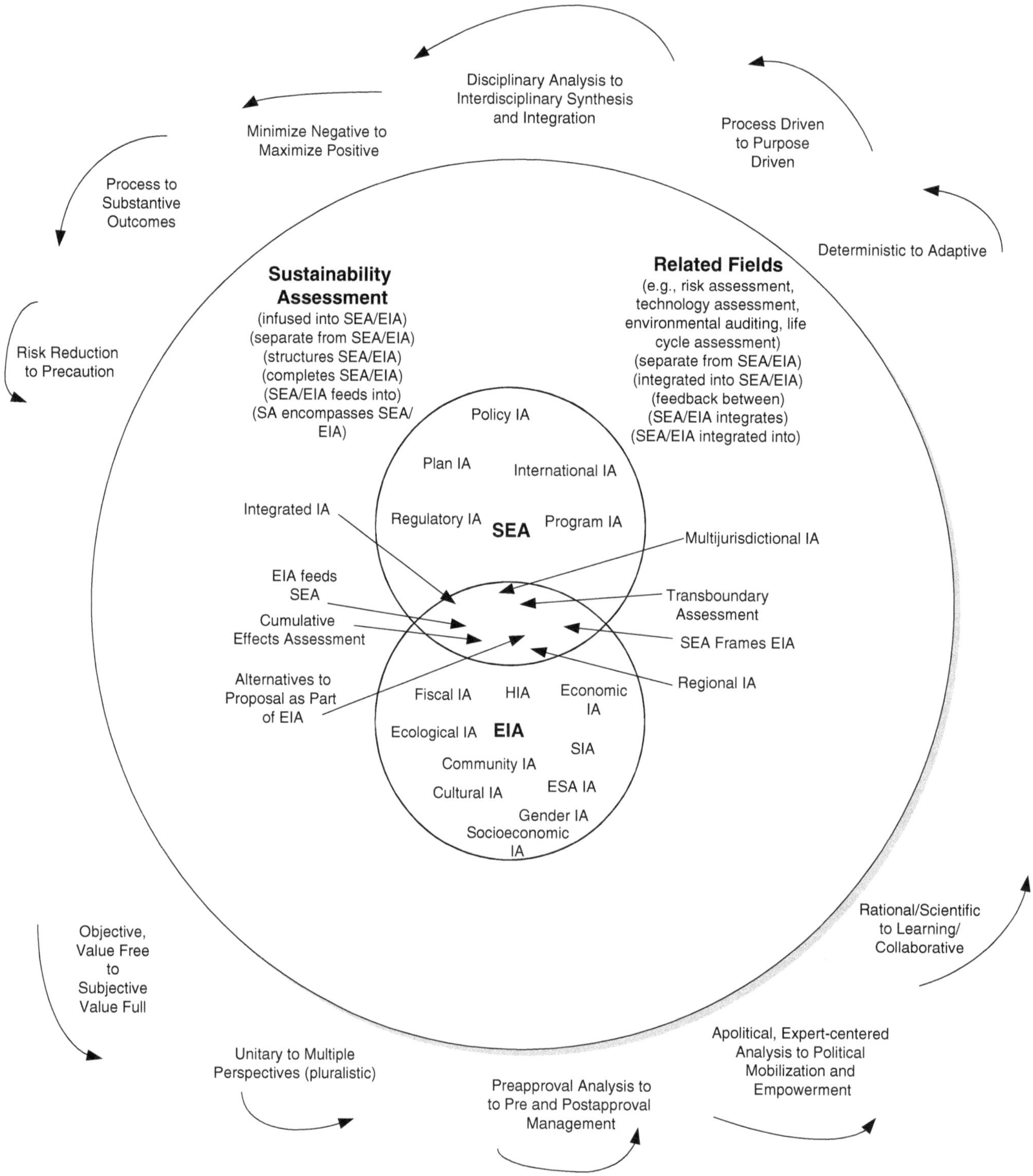

Figure 3.8 Greater influence by reorienting, linking, and integrating IA types.

should and should not be used to avoid the need, desire and, as appropriate, ability to circumvent IA requirements and processes. Preferably, initiatives to counter the two perspectives, where warranted, can be progressively integrated within alternative research and application strategies, explored from multiple perspectives, for making IA more influential of both stakeholders and decision making. Also, preferably such strategies should draw upon and be tested in collaborative stakeholder forums.

The influence-related concepts point to the need for more effective links between IA theory and practice, for better IA follow-up and auditing, and for formulating and applying IA effectiveness analyses from multiple perspectives. Both the procedural and substantive effectiveness of IA institutional arrangements and IA processes and methods need to be more systematically evaluated. Key applied research priorities include the enhanced design of IA tracking systems; the more effective integration into IA practice of environmental and sustainability objectives, criteria, thresholds, and trade-off rules; the formulation, application, and testing of procedural and substantive ethical standards; an enhanced understanding of the existing and potential role of IA in the exercise of power in decision making; and the current and potential role of professional accreditation in IA practice.

The influence-related frameworks underscore the need to more effectively draw upon insights and lessons, appreciating differences, from related fields of practice. Better practice in adapting IA to varying contexts is crucial to making IA more relevant and influential. Closing the gulf between IA theory and practice must be a priority. Much more attention needs to be devoted to exploring the complex web, both existing and potential, connecting IA institutional arrangements, IA practice, decision making, and stakeholders. More attention needs to be devoted to designing and evaluating the relative merits of alternative IA systems—both individual components and overall systems. Systematic analyses need to be undertaken of the effectiveness of alternative frameworks and approaches for linking and integrating SEA and EIA, SEA types, EIA types, IA and related fields, and SA and EIA/SEA. Alternative frameworks and procedures also need to be explored for shifting the orientation of IA practice more toward the adaptive, the purposeful, the substantive, the integrative, the positive, the precautionary, the value-full, the pluralistic, the political, and management.

The preceding initiatives need to be integrated within coherent, transparent, broadly supported, and reliably funded applied IA research strategies. Greater IA influence entails particular regard to identifying and overcoming the barriers that inhibit more effective and influential IA practices. It also necessitates integrating multiple perspectives.

3.5 INSTITUTING AN INFLUENTIAL IA PROCESS

3.5.1 Management at the Regulatory Level

Each of the four jurisdictions (the United States, Canada, Europe, Australia) has instituted a range of measures potentially conducive to greater IA decision-making influence. Table 3.1 provides an example list of measures introduced in the four jurisdictions that potentially have a bearing on decision-making influence.

As is immediately evident from Table 3.1, such measures, at best, facilitate the potential for, rather than ensure, greater decision-making influence. The potential for greater decision-making influence is, for example, enhanced if decision making occurs within rather than prior or subsequent to the IA process (e.g., mandatory scoping, early public involvement while choices are still open, the bounding of political and administrative discretion, various tiered IA requirements are broadly applied to different proposal types and decision-making levels); if the use of exemptions is fully substantiated and not abused; if requirements ensure that the scope of choices is broad rather than narrow (e.g., reasonable alternatives as collaboratively determined, broad definition of environment and effects including cumulative and transparent effects); and if the IA process is directed and bounded by explicit environmental principles, priorities, mandates, imperatives, and limits (especially with reference to sustainability, biodiversity, and climate change). Decision-making influence also can be facilitated if specific decision-making considerations, which must be considered during each IA process stage, are specified; if IA specialists must be accredited; and if the methods employed to evaluate effects and compare choices and the criteria and procedures used to determine significance must be explicit, substantiated, and consistent with good practice standards.

The role of the public in influencing and shaping decision making can be aided if ample public notice is provided prior to each decision; if all pertinent documents and analyses are readily available in a timely manner; if public involvement opportunities, with sufficient time and resources (e.g., participant funding) for meaningful involvement, are provided prior to each process decision; if the range of interested and affected public(s) is not restricted; if the potential for meaningful public involvement is not inhibited by formal and legalistic participation procedures; if ready and timely electronic access is provided to all pertinent documents; if access to mediation is provided for and actively encouraged; if documents and analyses are presented in a manner conducive to public understanding and involvement, consistent with the value and interests of each party (e.g., explicit consideration of distributional issues, consultation designed to meet the needs and characteristics of each stakeholder); if proactive public participation, including delegation of public participation responsibilities, is instituted for traditionally excluded parties (e.g., aboriginal peoples); if the basis for decisions must be substantiated (taking into account explicit decision-making criteria and public comments and employing explicit and open decision-making procedures); if the views of interested and affected parties must be made explicit together with their role in decision making, if there is the right to appeal decisions (i.e., access to justice); and if opportunities are provided for the delegation to and collaboration/harmonization with lower decision-making levels. In the latter case, IA substitution/delegation should only be permitted if public participation performance standards are not compromised; explicit criteria, standards and procedures, and decision rules are applied; adequate resources are provided; decisions are transparent and subject to appeal; and substitution/delegation is subject to independent performance evaluation.

Decision making is potentially more transparent and open to external influence if interconnections with related requirements (including international obligations) are made explicit and are systematically considered; if requirements are adjusted to suit decision-making levels, IA types, and

Table 3.1 Positive and Negative Regulatory Examples Regarding Decision-Making Influence

United States	Canada	Europe	Australia
(+) NEPA applied to a broad range of major actions (+) Detailed and explicit scoping requirements (+) Detailed and explicit significance determination procedures and criteria (+) Requirements to substantiate methods (+) Extensive provisions to ensure review process focused, timely, and efficient (+) Detailed intergovernmental, agency, and public notification, coordination, and involvement requirements (+) Proposed measures to enhance sustainability capacity of EPA (+) A range of measures to modernize and reinvigorate NEPA (+) Retrospective regulatory review initiative (+) Pilot project to facilitate innovation in IA review (+) Enhanced Internet access (+) Requirement to address reasonable alternatives (+) Explicit consideration of environmental justice (+) Detailed document content requirements (+) Initiatives to more effectively address sustainability and climate change (−) Limited application in practice of NEPA to strategic level (−) Emphasis of procedure over substance (−) Concerns about increasing costs and timing associated with fears of litigation (−) Concern that public role largely one of opposition rather than active participation	(+) Application broadened to include Crown corporations, projects on native reserves, and airport authority projects (+) Focus on major projects with potentially significant adverse effects (+) Greater provincial influence with substitution/delegation opportunity (+) Greater clarity in role definition, especially coordinative responsibility of CEA Agency (+) Follow-up requirements and enforcement penalties increase the potential for project modifications (+) Auditing and quality assurance provisions (+) Participant funding for all projects (+) Emphasis on aboriginal consultation (+) Enhanced Internet access (−) Opportunity for involvement only after major decisions made (−) New requirements leave a large degree of ministerial and cabinet discretion (−) Tight time limits (−) Exclusion of all but major projects and narrow and selective definitions of environment, effects, and alternatives (−) Explicit opportunity to approve even when significant adverse environmental effects (−) Lack of a link to strategic level (−) Narrow definition of interested party (−) Infrastructure project exemptions (−) Delegation/substitution could inhibit influence if process less open (−) Elimination of National Round Table on the Environment and Economy (−) Process for reviewing cabinet submissions limited and largely closed	(+) Proposed Project Directive (PPD): more explicit and transparent screening (+) PPD: mandatory scoping; explicit requirements (+) PPD: explicit significance criteria that must be taken into account and show to have been considered (+) PPD: requirement to provide explicit reasons for decisions and how environmental considerations integrated (+) PPD: detailed requirements regarding information to be provided to public (+) PPD: requirement to consider reasonable alternatives (as part of scoping and as defined by competent authorities) (+) PPD: mandatory monitoring (+) PPD: comprehensive definition of effects, including cumulative effects (+) PPD: focus on environmental priorities (e.g., biodiversity, climate change) (+) PPD: quality control mechanisms, including accreditation (+) PPD: requirement to summarize comments received (+) PPD: focus on major projects (+) PPD: measures to facilitate coordination with related assessments (+) Broad application of SEA directive; SEA report must be taken into account, influence demonstrated, mandatory scoping, explicit significance criteria, reasonable alternatives, and reasons for alternatives selection, required consideration of transboundary and cumulative effects, and required monitoring (+) SEA procedures for trade agreements and EU policies and regulations (+) Extensive studies of effectiveness (−) PPD: possibility of less influence for nonmajor projects (−) PPD: potential inhibiting role of time limits (−) PPD: no enhancements in access to justice (−) Tiering and coordination mechanisms not well defined	(+) The explicit identification of matters of national environmental significance (+) The explicit identification of principles of ecologically sustainable development; minister required to take into account (+) Cross referencing of international environmental obligations (+) Scope, documentation, and public notification and comment requirements for various IA documents (+) Postapproval decision making addressed through annual compliance audits and environmental management plan requirements for selected IA documents (+) Acceptance by government of proposals to strengthen use of SEA; undertake regional environmental planning, and ecosystem management; make IA accreditation criteria, standards, and procedures more open and explicit (including review and performance auditing and oversight); provide reasons for significant decisions; introduce measures to make decision making more open and focused; make greater use of public inquiries and joint assessment panels; provide enhanced electronic access, ensure longer public comment periods; and strengthen monitoring and auditing (−) Stipulation that social and economic matters must be considered and decisions placed within context of country's social and economic needs

contextual characteristics; if major, especially controversial projects are subject to open and independent hearings/reviews (providing governments must explicitly respond to all findings and recommendations); if decision making is extended into the postapproval phase (e.g., mandatory follow-up requirements, noncompliance penalties, independent third party auditing); if lessons conducive to enhanced public and agency influence are audited and integrated into IA requirements, guidelines and procedures; if the roles of all major parties are clearly defined (especially coordinative roles); if impact significance criteria and procedures are included in requirements; and if the application of IA requirements is informed by independent effectiveness of external involvement analyses.

The measures listed above provide a cross section of potential approaches for enhancing IA decision-making influence. Care must be taken not to "graft on" measures appropriate in one setting to another setting in which the measures may be entirely inappropriate. There is a possibility that measures applied in another jurisdiction may appear, on the surface, to be highly desirable. But these measures may not have been scrutinized for effectiveness and unintended side effects and/or stakeholder perspectives regarding desirability and effectiveness could vary dramatically. Nevertheless, a selective overview of measures employed in the four jurisdictions does provide an initial sense of the possibilities for enhancing IA decision-making influence through regulatory reform. None of these measures are likely to be effective if a political commitment is lacking and if the resources provided are inadequate. Care should be taken to minimize bias in favor of proponents, rational-technical experts, entrenched interests, and bureaucratic elites. Good practice approaches to facilitating enhanced IA decision-making influence are addressed in greater detail in Section 3.6.

3.5.2 Management at the Applied Level

Figure 3.9 depicts an example of an influential IA process. The figure and the process description that follow address the concerns raised by the two perspectives and draw upon the concepts, frameworks, research priorities, and good practices presented for making IA requirements and processes more influential.

Core Process Description The core elements of the example influential IA process are similar to the SEA and EIA processes depicted in Chapter 2. A screening process determines if the SEA or the EIA process will be undertaken. The process is designed through a scoping procedure that sets the context; determines the process and document purpose; identifies key issues, problem areas, and stakeholders; bounds the process; establishes the principles, objectives, and imperatives that will guide and structure the process; identifies potential alternatives; and indicates any linkages to related public and private actions (Ross et al., 2006). Scoping at the SEA level includes characterizing baseline conditions. Scoping at the EIA level also identifies valued environmental components (VECs), and frames the process within broader policies, plans, and programs. It then goes on, in a separate step, to characterize proposal characteristics and baseline conditions. Scoping is sometimes detailed in a draft and final document, and can be subject to public and agency review. The assessment stage, with both SEA and EIA processes, identifies, screens, and evaluates alternatives, and then optimizes (through mitigation and enhancement) the preferred alternative (O'Faircheallaigh, 2009). Assessment, at the EIA level, is undertaken at a greater level of detail. A systems perspective is maintained throughout (Bond and Morrison-Saunders, 2011; Hanna, 2005). Individual and cumulative effects are fully assessed (Benson, 2003; Hanna, 2005). The SEA and EIA analyses are documented in draft and final reports, and subject to agency and public review. The reports inform, guide, and substantiate decision making. The proposed action is then implemented (if approved), and subject to monitoring and follow-up. Adequate resources are allocated to impact management and to the assessment of effectiveness (O'Faircheallaigh, 2009). Links, where appropriate, are made to related decision-making areas.

Process Refinements Several refinements have been made to the core IA process to make it more relevant to stakeholders, decision makers, and IA objectives. The IA process, at both the SEA and the EIA level, is purpose-driven (Cashmore, 2004). Substantive objectives, principles, and imperatives drive and structure the process (Noble and Storey, 2005). The net environmental contribution of the proposed action is assessed, both as a basis for decision making, and as a means of assessing the intended and unintended consequences of implementing the proposed action. Decisions and decision makers are not simply informed. IA and decision making are integrated. Decision making is transparent (encompassing both IA and non-IA considerations), applied (demonstrated exercise of power), and fully substantiated (Cashmore et al., 2010). A concerted effort is made to systematically link and integrate SEA and EIA outputs to related decisions and actions. At the SEA level, related policies, plans and programs, and other SEAs are linked and integrated. At the EIA level, a particular effort is made to transcend individual disciplines and EIA subtypes (e.g., SIA, EcIA, HIA) (Hanna, 2005).

The influential IA process incorporates more continuous forms of stakeholder involvement from the outset (Sinclair et al., 2008). It adapts and applies good practice knowledge and methods. It shares lessons and insights. It systematically assesses effectiveness. It integrates data and knowledge into broader databases. The SEA process bounds, directs, and frames the EIA process (Benson, 2003). As a result, the EIA process can more readily address need, alternatives to the project, cumulative effects, transboundary effects and inter-jurisdictional connections. The EIA process, in turn, feeds

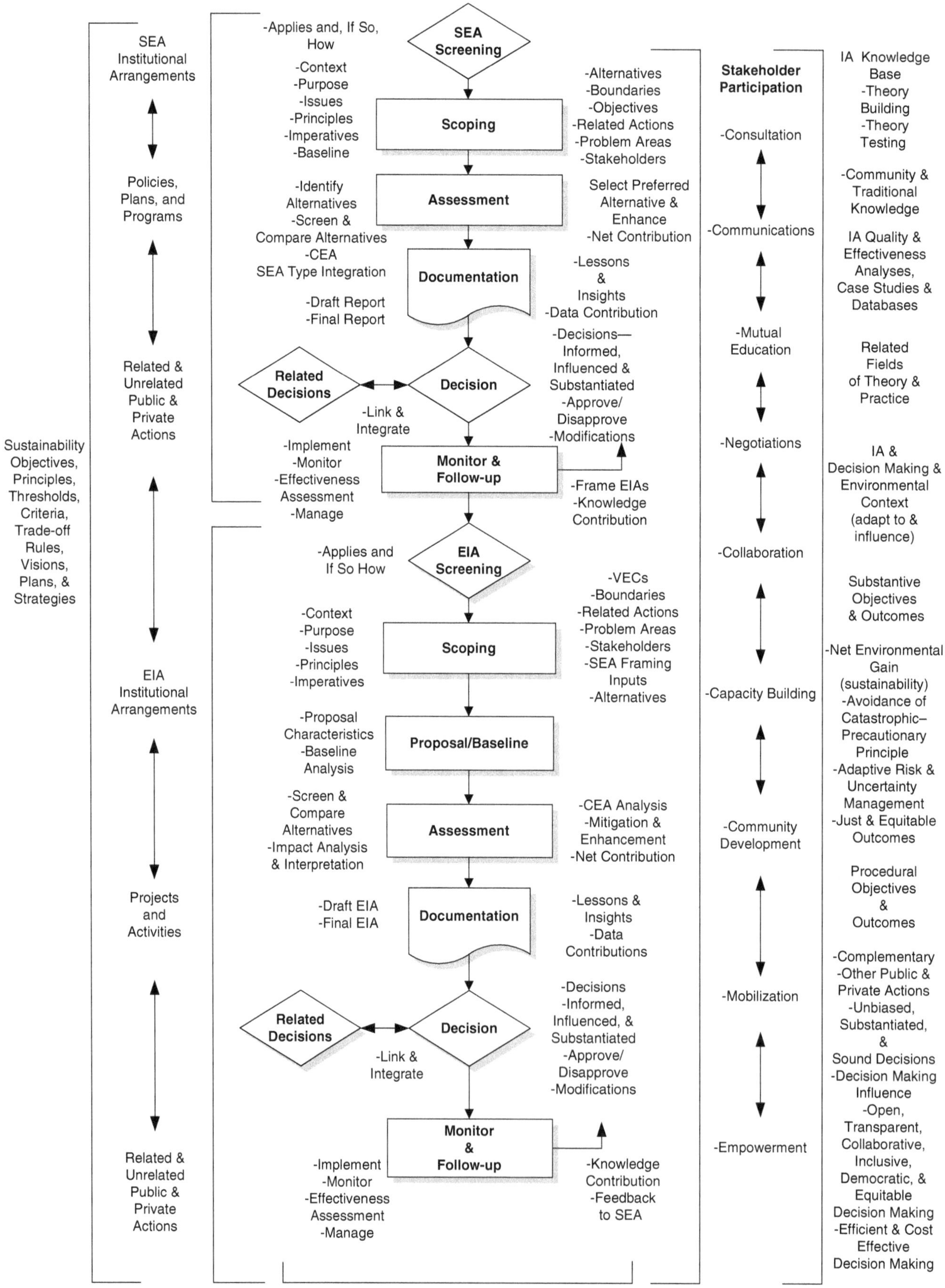

Figure 3.9 Example of an influential IA process.

back data and analysis to SEA processes. Differences among IA types are taken into account.

Framing the Process An influential IA process proactively identifies and seeks to realize environmental ends. Both SEA and EIA are infused by sustainability values, visions, objectives, principles, criteria, thresholds, and trade-off rules derived from sustainability plans and strategies (Benson, 2003). SEA and EIA scope and contents also are bounded and structured by institutional arrangements. Major public and private actions, both individually and collectively, with the potential for significant environmental effects, are addressed by SEA/EIA institutional arrangements and/or some combination of SEA/EIA institutional arrangements and complementary regulatory and administrative measures. The overall IA system is designed and managed to facilitate the realization of tangible environmental objectives. The credibility of such combinations of measures is likely to be greater when such measures, and related decision making, are transparent, systematic, substantiated, and inclusive of stakeholder interests and perspectives.

An influential IA process is more credible and relevant to potentially interested and affected parties when ample provision is made for stakeholder involvement at every step in the IA process (Doelle and Sinclair, 2006). Stakeholder acceptance of the process is more likely if ample provision is made for all stakeholders to be actively involved, and if the form of involvement extends beyond one-way consultation, communications, and education to encompass more interactive forms of involvement such as negotiations, collaboration, capacity building, community development, mobilization, and empowerment (Doelle and Sinclair, 2006; Petts, 2003).

In order to avoid the dismissal and circumvention of IA requirements and processes on the grounds that they are little more than empty "paper pushing" exercises, an influential IA process is guided and structured by explicit, substantive, and procedural objectives. An IA process is more likely to be considered influential if tangible environmental performance standards such as net environmental gain, the application of the precautionary principle, adaptive risk and uncertainty management, and just and equitable outcomes (especially for vulnerable populations and environmental features) are applied. Procedural objectives such as demonstrably complementary to other public and private actions and forms of environmental management; unbiased, substantiated, and sound decisions; demonstrated stakeholder decision-making influence; transparent, collaborative, inclusive, democratic, and equitable decision making; and efficient and cost-effective decision making, which bound and structure the IA process, also are likely to increase the credibility and influence of the IA process from multiple perspectives.

The influence of the IA process will quickly dissipate, from the perspective of many stakeholders, if the level of IA practice does not evolve and improve in conjunction with the IA knowledge base. There has to be a clearly evident "learning curve." IA knowledge has to be systematically adapted and integrated into the IA process (i.e., theory testing and refinement). The IA process, in turn, has to contribute to the IA knowledge base (i.e., theory building), through, for example, IA quality and effectiveness analyses, case studies, databases, and process and document performance rating systems (Tzoumis, 2007).

The influence of the IA process is further enhanced when community and traditional knowledge are effectively integrated; when effective use is made of knowledge and methods derived from related fields of theory and practice; and when the IA process knowledge base and methods are effectively adapted to the physical, ecological, social, economic, cultural, institutional, and decision-making context. Concomitantly, as an instrument for positive change, an influential IA process actively seeks to positively influence the decision making and environmental context.

Stakeholder Perspectives The influential IA process, as characterized in Figure 3.9, is unlikely to be uniformly embraced by all stakeholders. Influence, from the perspective of proponents and other private sector parties, is likely to be viewed largely in terms of an efficient and cost-effective planning and decision-making process, with explicit, consistent, focused, timely, and not unduly burdensome, requirements and procedures. Decision makers will tend to be looking for sound, succinct, and understandable decision-making advice and outcomes, consistent with their values, likely to benefit their constituents, and unlikely to engender a storm of criticism and controversy that could reflect back on them. Regulators will tend to look favorably on an IA process that is demonstrably compliant with and supportive of the mandate and mission of their agency, and that facilitates the expeditious execution of their IA and non-IA responsibilities. IA practitioners will tend to favor an IA process that is consistent with regulatory requirements and good practice performance standards, results in procedural and substantive outcomes supportive of their values and professional ethical standards, and can be undertaken within available resources.

Members of the public will tend to consider an IA process relevant and influential when it is directly tied to decision making that affects their concerns and interests; when it facilitates the achievement of community goals and aspirations; when potential regional and local adverse effects are prevented, minimized, and compensated for; when IA-related decision making is open, transparent, inclusive, collaborative, and democratic; when environmental outcomes are positive and equitable; when explicit and effective use is made of community and traditional knowledge; when public concerns, preferences, and interests are respected and integrated into the process; and when decision-making procedures and outcomes facilitate community development, mobilization, and empowerment. Members of environmental groups will tend to judge the influence of an IA process largely in terms of the extent of decision-making influence (i.e., will it be the primary basis for decision making?) and the extent to which it facilitates the achievement of tangible environmental outcomes (both in absolute terms and relative to alternative actions and decision-making and environmental management instruments). Members of the academic community will likely focus on the potential role of the IA process, relative to other environmental management instruments, in contributing to human knowledge, and in advancing procedural (e.g., more open, collaborative, and democratic decision making) and substantive imperatives (e.g., sustainability).

These varying influence performance standards are not necessarily conflicting. It is possible to design and manage an IA process that effectively integrates and balances different stakeholder perspectives. However, consensus is not always possible or desirable. If the ultimate standard is positive environmental outcomes, some proposed actions will need to be rejected or severely modified, and some IA processes will necessarily be more costly, time-consuming, and onerous than some parties desire. At the same time, a highly complex, lengthy, and burdensome IA process is not necessarily conducive to a higher level of environmental quality. The net result over time, over numerous IA processes, tends to be the withdrawal or rejection of environmentally desirable proposals, the continuation and expansion of existing environmentally less desirable activities, a greater propensity to circumvent the process, and streamlining "reforms" to IA institutional arrangements that severely undermine IA effectiveness.

Steering a path that effectively links and balances varying perspectives concerning the essential attributes of an influential IA process is far easier when the major stakeholders respect the views, concerns, and interests of one another; when they cooperatively and collaboratively work together to facilitate the achievement of common ends; and when the IA process is designed and managed to facilitate cooperation and communications among the major parties. The likelihood of achieving a common standard of influence is greatly diminished if the views, interests, and potential contributions of other parties are summarily dismissed; if data, analyses, and interpretations are biased, misrepresented, or unsupported; if the competence of IA and other pertinent practitioners is perceived to be doubtful; if the process is poorly adapted to the context; and if one or more parties seek to or do circumvent the process.

Process and Practice Variations by IA Type The desire to influence decision making is a recurrent theme in all forms of IA practice, as summarized in Table 3.2. Examples of decision-making-related themes that cut across IA types include the desire to broaden the decision-making agenda to encompass environmental concerns and a broader range of choices; to alter how proponents, governments, and other stakeholders go about making and implementing decisions; to make a difference in both how decisions are made and the environmental outcomes that flow from those decisions; and to enhance public access to, involvement in, and influence over those decisions that may directly or indirectly affect their lives. Increasingly all IA types emphasize the infusion of sustainability aspirations, principles, and perspectives into decision making, albeit with varying interpretations. All IA types stress the importance of follow-up and the auditing of effectiveness as means for facilitating enhanced decision-making influence. All recognize the need for capacity building if the decision-making role of parties, traditionally excluded from decision making, is to be facilitated. There also are many differences among IA types in the manner in which they address interconnections between IA practice and decision making.

Sustainability Assessment (SA) SA is concerned with more than influencing decision making by broadening the range of environmental considerations on the decision-making agenda. SA aims to reorient decision making toward the realization of sustainability ends, to ensure adherence to sustainability limits and principles, and to move decision-making outcomes away from unsustainable practices. It strives to inculcate an environmental ethic into decision making, to foster more holistic environmentally oriented decision making, to facilitate links and integration among all forms of environmental requirements and decision making to extend the temporal (e.g., future generations) and spatial (e.g., a global and a local perspective) boundaries of decision making, to broaden the range of choices seriously considered in decision making, to reorient decision making away from simply minimizing the negative to realizing mutually reinforcing gains on multiple fronts, to facilitate more bottom-up participation and influence in decision making, and to favor more cautious and adaptive decision-making practices.

Strategic Environmental Assessment (SEA) The relationship of SEA practices and decision making is an especially close one. SEA and decision making more than overlap—the boundaries are blurred and often indistinguishable. With SEA it is not so much IA informing decision making but rather SEA as a tool for redefining and reforming both formal and informal institutional structures and procedures that directly and indirectly influence the environment. Understanding, adapting to, and reshaping context is a central attribute of influential SEA practice. Particular emphasis is placed on such themes as identifying key decision-making leverage points, fostering multilevel and directional integration, proactively facilitating organizational learning, adapting to and influencing an inherently uncertain and continuous decision-making environment, the need to overcome institutional barriers and resistance, the importance of facilitating collaboration and conflict-resolution, and SEA as a tool for enhancing decision-making capacity.

Environmental Impact Assessment (EIA) At the project EIA level the relationship to decision making is more structured. Emphasis is placed on effectively informing

Table 3.2 Influential IA Practice Characteristics by IA Type

Influential SA Practices	Influential SEA Practices	Influential EIA Practices
Applies directly, or through other IA or other instruments, to all public and private actions that might have significant effects on sustainability prospects Treats as umbrella for integrating other IA types, providing not serving as vehicle for promoting prevailing economic agenda Establishes in law requirement that decisions demonstrably consistent with and supportive of sustainability Uses broad sustainability visions, goals, strategies, principles, objectives, criteria, and indicators to guide the process Uses sustainability principles, criteria, thresholds, and decision-making rules to bound and structure the process; explicitly links to environmental and regional social limits and capacities Addresses the environment holistically, including explicit references to cumulative effects management, and intra- and intergenerational equity Incorporates explicit links to international laws, conventions, and other global governance mechanisms Explicitly links IA levels and types, and to institutional and organizational sustainability and laws, strategies, policies, and plans, including identification of conflicts Treats contribution to sustainability as the main test of purposes, options, and practices Emphasizes enhancement in all forms of IA Seeks mutually reinforcing gains on all fronts in preference to significant, especially permanent, losses, or trade-offs (e.g., natural capital substituted by human capital) Provides criteria and trade-off rules as the basis for separating the sustainable from the unsustainable; for assessing relative sustainability contributions; and for assessing sustainability decision quality Explicitly favors caution and adaptation and requires independent monitoring and reporting Assesses SA institutional changes and reforms against good sustainability governance framework, adapted to organizational and government context Favors bottom-up participation and decentralization; broadly engages public and communities Ensures open to public scrutiny, public participation, and public legal action to compel compliance	Frames with formal, transparent, consistent, and explicit SEA institutional requirements and learns from informal SEA experiences Seeks to enhance understanding of political, institutional, and cultural context; designed to fit within and to progressively influence Ensures infrastructure in place to provide data necessary to support SEA Employs multiple spatial and temporal scales Is guided by and infused with holistic sustainability and environmental values, visions, goals, objectives, and criteria Systematically fosters horizontal integration Systematically addresses interrelationships and fosters integration among overlapping, complementary, and competing objectives, criteria, knowledge systems, and interests Defines alternatives broadly to encompass both alternatives to meet a need or problem and alternatives to an existing policy, plan, or program Links SEA to emergent, informal strategies, and formal decisions at all levels; identifies and takes advantage of key leverage points/policy windows Treats SEA as tool for proactive and strategic interaction with decision making rather than as a means of reporting largely negative environmental effects Shifts emphasis beyond simply effects to encompass environmental issues, concerns, and values in effort to make more central to decision making Tiers various SEA types and establishes and strengthens multidirectional links; ensures outputs directly linked to downstream decision inputs and EIA activities Treats SEA as an inherently uncertain, continuous process and a catalyst for organizational learning Provides continuous information to decision makers Clearly defines decision-making roles and responsibilities; seeks planning ownership of SEA Uses follow-up to determine the value added from SEA to and to inform theory and guidance Places more emphasis on scoping, collaboration, negotiation, persuasion, and role of SEA as social/rational learning tool; less emphasis on prediction and evaluation Makes full provision for public access and public participation; fosters collaborative/democratic approach Seeks to overcome institutional and bureaucratic barriers—includes reforming	Complies with all EIA requirements, carefully interprets guidelines and legal precedents, and systematically draws upon IA knowledge Roles, tasks, and responsibilities clearly defined Process designed and managed to be efficient and cost-effective Draws upon IA procedural design choices and carefully matches to context Basis for strengths and limitations of all data sources, boundaries, assumptions, analyses, interpretations, and conclusions, explicit and fully explained and substantiated Process focuses on key issues and on potentially significant individual and cumulative effects Need and alternatives to the proposed action, and alternative means of carrying out the proposed action openly assessed Employs rigorous impact prediction, CEA, and impact management methods Uses holistic, interdisciplinary perspective when establishing boundaries and when establishing causal networks Fully integrates community and traditional knowledge, and public concerns, preferences, values, and aspirations Allocates sufficient time and resources for effective and meaningful participation by all interested and affected parties Provides public consultation from the outset and throughout the process; provides additional resources to marginalized groups and indigenous peoples to participate, and to conduct their own analyses and consultation Explicitly addresses procedural and substantive equity and fairness issues Makes a concerted effort to facilitate community and public learning, capacity building, collaboration, coalition building, and empowerment Defines the environment and effects broadly Fully integrates substantive, sustainability goals, thresholds, and trade-off rules Avoids and minimizes negative effects, enhances and creates positive effects, creates and explores best, environmentally practical alternatives, and seeks to realize community aspirations Addresses uncertainty concerns through avoidance, management at source, the precautionary principle, the systematic identification and management of uncertainties, a learning adaptive

(*continued*)

Table 3.2 (*Continued*)

Influential SA Practices	Influential SEA Practices	Influential EIA Practices
Institutes sustainability assessment capacity building, including provision for sustainability advisors Assesses sustainability decision-making effectiveness and identifies jurisdictional gaps	and institutionalizing environmental governance structures processes Enhances institutional capacity to undertake and implement SEA and democratic effectiveness Monitors environmental and nonenvironmental changes, goals achievement, and implementation performance Utilizes professional and institutional means to enhance SEA capacity to influence decision makers Provides data on SEA lessons, successes, obstacles, pitfalls, and benefits, with particular reference to effective implementation (i.e., more SEA self-reflection)	management approach, and the systematic integration and application of uncertainty, complexity, and risk avoidance and management methods

Influential EcIA Practices	Influential SIA Practices	Influential HIA Practices
Integrates biodiversity and ecological IA into SEA and EIA requirements and institutional arrangements Links to international biodiversity and protected area and species treaties, conventions, and agreements Links to and places within the context of international, national, and local ecological and conservation policies and good practices Places within context of ecological management plans Liaises and cooperates with other jurisdictions, experts, and nongovernment organizations; works toward common standards and guidance and share lessons, insights, and practices Facilitates stakeholder involvement in decision making and comanagement Focuses on ecosystem relationships and processes, enhances benefit sharing, uses adaptive management practices (at the appropriate scale and decentralized where practical), cooperatively and collaboratively involves interested and affected parties, and adopts a long term/regional systems perspective Seeks to further ecological sustainability, maintain and enhance natural capital, and ensure no adverse effect on sustainable use of biological resources Seeks to avoid irreversible losses, enhance, restore, and add to biodiversity at the ecosystem, species, and genotype levels Ensures the sustainable use of biodiversity resources, and ensures fair and equitable sharing of biodiversity benefits	Ensures IA requirements include positive and negative, direct and indirect social environment and effects Advocates for full weighting of social and psychological impacts Applies SIA early and fully in planning and decision-making process Encompasses social, economic, cultural heritage (including spirituality), political, and health effects Seeks a holistic, social–cultural–historical understanding of society, of social and economic processes and systems, and of potentially affected local and regional populations and communities (e.g., community profiles) from outset of process Recognizes that SIA is inherently political, and seeks to decentralize decision-making power Seeks to contribute to social learning, capacity building, socioeconomic empowerment of the least powerful, and the realization of social potential Seeks to identify and counter institutional constraints to SIA Seeks to bring local knowledge into decision-making process and to demonstrate potential role of SIA in enhancing participation, in facilitating mutual learning, and in alleviating tensions Focuses on key elements of human environment and provides quality information for use in decision making Recognizes that SIA is often hampered by institutional, financial, and professional constraints; makes proactive effort to prevent and offset	Clarifies legal standing of HIA as part of SEA/EIA and, where appropriate, independent from SEA/EIA Seeks to obtain political and government support for HIA Clearly and simply demonstrates links from HIA to policy issues and problems and HIA utility Seeks to institutionalize HIA Clarifies criteria for initiating, conducting, and completing HIA Sponsors HIA demonstration projects Institutes sustained dialogue among researchers, practitioners, affected populations, and policy makers Creates single accessible source of information about HIAs; ensures ongoing resources Engages decision makers and other stakeholders Develops and applies frameworks for assessing costs and benefits of HIA Seeks to mainstream HIA so triggered as part of routine decision making Devotes particular attention to avoidable, involuntary, adverse, irreversible, and catastrophic effects Systematically addresses health hazards and risks, health promoters and opportunities, potential health impacts, their pathways and potential outcomes Explicitly acknowledges limitations of analysis and methodology, and associated implications Involves and engages health experts, and ensures coordination between IA and health practitioners; clarifies roles

Table 3.2 *(Continued)*

Influential EcIA Practices	Influential SIA Practices	Influential HIA Practices
Distinguishes among biodiversity values, social/community values, and economic values associated with ecological features Makes explicit reference to relevant aspects of ecological structure and function upon which features depend Focuses analysis, where possible, using indicator species or valued ecosystem components Identifies and evaluates ecological resources likely to be affected, biological changes likely to be affected, valued ecological resources and features, and significant ecological features Employs best practical options (mitigation) for maintaining biological diversity (including generous offsets) Assesses impacts at bioregional, ecosystem, habitat, community, species, population, and below species levels Applies ecological standards, guidelines, and objectives, where practical, to support significance decisions; protects sites and features designated for nature conservation, and rare, endangered, and vulnerable habitats and species Institutes biodiversity monitoring program; ensures commitment to and adequate resources for enforcement and implementation Cooperates with biodiversity partnerships and information networks	Identifies key public social and cultural issues, respects fundamental human rights, promotes health and safety; avoid and minimize impacts on physical and cultural heritage resources Devotes particular attention to designated sites, and to concerns of indigenous, ethnic, and cultural groups Acknowledges the legitimacy of public risk perception; seeks intergenerational, intragenerational, and gender equity; seeks to recognize and preserve diversity and fully integrate local and traditional knowledge, concerns, perspectives, and experiences Seeks to end the marginalization of indigenous peoples regarding development in traditional lands, and provide for control of SIA by indigenous peoples Systematically addresses the distribution of impacts among different groups in society, ensures that environmental justice and gender issues are fully described and analyzed, and makes impacts on vulnerable groups a priority Seeks to help affected populations and communities understand, participate in, and cope with potential impacts, to use SIA and proposed actions as catalysts for achieving positive community visions and goals (social sustainability), and to attain broad public acceptance and support Emphasizes critical importance of public participation and central role of SIA practitioners (subject to ethical boundaries) in consulting and negotiating with stakeholders Seeks to develop and apply a simple and workable SIA follow-up model; ensures government agencies have institutional and financial capability to implement Adjusts approach to suit regulated and deregulated contexts	Integrates public and decision-maker concerns and perspectives about potential health effects Seeks to reduce inequities that result from avoidable differences in health determinants and/or outcomes Seeks to reduce the burden on health sector services, to strengthen health services, and to safeguard health and well being Provides monitoring and evaluation guidance and ensures health concerns are fully integrated into follow-up Involves communities in a participatory manner; fosters stakeholder/community buy-in Seeks to understand how decision makers may react to HIA and explores means to affect Seeks to demonstrate role of HIA in making decision makers more accountable to constituencies Links to broader efforts to promote public health across multiple sectors Addresses crosscutting health issues and sustainability repercussions Undertakes process and performance evaluations of completed HIAs Contributes to HIA capacity building Practitioners share experiences, lessons, insights, promising practices, and recommendations for improvement

Sources: Ayre and Calloway (2005), Ali et al. (2008), Atkinson and Cooke (2005), Becker et al. (2005), Benson (2003), Bhatia (2007), Bhatia et al. (2010), Bina (2007), Binder et al. (2010), Bond (2004), Bond and Morrison-Saunders (2011), Bonifazi et al. (2011), Buchan (2003), Burdge (2003a,b), CEAA (1996a, b), Chaker et al. (2006), Cherp et al. (2007), Clark et al. (2011), Cole (2004), Cole and Fielding (2007), Connor and Dovers (2004), Dalal-Clayton and Sadler (2004), Dannenberg et al. (2006), Doelle and Sinclair (2006), Dora (2004), Dovers (2005), Edwards (2005), Égré and Senécal (2003), Edelstein (2003), Edwards (2005), Elliott and Francis (2005), Erlanger et al. (2008), Fischer (2005), Fischer and Gazzola (2006), Fischer et al. (2010), George and Kirkpatrick (2008), Gibson (2006a, 2011), Hanna (2005), Hacking and Guthrie (2008), Harris and Spickett (2011), IAIA (1999, 2002a,b, 2003, 2005, 2006a,b, undated b), ICPGSIA (2003), IEEM (2006), Jiliberto (2011), João and Mcclauchlan (2011); João et al. (2011); Jones and Slinn (2008); Kemm (2005); Kiewiet and Vos (2007); Kirchhoff et al. (2009); Kirkpatrick and George (2006), Knaus et al. (2006), Kreigar et al. (2003), Lane et al. (2003), Larsen et al. (2012), Lawrence (2009), Lobos and Partidário (2010), Manou and Papathanasiou (2009), Momtaz (2003), Morgan (2011), Morrison-Saunders and Fischer (2010), Morrison-Saunders and Hodgson (2009), Ng and Hui (2007), Noble (2000b, 2004b, 2008, 2009b), Noble and Bronson (2006), Nobel and Storey (2005), OECD (2006), O'Faircheallaigh (2009), Ortolano (2008), Partidário (2007), Partidário and Arts (2005), Partidário and Wilson (2011), Petäjäjärvi (2005), Peterlin et al. (2008), Petticrew et al. (2007), Pisani and Sandham (2006), Pope et al. (2004), Pope (2006), Pope and Dalal-Clayton (2011), Pope and Grace (2006), Pope and Klass (2010), Pritchard (2005), Quigley and Taylor (2003), Scanlon and Davis (2011), Scott (2011), Sharma (2010), Shepherd (2008), Sinclair et al. (2008), Stoeglehner et al. (2009), Slootweg and Kolhoff (2003), Swangjang et al. (2004), Tang (2010), Taylor et al. (2003, 2004), Thérivel (2010), Treweek et al. (2005), Tugwell and Johnson (2011), Vanclay (2003, 2006, 2010), Wale and Yalew (2010).

and influencing decision making regarding proposed or potential projects at a point when decision making is open to the consideration of reasonable alternatives and to the integration of a broad range of environmental effects. Influential EIA practice seeks to move beyond IA as either a post hoc rationalization or as a bureaucratic barrier to be overcome. Influential EIA practice entails clearly defined roles and responsibilities; the integration of IA into decision making from initial problem/opportunity definition right through to postapproval follow-up; the systematic analysis and interpretation of options and individual and cumulative effects; explicit and fully substantiated assumptions, methods, interpretations, conclusions, and recommendations; the use of proactive measures and resources to facilitate public understanding, involvement, and influence prior to every decision; and an adaptive approach that effectively identifies and manages risks and uncertainties. Influential EIA practice provides information, analyses, and interpretations in a form that is readily understandable; is appropriate to the context and audience; and is directly linked to potential decisions (including how they might be implemented). It is focused on key stakeholder issues; efficiently utilizes available resources; effectively draws upon both technical/scientific and community/indigenous knowledge; appreciates and integrates the perspectives, interests, and values of each party; treats IA as a mechanism for fostering collaboration and organizational/social learning and capacity building; and explicitly and systematically addresses procedural and substantive equity and fairness issues.

Ecological Impact Assessment (EcIA) Influential EcIA practice places particular emphasis on links to international, national, and local ecological and biodiversity agreements, policies, requirements, and initiatives; on cooperation, joint planning and comanagement with other jurisdictions and environmental NGOs; on the integration and application of ecological standards, guidelines, and objectives; and on the systematic adaptation and application of ecological knowledge and perspectives at multiple ecological and decision-making levels.

Social Impact Assessment (SIA) Influential SIA practice recognizes the inherently political nature of IA and decision making, and the critical importance of public participation. It ensures that contextual variations are fully reflected in decisions and in the outcomes from decisions. It proactively seeks to decentralize decision making; fully integrates social, cultural, heritage, and psychological effects and concerns; systematically addresses interconnections among ecological, social, and health effects; facilitates social learning, capacity building and empowerment; overcomes institutional constraints to SIA; respect human rights; fully considers the distribution of effects among population groups; ameliorates and overcomes procedural and substantive inequities--especially for the least advantaged; incorporates community and indigenous knowledge, perspectives, and experiences; and attains broad public acceptance and support, consistent with the realization of social potential and aspirations.

Health Impact Assessment (HIA) Influential HIA practices focus on obtaining political and government support for the institutionalizing of HIA; on directly linking HIA processes and outcomes to policy issues and problems; on demonstrating (e.g., through demonstration projects and effectiveness analyses) the benefits to all stakeholders of HIA; on facilitating dialogue among researchers, practitioners, affected populations, and decision makers; and on ensuring a sound and readily available (to all parties) HIA knowledge base. HIA practice, if it is to be influential, must explicitly and systematically address and manage the health concerns and priorities of all parties; clearly acknowledge uncertainties and limits; promote health rather than just seek to minimize adverse health effects; reduce inequities; foster links to health practitioners; focus particular attention on avoidable, involuntary, adverse, irreversible, and catastrophic health risks (especially to the most vulnerable); demonstrate benefits to the delivery of health services; facilitate stakeholder and community "buy-in" and support; and contribute to HIA capacity building.

3.6 CONTEMPORARY CHALLENGE—IA INFLUENCE—GOOD PRACTICE GUIDANCE

Enhancing the influence of IA in decision making is a complex task involving a variety of strategies and tactics. Tables 3.3 and 3.4, respectively, use criteria to structure the presentation of potential good practices for making IA more influential at the regulatory level and applied levels.

These criteria and practices represent, at best, a list of possible measures. Careful consideration should be given to whether they are appropriate to the decision-making context and, if so, what adaptations might be needed. Individual measures need to be knit together into a coherent strategy. Such a strategy should be the product of an open and collaborative process involving all interested and affected parties. It also should draw upon experiences elsewhere (again with appropriate contextual adaptations) and IA literature. Adjustments during the course of implementation will always be necessary. An open and independent audit of the effectiveness of the measures, both individually and collectively, is essential.

3.7 SUMMING UP

In this chapter we describe an influential IA process, a process directed toward making IA practice more integrated with and influential over decision making. Four stories describe applied, IA influence-related experiences in which (1) the major parties decide to circumvent the IA process on the grounds that the proposal is too "important"; (2) SEA is used as an instrument for enhancing public policy-making

Table 3.3 Examples of Regulatory Level Good Practices: Making IA More Influential

Criteria	Practices
Enhancing the understanding of decision making	Sponsor applied research on obstacles and pitfalls and on what makes different planning systems effective Include provisions for undertaking of an institutional analysis (e.g., objectives, legal and regulatory framework, informal rules, decision-making processes, implementation, resources, interactions) Seek to understand what makes IA effective in different planning systems Treat IA as open learning process, and decision making as process of learning and negotiations among multiple actors Explore strategies for overcoming institutional resistance to instilling environmental values
Constraining and substantiating decision making	Require reasons for all IA decisions Require decision makers to follow IA recommendations or provide reasons (consistent with purpose of legislation) for not doing so Seek senior management commitment Ensure authorities have competence and duties to take environmental matters into account Require agencies to avoid or minimize proposal's negative environmental effects Require that final decisions take into account public views and constraints
Ensuring decision making is easier and better informed	Clarify IA aims and accomplishments Require summary documents Emphasize, in requirements and guidelines, readability, reduced length, and highlighting of significant impacts Clarify objectives Insist on transparent planning and decision making Tailor communications to decision-making needs and preferences
Reducing the barriers to IA decision-making effectiveness	Reform institutions to integrate sustainability and environmental values into mandate Contribute sufficient resources to implement IA Seek to decentralize decision making Seek to identify and then enact measures to reduce self-sufficiency and exclusion of strong institutions and interest groups that tend to dominate bureaucracies; may require legal provisions and networking
Making decision making more explicit, consistent, and up-to-date	Explicitly identify agency roles, responsibilities, and accountability, keyed to decision points Incorporate SEA and EIA performance (measures of success—quality control) tracking measures by agency; including national repository, available online Sponsor development of quality and effectiveness review protocols Establish working group to investigate need for further improvements Clarify and make explicit such matters as contacts, languages, timing, notification, public consultation, interpretation of terms, document contents, and follow-up
Making decision making more transparent, open, inclusive, democratic, and empowering	Make all completed assessments public; central registry Define public broadly, define effective notice and acceptable processes for engaging public at key decisions Rectify inadequate timelines Require demonstration of public role in decisions Provide opportunities for appeal of process or decision outputs Promote in legislation and guidelines, multistakeholder and public participation, cooperation, negotiations, and dispute resolution Establish participant funding and IA capacity building programs Require documentation of agency consultation procedures Devote more attention to subsidiarity Strengthen links among planning, decision making, and participation Enhance democratization, reframe decision making to shift balance of power toward marginalized groups
Making decision making more environmentally substantive	Give IA statutory, substantive purpose, principles, and objectives (by government level); require decisions to be consistent with Seek political commitment to objectives of legislation Devote greater attention to advancing systems and practices for ensuring substantive outcomes Establish clearing houses/repositories for environmental information and data Strengthen science–policy links (e.g., through knowledge brokering) Promote and incorporate environmental values; seek to integrate into institutions

(continued)

Table 3.3 (*Continued*)

Criteria	Practices
	Increase weight given to environmental resources and capacities in IA systems
	Institute higher approval test—enhancements and net sustainability gains
Demonstrating the benefits of IA	Require and encourage top level political and bureaucratic buy-in
	Require IA performance evaluation
	Sponsor applied research—IA effectiveness and added value
	Emphasize effectiveness in monitoring requirements
	Solicit and facilitate public sector "ownership" of IA outcomes
Linking IA and to related decision making	Formalize tiering of policies, plans, programs, and projects
	Link SEA to downstream decision inputs and decision activities
	Promote interagency cooperation and coordination
	Explore opportunities for greater unity and comparability among substantive and procedural IA and environmental requirements
Linking IA to and integrating IA with policy/plan/program and project decision making	Integrate with policies, plans, programs, and projects, wherever practical
	Focus on planning and decision-making links
	Investigate barriers to ecological, social, and health integration, and how might be addressed
Enhancing the organizational and institutional capacity to undertake IA	Provide for IA and IA-type training and capacity building (e.g., technical training, awareness-raising workshops, support for institutionalizing of evaluation systems, networking, guidelines)
	Enhance regulatory, financial, and personal capacities of IA system, authorities, and other participants to undertake and effectively participate in IA and ameliorate technological and logistical constraints
	Institute management systems to apply requirements and support implementation (e.g., compliance outreach, training and organizational support, education of staff, contractors and decision makers)
	Sponsor institutional analysis; ameliorate weaknesses that inhibit effective IA
Enhancing the fit between IA and the decision-making context	Emphasize need to design to suit problem and institutional context
	Consider planning culture differences and implications
	Sponsor adaptation of quality and effectiveness review protocols to match decision-making context
	Adapt IA policies and requirements in conjunction with changing institutional environment and societal reform
Enhancing the professional status of IA practitioners	Consider mandatory registration of IA professionals (by IA type), IA training, codes of conduct, and methodological guidelines

Sources: Benson (2003), Bond (2004), Bond and Morrison-Saunders (2011), Booth and Skelton (2011a,b), Campbell (2003), Cashmore (2004), Cashmore et al. (2004), Bonifazi et al. (2011), Bredariol and Marini (2003), Briffet et al. (2003), Burdge (2008), Buuren and Nooteboom (2009), Chaker et al. (2006), Cherp and Antypas (2003), Craik (2008), Dimento and Ingram (2005), Doberstein (2003), Doelle and Sinclair (2006), Eccleston (2008), EC (2008b, 2009d, 2010, 2011b), Elling (2007, 2011), European Court of Auditors (2010), Evaluation Partnership (2007), Fischer (2005, 2006); Fischer et al. (2009); Fuggle (2005a,b); Gazzola (2008); Hanna (2005); Hanusch and Fischer (2011); Hansen and Wolff (2011); Hegmann and Yarranton (2011); Heinma and Põder (2010), Hildén et al. (2004), Hinte et al. (2007), IAIA (undated a), ICCL (2001), Jackson and Dixon (2006), Jha-Thakur et al. (2009), Jiricka and Pröbstl (2008), Jiliberto (2011), Jiricki and Pröbstl (2009), João (2007), Jones and Slinn (2008), Karjalainen and Järvikoski (2010), Kessler and Abaza (2006), Keys et al. (2011), Kørnøv and Dalkmann (2011), Kolhoff et al. (2009), Kolkman et al. (2007), Kørnøv and Thissen (2000); Larsen et al. (2012); Lobos and Partidário (2010); Lyhne (2009); McCluskey and João (2011); Noble (2004a,c, 2009a,b); Noble and Bronson (2006), Noble and Storey (2005), Nykvist and Nilsson (2009), OECD (2006), O'Faircheallaigh (2009, 2010), Page (2006), Partidário and Coutinho (2011), Persson and Nilsson (2007), Pisani and Sandham (2006), Pischke and Cashmore (2006), Pölönen (2006), Pullin and Knight (2003), Retief (2007b), Retief et al. (2008), Ross et al. (2006), Runhaar and Driessen (2007), Runhaar et al. (2010), Sánchez and Gallardo (2005), Scott (2011), Sheate (2011), Sheate and Partidário (2010), Smythe and Isber (2003), Söderman and Kallio (2009), Soneryd and Weldon (2003), Stern et al. (2009), Stern and Predmore (2011), Stoeglehner et al. (2009), Thérivel (2010); Tinker et al. (2005), Tzoumis (2007), Vicente and Partidário (2006), Weaver et al. (2008), Weston (2011), Wirutskulshai et al. (2011), Zhu et al. (2010).

influence; (3) various approaches are explored in one jurisdiction (the Netherlands) for making IA more relevant and influential; and (4) the key environmental issue (i.e., whether the proposal should proceed) was not addressed by the EIA, and the EIA process and documents had a negligible influence on decision making.

Three negative perspectives represent the problem. The first perspective concerns the view that IA is a simple process readily mastered with only a modicum of methodological knowledge and experience. The second perspective views IA as more trouble than it is worth. The third perspective argues that proposed actions directed towards environmental ends and environmentally conscious policies, plans, programs, and projects already achieve IA-related ends without the costs and biases associated with IA-related means. Together these perspectives result in a tendency to circumvent or avoid IA processes and/or they undermine the effectiveness of IA processes. The direction entails determining the legitimacy of these perspectives, offsetting potentially valid aspects of these perspectives, and defining more relevant IA requirements and processes.

Table 3.4 Examples of Applied Level Good Practices—Making IA More Influential

Criteria	Practices
Enhancing the understanding of decision making	Seek to enhance understanding of "real" decision-making (e.g., customs, tools, formal and informal rules, yardsticks, norms, priorities, competencies, cultures, sources, and culture of resistance to and disownment of change, interests, incentives, frames of reference, how power directly and indirectly exercised, cognitive and resource limits, distribution of decision making among actors) Seek to enhance understanding of beliefs, roles, values, biases, preferences, norms, ambiguity, experiences, convictions, interests, and needs of decision makers and of other stakeholders Promote discourse analysis and discourse reflection by stakeholders Identify all parties, interests, and views associated with proposed action and alternatives; analyze conflicts Identify current and potential roles of IA in decision making Assess varying perceptions of IA practitioners and decision makers
Constraining and substantiating decision making	Ensure reasons consistent with requirements, purposes, principles, and objectives Substantiate choice and application of all assumptions and methods Make IA preparation roles and responsibilities explicit Make implementation roles and responsibilities explicit Ensure that public views not constrained or narrowed Provide and apply clear decision-making criteria; link to legal requirements, objectives, and ethical principles
Ensuring decision making is easier and better informed	Focus on main messages and results in nontechnical form adjusted to readers Seek to enhance communications to decision makers (e.g., bridging problems and solutions) and among stakeholders Provide roadmaps and meet needs of readers Initiate early consultation with decision makers Provide evidence-based framework for decision steps Provide sufficient, reliable, and usable information for planning and decision making Identify and remedy communications deficits Focus on critical factors that could make a decision-making difference
Reducing the barriers to IA decision-making effectiveness	Concentrate on performance effectiveness Foster ownership of outcomes Seek to facilitate dialogue, adjust perceptions, and identify shared interests and meanings Seek to understand basis for disputes, help parties reframe and work toward settlement; may require role for dispute resolution specialists Seek to positively influence decision makers' capacity to accept by changing mind sets and motivations at senior levels Involve decision makers as active participants from process outset Demonstrate added value of IA to decision making
Making decision making more explicit, consistent, and up-to-date	Use decision tools and frameworks to facilitate more systematic and consistent decision making Apply criteria for assessing IA expert judgments and decision consistency Clearly define roles and responsibilities of assessors and planners Provide more best practice examples and lessons and knowledge sharing forums IA practitioners, organizations, and professional bodies should raise bar and demand better quality assurance Ensure consistence with good practice standards and quality and effectiveness review results
Making decision making more transparent, open, inclusive, democratic, and empowering	Seek to influence decision making through participation and involvement in structured decision making Seek to integrate environmental inputs into critical decision windows Assess direct and indirect democratic effectiveness Facilitate community empowerment; facilitate bottom-up participation and influence Institute early in process and undertake throughout action life cycle Treat IA as a systematic negotiating tool to identify the best option that meets shared stakeholder objectives Facilitate two-way communication and collaboration among stakeholders from outset of process Facilitate coordination between government and nongovernment Foster bottom-up participation and influence; consider and assess democratic effectiveness Respect treaty and aboriginal rights Eliminate bias against community and aboriginal views and knowledge Extend public involvement into decision making and into postapproval decisions Broaden temporal and spatial boundaries

(continued)

Table 3.4 (*Continued*)

Criteria	Practices
Making decision making more environmentally substantive	Focus on achievement of environmental and sustainability vision and goals (e.g., positive sustainability contribution); follow-up whether goals achieved Focus on integration of environmental values Institute new forms of sustainability plans and appraisals Identify and apply explicit sustainability, environmental, and substantive ethical principles and standards Undertake environmental advocacy when highly normative and politically constrained issue Shift IA orientation to promoting substantive environmental concerns, especially sustainability
Demonstrating the benefits of IA	Undertake applied research—cases that demonstrate added value Seek stakeholder buy-in Link IA outcomes to public interest Use follow-up to demonstrate value added by IA for all stakeholders Emphasize effectiveness in monitoring Demonstrate how IA consistent with and supportive of interests, values, and preferences of decision makers and senior administrators
Extending decision making past approvals	Employ strong enforcement mechanisms Ensure sufficient resources to impact monitoring so can evaluate impacts and mitigation effectiveness Independently review and audit decision making Employ checklists and audit protocols to ensure mitigation measures successfully implemented Link to and integrate with environmental management systems Assess effectiveness in terms of contribution to decisions, procedural quality of process, and quality of stakeholder participation
Linking IA to related decision making	Link to national and institutional environmental and sustainability policies and strategies Link to state-of-the-environment reporting
Linking and integrating IA with policy/plan/program and project decision making	Integrate institutional and political factors Comply with legal requirements Network between assessors and planners
Enhancing the organizational and institutional capacity to undertake IA	Foster organizational learning Assess legal, policy, and development control implications Treat IA as catalyst for organizational learning Make institutional analysis part of IA process Learn from effectiveness/performance assessments Seek to instill IA knowledge and skills in sectors responsible for development planning Seek to foster capacities of key stakeholders
Enhancing the fit between IA and the decision-making context	Systematically adapt to decision-making context Seek opportunities to positively influence context Adapt quality and effectiveness review protocols to match context Consider contextual uncertainties and decision-making implications
Enhancing the professional status of IA practitioners	Participate in environmental advocacy, especially when highly normative and politically constrained issue Practitioners should actively seek to enhance IA quality and effectiveness Foster professional networks and adherence to good practice standards

Sources: Benson (2003), Bond (2004), Bond and Morrison-Saunders (2011), Booth and Skelton (2011a,b), Cashmore (2004), Cashmore et al. (2004), Bonifazi et al. (2011), Bredariol and Marini (2003), Briffet et al. (2003), Burdge (2008), Buuren and Nooteboom (2009), Campbell (2003), Chaker et al. (2006), Cherp and Antypas (2003), Craik (2008), Dimento and Ingram (2005), Doberstein (2003), Doelle and Sinclair (2006), Eccleston (2008), EC (2008b, 2009d, 2010, 2011a,sb), Elling (2007, 2011), European Court of Auditors (2010), Evaluation Partnership (2007), Fischer (2005, 2006); Fischer et al. (2009); Fuggle (2005a); Gazzola (2008); Hanna (2005); Hanusch and Fischer (2011); Hansen and Wolff (2011); Hegmann and Yarranton (2011); Heinma and Põder (2010), Hildén et al. (2004), Hinte et al. (2007), IAIA (undated b), Jackson and Dixon (2006), Jha-Thakur et al. (2009), Jiricka and Pröbstl (2008), Jiliberto (2011), Jiricki and Pröbstl (2009), João (2007), Jones and Slinn (2008), Karjalainen and Järvikoski (2010), Kessler and Abaza (2006), Keys et al. (2011), Kørnøv and Dalkmann (2011), Kolhoff et al., (2009), Kolkman et al. (2007), Kørnøv and Thissen (2000); Larsen et al. (2012); Lobos and Partidário (2010); Lyhne (2009); McCluskey and João (2011); Noble (2004a,c, 2009b); Noble and Bronson (2006), Noble and Storey (2005), Nykvist and Nilsson (2009), OECD (2006), O'Faircheallaigh (2009, 2010), Page (2006), Partidário and Coutinho (2011), Persson and Nilsson (2007), Pisani and Sandham (2006), Pischke and Cashmore (2006), Pölönen (2006), Pullin and Knight (2003), Retief (2007b), Retief et al. (2008), Ross et al. (2006), Runhaar and Driessen (2007), Runhaar et al. (2010), Sánchez and Gallardo (2005), Scott (2011), Sheate (2011), Sheate and Partidário (2010), Smythe and Isber (2003), Söderman and Kallio (2009), Soneryd and Weldon (2003), Stern et al. (2009), Stern and Predmore (2011), Stoeglehner et al. (2009), Thérivel (2010); Tinker et al. (2005), Tzoumis (2007), Vicente and Partidário (2006), Weaver et al. (2008), Weston (2011), Wirutskulshai et al. (2011), Zhu et al. (2010).

Perspective 1 (What Could Be Simpler?) envisions the IA process as a simple, rational planning procedure. All that is required to effectively manage the IA process, it maintains, is a cursory knowledge of IA methodology, familiarity with IA requirements and guidelines, and some experience with comparable proposals and processes. The IA knowledge base is viewed as limited and largely static. Little knowledge from related fields is considered relevant.

Perspective 2 (It Can't, Won't, or Shouldn't Be Done) asserts that IA cannot, will not, or should not be institutionalized and applied as a decision-making aid. A variety of overlapping and mutually reinforcing reasons have been offered up to support this conclusion. IA, it is argued, involves inherently artificial boundaries; is largely atheoretical; lacks focus or tends to focus on the wrong things; seeks (unsuccessfully) to predict and manage long-term change; is confounded by irreconcilable value changes; is inherently biased and subjective; tends to be excluded from or peripheral to decision making; inequitably allocates resources; entails requirements and processes that are neither necessary nor effective; and makes no or a minimal, substantive environmental contribution.

Perspective 3 (We Already Do That) starts from the premise that the major purpose of IA (i.e., more environmentally sound decision making) is or can be achieved through environmentally driven and shaped policies, plans, programs, and projects, without the imposition of "action-forcing" IA requirements. Proponents of this perspective point to the many public policies, plans, and programs designed and implemented to achieve environmental aspirations—initiatives launched and undertaken without the necessity of grafting on IA requirements. They maintain that there already is a long tradition of infusing environmental perspectives and knowledge into public and private decision making. They also argue that IA is wasteful, redundant, and unnecessary; distorts planning and decision making; has a negligible planning role; and represents little more than a bureaucratic hurdle.

Although the three perspectives represent a partially valid criticism of some IA practice, the arguments are overstated, many positive examples do not display the characteristics cited, and there is considerable scope for improvement to the extent that IA practice falls "short of the mark." Contrary to Perspective 1, although basic IA processes are simple, there are a huge array of process design choices available, choices that should be drawn upon for complex decision-making situations, to match process to context, and to adapt to changing conditions and needs. The IA knowledge base is extensive, rapidly expanding, and complex. A thorough understanding of the IA knowledge base, and the knowledge base of related fields, is essential to good IA practice. Considerable experience and care is needed to adapt knowledge to practice, to transfer experiences from comparable proposals, and to adequately interpret IA requirements, guidelines, and advice.

There also is considerable potential for ameliorating the shortcomings cited by adherents to Perspective 2. Boundaries can be effectively and systematically determined and substantiated. IA theory building and testing is both possible and highly desirable. IA practice can be focused in a supportable manner. Techniques are available for effectively predicting and managing long-term environmental changes. Value differences can be both embraced and ameliorated. Bias can be minimized. Subjective interpretations can be clearly and consistently applied and supported. IA requirements and processes can be reformed to encompass all relevant decision making. Resource inequities can be offset. IA requirements and processes are both necessary and potentially effective. IA can and has made positive contributions to the achievement of environmental objectives. Considerable scope for improvement remains to further ameliorate the negative tendencies of some IA practice, to build upon positive examples, and to make more effective use of available knowledge and experience.

Perspective 3 overstates the extent to which the planning of policies, plans, programs, and projects, without the integration of IA requirements, fully and adequately addresses environmental concerns. It also overstates the deficiencies and limitations of IA as a field of theory and practice. Advocates of Perspective 3 tend to incorrectly assume that (1) environmental proposals have negligible adverse environmental impacts (substantial unintended adverse environmental effects often occur); (2) environmental considerations and aspirations are enough (often intentions fall well short of accomplishments); (3) action-forcing requirements are unnecessary (IA requirements ensure minimal performance standards are met and IA guidelines can facilitate more consistent good practice); and (4) impact assessment is inherently biased and has nothing to contribute methodologically (IA biases have been largely transcended in recent years and the field has much to offer regarding such matters as the prediction and interpretation of environmental effects, the generation and evaluation of alternatives, impact management, and stakeholder involvement).

Several concepts, integrative frameworks, and knowledge base and research priorities, pertinent to designing and managing influential IA regulatory and applied processes and systems, are described to establish a foundation for good practice approaches to facilitating decision-making influence.

There are positive and negative examples of influence-related IA requirements and initiatives in each of the four jurisdictions. Influence at the IA regulatory level includes the IA regulatory system, the IA administrative system, and links between IA types and decision making. The state of IA practice has advanced to the point that good practices can be identified in each of these areas. Contextual adjustments and a collaborative approach are, however, necessary.

An example influential IA process is depicted. The process, in common with most IA process characterizations, includes such activities as screening, scoping, alternatives generation and evaluation, baseline and effects analysis and interpretation, documentation, decision making, monitoring, and follow-up. Unlike typical IA process descriptions, the influential IA process also links and integrates SEA and EIA types and decision-making levels, systematically draws upon and contributes to the IA and related fields knowledge base, and fully integrates all stakeholder perspectives and interests from the outset, and throughout the process. In addition, the process is guided and structured by explicit sustainability and other environmental values and objectives, stakeholder interactions are guided by procedural objectives, and direct and explicit links are made to related decisions and to IA institutional arrangements. Also, a systems perspective is maintained throughout; a proactive effort is made to inform, influence, and substantiate decision making; care is taken to ensure that the process suits the context; and community and traditional knowledge are fully and effectively integrated into the process. Given the likelihood of varying perceptions of influence among IA stakeholders, an influential IA process must necessarily steer a careful path if it is to effectively link and balance these varying perspectives. It is far easier to achieve such a balance when the major stakeholders respect one another; when they collaboratively work together for common goals; and when the process facilitates cooperation and communications.

Good practices for making IA more influential at the regulatory and applied levels, at the SEA and EIA levels, and for individual IA types (SA, SIA, EcIA, HIA) are presented. These possible measures need to be adapted to context and integrated into a coherent strategy. The strategy should emerge from an open and collaborative process involving all interested and affected parties. The strategy also should draw upon experience elsewhere and IA literature, be adjusted through implementation, and independently audited for effectiveness (as interpreted from multiple perspectives).

Chapter 4

How to Make IAs More Rigorous

4.1 HIGHLIGHTS

This chapter tests the premise that IA processes, documents, and methods should be more scientifically rigorous. Chapter 1 identifies the lack of scientific rigor as a recurrent shortcoming of IA practice. This chapter seeks to address that shortcoming.

- The analysis begins with two applied anecdotes. The stories describe applied experiences associated with efforts to make IA practice more rigorous.
- The analysis in Section 4.3 then defines the problem, which is the inadequate and ineffective use of scientific principles, knowledge, and methods in the IA process. We provide practical advice for making IA documents and processes, at both the SEA and project EIA levels and for various IA types, more rigorous, while allowing for the limits of science in IA practice.
- In Section 4.4 we identify relevant principles and assumptions commonly ascribed to analytical science. Debates concerning analytical science are explored both in general and for applied fields such as IA. These analyses provide the basis for defining a rigorous IA process.
- In Section 4.5 we detail how a rigorous EIA process could be implemented at the regulatory and applied levels. In Section 4.5.1 we infuse a "scientific" perspective into IA regulatory requirements and guidelines, in Section 4.5.2 we integrate a "scientific" perspective into applied processes, and in Section 4.5.3 we address IA theory building at the SEA and project EIA levels, and for various IA types (SA, EcIA, SIA, and HIA).
- In Section 4.6 we address the contemporary challenge of good practice follow-up approaches. We address both good practice ends and means and present an array of regulatory and applied level examples of follow-up good practices.
- In Section 4.7 we highlight the major insights and lessons derived from the analysis.

4.2 INSIGHTS FROM PRACTICE

4.2.1 Poor Practice Rides Roughshod Over Sound Theory

Located in the Peace Region of British Columbia (BC), Canada, the Hermann Coal Mine Project was proposed at a time of heightened concern by resident indigenous groups over the compromised health of the local ecosystems. The region was (and likely still is) the quintessence of cumulative impacts. Examples of the intensity and diversity of development projects approved as of 2012 include approximately 18 000 oil and gas well sites, 10 000 industrial facilities, 6000 forestry cut blocks, 5 coal mines, 6 wind farms, 2 large-scale hydroelectric dams, 60 000 km of roads, and 100 000 km of pipelines. None of the existing projects, however, included a SIA that adequately examined project-specific (let alone cumulative) impacts upon the cultural values held by the indigenous groups.

In October 2006, the BC Environmental Assessment Office (EAO) released the draft terms of reference (TOR) for the EIA of the Hermann Mine to the public for their review and comment. Since a draft TOR is modified in some cases (before finalization), in order for the EIA to respond properly to the concerns of the public, an indigenous group took the opportunity to provide constructive, science-based comments with the intention of making the SIA more culturally inclusive.

Considerable emphasis was placed upon the theoretical orientation. The proponent's use of the "modernization paradigm" as the basis for the SIA was viewed by the indigenous group to be highly problematic, since it assumes (albeit erroneously) that industrial development is inherently beneficial to them and that their culture will inevitably be assimilated into mainstream society such that their land-based economy will be replaced with one that is industrial based. They also noted that the proponent's theory led to the selection of valued social components (VSCs), specifically, housing, transportation, services, labor supply, and community health conditions, which would not accurately assess impacts to their cultural values. Overall, the framework did not adequately take into account the essence of their cultural traditions, customs, and land-use practices, or of the cumulative impacts to such values.

Impact Assessment: Practical Solutions to Recurrent Problems and Contemporary Challenges, Second Edition. David P. Lawrence.

An appropriate remedy, in their view, was the application of the "subsistence/adaptation paradigm" as a way to properly ground the SIA. Because that theory places a significant emphasis on the ability of indigenous groups to preserve cultural lifestyles and the ability to mix the traditional economy with the mainstream economy in situations where integration maintains or enhances cultural integrity, the selection of VSCs reflecting this theory would echo their cultural values. They recommended the SIA use the following VSCs: livelihoods and economic vitality, population structure and health, social stability and community well-being, education and training, and cultural well-being.

Irrespective of the fact that the approach put forward by the indigenous group was in tune with socioscientific standard of SIA research, the proponent stated that it was not going to revise the scope of work, and thus refused to modify the draft TOR. Its reason was simple: The company planned to submit the EIA report by mid-November 2006 (mere weeks after the public comment period ended), and thus did not have the time to change its design and methodology. This shed light on the proponent's predisposition; that is, it likely had no intention of seriously considering or integrating input into the TOR if such a change increased the risk to its project. Not surprisingly, the EAO and the indigenous group disagreed with the proponent's response, which was, at a minimum, contrary to the principle of soliciting input in good faith. The feedback from the indigenous group was considered by the EAO to be scientifically credible, and as such, the proponent was forced to include the Indigenous approach and chosen VSCs in the final TOR.

Without collecting the relevant primary or secondary data to address the data gaps, the proponent submitted its EIA report two months later. The EAO (unexpectedly) accepted the report even though it failed to fulfill the requirements of the TOR; indeed, the document admitted that the assessment was "incomplete" and the proponent did not have the necessary data to assess impacts to culture. None of the indigenous groups took the decision lightly. They boycotted the process and insisted that the timelines be suspended until the impacts to their cultures were adequately assessed. The EAO initially declined to suspend the process, but did so after the proponent (albeit reluctantly) agreed.

Neither the data gaps nor the assessment of the potential impacts to cultural values were, by the end of the suspension period, reconciled. The Hermann Mine was subsequently approved by BC although it lacked a scientifically rigorous and culturally appropriate EIA. What this case revealed was that SIA theory may have come a long way since its inauguration in Canada approximately 40 years ago, but its current practice remains fixed in the domain of public policy (based on heartland–hinterland ideologies) rather than sound scientific analysis.

BRUCE R. MUIR

CTQ Consultants Ltd., Kamloops, British Columbia, Canada

4.2.2 Better Science = Better Decisions: the Case of Igamba Falls

The Kigoma region of western Tanzania is desperately short of power; less than 3% of households are supplied with electricity. Electricity is supplied by the Tanzania Electricity Company using old diesel-powered generators that are no longer capable of operating at full capacity. As the diesel fuel has to be transported by rail from the port of Dar es Salaam, 1200 km (750 miles) away, both the supply of fuel and the electricity generated are erratic and expensive. The region is not linked with the national transmission grid. This inadequate supply of electricity greatly hinders industrial development and regional economic growth. The Government of Tanzania recognized in the mid 20th Century that improved electrification of the Kigoma region will have many beneficial impacts, and has been actively looking for ways to provide it with an adequate, reliable, and affordable supply of electrical power.

Since the early 1980's the Tanzanian government has been working closely with the Joint UNDP/World Bank Energy Sector Management Assistance Program. An early initiative was a 1982 feasibility study of the potential for mini hydropower development in the region. This study explored the possibilities of hydropower development on the Malagarasi River at the 20 m high Igamba falls about 60 km southeast of Kigoma. Although hydropower generation at the falls was technically feasible, this early study concluded that the Igamba Falls project was uneconomic, even with funding from the World Bank, and was thus not pursued.

During 2004–2005, a further study by a different international consultancy undertook a prefeasibility study of a modified plan for hydropower generation at the Igamba Falls. This study collected extensive topographic, environmental, social, geological, and hydrological data. Sections of the report dealt with aquatic environment, water quality, sediment load, terrestrial ecology, land use and vegetation cover types, habitat assessment, geology and soils, landscape and land use, settlement patterns and land tenure, demography, sources of income, currently protected areas, and a proposed protection area. Particular attention was given to several features of the project site. The unusual geology—highly fractured calcareous sandstone that allows a lot of water to flow underground through fissures and solution channels—caused concern and required an explorative drilling program. The barrier that the falls posed to aquatic migrations, and thus to potential species separation, was recognized and led to a comparison of fish being caught above and below the falls by indigenous fishermen. Although the site impinges marginally on the Masito Conservation Area—known for its chimpanzee population—the fact that the project requires a bridge across the river, was recognized as a problem as this would facilitate human access to the conservation area. The study proposed measures to overcome these negative impacts. Nevertheless, the economic and financial analyses indicated that the project would not be viable unless it could be financed from a "soft loan" repayable at low interest over 30 or more years.

During 2005 a team of international scientists, led by Dr Ellinor Michel of the British Natural History Museum, undertook a three-week survey of the 450,000 ha Malagarasi-Moyovosi wetlands. This biodiversity and limnological study was completely unrelated to the hydropower project and was undertaken because the Malagarasi drainage in western Tanzania is among the world's most important, yet least studied, wetland regions. Detailed biological sampling took place at 40 sites, one of which was external to the wetlands and in the gorge through which the Malagarasi River discharges into Lake Tanganyika: serendipitously, the site of the proposed Igamba Falls hydroproject was surveyed because a track to the river had been opened to allow access for hydroproject vehicles.

Sampling of the Igamba Falls led directly to the statement in the report: "Our aquatic surveys of the remote Igamba Falls ... revealed a number of new fish and mollusk species, making this site not only of special interest, but also of special concern as it is under discussion for hydropower installation.... Our aquatic sampling there was of very limited duration and we strongly advocate detailed biodiversity work be continued in the region."

In 2008 the Tanzanian Government (GoT) entered into a compact with the Millennium Challenge Corporation (MCC), created in 2004 by the Congress of the United States of America. The compact allows for grants to projects in the energy sector that will reduce poverty through economic growth, increased productivity, and raised incomes. This compact opened the way for immediate action on the previously planned and studied 8 MW run-of-river hydropower plant at the Igamba Falls.

The earlier environmental and social impact assessment (ESIA) was quickly updated by an international consultant, but was found inadequate by the National Environment Management Council in Tanzania and by the MCC. A further ESIA was consequently commissioned from a different international consultant to meet MCC environmental and social guidelines and GoT requirements. This contract required detailed feasibility studies, a baseline aquatic survey at Igamba Falls, as well as a comprehensive cumulative impact assessment for the project. But, unaware of the results of the Michel et al. survey, the terms of reference failed to specify the extensive surveys and analyses necessary for a scientifically sound examination of project impacts on endemic species, in particular the new fish and snail species that had been found.

In early 2009, the MCC became aware of the significant biodiversity questions raised by Dr. Michel and her colleagues and recognized that specialized expertise would be necessary to address the issues raised and contribute to the project's ongoing environmental assessment. A further study by internationally acclaimed scientists was therefore commissioned to allow assessment of the biological significance of the three new species that had been previously discovered at the Igamba Falls to determine whether these species exist anywhere else and whether mitigation measures could ensure that they would survive the construction and operation of a hydropower facility.

This supplementary study confirmed that the Igamba Falls are unique. Their geology and geomorphology are most unusual, the calcareously bound sandstone is fractured, and scour holes and solution channels are ubiquitous. Subsurface water flows through horizontal strata and vertical joints creating a unique habitat. Water chemistry in the Igamba Falls area differs from that upstream and downstream, and biological productivity is substantially different from any other site on the Malagarasi River. Consequently, the flora and fauna of the falls are unique. The study also found that that the physical, biological, and chemical character of Igamba Falls would be fundamentally changed by the hydropower project, and that mitigation measures needed much more study.

An independent advisory panel (IAP) of international and Tanzanian experts was appointed to review all relevant studies and to suggest a way forward. The IAP found that the Igamba Falls did require special consideration due to their unique ecological, geological, and hydrological characteristics. It identified many significant impacts that would occur should the project go ahead as planned. Consequently, steps and further studies to protect the unique habitats and their endemic flora and fauna were recommended.

Despite a strong desire to fund the hydropower project to promote regional development and alleviate poverty, the MCC was faced with the requirement in its founding legislation that it may not finance projects that are "likely to cause a significant environmental, health, or safety hazard." Consequently, the MCC reached the conclusion that it could not fund the Igamba Falls hydropower project. Some viewed this decision as "snails being more important than power and development," particularly as none of the species have economic or cultural value. Other commentators concluded that development and poverty reduction were being sacrificed on the altar of biodiversity.

The Igamba Falls project contains several lessons for impact assessment professionals: First, the need for utmost caution and unusually thorough scientific investigations whenever projects are being planned in locations that have highly unusual physical features likely to give rise to rare habitats and endemic species. Second, no matter how well the environmental and social assessments are planned, be prepared for serendipitous discoveries and do not jump to premature conclusions. Scientific studies unrelated to formal ESIAs can, and should, inform project assessments. Third, studies of noncharismatic and noncommercial species are as important as studies of popular or commercially important species. Biodiversity assessments are an important aspect of environmental investigations. Finally, rapid assessments by highly qualified, experienced investigators can produce results that were overlooked by less experienced persons, even when working more slowly.

Postscript: The MCC has subsequently worked with the GoT to provide other options to meet Kigoma's electricity needs. These include repairs to generators and increased use of solar energy. The MCC also financed, at the request of the GoT, feasibility and environmental studies of another potential downstream location for a different Malagarasi hydroproject. That option would permit other donors, if the environmental impacts of the downstream location can be mitigated, to invest in the project and link the power generated into the Tanzanian grid.

RICHARD FUGGLE

Environmental Evaluation Unit, University of Cape Town, Cape Town, South Africa

4.3 DEFINING THE PROBLEM AND DECIDING ON A DIRECTION

The preceding stories demonstrate that there is potential role for the sciences in the IA process. The first story illustrates how the lack of sound, contextually appropriate, social science, especially if not effectively married with community and indigenous knowledge, can undermine IA credibility and effectiveness. The second story demonstrates the importance of a cautious approach to applied science in uncertain and sensitive environments, the need to extend the consideration of biodiversity to encompass noncharismatic and noncommercial species, the potentially significant decision-making role of biodiversity, and the value of highly qualified and experienced scientific investigators.

Critics of the prevailing and, in their judgment, too limited role of science in IA practice argue that IA processes, documents, and methods are too often ill defined, biased, subjective, and excessively descriptive (Whitney, 1986). They generally point to the limited foundation of sound scientific theory and knowledge (Dimento and Ingram, 2005; Greer-Wooten, 1997). They note that objectives are usually poorly stated. Study designs and standards of inquiry, they suggest, tend to be weak to nonexistent (Whitney, 1986). The methods applied, they point out, are far more simplistic than those formulated and applied by researchers (Lee, 2006). Commonly, they argue, spatial and temporal boundaries are either not defined or are defined too narrowly (Galbraith et al., 2007; Greer-Wooten, 1997). They indicate that the methods for characterizing environmental conditions, predicting changes with and without the proposal, and for managing effects are frequently vague, overly descriptive, poorly substantiated, and inconsistent with the scientific standards and protocols (Beanlands and Duinker, 1983; Greer-Wooten, 1997).

The critics tend to express particular concern with the limited attention devoted to the variability of natural phenomena, to social impacts, to environmental and impact interactions that transcend disciplinary boundaries, to cumulative effects, to comparable proposals and environments and to the postapproval monitoring and auditing of environmental conditions, the accuracy of impact prediction and the effectiveness of management measures (Burdge, 2004; Devlin and Yap, 2008; Hansen and Wolffe, 2011; Morgan, 1998; Noble, 2009b; Whitney, 1986). They further contend that qualified scientists are insufficiently involved in conducting and peer reviewing IA documents, methods, and procedures (Brown, 1986). These deficiencies result in, the critics conclude, unreliable predictions, avoidable uncertainties, an unsound decision-making basis, limited decision-making influence, the diminished credibility of science and scientists, a negligible contribution to the accumulation of knowledge, and a degraded environment (Dimento and Ingram, 2005; Morgan, 1998; Morrison-Saunders and Sadler, 2010; Whitney, 1986).

These views are not uniformly shared. Many argue that there are numerous dangers associated with a more scientific IA process. Some even suggest that a scientific IA process is inherently inappropriate. Between these two poles is a considerable middle ground occupied by those who would selectively apply, adapt, temper, and modify analytical scientific methods and principles. Further complicating the issue is a plethora of alternative conceptions of the nature and role of science as it is and as it could be applied for planning and decision-making purposes.

The journey from a desire for a more rigorous IA process to its application, therefore, involves intermediate steps. First, an overview of the principles and assumptions commonly ascribed to analytical science needs to be provided as a point of departure. Then debates regarding if and how analytical science might be modified in general and for applied fields such as IA need to be explored. The guidance for implementing a rigorous IA process addresses management at the regulatory and applied levels. Theory building variations between the SEA and project EIA level, and among IA types, are explored. Particular consideration is given to good follow-up practices. Ideal IA process criteria are used to identify the positive and negative tendencies of a scientific IA process.

4.4 SELECTING THE MOST APPROPRIATE ROUTE

The debates surrounding the applied role of science tend to treat classical analytical science as the "touchstone." Analytical science is viewed by some as a role model for applied fields such as IA. Others focus on its characteristics in making a case either against the use of science or in favor of substantial adaptations. Table 4.1 lists examples of terms that commonly crop up in the debate. Table 4.2 identifies characteristics often ascribed to analytical science. These characteristics sometimes operate in dramatic tension. Almost all analytical science characteristics are intensely debated. Any exploration of science in IA must inevitably

Table 4.1 A Few Key Terms in Analytical Science

Applied theory	• Findings applied to the solution of problems • Focus on facilitating decision making
Deduction	• Logical rules determine general premises, hypotheses, or theories • Conclusions about particulars follow from general premises • Testing approach often referred to as hypothetico-deductive method
Empiricism	• Research orientation that emphasizes facts, observations, and experiences over theory and conceptual reflection
Grounded theory	• Grounded in data obtained by research (contrasts with formal abstract theory)
Induction	• Process leads from particular facts and observations to general conclusions
Knowledge	• True belief acquired by a reliable method
Normative theory	• Hypotheses or other statements about what is right or wrong, desirable or undesirable, just or unjust in society
Positivism	• Rejects metaphysical speculation in favor of observation and experimentation as the preferred source of knowledge about the world • Aims to construct general laws or theories which express relationships between phenomena
Principles	• General rules for constructing models
Reductionism	• Belief that all phenomena can be reduced to a few laws and principles
Rigor	• Strict precision; exactness
Science	• A systematic series of empirical activities (methods) for constructing, representing, and analyzing knowledge about phenomena being studied • A set of normative commitments shared by a community of scholars • An occupation (scientists seeking to establish a body of knowledge) • Knowledge (tested facts and theories) • Applied to human needs and purposes (when applied rather than basic or pure science)
Scientism	• The claim that the positivist method is the only true method of obtaining knowledge
Theory	• A collection of hypotheses and predictions amenable to experimental testing • Organizes our concepts of and understating of the empirical world in a systematic way • A guide for defining what type of observations need to be made to understand phenomenon • A guide for interpreting observations

Sources: Bird (1998), Cashmore (2004), Giere (1999), Gower (1997), Patterson and Williams (1998), Rothman and Sudarshan (1998), Wilson (1998).

touch on some of these "science wars" debates (Giere, 1999). The following subsections highlight the implications of the "science wars" for IA theory and practice.

4.4.1 Absolute Truth Versus Relativism

The absolute truth versus relativism debate suggests that IA practitioners need not be constrained to a choice between the quest for absolute truth and the "anything goes" perspective of relativism. IA practitioners can draw upon the insights and methods of the natural and social sciences to enhance understanding, facilitate explanations, and contribute to improved decision making. As illustrated in Figure 4.1, IA practice can make a small contribution to middle-range theory building and successively greater contributions to micro theory building and pre-theory. These contributions will undoubtedly vary greatly in their quality, accuracy, scope, simplicity, and fruitfulness (Gower, 1997). They may not fully satisfy the "standards" of analytical science. A "tidy" hierarchy of mutually consistent and supportive theories is unlikely to emerge. A plurality of overlapping and often competing theories is the more likely result. But these efforts can, when due allowance is made for contextual variations, improve our understanding of and our role within the environment.

Much IA practice is either atheoretical or at the pre-theory level. Nevertheless, carefully formulated, applied, shared, and refined IA models, designs, concepts, frameworks, precedents, lessons, and tightly circumscribed generalizations still can lift IA practice beyond an exclusive focus on individual proposals and settings. A core body of good-practice IA knowledge is emerging. This process has accelerated in recent years with the proliferation of IA quality and effectiveness analyses, and with substantial theory building initiatives for individual IA types. The construction of an IA knowledge base cannot and should not be limited to the application and adaptation of a narrowly defined range of scientific perspectives and procedures. The claims and contributions of other scientific and nonscientific perspectives and of other modes of reasoning also must be recognized and accommodated (Healey, 1997). It will often be helpful to view IA related issues from multiple perspectives (sometimes this means alternative worldviews) and to apply multiple scientific and nonscientific paradigms and methods (Patterson and Williams, 1998). It is especially important that nonscientists be given a voice, expressed,

Table 4.2 Characteristics Commonly Ascribed to Analytical Science

Objective	• Values separable from facts • Science should not be subject to preconceptions (unbiased, impartial) • Science and scientists can and should be value-free
Independent	• Independent from subject, values, moral and political commitments, and environment (disinterested observer) • Judged on academic grounds; no reflection on individual
Reducible	• Deducible from the smallest number of possible axioms • Search for laws and law-like generalizations (ideal universal) • Largest amount of information with least effort (elegant)
Heuristic	• Builds on knowledge base (each addition contributes to enhanced understanding) • Continually testing and improving • Joined by theory
Methodological	• Accepted procedures for observing phenomena • Importance of observation, data, and evidence • Rigorous (precise), reliable, and standardized methods of investigation • Preference for quantitative results and experimental methods (most satisfactory form of evidence) • Reliability enhanced when multiple methods applied
Technological	• Reciprocal relationship between science and technology • Great advances in sciences often related to new tools • Omnipresence in science of machines, instruments, and experimental setups
Prone to positivism and scientism	• Science preferred source of knowledge (sometimes argued only valid source of knowledge)
Natural science model	• Physics as model for natural sciences • Natural sciences as model for social sciences • Scientific model for humanities • Scientific (normative) model for planning and decision making (e.g., IA, policy science, scientific management, scientific planning, scientific politics)
Explanatory	• Proper roles for science—measurement, observation, explanation, and prediction • No "ought" in science • Facilitates understanding • Conducive to decision making (knowledge base)
Verifiable	• Explicit assumptions and procedures • Traceable and replicable procedures • Possible to determine if correct (verifiable) or not (falsification) (possible to be weakly or strongly verifiable) • Prizes observation and measurement as primary means of explaining phenomena in comparable situations
Real	• Sometimes succeeds in stating the truth or a good approximation of • Knowledge is experienced based (meaning grounded in observation) • Objective world; can be observed and recorded in an objective manner
Collective	• A collective activity of discovery • Knowledge is shared (open, iterative) • Advanced by constructive discussion, analysis, and criticism (organized skepticism) • Must satisfy standards of peers
Pluralistic	• Multiple paradigms, theories, and concepts • Heterogeneous—many divisions within field • Multiple perspectives on definition, practice, and application of science
Consilient	• Underlying methodological unity • Fundamental laws and principles underlie every branch of learning (unity of knowledge) • Trend toward bridging of fields (interconnections, consistencies, middle ground, transcending concepts)
Certain	• Sufficient knowledge to measure and predict with accuracy • Manageable uncertainty
Causal	• Events have determinate causes • Causes precede events
Complex	• World viewed as a set of complex problems • Large number of variables and interrelationships • Still amenable to scientific methods
Beneficial or benign	• Can provide sound basis for decision making • Major contributions to society and environment

Sources: Barrow (1998), Cashmore (2004), Curtis and Epp (1999), Dawkins (1998), Erckmann (1986), Gower (1997), Greene (1999), Lemons and Brown (1990), Miller (1993), Orwell (2007), Patterson and Williams (1998), Pickering (1995), Porritt (2000), Rothman and Sudarshan (1998), Wilson (1998).

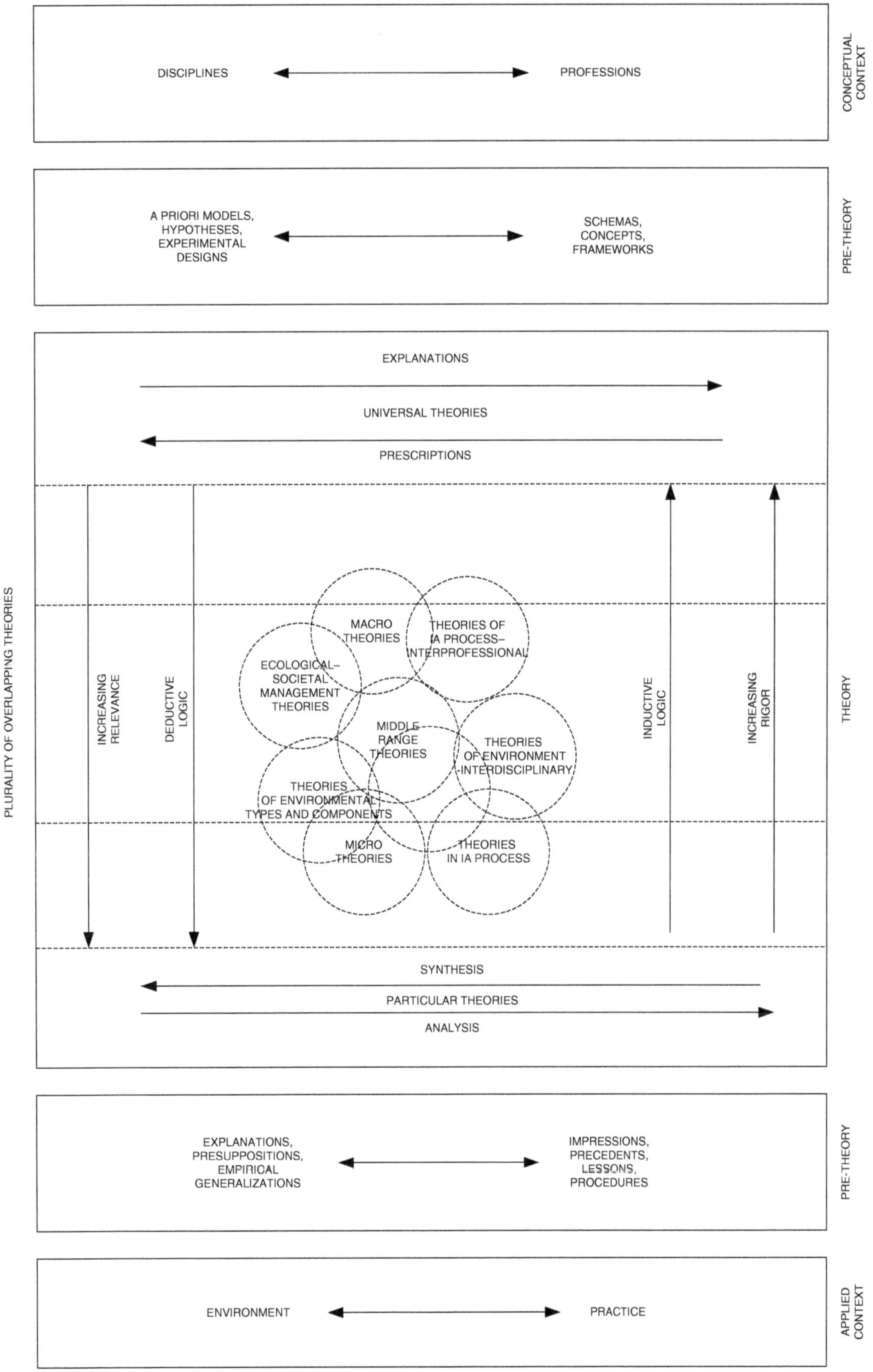

Figure 4.1 IA theory levels. Adapted from Lawrence (1997c).

if possible, in well-structured public interest arguments (Parkin, 1996).

4.4.2 Rigor Versus Relevance

The rigor versus relevance debate in IA is largely a false dichotomy. Natural and social scientific knowledge is a valuable resource for IA practitioners. It contributes to understanding and facilitates both significance interpretations and management actions (Healey, 1997). Appreciating the practical limits of IA practice, empirically adequate impact predictions are still conducive to better decision making and to better postapproval environmental management. Boundaries, assumptions, and models can and should be explicit and substantiated. Pilot studies and reviews of comparable proposals and environments can contribute to more accurate predictions and to more effective management actions. Monitoring can test the accuracy of impact predictions. Peer review can test methods and methods application. The substantive and methodological knowledge acquired through IA practice can be more broadly shared. Applied science, conducted under the auspices of IA practice, can be targeted to problems, opportunities, knowledge gaps, and uncertainties.

There can, moreover, be different degrees, standards, and forms of rigor. There are alternatives to analytical science (such as new or holistic science, chaos and complexity theory, regulatory or applied science), which, to varying degrees, relax and adapt the standards of analytical science. Rigor can be selectively applied. The issue then becomes which scientific standards are appropriate and reasonable, given the constraints and objectives of IA practice, rather than whether IA can or should be either rigorous or relevant. There is, however, a danger (some would say a "slippery slope") with selectively abandoning and adapting scientific standards. At some point one has ventured so far beyond the realm of applied science, that notwithstanding scientific "labels," what remains is no more than unsupported speculative thinking (Miller, 1993). Appreciating such risks, it still seems possible for IA practice to effectively blend both rigor and relevance. But any mixing of rigor and relevance will need to be carefully, systematically, and explicitly defined and substantiated.

4.4.3 Objectivity Versus Subjectivity

IA practice inevitably combines the objective and the subjective. Clearly, IA practitioners are not value-free. Much of IA practice is subjective. This does not mean that those advocating objectivity in science and in IA do not have valid points. Underlying the appeal for objectivity is a concern that if independence and transparency are simply abandoned, IA practitioners, documents, and procedures will be biased (often implicitly) against other stakeholders and against the environment whenever either conflicts with proponent interests. Some commentators will naturally conclude that such biases are inherent to IA. This is undoubtedly true to some degree, but bias can be reduced and environmental values can be applied.

Bias can be ameliorated if subjective judgments regarding assumptions, constraints, choice and application of methods and procedures, interpretations, and uncertainties are explicit, unambiguous, and substantiated (Beder, 1993; Rothman and Sudarshan, 1998; Mostert, 1996). Reports and methods can be peer reviewed. Professional codes of practice can be strictly adhered to and IA professionals can be accredited. Sometimes it is preferable if the proponent does not prepare some (e.g., interest groups generate own data) or all (e.g., IA reports and/or peer review reports prepared for the third party) IA documents (Beder, 1993). A tradition of openness and honesty in reporting can be fostered (Lee et al., 1992). Raw data and input reports can be made available (Beder, 1993). "Whistle blowers" can be encouraged and protected (Beder, 1993). All interested and affected parties can be freely and openly consulted and involved in interpretations and decisions (Rorty, 1991; Mostert, 1996).

Norms, values, and the interests of all parties can be systematically integrated into the IA process (Parkin, 1996; Mostert, 1996). Explicit environmental values, objectives, criteria, and ethical standards can be identified and applied (Mostert, 1996). Differences and tradeoffs among value-based positions can be systematically and explicitly explored and substantiated. The role of scientists and nonscientists in contributing to judgments will vary by IA process activity. Scientists could assume a greater role in impact analysis and monitoring. Nonscientists could take the lead in identifying issues and in impact evaluation and decision making (Morgan, 1998). This blending and blurring of the objective and the subjective does not mean that realism in science and in IA need be abandoned (Rorty, 1991). Although scientists and scientific methods are subjective (especially as applied in IA practice), this does not mean that there is no "objective truth." Predictions can be more or less accurate. Environmental management and IA practice can be more or less effective (Giere, 1999; Gower, 1997). What is required is that the objective, subjective, and objective/subjective elements of IA practice be transparent, substantiated, jointly determined (with interested and affected parties), and conducive to environmental enhancement.

4.4.4 Beneficial Versus Detrimental to Environment

IA is intended to benefit the environment or, at least minimize detrimental effects on the environment. Scientific IA approaches, in common with science, have been criticized for contributing to environmental degradation by reinforcing an aggressive, exploitive, reductionist, and arrogant worldview (Bowler, 1992; Porritt, 2000). The net result, it is argued, is a nonsustainable economic and social culture (O'Riordan, 1995).

IA practitioners need to guard against these negative tendencies. They can moderate or replace subtle assumptions regarding such matters as the levels of certainty and control, the beneficial links between science and technology, and the preeminence of scientists and scientific knowledge. The claims of holistic scientific theories regarding greater respect for the environment also can be scrutinized and tested (Bowler, 1992). Scientific principles and methods can be guided by and integrated with environmental ethical principles, imperatives, and standards (see Chapter 10).

4.4.5 Beneficial Versus Detrimental to Democracy

A common benefit attributed to IA is more open, transparent, informed, and democratic decision making. However, an overreliance on and uncritical acceptance of "expert" knowledge and interpretations (even in inherently subjective areas) can inhibit democratic debate (Bowler, 1992). Political values, perspectives, and interests can be implicitly subsumed within "objective expert" opinions. Science as expressed through IA can become a tool for political persuasion (Ozawa, 1991). The interests that sponsor applied IA research can distort priorities and findings.

These negative tendencies can be offset and ameliorated. The contributions of scientists can be acknowledged, without scientists acting like or being treated as all-knowing seers (Rorty, 1991). Community initiated research and interactions between scientists and the community can be encouraged (Henman, 1997). Other knowledge sources, such as traditional knowledge, can be integrated into decision making. Greater consideration can be given to subfields such as civic science (Porritt, 2000). An increased effort can be made to blend scientific IA practice with ethical democratic principles, perspectives, and imperatives (Bowler, 1992).

4.4.6 Espoused Versus Applied Science

The major discrepancy between the theory and practice of science, in an applied field such as IA, does not mean that the espoused model of science has nothing to offer IA practitioners. IA practice, except to its fiercest critics, is acknowledged to have generally improved over the past two decades. Arguably, improvements would have been greater had there been more of an effort to systematically define and build upon a core body of good practice, albeit with contextual adjustments. A systematic exploration of IA as applied (an inductive analysis) is likely to reveal patterns of good and bad IA process management. Although these patterns are largely at the level of pre-theory (as illustrated in Figure 4.1), they could lead over time (if tightly circumscribed within contextual limits) to micro-theory building. Similarly, IA concepts and frameworks for IA process management can be refined and tested in practice (i.e., a deductive analysis). Instead of choosing between espoused and applied science and IA, practitioners could iteratively move between the two in a progressive (albeit disjointed) process of knowledge accumulation, derivation, and application.

4.4.7 Predict and Control Versus Manage and Adapt

It is an overstatement to suggest that science as applied in IA is ill equipped to deal with risks and uncertainties or that complexity is the "order (or perhaps more appropriately 'disorder') of the day" in IA practice. Uncertainty can be partially addressed in IA by integrating techniques such as human health and ecological risk assessment. Gaps, uncertainties, and value assumptions, together with their implications, can be explicitly identified and explored (Lemons and Brown, 1990). Conservative assumptions can be applied. Multicriteria decision aids can facilitate the management of complex knowledge (Kain and Söderberg, 2008). Greater emphasis can be placed on minimizing type II errors (effects when none are predicted) (Reckhow, 1994). There could be situations where applying the precautionary principle is warranted (e.g., new technologies, processes or chemicals, catastrophic potential), where adaptive management is appropriate, or where elements of holistic and sustainability science are helpful.

Care should be taken not to overstate predictive and control limits or to abandon potentially valid and useful elements of analytical science. Holistic and sustainability science, moreover, may produce intriguing concepts and frameworks (Bond and Morrison-Saunders, 2009, 2011; Faber et al., 2010). But these concepts and frameworks may bear only a passing resemblance to reality (i.e., patterns imposed on rather than tested against or derived from the surrounding world) (Miller, 1993).

4.4.8 Analysis Versus Synthesis

Analysis is a central attribute of IA practice. Potentially affected components and functions of the environment must be determined. Impacts must be identified, predicted, interpreted, and managed. However, neither science nor IA end or should end with analysis. IA has always been integrative. Overall conclusions regarding both preferred alternatives and proposal acceptability must be reached to establish a sound decision-making basis. More attention is now devoted in IA theory and practice to interrelationships among environmental components, as reflected, for example, in ecological and socioeconomic models. Practical approaches have been formulated and applied for addressing cumulative effects; for considering economic, social, and biophysical interconnections (interdisciplinary rather than multidisciplinary analysis and synthesis increasingly under the umbrella of sustainability); and for integrating individual measures within management strategies (Caldwell, 1988). Also, as described in Chapter 2, greater attention is being devoted to interrelationships between IA, among IA types and levels, and with other forms of environmental management.

IA practice constraints necessitate professional judgment, adaptation, and creativity. Holistic and sustainability science could be especially useful when "wicked," trans-scientific, and "messy" problems must be managed (Kates, 2000; Miller, 1993). Analytical science could be better suited to situations, where problems can be readily circumscribed, where much is known, and where available methods appear adequate (Miller, 1993). For most IA problems, it could be better to maintain the analytic and holistic components in a "dramatic tension." Analytical methods and perspectives can be tempered by judgment, systems thinking, and a willingness to adapt and create. Holistic approaches need to be derived from or tested by empirical evidence obtained, where practical, by the judicious use of scientific methods and protocols.

4.4.9 Explanation Versus Prescription

IA is both explanatory (what effects are likely to occur?) and prescriptive (how are negative effects to be avoided and managed and positive effects to be enhanced?). The prescriptive IA role (i.e., to advance environmental values) is consistent with a management orientation. IA practice is not dissimilar to applied scientific research. Both are sponsored by government and industry and are directed toward social, environmental, and business purposes. IA-related research, although perhaps more tightly circumscribed, is still a form of scientific research. It can still, with appropriate qualifications, apply scientific principles and methods.

Many sociological, economic, and political science theories are normative. They extend beyond explanation. They also prescribe how institutions and society at large could more efficiently and effectively operate (Rorty, 1991). Applied fields such as management and planning also have combined the prescriptive and the explanatory under the umbrella of scientific planning and management, albeit with mixed results. IA can draw upon both the positive and negative lessons acquired in these fields.

4.4.10 Unified Science Versus a Plurality of Sciences

The divisions within science are considerable notwithstanding many integrative efforts. The same heterogeneous pattern of competing and overlapping theories and frameworks is repeated in applied fields such as planning, environmental management, and IA. A core body of IA knowledge and methods has yet to emerge more than tentatively. The debates surrounding whether science is, or should be, unified or pluralistic (including the middle-ground positions) indicate that defining good IA practice will not be a simple task. Being aware of the divisions within and among the natural and social sciences, as well as efforts to transcend divisions, could help interpret and place in context IA divisions and integrative initiatives.

An open and tolerant, albeit critical, posture to new theories, concepts, and frameworks is likely to be more conducive to insights and applications of value to IA practice than a strict application of analytical scientific protocols. Oftentimes, it will be necessary to select a mix of tools from both science and IA, appreciating their characteristics, strengths, limitations, and interconnections. These tools can then be adapted and applied to match proposal and environmental characteristics. No easy task! But perhaps one that is becoming less difficult as lessons and insights are increasingly emerging from IA quality and effectiveness analyses.

4.5 INSTITUTING A RIGOROUS IA PROCESS

4.5.1 Management at the Regulatory Level

The four jurisdictions (the United States, Canada, Europe, and Australia) have all instituted measures potentially conducive to greater IA rigor, as listed in Table 4.3. These measures illustrate a range of possible regulatory approaches to making IA practice more rigorous. Even collectively they fall far short of good practice standards. They do, however, represent a departure point for a more in-depth exploration of the subject.

They point to the value of an interdisciplinary approach and the need to document and substantiate methods and

Table 4.3 Positive and Negative Regulatory Level Examples Regarding IA Rigor

United States	Canada	Europe	Australia
(+) Emphasis on interdisciplinary approach, on analytic rather than encyclopedic documentation and rigorous evaluation of alternatives (+) Required to identify methods and sources (+) Extensive IA-related applied research and guidelines (e.g., CEA, mitigation and monitoring) (+) Presidential Memorandum on Scientific Integrity	(+) Power of Minister to issue guidelines and codes of practice (+) Power of Minister to establish research and advisory bodies (+) Requirement of follow-up program for verifying accuracy and determining effectiveness (+) Stronger agency role in promoting and monitoring compliance (+) Power to verify compliance or prevent noncompliance; includes orders, injunctions, prohibitions, and offenses (fines \$100,000 to \$400,000)	(+) Proposed Project Directive (PPD)—required to take into account current knowledge and methods of assessment (+) PPD—mandatory monitoring (to monitor significant adverse environmental effects, to assess implementation and effectiveness of mitigation and compensation measures, and to identify unforeseen adverse effects)	(+) Mitigation requirements (feasibility and effectiveness, proposed safeguards, statutory or policy basis, costs, environmental record of proponent) (+) Monitoring requirements (environmental management plan, outline, responsibilities, biodiversity monitoring) (+) Minister can require an environmental audit

Table 4.3 *(Continued)*

United States	Canada	Europe	Australia
(+) Oversight of IA system by combination of CEQ (advises President, resolves interagency disputes; policy and legislative recommendations, regulations, and guidelines; reviews agency procedures), EPA (review of environmental impacts, rating of adequacy of environmental impact statements, operational duties associated with EIS filings), and courts (interpretation and enforcement, largely procedural) (+) CEQ retrospective regulatory review plan; retrospective reviews of existing regulations (+) Final guidance specifying when there is a need to monitor environmental mitigation commitments (+) CEQ NEPA pilot program—solicitation to agencies and public to nominate projects employing innovative approaches to completing environmental reviews more efficiently and effectively; nomination and selection process (+) Information quality, objectivity, integrity, and utility guidance (US OMB, 2002) (+) Monitoring guidance (+) Annual NEPA reports (+) US DOE—quarterly report—lessons learned (−) Emphasis of procedure over substance (−) Limited application at strategic level	(+) EA Agency objects—promoting or conducting research on EA and development of EA techniques (including testing programs); promoting EA; promoting, monitoring, and facilitating compliance with Act; and promoting and monitoring EA quality (+) EA Agency may undertake studies or activities or conduct EA research; advises people re. EA and provides training opportunities, courses, and resources (+) History of sponsored applied research (+) Now legally enforceable decision statement (+) Cumulative effects provisions (+) Follow-up and CEA guidance (+) Quality Assurance Program (−) Research program has not yet been established; power to promote and conduct research does not mean that it will be re-established in a comparable form (−) Elimination of National Round Table on Economy and Environment (−) Removal of requirement to assess natural resource capacity (−) Issue of EA capability of Canadian Nuclear Safety Commission and National Energy Board (−) Issues of rigor of SEAs and large number of projects no longer subject to Act (−) Issues of maintenance of rigor with narrow definition of environment, narrow and selective definition of effects and alternatives and administrative/political discretion (−) Issue of maintenance of rigor when substitute or equivalent IAs are applied (−) Largely passive approach to regional studies (−) Lack of tiering	(+) PPD—required to identify monitoring parameters and duration (proportionate to nature, location, and size of project and significance of environmental effects) (+) PPD—periodic progress reports on implementation (+) PPD—required to describe forecasting methods and identify main uncertainties (+) PPD—competent authority to verify up-to-date information re. mitigation (+) PPD—requirement that reports prepared by or verified by competent and accredited experts (+) PPD – broad definition of environment and effects (+) PPD—requirement to address accumulation of impacts and reasonable alternatives (+) History of extensive applied research and a diverse array of quality and effectiveness reviews (+) Quality requirement for SEA Directive (+) SEA Directive—required monitoring (+) Oversight of IA systems provided by a combination of boards, courts, auditors, and independent effectiveness reviews (−) PPD—uncertain whether quality enhancement efforts consistent with time limits (−) History of considerable variability in IA quality and effectiveness	(+) Auditor General undertakes annual compliance audits (Auditor General 2002–2003) (+) Scientific advice provided through the Threatened Species Committee, the Biological Diversity Advisory Committee, and the Australian Heritage Committee (+) Planned National Centre for Cooperation on Environment and Development (to provide neutral forum within which industry, scientists, NGOs, and governments can cooperate in developing environmental standards, guidelines, and procedures) (+) Annual internal operations reviews of the EPBC Act (±) Recommendation by independent review of the EPBC Act for establishment of an independent environment commission to advise government on project approvals, strategic assessments, bioregional plans, and other statutory decisions; government did not accept recommendation (+) Government agreed to publish reasons for significant decisions (+) Government agreed to bring together and rationalize compliance and enforcement powers and responses (+) Detailed provisions for identifying and monitoring biodiversity (+) Consultation draft on Environmental Offsets Policy (−) Public can only take court action on procedural grounds (−) Limited attention to social sciences and concerns

sources. They underscore the importance of systematically assessing alternatives, the value of scientific procedural standards (especially as part of mitigation and monitoring requirements), the need for cumulative effects assessment requirements, and the pivotal role of government-sponsored IA scientific policy direction. They illustrate the roles that applied research, effectiveness reviews, pilot projects, training, and guidelines can assume in raising the level of IA practice. They demonstrate the potential benefits of professional accreditation requirements for individuals and organizations preparing and reviewing IAs, of independent oversight of IA legislation and regulatory compliance (including the potential for substantive legal challenges), of independent scientific advisory bodies, and of forums to facilitate collaboration between scientists and nonscientists. They point to the need to ensure and maintain scientific and technical integrity and capability, to broadly define the environment and effects, and to require the consideration of current knowledge (scientific and traditional) and good practice standards. They demonstrate the need to fully describe and substantiate forecasting methods; to explicitly identify uncertainties and related implications; to ensure that the basis for interpretations and decisions is open, transparent, and substantiated; and to systematically address interconnections among IA levels and between procedural and substantive environmental requirements.

The consistent identification and application of these types of principles and measures at the regulatory level could establish a foundation for a greater emphasis on sound science in IA practice. There are dangers when extending beyond general science-related principles and requirements, especially when seeking to integrate complex and sometimes conflicting social scientific protocols and suggested practice-based standards. Even the general scientific principles already incorporated into IA requirements are debatable and can be interpreted in multiple ways. It is essential to proceed with caution when instituting science-based IA requirements. It would be easy to fall into the trap of insisting on science-based requirements—requirements which are hotly debated in science, and which could constrain as much as benefit IA practice. A more flexible, performance-based approach could capitalize on the benefits of science while appreciating its limits for applied fields such as IA.

4.5.2 Management at the Applied Level

Figure 4.2 is an example of a rigorous IA process. Figure 4.2 and the process description that follows incorporate many scientific IA elements. IA process managers and participants can "pick and choose" relevant or appropriate elements. A rigorous IA process treats IA as an experiment. Predictions are hypotheses to be tested. The process is driven by independent, skilled, and qualified natural and social scientists. The scientists strive to ensure a rigorous, objective, and open process consistent with scientific principles and protocols. Scientific findings and interpretations are assumed to provide a sound decision-making basis.

Start-up The start-up to a rigorous IA process involves several, highly interrelated activities. The problem or opportunity to be addressed must be determined. Objectives, which represent the purpose of the experiment, must be formulated. Constraints and assumptions for bounding the experiment need to be established. It is necessary to identify the initial proposal characteristics that trigger the process. Preliminary methods for collecting, analyzing, integrating, and interpreting data must be identified. The need, an extension of the problem or opportunity, must be determined. The appropriate aspects of the context must be identified. Initial hypotheses, the preliminary alternative explanations suitable for testing, must be formulated. The experimental design, a research program for testing the hypotheses, must be prepared.

Experimental design principles are incorporated, where practical, into the rigorous IA process. Initial hypotheses of interest are formulated. The hypotheses are explanations of the likelihood and magnitude of potential impacts. Hypotheses are formulated for alternative proposal characteristics (options including the null hypothesis) and for alternative impact predictions for the proposal. The experiment is designed to suit the context, to achieve the objectives, and to operate within the constraints and assumptions. Graphical and statistical testing procedures using appropriate sampling frames support the analyses and the interpretations. The experiment is designed to minimize both type I (concluding that there are effects when none exist) and type II (concluding there are no effects where effects have occurred) errors.

Constraints (e.g., limits to knowledge) and assumptions (e.g., temporal and spatial boundaries, values) are explicit and substantiated. The study design addresses the implications of uncertainties and of contextual factors. A high level of rigor (the experimental methods and test protocols of analytical science) is applied when cause–effect relationships are simple and clearly structured. A lower level of rigor (more selective and superficial quasiexperimental and judgmental activities) is applied for more complex, larger scale, longer time horizon cases characterized by high levels of variability and low levels of predictability and control (Patterson and Williams, 1998; Beanlands and Duinker, 1983).

Multiple scientifically defensible methods structure the data collection and analysis, link and integrate impact predictions, and facilitate interactions among scientists (Bird, 1998). The methods are reliable and consistent (over a range of spatial and temporal scales). The startup activities are progressively refined as new data are incorporated into the IA process.

Analysis The analysis is highly structured and guided by explicit theoretical models (Lima and Marques, 2005). Multiple, reliable, and preferable precise environmental

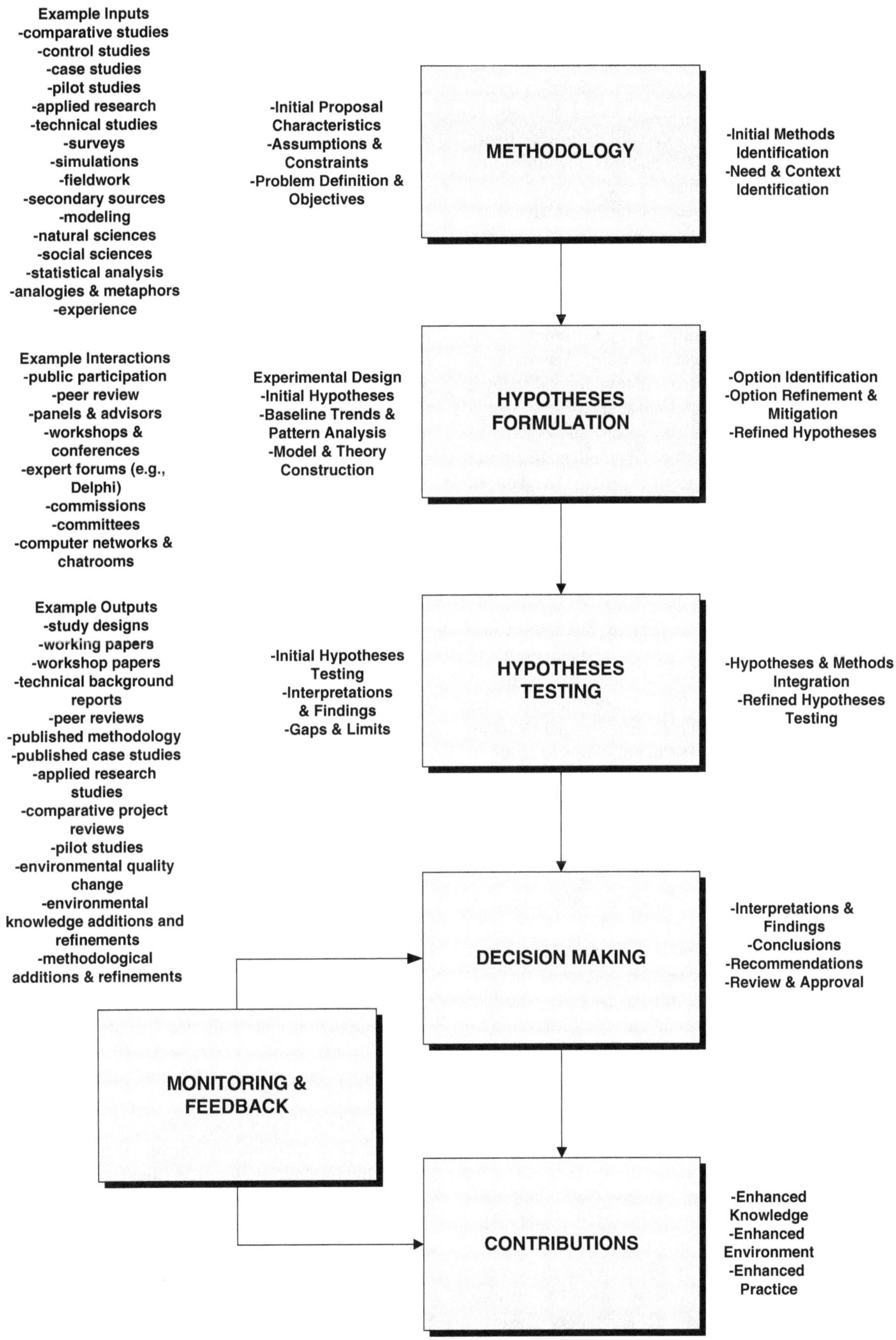

Figure 4.2 Example of a rigorous IA process. Adapted from Lawrence (2005a).

and social criteria and indicators are selected—criteria and indicators that can reveal (ideally statistically significant) changes in ecological and socioeconomic conditions. The level of detail (e.g., individual organisms, species, populations, communities) at which change occurs and can be reliably detected and predicted is carefully selected (Beanlands and Duinker, 1983). The analysis identifies and focuses on sensitive and significant social, economic, and ecological components, processes, and functions. Historical and emerging cumulative effects are identified (Cashmore, 2004).

Data, which can be readily aggregated or disaggregated, are collected over a range of temporal and spatial scales. Ideally, the IA database naturally extends from regional social, economic, and ecological indicators. The regional and greater statistics establish a context for subsequent impact interpretations. The baseline analysis is dynamic (trends and patterns over time and space) and conducive to the prediction and management of cumulative effects and systems level understanding (Seitz et al., 2011). The data are suitable for testing impact prediction accuracy.

Testable hypotheses are refined based on insights obtained from the baseline analyses. Predictions or implications are deduced for each competing hypothesis (Curtis and Epp, 1999; Greig and Duinker, 2011). A premium is placed on predictive precision (especially predictions that are easiest to test by observation and experimentation) and causal hypotheses (correlations between two kinds of events) (Bird, 1998; Wilson, 1998). Causal networks are identified and traced (Lord, 2011; Perdicoúlis and Glasson, 2006; Perdicoúlis et al., 2007). Predictions are first likely to be broad approximations. They are refined as more detailed data are integrated into the analysis (Greene, 1999). The object or process being studied is separated from its context to control confounding variables (Miller, 1993). Predictions address the magnitude, frequency, extent, and likelihood of potential effects. The bases for predictions are explicit.

Synthesis Interrelationships among environmental components and among potential impacts are addressed through models and occasionally preliminary theories. Multiple model types (conceptual, mathematical, physical, biological, social, economic, human health, and ecological risk) are applied. The models convey a systems level perspective on natural and socioeconomic conditions and provide a framework for predicting changes, including cumulative effects. They also are useful for testing options and mitigation methods (alternative hypotheses).

The models are derived from baseline spatial and temporal patterns and trends and from reviews of comparable, control, and pilot studies. They also draw upon natural and social science literature. They focus on critical environmental components and interrelationships. The refined hypotheses, emerging from the model and theory construction exercises, trace webs of causal connections beginning from the proposal, extending through various levels of direct and indirect effects, and ending with systems level effects. The models and theories are described and justified.

Models, theories, and hypotheses initially tend to be narrowly defined, generally within individual disciplines. A second round of model and theory building is usually necessary to address interconnections among the models and theories. Ideally, integrated assessment models or overarching theories can be constructed (Ravetz, 1998; Greene, 1999; Bird, 1998). Interconnections across consistent and mutually supportive theories and models are preferred. More commonly, interconnections and inconsistencies are only partially and subjectively addressed (Rothman and Sudarshan, 1998).

Hypothesis Testing The real measure of a rigorous IA process occurs when the validity of alternative hypotheses, based on explicit theoretical models, are rigorously tested (Curtis and Epp, 1999; Rossou and Makan, 2007). Hypothesis testing involves collecting empirical evidence and then modeling environmental conditions. An interrupted time series design for hypothesis testing tends to be preferred. This entails periodic tests, measurements, and observations of relevant variables at equally spaced intervals. The proposal is introduced at a predetermined interval. Hypotheses are tested (falsified) both prior (using comparative, control, case, and pilot studies) and subsequent (through monitoring) to approval (Burdge, 2004; Petticrew et al., 2007).

Hypothesis testing applies to the impacts predicted for the proposed action and to the impacts that could result from options and mitigation measures. Options tend to be excluded where severe impacts are likely based on reliable impact predictions and where there is a high degree of uncertainty and potentially severe consequences. Option comparison relies on a combination of social and natural scientific (preferably quantitative) indicators and substantiated scientific interpretations.

Gaps, limits, and uncertainties are explicitly identified together with potential implications. Uncertainties can be considerable given the complexity of the phenomena and the time and resource limits usually inherent to IA practice. Consequently, there is a tendency to incorporate procedures and assumptions that minimize the likelihood of underestimating the incidence and severity of adverse impacts. These procedures are explored more fully in Chapter 11.

Interpretations Extensive use is made of statistical tests of significance to facilitate interpretations in rigorous IA processes. Interpretations are often influenced by uncertainties regarding data reliability, the potential for systemic bias and nonlinear relationships. Both quantitative and logical techniques are used to analyze the evidence and to reach judgments based on the weight of evidence.

Interpretation is the creative component of science. Scientific interpretations encompass both espoused (reliance on statistical and other quantitative analyses) and applied (a systematic, creative, and collective endeavor) dimensions.

In the latter case, the boundaries between scientific and nonscientific IA processes overlap.

Approvals and Postapprovals Scientific findings and interpretations provide the primary decision-making basis. A scientific experimental design guides and structures the monitoring and follow-up program. Monitoring tests the accuracy of impact predictions and the effectiveness of mitigation measures. It can facilitate the enhancement of environmental and IA quality (Burdett, 2008a; Noble and Storey, 2005). Monitoring data can lead to new and refined hypotheses that are, in turn, tested through further monitoring (Storey and Noble, 2005). Follow-up assesses the validity and effectiveness of experimental design features, data collection and analysis procedures, model design and application, interpretations, environmental outcomes, and the overall rigorous IA process (Burdett, 2008a; Fuggle, 2005a,b; Noble, 2009b; Persson and Nilsson, 2007). It also can enhance environmental monitoring. Results can be incorporated into regional environmental databases (Gachechiladze et al., 2009).

Inputs, Outputs, and Interactions A rigorous IA process is far from closed. It extends from and contributes to natural and social scientific knowledge (Brown, 1986; Greig and Duinker, 2011). Links between SEA/EIA and CEA are systematically explored (Gunn and Noble, 2011). It draws upon interdisciplinary knowledge and on the experiences and insights of applied research scientists and IA practitioners (Caldwell, 1988). Model construction and hypotheses testing (prior to approvals) requires the systematic use of comparative, control, case, and pilot studies. Knowledge gaps are addressed through targeted research. Baseline analyses apply such scientific tools as surveys, field investigations, modeling, and computer simulations (Barrow, 1998). Statistical analyses aid significance interpretations. Analogies and metaphors often help structure the analysis but need to be confirmed through experimentation (Rothman and Sudarshan, 1998).

A scientific IA process is a collective endeavor. A team of natural and social scientists manage the process and provide specialist environmental knowledge and methods. Other scientists are involved as government reviewers and as specialist advisors and peer reviewers. Peer reviewers can assess the correctness of procedures and the plausibility of findings and conclusions (Hirschmann, 1994). Interactions among scientists occur through committees, workshops, commissions, panels, expert forums, and gateway websites (Dannenberg et al., 2006; Swor and Canter, 2011). Links to the broader scientific community are maintained by circulating and publishing (where practical) research findings and by means of human and computer networks (Barrow, 1998). Broader agency, political, and public participation occurs prior to major decisions and in the review of documentary outputs. Knowledge is freely exchanged and transferred (Sheate and Partidário, 2010). The public and politicians assume a more prominent role in defining the problem and objectives, in contributing to significance interpretations, in suggesting options, in participating in follow-up, and in tempering and testing conclusions and recommendations (Devlin, 2011; Hunsberger et al., 2005).

Rigorous IA processes tend to have numerous interim documentary outputs (e.g., study designs, working papers, applied research studies, pilot studies, technical reports, comparable project reviews, workshop reports, background studies, peer reviews). The findings from scientific and technical support studies are fully integrated into core documentation. IA process documentation extends into postapproval with the preparation and circulation of monitoring results. Consistent with good scientific practice assumptions, methods, and findings are fully and fairly represented. They also are independently evaluated and subjected to rigorous criticism. A rigorous IA process is expected to add to the environmental and IA knowledge base, to maintain and enhance IA quality and effectiveness, and to contribute to environmental protection and enhancement (Donnelly et al., 2008; João, 2007).

Adaptations and Variations There are multiple perspectives in the scientific community regarding what constitutes "the" scientific method or indeed whether a plurality of methods is both necessary and desirable. The same is the case for a rigorous IA process. The process, described in the preceding subsections, largely conforms to the tenets of analytical science. Several important modifications have been made, most notably the greater emphasis on integration, the infusion of environmental values and objectives, and the direct links to decision making.

This process could be tempered or replaced by, for example, a holistic, a management, a complexity, an applied, a conservation, or a civic scientific IA process (Cashmore, 2004; Levins, 2003). A rigorous IA process also varies by IA level and type (Nilsson et al., 2009) (see Section 4.5.4). These other rigorous IA processes are treated here as tempering considerations or at best variations rather than as alternatives to the analytical model. This is partly because they are not nearly as fully developed and have been applied to a much more limited extent. Also, if viewed as a replacement to analytical science, there is the question of whether the process that emerges is still primarily "scientific" in orientation.

Viewed solely in a tempering capacity, holistic or new science points to the value of creativity, intuition, imagination, judgment, and flexibility and to the potential roles of multiple trans-scientific and systems perspectives and frameworks (Miller, 1993; Porritt, 2000). Management science demonstrates the need to place IA-related science more firmly in the context of decision-making priorities, requirements, and limits. Complexity science and the precautionary principle underscore the central position of uncertainty, especially in applied fields such as IA (Levins, 2003; Tickner, 2003e). It also systematically addresses the implications of uncertainty for the ability to predict and manage and for the design and adaptation of planning processes and organizations (Patterson and Williams, 1998; Rothman and Sudarshan, 1998). Civic science points to the need for and value of having interested and affected members of the public assuming a less peripheral role in the IA process. Civic science argues that the public should assume a valuable role in shaping the process and in contributing to such tasks as data collection, analysis, and interpretation (Henman, 1997; Ozawa, 1991; Porritt, 2000).

The multiparadigm view of science is intriguing. It forms part of the foundation for the structure of this book. It offers many insights regarding potential ways of linking, integrating, and transcending multiple, overlapping models within and among disciplines and fields of practice. It also clearly demonstrates the many pitfalls and potential losses associated with pushing the integration process too far. To avoid repetition, the themes and insights derived from an exploration of multiparadigm approaches to science have been incorporated into Chapter 12 (which addresses interrelationships among the IA processes).

4.5.3 Theory Building

Science and IA have a complex relationship, as illustrated in Figure 4.3. The IA process described is driven, framed, and structured by analytical science. However, because of the practical realities of an applied field such as IA, its aspirations, to be realistic, are more in the nature of pre-theory—aspiring toward particular and middle-range theory building. Also, scientific IA processes and practices are tempered to integrate the relative, the relevant, the subjective, the practical, the applied, contextual adaptations, the need to connect and transcend disciplines and decision-making levels, the prescriptive, and the multiperspective nature of IA related practice. SEA and EIA practice (including the overlapping areas that encompass integrated IA, CEA, and various forms of tiered planning and decision making) are both extended and refined (by more technical and scientifically oriented substantive IA types), and challenged and restructured (by more transformative, political, holistic, and collaborative IA types). Individually SEA and EIA process types can, depending on the type or type combination, refine/reinforce or challenge/restructure rigorous IA practice. Theories, frameworks, and models derived from the natural and social sciences tend to extend and refine SEA and EIA science-based approaches. More integrative forms of science, such as sustainability science, are more prone to challenge and restructure IA theory and practice. Follow-up and auditing, as detailed in Section 4.6, assumes a critical role in applying, testing, adapting, enhancing, refining, and sharing scientific IA practice.

Numerous general measures are possible to facilitate IA theory building. A scientific integrity policy can be instituted. Good practice guidance can be provided regarding best practical science and traditional knowledge. Guidance materials can be based on monitoring and auditing outcomes. Greater emphasis can be placed on environmental outcomes and cumulative effects, and on uncertainty, complexity, surprise, and whole systems. Infrastructure enhancements can be instituted, such as the performance tracking of IA documents, a national IA database for the public and practitioners, and annual state-of-environment reporting (which integrates monitoring outcomes). Comparative studies of IA legislation, regulations, institutional arrangements, and context can be sponsored. IA performance evaluation can be linked to research. IA knowledge and experiences can be shared. Best practical science standards can be applied. Greater use can be made of comparative studies.

A proactive effort can be made to contribute to IA theory building and good practices. Care can be taken to delineate the rationale, limitations and implications of methods, to apply appropriate scales, time horizons, and level of detail, and to make appropriate contextual adjustments. IA theory-building efforts are facilitated if the environment and effects are broadly defined, if interconnections among IA types and cumulative effects receive particular attention, if assessment studies and applied IA research are broadly available, and if a proactive effort is made to connect and transcend organizational barriers. Drawing upon and contributing to interdisciplinary science, the integration of citizen knowledge and the maximizing of learning potential should be priorities. Knowledge limits should be explicitly acknowledged, the roles and responsibilities of specialists and advisors should be specified, key references should be integrated, the effectiveness of substantive outcomes should be assessed, and a clear theoretical framework should be applied.

4.5.4 Variations by IA Type

Table 4.4 details examples of theory-building possibilities for different IA levels (SEA and project EIA) and for various substantive IA types (SA, EcIA, SIA, and HIA). In general terms, these measures address such matters as scientific integrity, sponsored research and research priorities, methodological standards and protocols, good practice guidance, knowledge sharing and contribution procedures, supportive databases, study scope and design, transparent and substantiated methodology, connections to related bodies of knowledge, and contextual adaptations.

Strategic Environmental Assessment (SEA) At the SEA level, effective theory building starts from a particular need to test the suitability and effectiveness of alternative SEA institutional arrangements and approaches, to demonstrate the added value of SEA, and to forge more effective links between SEA and policy/plan making and decision making. It requires seeking a better understanding of organizational and institutional procedures and barriers and how they might be influenced and altered, demonstrating tangible contributions to environmental enhancement, and enhancing the level of SEA practice and the credibility of SEA

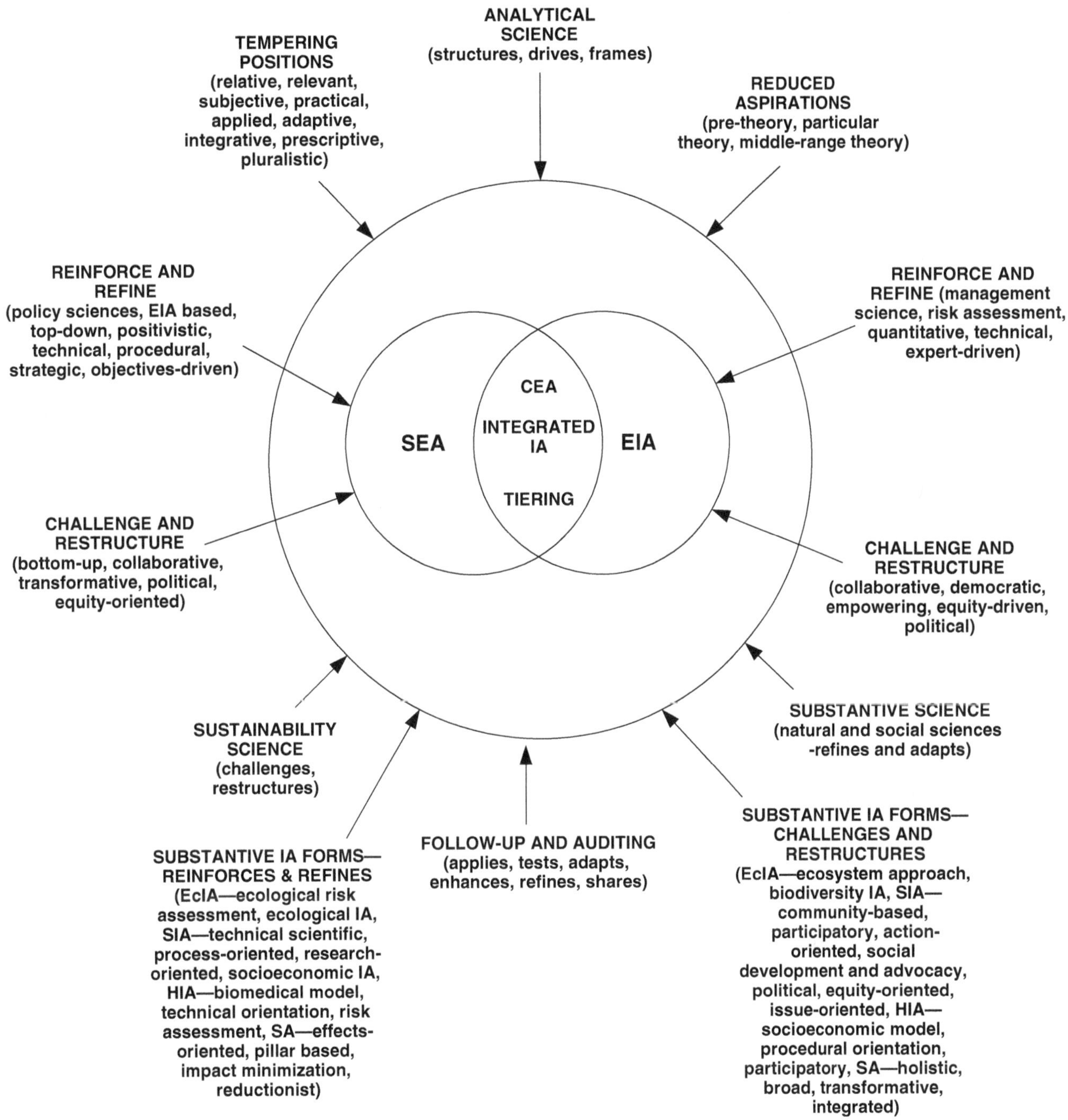

Figure 4.3 Science and IA.

practitioners. It also involves better differentiating between project-level and SEA good practices; distinguishing SEA good practices for and across varying decision-making levels, sectors and settings; more effectively integrating cumulative effects and manage uncertainties; and facilitating organizational and institutional learning, knowledge exchange, and capacity building.

Environmental Impact Assessment (EIA) At the project EIA level, theory building entails establishing more effective links between the scientific and EIA communities and knowledge systems; making more effective use (with appropriate adaptations) of scientific methods, models, and standards; enhancing the role of environmental databases in supporting EIA and the contribution of EIA to such systems; and more systematically undertaking EIA-related research targeted at key knowledge gaps and uncertainties. It also seeks to raise the quality of scientific practice and practitioners in EIA (especially regarding follow-up and auditing procedures); to more effectively test and refine EIA models,

Table 4.4 Rigorous IA Practice Characteristics by IA Type

Rigorous SA Practice	Rigorous SEA Practice	Rigorous EIA Practice
Clearly and strictly defines sustainability and sustainability purposes, extends time horizons, broadens spatial horizons, and includes positive and negative effects Provides guidance regarding means of integrating sustainability into each IA activity Links to sustainability priorities and targets at all levels—international, national, regional, and local—specifies what to sustain, why, and progress toward sustainability Focuses on links to sustainability objectives and reference points, genuinely sustainable decision making and environmental outcomes, equity concerns, and barriers to sustainability integration and strategies for overcoming Seeks to institutionalize sustainability and contribute to sustainability capacity building Links to other sustainability plans, strategies, programs, objectives, and standards Registers SA professionals Sponsors research on approaches for integrating and transcending pillars and disciplines, sustainability research systems and science, and SA as a platform to facilitate knowledge exchange and transfer Adopts postnormal science approach (uncertain facts, value disputes, and urgent) and holistic and dynamic perspective (constant improvement) Is creative, adaptive, and precautionary Draws upon and contributes to sustainability research and interweaves positive and negative, ends and means, and limits and opportunities Avoids reductionist methods and integrates good practice SA principles and procedures	Provides guidance linking SEA to policy and plan making Provides guidance for testing the added value of SEA policy, plan or, program development and downstream assessment Focuses on proactive and integrative SEA approaches, analyzing how organizations and institutions function, assessing the quality and effectiveness of SEA outputs, facilitating uncertainty management and organizational learning, overcoming obstacles and pitfalls to SEA, and enhancing decision-making influence Institutes and refines SEA/EIA tiering system and multiscale and multidirectional analysis Contributes to SEA capacity building Registers SEA professionals Sponsors research on effectiveness of SEA legal frameworks and alternative SEA approaches, frameworks, and models in varying decision settings Undertakes case studies and methodological comparisons across a range of sectors, jurisdictions, and SEA levels Explores data issues and spatial and temporal scales differences and implications Seeks to strengthen and making science–policy links transparent Integrates SEA and CEA Institutes quality assurance checklists, facilitates SEA knowledge exchange and transfer, refines the SEA toolkit, and assesses differences between what should and did happen	Requires independent monitoring and auditing of implementation, compliance, and auditing Establishes realistic standards for quality of EIA scientific practice Focuses on substantive purposes, causal processes, and mitigation effectiveness Seeks more reliable science to better inform EIA, better science–politics links, enhanced scientific support outside EIA community by tapping into public pressure, and encouraging regulators and targeted research Institutes science advisory boards, research and centers of excellence, and environmental data clearinghouses Contributes to EIA capacity building Registers EIA professionals Sponsors research on more effective ways of integrating EIA and CEA Seeks to create, test, and refine predictive models Applies experimental design where practical Treats science roles as applied, experimental, and naturalistic Emphasizes sound methodology, evidence standards, and concise analysis Uses EIA to generate testable hypotheses and monitoring data Identifies, refines, and tests causal network analysis

Rigorous EcIA Practice	Rigorous SIA Practice	Rigorous HIA Practice
Incorporates ecological and biodiversity principles, concepts, and techniques Provides guidance regarding integration of ecological concerns into each IA activity and concerning ecosystem approach and biodiversity principles and practices Links to ecological priorities and targets at all levels—international, national, regional, and local	Includes social–psychological impacts and social responses to impacts Provides guidance regarding integration of social concerns into each IA activity, social changes at micro (e.g., individual, family, business) and macro (community) scales, and social change processes and social impacts	Broadly defines health and health determinants Provides guidance regarding integration of health concerns into each IA activity; data types and sources for HIA application, HIA methods, models, and follow-up, indigenous community research in HIA, and HIA follow-up

Table 4.4 (*Continued*)

Rigorous EcIA Practice	Rigorous SIA Practice	Rigorous HIA Practice
Focuses on potential damage to and protection of unique, endemic, threatened, and declining species, habitats and ecosystems, holistic biotype/ecological function approaches, ecological sustainability, biodiversity and ecological follow-up, and barriers to ecological integration and strategies for overcoming Seeks to institutionalize EcIA Links to ecological risk assessment and to conservation and ecological planning and management Supplements holistic databases to systematically address ecosystem functions and ecological connections and networks Facilitates biodiversity partnerships and information networks and EcIA training and capacity building Registers EcIA professionals Sponsors research to address refinements to ecological scoping and follow-up frameworks and methods, enhancements to biodiversity and EcIA impact prediction models and tools, and habitat loss and fragmentation methods Explicitly recognizes natural system uncertainties and complexities, fully represents ecological levels and integrates biodiversity models and frameworks Defines EcIA roles at SEA and project EIA levels Draws upon and contributes to GIS modeling, biodiversity indicator, and ecosystem process research Promotes awareness and analysis of ecological effects Integrates good practice EcIA principles and procedures	Focuses on social sustainability, barriers to social integration and strategies for overcoming, means of combining and integrating SIA methods and perspectives, culturally appropriate social benefits, opportunities and capital enhancement, and social justice and distribution of effects (especially for most vulnerable) Seeks to institutionalize SIA Incorporates mechanisms for facilitating community empowerment Contributes to SIA training and capacity building Registers SIA professionals Links to social research systems Links to other social plans, strategies, programs, objectives, and standards Sponsors research on SIA approaches and methods adapted by proposal, IA and context type, and means of integrating SIA and indigenous knowledge, perspectives, rights, and positions Seeks to base SIA on sound and replicable scientific methods and concepts Adopts hypothesis testing approach for follow-up, where appropriate Ensures sound understanding of social and cultural context Defines SIA roles at SEA and project EIA levels Promotes awareness and analysis of social effects Integrates good practice SIA principles and procedures	Focuses on health sustainability, barriers to health integration and strategies for overcoming, the effective blending of health and social sciences for addressing complex causal links, especially health determinants, HIA and decision-making links, more robust magnitude and distribution prediction methods, and more effective approaches for engaging health authorities and experts Seeks to institutionalize HIA Applies a longitudinal demographic surveillance system to facilitate health monitoring and evaluation Provides institutional support for HIA Links to other health plans, strategies, programs, objectives, and standards Contributes to HIA training and capacity building Registers HIA professionals Defines HIA roles at SEA and project EIA levels Promotes awareness and analysis of health effects Seeks to strengthen scientific evidence in support of causal links Integrates good practice HIA principles and procedures Fully engages health professionals and experts Fully integrates community health perspectives and knowledge Refines methods application based on follow-up of both decision-making influence and health outcomes

Sources: Alshuwaikhat (2005), Ahmadvand and Karami (2009), Barth and Fuder (2002), Becker et al. (2005), Bhatia (2007), Bhatia and Seto (2011), Bhatia et al. (2010), Bond (2010), Bond and Morrison-Saunders (2009, 2011), Burdge (2003a,b, Burdge (2004), Cashmore (2004), Chaker et al. (2006), Cherp et al. (2011), Cranor (2003), Dannenberg et al. (2006), De Ridder et al. (2010), Dimento and Ingram (2005), Doelle and Sinclair (2006), Donnelly et al. (2008), Dovers (2005), Duffy (2008), Elliott et al. (2004), Égré and Senécal (2003), Faber et al. (2010), Fischer (2003, 2005, 2007b), Fischer et al. (2010), Forsyth et al. (2010), Gasparatus et al. (2007), Gachechiladze et al. (2009), Geneletti (2003), Gibson (2006a,b, Gibson,2011), Gontier et al. (2006), Govender et al. (2006), Greig and Duinker (2011), Grinde and Khare (2008), Gunn and Noble (2011), ten Hallers-Tjabbes (2003), Hanna (2009a), Hansel and Aylin (2003), Hansen and Wolff (2011), Harris-Roxas and Harris (2011), Hermans and Knippenberg (2006), Hunsberger et al. (2005), IAIA (2005, undated b), ICPGSIA (2003), IEEM (2006), Jha-Thakur et al. (2009), João (2007), João and Mclauchlan (2011), Jones and Slinn (2008), Kain and Söderberg (2008), Kates (2000), Kemm (2005), Kemm and Parry (2004a,b), Ketelsen (2003), Khera and Kumar (2010), Kobus (2005), Kumagai et al. (2006), Kwiatkowski (2011), Kwiatkowski et al. (2009), Lane et al.(2003), Lavallée and André (2005), Levins (2003), Lima and Marques (2005), Lobos and Partidário (2010), Lord (2011), Mandelik et al. (2005), McCaig (2005), Moles et al. (2008), Morrison-Saunders and Bailey (2003), Nilsson et al. (2009), Noble (2009a,b), Noble and Bronson (2006), Noble and Storey (2005), O'Faircheallaigh (2009), Partidário (2007), Partidário and Arts (2005), Perdicoúlis and Glasson (2006, 2009), Perdicoúlis et al. (2007), Persson and Nilsson (2007), Petticrew et al. (2004), Petticrew et al. (2007), Pisani and Sandham (2006), Pope and Grace (2006), Pope and Klass (2010), Quigley and Taylor (2003), Retief et al. (2008), Retief (2007a,b), Rossou and Makan (2007), Rotmans (2006), Rowan and Streather (2011), Runhaar (2009), Sadler (2005b), Scanlon and Davis (2011), Schirmer (2011), Seidler and Bawa (2003), Sheate and Partidário (2010), Seitz et al. (2011), Shepherd (2008), Sherrington (2005), Singh et al. (2009), Slootweg (2005), Söderman and Saarela (2010), Storey and Jones (2003), Storey and Noble (2005), Taylor et al. (2004), Thérivel et al. (2009), Tickner (2003c,e), Tzoumis (2007), US EPA (2011), Weiland (2010), Winkler et al. (2011), US EPA (2011), Utzinger et al. (2005), Vanclay (2010), Wlodarczyk and Tennyson (2003), Zhu et al. (2010), Ziller and Phibbs (2003).

methods, and mitigation measures; and to more effectively integrate substantive environmental concerns and priorities, such as cumulative effects and climate change, in a manner that results in tangible improvements in environmental quality.

Substantive IA Types Theory-building priorities, for the explicitly substantive forms of IA (SA, EcIA, SIA, HIA), only partially overlap with those identified for SEA and project-level EIA. Collectively, these IA types are concerned with more effective links across disciplines, decision-making levels, and IA types without compromising the integrity and ethical and disciplinary standards of individual specialties, with narrowing the gulf between theory and practice, with more effectively managing uncertainty and complexity, and with more effectively operating within political/administrative decision-making systems and IA institutional structures. They also strive to enhance the integration of theory and an appropriate mix of methods, models, and frameworks while still allowing for contextual variations; to forge better links to, contributions to, and integration with broader knowledge systems; to facilitate an enhanced understanding of interactions between human activities and the environment; and to make a tangible contribution to the enhanced state of the environment.

Sustainability Assessment (SA) SA theory building seeks to interconnect and transcend individual disciplinary sciences. This necessitates redefining IA practice, at the regulatory and applied levels, with clear sustainability objectives and benchmarks, broadened spatial and temporal boundaries, a broader definition of environment and effects, a greater focus on equity concerns, and a concerted effort to integrate sustainability into each IA activity. Tangible progress toward sustainability, as demonstrated through decision making and environmental outcomes, is required. Organizational reforms conducive to sustainability capacity building are essential. Effective links to related sustainability initiatives and objectives, to sustainability research systems, and among organizations and individuals with sustainability knowledge must be established. Enhancing the status of SA professionals would be helpful. A particular effort needs to be made to transcend individual disciplinary and other boundaries, to employ a holistic and dynamic perspective, to anticipate and more effectively manage risks and uncertainties, to generate creative mutually beneficial opportunities, and to incorporate and contribute to sustainability theory and good practices.

Ecological Impact Assessment (EcIA) EcIA theory building can be more effectively framed by regulatory requirements and guidelines that fully and systematically integrate ecological and biodiversity principles, concepts, objectives, and methods. A particular effort needs to be made to protect rare and highly valued species, habitats, and ecosystems; to make effective use of ecological and biodiversity techniques (especially as part of follow-up); to overcome barriers to the integration of ecological concerns; and to facilitate the achievement of ecological sustainability. An ecological perspective and sensitivity to biodiversity concerns must be integral to public and private organizational and institutional decision making if EcIA theory building is to be effective. This means holistic ecological databases; effective links to related fields of theory and practice; biodiversity and ecological partnerships and networks; and EcIA training, capacity building, and certification. Effective applied ecological and biodiversity research regarding such matters as ecological scoping, prediction, follow up, and uncertainty management also is essential. In addition, good EcIA practices, differentiated by SEA and project level and by ecological level, must be consistently tested, refined, applied and, wherever practical, supplemented.

Social Impact Assessment (SIA) Effective SIA theory building requires institutional arrangements (requirements, guidelines, infrastructure) that broadly define social impacts (e.g., social, cultural, sociopsychological, heritage); address social impacts at micro and macro levels; and encompass social change processes, social impacts, social responses to impacts, social justice, and ultimately social sustainability. Particular attention needs to be devoted to the distribution of social effects (especially regarding effects on the most vulnerable); to the realization of social benefits, opportunities, and capital enhancement; and to the facilitation of community empowerment. The organizational and institutional barriers to effective SIA theory building are considerable. Some of the steps needed to ameliorate those constraints include an ongoing effort to combine and integrate SIA methods and perspectives; better links to social research systems and to related social interventions, SIA training, capacity building, and the certification of SIA professionals; the systematic differentiation of SIA approaches by proposal, IA, and context type; the more effective integration of community and indigenous perspectives and knowledge; and the adaptation, application, and extension of good practice SIA principles and procedures—especially with regard to follow-up.

Health Impact Assessment (HIA) Effective HIA theory building is conditional on IA institutional arrangements that provide for a broad definition of health and health determinants; effective guidance regarding such matters as data types and sources; prediction methods, models, and follow-up; enhanced health data systems; strategies for more fully engaging health professionals and experts; links to related health initiatives; HIA training and capacity building; and the certification of HIA professionals. The HIA state of practice can be furthered by such theory-building initiatives as methodological refinements; additional effectiveness case studies; the integration of community health and indigenous perspectives and knowledge; and applied research regarding causal links, prediction methods, follow-up procedures, and decision-making influence. A particular effort is needed to remove barriers that impede the integration of health concerns into decision making, to better define HIA roles at the SEA and project EIA levels, and to further the cause of health sustainability.

4.6 CONTEMPORARY CHALLENGE—GOOD PRACTICE IA FOLLOW-UP

Good practice follow-up, as illustrated in Figure 4.4, draws upon a diverse array of regulatory and applied procedures and methods. It also necessitates clearly defined procedural

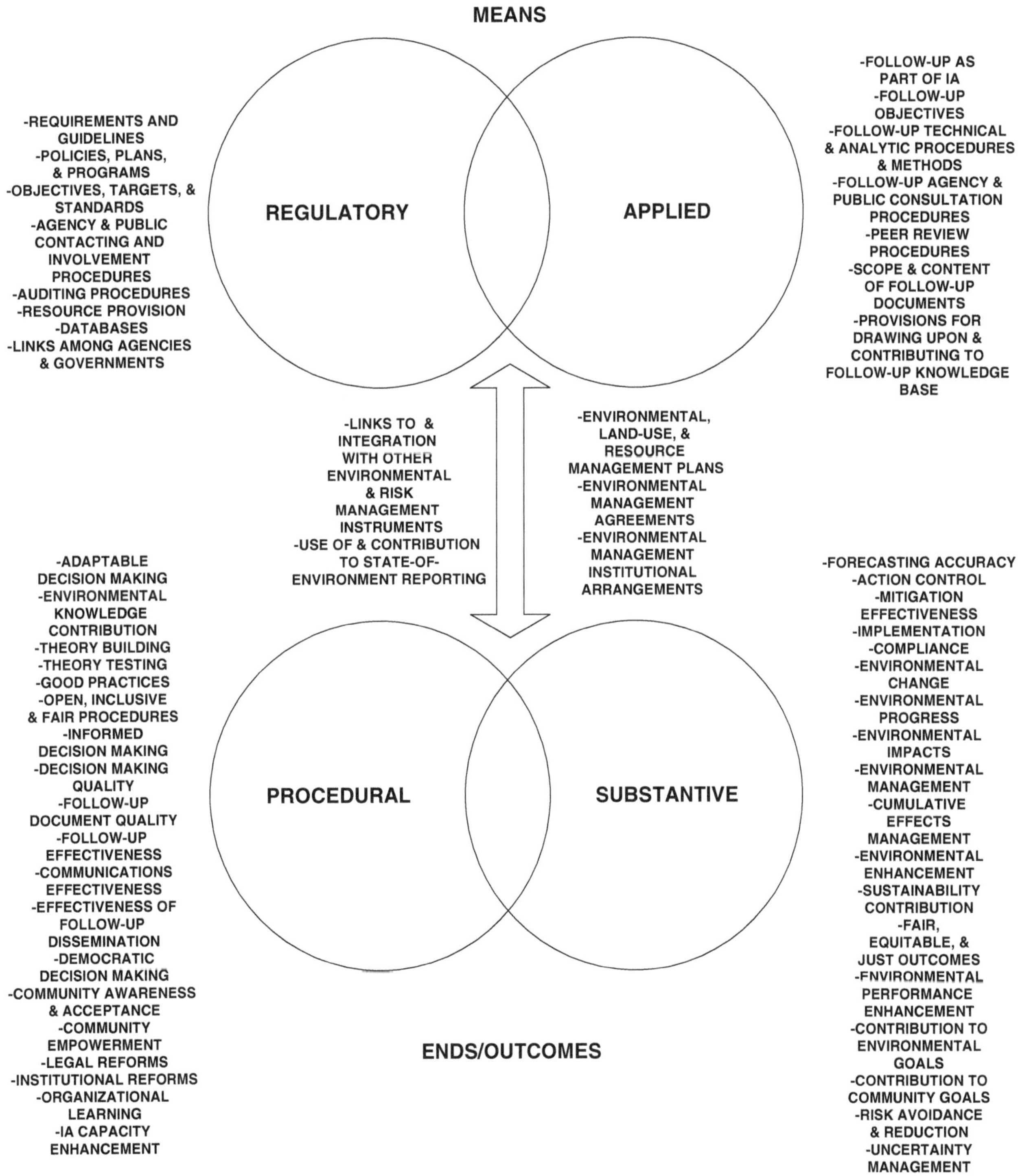

Figure 4.4 Examples of good practice follow-up.

and substantive objectives—objectives that are defined at the outset and tested for effectiveness throughout the course of implementation. These individual ends and means should be knit together into a coherent and fully substantiated impact management plan. The impact management plan, in turn, should draw upon and contribute to related environmental management instruments; state-of-the-environment reporting; and environmental, land-use, and resource management plans and strategies. IA follow-up, to be effective, must be framed within and formalized by action-forcing institutional arrangements and implementation mechanisms such as environmental management agreements.

The tendency has been to define follow-up narrowly (e.g., only biophysical impacts, only for the purpose of assessing forecasting accuracy and determining compliance and mitigation effectiveness). Good practice IA follow-up broadly defines the environment and effects, and encompasses a broad array of procedural and substantive objectives. Procedurally, IA follow-up can, for example, seek to ensure that decision making and decision-making documents are adaptable, fair, inclusive, transparent, and informative. Additionally, it can further legal and institutional reform; enhance the potential for fully substantiated and higher quality decisions; facilitate democratic decision making and community empowerment; contribute to IA capacity enhancement, organizational learning, and community awareness and acceptance; and assist in environmental and IA theory building and testing. In a substantive sense, in addition to assessing forecasting accuracy, compliance, and mitigation effectiveness, good practice IA follow-up can avoid and ameliorate adverse individual and cumulative environmental impacts; enhance the environmental performance of proposed actions; contribute to positive environmental change, community goals, environmental goals, and sustainability; reduce environmental risks and uncertainties; and increase the likelihood of just and equitable outcomes. Good practice IA follow-up clearly defines procedural and substantive objectives, adjusts objectives as needed through the course of implementation, and systematically tests if and the extent to which objectives are achieved.

Table 4.5 highlights a wide range of suggested IA follow-up good practices. The good practices encompass both the regulatory and applied levels. Good practice follow-up at the regulatory level includes possible measures addressed through requirements and guidance; the sponsorship of research and good practices; institutional reforms and practices; and the auditing of IA documents, procedures, and the overall IA system. Applied level IA follow-up good practices pertain to follow-up process design, the scope of the follow-up program, the analytical methods employed, consultative procedures, and external connections and theory building. To be effective, these individual measures need to be effectively integrated into a coherent and complementary set of follow-up institutional arrangements and applied objectives and procedures.

Table 4.5 Examples of Good IA Practices—Follow-up

Regulatory Level	Applied Level
Requirements and Guidance	*Process Design*
Require mitigation, follow-up, and reporting	Institute monitoring before approvals (baseline)
Require action and environmental monitoring	Design to suit proposed activities, potential effects, decision-making level, culture, and IA type
Require consideration of accidents, malfunctions, and natural disasters and risks (including climate change)	Identify and substantiate follow-up purpose, goals, and objectives, including clear commitments list
Require consideration of social, health, ecological, cumulative, transboundary, and sustainability effects	Identify follow-up roles, tasks, and responsibilities; change proponent must accept responsibility for implementing follow-up, and roles should be distinguished to avoid conflicts-of-interest
Require public and agency consultation	Ensure follow-up system is timely, adaptable, and action oriented
Require publication of follow-up results	Ensure actions effectively satisfy follow-up program goals
Require that decision making consider follow-up results	Evaluate, as applicable, policy, plan, program, and project performance; undertake necessary remedial actions
Require modifications to action based on monitoring results, where warranted	*Scope*
Require legally binding approval conditions; ensure enforceable	Identify and substantiate staged scope of follow-up (e.g., draft and final programs at varying levels of detail with provision for ongoing adjustments)
Require regular independent review of IA system (legislation, regulations, policies)	Sustain follow-up over entire life of activity (e.g., design, construction, operations, decommissioning)
Allow for compensation/environmental offsets	Broadly define environmental (physical, ecological, social, heritage, economic) effects
Stipulate duty of proponents to avoid and minimize harmful environmental effects	Monitor environmental quality, change, and progress
Permit legal appeals and provide significant penalties for noncompliance	Monitor adverse, beneficial, direct, indirect, cumulative, and sustainability effects
Establish follow-up performance standards	Monitor effectiveness; modify actions based on monitoring results
	Employ ongoing scoping to modify follow-up program in response to changing context and to long and short-term environmental changes

Table 4.5 (*Continued*)

Regulatory Level	Applied Level
Ensure government has legal authority to require independent audits of decision-making effectiveness and environmental outcomes Issue regularly updated, follow-up guidance materials; modify based on follow-up results *Research Sponsorship and Good Practices* Sponsor follow-up research Sponsor follow-up effectiveness reviews of IA systems and of IA utility Ensure that IA is followed up Integrate substantive objectives and principles into IA requirements Emphasize the role of follow-up in determining substantive outcomes Integrate follow-up results into IA quality control and environmental databases, education, and capacity building *Institutional Arrangements and Auditing* Establish independent oversight of follow-up Establish and refine links between IA and other forms of environmental planning and management Ensure adequate resources for follow-up and enforcement (including inspections, inspectors, education, and capacity building) Enhance environmental databases Institute quality assurance Institute tracking system for IA and follow-up document performance Institute annual compliance audits Institute public follow-up registry Provide for compliance monitoring (inspections, regulatory permits, agreements)	*Methodology* Ensure mitigation/compensation and follow-up are practical, cost-effective, verifiable, manageable, independent, and enforceable Clearly define and substantiate follow-up performance criteria (e.g., standards, targets, and indicators) and thresholds/criteria (e.g., when remediation needed); should be rigorous and reflect best practice Substantiate spatial study areas, temporal boundaries (frequency and duration), and monitoring areas Ensure monitoring information and outcomes are easily measured and unambiguous Identify and substantiate information sources, sampling designs, and procedures for filling gaps Identify and substantiate monitoring, evaluation, management, and communications methods Apply a precautionary approach; ensure capacity to provide early warning about irreversible trends Explain and substantiate results Subject follow-up to peer review Ensure adequate resources for follow-up *Consultation and Collaboration* All parties should clearly commit to follow-up Ensure monitoring sufficiently frequent to be useful to stakeholders without burdening implementation All parties should seek to openly cooperate in follow-up without prejudice Decisions and actions resulting from follow-up should be fair, transparent and communicated directly to stakeholders Integrate community and indigenous perspectives and knowledge Inform and actively engage local communities and other interested and affected parties in follow-up (including provision for community-based follow-up) Establish and actively participate in regional monitoring groups and multistakeholder bodies Recognize limits and pitfalls of negotiated environmental agreements Provide feedback on follow-up process and outcomes Ensure effective dissemination of follow-up results *External Connections and Theory Building* Link to and build on existing monitoring efforts Demonstrate compliance with pertinent regulatory requirements and public policies and objectives Demonstrate compliance with approval conditions Identify follow-up lessons and adjust as needed Integrate follow-up results into all stages of planning system Adhere to good practice standards Promote continuous learning from experience to improve future practice, including drawing upon and contributing to national and international networks Contribute to theory-building and testing Seek to demonstrate follow-up benefits

Sources: AGC (2004, 2008), Australian Government (2011d,f), Barth and Fuder (2002), Burdett (2008a), Burdge (2004), Canter (1993a), Canter and Atkinson (2011), Cashmore (2004), CEAA (2007d,e,f,g, CEAA,2009b, CEAA,2011a,b), Cherp et al. (2011), Craik (2008), Crawford et al. (2010), Devlin (2011), Eccleston (2008), EC (2002, 2010, 2011a), Emilsson et al. (2004), Evaluation Partnership (2007), Fuggle (2005a), Hanusch and Glasson (2008), Hayes and Morrison-Saunders (2007), Herring (2009), Hunsberger et al. (2005), Kemm and Parry (2004a,b), Law et al. (2005), Lee (2006), Lundberg et al. (2010), Marshall (2005), Marshall et al. (2005), Morrison-Saunders et al. (2003), Morrison-Saunders and Arts (2005), Morrison-Saunders and Sadler (2010), Nilsson et al. (2009), Noble (2009a), Noble and Birk (2011), Noble and Storey (2005), Partidário and Arts (2005), Persson and Nilsson (2007), Pölönen (2006), Ramos et al. (2004), Ridgway (2005), Sánchez and Gallardo (2005), Slotterback (2008), Smit and Spaling (1995), Swor and Canter (2011), Thérivel and Ross (2007), Thompson (2000b), Tinker et al. (2005).

4.7 SUMMING UP

In this chapter, we test the premise that IA processes, documents, and methods should be more scientifically rigorous. We present two practice-based stories where the role of science in the IA process is an issue. We provide the conceptual underpinning for a rigorous IA process. We describe a scientific IA process as it might be applied at the regulatory and applied levels. We address IA theory building at the SEA and project EIA levels. We provide examples of IA theory building for various IA types. We address the contemporary challenge of good practice follow-up.

The two stories demonstrate that there is a potential role for the sciences in the IA process. The first story illustrates how the lack of sound, contextually appropriate, social science, especially if not effectively married with community and indigenous knowledge, can undermine IA credibility and effectiveness. The second story demonstrates the importance of a cautious approach to applied science in uncertain and sensitive environments, the need to extend the consideration of biodiversity to encompass noncharismatic and noncommercial species, the potentially significant decision-making role of biodiversity, and the value of highly qualified and experienced scientific investigators.

The effort to formulate a rigorous IA process begins with an overview of the major criticisms advanced by critics. The thrust of these arguments is that IA should be treated as a form of applied research, consistent with prevailing standards and protocols of analytical science. This view is not uniformly shared. Some suggest that scientific standards are inappropriate. Others argue that the standards should be tempered.

Most commentators use analytical science as the touchstone in advancing their arguments. Consequently, analytical science is treated as the departure point. Key analytical science terms are defined. Examples of characteristics commonly ascribed to analytical science are listed. In attempting to establish a foundation for a rigorous IA process, it quickly became evident that the discussion surrounding the role of science in IA is part of a much larger, protracted, and often heated series of debates. A highly selective and simplified version of these debates is presented. Ten sets of opposing positions are presented, together with middle-ground positions. The debates concern such matters as whether science (and by extension a rigorous IA process) should strive for absolute truth, whether it should be rigorous or relevant, whether it is objective or subjective, whether it is beneficial or detrimental to the environment, and so on. IA process management implications are identified for each debate. The arguments are all potentially instructive for IA practitioners. In most cases IA process management will likely occupy a middle-ground position, but one tempered by a need to move closer to one position or the other depending on the IA activities involved and on contextual characteristics.

Regulatory scientific IA process management to this point has largely consisted of identifying general science-related principles for application in IA practice. Examples of rigor-related initiatives derived from the four jurisdictions are cited.

A detailed depiction is presented of a rigorous IA process. The IA process is treated as an experiment, consistent with analytical science. The focal point of the start-up activities is an initial set of hypotheses and an experimental design. The initial hypotheses are preliminary alternative explanations suitable for testing. The experimental design is a research program for testing the hypotheses. The latter integrates such matters as problem definition, objectives, context, need, and methods.

The baseline analysis involves selecting and applying multiple and reliable environment criteria and indicators. Preferably, the indicators can be aggregated or disaggregated. They provide a dynamic picture of trends and patterns. The analysis focuses on sensitive and significant environmental components, functions, and processes. Predictions are deduced from the hypotheses. The predictions pertain to effects from the proposed action and from options and mitigation measures. Interrelationships are addressed through conceptual and quantitative models and theories. The models and theories trace patterns of causal connections. A second round of model and theory building is usually necessary to address interconnections among models and theories. Hypotheses are tested prior to approvals using comparative, control, case, and pilot studies and after approvals using monitoring and follow-up studies. Gaps, limits, and uncertainties are explicitly identified together with implications. Interpretations, preferably supported by statistical analyses, are explicitly identified and substantiated. The scientific findings and interpretations are assumed to provide a sound decision-making basis. Postapproval activities involve both further hypothesis testing and contributions to the scientific and IA knowledge base.

The IA process extends from existing natural and social science knowledge. Ample use is made of scientific methods. Independent, skilled, and qualified scientists drive the process. The scientists strive to ensure a rigorous, open, and objective process, consistent with scientific principles and protocols. Peer reviewers assess the findings and procedures. Scientists interact and maintain contact with the broader scientific community. Stakeholders are involved prior to decisions and in reviewing documentary outputs. Documentation is consistent with good scientific practice. Research findings and methodological innovations are circulated and published, wherever practical. The process is adapted, as needed, by drawing upon alternative scientific paradigms—for example, holistic, management, complexity, and civic scientific approaches.

The relationship between science and IA is complex. The application of analytical applied science standards to SEA and project-level EIA can, depending on the context and prevailing perspectives, be tempered, reduced, reinforced, challenged, refined, or structured. The relationship also varies by IA level and type and is influenced by science types (e.g., sustainability science) and related fields of theory and practice.

Examples of IA theory-building, in general, at the SEA and project EIA level, and for substantive IA types (SA, EcIA, SIA, and HIA) are presented. Cross-cutting themes are identified. Applied theory-building characteristics for each IA type are identified.

Examples of good IA practices regarding follow-up are presented both at the regulatory and applied levels. The analysis addresses both applied and regulatory means and procedural and substantive ends and outcomes. Regulatory level good practices address measures pertaining to requirements and guidance, the sponsorship of research and good practices, and institutional arrangements and auditing. Applied level good practices concern process design, scoping, methodology, consultation and collaboration, and external connections and theory building.

Chapter 5

How to Make IAs More Rational

5.1 HIGHLIGHTS

This chapter addresses the question of how IA processes, documents, and methods can become more rational. The suggestions to make IA more rational are informed and tempered by the debate surrounding rationality both within and external to IA.

- The analysis begins with three applied anecdotes. The stories describe applied experiences associated with efforts to draw upon the strengths and temper the weaknesses of rationality. The analysis in Section 5.3 then defines the problem, which is either (depending on one's perspective) an IA process that is insufficiently rational, consistent, and systematic or an IA process that is too technical, rational, and autocratic. We raise the possibility that both positions are overstated and that there might be some fertile "middle ground."
- In Section 5.4 we provide a context for the rational IA process. We define rationality and identify various rationality forms. We describe the major characteristics of a typical rational planning process. We highlight attributed strengths and limitations. We briefly describe various adaptations to and alternatives to the typical process. We summarize how the debate has played out in IA literature.
- In Section 5.5 we detail how a rational IA process could be implemented at the regulatory and applied levels. In Section 5.5.1 we address how rational IA processes can be facilitated and structured at the regulatory level. In Section 5.5.2 we demonstrate how a rational IA process might be expressed at the applied level. We provide, in Section 5.5.3, examples of applied level rationality good practices for various IA types (SEA, project-level EIA, EcIA, SIA, HIA, and SA).
- In Section 5.6 we address the contemporary challenge of siting locally unwanted land uses (LULUs). We provide an overview of generic approaches, provide examples of good practices, and describe an example of a LULU siting process.
- In Section 5.7 we highlight the major insights and lessons derived from the analysis.

5.2 INSIGHTS FROM PRACTICE

5.2.1 The Willing Host Approach to Siting LULUs: a Reasonable Alternative to Technical Rationality

There is an emerging "reasonable" approach to the siting of locally unwanted land uses, which is not driven exclusively by technical and scientific rational analysis. The approach entails seeking out a Willing Host. Often these controversial facilities are beset with well-known historical examples of poor management as well as being rooted in misinformation perpetuated by opponents. Particular examples include waste management facilities, prisons, and nuclear waste facilities. There are examples of all three of these types of facilities being addressed through "willing host" processes in North America.

The Willing Host approach raises a number of issues. Individuals wonder about the extent of the environmental review that occurs when a Willing Host comes forward. Other issues focus on equity and fairness. However, it has been our observation that the Willing Host approach is definitely not a shortcut as a rigorous, fully detailed, thorough environmental assessment is still carried out. The approval agencies have no choice but to ensure that all environmental requirements are met for the location of the facility, whether or not it is a Willing Host. Indeed, one can argue that the Willing Host approach is another way of achieving the social license to operate once the environmental requirements are met.

If a Willing Host is identified and can meet all environmental assessment requirements, there are still potential problems. The problems may not relate specifically to the Willing Host community but to neighboring communities, the broad regional area, and all of the communities that are along any transportation routes.

In terms of the Willing Host approach, there are three examples to consider.

1. NWMO (Nuclear Waste Management Organization), a federal Canadian government agency, is charged with the responsibility to find one or more disposal sites for high-level nuclear waste. The NWMO Willing Host approach is a systematic six-step process as shown below.

Impact Assessment: Practical Solutions to Recurrent Problems and Contemporary Challenges, Second Edition. David P. Lawrence.

Step 1: The NWMO initiates the siting process with a broad program to provide information, answer questions, and build awareness.

Step 2: Communities identify their interest in learning more, and the NWMO provides detailed briefing.

Step 3: For interested communities, a preliminary assessment of potential suitability is conducted.

Step 4: For interested communities, potentially affected surrounding communities are engaged and detailed site evaluations are completed.

Step 5: Communities with confirmed suitable sites decide whether they are willing to accept the project, and propose the terms and conditions on which they would have the project proceed.

Step 6: The NWMO and the community with the preferred site enter into a formal agreement to host the project.

The completion of Step 6 then initiates the formal regulatory review process.

A community that embarks on this process can drop out at any point in time. A successful Willing Host must meet the basic geological requirements as well as have a thorough review of the overall social, economic, and environmental impacts and benefits. Currently there are 10 communities in Canada (7 in Ontario and 3 in Saskatchewan) that are going through the NWMO Willing Host process. A similar Willing Host process was very successful in finding a long-term, high-level nuclear disposal facility in Sweden.

2. The government of British Columbia (BC) recently had a competition for siting a medium security prison in Central BC The successful community was the Osoyoos Indian Band.

   The BC government invited 15 local governments to provide bids for the new prison facility to be built somewhere in the Central Okanagan. The facility would provide 500 construction jobs and 240 permanent, full-time jobs. A number of criteria were set out, including a large enough piece of serviced land near major highways and no more than 70 min from the various courthouses.

   Of the 15 local communities, only five took up the challenge and one eventually dropped out due to residents' opposition.

   The successful site is in a fully serviced industrial park owned and managed by the Osoyoos First Nation. The key to being successful in this competition is that Chief Clarence Louie provided the BC government with the results of a community referendum showing that the band members were supportive of the proposal.

3. The third example is that of a private entrepreneur to partner with a community or First Nation to develop municipal waste and/or construction waste sites in Ontario. This proponent of these waste facilities is looking for communities that are willing to consider entering into an agreement to provide such facilities. Basic requirements include access to a rail line and major highways. Although the technology for such waste sites is proven, there are still public concerns and a public stigma to such facilities. As a result, a Willing Host community provides the best opportunity for a successful siting and approval.

There are really two levels of activities with the Willing Host process.

First level is the full range of environmental assessment work and consultation that occurs with the Willing Host community. In most cases, Willing Hosts are small communities who are either stagnating or experiencing declining population and are looking to restart or regenerate their economies.

The second level is at the regional scale where neighboring communities, particularly those along transportation routes, believe that they will bear some of the burden but not necessarily get any of the benefits. This raises an issue in terms of ultimately making a choice between finding a Willing Host that meets the EA requirements and accepting the facilities while having to deal with transportation route for communities and/or nearby adjacent communities who for various reasons are opposed to the facility. This raises questions as to the weights one ultimately gives to the various positions in making the decision. Is it enough to find an acceptable Willing Host community to host a controversial facility? How far beyond the Willing Host community should you consider in decision making?

As the Willing Host approach is tested more and more, these questions will be challenges to address, and it is expected that they will be resolved through the application of the "reasonable man/woman approach."

PETER HOMENUCK

DPRA Canada, Toronto, Ontario, Canada

5.2.2 Collaborative Community-Based Reasoning

It has been suggested that solutions to some of the problems with strategic environmental assessment (SEA) practice may, in fact, depend on how it is implemented. As Connelly and Richardson (2005, p. 397) note, there are evident weaknesses in the technocratic approach to SEA and they contend that perhaps "good SEA is participative."

They point out that participatory, deliberative approaches to EA are asserted as the new orthodoxy. The social learning that occurs through such approaches helps in the shared understanding of values and reflects the greater underlying need for open deliberation in SEA as compared to EIA (Partidário, 2006). Given these points, the purpose of our work was to consider whether a community-based participative approach to strategic assessment could work and, if so, what lessons might be gained from using this approach in the development context (Sinclair et al., 2009).

Our starting point was community-based environmental assessment (CBEA), which has been adapted in an innovative way to smaller, community-based projects that utilize natural resources for basic livelihood needs (Spaling, 2003; also CIDA, 2005). Typical projects include boreholes, gravity water systems, small reservoirs, agroforestry, fishponds, and construction of latrines, clinics, schools, and small bridges.

Since these projects interact directly with biophysical systems, many already stressed, there is potential for resource degradation through over-extraction, land clearing, soil erosion, contamination, and other forms of exploitation. Application of EA to these projects is emerging as a way to facilitate management of local resources and ensure continued project benefits (Spaling, 2003, p. 152).

In community-based approaches to EA, a participatory forum facilitates a process of communal dialogue and collective decision making that includes the development of goals, the sharing of knowledge, negotiation and compromise, problem posing and problem solving, the evaluation of needs, and the definition of goals (Pallen, 1996; Neefjes, 2001; Spaling, 2003). This process helps communities clarify values, be more adaptive and proactive, respond to change, develop an appreciation for the human-/ecological interface, set personal and communal goals, and participate in a process where they are heard (Keen and Mahanty, 2006; Meredith, 1992).

Using a qualitative research design, we applied a CBEA approach to the strategic assessment of the the *Instituto Costarricense de Electricidad* (ICE) (Costa-Rica's publicly-owned electrical and telecommunications company) watershed management agricultural program (WMAP). This program addresses erosion and contamination problems caused by conventional farming practices (i.e., planting homogeneous crops, heavy reliance on chemicals, and regular tilling) in watersheds where ICE has hydro projects. We worked with farmers from two watersheds, Reventazón and Sarapiquí, who are collaborating with two different ICE WMAP teams.

The design of the community-based strategic environmental assessment (CBSEA) was highly participatory, utilizing common participatory rural appraisal tools, which promoted dialogue and sharing in a safe environment, focused on a single, small community program, and involved locals in event planning. The CBSEA process itself was broken down into four components that captured the common elements of SEA frameworks defined in the literature (Thérivel and Brown, 1999):

1. *Determining the purpose and presenting the program.* In the first workshop, the CBSEA process was explained to participants and ICE explained their proposed WMAP Phase II.
2. *Assessing the program and identifying alternatives.* In the second workshop, community participants collaboratively assessed the potential effects of the proposed ICE WMAP Phase II.
3. *Identifying real and potential impacts of the proposed program components including the identified alternatives.* In the third workshop, community participants identified real and potential social, economic, and environmental impacts that the components within the new modified proposed program would have if they were implemented.
4. *Sharing CBSEA results with the proponent.* In the fourth workshop, communities and related institutions [e.g., MAG, Ministry of the Environment and Natural Resources (MINAE), National Institute for Learning (INA)] came together to discuss the CBSEA process and the results from the process.

The case study revealed that people with little to no experience with EA were able to effectively participate in a participatory and structured SEA process. They were able to articulate preferred options, potential effects, and mitigation approaches with relative ease, especially those for the physical works components of the proposed program. The results also revealed that through this process participants were able to think more critically, even about some of the impacts of activities they were already undertaking on their farms. In the end, they learned and a more robust preferred program was the outcome.

The case study also revealed that a participatory CBSEA approach can accrue many benefits including meaningfully engaging community members in a natural resources management decision-making process, facilitating a more comprehensive assessment of incoming programs, individual and social learning outcomes, and facilitating a transition toward sustainability. These conclusions indicate that there can be great value in doing a SEA at the community level, at least in relation to programs.

A. JOHN SINCLAIR

Natural Resources Institute, University of Manitoba, Winnipeg, Manitoba, Canada

LAURA SIMS

Faculty of Education, Université de St. Boniface, Winnipeg, Manitoba, Canada

HARRY SPALING

Geography and Environmental Studies, King's University College, Edmonton, Alberta, Canada

5.2.3 A Role for Rational Analysis in Providing SEA Technical Support

A SEA was carried out for a spatial plan in Italy. One of the main tasks of the SEA was to contribute to the identification of possible locations for new urban development areas. A structured process was undertaken based on two main stages. First, a preliminary screening of possible locations was performed. Subsequently, these locations were compared to select the most suitable ones.

The first stage was conducted by spatial multicriteria analysis (SMCA). A criteria tree was developed that contained four main categories of criteria: ecosystems, water resources, quality of the urban environment, and energy and climate. Additionally, constraints based on existing laws and regulations were applied to rule out areas that could not be considered for urban development. Specific criteria were then linked to each category, and relevant maps were constructed in a Geographic Information System (GIS) environment. These criteria included, among others, ecological networks and ecosystem fragmentation, potential water pollution loads, accessibility to urban green areas and parks, landscape quality, noise, electromagnetic radiation, solar electricity potential, and availability of public transport. Some of the criterion maps were based on existing data, whereas others required GIS modeling that was performed in the preliminary stages of SEA. The maps were then normalized and aggregated, leading to an overall land suitability map for urban expansion. This map was used to extract a preliminary set of potential sites. This was done by applying two types of thresholds: one related to the suitability value, and the other related to the overall size of contiguous and suitable cells. Potential sites were those characterized by high suitability and by a relatively large area (the area threshold was provided by plan's regulations aimed at preventing excessive urban sprawl and fragmentation).

This process resulted in the preliminary selection of 15 sites. For each of these sites, the average performance value of the criteria previously listed were computed and used to run a set of additional multicriteria analyses, by changing the importance weight assigned to the four criteria categories. As a result, different perspectives were generated, each showing the ranking of the 15 sites, obtained by assuming different level of importance attached to criteria related to ecosystems, water resources, quality of the urban environment, and energy and climate, respectively. A final analysis was conducted to assess the sensitivity of the sites to changes in the criteria weights.

About a third of the sites ranked stably at the top of the rankings, regardless of the weights. The remaining sites had a performance more strongly connected with the weights. All these analyses were summarized in a short executive report, and detailed in a more technical appendix. The report was then provided to planners and policy makers who used it to interact with the general public according to the required steps of the planning procedure, and to select the final location of urban expansion areas. In this study, SEA played a very technical role, mainly consisting of providing evidence and a structured and documented process to support planning decisions.

DAVIDE GENELETTI

Department of Civil and Environmental Engineering, University of Trento, Trento, Italy

5.3 DEFINING THE PROBLEM AND DECIDING ON A DIRECTION

The three stories offer different perspectives on the potential roles of reasoning and technical analysis in the IA process. The first story shows how the Willing Host approach to siting LULUs offers a potentially creative procedure for introducing reason into complex and contentious IA problem solving. The second story describes a CBSEA reasoning approach that adheres to the major rational planning steps while still being highly collaborative, and able to effectively draw upon sound rational–technical analysis. The third story describes an IA process, which relied heavily upon technical quantitative analysis and was undertaken in a manner that supported and informed rather than supplanted the decision-making role of planners, policy makers and the public working together.

The three stories demonstrate that technical–rational analysis and collaborative reasoning have the potential to be complementary in an appropriately structured, dynamic, and inclusive IA process. They illustrate that rationality can be either a positive or a negative force in an IA process, depending on how it is applied, the rationality perspectives that it represents, and the fit between process and context. They suggest that there is a potential role for rationality in the IA process but that role tends to be more effective when it assumes a support role (rather than driving the process), when there is a high degree of community collaboration and influence, when procedural and substantive equity issues are explicitly addressed, and when proactive steps are taken to offset the negative propensities of rational IA processes.

If IA processes are "rational" they clearly define the problem. They explicitly identify goals and objectives. Alternatives are systematically identified and evaluated against goals and objectives. The planning process is integrated, from early on, into organizational planning and decision making. The preferred alternative is implemented. The achievement of goals and unintended consequences is monitored and managed. Up-to-date and appropriate technical methods are systematically integrated into the process. Formal checks of document quality are instituted.

Historically, IA theory and practice, for much of its existence, has been structured around a rational, information provision model (i.e., better information leads to better decisions) (Bond and Pope, 2012). Notwithstanding the rational assumptions and aspirations of IA requirements and processes, there is widespread doubt regarding whether IA is or should be rational. IA effectiveness analyses provide the first clues as to whether IA processes are rational. The environmental aspirations expressed in IA legislative goals are not being realized to the extent hoped for (Clark, 1997). Effectiveness ratings for problem definitions, objectives formulation, and preparing adequate terms of reference are low (Sadler, 1996; Spooner, 1998). Problems, root causes, and high and low objectives are often not satisfactorily identified (Lee and Kirkpatrick, 2006). The identification and evaluation of reasonable alternatives is a recurrent weakness in IA guidelines and documents (Barker and Wood, 1999; Eales and Sheate, 2011; ERM, 2000; Sadler, 1996; Sadler and Jurkeviciute, 2011; Spooner, 1998). Options tend to be too narrowly defined, option analysis is often problematic, and choices tend to be poorly justified (Galbraith et al., 2007; Lee and Kirkpatrick, 2006). The general tendency is to treat the alternatives analysis as a *pro forma* exercise (Sadler and Jurkeviciute, 2011). Methods selection and application could be greatly improved as could document quality (Sadler, 1996; ERM, 2000).

IA processes, documents, and methods in practice fall well short of rationality ideals. Most IA texts and much of IA literature seek to correct this "deficiency" (Canter, 1996; Gilpin, 1995; Glasson et al., 2005; Hanna, 2009b; Morris and Thérivel, 2009; Morgan, 1998; Noble, 2009b; Smith, 1993; Westman, 1985). IA requirements, they suggest, should include objectives. Proposals and alternatives, they argue, should be systematically evaluated against objectives. Specific measures are described to facilitate more effective procedural guidance and control, especially regarding the treatment of alternatives (Ortolano, 1993). Explanations are provided concerning how environmental considerations can be integrated into project management and into organizational decision making.

These commentators see value in a rational, systematic, normative approach, albeit one that is more environmentally substantive, that is tailored to the decision-making level and other contextual considerations, and that proactively seeks to effectively influence decision making (Fischer, 2003, 2005). These commentators worry that a flexible, adaptive IA approach will be worse for the environment. They stress the value of an explicit substantive, interconnected hierarchy of environmental/sustainability ends (both ultimate and milestone) to guide the IA process and to provide the basis for evaluating options (Donnelly et al., 2006, 2007; Fischer, 2003; Hacking and Guthrie, 2006).

A diversity of largely technical and scientific methods are identified, described, and compared. The methods apply to different IA activities, impact types, and environmental components. They also address interdisciplinary analysis and study team coordination. Quantitative and computer-based systems and multicriteria decision aids often receive particular attention (Geneletti, 2005; Bruner and Starkl, 2004; Hassan, 2008; Julien, 1995; Kain and Söderberg, 2008). The net result is assumed to be a more rational, environmentally substantive, scientifically and technically sound, comprehensive, and objective decision-making basis (Culhane et al., 1987; Fischer, 2003). Rational decisions and the achievement of the environmental objectives of IA requirements are assumed to flow naturally from the greater and more effective application of the recommended procedures and methods.

Some critics, in contrast, maintain that IA practice either is or strives to be too rational. Rational processes are seen as politically naïve and ineffective in their failure to appreciate how decisions are made within organizations and how power is wielded. As a consequence, they tend to have limited success in influencing decision making (Bond and Morrison-Saunders, 2011; Weston, 2010). They are considered impractical because of their inability to recognize cognitive and decision-making limitations and to effectively operate across disciplines and professions. They are considered inflexible and poorly equipped to deal with uncertainty and conflict (Boothroyd and Rees, 1985). They are viewed as oblivious to the implications of operating in a multiactor, multi-interest sociopolitical environment (Greer-Wooten, 1997). They are characterized as reductionist (Bond and Morrison-Saunders, 2011). They are described as not appreciating the ubiquitous role of competing values and interests in decision-making (Weston, 2010). They are described as failing to change social values or to facilitate social learning (Weston, 2010). They are condemned for excluding extrarational forms of knowledge and for being biased against the qualitative (Hodge, 2004). They are labeled as technocratic, autocratic, and antidemocratic on the grounds that they tend to be "top-down" and marginalize the public's role in the decision-making process (Binder et al., 2010). These critics argue that the rational IA process should be abandoned and replaced with more political, interpretative, intuitive, contextual, dynamic, holistic, iterative, collaborative, transdisciplinary and value-full planning and decision-making models and procedures (Bond and Morrison-Saunders, 2011; Binder et al., 2010; Boothroyd and Rees, 1985; Craig, 1990; Greer-Wooten, 1997; Torgerson, 1981; Weston, 2010).

Another group of commentators focus on how rationality is defined and how it is applied. They suggest that ecological, practical, and communicative rationality forms are more directly relevant to IA practice than are analytical rationality forms (Bartlett, 1997; Craik, 2008; Stoeglehner et al., 2010). They maintain that rationality tenets and assumptions should be relaxed and replaced with a more practical, political, social, adaptive, precautionary, substantive, and collaborative "reasoning" process (Barrow, 1997; Doberstein, 2004; Greer-Wooten, 1997; Kørnøv and Thissen, 2000). They argue for greater attention to and understanding of how decisions are actually made (Bond and Pope, 2012). They suggest viewing IA as an expression of discursive and democratic democracy (e.g., planning as a form of persuasive storytelling) (Richardson, 2005). They advocate the increased use of inclusive and dialogue-based participatory tools and a broadened conception of rationality to encompass, for example, indigenous perspectives (Crawford et al., 2010; Wikland, 2005). They argue for placing rationality in a broader context (i.e., vary rationality standards to match contextual conditions, constraints, and uncertainties) for bringing together the best of the qualitative and the quantitative, and for adjusting the level of detail to suit the scale and type of proposal (Dalal-Clayton and Sadler, 1998; Hodge, 2004). They suggest that IA practice should encompass, at each stage, multiple rationalities, consistent with the multiple values and value conflicts that permeate IA practice (Richardson, 2005). They acknowledge that the rational model can work well when goals are shared (i.e., a persuasive basis for justifying actions) (Craik, 2008). But they suggest that a more overtly political pluralist model is more appropriate when goals and interests conflict (Craik, 2008;

Richardson, 2005). They recommend selectively applying rationality.

Rationality is considered especially appropriate when there is a high degree of certainty and control and a low degree of conflict (Kørnøv, 1998). Commentators favor modifying and adapting rationality to accommodate knowledge and insights derived from experience, intuition, emotion, and imagination. They point to the need for decision-making to combine the objective and analytical with the subjective and integrative (Kørnøv and Thissen, 2000). These commentators see value in rationality for IA, but only a form of rationality that is carefully defined, tightly circumscribed, selectively applied, and integrated both with other forms of knowledge and into pragmatic, sociopolitical planning, and decision-making processes (Craik, 2008; Culhane, et al. 1987; Partidário, 1996).

As is evident from the above, there are at least three definitions of both the problem and the preferred direction. Applying valid rationality strengths, minimizing legitimate rationality deficiencies, and drawing upon alternative rationality definitions and applications can help integrate these problem definitions.

5.4 SELECTING THE MOST APPROPRIATE ROUTE

5.4.1 Definitions and Distinctions

Rationality has been a central theme of Western thinking since the Renaissance (Alexander, 1986). Definitions of rationality encompass many elements, as illustrated in Figure 5.1. Ideally, a rational IA process displays such *attributes* as purposeful, sensible, orderly, lucid, logical, coherent, transparent, explicit, replicable, consistent, reflective, reasoned, verifiable, and objective. It strives for accuracy, reliability, and accountability. It is technically sound and scientifically rigorous (see Chapter 4). It seeks to minimize such *irrational* factors as errors, falsehoods, contradictions, and incoherence. It tries to avoid nonsensical, irrelevant, ad hoc, illogical, unreflective, and inadequately supported arguments. Emotions, feelings, experiences, intuition, imagination, wisdom, habits, traditions, faith, and subjectivity are excluded or downplayed because they are nonrational or *extrarational.*

Rational attributes are *expressed* in thoughts, actions, opinions, judgments, interpretations, criticisms, reflections, decisions, conclusions, and recommendations. They are evident in deliberations and dialogue. Rational expressions are *supported* by reasoned intellectual analysis, logic, and argumentation. They are *informed* and *guided* by ideas, information, knowledge, and evidence. They are *justified* by sound and systematic application of technical and scientific methods, as conducted by qualified specialists. The expression of rationality in an IA process occurs more fully when such *preconditions* as openness, honesty, and trust prevail. Also essential is a nonoppressive environment where interested and affected parties are willing to participate in the process, engage in reasoned dialogue, and commit to a broader human or ecological purpose.

Rational expressions are *fulfilled* in problem solving, in opportunity seeking, in planning and in decision making. They also are *applied* in management, in lawmaking, in organizational design, and in communications and public involvement procedures. The expression and application of rationality is dependent on *context*. It varies for individuals, groups, and organizations. It operates differently at the multiorganizational and societal levels.

As indicated in Table 5.1, there are many rational *forms* relevant to IA practice. Purpose, need, and objectives are often defined before the IA commences, consistent with instrumental or technical rationality. IA requirements and practices are structured around a process, in accordance with procedural rationality. Sometimes IA legislation and practice identify and explore purpose and direction questions, as is the case with purposive (or value or normative) rationality. Occasionally IA, in common with substantive rationality, considers both alternative ends and means. IA both integrates individual effects (analytical rationality) and considers systems levels concerns and impacts (systems rationality). IA addresses and adopts social (social rationality), economic (economic rationality), political (political rationality), and ecological (ecological rationality) perspectives. IA is generally an "action-forcing" legal environmental management instrument (legal rationality). IA considers market implications (market rationality). It involves a highly participative process (communicative rationality). IA is constrained, focused, and pragmatic consistent with bounded, practical, and strategic rationality forms, respectively.

5.4.2 Core Characteristics

The rational planning and decision-making process, drawing heavily upon utility theory, began (at least in the postwar period) with the suggestion that decision-makers agree on goals, identify available alternatives for achieving goals, evaluate the consequences of alternatives, and select the alternative that comes closest to achieving the goals (Banfield, 1955; Nilsson and Dalkmann, 2010; Simon, 1976). Implementation of the preferred alternative was assumed. This process was refined through the 1960s and 1970s, as illustrated in Figure 5.2. The revised process begins with a problem, need, or opportunity to be addressed. An appropriate constellation of values (the public interest) was determined (Davidoff and Reiner, 1962). General values were distilled into goals, principles, objectives, and criteria—progressively more precise measures of progress toward the public interest. Goals, objectives, and criteria were ranked (Boyce, 1971). Methods for assembling and analyzing data, operating within resources, responding to pertinent constraints and opportunities, determining present and predicting future conditions, and deriving, screening, and comparing alternatives were formulated (Alexander, 1986; Friedmann, 1987). Uncertainties and variations in

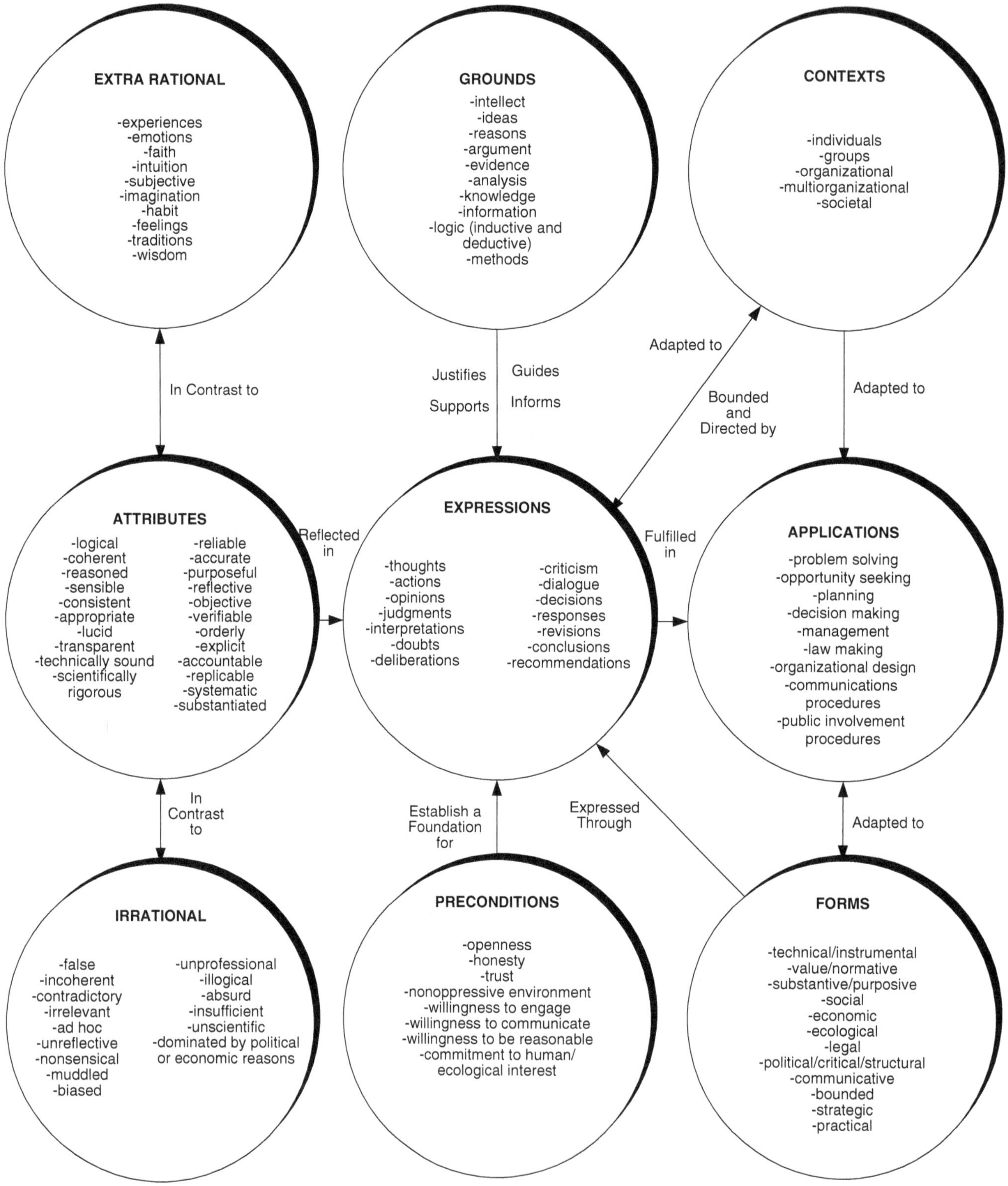

Figure 5.1 Rationality definition.

preferences were taken into account. More details were subsequently added concerning how the preferred alternative would be implemented. Allowance was made for interactions among process steps (scanning forward and feedback loops) (McLoughlin, 1969). Provision was also made for public and agency involvement, often prior to decision making and occasionally as inputs to each stage in the process.

Table 5.1 Potentially Relevant Forms of Rationality

Forms	Key Characteristics
Instrumental/technical/scientific rationality	Search for the best possible means (how to do things) for given ends (what could be achieved) Stress on efficient, logical and systematic goals achievement Based on causal explanations In accordance with scientific rules of assessment Emphasis on efficiency, measurement, and analysis Goals and objectives determined externally (teleological)
Procedural rationality	Rationality of the process The procedures used to choose actions Acceptance or rejection of a claim based on procedures or rules followed
Purposive/value/normative rationality	Rationality of ends Based on moral judgments Synthetic Evaluation and choice among goals
Functional rationality	Rationality inherent in the functioning of systems, societies, or organizations Clearly defined and calculable goals
Substantive rationality	Rationality of ends and means (deontological) Rationality of the outcome of the process Applies to individual decisions or actions
Analytical rationality	Understanding by breaking things into parts and by studying differences and links Additive (sum of parts)
Systems rationality	Understanding in terms of purpose and relevance Order flows from sense of whole
Social rationality	Seeks integration in social relations and social systems Makes social action possible and meaningful Assumes social formation prior to individual; identity from group; reason exercised for group
Economic rationality	Utilitarian Entails the maximum achievement of plurality of goals Underlain by principle of efficiency Assumes orderly measurement and aggregation
Political/critical/structural rationality	Rationality of decision-making structures In accordance with political rules of democratic decision making Preserves and improves decision structures Emphasis on practical capability for facing societal problems Requires an open, honest, informed debate Concerned with identifying and redressing structural inequities
Legal/administrative rationality	Reason inherent in clear, consistent, and detailed formal rules for preventing disputes and for providing solutions In accordance with administrative rules of policy preparation and implementation Reason inherent within context of organizational and institutional structures and procedures
Market rationality	Unconstrained pursuit of self-interest by individuals and organizations
Ecological rationality	A rationality of living systems An order of relationships among living systems and their environment
Communicative rationality	Organized dialogue to promote democracy and personal growth Concerned with the quality of the communications Stress on mutual understanding and counteracting of communications barriers and distortions
Bounded/limited/instrumental rationality	Search for satisfactory solution (good enough) Not all alternatives known or consequences considered; alternatives considered sequentially Not all preferences evoked Procedural Criterion—effectiveness, limited to understanding the impacts cognitively Contingent on environmental conditions
Aesthetic/expressive rationality	Criterion—truthfulness With an art system
Ethical/moral rationality	Criterion—rightness Within a moral system
Practical rationality	Starts with real, everyday life Pragmatic
Strategic rationality	Selective and contingent Adapted to local context and specific situation

Sources: Alexander (1986, 2000), Bartlett (1989), Becker et al. (2005), Braybrooke and Lindblom (1963), Elling (2011), Etzioni (1967), Forester (1999), Friedmann (1987), Habermas (1993), Healey (1997), Jiliberto (2011), Kørnøv (1998), Richardson (2005), Sager (1994), Verma (1998).

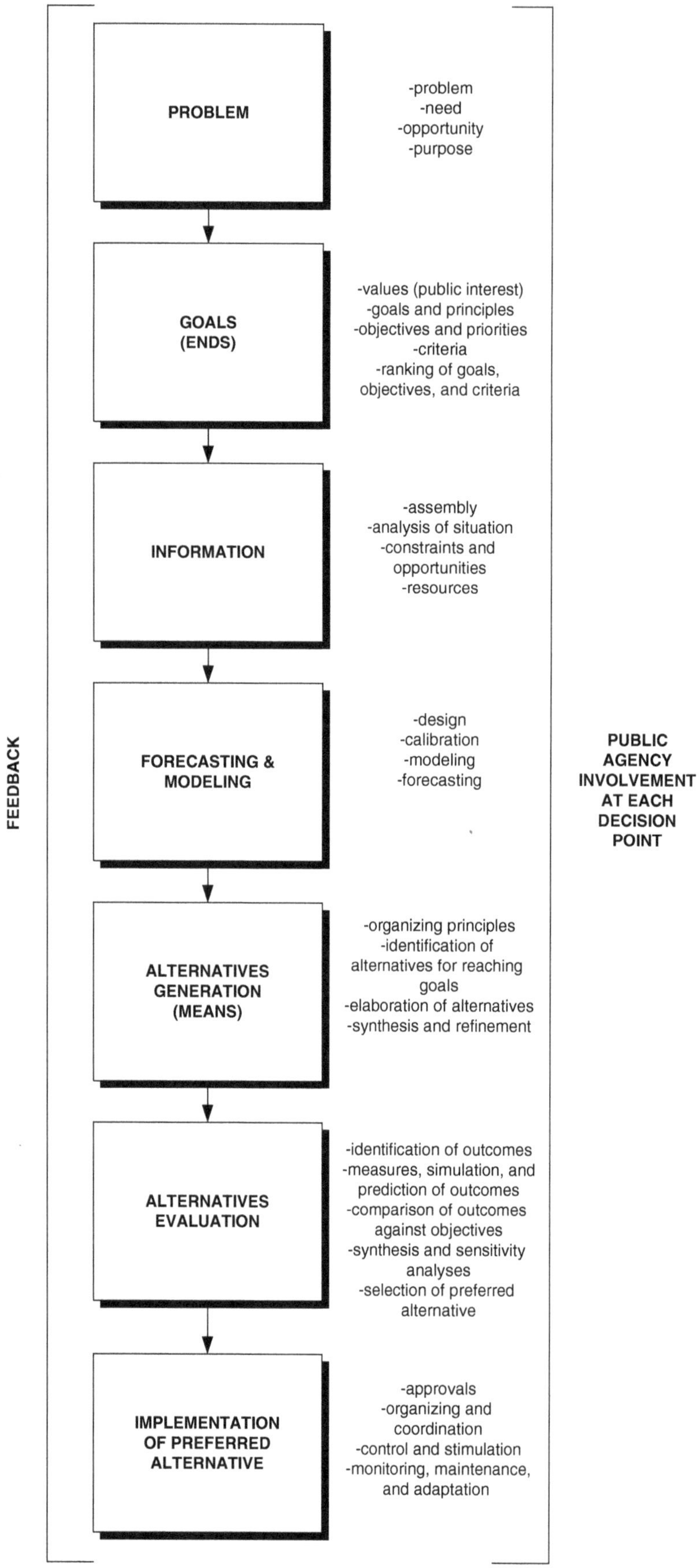

Figure 5.2 Rational process.

Table 5.2 Ascribed Rational Process Assumptions

General

- Reason is systematically applied (central to problem)
- The process is independent of the problem and of the context (process can be universally applied)
- Adequate time, skill, and resources
- Comprehensiveness—complete information
- Unitary actors
- Individual is purposeful and intentional
- Collective choices guided by aggregated individual preferences
- IA specialist as technician (objective, apolitical, unbiased)
- Sequential, analytical process
- Hierarchical decision-making structure
- People act rationally; their actions make sense with respect to the aims and the means they select to achieve their ends
- A pluralistic society (all competing interests have access to power)
- Collective rationality is the aggregation of individual rationality (utilitarian)
- Downward direction of control

Goals (Problems, Goals, Objectives)

- A well-defined and independent problem (susceptible to analysis and diagnosis)
- People have preferences and act in accordance with them
- Goals and objectives can be identified and articulated
- There is unitary public interest (value, goal, and objective consensus is possible)
- Goals and objectives guide process (also basis for evaluating alternatives)
- Specialists are value-neutral and can determine the public interest

Information

- Supremacy of technical and scientific knowledge (as determined by independent specialists)
- Complete and fully accessible baseline information
- Manageable uncertainties

Forecasting and Modeling

- Well defined action space (all relevant variables)
- Probability of occurrence can be predicted based on available data
- Predictable and controllable environment
- Stable society

Alternatives Generation (Alternatives, Plans, Strategies)

- Well-defined alternatives
- All reasonable alternatives available
- All alternatives examined
- Best alternative can be identified

Evaluation of Alternatives

- Well-defined and known outcome space
- All relevant consequences of each alternative can be determined
- Preferences are transitive (goals and objectives can be ranked)
- Alternatives can be assessed against goals, objectives, and criteria
- Predicted consequences and value preference differences can be amalgamated to select a preferred alternative (using formal evaluation methods)

Table 5.2 *(Continued)*

- Action chosen is based on hierarchy of preferences that gives the greatest benefit
- Uncertainties in predictions and value preferences can be addressed (e.g., using sensitivity analyses)
- Decisions based on evaluation of alternatives conducted by specialists

Implementation

- Full approval is obtained
- Proposed decision and decision implementation is unambiguous
- Rational information will improve decision making
- The environment is controllable and is controlled
- Possible to monitor all key variables and to make appropriate adaptations

Interrelationships

- Facts and values can be separated (objective from subjective)
- Ends and means can be separated and interrelationships between ends and means are clearly defined
- Independence of probabilities and utilities (what is expected is unaffected by what is wanted)
- Independence of analysis and evaluation
- Emphasis on consistency, transparency, and transitivity
- Separation of analysis/evaluation from implementation (technical from political)

Sources: Arts et al. (2011), Alexander (1986), Banfield (1955), Boyce (1971), Boyer (1983), Damasio (1994), Davidoff and Reiner (1962), Feldman and Khademian (2008), Forester (1984), Friedmann (1987), Harper and Stein (1992), Healey (1997), Kørnøv and Thissen (2000), Mintzberg (1994), Nilsson and Dalkmann (2010), Sager (1994), Simon (1976), Smith (1993).

Numerous assumptions have been ascribed to the rational process, as detailed in Table 5.2. It tends, for example, to be assumed that problems are well defined, that the environment and available choices are predictable and controllable, that a unitary public interest can be defined, and that decision makers are rational. It is expected that they will select and implement preferred alternatives based on the comprehensive and objective analyses of technical and scientific specialists. The rational process is seen as systematic, largely sequential, and optimizing (i.e., all alternatives considered and best alternative selected). These assumptions should be approached with caution. A part of the procedure for formulating a rational IA process involves determining which ascribed assumptions are intrinsic to the rational process and which either apply only to specific rationality forms or could be relaxed, adjusted, or abandoned.

5.4.3 Attributed Strengths and Limitations

The rational process has been described as simple, explicit, logical, consistent, systematic, and adaptable (Caldwell, 1991; Sager, 1994). It helps to clarify future directions, establish priorities, and explore potential courses of action (Faludi, 1986). It provides a clear, coherent, comprehensive, unbiased, and defensible basis for decision making (Briassoulis, 1989; Caldwell, 1991; Healey, 1997). It systematically integrates scientific and technical knowledge. The role of specialists in the process is clearly defined and legitimized (Benveniste, 1989). Although the process objectives may not be fully realized, it is still considered beneficial to seek to be unbiased, comprehensive, consistent, and systematic (Briassoulis, 1989; Faludi, 1986).

The rational process has been attacked on a host of fronts. It is labeled as unrealistic, ineffective, incomplete, and inappropriate. It is considered *unrealistic* because human cognitive limits are not adequately considered (Webber, 1983). It fallaciously assumes that problems are well structured; goals and beliefs are clear and unambiguous; adequate, largely quantitative, environmental information is available; existing environmental conditions can be extrapolated into the future; all alternatives are available; all consequences can be determined; decision making is rational; and the preferred alternative can and will be implemented (Bond and Morrison-Saunders, 2011; Forester, 1989; Mintzberg, 1994; Healey, 1997; Webber, 1983). It incorrectly assumes that all actors in the process are rational and that adequate resources are available to support a comprehensive analysis (Briassoulis, 1989; Forester, 1989). It fallaciously assumes complete information, unitary actors, well-defined problems, consistent objectives, interests and values, minimal and manageable uncertainties, and clearly defined hierarchical decision-making mechanisms (Arts et al., 2011; Feldman and Khademian, 2008). It fails to recognize the extent to which ends and means, reason and emotion, and analysis, evaluation, and implementation are necessarily interwoven (Damasio, 1994). It makes the impossible assumptions that all possible alternatives can be identified and compared, and that the "best" alternative can be selected (Morgan, 2012).

The rational process tends to be *ineffective* because it does not consider practical, "common sense" considerations and solutions (Saul, 1992). It fails to focus the limited, available resources (Benveniste, 1989). The rational process concentrates on internal analyses, assuming that external environmental conditions are stable and can be controlled (Benveniste, 1989). As a result, it is not well adapted to contextual characteristics and does not respond promptly to changing circumstances (Mintzberg, 1994). It is especially ineffective on the political front (Richardson, 2005). It fails to take into account the nonrational, politically driven world in which it operates (Jay et al., 2007; Thérivel, 2010). It has difficulty in explaining such issues as power, conflict, trust, solidarity, inequality, communications, and legitimacy (Nilsson and Dalkmann, 2010; Richardson, 2005). It fails to take into account cognitive and resource limits, behavioral biases, ambiguity, variable preferences and norms, and the distribution of decision making among actors (Kørnøv and Thissen, 2000). Sometimes it is used to rationalize and legitimize political decisions (Nilsson and Dalkmann, 2010). Instead of grappling with these issues, it simply assumes implementation. Consequently, it fails to consider and address bureaucratic, political, and structural implementation obstacles and opportunities (Alexander, 1986; Thérivel, 2010).

The rational process is *incomplete*. It lacks social and environmental content or substance (Boyer, 1983). It seeks to attain goals (as a generic concept) but it is not driven, guided, and bounded by specific tangible social and environmental values, preferences, ethical principles, and imperatives (Beauregard, 1987; Nilsson and Dalkmann, 2010). Although it is inherently value-laden, it masks value-based decisions by portraying decision making as value free and objective (Nilsson and Dalkmann, 2010). It is conducive to systematic analysis but lacks a holistic perspective (Mintzberg, 1994). The image of people as rational decision makers is especially constraining. No provision is made for the contributions of extrarational insights, knowledge, experiences, wisdom, and methods (Alexander, 2000; Friedmann, 1987; Healey, 1997). The implications for process design and management of the subjective, social, and political nature of decision making are not addressed (Saul, 1992; Webber, 1983). The likelihood that perspectives and interests will clash is not considered (Boyer, 1983). The net result is a highly circumscribed and artificial view of people, how they reason, how they interact, and how, collectively, they reach and implement decisions.

The rational process is *abstract*. No effort is made to fit the process to the context. Consequently, process and context are often poorly matched. The rational process is especially *inappropriate* in situations characterized by high levels of complexity, uncertainty and conflict (Briassoulis, 1989; Damasio, 1994; Healey, 1997). The rational process and the technical and scientific "experts" who support the process are presumed to be objective, unbiased, and value-free. In truth, both the experts and the process are prone to numerous, often hidden, biases (Boyer, 1983; Mintzberg, 1994). "Objective" technical and scientific knowledge and methods are valued over subjective knowledge (Poulton, 1990). Analysis is favored over synthesis. Efficiency takes precedence over effectiveness. The process is more mechanistic than humanistic or ecocentric. Experts are the primary custodians of knowledge. Professional "mystifications" and rationalizations can inhibit public understanding and involvement (Forester, 1989; Saul, 1992). They also can lead to contempt by the specialists for the people (Saul, 1992). Often "depoliticized," expert-driven processes become autocratic. They can mask political purposes. They tend to reinforce the existing distribution of power (Benveniste, 1989). Sometimes they compound existing or even create new inequities. They are of little value in identifying or resolving fundamental value disputes (Campbell, 2003). The rational process, according to many of its critics, has a propensity to be highly undemocratic.

The foregoing ascribed strengths and limitations are not necessarily inherent to the rational process. They could simply be tendencies. Positive tendencies can be reinforced. Negative tendencies can be offset. Still, it is prudent to take these tendencies into account when designing and managing IA processes with rational elements.

5.4.4 The Response

Several responses to the identified shortcomings seek to make the rational process more realistic and effective. Incrementalism advocates a bounded or limited rational process where satisfactory (rather than ideal) decisions are made in a continuous, sequential, informal, and interactive bargaining process in a highly constrained and uncertain environment (Braybrooke and Lindblom, 1963; Feldman and Khademian, 2008; Lindblom, 1965). A few alternatives are sequentially assessed based on the test of agreement. Bounded rationality entails a rational decision-making process within the boundaries of the limited capacity of people to be value free and objective (Nilsson and Dalkmann, 2010; Simon, 1957). People satisfice rather than maximize (Simon, 1976). Mixed scanning envisions a two-tier planning process with incremental problem solving at the operational level and strategic level policy making to address major changes and issues (Etzioni, 1967, 1986; Nilsson and Dalkmann, 2010).

Effective planning concentrates on building and applying practical political skills to facilitate implementation and to manage uncertainties (Benveniste, 1989). Strategic planning is selective and issue, action, and implementation oriented. It is bounded within real decisions (Nilsson and Dalkmann, 2010). It systematically scans external and internal environmental conditions to maximize opportunities and minimize threats (Mintzberg, 1994). Contingency planning seeks to match procedural characteristics and environmental conditions (Alexander, 1986). The organizational behavior model focuses on understanding and progressively reforming organizational and institutional behavioral patterns (Feldman and Khademian, 2008). Theory-in-action and reflection-in-action explore how practitioners pragmatically design, reflect on, reframe, and implement policies in practice (Schön and Rein, 1994; Schön, 1983). Strategic choice involves a collaborative, highly iterative problem-structuring process that continuously manages uncertainties (Friend and Hickling, 1997).

Other responses make the rational process more deliberative, substantive, collaborative, and democratic (Morgan, 2012). Advocacy planning, extending from the legal model (legal rationality), focuses on the needs of the poor in a pluralistic society (Davidoff, 1965). Social learning and related organizational development and societal guidance concepts offer more humanistic, organic, interactive, and adaptive planning and organizational models (Friedmann, 1987). Critical planning and related concepts such as social justice, social mobilization, equity planning, progressive planning, radical planning, and structural planning seek to identify and redress social injustices and power inequities (Forester, 1989; Friedmann, 1987; Harper and Stein, 1992; Morgan, 2012; Rawls, 2001). Substantive planning processes attempt to realize and operate within tangible humanistic, ecological, communitarian, and sustainability principles, limits, and imperatives (Beatley, 1995; Etzioni, 1995; Friedmann, 1987). Communicative and collaborative approaches integrate reasoned, ethical, and practical discourse and argumentation into interactive and value-full collaborative forums (Goldstein, 1984; Forester, 1999; Healey, 1997). They also minimize communication distortions; facilitate participation, consensus building and conflict resolution; and justify moral norms (Campbell, 2003; Habermas, 1993; Innes, 1995; Sager, 1994).

The debate surrounding the rational process has cycled through multiple iterations. No consensus has emerged nor is likely to given the clash of perspectives and interests. Many of these perspectives are integrated into the rational IA process presented in this chapter. Approaches that cannot be fully incorporated into a rational IA process are integrated into other IA processes presented in subsequent chapters.

5.4.5 IA and Rationality

Some IA literature, especially the sources that advocate the wider application of scientific and technical methods, are either oblivious to the debates surrounding rationality or come down firmly in the technical analytical camp (Canter, 1996; Gilpin, 1995; Morris and Thérivel, 2009). They tend to maintain that the process should be comprehensive, scientific, rational, and objective. They generally focus on the appropriate application of technical, often quantitative, methods by specialists (Geneletti, 2005; Julien, 1995).

Many IA process characterizations (as described in Chapter 2) truncate, perhaps not consciously, the rational process by moving directly to criteria application to a proposed action and to "reasonable" alternatives. This tends to occur because the process is triggered only after a proposal is well defined. Project-level EIA, in particular, tends to be reactive, focusing on the negative, rather than on consistency with values (i.e., a value-rational view) and on the achievement of aspirational objectives (Hansen and Kørnøv, 2010; Pope et al., 2005). Some SEA forms, especially objectives-led integrated assessment, are less prone to this flaw. Greater attention could be devoted to problem structuring, to formulating goals and objectives (substantive rationality), to formulating alternative goals and objectives (purposive or value rationality), and to procedures for generating alternatives. SEA process depictions give more attention to "front-end" activities such as problem definition, goal setting, and alternatives formulation.

IA has partially benefited from rationality positive features and avoided some rationality limitations. IA, in common with social and ecological rationality, is driven and shaped by an environmental and social ethic (Bartlett, 1997; Craik, 2008). Process and substance are married, increasingly in an effort to further both sustainability and social/ecological justice (Doberstein, 2004; Sadler, 1996). Frameworks have been developed and applied for linking and integrating substantive IA concerns (Hacking and Guthrie, 2008). IA operates within limits (scoping), is focused (reasonable alternatives that satisfy need, significant effects), appreciates the needs for synthesis (cumulative effects), adaptively manages risks and uncertainties, and extends beyond decision making (monitoring and auditing) (Barrow, 1997; Craik, 2008; Glasson et al., 1999; Holling, 1978; Schmidt, 2011a). IA requirements and process descriptions demonstrate that it is impractical to identify all alternatives, to select the "best" alternative and to assess all consequences (Bartlett, 1997; Culhane et al., 1987; Kørnøv and Thissen, 2000). They appreciate the potential contributions of the extrarational (traditional knowledge, for example). They also recognize the value-full, social, collaborative, and political nature of the IA process (Interorganizational Committee, 1994).

SEA has tended to move the farthest from the rational–technical ideal (i.e., from positivism to postpositivism) (Bond and Pope, 2012). Early SEA procedural characterizations often mirrored rational-positivistic project-level EIA processes (Tetlow and Hanusch, 2012). Over time, the emphasis on consequences and environmental impacts has been replaced with vaguer concepts such as environmental issues and aspects. Instead of simply informing decision making with technical–rational information, SEA now seeks to broaden (to redefine problems and to include environmental aspects), and to directly interact with, and influence decision making (i.e., SEA as a social struggle over problem definition and future choices) (Jiliberto, 2011; McCluskey and João, 2011). The postpositive SEA perspective presumes many actors with multiple, often conflicting, goals (Adelle and Weiland, 2012). SEA has become increasingly more strategic (more policy based and plan-shaping), integrative (with planning and decision making), proactive, decision-centered, continuous, adaptive, collaborative, learning oriented, creative, and overtly political (Lobos and Partidário, 2010; McCluskey and João, 2011; Partidário, 2007; Tetlow and Hanusch, 2012). SEA purposes (especially sustainability); contextual adaptations; the role of participation, collaboration, and negotiation activities (especially during scoping); and the facilitation of social learning and institutional reform are now stressed to a greater degree than prediction and evaluation techniques (Bina, 2007; Tetlow and Hanusch, 2012). At the same time, as these orientation changes, there continue to be calls for more technical-quantitative (often using multicriteria analysis) analyses for assessing and ranking alternatives and for conducting cumulative effects assessment (Geneletti, 2005). These changes, at the conceptual level, are more evident at the policy level. At planning and program levels, SEA practice tends to be strongly influenced by the rational EIA model (Lobos and Partidário, 2010).

Some rationality strengths are evident in rationality-based IA practice. Rational approaches are especially effective during information gathering and analysis and in structuring follow-up (Morrison-Saunders and Sadler, 2010; Noble and Birk, 2011). Systematic rational procedures can help to transparently compare alternatives (Geneletti, 2005). They can, when they take the form of a joint and inclusive reasoning process that systematically explores both ends and means, ameliorate and transcend value-based political debates (Lemonick, 2010). Such approaches can effectively integrate varying stakeholder values, perspectives, and preferences (Hansen and Kørnøv, 2010). Different evaluation parameters and methods, uncertainty factors, and procedures for systematically exploring indirect and cumulative effects can be included in such analyses (Geneletti, 2005). Rational IA approaches have been credited with enhancing project design, informing decision making, and contributing to institutional involvement and public involvement (Jay et al., 2007; Noble, 2009b). The rational approach still reflects much of IA practice (Noble, 2009b).

Some rationality limitations also are evident in rationality-based IA practice. Rational IA approaches tend to be at odds with the realities of decision making (Jay et al., 2007). Technical–rational IA approaches have little to offer when decisions involve value-based trade-offs and conflicts among competing interests (Craik, 2008; Richardson, 2005). They fail to consider social, political, or cultural ambiguities and rarely lead to organizational behavior improvements (Ebrahim, 2008). The reductionist, analytic nature of rational IA practice tends to inhibit the consideration of cumulative and transboundary effects (Devlin and Yap, 2008). Rational IA approaches tend to inhibit two-way communications and dialogue-based participation (Wikland, 2005).

Many rationality debates are mirrored in IA literature (Morgan, 2012). There are lively discussions concerning whether SIA should be technical, political, or collaborative (Bartlett, 1997; Craig, 1990; Greer-Wooten, 1997; Lockie, 2001; Wikland, 2005). Rational–technical and adaptive/ecological approaches are compared and contrasted (Boothroyd and Rees, 1985). There are debates regarding whether reason and rationality should guide the process or be replaced by a more transformative approach directed toward raising political awareness and changing values, world views, and institutional behavior (Bina et al., 2011; Torgerson, 1981). There are discussions regarding whether IA practice should be apolitical or should allow political considerations to enter into decision making from the outset (Craik, 2008; Richardson, 2005). The validity of emotions and experiences as a decision-making basis is raised as an issue. There are discussions regarding whether IA should be comprehensive and rigorous or is necessarily practical and constrained (Kørnøv, 1998). Efforts have been made to integrate additional rationality types (e.g., aesthetic, ethical) into IA practice (Elling, 2007). SEA characterizations range from a close parallel to the rational process to processes that share many of the characteristics of strategic planning, the strategic choice method, mixed scanning and effective planning (Glasson et al., 1999; Partidário, 1996; Kørnøv and Thissen, 2000).

IA literature and practice could benefit from a closer scrutiny of the rationality debates in other fields (e.g., critical theory) (Elling, 2007; Richardson, 2005; Weston, 2010; Wikland, 2005). Particular consideration could be given to the potential benefits associated with a broader definition of rationality (see Table 5.1). IA practitioners could assess whether the assumptions and limitations ascribed to rational processes apply to current and proposed IA processes. The systematic integration of different rationality forms into the IA process could help guard against some of the excesses of technical–analytical rationality. Practitioners and other process participants could reflect on how they individually and collectively apply reason to build theory in practice. The efforts to foster reasoned, practical, and ethical discourse and discourse reflection could be especially appropriate for collaborative IA processes (Runhaar et al., 2010). Practitioners could consider the role of the extrarational in planning and decision making. They could seek a better match between process and context by appreciating the contingent nature of rationality. SEA practitioners could learn from the experiences of strategic planning and the strategic choice method.

The preceding overview suggests that the uncritical application of a rational IA process is bound to be problematic. Some measure of reform in the marriage of IA and rationality is essential. What remains to be determined is the nature of the reforms and whether those reforms should involve modest adjustments or a more fundamental reformulation. Figure 5.3 identifies some of the major perspectives regarding both more modest and more fundamental reforms to the rational IA process model. At the modest change end of the spectrum, the rational IA process is sufficiently flexible that it can be made more practical, participative, substantive, influential, value-full, and integrative. It also can be adjusted to be less absolute and certain. At the major change end of the spectrum, technical–rational analysis is shifted "to the sidelines," assuming, at best, a support role. Reason is fundamentally redefined and reoriented to be compatible with and supportive of various alternative perspectives regarding the rationale for and nature of IA-related decision making. Arguably, an IA process is "reasonable" to the extent that it, for example, facilitates fundamental political/economic reform, contributes to an environmental transformation, influences and shapes decision making, is context-based, embraces and embeds a precautionary perspective, is shaped by holistic perspectives rather than analytic methods, and contributes to a collaborative, "bottom-up" planning approach. Chapters 6–11 address many of these themes in greater detail.

As illustrated conceptually in Figure 5.3, there is some potential for combining elements of both modest and major reforms. The outcome from these debates, in terms of the nature and role of IA practice at both the regulatory and applied levels, is far from clear. Based on historical experience, rationality will continue to assume a prominent role in IA practice, especially at the project level, albeit partially reformed. The extent to which the conceptual debates regarding reason and rationality will percolate down to the applied level is yet to be determined.

5.5 INSTITUTING A RATIONAL IA PROCESS

5.5.1 Management at the Regulatory Level

The four jurisdictions all include elements of rational planning in their IA requirements and guidelines, as highlighted in Table 5.3. An overview of the four jurisdictional approaches to the treatment of alternatives points to the value of a clear set of substantive and procedural policies, objectives, and principles, embedded in the IA legislation, to serve as a basis for identifying, screening, and comparing alternatives. It illustrates the need for clear requirements and guidelines concerning the treatment of purpose and need, the range and type of alternatives to consider for various classes of proposed actions and settings (refined through proposal-specific guidance), the criteria to employ when screening alternatives, and the range of factors to consider when comparing alternatives. It underscores the need to explain and substantiate the methods used to evaluate alternatives, the need to integrate mitigation into the alternatives analysis, and the critical role that agencies and the public should assume in identifying, screening, and comparing alternatives. It demonstrates the importance of substantiating the basis for decisions related to alternatives, and the need to extend the assessment of alternatives into impact management activities.

Experience in the four jurisdictions illustrates the importance of ensuring that the evaluation and the preferred alternative(s) are consistent with the realization of the IA legislation objectives and principles, are supportive of other substantive environmental requirements and policies, and provide for the tiering of IA levels (such that strategic options are fully considered and frame the consideration of project-level options). It demonstrates the need to fully consider the implications of uncertainties, fully assess distributional differences among alternatives (over time, over space, among population groups—especially the most vulnerable), and require the consideration of an environmentally preferred alternative. It points to the need to fully explore alternatives when environmental sensitivity, risks to human health, the potential for catastrophic consequences, resource depletion, severe climate change impacts, and other environmental policy priorities are significant issues. It demonstrates the importance of requiring and facilitating public and agency involvement in option evaluation, and of clearly documenting the contribution and roles of agencies and the public in the evaluation process.

The approaches adopted by the four jurisdictions to address alternatives also illustrate a range of potential pitfalls in the treatment of alternatives. Examples include the problems that arise when the proponent alone determines the alternatives, when the public and agencies are excluded from

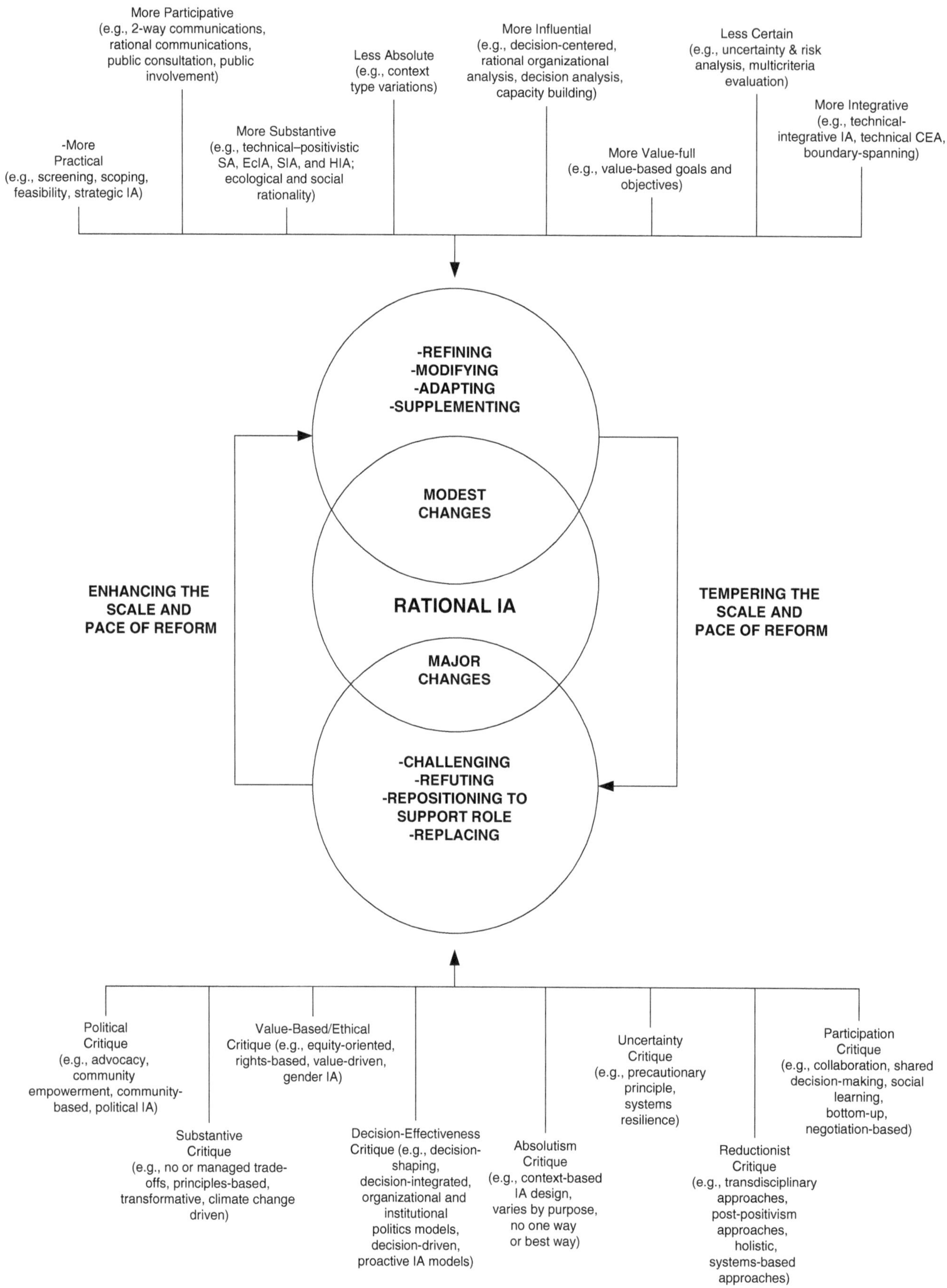

Figure 5.3 Rationalism in IA—choices for reform.

Table 5.3 Positive and Negative Regulatory Level IA Examples Regarding Rationality

United States	Canada	Europe	Australia
(+) NEPA refers to overall environmental purpose and national environmental policy (+) Required to address purpose (condition change desired) and need (if purpose met, existing conditions that call for improvement) (+) The purpose and need should not inappropriately narrow the range of reasonable alternatives (+) CEQ guidance on purpose and need and on selection and evaluation of alternatives, in various sectors and for various proposal types (+) All reasonable alternatives must be examined for EAs (at a lower level of detail) and EISs, including the no-action alternative (mandatory for EISs), the agency's preferred alternative, the environmentally preferred alternative, and all reasonable alternatives (including those not within the jurisdiction of the lead agency, which would avoid and minimize adverse environmental effects) (+) A particular need to assess reasonable alternatives is indicated when there are resource conflicts (+) The no-action alternative is generally viewed as current management practices and includes the actions of others (+) The environmentally preferred alternative is expected to promote NEPA's national environmental policy (+) Reference is made to the careful consideration of alternatives requested by groups and the public (+) Particular need to explain rationale when there is only one reasonable alternative (+) References made to a broad array of possible alternatives (e.g., alternative locations, alternative technologies, alternative transportation means, environmental release choices, cost-effective waste minimization and pollution prevention activities, economically beneficial landscape practices)	(+) Identifies substantive purposes for Act (e.g., sustainability, encourage study of cumulative effects, precautionary approach) (+) Identifies procedural purposes for Act (e.g., cooperation and communications with other governments and with aboriginal peoples) (+) CEA Agency objectives include promoting EA in a manner consistent with purposes of Act and promoting and monitoring EA quality (+) History of panel use of purposes of Act to evaluate proposed undertakings (e. g., sustainability test) (+) Factors to consider include purpose of designated project and alternative means of carrying out project (technically and economically feasible) (CEAA 1998a) (±) Guidance for alternatives analysis; question of continuing validity under new Act (+) Enhanced monitoring requirements could facilitate consideration of management options during implementation (+) SEA requirements—evaluation of alternatives and principal environmental considerations should be fully integrated into option analysis (−) New Act eliminates reference to alternatives to proposed action (−) Narrow, selective, and discretionary definition of environment and effects likely to inhibit systematic generation and evaluation of alternatives	(+) Proposed EIA Project Directive (PPD)—required to assess reasonable alternatives; includes baseline scenario (likely evolution of existing state of the environment) (+) PPD—views assessment of reasonable alternatives as a means of improving process quality and of allowing integration of environmental considerations early in project design (+) PPD—required to consider alternatives to the proposed project and extent considered (+) PPD—required to justify choice of preferred alternative (+) PPD—scoping requirements provide for agency and public input before decisions made on preferred alternative (+) PPD—cross-references to substantive environmental requirements (e.g., biodiversity, climate change, disaster risks) facilitate consideration in analysis of alternatives (+) PPD—broad definition of effects and environmental sensitivity facilitates alternatives evaluation (+) PPD—mandatory monitoring could facilitate consideration of management options during implementation (+) SEA Directive—refers to main objective and relationship to other plans and programs; can be applied in alternatives analysis	(+) Legislation includes specific environmental, social, sustainability, and cultural objectives and ecologically sustainable development (ESD) principles (+) Detailed requirements and guidance focused on matters of national environmental significance (+) Project-specific guidelines refer to compliance with objectives of legislation and with ESD principles (+) Requires the identification of alternatives, the assessment of impacts, the determination of mitigation measures and a clear basis for the preferred alternative (+) Reference (draft PERs and EISs) to consequences of not proceeding, complete description of impacts, and sufficient clarity regarding basis for preferred alternative; more detail in project-specific guidelines (+) Project guidelines refer to comparing short, medium and long-term advantages and disadvantages, reasons for preference and compliance with legislative objectives and ESD principles (±) Reference to alternative locations and feasible alternatives to the proposed action considered but not proposed (including no action if relevant); leaves choice of alternatives at discretion of proponent and provides minimal guidance

(continued)

Table 5.3 (*Continued*)

United States	Canada	Europe	Australia
(+) Requirements to rigorously explore and objectively evaluate all reasonable alternatives—required to specify how it will achieve NEPA's goals, consider mitigation, provide a clear rationale for choice, use analytic rather than encyclopedic approach (in comparative form), include relevant information (including data gap implications), ensure scientific and professional integrity when selecting methods, provide substantial treatment to alternatives considered in detail, include unavoidable adverse impacts and irreversible and unavoidable resource commitments, and avoid post hoc rationalizations (±) Not necessary to develop alternatives when no significant impacts anticipated; could be problematic if basis for significance determination flawed (±) Screening role of limitation to reasonable, technically, and economically practical and feasible alternatives; reasons for alternatives screened and basis for choice required; issue of how systematically undertaken and substantiated (−) There is no requirement to specify project goals or objectives (−) The NEPA national environmental policy and objectives are very general (−) Compliance with NEPA in practice has been predominantly procedural	(−) Issue of whether EAs undertaken under substitution/equivalency provisions will systematically generate and evaluate alternatives (−) Closed evaluation of alternatives; decisions regarding preferred alternative already made by the time that project is designated (−) Limitation of application of Act to major projects excludes consideration of strategic alternatives and means no IA requirements related to consideration of alternatives for moderate and small projects	(+) SEA Directive—required to consider reasonable alternatives and document in environmental report reasons for selecting preferred alternative and how assessment is undertaken (−) PPD—reference to consideration of reasonable alternatives early in project design suggests that project is predetermined and choices only relate to implementation choices (−) PPD—general nature of purposes of legislation inhibits systematic test of and extent to which alternatives contribute to the realization of substantive environmental goals (−) PPD—focus on major projects could inhibit systematic consideration of alternatives for small projects (±) SEA Directive—no-action alternative to be considered, where appropriate	(±) Recent review recommends that Minister be given power to request information on alternatives; government agreed to change and indicated would encourage early consideration of options and produce guidelines on prudent and feasible alternatives; encouraging is not the same as requiring (+) Accepted recommendation; Minister to provide reasons for all significant decisions (e.g., approval or not)

the process until after the alternatives evaluation has been completed, if the substantive mandate of IA legislation is too weak or general to serve as a test of proposal and option acceptability, and if adequate links are not made to related substantive environmental requirements. Other potential pitfalls include not requiring the assessment of the no-action and strategic alternatives (facilitated through tiering); unduly restricting the range of potential alternatives (e.g., alternative means only, only alternatives within the agency's jurisdiction); leaving open the possibility that the basis for screening alternatives is so general that options can be summarily dismissed with a negligible rationale; not insisting that proposals and options that run counter to environmental policies, standards, and limits be screened out; artificially constraining the comparison of alternatives by narrow definitions of the environment and effects; not requiring that the methods used to screen and compare alternatives are transparent, substantiated, systematic, and inclusive; not requiring that the implications of mitigation and uncertainties for the alternatives analysis be considered; not extending the alternatives analysis into the follow-up phase; and not requiring substantiation for political approval steps.

In terms of future directions, IA requirements and guidelines, drawing upon the rationality debates, could seek to foster the conditions necessary for reasoned, ethical, and practical procedures for formulating and evaluating alternatives. More attention could be devoted to problem and goal definition activities (value and substantive rationality), to matching process and context (contingency planning), and to facilitating and accommodating ecological, social, political, and communicative forms of rationality. Allowance should be made for alternative goals and objectives when there are value conflicts. A range of specific environmental goals and principles within IA legislation can provide a consistent litmus test for proposal acceptability, providing the loop is closed by stipulating that each alternative must be assessed against the goals and comply with the principles. Such provisions also could facilitate alternatives comparison. Care should be taken not to prescribe particular evaluation methods at the regulatory levels (e.g., the propensity to insist that only quantitative evaluation procedures are appropriate). It is, however, reasonable to emphasize that the evaluation be unbiased, include performance criteria that evaluation methods must satisfy, and provide guidance regarding good practice standards for generating, screening, and comparing alternatives (Lawrence, 1993). IA requirements could be reviewed to determine if they exhibit the ascribed rational process assumptions and could potentially be conducive (however unwittingly) to the ascribed rational process negative tendencies. Requirements should be sufficiently flexible to retain and foster reasoning in IA processes without precluding the potential contributions of variations of and alternatives to rationality.

5.5.2 Management at the Applied Level

Figure 5.4 is an example of a rational IA process. Figure 5.4 and the process description that follows integrates and is built upon suggested rational process elements as advanced in IA and related literature. IA process managers and participants can "pick and choose" the relevant and appropriate elements.

Start-up The process begins by characterizing the problems and opportunities and analyzing the need to solve the problems and take advantage of the opportunities. These analyses make it possible to identify the purpose for the process and the purpose for any proposed actions. A study design is then prepared to describe how the purpose is to be realized and the needs met. Management and specialist teams are assembled to fulfill the study requirements. Broad goals, objectives, principles, and priorities are formulated to guide the process. They also provide a preliminary vision of conditions, as they might exist, if the problems are solved and the opportunities met. The goals flow directly from the problems and opportunities. The objectives refine the goals. The principles are broad performance standards. The priorities are system characteristics most directly and immediately relevant to problem resolution and opportunity realization. The goals and objectives are specific and substantive. They address such concerns as sustainability, environmental quality, social and environmental justice, biodiversity, heritage, resource conservation, and energy efficiency.

A public scoping program, a form of practical rationality, focuses the process on key public and agency concerns and issues, major stakeholders, potentially significant impacts, shared interests, potential perspective, value and interest differences, and likely alternatives. Alternative goals, objectives, and principles are formulated, where necessary, to address major perspective, value and interest differences (i.e., purposive or value rationality). A preliminary list of potentially applicable methods is compiled. These methods could support such activities as alternatives formulation and evaluation, data collection and analysis, impact prediction and interpretation, and public communications and participation. Both quantitative and qualitative methods could be identified. An environmental overview ensures that the goals, methods, and scoping activities are relevant and appropriate to the situation (i.e., the matching of process and context). The basic characteristics of the proposed action are identified. The proposed action is shaped to meet the need, fulfill the goals and objectives and be consistent with the principles and priorities. Temporal, spatial, and jurisdictional boundaries for the IA are established. Consistent with strategic rationality, the external and internal environments are scanned to determine constraints, opportunities, and limits.

Alternatives to the Proposed Action The major proposed action characteristics are described. Potential impact sources, stemming from the action, are identified. Aspects of the action, where choices exist, are highlighted. Potentially reasonable alternative ways of meeting the objectives and satisfying the principles are identified. If, for example, the need is transportation related, consideration might be given to such alternatives as no change, deferment, land-use planning changes, demand control procedures, growth management, and alternative modes. Clearly defined and fully substantiated exclusionary criteria are formulated to ensure a consistent basis for rejecting unacceptable alternatives. A screening procedure is formulated. Data are collected and complied to support the screening analysis. The exclusionary criteria are applied to the alternatives to the proposal. Alternatives are only rejected if they clearly meet or exceed exclusionary criteria, taking into account mitigation potential and uncertainties.

The alternatives to the proposed action, remaining after the screening analysis, are compared. This analysis is undertaken at a broad level of detail consistent with the diverse nature of the alternatives. Supplementary data collection occurs to support the comparative analysis. Where practical, impacts are scaled. This ensures a consistent approach to impact magnitude. Comparative evaluation methods are formulated appropriate to the available data, the level of detail, and the nature of the alternatives. Comparative evaluation criteria are formulated. Care is taken to ensure that the criteria make it possible to address whether and the extent to which the alternatives contribute to the achievement of the goals and objectives and are consistent with the principles and priorities. Objectives and criteria are ranked and, where necessary to support the evaluation methods, weighted. Alternative rankings address value, interest, and objective variations. Both qualitative and quantitative evaluation methods are employed to combine the criteria rankings and the scaled impact data. Multiple sensitivity analyses address areas of uncertainty, the implications of mitigation and enhancement measures, and variations in criteria rankings. Carefully reasoned arguments justify the selected

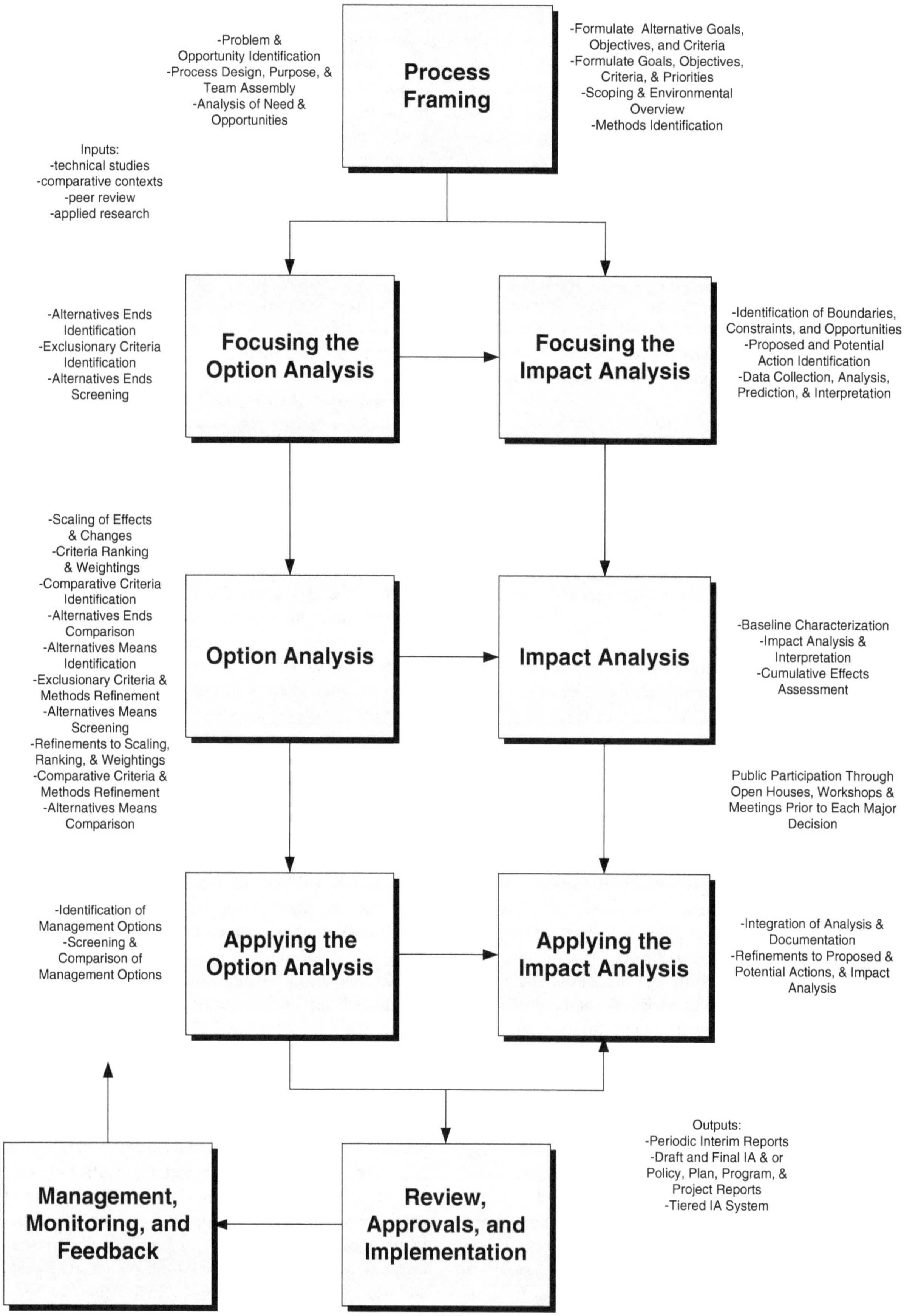

Figure 5.4 Example of a rational IA process. Adapted from Lawrence (2005a).

alternative. The implications of any residual uncertainties are explored and explained.

A further data collection and analysis round provides the basis for identifying, analyzing, predicting, and interpreting impacts potentially associated with the selected alternatives. Both technical and nontechnical forms of knowledge and experience are considered. Baseline environmental conditions are characterized. Likely future environmental conditions, assuming no proposed actions, are predicted. Sensitive and significant social and ecological components, processes, interactions, and systems are identified. The characteristics of the proposed action are refined. Potential individual impacts are identified and then predicted. Criteria are applied to ensure a consistent approach to characterizing the magnitude and distribution (over time, over space, and among social groups) of potential impacts. Quantitative (e.g., quantitative models), semiquantitative, and qualitative methods (e.g., conceptual models, scenarios) characterize baseline conditions and potential future impacts. Impact predictions are refined after incorporating mitigation and enhancement measures into proposed action characteristics. Key environmental interrelationships and patterns of direct and indirect impacts are determined. Uncertainties associated with both baseline conditions and predicted impacts are identified and their implications explored. Criteria are applied to ensure the consistent treatment of impact significance. The impact significance criteria take into account such matters as impact magnitude, impact distribution, public and agency priorities, mitigation potential, and the levels and types of risks and uncertainties. A thorough rationale is prepared for all findings, interpretations, and conclusions.

Alternative Means of Carrying Out the Proposed Actions The "alternative means analysis" employs the same steps and methods as the alternatives to analysis but at a greater level of detail. If the preferred alternative, emerging from the alternatives to analysis, for example, is a highway system, alternative means could include route alternatives, alternative vertical and horizontal alignments, alternative intersection locations, and alternative intersection designs. Supplementary data collection and analysis are undertaken during this stage to support the cumulative effects analyses. Refinements are first made to the individual impact analysis to take into account changes to the proposed action characteristics. The cumulative effects analysis then addresses multiple additive and nonadditive effects on individual environmental components and on broader social, economic, and ecological systems. The effects result from the proposed action in conjunction with other historical, current and likely future actions and activities. The cumulative effects analysis focuses on such concerns as temporal and spatial crowding and discontinuities; indirect, growth inducing, and threshold effects; biomagnification; and feedback effects. As with the individual impact analysis, effects are quantified to the extent practical. Explicit criteria are applied to ensure the consistent treatment of impact magnitude and significance. Areas of uncertainty and related implications are highlighted. Adjustments are made to proposed action characteristics to avoid and minimize potentially significant cumulative effects and uncertainties. Actions with other parties are coordinated. A clear rationale is provided for all findings, interpretations, and conclusions.

Approvals and Postapprovals A management program is prepared once the proposed action characteristics are largely determined. The management program integrates and coordinates mitigation, enhancement, monitoring, feedback, contingency, and auditing procedures and methods. There are likely to be some options concerning management program elements. Management options are screened and compared using methods comparable to those associated with the alternatives analyses. Responsibilities and commitments are detailed. Some fine-tuning of the proposed actions occurs to further enhance benefits and to minimize adverse effects.

The analyses are integrated, refined, and consolidated into draft and final IA documents. Points of confusion are clarified. Concerns and objections are thoroughly considered and addressed. Further refinements are made to the proposed actions based on inputs received during review and approval and (if the proposed action is approved) as a result of monitoring and feedback. Obstacles to implementation are anticipated. Once and if approval occurs, a systematic effort is made to facilitate implementation. Methodological improvements are made for subsequent application based on the auditing of the IA process.

Inputs, Outputs, and Interactions The IA process is supported by technical studies, reviews of comparable proposals and environments, peer reviews, and applied research. The public is involved in identifying concerns and preferences; in reviewing analyses and preliminary findings and conclusions; in identifying alternatives; in expressing opinions regarding criteria rankings, acceptable and preferred alternatives, significant impacts, and conclusions; and in responding to interim and draft documents. The public participates through open houses, workshops, and meetings prior to major decisions. The communications and consultation methods are jointly formulated and adapted with interested and affected parties. A proactive effort (e.g., participant funding, additional resources) is made to involve groups and organizations less likely or able to participate in the process. Care is taken to minimize communications and involvement distortions and inequities, consistent with communicative rationality principles. Close and frequent contact is maintained with regulatory review agencies.

Periodic interim reports are released as the process unfolds. Draft and final IA summary and detailed reports are broadly circulated. All documents are designed to be lucid, unbiased, traceable, technically sound, scientifically rigorous, accurate, and consistent. Inputs received from agencies and the public are recorded and addressed. Changes made to documents as a result of inputs received are clearly specified. A clear rationale is provided for suggested changes not made.

5.5.3 Applied Rationality Good Practices by IA Type

Section 5.5.2 presents a generic description of a rational IA process. However, a rational IA process, at a strategic level, as compared to a project EIA level, and a rational IA process for various substantive IA types (e.g., EcIA, SIA, HIA, SA) can be quite different. It, therefore, follows that rationality good practices also would vary by level and IA type. Table 5.4 summarizes suggested rationality regulatory and applied good practices by decision-making level and type. The general rationality good practices exhibit a broad definition of ends, alternatives and effects, a close connection to decision making, a high degree of stakeholder involvement, the

Table 5.4 Rational IA Practice Characteristics by IA Type

Rational SA Practice	Rational SEA Practice	Rational EIA Practice
Includes sustainability as basic purpose of IA legislation; provides guidance for integrating sustainability concerns in IA purpose and need Requires that sustainability concerns be explicitly addressed in procedures for identifying alternatives Requires the identification of sustainable alternatives Provides guidance for generating environmentally sustainable alternatives Requires explicit criteria, procedures, and reasons for screening out unsustainable alternatives Requires that evaluation procedures for comparing alternatives include sustainability criteria Requires that reasons for selecting preferred alternative(s) take into account sustainability criteria Provides guidance for integrating substantive environmental concerns, including sustainability, into alternatives analysis Clearly defines sustainability meaning (preferably strong sustainability) Defines need and purpose in light of explicit sustainability objectives, principles, criteria, and indicators/targets Addresses interactions among objectives Considers systems as a whole; integrates economic, environmental, and social aspects; requires multi to inter to transdisciplinary approach Engages community in developing sustainability vision Challenges entrenched policy beliefs and traditions Seeks to build desirable and resilient future Includes global, regional, and local sustainability Seeks creative opportunities and positive contributions Seeks to maximize environmental, social, and economic objectives Seeks mutually reinforcing gains and greatest overall benefit rather than balancing	Requires identification of purpose, need, and opportunities to be served by proposed policy, plan, or program Requires identification of objectives for proposed policy, plan, or program Requires identification of links to other policies, plans, and programs Provides guidance for addressing purpose and need for policies, plans, and program Requires the consideration of alternatives including the "no change" alternative Requires that alternatives considered include all reasonable means for achieving purpose and need for proposed policy, plan, or program Identifies alternatives or alternative types that must be considered for specific policy, plan, or program types Provides guidance for generating alternatives for policies, plans, and programs (e.g., typical sector and plan type alternatives) Requires explicit reasons for policy, plan, and program choices rejected and preferred Requires consistent and explicit evaluation criteria and procedures for screening and comparing policy, plan, and program choices Requires that alternatives analysis addresses differences in contribution to achievement of environmental (broadly defined) and sustainability objectives Requires that alternatives analysis address differences in potentially significant positive and negative environmental consequences including cumulative effects Requires the integration of mitigation into the alternatives analysis when there is potential for significant adverse effects Provides guidance for screening and comparing alternatives for various policy, plan, and program types Treats role of SEA as mainstreaming environmental considerations into decision making Emphasizes SEA role in facilitating innovative and creative strategic decision making	Requires consideration of need for undertaking (public interest) Requires explicit identification of purpose and objectives for proposed undertaking Requires identification of links to other projects Provides guidance for addressing purpose and need for project-level EIAs Requires the consideration of alternatives to the proposed action for major or complex projects or projects with potentially significant adverse effects, including the "no-action" alternative, alternative locations, and variations in the pace and scale of the proposed project Requires the consideration of alternative means for all proposed projects Requires that alternatives analysis address choices most likely to prevent and reduce the likelihood and severity of significant adverse individual and cumulative environmental effects Identifies alternatives or alternative types that must be considered for various classes of projects Provides guidance for generating alternatives (e.g., typical project type alternatives) Requires explicit reasons for project alternatives rejected and preferred Requires that alternatives analysis, including criteria, is consistent with definition of environment and effects in EIA legislation Requires evaluation of consistency with pertinent public policies, plans, programs, and standards Requires that alternatives comparison analysis explicitly consider mitigation and enhancement potential

Table 5.4 (*Continued*)

Rational SA Practice	Rational SEA Practice	Rational EIA Practice
Seeks sustainable outcomes and demonstrated influence on decision making not just sustainability-oriented processes Seeks resilience of socioecological systems Alternatives should be informed by sustainability issues and objectives and consistent with fundamental sustainability requirements Identifies and applies inviolate limits to identify and screen alternatives Protects the future; includes long-term choices Avoids undesirable trade-offs Favors options incorporating adaptive design Contribution to sustainability should be the main test Applies exclusionary sustainability criteria Seeks to make effective use of bottom-up integrative and transdisciplinary evaluation methods Applies explicit sustainability criteria in comparing alternatives; including inter/intragenerational equity Explicitly applies sustainability trade-off rules (e.g., net gains, avoidance of significant adverse effects, protection of future, explicit justification, open process) Factors in long time horizons Seeks multiple, mutually reinforcing, and lasting gains while avoiding significant adverse effects Provides compensation or offsets Interweaves ends and means Assesses overarching consistency of sector targets Employs frameworks and methods that demonstrate progress toward sustainability Employs methods that link and integrate substantive environmental concerns Treats trade-offs as last resort; identifies and applies trade-off rules Preferred alternative should be designed for resilience and adaptability in face of risks and uncertainties (e.g., precautionary approach)	Develops environmental objectives within the planning process Seeks to redefine problems in a manner that encompasses environmental aspects Explicitly identifies links between policy and SEA objectives Views SEA as a form of organizational and social learning Identifies a range of alternative strategic options for meeting the objectives Identifies options that meet general public concerns Draws upon community-based approaches Choice of alternatives takes into account decision context, decision tier, and stakeholder proposals Screens options that threaten critical, irreplaceable, and most vulnerable system elements Structures alternatives on basis of such distinctions as the testing of need/demand, how need to be met (model/process), and location (where) Systematically links objectives, targets, and indicators to environmental receptors Considers multiple temporal (e.g., generational, decisional) and spatial (e.g., global, macroregion, region, municipality, site) scales for various SEA types, and for project-level EIA Focuses on factors that could make a decision-making difference Employs resilience indicators and criteria; favors options that facilitate management resilience Integrates participatory and community-based approaches into SEA evaluation processes	Requires that alternatives analysis address differences in potentially significant negative environmental consequences, including cumulative effects Provides guidance for screening and comparing alternatives for various project types Clearly defines underlying goal or objective Ensures need is not so narrowly defined that excludes reasonable alternatives Identifies values and interests of key parties Identifies values in which IA performance should be anchored (needs) Includes pace and scale of alternatives Includes a range of reasonably foreseeable expansions and future developments Focuses on reasonable alternatives that best satisfy underlying need Seeks alternatives to problematic activities Evaluates options against needs and societal capacities Evaluates options taking into account distribution of effects and consistency with values and interests of key parties; favors most vulnerable Integrates consideration of indirect and cumulative effects Integrates community responses (e.g., social vitality, economic viability, political efficacy) Integrates mitigation potential into alternatives analysis; tracks mitigation implementation Assesses robustness of evaluation (e.g., sensitivity analyses) results; takes risk aversive decisions

Rational EcIA Practice	Rational SIA Practice	Rational HIA Practice
Includes objectives in IA legislation Provides guidance for integrating ecological environmental concerns into IA purpose and need Requires that ecological concerns be explicitly addressed in procedures for identifying alternatives Requires consideration of alternatives when there are potentially significant ecological effects or resource conflicts	Includes social objectives in IA legislation Provides guidance for integrating social concerns into IA purpose and need Requires that social concerns be explicitly addressed in procedures for identifying alternatives Requires the consideration of alternative when there are potentially significant social effects Requires the identification of the environmentally (including social) preferred alternative	Includes health objectives in IA legislation Provides guidance for integrating health concerns into IA purpose and need Requires that health concerns be explicitly addressed in procedures for identifying alternatives Requires the consideration of alternatives when there are potentially significant health effects

(*continued*)

Table 5.4 (*Continued*)

Rational EcIA Practice	Rational SIA Practice	Rational HIA Practice
Requires the identification of the environmentally (including ecological) preferred alternative Provides guidance for generating ecologically desirable alternatives Requires explicit criteria, procedures, and reasons for screening out unacceptable alternatives from an ecological perspective Requires that evaluation procedures for comparing alternatives include explicit ecological criteria Requires that reasons for selecting preferred alternative(s) take into account ecological criteria Provides guidance for integrating ecological environmental concerns, including ecological sustainability, into alternatives analysis Ensures consistency with ecological rationality principles Focuses on biodiversity conservation; builds in biodiversity objectives Integrates EcIA principles and good practices Includes principle of "no net biodiversity loss" Integrates international, national, regional, and local biodiversity priorities and targets Screens alternatives that result in irreversible biodiversity loss Seeks alternatives that minimize biodiversity loss Seeks alternatives that make positive contribution to biodiversity Assesses in terms of potential to contribute to biodiversity conservation and ecosystem services Assesses differences in biodiversity risks and opportunities Assesses differences in degree of threats to species and habitats Integrates uncertainty factors into comparison of biodiversity differences (e.g., sensitivity analyses)	Generates socially substantive alternatives Requires explicit criteria, procedures, and reasons for screening out socially unacceptable alternatives Requires that evaluation procedures for comparing alternatives include explicit social criteria Requires that reasons for selecting preferred alternative(s) take into account social criteria Provides guidance for integrating social environmental concerns, including social sustainability, into alternatives analysis Defines rationality to include organizational learning and ability to positively respond to conflict Designs to overcome tension between political and technical; ensures procedural justice Integrates SIA principles, good practices, and social performance standards Identifies social justice and community cohesion issues Includes human rights and free, prior, and informed consent principles Includes all values and interests of parties in goals Integrates social scientific and indigenous knowledge Identifies alternatives that go beyond preventing negative (e.g., building social capital, capacity building, good governance, community engagement, social inclusion) Links values and interests to proposed alternatives Public acceptance should be a strategic priority SIA and community outreach should contribute to the identification of alternatives Ensures environmental and social justice issues are fully described and analyzed Integrates social risk assessment Recognize critical role of perceived impacts Clearly identifies who wins and loses; emphasizes vulnerability of underrepresented and disadvantaged populations; and emphasizes enhancement of lives of vulnerable and disadvantaged Seeks consensus with community Seek socially just distributional outcomes Seeks to facilitate poverty alleviation through sustainable development	Requires the identification of the environmentally (including health) preferred alternative Provides guidance for generating preferred health alternatives Requires explicit criteria, procedures, and reasons for screening out unacceptable alternatives from a health perspective Requires that evaluation procedures for comparing alternatives include explicit health criteria Requires that reasons for selecting preferred alternative(s) take into account health criteria Provides guidance for integrating health concerns, including health sustainability, into alternatives analysis Recognizes that health-related policy making is generally incremental Explicitly identifies health-related values, objectives, and criteria Clearly and broadly defines health Fully engages health professionals and community Integrates HIA principles and good practices Fully engages health experts and community in alternatives identification Focuses on alternatives that enhance ability of decision makers to make precautionary decisions Gives greater emphasis to broadly defined, alternatives (e.g., knowledge, institutional, and goal alternatives) Fully and proactively integrates health concerns into alternatives evaluation Emphasizes decision-making effectiveness, equity and differences in health outcomes among alternatives and across populations Provides, in evaluation, for negotiations among interested parties Takes into account social and behavioral aspects of health

Sources: Becker et al. (2005), Binder et al. (2010), Bond (2010), Bond et al. (2012), Burdge (2004), Croal et al. (2010), Dalal-Clayton and Sadler (2004), Desmond (2007), Donnelly et al. (2006, 2007), Donnelly et al. (2008), Dovers (2005), Eales and Sheate (2011), Égré and Senécal (2003), Elling (2007), Esteves et al. (2012), Fischer et al. (2010), Geneletti (2003, 2005), Genter et al. (2008), Gibson (2006a, 2011), Hacking and Guthrie (2008), Hassan (2008), Hansen and Kørnøv (2010), Harris-Roxas et al. (2012), ICPGSIA (2003), Jiliberto (2011), João and Mclauchlan (2011), Karjalainen and Järvikoski (2010), Khera and Kumar (2010), Kobus (2005), Lane et al. (2003), Kemm and Parry (2004a,b), Kolkman et al. (2007), Lemonick (2010), McCluskey and João (2011), Meynell (2005), Partidário (2007), Partidário and Coutinho (2011), Pope and Grace (2006), Pope et al. (2005), Schmidt (2011a,b), Sheate (2011), Sinclair et al. (2009), Steinemann (2001), Stoeglehner et al. (2010), Smith (2007), Slootweg et al. (2010), Slotterback (2008), Thérivel (2010), Thérivel and Partidário (1996), Tickner (2004), Treweek et al. (2011), Vanclay (2006), Youngkin et al. (2003), Wiek and Binder (2005).

reasoned substantiation of all choices, and a proactive effort to making the process more environmentally substantive.

Strategic Environmental Assessment (SEA) At the SEA level, rationality good practice is concerned with mainstreaming environmental concerns into decision making. Increasingly, it is largely about the reasoned marriage of ends and means, within an organizational/social learning, iterative, creative and adaptive, context-dependent, decision/policy oriented, holistic, and participatory framework. In recent years, it has become much less about the systematic application of technical/scientific models and methods, generally borrowed from EIA theory and practice. This pattern is especially evident at the policy level.

Environmental Impact Assessment (EIA) Project-level EIA rationality good practices seek to broadly define ends (e.g., need, goals, objectives, and criteria) and means (e.g., alternatives, mitigation/enhancement measures). This is a departure from earlier process formulations, which tended to downplay ends and focus instead on ameliorating the negative effects of a predefined proposed action. Contemporary EIA-level rationality good practice tends to take the form of reasoned, inclusive, and substantiated argumentation rather than technical, expert-driven analysis, although the latter still assumes an important support role. EIA-level rationality also devotes more attention to such matters as nonstructural alternatives, indirect and cumulative effects, the distribution of consequences (especially with regard to vulnerable populations), risk and uncertainty management, and links between effects and ecological/community capacity.

Ecological Impact Assessment (EcIA) Rationality in EcIA combines the rational model (e.g., objectives, alternatives, support by experts and scientific methods) with substantive ecological knowledge, principles, and priorities. It is especially focused on such themes as maintaining and enhancing biodiversity, and avoiding and minimizing threats to rare and endangered species and habitats. It counterbalances a technical/scientific orientation with a holistic perspective, and an emphasis on uncertainty management and adaptability.

Social Impact Assessment (SIA) SIA is more ambivalent about rationality. It explicitly recognizes the tension between the technical and the political, and the tendency of technical/social scientific SIA approaches to inhibit collaborative public involvement. Nevertheless, SIA rationality good practices explicitly identify ends, suggest alternatives, and screen and compare alternatives in terms of ends achievement. However, SIA rationality good practice tends to ameliorate the negative tendencies of technical–rational analyses by, for example, explicitly integrating social values and ethical procedural and substantive principles (e.g., human rights, social justice); emphasizing options that meet social needs and aspirations (e.g., social capital, good governance, social inclusion); stressing the value of and need for community knowledge, perceptions, involvement, influence, support, and acceptance; and focusing on the distribution of benefits and adverse effects (with particular regard to the vulnerable and disadvantaged).

Health Impact Assessment (HIA) Rationality in HIA wavers between a "health sciences"–technical–quantitative rational model and an inclusive–collaborative–qualitative–social science reasoning model to the integration of health concerns into decision making. The former approach is more top-down, narrowly defined, and expert driven. The latter approach is more bottom-up, broadly defined, and community driven. Some efforts have been made to combine elements of both approaches. Such integrated approaches offer the potential to take advantage of the positive tendencies of rational–technical analysis while offsetting its negative tendencies. Both approaches seek to fully integrate health concerns into the determination of ends (e.g., health-related values, objectives, and criteria) and into the generation and evaluation of means (e.g., health outcomes, the distribution of health effects). They also seek to better engage both health professionals and the community in the process.

Sustainability Assessment (SA) Rational SA practice tends to maintain the positive and negative tendencies of rationality in a sort of dramatic tension. Much is made of the need to explicitly identify sustainability objectives, criteria, and targets, to broadly define alternatives, and to systematically screen (e.g., against sustainability thresholds) and compare (e.g., trade-off rules) alternatives in terms of if, and the extent to which, they support sustainability ends. At the same time, SA good practices seek to offset the negative tendencies of technical–rational approaches (e.g., reductionist, inhibits public involvement, lacking substance, overestimates certainty, limits decision-making influence, value free) by, for example, emphasizing holistic, transdisciplinary synthesis, stressing the need for collaboration among interested and affected parties, pointing to the importance of resilience and precaution, identifying and applying explicit criteria and decision rules that bound and direct decision making, and integrating explicit substantive environmental values, ends, limits, and perspectives.

5.6 CONTEMPORARY CHALLENGE—SITING "LOCALLY UNWANTED LAND USES"

5.6.1 The Challenge

One of the most vexing problems in contemporary IA practice is the siting of locally unwanted land uses (LULUs) (examples include pipelines, power plants, waste treatment/disposal facilities, and mines), where there is a broad public interest (at least from the perspective of most senior governments and proponents); the prospect of increased local employment, purchases, and payments (although often not equitably distributed); a concentration of negative environmental and social effects and risks in the vicinity of the proposed facilities; and serious doubts (at the local level and from the perspective of broader public interest groups) regarding the long-term sustainability of such undertakings.

The conventional approach to the siting of such facilities has been largely rational and technical. Areas and sites have been systematically analyzed for environmental suitability (both for minimizing constraints and for taking advantage of opportunities), and for the effective and efficient provision of services. Options have been systematically identified, screening, and compared. Ample provision is generally made for public and agency involvement (largely in the form of information and consultation) prior to key decision points in the siting process. This "track record" of rationally driven approaches (highlighted as the environmental suitability siting approach in Table 5.5) has been mixed at best. Intense regional, local, and often and international national opposition has been commonplace. Need is challenged, especially on sustainability grounds. Severe doubts are raised regarding whether the frequency and severity of environmental and social impacts and risks have been underestimated. Fairness, both in a procedural and substantive sense, and in terms of the distribution of facilities and costs and benefits over space and time, tends to be a major area of concern. Another recurrent concern has been the limited influence that other affected and interested parties, especially local communities, have over either the process or its outcomes. These parties argue that they need and deserve much more influence and control over the siting process and regarding if, where, in what form and pace such facilities should be established and operated. These equity and community control concerns have led to the gradual emergence of alternative siting processes (as highlighted in Table 5.6 and as described in Section 5.2.1).

Table 5.5 Major Siting Approaches and Subsets

Environmental Suitability	Social Equity	Community Control
Constraint Minimization • Area screening/identification • Site screening/identification • Site comparison	*Fairness: Distribution of Facilities* • Unfair locations • Fair locations	*Locational Control* • Voluntary communities/local veto • Voluntary site location • Voluntary sites • Voluntary access routes • Combinations of above
Opportunity Maximization • Physical suitability • Existing facilities • Site rehabilitation • Compatible land uses • Available lands • Performance standards—private bids	*Fairness: Distribution of Costs and Benefits* • Avoid imbalances • Redress imbalances • Redistribute to more than redress imbalances	*Procedural Control* • Citizen siting authority • Siting partnerships • Citizen advisors • Use of third parties
Service Maximization • Service to facility • Service from facility	*Procedural Fairness* • Participant/intervenor funding assistance • Conflict resolution/consensus building • Community involvement	*Facility Control* • Needs analysis • Policy/program options • Alternative technologies • System characteristics • Facility characteristics • Facility management
Variations and Combinations • Evaluation methods • Project types • Proponent type • Impact management methods • Environmental context • Siting requirements • Combinations	*Variations and Combinations* • Procedural fairness and location distribution fairness • Location distribution fairness and fairness in distribution of benefits and costs • Procedural fairness and fairness in distribution of benefits and costs	*Variations and Combinations* • Locational/procedural control • Procedural/facility control • Locational/facility control • Locational/procedural and facility control

Source: Adapted from Lawrence (1996).

Table 5.6 Examples of Good Practices—Siting Locally Unwanted Land Uses

Context (framing)	View proponents and proposed actions as proactive, potential catalysts/bridges for lasting, desirable, and resilient social and ecological future through the provision of improved services, infrastructure, building, employment, local development initiatives, and other community and ecological benefits Place within the context of and complement existing policies, plans, and programs (e.g., sector, regional, land use, conservation); supplement, as needed, to fill gaps (e.g., SEA) before process commences Clarify intergovernmental revenue-sharing policies and protocols and proponent and government's social, economic, and ecological responsibilities, policies, and obligations Make effort to address preexisting environmental concerns and relationship difficulties among key parties Ensure appropriate policies, legal framework, approval procedures, and mechanisms in place (e.g., sustainability test, requirement to enhance positive effects, authority to address such issues as cumulative effects, equity issues, legacy and bridging impacts, pace and scale of development); reform as needed Ensure appropriate guidance materials in place (e.g., collaborative consultation, ethical standards) Draw upon lessons and insights regarding other social dilemma situations and intractable environmental disputes Seek enhanced understanding of nature and basis for both social and private determinants of opposition in comparable situations
Start-up	Seek voluntary opportunities (e.g., voluntary communities and routes) Actively seek "buy-in" by all parties of process and outcomes (i.e., free, prior, and informed consent); seek financial and nonfinancial partnership among proponent, government, and communities Undertake comprehensive ecological, social, and economic profile of region; ensure sound understanding of regional and local context and issues Assess baseline ecological and social carrying capacity (focus on enhancing) and cumulative effects (focus on avoiding and ameliorating) Predict a range of potential and likely baseline futures, identify a range of reasonably foreseeable expansions and future developments, and identify all planned, proposed, and likely future activities affecting the same environment Ensure appropriate range of technical/scientific and procedural specialists; ensure good proportion of local staff and sustained involvement of SIA and public involvement practitioners Clearly and, where necessary, legally define all key terms
Scoping	Design approach, jointly with interested and affected parties to suit context (e.g., prior experiences, cumulative effects, local values, and priorities) Embed framing and start-up commitments in scoping Identify and characterize (e.g., nature, distribution, extent shared or conflicting) local concerns, preferences, interests, and values from outset Clearly define meaningful consultation, empowerment, contextually appropriate benefits, and what constitutes success for each party Ensure process is thorough and efficient Ensure early consideration of positive impacts, benefits, and beneficiaries
Guidance	Jointly determine values, goals, objectives, criteria, and limits to guide and bound process Guide by ethical principles (e.g., inter and intragenerational equity, internalization of costs, polluter pay, protection and promotion of health and safety, precautionary, multisectoral integration, duty to consult, integration of human rights—personal, property, indigenous peoples) Maximize sustainability opportunities Seek social, economic, and community development; improved health and well-being; improved biodiversity; restored ecosystem and landscape character; and protected and respected cultural heritage resources Design process to be consistent with IA and environmental management best practice and with social performance standards
Procedural screening (unacceptable)	No infringement of human and indigenous rights No use of violence, intimidation, harassment, or undue force No bias or withheld information (lack of transparency) No use of benefits to "cover up" negative, knowledge gaps, lack of data, political or institutional barriers, or to "win favor" for harmful projects No one-way communications or "token" consultation Insufficient time or resources to meaningfully participate No procedural inequities or unfairness No procedures inappropriate to context

(continued)

Table 5.6 (*Continued*)

Substantive screening (unacceptable)	No compromise of health or safety standards No compromise of environmental standards or IA requirements Consistent with government policies and targets (consistent with public interest) No unsustainable options No severe risks to protected areas and species No net significant adverse effects (environmentally unacceptable) No significant uncertainties regarding potential for catastrophic long-term consequences Scale, pace, or location of project not to exceed or threaten to exceed ecological, social, or institutional carrying capacity, including consideration of cumulative effects
Comparative process/IA	Institute collaborative process including early and ongoing involvement and meaningful participation and reassurance Ensure open and transparent process with broad scope of community involvement (e.g., use of interactive community forums) and particular effort to involve underrepresented and disadvantaged populations Ensure procedurally equitable and fair process Search for synergistic, enduring opportunities to meet both proponent goals and broader benefits to community and environment; seek to match project benefits and community needs and aspirations Anticipate and deal with issues before they become problems; seek best possible compromise Give due consideration to all reasonable alternatives (including no action, environmentally preferred, all alternatives proposed by interested and affected parties, scale and pace options, alternative goals, institutional alternatives); interweave ends and means and explicitly justify trade-offs Seek consensus in choice of preferred alternatives, mitigation/enhancement measures, local benefits, and impact management strategy Assess environmental and social sustainability and capital, with and without project and, if practical, against control communities Give full consideration to locally appropriate mitigation and enhancement (enhancement hierarchy—project, local area, wider area) Systematically identify and address uncertainties in a manner consistent with precautionary principle Test robustness of alternatives and effects assessment against varying project and baseline scenarios
Preferred procedural outcomes	Project is acceptable to market, is politically acceptable (at all levels), and is broadly acceptable to and supported by community and public; free, prior, and informed consent (clearly and consistently defined) Complements and consistent with relevant policies, plans, programs, and projects Facilitates critical social learning Political efficacy and participatory democracy strengthened Decision making decentralized (subsidiarity) Shared and collaborative decision making Equitable and fair decision making Communities have capacity and resources to maximize project benefits Facility comanagement (proponent, government, affected communities/publics)
Preferred substantive outcomes	Contributes to ecologically, economically, and socioculturally sustainable environment, with dependencies recognized, mutually reinforcing gains, trade-offs a last resort, trade-off rules, and future protected Net biodiversity benefits or enhancements, restored biodiversity, improved biodiversity security, improved ecosystem services, and improved ecological resilience Contributes to equitable/just environment, over space, over time, and among populations and communities, with emphasis on enhancement of marginalized groups and correction of past injustices Community cohesion, capacity building, empowerment and realization of human and social potential furthered, dependence reduced, and social and ecological diversity and resilience maintained and enhanced Provides catalyst for realization of community aspirations and human potential and helps build social and human capital
Formalizing decisions	Impact management and benefits agreements with affected communities (e.g., employment opportunities, community investment, environmental restrictions, revenue sharing, dispute resolution, social and cultural programs, local content requirements—participation of local people in work force and supply chain) Joint determination, with communities, of community investment opportunities; considerable local discretion in how payments spent

Table 5.6 (*Continued*)

	Transition planning agreements and plans and initiatives to build individual, community and entrepreneurial capacities, sufficient transition funding and post project legacy funding; funding structured to smooth "boom and bust" cycle and extend beyond project life
	Control of SIA by indigenous people, where applicable
	Intergovernmental revenue sharing and implementation responsibility agreements
Implementation and follow-up	Ensure impact management goes well beyond compliance and physical effects management (e.g., cumulative effects management, anticipating and recognizing emerging concerns, socioeconomic positive and negative effects, benefits enhancement)
	Ensure environmental changes, and positive and negative direct, indirect and cumulative effects independently monitored and adjusted as needed
	Seek to extend reach of benefits, ensure equitable sharing, and link socioeconomic IA to sourcing strategies
	Incorporate contingency planning and design adaptive management strategies
	Clearly define follow-up responsibilities by party; hold accountable
	Make provision for interested and affected parties to help select indicators and measures of social and environmental impacts and benefits and determination of appropriate responses
	Integrate impact management objectives identified by community and ensure sufficient resources so that communities can monitor and manage social impacts themselves; provide guidance and assistance as needed
	Incorporate credible and effective grievance mechanisms (principles, process, and resources) (including possible use of ombudsman)
	Ensure sufficient resources for post approval impact management, including provision for independent oversight and full community involvement in monitoring and management
	Work with communities and government to institute, maintain, and enhance necessary institutional and organizational changes
	Ensure all levels of government have institutional and financial capability to undertake follow-up responsibilities; facilitate capacity building as needed
	Undertake independent procedural and substantive effectiveness audits
	Share lessons, best practices, experiences, and insights
Knowledge support	Undertake broad ranging and integrated environmental suitability and impact assessment (ecological; social–psychological; health; sustainability; direct, indirect, and cumulative; positive and negative) analyses
	Assess distribution of costs and benefits, including environmental justice issues
	Assess indirect, cumulative, legacy, and bridging effects; assess impact management preparedness
	Analyze risks and uncertainties from a precautionary perspective
	Fully integrate community and indigenous knowledge
	Fully integrate scientific and technical knowledge
	Undertake social development needs and infrastructure and service capability analyses
	Undertake decision-making effectiveness and community goals achievement analyses
	Systematically draw upon experiences elsewhere and good practices
Procedural/community support	Provide resources for independent review of technical/scientific analyses
	Use a combination of methods (e.g., surveys, literature review, interviews, meetings), appropriate to context, to determine public attitudes, preferences, and environmental perceptions (including perceived risks)
	Integrate local knowledge and experience
	Integrate traditional knowledge and respect intellectual property rights
	Provide for early and ongoing public participation, including community outreach
	Allow for third-party assistance (e.g., facilitation, mediation)
	Provide translation as needed
	Ensure procedural fairness; necessary if process legitimacy is to be accepted
	Provide necessary resources (e.g., funds, training) for meaningful participation, including sufficient time and money to meaningfully participate
	Undertake stakeholder participation effectiveness analysis
	Undertake independent research of effectiveness of impact and benefits agreements

Sources: Ahmadvand and Karami (2009), Armour (1990a), Bond et al. (2012), Booth and Skelton (2011b), Burdge (2004), Campbell (2003), Cavatassi and Atkinson (2003), Edelstein (2003), Égré and Senécal (2003), Esteves and Barclay (2011), Esteves and Vanclay (2009), Esteves et al. (2012), Fischer (2003), Galbraith et al. (2007), Gibson (2006a, 2011), Harris et al. (2003), Harris-Roxas et al. (2012), IAIA (2003, undated b), ICPGSIA (2003), IFC (2009), João et al. (2011), Karjalainen and Järvikoski (2010), Lane et al. (2003), Noble (2009b), O'Faircheallaigh (2009), Orenstein et al. (2010), Rajvanshi et al. (2011), Ross and McGee (2006), Rowan and Streather (2011), SPTF (undated), Storey and Jones (2003), Tamburrini et al. (2011), Vanclay (2003), Walker (2003, 2010), Weaver et al. (2008), Wlodarczyk and Tennyson (2003), Wolsink (2010), Youngkin et al. (2003).

5.6.2 Generic Siting Approaches

As illustrated in Figure 5.5, the subsets within these three major siting processes and the siting processes themselves overlap to a considerable degree. These overlaps are critical inasmuch as any one approach will and has been criticized for ignoring or undervaluing the legitimate concerns that the other approaches represent. The question then becomes which combinations of approaches are best suited to the siting of "locally unwanted land uses". Some recent siting initiatives have tended to retain the rational–technical approach as the core, but then have addressed equity concerns through an increased emphasis on local benefits (often formalized in local benefits agreements) and local control concerns through a greater community role in facility design and operation (often formalized in impact management agreements). Although ameliorating some equity and community influence concerns, intense opposition has tended to continue on the grounds that issues of need and sustainability have been glossed over and many environmental and community concerns have not been addressed or addressed adequately. Positions for and against such facilities have tended to remain polarized. The question that remains then is—are there other approach combinations available that come closer to

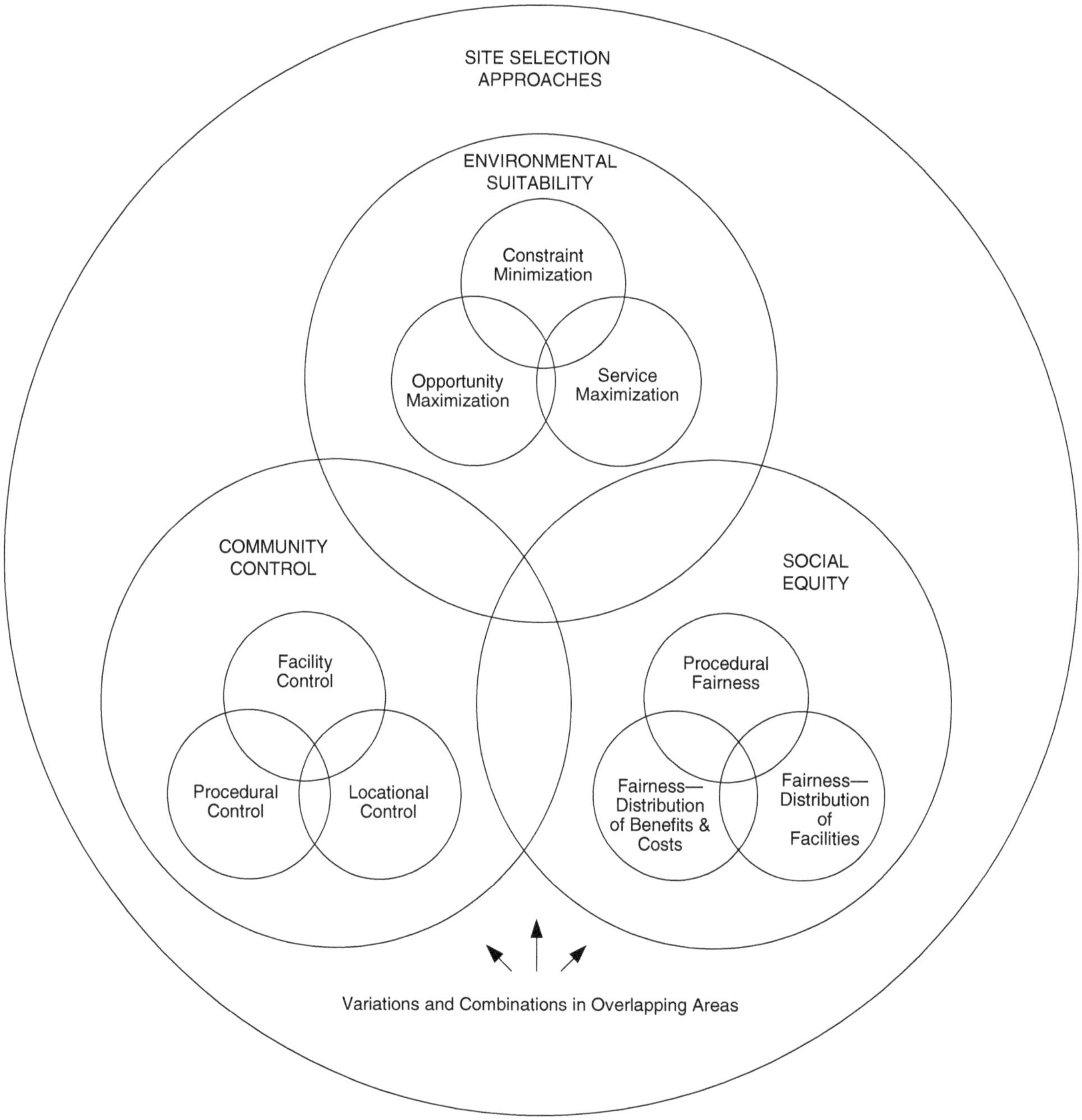

Figure 5.5 Site selection approaches. Adapted from Lawrence (1996).

adequately addressing the concerns and preferences of all interested and affected parties?

5.6.3 An Example of LULU Siting Process

Premises A possible departure point is the good practices suggested in IA literature for siting LULUs. Table 5.6 provides a consolidated list of such practices. An example of a process based on such practices is described below and is highlighted in Figure 5.6. The process begins from four premises: (1) need must be defined broadly (i.e., more than a market opportunity, demonstrably in the long-term public interest); (2) the proposed action is more than a remedy to a need, it also must represent an opportunity and a catalyst for meeting the long term needs and aspirations of a constellation of interested and affected parties, especially the environment; (3) the manner in which the siting process is conducted (i.e., the means) is as important as the needs and aspirations that the undertaking seeks to fulfill; and (4) elements of the community control/social equity approach should drive the process with the rational–technical approach assuming a partial support role, in conjunction with a range of other nontechnical procedural and substantive support mechanisms.

Framing All too often LULU siting processes commence (and often end) before certain framing elements are in place. There should, for example, be a well-defined array of environmental and land use policies, plans, and programs (and related SEAs) in place that establish the strategic context within which the proposed action will be established. The necessary laws, regulations, and guidelines (and related infrastructure) need to be in place. The division of responsibilities and revenues among the key parties needs to be clearly defined. Proactively addressing preexisting environmental problems and inequities in the geographic areas under consideration should be a priority. A concerted effort should be made to draw upon the knowledge base established with comparable projects in comparable environments, with a particular emphasis on understanding and appreciating the basis for opposition to LULUs. The "wheel does not have to be reinvented." Addressing such matters retroactively, through a project-specific EIA, is almost always doomed to failure. Some significant capacity building may be necessary before all parties can actively and fully participate in the process.

Start-up Start-up for a LULU siting process ideally commences with "volunteers" (i.e., areas, communities, and sites potentially willing to "host" a proposed facility). Community acceptance/support always at the end, and preferably from the outset, can go a considerable distance in alleviating the community control concern. Even when a volunteer siting approach is not practical (e.g., fixed pipeline route) or there are no volunteers, the aim always should be to progressively build a financial and nonfinancial partnership among the interested and affected parties, consistent with the "free and informed consent" principle. This may involve formal agreements (e.g., addressing local benefits and impact management)—agreements that are progressively refined through the process, and which are conditional on the final outcomes from the process. Such initial agreements should not be viewed as final, as providing a basis for excluding parties from the process (i.e., uninformed consent), or as a means of compromising substantive or procedural standards. Start-up also entails establishing a comprehensive profile of the region; determining the baseline ecological and social carrying capacity; ensuring a sound understanding of the regional and local context and issues; assessing the implications of preexisting cumulative effects; projecting likely future baseline conditions (preferably with a range of scenarios encompassing varying assumptions); identifying a range of facility-related options, assumptions, and potential future developments (including scale and pace of development choices that the proponent may be reluctant to consider); and identifying other planned, proposed, and likely facilities and activities in the area (to set up the cumulative effects assessment). The necessary range of skills and experiences should be determined, key terms should be defined (to avoid confusion down the line), and an initial study team assembled.

Scoping Scoping is a critical activity if resources are to be effectively and efficiently allocated, if the process is to proceed expeditiously, if critical concerns and issues are to receive the attention they deserve, and if the major parties that could be affected by the proposal are to be fully and effectively involved in the process. Consistent with the community control aspects of the approach, the concerns, preferences, interests, and values of all interested and affected parties should be fully identified and characterized. A particular effort should be made to clearly define what constitutes success, meaningful participation, and empowerment for each party. Consistent with the social equity aspects of the approach, early consideration should be given to potential positive effects, current and anticipated inequities, potential and preferred benefits and beneficiaries, and contextually appropriate benefit enhancement choices.

Siting Process Guidance The outputs from the framing, start-up and scoping activities provide the foundation for siting process guidance activities and documents. Guidance directs and bounds the siting process. The direction (e.g., values, goals, objectives, ethical principles, priorities, and criteria) and boundaries (e.g., ethical limits, beyond the mandate of the parties) for the process should be jointly determined by the interested and affected parties. The maximizing of sustainability should be an integrating theme. Consistent with the mutually reinforcing benefits approach to siting, the guidance activities should seek to structure and direct a process that seeks synergistic, enduring opportunities for all parties, encompassing such concerns as social,

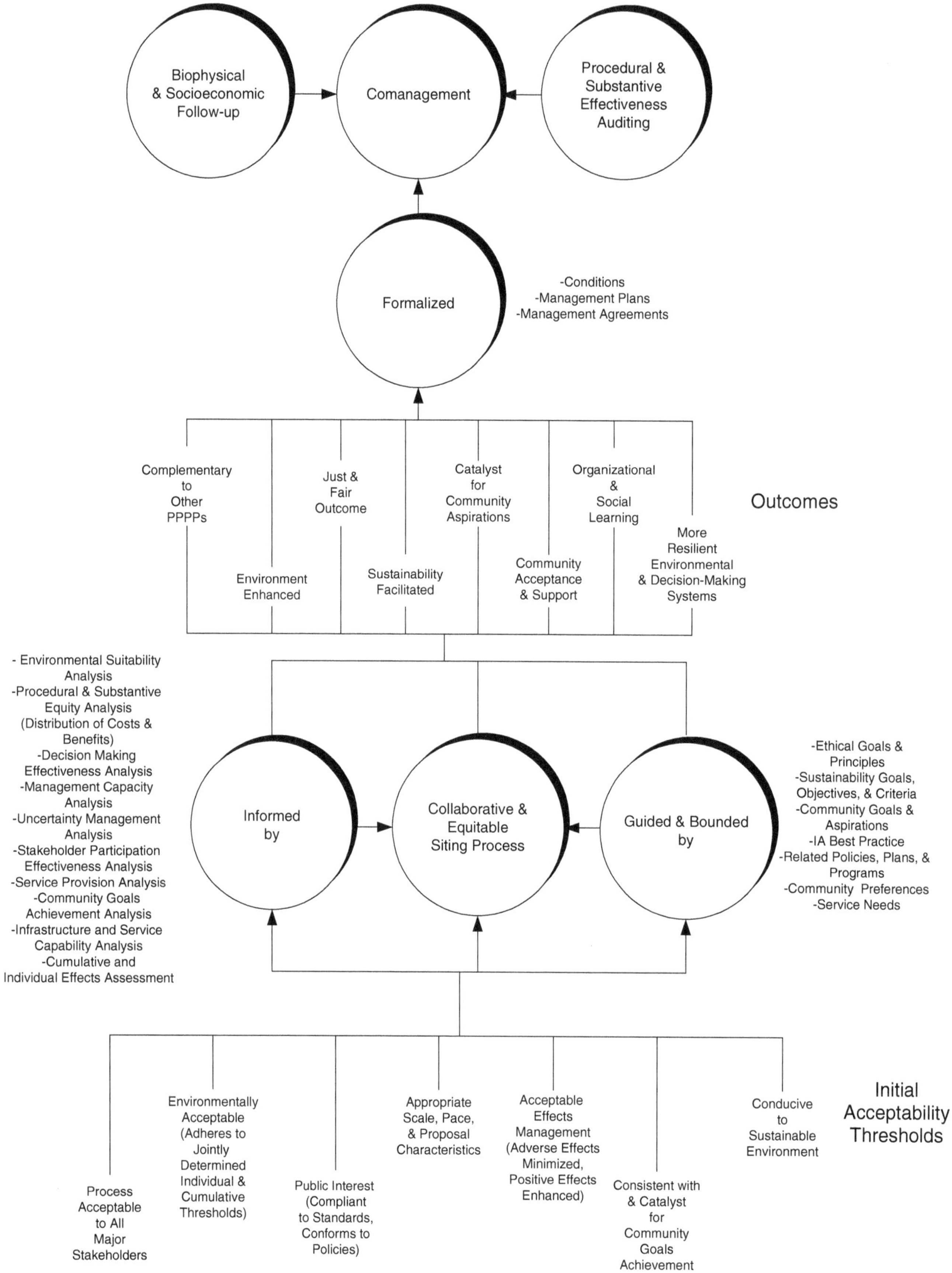

Figure 5.6 A conceptual model for siting LULUs.

economic, and community development; improved health and well-being; improved biodiversity; restored ecosystem and landscape character; and protected and respected cultural heritage resources. Again, recognizing the importance of drawing upon experience elsewhere, particular care should be taken to consider IA and environmental management best practices and social performance standards.

Screening The early and explicit screening of unacceptable procedural and substantive choices is critical to ensuring that the process remains "on track" and potentially acceptable to all the major parties. Clear substantive thresholds of acceptability should be defined and consistently applied. Potential examples include unsustainable options, severe risks to protected areas and species, major uncertainties with the potential for catastrophic consequences, threats to social or ecological carrying capacity, significant adverse individual or cumulative effects, and effects that are likely to contravene or be inconsistent with regulatory standards or public policies. Unacceptable procedural practices (e.g., exclusion from process, insufficient time or resources to participate effectively, intimidation, lack of transparency) also should be explicitly identified. A particular effort should be made to avoid the inappropriate use of benefits to "cover up" negative effects, knowledge gaps, lack of data, or political and institutional barriers, or to "win favor" for harmful projects. Consistent with the community control–social equity thrust of the approach, the line between acceptable and unacceptable procedural and substantive outcomes should be jointly determined by the major parties, and should emphasize avoiding and ameliorating inequities.

Comparative/IA Process The comparative/IA part of the process should focus on identifying and elaborating on a mutually supportive set of enduring benefits that meet both project goals and facilitate the realization of broader community and environmental benefits and aspirations, with a particular emphasis on sustainability. Due consideration should be given to all reasonable alternatives (e.g., no-action, environmentally preferred, alternative goals, all public proposals, institutional alternatives, pace and scale options). Adverse, potentially significant effects should be avoided and mitigated (using locally appropriate measures). Positive effects and benefits should be enhanced (again in a locally appropriate manner). Uncertainties should be systematically identified and addressed in a manner consistent with the precautionary principle. The primary tests of all options and potential mitigation/enhancement measures should be sustainability, community support, outcome fairness, adaptability, consistency with substantive goals, appropriate to context, and supportive of community needs and aspirations. The process should be open, transparent, inclusive, procedural fair, collaborative, and issue-oriented. A special effort should be made to include and address the concerns and interests of disadvantaged and underrepresented populations.

Outcomes The procedural and substantive outcomes from the process should be jointly determined and supported by the major parties. Examples of potentially preferred substantive outcomes include sustainability; net biodiversity benefits; a more equitable and just environment; potential adverse effects prevented, avoided and ameliorated; positive effects generated and enhanced; community aspirations realized; human potential furthered; positive legacy; reduced inequities; greater resilience; and enhanced community capacity, cohesion, and empowerment. Examples of potentially preferred procedural outcomes include market acceptability, community acceptance and support, political support, democracy strengthened, policy consistency, decentralization, and facility comanagement. The outcomes provide the basis for final proposal-related decision making.

Formalizing Outcomes Both procedural and substantive outcomes from the process are formalized in agreements, appreciating that agreement formalization is progressive and parallels the overall siting/IA process. These agreements could address such matters as employment, community investment, revenue sharing, dispute resolution, social and cultural programs, local content, environmental management, division of responsibilities, reporting requirements, capacity building, transitional planning, authority delegation, funding, and comanagement.

Implementation and Follow-up Implementation and follow-up, consistent with the emphasis on community control and social equity, should take the form of comanagement, and should proactively seek to avoid and reduce inequities. It also should be fully informed by technical/scientific analyses and nontechnical knowledge. The parties should work together to select appropriate positive and negative; and direct, indirect, and cumulative measures and thresholds. Independent monitoring and oversight is generally preferred. Responsibilities should be clearly defined. Funding and other resources, organizational reforms, and capacity building will be necessary to ensure the effective participation of all interested and affected parties. Full and ongoing community collaboration is essential, up to and including the delegation and funding, where possible, of SIA monitoring and management. The overall approach to impact management should be guided and directed by sustainability and other substantive environmental imperatives and social equity concerns. It also should be transparent, adaptive, and inclusive. Clearly defined and jointly supported grievance procedures should be instituted. Implementation and follow-up activities should be subject to independent procedural and substantive audits, which, in turn, should contribute to IA, public participation, and environmental management theory building.

Support A LULU siting process, such as the one described above can only be effective if it is fully supported by technical and nontechnical knowledge and by ongoing public and agency communications, consultation, collaborations, negotiations, and delegation. Technical and scientific analyses, while vital, should not be viewed as the sole or preeminent knowledge source. Community and indigenous knowledge is equally, and in some cases, more important. These knowledge sources should address such matters as options evaluation; environmental suitability and impact assessment (broadly defined); the magnitude and distribution of costs and benefits; risk and uncertainty analysis (from a precautionary perspective); mitigation, enhancement, and impact management; social development needs and infrastructure and service capability and capacity building; decision-making effectiveness; and goals achievement. The knowledge base for the process also should not be limited to the information and knowledge generated within the process. Ample use should be made, with appropriate contextual adjustments, of comparative experiences, knowledge, and good practices. An effective LULU siting process must be inherently collaborative. This necessitates ample and appropriate procedural support encompassing such matters as independent peer review, an extensive array of public participation methods and specialist advice, procedures for integrating community and traditional knowledge, procedures for determining and applying procedural ethical principles, third-party assistance (e.g., facilitation, mediation), and translation services. A particular effort should be made to enable (with appropriate resource and capacity building support) interested and affected parties (especially indigenous peoples and local communities) to undertake their own consultation and knowledge-building activities.

Future Steps The example process represented in Figure 5.6 and described above, and the good practices presented in Table 5.6, are far from a formula for siting LULUs. Some disputes are simply intractable, regardless of how they are approached. Many of the measures are untested in practice. What is successful in one context could be singularly unsuccessful in another. Definitions of success will vary, sometimes dramatically. Individual measures that generally work well may operate at cross-purposes when combined with certain other measures. What is, however, evident is the considerable gulf between good practice performance standards and the current state of practice. What is needed is the systematic effectiveness testing of various combinations of suggested good practice LULU siting approaches in a range of contexts. Rational analyses could assume a worthwhile, albeit partial, support role in such endeavors.

5.7 SUMMING UP

This chapter addresses the question of whether and how IA processes, documents, and methods can become more rational.

The three stories offer different perspectives on the potential roles of reasoning and technical analysis in the IA process. The first story shows how the Willing Host approach, to siting locally unwanted land uses, offers a potentially creative procedure for introducing reason into complex and contentious IA problem solving. The second story describes a community-based SEA reasoning approach that adheres to the major rational planning steps while still being highly collaborative, and able to effectively draw upon sound rational–technical analysis. The third story describes an IA process, which relied heavily upon technical quantitative analysis but also which was undertaken in a manner that supported and informed rather than supplanted the decision-making role of planners, policy makers, and the public working together. The three stories demonstrate that there is a potential role for rationality in the IA process but that role tends to be more effective when it assumes a support role (rather than driving the process), when there is a high degree of community collaboration and influence, when procedural and substantive equity issues are explicitly addressed, and when proactive steps are taken to offset the negative propensities of rational IA processes.

The problem can be posed in three ways: (1) IA processes in practice are insufficiently rational; (2) IA processes in practice are too rational; and (3) the ways in which rationality is defined and applied in IA processes need to be modified. There are valid arguments in support of all three positions. The problems are addressed by exploring the potential to apply rationality strengths, minimize rationality deficiencies, and draw upon alternative rationality definitions and applications.

Rationality attributes (e.g., logical, consistent, systematic) are identified and contrasted with irrational and extrarational decision-making factors. There are many rationality forms relevant to IA process management. Rationality expressions and applications vary depending on context. The rational process, as commonly described, involves defining a problem, identifying goals, collecting and analyzing information, forecasting and modeling future conditions, generating and evaluating alternatives, and implementing the preferred alternative. Public and agency involvement tends to take place prior to major decisions in the process. A great many assumptions have been ascribed to rationality. Ascribed rationality assumptions should be carefully considered because they are often implicit in IA processes.

Many strengths and limitations are attributed to the rational process. These strengths and limitations are tendencies that can be offset or reinforced. Several responses to the identified shortcomings bound and focus the process in an effort to make it more realistic and effective. Others selectively adapt and combine social, political, legal, ecological, and communicative rationality forms to make the process more substantive and democratic.

Rationality strengths and limitations are evident in rationality-based IA practice. IA has avoided some rationality limitations partly because of IA process characteristics (e.g., environmental ethic, scoping, cumulative effects assessment) and partly through deliberate efforts to offset negative tendencies. This is especially the case for SEA. Many rationality debates are mirrored in IA literature. IA literature and practice could derive additional benefits from a closer examination of the rationality debates both within and external to IA literature and practice.

Each of the four jurisdictions address, in different ways, purpose and need, the generation of alternatives, the screening of alternatives, and the comparison of alternatives. The range of rationality approaches encompassed by the four jurisdictions, while instructive, fall well short of good regulatory practice, in general, for different IA levels (e.g., SEA, project-level EIA), and for various IA types (e.g., EcIA, SIA, HIA, and SA). A greater effort could be made to draw upon these good practices.

The example rational IA process characterizes the problem or opportunity, analyzes need, determines the process and proposal purpose, and assembles a study team. Goals, objectives, principles, and priorities are determined. Alternative goals, objectives, and principles are formulated, where warranted. Methods are identified. An environmental overview is conducted. A scoping program is formulated and applied. Boundaries for the process are identified. Internal and external constraints and opportunities are identified. The proposed actions are described. Potentially reasonable ways of meeting the objectives and satisfying the principles are assessed. Unreasonable alternatives are excluded by applying exclusionary criteria. Reasonable alternatives are compared using evaluation methods, which combine scaled effects with criteria rankings and weightings. Uncertainties and variations in preferences are addressed through sensitivity analyses. Mitigation potential is integrated into the analysis. The analysis is supported by methods refinements and by data collection, analysis, prediction, and interpretation. The alternative means of carrying out of the proposed actions also are assessed. The same basic steps are followed but at a greater level of detail. Baseline conditions are characterized and individual and cumulative impacts, stemming from the proposed action, are identified, predicted, and interpreted, in parallel with the alternatives to and alternatives means analyses.

An impact management program refines and facilitates the implementation of the proposed actions. Options associated with the management program are generated and assessed in a manner comparable to the alternatives analyses. A clear and consistent decision-making basis is established, taking into account agency and public comments and suggestions. If approved, monitoring and auditing programs are undertaken. The monitoring program minimizes adverse impacts and maintains or enhances benefits. The auditing program facilitates methodological refinements. The process is supported by technical studies, reviews of comparable proposals and environments, peer reviews, and applied research. The public identifies concerns and suggestions and responds to analyses and documents. Communications and involvement distortions and inequities are minimized. Periodic interim reports are released. The draft and final report are broadly distributed. The documents provide a clear, unbiased, systematic, and accurate decision-making basis.

Good practice guidance is described for making IA more rational at the SEA and project EIA levels, and for EcIA, SIA, HIA, and SA. Each IA type seeks, albeit in different ways, to draw upon the positive tendencies and offset the negative tendencies of rationalism. There is considerable potential for mutual learning. At the same time, differences in approach, perspectives, and values need to be respected.

The siting of locally unwanted land uses represents one of the most vexing problems in contemporary IA practice. The "track record" of conventional technical rationality-driven approaches has been mixed at best. Intense opposition is commonplace, and is often based on valid procedural and substantive criticisms of the process and its outcomes. A greater effort needs to be made to integrate aspects of social equity and community control siting approaches with the more technical aspects of environmental suitability analysis. Rational–technical analyses tend to be more effective, in the siting of LULUs, when they serve a support role to more social equity–community control approaches. Experience in the field has evolved to the point that important insights, appreciating the need for contextual adjustments, can be gained by drawing upon good practice guidance for framing the process, for individual process activities (e.g., start-up, scoping, screening, comparison, formalizing decisions, implementation, and follow-up), and for supporting the process (both knowledge support and procedural/community support). Sufficient experience also has been acquired to suggest possible procedural and substantive screening thresholds and possible preferred procedural and substantive outcomes. Examples of good practice guidance for siting LULUs are presented. More applied research is needed to determine which combinations of practices are most appropriate to which combination of contextual characteristics.

Chapter 6

How to Make IAs More Substantive

6.1 HIGHLIGHTS

This chapter is concerned with designing and managing regulatory and applied IA processes to better integrate environmental perspectives, values, and knowledge.

- The analysis begins with two applied anecdotes. The stories describe applied experiences associated with efforts to make IA practice more substantive.
- The analysis in Section 6.3 then defines the problem. The general problem is a shortfall between IA environmental aspirations and achievements. The more specific problem is the role that the IA process assumes in widening or narrowing that gap. We describe how to make the IA process more conducive to integrating environmental perspectives, values, and knowledge and to furthering environmental objectives.
- In Section 6.4 we provide an overview of a range of ecological, social, and sustainability concepts. Attributes pertinent to IA process management are highlighted. Methods that could facilitate the integration process are also briefly described.
- In Section 6.5 we detail how an environmentally substantive IA process could be implemented at the regulatory and applied levels. We address implications and future directions, and identify generic regulatory approaches for integrating IA and substance. We then demonstrate how an environmentally substantive IA process might be expressed at the applied level. We provide good practice examples at the SEA and project EIA levels and for SAs, EcIAs, SIAs, and HIAs.
- In Section 6.6 we address the contemporary challenge of horizontal integration. We consider if and how to better integrate various substantive environmental (e.g., ecological, social, economic) concerns into IA requirements and processes.
- In Section 6.7 we highlight the major insights and lessons derived from the analysis.

6.2 INSIGHTS FROM PRACTICE

6.2.1 Assessing Significant Socioeconomic Impacts in EIAs for the Decommissioning of UK Nuclear Power Stations

The 1997 amendments to the EC EIA Directive included, among other things, the addition of the decommissioning of nuclear power stations and other nuclear reactors for mandatory EIA. This was the first specific identification of the need for assessment for the decommissioning of major projects. The United Kingdom has over 20 nuclear reactors, many commissioned in the 1960s/early 1970s and coming to the end of their operational life. Decommissioning operations are now underway for many power stations, mainly with Magnox reactors of approximately 400 MW. The operations include three main stages stretching over 100 years.

The EIAs have been largely coordinated by Magnox Electric and often involve a number of consultancies with particular specialist expertise. The Impact Assessment Unit (IAU) at Oxford Brookes University has undertaken the socioeconomic assessment input for many of the decommissioning project EIAs. The requirements of the EC Directive have produced a range of guidance and standards, with a particular focus on the biophysical, including safety issues associated with the management of radioactive waste and facilities, but guidance is much more limited for socioeconomic issues. Yet, perhaps surprisingly, socioeconomic impacts have emerged as particularly high ranking impacts in the assessments. This is a reflection of how, in decommissioning, many of the impact effects of development are reversed. Thus, decommissioning will lead to landscape and visual improvements, and to ecology, water quality, and air quality benefits. But, for socioeconomic benefits, there are important employment changes, with the pool of well-paid and previously very stable jobs falling by initially an order of 50–70%, then almost by 100%—before rebounding substantially, but alas not for 100 years! These losses have knock-on effects on the associated local and regional manufacturing jobs, demography, and economic and social services.

Impact Assessment: Practical Solutions to Recurrent Problems and Contemporary Challenges, Second Edition. David P. Lawrence.

The focus on socioeconomic impacts also raises interesting issues with regard to the criteria for assessing impact significance. There are no easily applicable "state of local society" standards against which the predicted impacts of a development can be assessed. While an increase in local unemployment may be regarded as negative, and a decrease in local crime as positive, there are no absolute standards. Thus, views on the significance of economic impacts, such as the proportion and types of local employment on a project, are often political and arbitrary. Nevertheless, it is sometimes possible to identify what might be termed threshold or step changes in the socioeconomic profile of an area. For example, it may be possible to identify predicted impacts, which threaten to either swamp or badly deplete local labor demand. It is valuable if the practitioner can identify possible criteria used in the analysis for a range of levels of impacts, which at least provides the basis for informed debate. The insert below provides an example from one of our IAU decommissioning studies. While this is an imprecise exercise, it can also provide a very useful basis for impact discussion.

Assessing the Local Impact Magnitude of Socioeconomic Impacts: Extracts from a UK Nuclear Power Station Decommissioning Project

Type of Impact	Local Context	Negligible Impact	Minor Impact	Moderate Impact	Major Impact
		Demographic Impacts			
Change in local population level of more than 2%	Population growth (2001–2009)	Change in local population of less than + or −0.25%	Change in local population of + or −0.25%	Change in local population of + or −1 to 2%	Change in population of more than 2%
		Direct and Indirect Employment Impacts			
Change in employment levels in local economy	Employment growth (ABI estimate 2001–2007)	Change of less than + or −0.25% on employment levels in the local economy	Change of + or −0.1% on baseline employment levels in the local economy	Change of + or −1 to 2% on baseline employment levels in the local economy	Change of more than + or −2% on baseline levels in local economy
Change in unemployment level in the local economy	Claimant % unemployment rates (June 2010)	Change of less than + or −2% in claimant unemployment	Change of + or −2.5% in claimant unemployment	Change of + or −5 to 10% in claimant unemployment	Change of + or −10% in claimant unemployment level
		Accommodation Pressures and Development			
Change in stock of local housing	Housing stock growth (2001–2008)	Change of less than + or −0.25% on baseline housing stock	Change of + or −0.25 to 1% on baseline housing stock	Change of + or −1 to 2% on baseline housing stock	Change of + or −2% on baseline housing stock

In the assessment of significance, the analyst must be aware of the importance of triangulation of information. Multiple perspectives on significance can be gained from many sources, including key local opinion formers (especially local councilors and officials), the local population (via surveys, focus groups, etc.), and the local press, which can be very powerful as an opinion former. All can help to assess the significance, perceived and actual, of the various socioeconomic impacts.

JOHN GLASSON

Department of Planning, Faculty of Technology, Design and Environment, Oxford Brookes University, Oxford, UK

6.2.2 Applying a Sustainability Test to a Hydrocarbon Megaproject

The ambition and innovative potential of environmental assessment regimes vary greatly. In jurisdictions where assessment has not advanced far beyond regulatory licensing, the focus of assessment is mostly limited to mitigating the adverse biophysical effects of the already designed and proposed undertakings. More demanding regimes take a broader approach. They define "environment" to include interactions among social, economic, and cultural as well as biophysical effects; encourage careful attention to these effects from the outset of the planning process; and provide for public engagement in critical review of purposes and comparative evaluation of alternatives. But even these regimes are mostly concerned about avoiding negative effects, rather than demanding lasting gains.

The leading edge exceptions are assessment processes that aim to ensure positive contributions to sustainability. These require proponents to establish that their project (or strategic undertaking) will provide overall benefits and leave the relevant communities and ecologies in a better shape. In Canada, since the mid-1990s, five major project assessment review panels have contributed to sustainability test. Three involved mines, one was about a major hydropower dam and one, the most advanced case example, centered on a natural gas development and transportation proposal.

The Mackenzie Gas Project was an industry consortium's proposal for a multicomponent venture estimated to cost about 16.3 billion Canadian dollars. The proposal, as filed for assessment review, included development of three gas fields and associated gas gathering infrastructure in the Mackenzie Delta area of the Northwest Territories, in addition to a 1200 km gas pipeline south through the Mackenzie Valley to connect with the existing continental pipeline network, and a natural gas liquids pipeline about halfway along the same route to an existing liquids pipeline. The area to be affected was huge, sparsely populated (a few dozen mostly small communities), and in some ways, both ecologically fragile and socioeconomically stressed.

Anticipating the proposal, the relevant federal, territorial, and aboriginal authorities agreed to a process merging their three separate assessment regimes. In 2004, they appointed a seven-person Joint Review Panel to evaluate the project in light of public hearings centered on the proponents' Environmental Impact Statement. The panel's terms of reference explicitly established "contribution to sustainability" as a fundamental principle for the assessment, along with respect for traditional knowledge, land claims and treaties, diversity, and the precautionary approach (IGC et al., 2004, p. 4).

The project represented not only welcome economic opportunities but also significant challenges. As a nonrenewable resource undertaking, it could not itself be sustainable. Any positive legacy would depend not only on avoidance of lasting damage, but also on planning for implementation and revenue use to build the foundations for sustainable livelihoods after the resource was depleted. Also, the proposal filed for panel review was only part of the story. The proponents designed the gas pipeline for potential throughput expansion to more than twice the flow expected from the three fields in the proposal. Some scenarios presented to the panel anticipated another doubling of throughput. While no one was in a position to know how much more development would be induced, everyone expected additional exploration, extraction, and transportation that would bring higher revenues, further opportunities, and more stresses on ecological, social, and administrative capacities. The big issues consequently involved cumulative effects and legacies. And the options before the panel went beyond whether or not to recommend approval of the project as filed. They also included what pace and scale of development would bring the greatest benefits and the least risk. In effect, the case was a strategic assessment of choices for regional futures in the guise of a project assessment review.

Near the outset, the panel set out its contribution to sustainability test, explaining that it would "evaluate the specific and overall sustainability effects of the proposed project and whether the proposed project will bring lasting net gains and whether the trade-offs made to ensure these gains are acceptable in the circumstances" (Mackenzie Panel, 2005).

After holding public hearings in 26 communities, the panel prepared a detailed final report. It applied an analytical framework based on 36 key issues in five core categories that were meant to cover the full suite of requirements for progress toward sustainability (Mackenzie Panel, 2009, especially Chapters 5 and 19):

- cumulative impacts on the biophysical environment;
- cumulative impacts on the human environment;
- equity impacts (fair distribution of benefits and risks);
- legacy and bridging impacts;
- cumulative impacts management and preparedness (capacities for managing the risks and opportunities).

The report's final chapter summarized the panel's evaluation, comparing impacts in each issue area for the null option (no project), the basic project as filed, and the range of project expansion and induced development scenarios distinguished by the volume of gas pipeline throughput (Mackenzie Panel, 2009, Chapter 19). The panel also determined, for the full range of options, how the likely impacts would be affected if the panel's recommendations were implemented effectively, and the potential interactions between the effects and the nature of the remaining trade-offs.

The panel concluded that the project, with some expansion beyond the immediately proposed activities, would contribute to regional sustainability, if accompanied by effective implementation of 176 recommendations (Mackenzie Panel, 2009, pp. 613–615). Some of the recommendations were conventionally focused on the conditions to be written into licenses and permits. But the most innovative recommendations called for steps by the federal and territorial governments to mitigate adverse cumulative effects and enhance lasting benefits. Most controversially, the panel urged active management of the pace and scale of development and use of nonrenewable resource development revenues and other opportunities to foster transition to "a more diverse, flexible and lasting basis for livelihoods in the region" (Mackenzie Panel, 2009, p. 602).

The panel's legacy was mixed. The breadth, rigor, and transparency of its sustainability-based analysis set an exemplary standard for future applications. But the panel lost credibility and good will by taking 2 years to write its report. The receiving governments formally rejected the panel's most important recommendations, including those about managing the pace, scale, and cumulative effects of development (Canada and the Northwest Territories, 2010). Nonetheless, on some matters, other jurisdictions have acted on rejected recommendations. The territorial government has, for example, initiated the establishment of a heritage fund to build a positive legacy from nonrenewable resource income. The proposed project received government approvals, but due to falling natural gas prices, it is not now economically viable and may not be resurrected for many years.

In the larger context, the Canadian federal government has passed a new assessment law that eliminates or narrows most of the federal involvement in environmental assessments, relying instead on a patchwork of provincial and territorial assessment regimes. The official rationale is that current economic imperatives justify the removal of impediments to resource exploitation projects. These changes reverse decades of environmental assessment evolution and will weaken the near term prospects for sustainability assessment in Canada. Eventually, however, the effects are likely to reveal again the underlying wisdom of decision making that seeks to deliver both stewardship and lasting well being.

ROBERT B. GIBSON

Environment and Resource Studies, University of Waterloo, Waterloo, Ontario, Canada

6.3 DEFINING THE PROBLEM AND DECIDING ON A DIRECTION

The preceding stories demonstrate that IA processes, to be effective, must be substantive. Specific substantive concerns, as illustrated in the SIA story, need to be incorporated into explicit, consistently applied, and appropriately adapted (to suit the decision-making context) significance criteria and judgments. A general commitment to sustainability is of little value in terms of making IA practice more substantive. As demonstrated in the SA story, the derivation and successful application of a sustainability test requires a sustainability-based regulatory framework coupled with clearly defined and locally adapted definitions, criteria, thresholds, and trade-off rules. These mutually supportive sustainability elements must collectively and explicitly mitigate cumulative effects and facilitate the realization of lasting benefits.

The relationships between the process and substance in IA practice, as illustrated in Figure 6.1, can be approached from many perspectives. The position can and has been taken that IA is a negative force in the environmental movement. Some label it as a deceitful procedure that legitimizes, without excluding or appreciably altering, environmental unsound projects, actions, and practices. IA is considered as an empty and a wasteful paper-processing exercise. Resources devoted to IA should, it is argued, be redirected to redefining agency missions and to tangible initiatives that directly advance the cause of environmental quality and sustainability. Alternative environmental management approaches (such as the ecosystem approach and adaptive environmental management) are sometimes suggested as tools that could replace IA. The point occasionally made is that IA presumes a degree of predictability and control that is so inconsistent with the knowledge base and the institutional structure that it is doomed to fail as an effective environmental management instrument. At best, it is concluded, IA is a procedural instrument with no substantive content. A fundamental reorientation away from process and procedure and toward substantive purposes is needed (Cashmore, 2004). Substance is added only when and if IA is linked to and placed within the context of substantive environmental management frameworks and tools.

The argument that IA is invariably either a negative force or that it serves no substantive purpose is not made now as often as it was in the early days of IA practice—at least not by those having knowledge of IA practice. Nevertheless, the arguments are useful because IA can and too often does reinforce the status quo and waste resources. It also can be of limited value if knowledge and control limitations are not recognized and addressed, if IA requirements are watered down and made so discretionary that they constitute little more than the illusion of action, and if the relationship between IA and other environmental management instruments is not considered. The substantive role of IA is shared with numerous other environmental management forms. Thus, making the IA process more substantive requires complementary roles among instruments and within broader frameworks.

It is difficult to reach firm conclusions concerning the IA process—substance problem because of uncertainties surrounding both the knowledge base for IA and the consequences from IA. IA has been advanced as an effective mechanism for integrating social and environmental values and perspectives into institutional practice. Although there is ample favorable (and unfavorable) anecdotal evidence, it is an overstatement to conclude that IA has either been highly effective or ineffective in bringing about such a transformation when proponents tend to view IA requirements as a hurdle rather than a useful planning and decision-making mechanism (Bartlett, 1989; Eales and Sheate, 2011). The tendency under most IA systems is to work through a process rather than toward a substantive objective such as sustainability (Galbraith et al., 2007). It is also unclear if the results without IA would have been substantially different. The gap between aspirations and outcomes, in terms of global environmental and sustainability concerns, is especially wide (Gibson et al., 2005).

A similar aura of uncertainty surrounds whether IA appreciably reduces adverse environmental impacts and whether it greatly enhances natural and social environmental conditions and benefits. Some benefits in terms of the enhanced consideration of environment in decision making (as compared with past neglect and failure) are evident (Cashmore, 2004). What is much less clear is the extent to which IA has contributed to substantive environmental goals such as sustainability. These uncertainties stem in part from a spotty, albeit improving, record of monitoring environmental impacts, and mitigation effectiveness (Clark, 1997). They also result from a mixed record in furthering IA environmental aspirations. Even when applied, the procedural orientation of IA means that there is no obligation to select less environmentally damaging alternatives or to avoid or minimize negative environmental effects, much less to realize net environmental benefits (Pölönen, 2006;

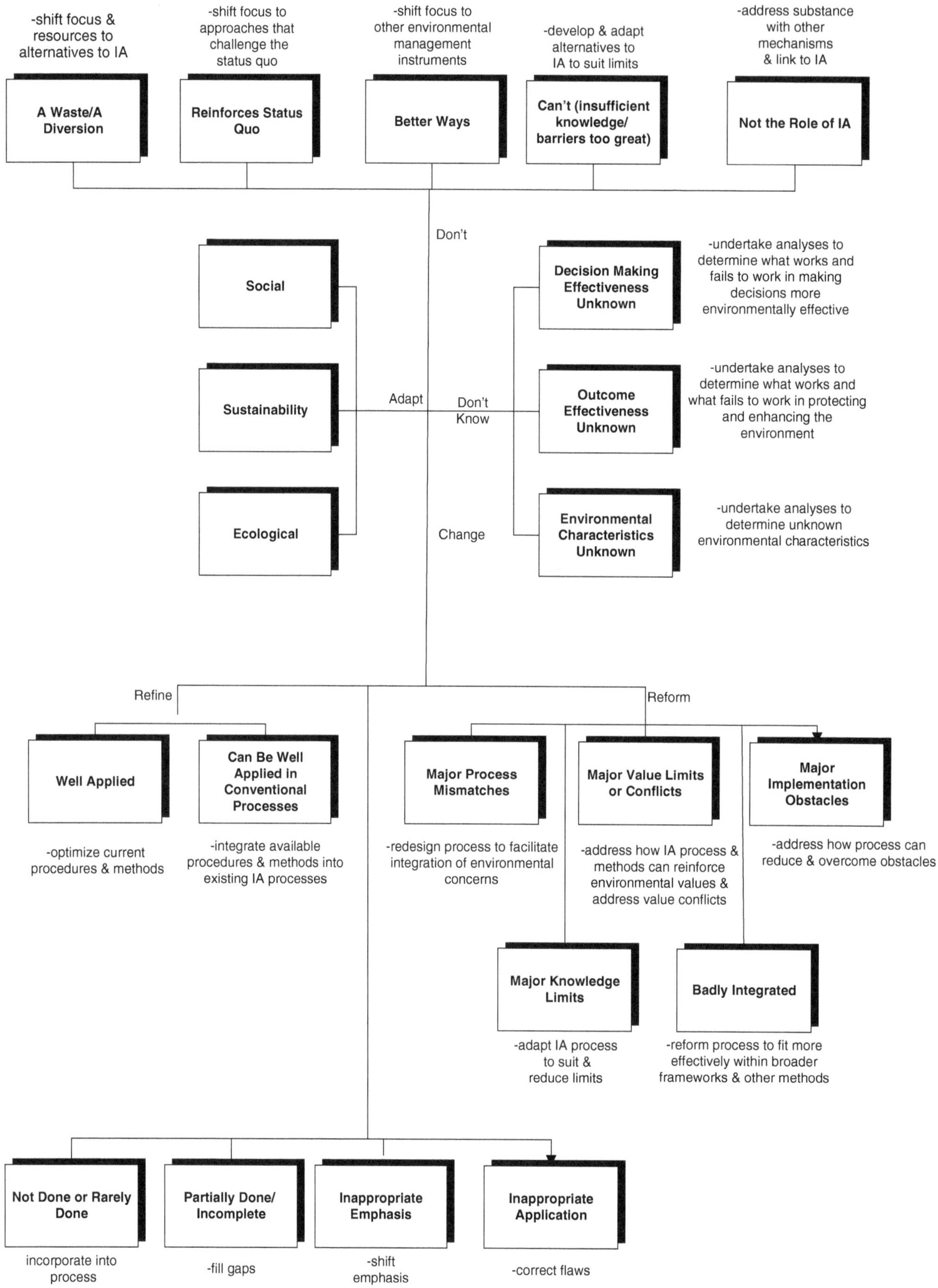

Figure 6.1 Perspectives on substance in the IA process.

Keys et al., 2011). IA effectiveness ratings concerning decision-making benefits, reduced social and ecological impacts, and contributions to sustainability, all leave considerable room for improvement (Sadler, 1996). It is difficult to reach firm conclusions regarding IA benefits because of the myriad of uncertainties surrounding the analysis, prediction, and management of ecological, social, and administrative systems (Dearden and Mitchell, 1998). These uncertainties are compounded when social and ecological systems are linked to assess cumulative and sustainability effects.

IA is often credited with contributing to a greater sensitivity to and accountability for environmental consequences by agencies (Andrews, 1997; Caldwell, 1997). It also is viewed as helping keep environmental issues before the public (Moore, 1992). Although there is a sense of overall progress, there is considerable variability in how well the environmental concerns are integrated into decision making and regarding the extent to which environmental quality is enhanced (Dennis, 1997; Gibson, 2011; Eales and Sheate, 2011). Decision-making influence has tended to be limited in part because of institutional and political resistance, and in part because of the failure to proactively demonstrate the merits of environmentally substantive planning and decision making (Gibson, 2011). Oftentimes, a holistic perspective or even a passing knowledge of sustainability is lacking among specialized line agencies (Kirkpatrick and George, 2006). The lack of a holistic perspective on the environment is especially evident in the latter stages of IA processes (e.g., impact management) (Kørnøv et al., 2005). Most IA texts assume that disciplinary and interdisciplinary knowledge and methods can be readily incorporated into conventional IA process activities and stages (Canter, 1996; Morgan, 1998; Petts, 1999). The point is commonly made that "tiering" (fitting project-level EIA within a SEA framework) greatly expedites the process (Morris and Thérivel, 1995; Vanclay and Bronstein, 1995).

Many commentators suggest that the marriage of process and substance will necessitate both procedural and substantive adjustments. Some argue that IAs commonly lack an ecosystem perspective; remain a weak tool for integrating biodiversity, cultural heritage, cumulative effects, and climate change concerns; and fail to explore how social impacts are socially constructed (Beanlands and Duinker, 1983; Bond et al., 2004; Brooke, 1998; Burdge, 2002; Greer-Wooten, 1997; Sadler and Jurkeviciute, 2011; Weiland, 2010). They point out that the link between the environment and sustainability is often poorly defined (Thérivel, 2010). They suggest the need for a greater effort to include ecological principles; to fully address biodiversity impacts; to more effectively integrate social concerns and knowledge; to apply nonpositivist social science approaches and perspectives; to consider the benefits of conservation; and to devote more attention to global warming, cumulative effects, and other sustainability-related concerns (Andrews, 1997; Burdge, 2002; Byron and Treweek, 2005; Clark, 1997; Kaufman, 1997; Lockie, 2001; Moore, 1992; Treweek, 1999). They emphasize the need to devote more attention to capacity building, to ameliorating technical and logistical constraints, and to overcoming institutional and political resistance (Hanusch and Fischer, 2011). They point to the need to implement IA policy goals, to monitor social and ecological effects, to employ more than a weak conception of sustainability, and to ensure bureaucratic accountability (Bronfman, 1991; Dearden and Mitchell, 1998, Eales and Sheate, 2011; Kaufman, 1997; Treweek, 1999). They stress the need for IA to more effectively address resource-related impacts, most notably agriculture, forestry, and fisheries (Duffy, 2006).

IA processes and practices should, various commentators stress, be more interdisciplinary, place-based, and adaptive (Dearden and Mitchell, 1998; US CEQ, 1997a). The robust monitoring of trade-offs among environmental components is considered critical (Burdett, 2008a). They point out that interconnections among social, economic, and ecological systems and impacts continue to receive scant attention (Franz and Kirkpatrick, 2007; IEEP, 2004). They highlight such methodological shortcomings as inadequate study designs, inappropriate temporal and spatial boundaries, inadequate data, inappropriate statistical techniques, critical impacts not predicted, and an insufficient decision-making basis (Alberti and Parker, 1991; Beanlands and Duinker, 1983; Freudenburg, 1986). Many methodological limitations are traced to a poor understanding of social and natural systems and of available social and ecological scientific concepts, models, and methods (Beanlands and Duinker, 1983; Burdge, 2002; Craig, 1990; Treweek and Hankard, 1998). These shortcomings imply an iterative IA process–substance relationship, with cycles of adjustments in an ongoing effort to match procedural and substantive characteristics.

Some commentators maintain that it is premature to reorient IA requirements and practices to meet substantive environmental ends. They argue that fundamental value shifts are a prerequisite to an IA process driven by ecological, social, and sustainability ethical principles (Euston, 1997; Kaufman, 1997). IA can play a supporting role in advancing such values and in addressing basic value conflicts, but not through conventional procedures (Benson, 2003). Much ecological and social knowledge, upon which IA depends, is fraught with uncertainties, especially when cause–effect relationships must be discerned and future environmental effects must be predicted (Dearden and Mitchell, 1998). Additional complexity is added when cumulative ecological and social system effects must be determined and conclusions reached regarding sustainability implications. The substantive effectiveness of IA is further severely inhibited by "balkanized" government environmental and resource responsibilities (Weiner, 1997). Effective partnerships must be established and maintained among agencies, with nongovernment organizations, and with the public (Dennis, 1997; IEMTF, 1995). These substantive value, knowledge, implementation, and institutional obstacles and challenges necessitate, it is argued, fundamental reforms to the IA process.

The problem is not the same for ecological, social, and sustainability concerns, although IA effectiveness reviews suggest that there is substantial room for improvement in all three areas (Denq and Altenhofel, 1997; Freudenburg, 1986; Sadler, 1996). Effectiveness ratings tend to be lower for treating social concerns and lower still for treating sustainability concerns. The nature of the concerns naturally varies. An IA process conducive to addressing ecological concerns will not necessarily be appropriate for incorporating social or sustainability concerns. Adaptations will be necessary for each type of concern as well as for interactions among concerns. An often intensive debate continues to surround the question of whether IA should be limited to ecological concerns or also cover social and economic impacts (Thérivel, 2010) (See Section 6.6).

There is no simple answer to the question—Which of these perspectives is valid or the most valid? A greater understanding of environmental conditions and of decision making and outcome effectiveness is required. Pending such knowledge, it must be assumed that all problem perspectives and solutions, depending on the circumstances and to varying degrees, are valid. Sometimes IA is more trouble than it is worth. Sometimes the substance–process relationship needs refinement, sometimes modification, and sometimes reform. Sometimes the issues and solutions vary depending on whether ecological, social, or sustainability concerns are being addressed. The direction then is to explore how IA process management can better address this constellation of interrelated problems and solutions.

6.4 SELECTING THE MOST APPROPRIATE ROUTE

This overview of substantive environmental concepts establishes the foundation for enhanced IA process management. The analysis is necessarily abbreviated and selective. Only ecological, social, and sustainability concepts are considered. IA effectiveness ratings for treating such concerns tend to be low. The choice of concepts is admittedly arbitrary. Only concept characteristics directly relevant to IA process management are identified. Key concept attributes are first described. Then IA process management implications are explored.

6.4.1 Ecological Concepts

Ecology is a branch of the biological sciences concerned with the relationships between organisms and their environments, including relationships with other organisms. EcIA is a formal process of identifying, quantifying, and evaluating the potential impacts of defined actions on ecosystems (Treweek, 1999).

Table 6.1 identifies and briefly describes several ecological concepts potentially relevant to IA process management. The recurrent themes exhibited in the concepts imply an IA process distinctly different from the conventional process assumed in most IA texts. All the concepts begin with an ecological systems perspective. They see planning and management as shaped by ecological visions, goals, and principles. Disciplinary boundaries are spanned and transcended. Less emphasis is placed on comprehensive disciplinary analyses and single function institutions. More stress is placed on selective, transdisciplinary synthesis and place-based coalitions of agencies and stakeholders. Temporal and spatial boundaries are extended to match natural patterns and rhythms. Multiple spatial and temporal horizons and boundaries are employed. Natural systems are seen as dynamic, self-organizing, complex, evolving, and uncertain. Planning processes, to match such characteristics, are viewed as necessarily open, adaptive, creative, collaborative, iterative, selective, and action-oriented. A life cycle analysis of the ecological footprint of proposed actions is undertaken (Hansen and Wolff, 2011).

Action-induced stresses can result in ecological thresholds being exceeded, notwithstanding resiliency. Severe prediction and control limits are noted. Thresholds are difficult to discern and often change. Major implementation barriers and obstacles are identified. The value of scientific and rational knowledge and methods is recognized. But the need to integrate extra-rational perspectives, values, and interests is also acknowledged. The distinction between natural and human (e.g., social, political, and economic) environments is seen as forced and inhibiting. Distinctions among environmental management instruments, of which IA is only one, are also seen as artificial. Preapproval analysis is no longer the preoccupation. Instead, continuous management approaches are advocated that extend through implementation and that rely heavily upon monitoring and adaptive management.

The scientific and rational IA processes (described in Chapters 4 and 5) display few of these characteristics. They can assume a valuable supportive role. However, they appear poorly suited to integrating ecological substance and the IA process. The assumption that process and substance are independent cannot be supported. A substantive IA process will be conducive to integrating ecological perspectives, knowledge, and methods only if it is designed and managed with a sound appreciation of the procedural implications of substantive characteristics. Judging from the characteristics of the concepts presented in Table 6.1, it will be almost impossible to imbue an ecological perspective into a comprehensive, rigid, closed, top-down, and lineal IA process that assumes implementation and a high degree of predictability and control.

Biodiversity IA and ecological IA have expanded dramatically in recent years, both at a conceptual and at an applied level. Concepts, such as the ecosystem approach and adaptive management, have been refined. An extensive array of good practice guidance is available (IAIA, 2005). Nevertheless, application in practice has been, at best, mixed. IA

Table 6.1 Examples of Potentially Relevant Ecological Concepts

Applied ecology (also ecosystem sciences)	Requires an ecological perspective, adherence to basic ecological concepts, and an appropriate interdisciplinary conceptual framework Ecological systems—self-controlled within constraints, evolving and complex; ecosystems part of a larger sociobiophysical system (human cultures and environments part of system) Importance of temporal and spatial boundaries, which reflect ecological processes Seeks to better understand self-organizing structures and processes (management challenge to protect self-organizing capacities—ecosystem integrity) Recognizes highly dynamic systems, extreme variation, and major predictive and management constraints; strong interest in scale, patterns, rhythms, and thresholds in biophysical systems Focuses on key variables, key processes, and ecosystem tolerance; importance of habitat and biological diversity Focuses on questions relevant to decision-making choices
Ecological impact assessment	Identifies, quantifies, and evaluates the potential impacts of defined actions on ecosystems and their components Tends to fit within major IA stages; scoping, focusing, and ecological monitoring critical Links ecological evaluation and EIA and SEA Objectives—maintenance of natural areas and their biological diversity and maintenance of social functions provided by natural areas Focuses on interactions between ecological stressors and receptors; a systems perspective Addresses the well-being of ecosystem services beneficiaries, ecosystem services, and direct and indirect drivers of ecosystem change Concerned with the state of the environment (e.g., biodiversity loss); addresses impacts on ecosystems and their components (valued ecosystem components) Use of applied ecological concepts, principles, and methods (e.g., surveys, taxonomic classification, GIS, modeling, statistical analysis, ecological evaluation, monitoring, landscape planning, ecological footprint analysis, carrying capacity analysis) Takes into account barriers, limits, and uncertainties (institutional, knowledge, methodological, natural variation) Closely linked to socioeconomic impacts, risk assessment, pollution control, and land use and resource planning
Environmental indicators	Methods and measures to monitor environmental status (trends and conditions) Important for framing problems and for determining solutions Physical, chemical, and biological variables used to construct environmental change indicators; incorporated into environmental statistics (state-of-environment); a decision-making input Example indicator categories—response indicators (overall biological conditions), exposure or habitat indicators (ecosystem exposure to pollutants and habitat degradation), and stressor indicators (human and natural processes that change exposure and habitat conditions) Can help monitor environmental problems; responses also depend on social and political considerations
Biodiversity-inclusive IA	The array of interacting, genetically distinct populations, and species in a region, the communities they comprise, and the variety of ecosystems of which they are functioning parts (composition, structure, and processes); relationships and interactions are critical Management objectives—the conservation of biological diversity, the sustainable use of components, the fair and equitable sharing of benefits from genetic resources, and the ecosystem approach Identifies and applies principles (e.g., no net loss, net conservation benefit, precautionary principle, use of local, traditional, and indigenous knowledge, public participation, equitable sharing, sustainable use) Importance of inventorying the present state of biological resources Undertakes biodiversity screening (e.g., biodiversity threats, potential impacts on protected areas and species) and scoping Seeks biodiversity friendly alternatives; maps and evaluates biodiversity conservation differences Particular concern with habitat loss and fragmentation. Components—regional ecosystem diversity, local ecosystem diversity, and genetic diversity Utilizes factors contributing to biodiversity decline—physical alteration, pollution, over harvesting, introduction of exotic species, natural processes, and global climate change Identifies direct and indirect drivers that cause biological change and conceptual links; lists ecosystem services provided (e.g., fish, timber, regulating as in water purification, biological control, cultural services) Applies biodiversity principles (e.g., big picture or landscape perspective, protect communities, and ecosystems; minimize fragmentation; promote native species; protect unique or sensitive environments; maintain or mimic natural ecosystems processes; maintain structural diversity and genetic diversity; restore ecosystems, communities, and species) Ensures multispecies emphasis; seeks to protect broader habitats and ecosystems that support biodiversity Stresses need to think in terms of comprehensive multiscale ecological networks and to adopt nested hierarchical conservation strategies

(*continued*)

Table 6.1 (*Continued*)

	Assesses the extent of biodiversity loss and reduction or loss of biodiversity function (use of biodiversity indicators) Potential role of GIS-based ecological models; importance of assessing indirect, secondary, and cumulative biodiversity impacts Estimates uncertainty factors affecting impact evaluation and effect on results Value of people-centered approaches to integrating biodiversity into the broader livelihoods sustainability context
Ecosystem approach (also ecosystem management)	Place-driven environmental protection strategy; uses of natural boundaries and ecological indicators Whole system and broad regional and temporal perspectives (multiple scales and time horizons); appreciates the dynamic nature of ecosystems Sets long-term ecosystem management objectives; conserves ecosystem function and structure; maintenance of ecosystem services a priority Based on ecosystem integrity and sustainability principles and values; incorporates such concepts as carrying capacity, resilience, self-organization, community diversity, and stability; and the precautionary principle Seeks to restore and sustain health, productivity, and biological diversity of ecosystems and quality of life (the humans part of environment); manage within functioning limits Stresses the need for dynamic, transdisciplinary, visionary (explicit ecosystem goals), proactive, adaptive, and participatory planning process Internalizes ecosystem costs and benefits Recognizes the importance of institutional arrangements (especially coordination and communications) and of integration with social and economic goals and context; provides incentives to promote biodiversity conservation and sustainable use
Environmental planning and management (also resource planning and management)	Largely rational planning process adapted to integrate environmental knowledge and methods; methods have roots in ecological and social sciences Encourages inclusion of ecological perspectives Stresses the need to reach consensus on environmental issues Multi to interdisciplinary; stress the needs for comprehensive approach Usually advisory and participatory Increasing recognition of the need to address environmental justice and equity issues
Integrated environmental and resource management and assessment	Advocates need for a more effective, integrated, and coordinated approach Holistic, regional, and ecosystem-based perspective; stresses the preservation of natural systems integrity Interconnective, goal-oriented, and strategic; involves both human and natural resources in ecosystem Devotes greater attention to social, political, economic, and institutional factors operating in an ecosystem (including opportunities and barriers stemming from institutional arrangements) and to links to sustainability Supported by integrated management systems (e.g., database management, GIS, expert systems) Recognizes the importance of stakeholder collaboration and that of conflict management Recognizes the importance of context and links to urban and regional planning
Adaptive environmental assessment and management	Iterative decision-making process; mimics the dynamic, cyclic, and surprise-ridden state of nature; decisions and assumptions revisited; long-term research, monitoring and management critical; seeks more resilient policy Generally involves a series of workshops facilitated by a core groups of experts; focuses on building and testing (usually computer) models as tools for generating and testing options; ongoing data acquisition Combines scientific information with a forum for interested and affected parties; a minimum regrets planning tool Emphasizes interdisciplinary communications and collaboration; integrates societal and ecosystem goals and values Carries IA into ongoing management; highlights the importance of monitoring and that of adaptive management in the face of uncertainty and complexity; an open and continuous learning processes—learning by doing
Traditional knowledge	A way of knowing and thinking about relationships of living beings (including humans) with one another and with the environment (a way of life) A cumulative body of knowledge and beliefs, handed down through generations by cultural transmission Relies on observation and knowledge of indigenous people Holistic—a form of environmental knowledge that integrates the social, ethical, cultural, technical, scientific, historic, ecological, and spiritual; emphasizes interrelationships; avoids scientific reduction Includes interrelationships among physical, biological, and human; humans as participants in environment rather than only as observers Fluid and flexible; importance of understanding how it operates in indigenous contexts; often misunderstood and misapplied

Sources: Alberti and Parker (1991), Armitage (1995), Bagri et al. (1998), Barrow (1997), Beanlands and Duinker (1983), Beatley (2000), Berkes (1993), CEAA (1996a,b), Coleman (1996), Dearden and Mitchell (1998), Geneletti (2002, 2003), Geneletti et al. (2003), Genter et al. (2008), Hansen and Wolff (2011), Gontier et al. (2006), Hanusch and Fischer (2011), Hegmann and Yarranton (1995), Hollick (1993), Holling (1978), Hooper et al. (1999), IAIA (2005), IEEM (2006), IEMTF (1995), Khera and Kumar (2010), Kozlowski (1990), Landsberg et al. (2011), Letsela et al. (2010), Lou and Rykiel (1992), Margerum (1997), Sallenave (1994), Sheate (2011), Shepherd (2008), Sherrington (2005), Slocombe (1993), Slootweg (2005), Smith (1993), Slootweg and Kolhoff (2003), Slootweg et al. (2006), Treweek, (1995, 1999), Treweek et al. (2011), US CEQ (1993), Wackernagel and Rees (1996), Wegner et al. (2005), Wieringa and Morton (1996), Wiles et al. (1999).

ecological and biodiversity institutional arrangements and analyses have often been limited and of poor quality (Mandelik et al., 2005; Manou and Papathanasiou, 2009). In common with scientific and rational IA processes, ecological IA tends to be expert-centered. It continues to be hampered by data limitations. Procedures for overcoming institutional and implementation barriers; for integrating ecological and biodiversity considerations into decision making; for managing uncertainty; and for facilitating adaptation, creativity, and collaboration are far from fully developed (Fischer, 2011; Hooper et al., 1999). The concepts encompass such a wide range of aspirations (several of which are pursued in other chapters) that they run the risk of becoming either overly general or overly complex. While appreciating and addressing these potential limitations and obstacles, these concepts demonstrably help make IA processes more substantive.

6.4.2 Social Concepts

SIA is the process of managing the social issues, consequences (intended and unintended, positive and negative), and invoked social change processes associated with planned interventions (Esteves et al., 2012; Vanclay, 2003). Social, as applied in impact assessment, is a broad and therefore a difficult-to-define term or concept. It is also defined and applied in different ways (Burdge, 2002). It has been described as a field of research, a discourse, a paradigm, an interdisciplinary and/or transdisciplinary social science, and a subdiscipline (Esteves et al., 2012). It encompasses the evaluation of all impacts on humans and the ways in which people and communities interact with their sociocultural, economic, and biophysical surroundings (IAIA, 2003; Jones and Slinn, 2008). As such, it is strongly linked to and arguably subsumes such fields as aesthetic, heritage, archaeological, cultural, community, demographic, development, economic, fiscal, gender, health, institutional, infrastructural, indigenous rights, political, poverty, psychological, resource, and social and human capital impact analyses (IAIA, 2003). SIA can refer to the distinguishing characteristics of people, communities, and society (e.g., demographic, cultural, institutions, customs, traditions, political systems). It can involve multiple levels of human aggregation (e.g., families, groups, organizations, communities, society). It can include perceptions, attitudes, norms, values, aspirations, and beliefs. It can pertain to patterns of association, interactions, and interdependencies. It can refer to health and social well-being, quality of life and living environment, economic and material well-being, cultural values and integrity, personal and property rights, and gender relations. It encompasses the changes in people's way of life, their culture, their community, their political system, their environment, their health and well-being, their rights, and their fears and aspirations (Vanclay, 2003; Youngkin et al., 2003).

Depending on the definition, SIA can apply theories, concepts, and methods from such disciplines as economics, anthropology, political science, psychology, history, philosophy, and archaeology, and such professions as land use planning, social planning, landscape analysis, health planning, risk management, resource management, public involvement, and environmental management. It is highly interactive with the physical (both natural and built), resources, and the ecological. It is both a field of study (e.g., social sciences) and a field of application (e.g., SIA). It is an evolving field that embraces a diversity of distinct, partially overlapping and partially conflicting concepts, models, theories, perspectives, and frameworks. There are numerous frameworks available for structuring social criteria.

IA texts generally treat social impacts as a distinct discipline, but, for the most part, as a subset of SEA or project-based EIA. Sometimes social and economic impacts are considered separately. Sometimes they are combined under the umbrella of socioeconomic impacts. Occasionally, health impacts and cultural, historical, and archaeological impacts are addressed separately. The track record of integrating SIA into agency decision making and into the assessment process has been, at best, mixed (Bronfman, 1991; Burdge, 2004). Social concerns tend to be a secondary consideration in IA requirements (Burdge, 2004). Too often, SIA has assumed a marginal decision-making role (Burdge, 2002; Lockie, 2001).

The major stages in the SIA process largely parallel those of the IA process. Sometimes SIA processes commence with public involvement. They often include a separate stage for predicting public responses to impacts (Finsterbusch, 1995; Interorganizational Committee, 1994). Social impacts are not the same as ecological impacts (Barrow, 1997). People react in anticipation of and adapt to change, oftentimes based on perceptions and attitudes (Edelstein, 2003; Égré and Senécal, 2003; Peterlin et al., 2008; Wlodarczyk and Tennyson, 2003). Human reactions vary greatly among individuals and groups and over time. Social phenomena are difficult to predict (Finsterbusch, 1995). Social units are not fixed structures. Social phenomena involve adaptive interactions. SIA involves both social change processes (intervening variables that may lead to impacts) and social impacts (intended and unintended consequences on the human environment from planned interventions) (Vanclay, 2002). The IA process can influence how people anticipate and adapt to change.

Table 6.2 identifies and briefly describes several social concepts potentially relevant to IA process management. In considering these concepts the dangers of preconceptions and implicit assumptions about the conduct of the process, about the choice and application of methods, about the perspectives of potentially interested and affected parties, and about the potential social impacts are immediately apparent. The analysis and interpretation of social impacts should be approached with caution. Assumptions should be carefully scrutinized. Ongoing adaptations would be required. The world and proposed actions should be seen

Table 6.2 Examples of Potentially Relevant Social Concepts

Technical/scientific SIA	Relies heavily on natural and social sciences Seeks to maximize net social welfare (utilitarianism) Employs an adapted rational planning process Addresses change with and without proposal against control study Sees SIA as a technical component of the planning and decision-making process Uses a reductionist and objective research mode (scientist as neutral observer) Particularly helpful in structuring monitoring
Political SIA (also social conflict)	Conflict over resources and interests central to social life An interest-based approach; decision-oriented, value laden, and political Social life is diverse; social order is based on manipulation and control by dominant forces Tends to be issue-oriented; emphasizes openness, rectifying inequities and empowerment Seeks to understand basis for conflicts and how conflicts escalate Seeks to manage or contain conflict sufficient to identify mutually acceptable actions Stresses the need to strengthen the local institutional base (i.e., capacity building and community development) May be a realistic approach when positions are polarized and "worldviews" conflict
Participative/community-based SIA	In the tradition of collaboration and mutual learning of researcher and community, leading to social action Knowledge and experiences of individuals most affected by proposed change as the basis for impact prediction Roles of SIA practitioner—facilitator of knowledge sharing, interpretation, and impact reporting Process value laden and political Focused on community concerns Often employs techniques such as interactive community forums and frameworks such as community response (social viability, economic viability, political efficacy) and community organization models Emphasizes building social capital, capacity building, good governance, community engagement, and social inclusion Seeks to turn the impacted people into beneficiaries
Positivistic social science	Modeled after natural sciences; hypotheses tested by carefully analyzing the "numbers" Researcher as detached, neutral, and objective Stresses the value of an experimental, objective research approach, which seeks to logically explain cause–effect relationships Although flawed in its assumption of objectivity, is still instructive in terms of systematic and explicit research procedures; helpful in detecting methodological bias but contains own, often implicit, assumptions Social sciences can be difficult to apply in SIA because of inconsistencies in units of analysis, theoretical models, and language; social scientific traditions tend to be critical and discursive rather than predictive and explanatory (lack of a reliable set of theories for predicting social impacts) Weak on anticipating changes in the unquantifiable
Functional, ecological, and systems theory	Assumes shared norms and values in society Assumes a stable, cohesive, consensus based, and orderly social system; based on reciprocity, cooperation, and recognition of authority Assumes that system units are functionally related; change seen as an outside disturbance to an otherwise harmonious system; change accommodated by subtle shifts in system parts Reflected in most rational and participatory IA processes
Interpretative social science	Adopts a practical approach; not value free; common sense a vital information source Seeks to understand how people manage their everyday lives and construct meaning in natural settings Recognizes that people experience social reality in different ways Sees the unique features of specific contexts as essential to an understanding of social meaning May be helpful in addressing community-level impacts
Critical social science	Sees social science as critical and action-oriented; a political, moral activity Research conducted to critique and transform social relations Focuses on identifying and rectifying distortions and inequities Argues that social reality has multiple layers (illusions, myths, distortions, false consciousness) Potentially useful for addressing community empowerment issues

Table 6.2 (*Continued*)

Exchange theory	Assumes that human behavior reflects peoples' attempts to maximize rewards (utility) for involvement Expects that people will only become involved (and will continue to be involved) if they will benefit (or are rewarded) for their involvement; interaction seen as an exchange of rewards Requires a careful analysis of rewards (e.g., monetary, prestige, power, appreciation) that a setting is offering and how the rewards and patterns of exchange will be affected by a given change Sometimes basis for expectation that level of community acceptance will increase with the level of local benefits and compensation offered; of dubious validity and can be ethically problematic Has been incorporated into some siting approaches and explains some behavior
Symbolic meaning	Focuses on the inferred meaning attached to actions rather than to actions themselves People learn meanings and symbols in social interactions; can also alter meanings through introspection (their own interpretations of situations) and through interactions Conflicts may be exacerbated by definitions of situations Definitions of the situation by groups and individuals highly relevant to IA practice
Social learning	An approach for linking social concerns and public participation Includes both cognitive enhancement (e.g., learning about problem, learning about the values and interests of others) and moral development (e.g., developing a sense of self-respect and respect for others, developing moral reasoning skills) A potentially useful procedure for integrating social and moral considerations into participatory planning approaches
Phenomenological sociology	Focuses on describing and studying one's own and others' experiences without preconceptions Importance of avoiding preconceptions about external causes and consequences Opposed to objectivism, positivism, the acceptance of unobservable matters, and unsupported speculative thinking Seeks to analyze and describe everyday life; assumes that people create the world rather than being formed by social forces Although highly theoretical, underscores the need to begin with as few preconceptions as practical and to start from public and other stakeholder experiences and perspectives

Sources: Becker et al. (2003, 2004), Burdge (1995, 2004), Craib (1984), Craig (1990), Halstead et al. (1984), Harris et al. (2003), IAIA (2003), ICPGSIA (2003), Lane et al. (2003), Lee (2000), Manring et al. (1990), Newman (1997), Puschchak and Farrugia-Uhalde (2009), Ritzer (1996), Ross and McGee (2006), Rossou and Makan (2007), Schirmer (2011), Storey and Noble (2005), Vanclay (2003, 2006), Webler et al. (1995).

through the eyes of and as experienced by potentially affected parties. If social impacts are, in part, socially constructed, this suggests a socially constructed IA process (i.e., collaboratively designed and managed with interested and affected parties). Perspectives, norms, perceptions, beliefs, and values will change over time and can vary greatly depending on individual and group characteristics and depending on the level of social aggregation. This suggests iteratively exploring social impacts from multiple perspectives and at multiple levels. The magnitude and nature of social impacts are partly dependent on the process. The process is both an end (e.g., to provide a sound decision-making basis) and a means (e.g., a way of facilitating community empowerment, of avoiding and ameliorating adverse social impacts, of contributing to social viability, and of generating and enhancing social benefits) (Ross and McGee, 2006; Vanclay, 2003). This implies a process that is sensitive to public perceptions and perspectives, that actively seeks to manage positive and negative impacts from the outset (i.e., impact management as a continuous function), and that facilitates the achievement of community objectives.

Meaning and value are socially determined and are adjusted through social interactions. Dialogue is central to social interactions. Distortions in dialogue can exacerbate social impacts. The IA process is a form of social interaction. Dialogue and community/indigenous knowledge are central attributes of the process (IAIA, undated, a). This suggests designing and managing the process to facilitate dialogue, to contribute to co-learning, and to minimize communications distortions (Lockie, 2001). It also points to the need to understand how the IA process, as a form of social interaction, fits within and potentially affects existing social interaction patterns. Social interactions and impacts are both political and ethical. This suggests an IA process consistent with procedural and ethical principles and standards (e.g., equity, enhancement of marginal groups, gender equity) (see Chapter 10), conducive to the realization of social aspirations (e.g., reduced dependency, capacity building, building of social and human capital), and facilitative of the attainment of political objectives (e.g., empowerment, subsidiarity, strengthen democracy) (see Chapter 8) (Sharma, 2010; Vanclay, 2003). It also can entail community or indigenous control of the SIA (O'Faircheallaigh, 2009).

A reorientation of the SIA process, consistent with the view that SIA is primarily a mechanism for facilitating constructive social and political interaction and change, could result in less emphasis on impact prediction; more stress on co-learning and impact management; and a process contingent on the free, prior, and informed consent of proposal-affected people, formalized through an impact and benefits agreement (Esteves et al., 2012; Lockie, 2001). A taxonomic IA approach (which assumes minimal interactions among impact categories) is highly inappropriate given the dynamic nature of social interactions and impacts. What is required instead is an IA process built around conceptual models, frameworks, and stories that explore and trace through patterns of interaction and available choices from multiple perspectives (Vanclay, 2002). SIA is particularly concerned with the differential distribution of impacts among different groups in society, especially the most vulnerable (Vanclay, 2003). The most effective SIA processes draw upon and effectively integrate elements from technical, participatory, and political SIA models and methods (Ahmadvand and Karami, 2009; Becker et al., 2004; Lane et al., 2003).

The social sciences can make an important contribution to designing and applying an IA/SIA process. SIA encompasses such core social concepts as culture, community, power, human rights, gender, justice, place, resilience, sustainability livelihoods, capital, and participation (Esteves et al., 2012). But it is not a simple case of directly applying social science methods and models. There are multiple, overlapping, and conflicting social scientific models and methods available. Social science theories, models, and concepts can be quite effective in characterizing social change processes and impacts (Lima and Marques, 2005). They tend to be less effective in predicting the direction and magnitude of social changes and impacts. They are especially problematic regarding qualitative but significant aspects of the social environment (Puschchak and Farrugia-Uhalde, 2009). There are numerous interpretations of the appropriate purposes for and conduct of applied social research. Moving from the theoretical and the explanatory to the prescriptive and the practical can be very difficult. Contextual adjustments are essential. A clearly articulated theoretical framework is essential for effective social impact monitoring (Rossou and Makan, 2007). It is also important to identify and appreciate the implications of knowledge, resource, and control constraints and obstacles.

Notwithstanding the strong consensus on good SIA practice, the often-marginal role of social considerations in IA processes suggests the need for SIA practitioners to proactively advocate and extend the role of SIA within and among organizations (Bronfman, 1991; Esteves et al., 2012). The impartial analyst role is insufficient. It also implies the need for clearer definitions, enhanced methods, more follow-up research, a concerted effort to enlarge the SIA knowledge base, and a reconsideration of the nature of the SIA process (Burdge, 2002; Lockie, 2001).

The social concepts, in common with the ecological concepts, can shape and can be influenced by the IA process. Process and substance are intertwined. It is becoming increasingly possible to identify performance standards, albeit with contextual adjustments, for a socially substantive and sustainable IA process (ICPGSIA, 2003; Vanclay, 2003). General procedural characteristics, more and less conducive to combining substance and process in the IA process, can also be identified. SIA practice (e.g., locally appropriate mitigation, community benefits, equitable distribution of benefits) can be further enhanced with a greater emphasis on social follow-up, especially when it is structured, simple and workable, broadly available, empowers local residents, and facilitates social sustainability (Brown et al., 2003; Burdge, 2003b; Esteves and Vanclay, 2009; Lavallée and André, 2005; Lima and Marques, 2005; Petäjäjärvi, 2005; Rossou and Makan, 2007; Storey and Jones, 2003; Storey and Noble, 2005).

6.4.3 Sustainability Concepts

The roots of sustainability or sustainable development, as a concept, have been traced well back into the nineteenth century and beyond. The definition most commonly used as a point of departure is that of the World Commission on Environment and Development (WCED)—"development that meets the needs of the present without compromising the ability of future generations to meet their own needs" (WCED, 1987, p. 8). There is a continuing, albeit diminishing, debate surrounding how broadly sustainable development or sustainability (sustainability is used here for the sake of brevity and to avoid the impression that sustainability is only associated with development) should be defined. It is increasingly accepted that sustainability should encompass such considerations as intrageneration inequities, spatial inequities, human aspirations, other species needs, public participation in decision making, ecological limits, and relationships among sustainability forms and sustainability instruments. Common to most definitions is a desire to maintain, over an indefinite future, necessary and desired attributes of the sociopolitical system and of the natural environment (i.e., what society and communities want to sustain) (Deakin et al., 2002; Hanna, 2009a; Robinson et al., 1990).

Most recent characterizations view sustainability as holistic, ambiguous, adaptable, normative (i.e., objective-driven), and integrative rather than reductionist, deterministic, rigid, baseline-driven, and value-free (Bond, 2010; Gibson, 2006a; Pope, 2006; Pope et al., 2005). Sustainability also tends to be characterized as a continuous process of improvement (providing sustainability outcomes also are delivered); as more than just a global environmental perspective (also encompassing local, organizational, and societal perspectives); as necessarily democratic, open, inclusive, and equitable; as principles based rather than pillar-based, and as both universal and context-dependent (Binder et al., 2010; Bond, 2010; Craik, 2008; Dalal-Clayton and Sadler, 2004; Faber et al., 2010; Gasparatos et al., 2007; Gibson, 2006a; Grinde and Khare, 2008; Pope et al., 2004; Pope, 2006; Pope and Dalal-Clayton, 2011).

Some key interrelationships among sustainability elements are highlighted in Figure 6.2. Definitions, thresholds,

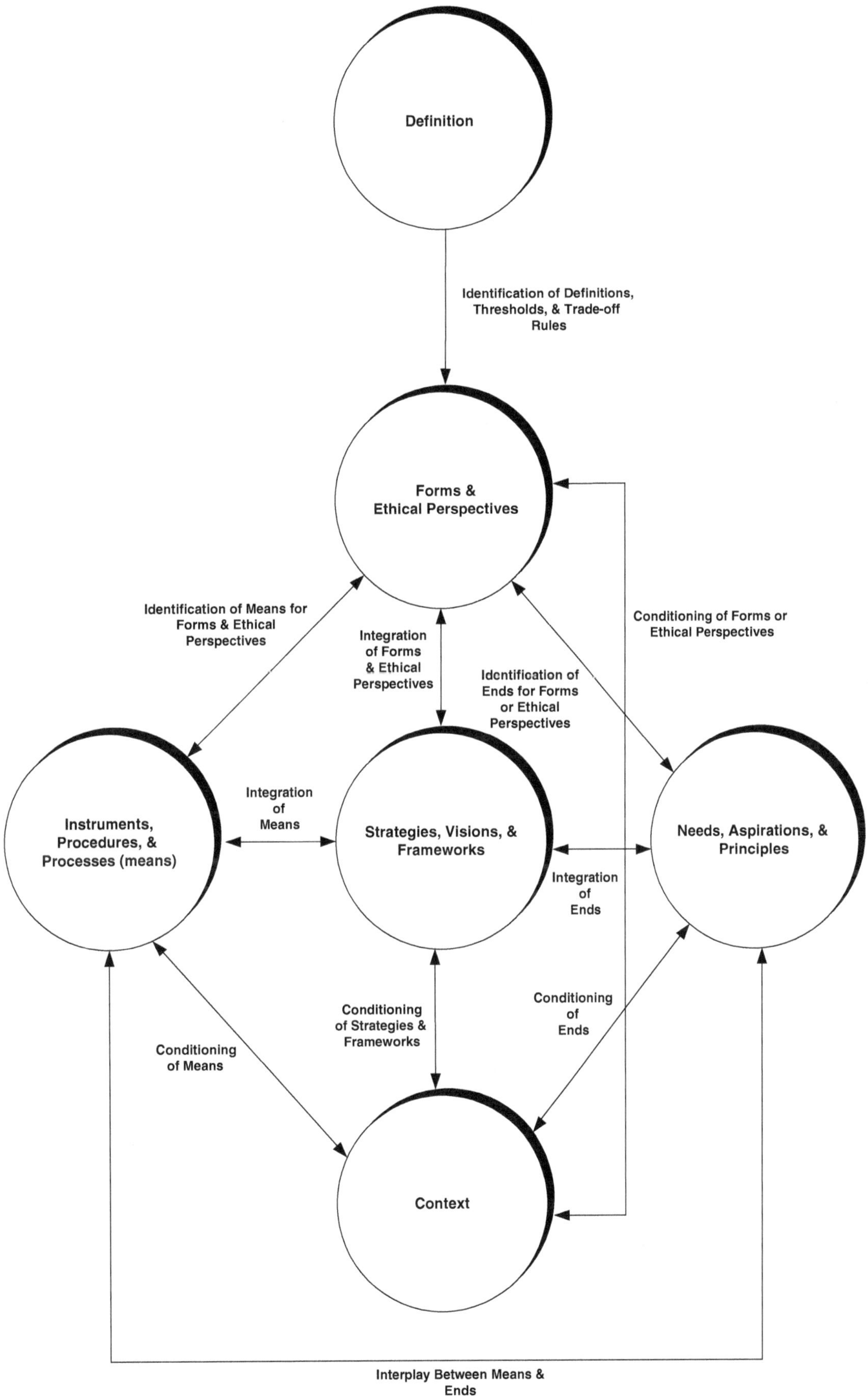

Figure 6.2 Sustainability elements. Adapted from Lawrence (1997b).

and trade-off rules are refined through sustainability forms and ethical perspectives. Sustainability forms concern overlapping and interdependent value systems (e.g., ecological, social, economic) (Franz and Kirkpatrick, 2007; Sadler, 1996). There is much debate concerning where the greatest emphasis should be placed and regarding how to best address interactions and interdependencies. Underlying these debates is a multiplicity of institutional, ideological, and academic sustainability perspectives and worldviews (Mebratu, 1998). These perspective differences are reflected in varying positions regarding such matters as the treatment of growth, weak versus strong sustainability (the latter is increasingly favored), the role of government and the market, and the importance of ecological limits and social justice (Bond, 2010; Constanza, 2000; Eales and Sheate, 2011). The net result is multiple versions of sustainability and multiple approaches to undertaking SA (e.g., minimizing negative vs. objectives-led vs. thresholds-driven, top-down vs. inside out vs. bottom-up, principle-based vs. triple bottom line or pillar-based, sustainability appraisal vs. sustainability IA vs. integrated IA, internal vs. external) (Binder et al., 2010; Fischer, 2007b; Henriques and Richardson, 2004; Pope et al., 2004; Pope, 2006; Robinson et al., 1990). Notwithstanding such differences, there are broad principles and imperatives common to many sustainability perspectives.

Sustainability forms and perspectives provide a basis for identifying sustainability ends (needs, aspirations, and principles) and sustainability means (instruments, procedures, thresholds, trade-off rules, and processes). Sustainability ends are both procedural (e.g., openness, fairness, participation) and substantive (e.g., ecosystem integrity, protection of biological diversity, enhanced quality of life, satisfaction of basic human needs, social justice). Sustainability ends can assume different forms (e.g., goals, principles), can represent both aspirations and thresholds that define unsustainability, and can operate at varying levels of detail (e.g., broad goals, specific objectives, priorities, criteria or indicators) (Hacking and Guthrie, 2006; Moles et al., 2008). Sustainability means are the mechanisms by which stakeholders work separately and together to move toward sustainability ends (e.g., green planning, legal instruments, indicators, thresholds and trade-off rules, financial incentives and penalties, institutional reforms, direct citizen mobilization, applied research, consumption and lifestyle choices, forums for joint planning and cooperation). Sustainability forms, ends, and means are brought together in sustainability strategies, visions, models, and frameworks. It is through such integrative mechanisms that core sustainability principles, themes, limits, decision rules, approaches, and methods are identified and applied (Brooke, 1998; Devuyst, 1999; Gibson, 2001, 2010). Part of application involves adapting ends, means, and strategies for different situations (i.e., sustainability as a situated concept) and clearly defined responsibilities (Binder et al., 2010; Kobus, 2005; Lawrence, 2009; Shearman, 1990). This, in turn, necessitates a regional/territorial understanding and definitions for sustainability (Péti, 2012).

The general interrelationships between sustainability and IA are illustrated in Figure 6.3. IA should be a proactive instrument for promoting sustainability (Sadler and Jurkeviciute, 2011). Sustainability and IA can be integrated at three levels—the conceptual level (theory and research), the regulatory level (sustainability-related IA requirements embedded within an overall sustainability governance regime), and the applied level (integrating sustainability concerns into IA practice) (Gibson, 2010; Gibson et al., 2005). Both IA and sustainability are concerned with maintaining and enhancing ecological, economic, and social environments. They reform, manage, and apply science and technology, institutional arrangements, and human environmental interventions (Pope and Grace, 2006). Both address interrelationships within and between environments and human activities. Sustainability can and should provide a means for redefining IA. IA can and should be an instrument for facilitating sustainability.

IA texts initially tended to assume that sustainability is an input to and an output from the conventional IA process. The essential features of the process remain unchanged. The relationships between sustainability and the IA process, however, are more complex, as illustrated in Figure 6.4 (Note: the numbers in Figure 6.4 are explained in the text). Sustainability offers the potential to extend and redefine the IA process within a fundamentally restructured institutional and societal setting (Pope and Grace, 2006). Project-level EIA and SEA, for example, identify, predict, and manage direct, indirect, and cumulative effects (1). Cumulative effects tend to be addressed incrementally (e.g., project effects in combination with the effects of related activities). Effects are projected into the future, usually assuming current trends persist. Sustainability, in contrast, starts with holistic images (often both desirable and undesirable) of the future and a systems perspective (Grinde and Khare, 2008) (7). It challenges conventional thinking and practice (Gibson et al., 2005). It begins from the whole and moves to the parts. It focuses on links and interdependencies. It adopts a long- and short-term perspective (Gibson et al., 2005). It does not assume that there is a single potential future or that trends define the future. It is embedded in complexity and surprise, understands the irreducibility of uncertainties and risks, adopts a precautionary approach, recognizes the importance of systems resilience and its determinants, proactively seeks creative innovation opportunities, and utilizes adaptive management and governance approaches (Gibson et al., 2005; Govender et al., 2006). It both extends from the present and traces connections back to the future. It pursues multiple, reinforcing gains (Gibson, 2010). It seeks a lasting sustainability legacy from planned, proposed, and potential change (Gibson, 2011). It meaningfully engages the broader community. It integrates IA types and decision making (horizontally across agencies and vertically among governments—both up and down) (Pope and Grace, 2006).

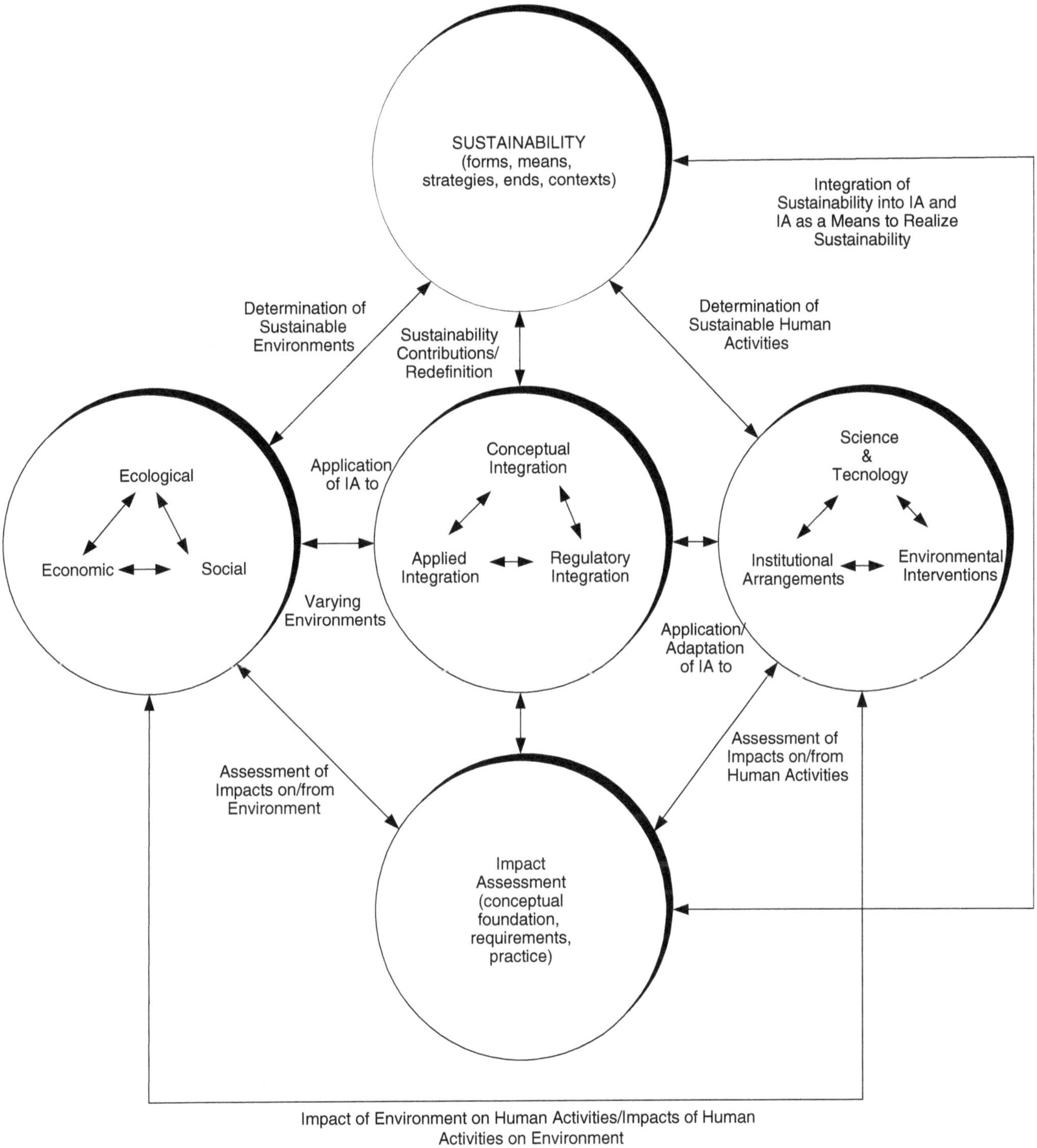

Figure 6.3 Integrating IA and sustainability. Adapted from Lawrence (1997b).

It is both shaped by (e.g., understanding of issues) and proactively strives to influence and reform policy and institutional contexts (Pope and Grace, 2006).

Conventional EIA and SEA approaches usually address significance in terms of the importance of individual environmental components, effects, or interactions (Hanna, 2005). Relationships between significance determinations and context are often poorly defined. Limited attention is given to systemic complexity and uncertainty. Sustainability visions (both ultimate state and milestones) and limits provide a context and a touchstone for significance interpretations and impact management actions (2) (Hacking and Guthrie, 2006; Sadler, 1996). Options, proposed actions, and effects can be assessed in terms of whether they respect inviolable limits (e.g., environmental or social carrying capacities or thresholds) (4) and in terms of their relative

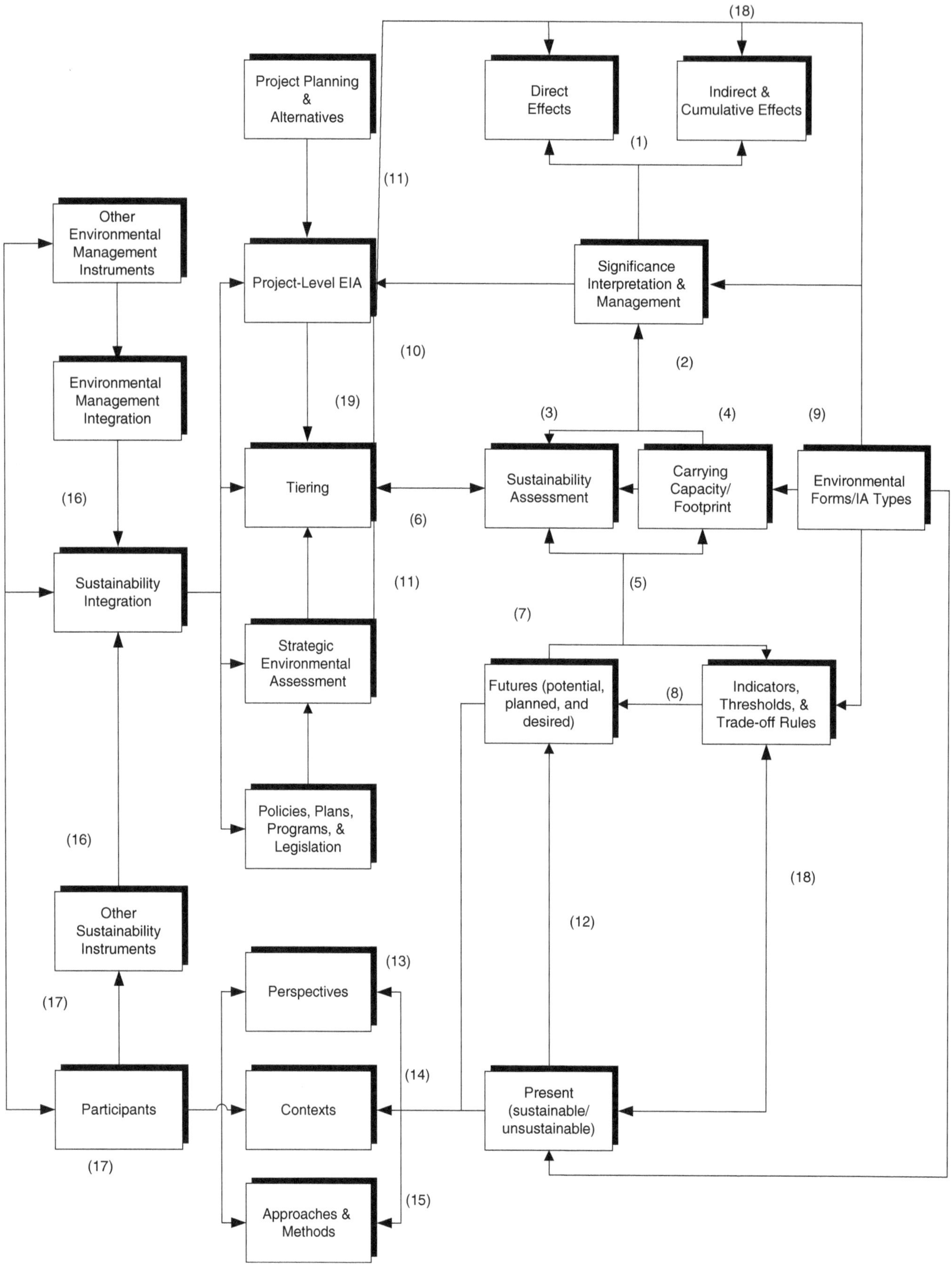

Figure 6.4 Sustainability in IA process.

contribution to sustainability visions, goals, targets, criteria, and principles (3) (Gibson et al., 2005; Noorbakhsh and Ranjan, 1999; Sadler, 1996). Sustainability analyses can be supported by qualitative and quantitative decision aid systems, models, and frameworks (Bruner and Starkl, 2004; Hassan, 2008; Hodge, 2004; Kain and Söderberg, 2008; Moles et al., 2008). Proposed actions or alternatives can be a catalyst for sustainability. A higher test is established for approvals (e.g., enhancements, net gains for sustainability) (Hanna, 2005).

Sustainability assessment could be considered the highest "rung" in the assessment ladder (6) (George and Kirkpatrick, 2008; Rotmans, 2006). It provides a context for SEA and links local actions to global concerns (Berke, 2002). Trade and other international agreements and broad strategies, which seek to integrate individual SEAs, could be subjected to a sustainability assessment (Elkin and Voiturez, 2009). Sustainability analyses could incorporate global and transboundary effects and priorities into integrative strategies, into lower order SEAs, and into project-level EIAs. Ultimately, SAs can represent a mechanism for integrating all IA forms and environmental management tools, both among (e.g., IA as a contribution to international governance) and within jurisdictions (Burdett, 2008b; George and Kirkpatrick, 2008). The database for EIA and SEA often has gaps and inconsistencies, especially as scales are broadened (e.g., to address cumulative effects) and as time horizons are extended. Sustainability criteria and indicators, which combine and supplement environmental, social, and economic indicators and which systematically cross and transcend pillar (social, economic, ecological) issues, can address these gaps (Gibson et al., 2005) (8). Ends and means are intertwined (Gibson et al., 2005). Sustainability principles, criteria, indicators, and trade-off rules, supported by participatory IA, also help determine whether ecological and social thresholds are being approached and whether progress is being made toward sustainability targets (18) (Fahy and Cinnéide, 2007; Hermans and Knippenberg, 2006; Jepson, 2001).

Conventional project-level EIA and SEA commonly treat social, economic, and ecological effects separately. Critical links among such effects are usually considered when addressing cumulative effects. Sustainability, in common with integrated assessment, recognizes that social, economic, and ecological systems are highly interdependent (Rotmans, 2006) (9). It uses holistic visions and integrative frameworks to address interdependencies from the outset (Bond and Morrison-Saunders, 2011). Sustainability decision rules often focus on interdependencies (e.g., economic growth within ecological carrying capacity). Sustainability can help focus project-level EIAs and SEAs (i.e., contributing to or undermining of sustainability) (10). EIAs and SEAs can be guided by sustainability visions, goals, and precisely defined principles, thresholds, criteria, and trade-off rules (Benson, 2003; Gibson et al., 2005). They can broaden spatial and temporal boundaries to address global and intergenerational impacts (Hacking and Guthrie, 2008). They can characterize baseline conditions in terms of sustainable and unsustainable activities and environments. The generation and evaluation of alternatives can focus on precautionary choices that are likely to be conducive to sustainability, offer the greatest overall benefit, and avoid undesirable trade-offs (Gibson et al., 2005) (11). Procedural and substantive ethical principles (e.g., inter- and intragenerational equity) can be prominently featured (Bond and Morrison-Saunders, 2011).

Proposed actions, rather than being viewed only as environmental intrusions (to be ameliorated to acceptable levels), can be treated instead, as potential sustainability opportunities or catalysts. Unsustainable alternatives can be screened from consideration (e.g., threat to carrying capacity). The remaining alternatives can be evaluated using sustainability decision rules (e.g., maintenance of natural capital, waste generation within assimilative capacity, renewable resources within regeneration rate, nonrenewable resources equal to substitution rate) (Gibson, 2006a; Goodland, 1993; Noorbakhsh and Ranjan, 1999; Sadler, 1996). Mitigation, compensation, and local benefits measures can maintain and enhance sustainability (e.g., in kind compensation for natural capital loss, fair distribution of benefits and risks) (Gibson, 2011). Before and after comparisons can be undertaken to determine sustainability-related changes (Thérivel and Minas, 2002).

The overall IA process can continually reflect back to the original vision and objectives (Bond and Morrison-Saunders, 2011). It can be supported by sustainability science, guided by sustainability procedural (e.g., keep options open, precautionary principle, a fair, accessible, efficient, and effective process) and substantive principles, informed by sustainability advisors, and supported by integrative frameworks, networks, models, and methods (Beatley, 1995; Gibson, 2001; Gibson et al., 2005; Hacking and Guthrie, 2008; Kates, 2000; Sadler, 1996; Scanlon and Davis, 2011; Slocombe, 1993). The process is open, transparent, iterative, inclusive, and learning oriented (Gibson et al., 2005; Singh et al., 2009; Täbara and Pahl-Wostl, 2007; Tuinstra et al., 2008). Proposed decisions must operate within key sustainability constraints, thresholds, and limits. They also must minimize the negative, optimize the positive, compensate for the loss of valuable features and benefits, and adhere to predefined trade-off rules (e.g., net overall and mutually beneficial, social, economic, and ecological gains, burden of proof on proponent, protection of the future) (Gibson, 2006a; Thérivel, 2010). SA can be broadened to subsume other IA forms, provide the institutional context for formulating and implementing SA good practice norms, make sustainability operational in an organizational setting, and contribute to international governance (Craik, 2008; George and Kirkpatrick, 2008; Kiewiet and Vos, 2007).

Sustainability recognizes that there are multiple perspectives concerning how the world is, how the world is likely to be, and how the world should be (13). These perspective differences result in many different pathways (i.e., an open-ended process) from the present to sustainable or unsustainable futures (Gibson et al., 2005) (12). Sustainability initiatives seek to identify the "overlapping consensus" among interested and affected parties that will provide a basis of action (17) (Rawls, 2001). A variety of approaches and methods can both define a sustainable future (e.g., assessment frameworks, visions, scenarios, models, participatory tools, indicator sets) and assess the contributions of individual proposals (e.g., apportionment techniques, sustainability indices, footprint analysis, multicriteria analysis) (14) (De Ridder et al., 2010; George, 1997; Jiliberto, 2004; Lee, 2006; Rotmans, 2006; Singh et al., 2009; Wackernagel and Rees, 1996; Wiek and Binder, 2005). Adjustments are also made for different settings and situations (e.g., a territorial/regional understanding of sustainability) (Kobus, 2005; Ng and Hui, 2007; Péti, 2012) (15). Multiple perspectives and methods are available for characterizing present and potential future conditions (i.e., methodological pluralism) (Gasparatos et al., 2007).

Project-level EIA and SEA operate largely independently from other sustainability instruments. Connections are made to other instruments but usually only after the process is well advanced, often during the review and approval stage. Sustainability initiatives recognize that many mutually supportive instruments are required (16). Efforts are made to ensure complementary visions, actions, and monitoring systems (12). The IA process could be modified to more effectively address the advantages and constraints associated with integrating sustainability concerns into both EIA and SEA (19). Ultimately, other IA forms should be transcended by and integrated within SA (Burdett, 2008b). These efforts could be broadened to embed, from the outset, SEA and EIA within the full network of sustainability instruments (16 and 19).

Many issues, obstacles, and dilemmas remain concerning how to best integrate sustainability and the IA process. The theoretical base for sustainability is still in need of refinement. Many questions are still being raised regarding how to best determine what is sustainable, over what area, and for how long (Briassoulis, 1999; Shearman, 1990). There are many debates concerning who integrates the various analyses and decides what is and what is not sustainable (Morrison-Saunders and Thérivel, 2006; Robinson et al., 1990). Apportionment procedures, how to consider uncertainties, and the treatment of compromises and trade-offs are difficult issues requiring further attention (Gibson, 2001, 2006). The inherent flexibility of sustainability can result in the adoption of alternatives that are "good enough," rather than sustainable (Bond and Morrison-Saunders, 2009). The fragmentation of disciplines, sectors, and institutions; weak vertical and horizontal integration; jurisdictional gaps; the fundamentally weak concept of sustainability most commonly applied; the lack of political and institutional will, understanding and SA competency; and the disconnect between attempts to govern the global environment and manage the global economy continue to hinder IA and sustainability integration efforts (Ayre and Calloway, 2005; Eales and Sheate, 2011; Kirkpatrick and George, 2006; Kobus, 2005; Tang, 2010). Sustainability challenges prevailing assumptions, institutions, and practices (Gibson et al., 2005) Accordingly, resistance to change and to the integration of SA into decision making tends to be the norm, with IA in any form (but especially SA) being viewed as an unwelcome hurdle rather than a useful planning mechanism (Eales and Sheate, 2011; George and Kirkpatrick, 2008; Voituriez et al., 2006). The causes of unsustainability are deeply embedded within prevailing institutions (Dovers, 2005). Reorienting and reforming the institutional and political context of IA is, at best, a long-term generational goal (Dovers, 2005).

Monitoring (especially of critical feedback loops and of decision quality), effective SA guidance, far greater emphasis on enhancement, and the robust auditing of SA experiences (which approaches and methods delivered and failed to deliver practical results, sustainability effectiveness) are critical to the advancement of the sustainability knowledge base (Burdett, 2008a; Grinde and Khare, 2008; Hacking and Guthrie, 2008; João et al., 2011; Morrison-Saunders and Hodgson, 2009; Pope and Klass, 2010; Thérivel et al., 2009). Care needs to be taken that more quantitative approaches to integrated IA and SA do not undervalue the qualitative, inhibit stakeholder involvement, and underestimate uncertainties (Duncan, 2008; Hodge, 2004). So far, effectiveness SA reviews range from discouraging to mildly positive (Thérivel et al., 2009; Wilson, 2010). Particular weaknesses include a failure to envision radically different futures; a reactive rather than a proactive approach to policy and option design; a tendency to rely on superficial check-list approaches (breadth over depth); and the less than holistic treatment of climate change, health, and gender impacts (Eales et al., 2005; Kessler and Abaza, 2006; Milner et al., 2005; Wilson, 2010). More attention needs to be devoted to capacity building, interagency coordination, and multi-stakeholder participation (Kessler and Abaza, 2006).

Some argue that sustainability is either not possible or is a "smokescreen" for "business as usual." Others suggest that more fundamental changes in values and behavior, coupled with fundamental institutional changes, are necessary before any discernible progress toward sustainability can be made (Connor and Dovers, 2004). On the bright side, the range of sustainability initiatives is enormous and the record of tangible improvements from these initiatives is considerable. Sufficient experience in undertaking sustainability assessments or appraisals has already been acquired and effectiveness factors (e.g., broad local involvement, early in the process, adequate resources) have been identified (Thérivel and Minas, 2002). An optimistic interpretation would be that IA practice can build on the successes while appreciating and addressing the constraints. It is an overstatement to suggest that IA has made more than a minor contribution to sustainability to this point. IA process reforms, along the lines described above, could increase that contribution.

6.4.4 Methods

Formulating and applying a substantive IA process requires numerous methods sensitive to ecological, social, and sustainability characteristics and objectives. The ecological, social, and sustainability concepts described in the previous subsections demonstrate the need to systematically address interrelationships. IA practice makes considerable use of network analysis, systems diagrams, and modeling to address interconnections and interdependencies. The systematic consideration of interrelationships finds its fullest expression in cumulative effects assessment (CEA). All four jurisdictions provide guidance for selecting and applying CEA methods. Methods for addressing interrelationships among disciplines are not as fully developed as those for considering interrelationships within disciplines.

IA is about decision making for the future. Substantive IA processes must consider long-term implications and explore pathways toward and back from sustainable futures. IA generally relies on projection and forecasting techniques when anticipating future conditions. Although helpful, such techniques provide only a partial picture of a potential future. They are also weak on social concerns, are lacking in vision, often underestimate uncertainties, and are prone to quantitative and conservative biases. Visioning, scenario writing, and story telling are better able to integrate qualitative, social, ecological, and political considerations. They can also provide multiple images of a desired future and that of varying routes to that future. Backcasting helps work back through decisions and actions from a desired future to the present.

The IA process should establish proximity to thresholds; assess progress toward ecological, social, and sustainability ends; and compare alternative courses of action. Ecological footprint analysis, carrying capacity analysis, and environmental indicators (see Table 6.1) can help assess status and choices. The process must adapt to and manage uncertainties and data gaps. Rapid rural appraisal, scenario writing, and adaptive environmental assessment (also see Chapter 11) are well suited to addressing uncertainties. A high level of community participation is essential for making IA processes more substantive. Visioning, story telling, participatory rural appraisal, the ecosystem approach, and social learning can all help in involving the interested and potentially affected parties.

6.5 INSTITUTING A SUBSTANTIVE IA PROCESS

6.5.1 Management at the Regulatory Level

All four jurisdictions (the United States, Canada, Europe, and Australia) explicitly combine procedural and substantive requirements, albeit in different ways and with varying degrees of success. Table 6.3 provides examples of how the four jurisdictions have sought to integrate procedural and substantive requirements. IA requirements can, for example, be made more substantive if IA legislation includes detailed environmental and sustainability policies, objectives, tests, and principles as a "touchstone" for generating and evaluating alternatives and as a framework for identifying, interpreting, and managing effects. The cause of more substantive IA requirements can be furthered by a broad definition of actions, proponents, alternatives, environment, and effects (including cumulative effects) by explicit and systematic links to related substantive environmental policies, plans, strategies, and requirements and through systematic tiering and horizontal integration institutional arrangements. It can also be facilitated by detailed substantive environmental guidance (e.g., biodiversity, human health, social heritage, climate change, significance determination factors); through triggers and special requirements for environmentally sensitive and significant areas; by means of detailed scoping, follow-up, and auditing requirements; and through applied environmental research and sponsoring initiatives. Care should be taken to ensure that environmental quality is not compromised by timing limits, IA substitution institutional arrangements, exemptions, and discretionary authority.

Arguably, the broad application of different forms of IA requirements to various classes of actions is consistent with the intent of IA to broadly integrate substantive environmental concerns into public and private decision making. The danger is that an all-encompassing set of requirements could consume a vast amount of resources, oftentimes in vacuous procedural requirements of minimal environmental benefit. A narrower range of proposed actions, it could be posited, with more tightly circumscribed requirements, could ensure that available resources are focused on those actions most likely to induce significant adverse effects. But such focused requirements ignore the intent of making decision making more environmentally substantive and often miss out on potentially significant environmental effects associated with vulnerable settings and cumulative effects potential. The middle ground between these two positions is more onerous IA requirements for major (with clear thresholds) actions and/or actions in highly sensitive/significant settings; simpler and less onerous requirements (e.g., screening procedures, class assessments) for routine decision making involving actions not likely to induce significant adverse effects; and a more discretionary, but transparent, procedure for actions that, depending on the circumstances (e.g., cumulative effects potential), could fall into either camp. This type of streaming approach to "IA triggering" is facilitated when there is an IA hierarchy such that higher level IAs requirements bind and shape lower level IAs.

On the surface, IA requirements that broadly define the environment (e.g., physical, ecological, social, economic, cultural, health, heritage, sustainability) and effects (e.g.,

Table 6.3 Positive and Negative Examples of Regulatory Level IA Examples Regarding Environmental Substance

United States	Canada	Europe	Australia
(±) National Environment Policy Act (NEPA) broadly outlines national environment policy (e.g., harmony between man and environment, health and welfare, balancing population and resource use, present and future generations); courts have ruled only a procedural duty (±) Broad range of actions (major actions funded, assisted, conducted, regulated, or approved by federal agencies); in practice, largely projects and, to a much lesser extent, plans and programs (±) Broad definition of environment and effects; positive and negative direct, indirect, and cumulative, ecological, social, economic, health, historic, archaeological, and health effects; social, health, and economic effects on own not an EIS trigger (+) References irreversible and irretrievable resource commitments and conservation potential (+) Context and intensity factors provided for significance determinations (+) Considerable substantive guidance (e.g., human health, biodiversity, protected areas, heritage resources, energy conservation, coastal zone management, floodplain management, forest health, environmental quality, pollution and waste prevention and control, cultural resources) (King, 2002) (+) Extensive array of executive orders (e.g., wetlands, invasive species, protected areas, migratory birds) (+) Applied research and coordination initiatives (e.g., ecosystem approach, adaptive management, coastal zone management)	(+) Broad purposes (e.g., promoting sustainable development, avoiding significant adverse environmental effects, cumulative effects) (+) Application broadened to include Crown Corporations, federally funded projects on reserve lands and national airport authorities (+) Factors include effects significance, accident and malfunction potential, and cumulative effects (+) Regulation triggers—selective project types in wildlife or migratory bird sanctuaries, offshore or in Yukon or Northwest Territories (+) Effects definition cross references specific environmental legislation (e.g., Fisheries Act, Species at Risk Act) (+) Treatment of substantive environmental concerns strengthened by mandatory and enhanced monitoring provisions (+) Recent panel decisions have applied a sustainability test and utilized the precautionary principle (+) Each federal agency required to prepare a sustainable development strategy and federal government has a green plan and a sustainability strategy (+) Oversight role by Commissioner of Environment and Sustainable Development (+) Physical and cultural heritage guidance (CEAA, 1998b) (+) Aboriginal Traditional Knowledge guidance (CEAA 2010a) (+) Sponsored research on sustainability-based EA (±) Makes provisions for regional studies but no trigger	(+) Proposed EIA Project Directive (PPD) addresses environmental issues such as resource efficiency, biodiversity, climate change, and disaster risks (+) PPD—linked to Soil Thematic Strategy and Roadmap to a Resource Efficient Europe, UN Convention on Biological Diversity, and Europe strategy for smart, sustainable, and inclusive growth (+) PPD—broad definition of effects (+) PPD—detailed requirements regarding environmental project characteristics, project location, and characteristics of potential impacts (+) PPD approach to substantive environmental concerns strengthened by mandatory scoping and monitoring (−) PPD—limited to major projects (−) PPD time limits could inhibit systematic consideration of environmental concerns (+) SEA Directive—seeks to provide a high level of environmental protection; sustainable development an objective (+) SEA Directive—requirement to identify existing environmental characteristics and problems; reference to value and vulnerability of area due to special natural characteristics or cultural heritage, exceeded environmental quality standards or limit values, and intensive land uses (±) SEA Directive—mandatory for plans/programs prepared for agriculture, forestry, fisheries, energy, transport, waste/water management, telecommunications, tourism, town, and country	(+) Aims of Environment Protection and Biodiversity Conservation (EPBC) Act—protecting environment, especially matters of national environmental significance, biodiversity conservation, protection of world and national heritage, and promotion of ecological sustainable development (+) Detailed conservation of biodiversity provisions, including principles of ecologically sustainable development (Australian Government, 2011a) (+) Specifies matters of national environmental significance; detailed requirements and guidance (+) Provisions for SEAs of policies, plans or programs (e.g., regional development pressures, high growth, multiple stakeholders, complex actions, cumulative effects) (+) SEA mandatory for fisheries and regional marine planning; applied to national environmental protection measures (+) Accepted reforms—greater use of regional environmental plans, strategic assessment, and conservation agreements (+) Environment—ecosystems and parts, including people and communities, natural and physical resources, qualities and characteristics of locations, places, and areas, and social, cultural, and economic aspects; includes indigenous heritage values (+) Direct and indirect impacts; references to short and long impacts and whether unknown, unpredictable, or irreversible significance; courts interpreted as including cumulative effects (+) In response to independent review, government agreed to include vulnerable ecological communities as a matter of

Table 6.3 *(Continued)*

United States	Canada	Europe	Australia
(±) NEPA goals consistent with sustainability; inhibited in practice by procedural orientation (+) Draft guidance document on climate change effects and greenhouse gas emissions (+) US EPA Center for Sustainability (+) NRC research report on incorporating sustainability into EPA principles and practices (+) Various initiatives to improve environmental, energy, and economic performance (+) Updated principles and guidelines for water and land-related implementation studies (+) Gulf coast ecosystem restoration (working group, road map for restoring ecosystem resilience and sustainability) (+) Review of mineral management service (MMS) NEPA procedures for managing natural gas, oil, and other mineral resources on the outer continental shelf (+) Federal sustainability initiative (+) Interagency Ocean Policy Task Force	(±) Focuses on significant adverse environmental effects (−) Narrow range of decision makers (only three), EA Agency, National Energy Board, (NEB) and Canadian Nuclear Safety Commission (CNSC) actions and designated projects only (from project list subject to screening and on a case-by-case basis) (±) Physical activity regulations—various physical projects in selected locations and/or on basis of scale thresholds and other project characteristics (−) Limited to major projects (−) Narrow definition of environment (biophysical only) and effects (areas of federal jurisdiction, broader for effects on aboriginal people; discretion to broaden) (Gibson, 2009, 2012) (−) Application constrained by exemption of infrastructure projects, initiation review period, time limits, and potentially by provincial substitution provisions (−) Removes the requirement of assessing the capability of natural resources to meet future needs (±) Biophysical definition of environment for cabinet submissions; SEA effects—positive and negative ecological, social, and economic effects	and land use planning; includes community-supported programs and plans; excludes national defense, minor, financial or civil emergency; not applied to policies (±) SEA Directive—broad range of environmental issues and effects (e.g., biodiversity, population, human health, fauna, flora, soil, water, air, climate, material assets, protected areas, vulnerability, cultural heritage, landscape, and interrelationships; does not require consideration of social and economic effects; only negative significant environmental effects; includes probability, duration, frequency, and reversibility, transboundary and cumulative effects, human health and environmental risks) (+) SEA Directive cross references sustainability action plans and strategies; required to identify environmental protection objectives (+) European Commission—sustainability assessment for all major policy proposals (sustainability assessment unit) and for trade agreements (+) Extensive substantive guidance and applied research (ECORYS Research and Consulting in Collaboration with IZA, 2010; EC, 2009b; IEEP, 2004; Teller and Bond, 2002; UNEP, 2002)	national environmental significance (+) Protection of matters of national environmental significance also addressed through conservation agreements, bilateral agreements and bioregional planning (+) Cross references to obligations under international environmental conventions and intergovernmental environmental agreements (+) National environmental strategies (e.g., biodiversity, forests, oceans, sustainable development, greenhouse gas emissions) (+) Australian government draft biodiversity policy and draft environmental offsets policy (+) Required to address links to related actions and to allow for federal or state policies, plans, or programs (−) Requirements and guidance regarding social, economic, and equity matters very general

direct, indirect and cumulative, positive and negative, short- and long-term effects) are more likely to provide a comprehensive and integrated assessment of the potential consequences of proposed actions and their alternatives. Such a comprehensive approach is also potentially more conducive to the determination of whether sustainability is being promoted or inhibited. However, as detailed in Section 6.6, there is a well-founded fear that under such an IA regime, socioeconomic benefits will overwhelm the assessment and biophysical concerns will receive much less attention and resources. Socioeconomic benefits, moreover, it is argued, already assume a dominant role in decision making outside the IA process (i.e., a form of double-counting). The counter to this argument is that a system that is limited to biophysical concerns will exclude entirely valid adverse social and economic effects; push IA even further to the periphery of decision making; result in an even less transparent and accessible decision-making process (outside IA requirements); and undermine any efforts to coherently address complex interactions among effect types, cumulative effects, and sustainability.

A somewhat less direct but still important way of integrating substantive environmental concerns into IA requirements is through explicit links to related environmental requirements and forms of environmental management. The linking and blending of these various instruments can, at least in theory, provide a systematic framework for integrating procedural and substantive requirements, vertically among governments and decision-making levels, horizontally across agencies, and among various environmental planning and management initiatives. This type of loosely connected network can, if connections and roles are clearly defined, be effective, efficient, and adaptable, while maintaining the independence of various environmental management and planning initiatives and requirements. The integration of requirements targeted to specific environmental components can help identify objectives and priorities, raise minimum standards, and establish coherent analysis and management approaches (Carter and Howe, 2006; Pritchard, 2005). The greatest danger is that any potential for a holistic perspective will be severely diminished. This is especially the case if IA requirements lack clear substantive objectives and selectively address environmental components and effects; if the links among instruments are not explicit with clearly defined roles; if higher level planning and decision making does not frame lower level planning and decision making; and if there are major substantive gaps, missing connections, unwarranted duplication, and contradictory objectives, roles, and requirements. In such cases, "the whole is decidedly less than the sum of the parts."

Optimistically, a loose network of requirements, guidelines, and initiatives will gradually coalesce into a broad strategy or into a mutually consistent, but loosely affiliated series of strategies. This strategy or strategies will, in turn, lead to a more central role for environmental considerations in decision making and to tangible environmental improvements in the direction of sustainability. IA would assume a pivotal role in this effort. The IA process would be progressively reformed to be more fully conducive to realizing substantive environmental ends. The pessimistic interpretation is that the present "patchwork quilt" will remain largely unchanged or will be narrowed in scope and application. Major gaps and inconsistencies will continue. The EIA and SEA requirements will be too general to make an appreciable difference. Progress (or not) toward sustainable futures will be unclear. The implications for IA process management will remain largely unexplored. Only time will tell which interpretation is more accurate in which jurisdiction.

Figure 6.5 highlights several ways in which IA substance and process can be integrated, based on the experiences in the four jurisdictions. Procedural IA requirements, structured within an IA hierarchy, can integrate substantive IA types, be linked to related substantive environmental requirements, and be placed within the context of sustainability strategies and plans. Substantive requirements can be built directly into IA legislation and regulations. IA requirements can be formally merged with other environmental requirements and/or informally linked to strategic and regional planning and management efforts. Substantive requirements can be comprehensive (e.g., ecological, social, economic, sustainability). They can focus on specific priorities (e.g., ecological sustainability, environmental justice). Procedural and substantive requirements can be addressed through tiering (e.g., regulatory/sustainability assessment, SEA, project-level EIA). An effective blending of the procedural and the substantive will incorporate elements of each of these approaches. The four jurisdictions include aspects of each approach, albeit in different ways and to varying degrees. Distinguishing between effective and less effective combinations would require systematic effectiveness analyses, tempered by jurisdiction-specific adjustments. Still, at least on the surface, directly integrating substantive requirements into IA legislation, regulations, and guidelines, coupled with procedural guidance conducive to realizing substantive objectives and requirements, would seem the most direct route to more substantively effective IA practice.

6.5.2 Management at the Applied Level

Figure 6.6 illustrates an example of substantive IA process. The figure and the process description that follows draw upon the concepts and methods presented in Sections 6.3 and 6.4. IA process managers and participants can "pick and choose" the relevant and appropriate elements.

Start-Up IA, and by extension the IA process, is an instrument for realizing tangible ecological, social, and sustainability objectives. The process begins with an overview of pertinent environmental and sustainability plans, strategies, programs, and other public, private, and multistakeholder initiatives. These initiatives could influence the current and future environmental and community conditions in the geographic areas potentially affected by proposed actions. They provide a context. They could be undermined, complemented, or unaffected by any proposed actions. An analysis of need and opportunities is undertaken. Need and opportunities are addressed both in the conventional sense (e.g., market opportunities, public service needs) and in the sense of identified sustainability problems (i.e., shortfalls between sustainability objectives and expected future conditions). Parties potentially interested and affected by potential actions are identified. The perspectives, concerns, and positions of these parties are identified. Varying worldviews regarding current environmental conditions and desired future conditions are determined. Known sustainability constraints and opportunities are identified.

Study design and scoping are informed by these background analyses. They focus the process on major issues,

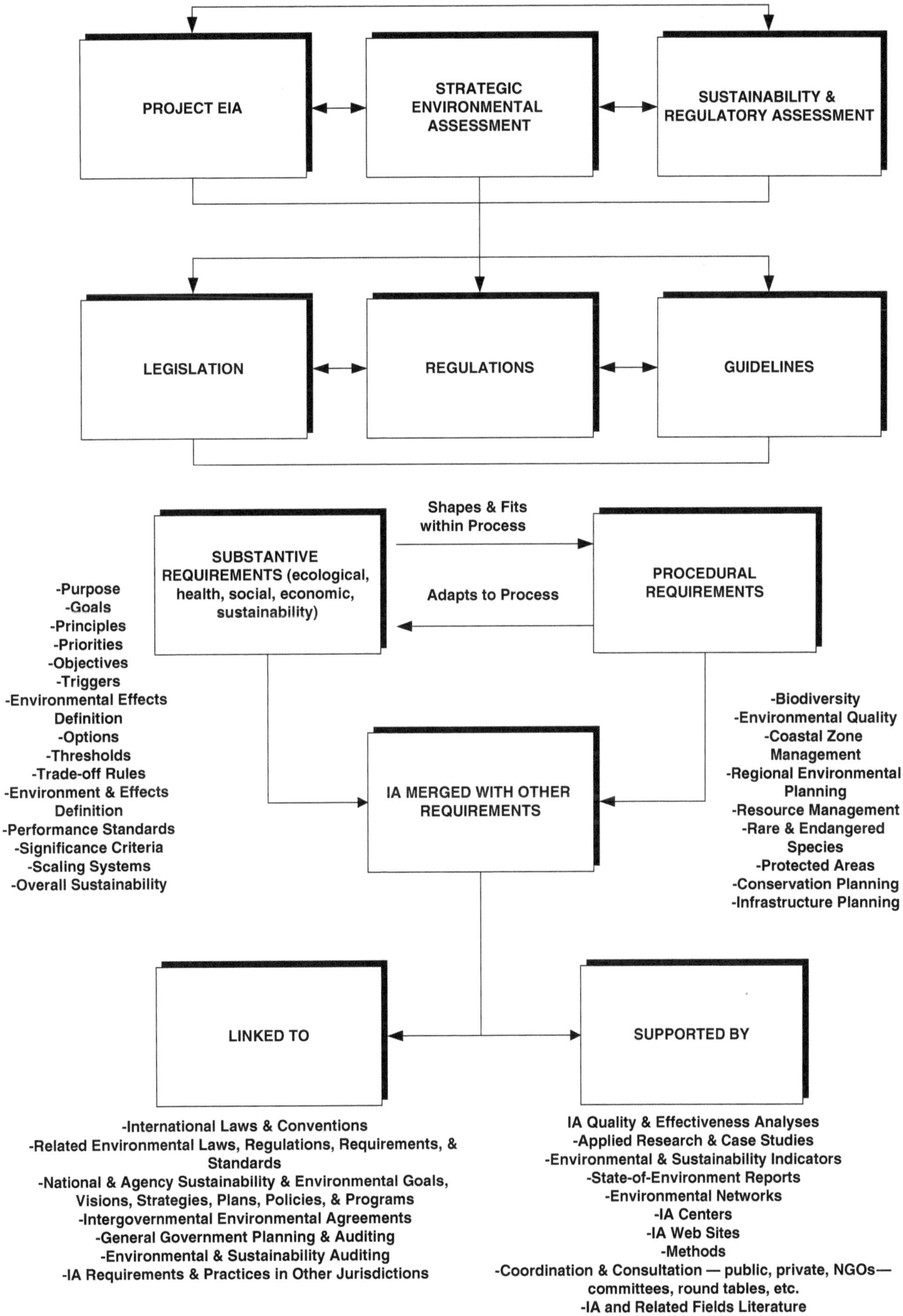

Figure 6.5 Examples of regulatory approaches to integrating IA and substance.

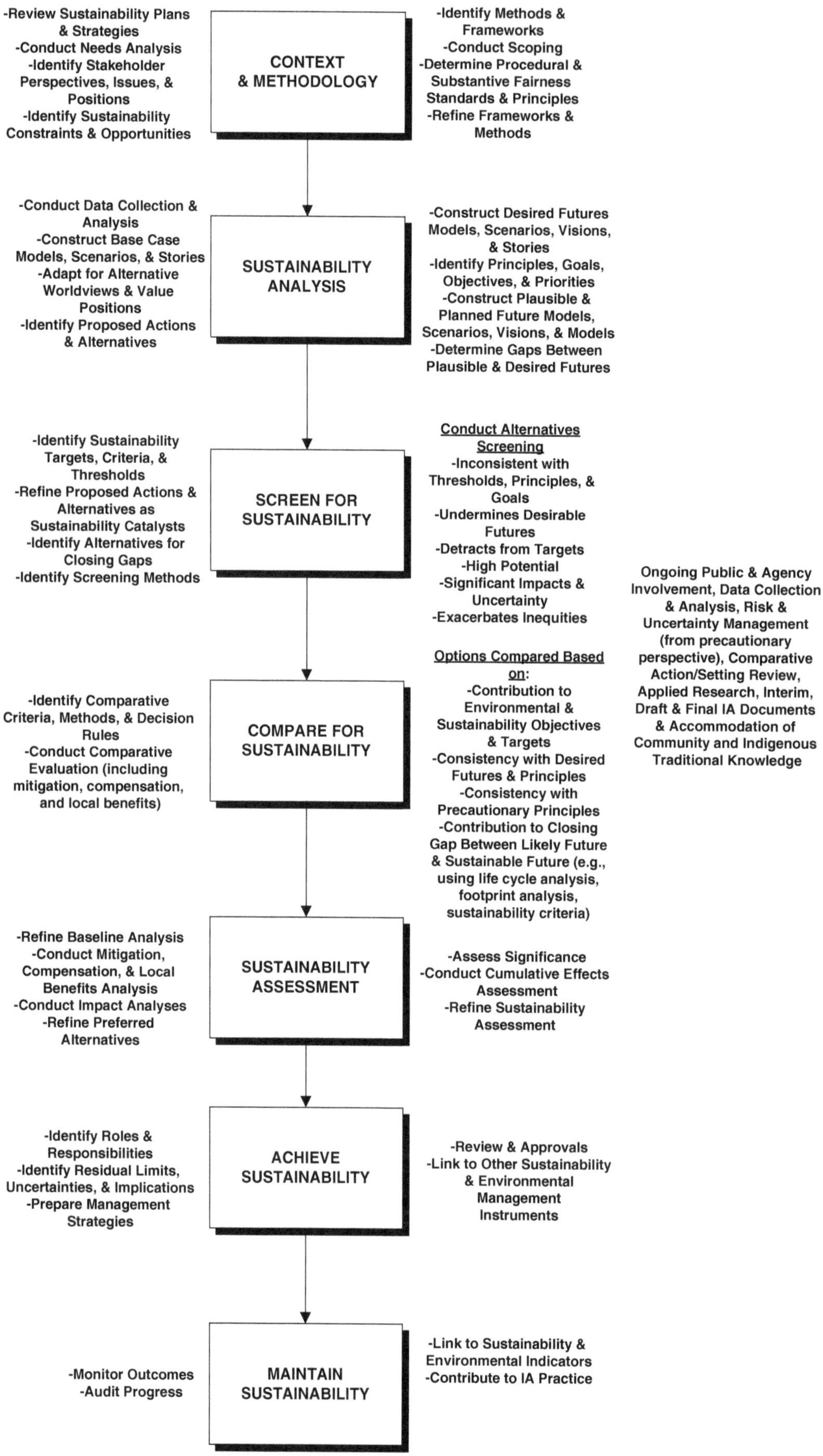

Figure 6.6 Example of a substantive IA process. Adapted from Lawrence (2005a).

needs, and participants within the context of relevant sustainability initiatives, constraints, opportunities, and perspectives. Frameworks for guiding the process and the types of methods likely to be applied in the IA process are identified. Procedural fairness standards and principles are jointly determined to facilitate the full and equal participation of interested and potentially affected parties.

Sustainability Analysis The sustainability analysis establishes the foundation for sustainability assessment. The sustainability analysis employs multiple qualitative and quantitative methods to characterize the present (sustainable and unsustainable activities and environments) and the future (plausible, planned, and desired)—all from multiple perspectives. Conventional IA processes tend to view proposed actions (and their alternatives) as intrusions upon social, economic, and ecological systems—the negative effects of which are to be minimized. In this case, the proposed actions and alternatives are envisioned as potential catalysts for a sustainable future. Methods and frameworks for undertaking the sustainability analysis are formulated and then progressively refined and adapted. A preliminary set of proposed actions and options are identified, refined, and adapted in an ongoing effort to meet the identified need while at the same time facilitating sustainability.

A series of base case (i.e., past and present conditions without proposed and possible actions or their alternatives) scenarios, models, and stories are first constructed. Conceptual and quantitative network diagrams and models explore spatial and temporal patterns of interrelationships between historical and current environmental conditions and activities. Alternative assumptions and inputs address uncertainties and explore varying interpretations. Scenarios and stories describe and explain how present conditions evolved from the past. These integrative tools are jointly formulated and applied with interested and affected parties. The analysis draws upon scientific and traditional data and knowledge. Different base case characterizations reflect alternative worldviews and value positions. Frameworks are constructed to explore overlaps and interconnections among disciplinary models and systems. Group consensus building and conflict resolution techniques identify common ground and residual areas of dispute among the characterizations. The analysis is undertaken at multiple levels (e.g., regional, community, local). It focuses on sensitive and significant environmental components and processes that could potentially be affected by proposed actions and alternatives. Activity patterns that appear sustainable and unsustainable, positive sustainability initiatives (that have and are making a difference), and sustainability constraints and opportunities (building on the start-up analysis) are identified. Multiple iterations are required to explore gaps, uncertainties, links, and varying interpretations. The outcome from these efforts is a small number of partially overlapping and partially interconnected base case characterizations. The core values and underlying assumptions of each are identified. Varying interpretations and perspectives are noted. Where practical, ecological, and social carrying capacities surpassed or in jeopardy are indicated. Major uncertainties and potential implications are highlighted.

The analysis next addresses future conditions, again in multiple ways and from multiple perspectives. Conventional forecasts identify trends in key conditions pertinent to sustainability, appreciating that trends are often a poor predictor of the future. Key attributes of the planned future, as envisioned in public policy and spatial planning documents, are highlighted. Plausible future conditions (pertinent to both sustainability and to proposed actions and their alternatives) are addressed using models, visions, scenarios, and stories. These plausible futures address how conditions (e.g., demographic, ecological, social, economic, political, institutional, technology) might evolve taking into account baseline conditions, forecasts (including those prepared by others), proposed activities in the area, sustainability instruments in operation, and the planned future. Visions, models, scenarios, network diagrams, backcasting, and stories also characterize desired, sustainable futures. The desired futures include both ends states (for various time horizons) and sequences of events (from the present to the future and back from the future to the present). The plausible and desired futures encompass multiple worldviews and value positions, alternative assumptions and interpretations, and varying perspectives regarding the nature and implications of uncertainties. Group processes are again used to search for common ground, to build consensus and to resolve disputes. A complete consensus does not emerge. Residual differences are highlighted for subsequent application in sensitivity analyses and in contingency measures. The analysis focuses on identifying and characterizing discrepancies between plausible, planned futures, and sustainable futures. Instances where plausible futures suggest that social or ecological thresholds could be surpassed are highlighted. The gaps and thresholds provide the basis for identifying sustainability principles, goals, objectives, and priorities. Major remaining uncertainties and potential implications are noted. The analysis is placed within the context of broader government and multistakeholder sustainability initiatives. It also builds on other efforts to characterize plausible and desired future conditions.

Sustainability Assessment Possible actions and their alternatives are reconsidered taking into account the gaps (including thresholds) between the plausible and the desired futures. Ways in which the gaps could be closed are assessed. The actions and the alternatives are modified, to the practical extent to assume the role of a catalyst for narrowing the gaps. The possibility of generating additional alternatives, which might better address the sustainability shortfalls, is also explored.

The sustainability principles, goals, objectives, and priorities are refined into more specific targets, criteria, and thresholds. Methods and exclusionary criteria for screening out unacceptable alternatives are formulated. Alternative screening criteria are formulated to reflect varying perspectives where a consensus cannot be reached among the interested and affected parties. Decision rules are formulated for addressing situations where the results from applying alternative screening criteria conflict. The screening criteria are applied to the proposed action and to the alternatives. Alternatives that are, for example, inconsistent with sustainability thresholds, principles, and objectives, undermine desirable futures, detract from targets, and result in major impacts and exacerbate inequities for socially disadvantaged groups are excluded from further consideration. A precautionary approach is adopted. Thus, marginal alternatives, characterized by major potential impacts and high levels of uncertainty, are also excluded.

Criteria, methods, and decision rules for comparing the remaining alternatives are formulated. The alternatives are compared for their contributions to environmental and sustainability objectives, for their consistency with desired futures and with sustainability principles and criteria, and for their potential contribution to closing the gaps between plausible and desired futures. The comparison of alternatives is supported by techniques such as life cycle and footprint analysis. Uncertainties and alternative interpretations are addressed with sensitivity analyses. A precautionary approach is applied. The implications of applying mitigation and enhancement measures are explored. Preferred alternatives are selected, supported by a clear rationale.

Baseline conditions, pertinent to the preferred alternatives, are characterized. Individual and cumulative impacts associated with the preferred alternatives are identified, predicted, and interpreted (in terms of significance). Mitigation and enhancement measures are introduced to prevent and offset adverse impacts and to enhance benefits. These activities largely mirror those associated with conventional IA processes. However, unlike conventional processes, the analysis focuses on using and refining the proposed and potential actions as sustainability instruments or catalysts. The impact analysis also builds toward an overall assessment of contribution to sustainability (e.g., contribution to or amelioration of global and transboundary problems, maintenance of ecological integrity, maintenance of natural capital, waste generation within assimilative capacity, reduced energy and material use, maintenance of environmental quality, biodiversity maintenance, amelioration of intergenerational and intragenerational inequities, pollution prevention, avoidance of risk to carrying capacity). Implementation, monitoring, and auditing roles and responsibilities are specified. Residual limits (e.g., knowledge, resource, institutional) and uncertainties, together with associated implications, are detailed from a precautionary perspective (e.g., minimum regrets).

Interested and affected parties are fully involved in all facets of the sustainability assessment.

Approvals and Postapprovals The process culminates in final conclusions regarding the acceptability, on sustainability grounds, of proposed and potential actions. A management strategy is included with links to other IA tiers and to other environmental management and sustainability instruments, networks, and strategies. Measures to ensure follow-up coordination, communications, cooperation, public involvement, and coalition/capacity building are instituted. Consideration is given to how to best overcome implementation barriers. The implementation measures could include institutional reform.

If proposed action(s) are acceptable, with or without approval conditions, environmental changes, impacts, and mitigation/enhancement are monitored. Monitoring results provide the basis for ongoing environmental and sustainability management. The environmental change monitoring results are incorporated into broader environmental and sustainability indicator systems. The IA process is audited. The lessons learned from the auditing analysis are broadly circulated as a contribution to IA practice.

Ongoing Activities The IA process is highly iterative, dynamic, and collaborative. It provides for both continuous (e.g., advisory committees) and periodic (e.g., workshops, forums, conferences, open houses) public and agency involvement opportunities. The workshops, conferences, and forums support generating, refining, and integrating visions, scenarios, models, and stories. They also contribute to generating and applying goals, objectives, criteria, principles, and decision rules. Surveys, interviews, meetings, and focus groups help identify perspectives, concerns, and initiatives. Periodic opportunities (e.g., open houses, meetings) are provided for broader public and government official (elected and nonelected) involvement. Provision is made for public and agency involvement during the postapprovals stage. Participant assistance is provided to ensure the effective involvement of all interested and affected parties. Specialists participating in the process are a part of and maintain contact with broader environmental management and sustainability networks.

Data and analysis, including traditional knowledge and reviews of comparable situations, are incorporated into each activity. Particular consideration is given to related sustainability experiences. Technical (e.g., model building, scenario generation) and procedural (e.g., consensus building, conflict resolution) advice and applied research are provided whenever needed. Impact and uncertainty management are continuous functions in the process. Numerous interim documents are generated as the process unfolds. The results of the process are consolidated into draft and final IA documents, which provide a complete decision-making basis. Documentation extends into postapprovals with the production and circulation of monitoring and auditing reports. The process is knit together by the central purpose and theme of facilitating substantive, sustainable environmental improvement.

Adaptations by IA Type Substantive IA processes vary by IA type and level. Table 6.4 outlines examples of good

Table 6.4 Substantive IA Practice by IA Type

Substantive SA Practice	Substantive SEA Practice	Substantive EIA Practice
Treats SA as a means of integrating all substantive IA types and levels Employs strong conception of sustainability; seeks sustainable outcomes milestones and ultimate state not just sustainability-oriented processes Establishes a workable sustainability concept in decision context Approaches sustainability from a systems perspective; principles/objectives rather than pillar-based Explicitly commits to sustainability objectives, criteria and trade-off rules Clearly identifies sustainability principles, purposes, issues, and indicators Applies no net loss principle; seeks positive sustainability contribution Avoids a reductionist approach; seeks to facilitate resilient socioecological systems Emphasizes climate change Monitors from a sustainability perspective; protects the future Uses models to assess regional sustainability Uses conceptual frameworks and indicator systems to assess progress toward sustainability Uses sustainability criteria in alternatives generation; seeks alternatives that offer greatest overall benefits and avoid undesirable trade-offs Ensures preferred alternative designed for resilience and adaptability Identifies critical decision windows that can influence sustainability outcomes	Places within substantive policy context Ensures SEA role in implementing substantive policies and requirements Uses biodiversity, social and sustainability checklists Identifies, constructs, and applies environmental objectives and indicator systems Identifies biodiversity social, health and sustainability friendly options Applies holistic approach Operates at multiple spatial levels and temporal horizons Considers cumulative threats Expands the use of SA, SIA, HIA, and EcIA in SEA and integration of each into SEA Seeks an enhanced understanding of ecosystems and of society, and related processes Considers distribution of effects (ecological, intragenerational, intergenerational, vulnerable populations) Assesses micro- and macroecological and societal changes independent from and resulting from impacting events Devotes more attention to ecological, social, economic, and health issues in predecision and follow-up stage Clearly explains models and assumptions if quantitative approach Enhances recognition of ecological, social, and health determinants Institutes ecological, social, and health disparity policy initiatives Institutes longitudinal surveillance systems Uses SA to identify SEA shortcomings; need for SEA to more fully engage sustainability concept and apply sustainability science Recognizes the irreducibility of risk and uncertainty	Identifies and evaluates ecological and social/health resources and features likely to be affected Identifies changes likely to affect valued ecological and social components Ensures appropriate scoping (e.g., temporal and spatial scales, choice of measurable indicators) Predicts, characterizes, mitigates/enhances, and assesses significance of impacts; monitors and manages (including use of direct and indirect environmental offsets and compensation) Emphasizes positive outcomes for biophysical and socioeconomic environment Utilizes participatory approaches (e.g., interactive community forum) to assess options and impacts Seeks a fair distribution of effects and risks; enhances distributional equity over space and generations Emphasizes social/ecological enhancements Seeks free, informed, and prior consent of affected publics Strengthens local project governance (e.g., impact and benefits agreements) Seeks to balance best available science, competing societal objectives, and local political considerations Utilizes GIS and surveillance systems Assesses cumulative impacts on biophysical and human environment Assesses equity and legacy implications (including pace and scale options); seeks lasting benefits; ensures active transition planning

Substantive EcIA Practice	Substantive SIA Practice	Substantive HIA Practice
Identifies biodiversity management objectives and principles (e.g., no net loss, net benefits—enhancement) Proactively considers biodiversity from the outset; includes ecosystem services, ecological connectivity, and landscape Ensures a clear understanding of environmental capacity thresholds Ensures consistency with international obligations (e.g., treaties, conventions) and national biodiversity policy Collaboratively involves ecologists	Integrates social concerns from the outset Obtains a good understanding of the likely affected communities (profiling); identifies community needs and aspirations Ensures effective public participation, collaboration, and mutual learning Integrates local knowledge Assesses distributive and poverty effects (e.g., intra and intergenerational equity, gender equity, human rights and justice issues, perception changes, sociopsychological effects, cultural	Integrates health disciplines Integrates and engages health professions Seeks industry-wide HIA standards Devotes greater attention to health inequities and to health and cumulative effects determinants Ensures that HIA is gender sensitive Assesses the effect of HIA on population health (e.g., retrospective evaluation) Addresses the need for better health information systems, knowledge of health impacts, and access to previous HIAs Ensures sufficient resources

Table 6.4 (*Continued*)

Substantive EcIA Practice	Substantive SIA Practice	Substantive HIA Practice
Determines ecosystem loss, compensates for biodiversity loss, and restores ecological processes (reverse harm) Takes an adaptive approach to building ecosystem resilience; integrates ecological mitigation and enhancement measures	heritage impacts, and cumulative effects on human environment) Applies social performance standards Ensures effective (structured, hypothesis-testing) social follow-up Emphasizes social sustainability Identifies social development needs	Broadly defines health determinants (e.g., gender) Facilitates HIA capacity building Broadly defines health alternatives (e.g., goals, knowledge, institutional)

Sources: Ahmadvand and Karami (2009), Athanas (2005), Australian Government (2011b,c), Becker et al. (2003), Benson (2003), Bond (2010), Bond and Morrison Saunders (2009, 2011), Bond and Pope (2012), Bond et al. (2004, 2012), Burdett (2008a,b), Burdge (2003b, 2004), Byron and Treweek (2005), CEAA (1996a,b), Croal et al. (2010), Dalal-Clayton and Sadler (2004), Dannenberg et al. (2006), Desmond (2007), Devuyst (2000), Donnelly et al. (2006, 2007), Donnelly and O'Mahoney (2011), Eales and Sheate (2011), Esteves and Vanclay (2009), Esteves et al. (2012), Slootweg et al. (2006), Fahy and Cinnéde (2007), Fischer (2003, 2011), Fischer et al. (2010), Franz and Kirkpatrick (2007), Gasparatos et al. (2007), Geneletti (2002, 2003), Geneletti et al. (2003), Genter et al. (2008), George (1999), Ghanimé et al. (2011), Gibson (2000, 2006a, 2009, 2011), Gibson et al. 2005; Gontier et al., 2006; Govender et al., 2006; Grinde and Khare 2008; Hacking and Guthrie 2006; Harris et al. (2003), Harris-Roxas et al. (2012), IAIA (2003, 2005), ICPGSIA (2003), IEEM (2006), Jiliberto (2004), João et al. (2011), Jones and Slinn (2008), Khera and Kumar (2010), Kobus (2005), Kolhoff and Slootweg (2005), Krieger et al. (2003), Kumagai et al. (2006), Landsberg et al. (2011), Letsela et al. (2010), Lord (2011), Mandelik et al., (2005), Morrison-Saunders and Arts (2005), Meynell (2005), Morgan (2003a), Noble and Bronson (2006), Péti (2012), Partidário (undated), Pischke and Cashmore (2006), Peterlin et al. (2008), Pope (2006), Pope and Grace (2006), Pope et al. (2004, 2005), Pritchard (2005), Quigley and Taylor (2003), Puschchak and Farrugia-Uhalde (2009), Rajvanshi et al. (2011), Rowan and Streather (2011), Sadler and Jurkeviciute (2011), Schirmer (2011), Sharma (2010), Sheate (2011), Slootweg and Kolhoff (2003), Söderman and Saarela (2010), Storey and Jones (2003), Storey and Noble (2005), Tetlow and Hanusch (2012), Thérivel (2010), Treweek et al. (2005, 2011), US CEQ (2010a), Utzinger et al. (2005), Villani (2011), Wale and Yalew (2010), Wegner et al. (2005).

practices for integrating substantive environmental concerns by IA level (SEA, project EIA) and by substantive IA type (EcIA, SIA, HIA, SA).

SA, EcIA, SIA, and HIA are inherently substantive. Key substantive issues with these IA types tend to pertain to which substantive concerns are especially critical, and how they can be most effectively introduced, linked, and integrated into a coherent whole.

Substantive EcIA practice proactively emphasizes biodiversity management principles and objectives, ecosystem services, and ecological connectivity and landscape issues from the outset. It stresses the importance of placing IA practice within the context of international and national policies and requirements, collaboratively involves ecologists, and focuses on environmental capacity issues. It seeks to determine, compensate for, and restore ecosystem loss; adopt an adaptive management approach to building ecosystem resilience; and integrate ecological mitigation and compensation measures.

Substantive SIA practice focuses on intergenerational and intragenerational equity, human rights and justice, perception changes, sociopsychological effects, cultural heritage effects, and cumulative human environmental effects. It emphasizes the importance of integrating social concerns from the outset. It stresses effective public participation, collaboration, and mutual learning. It seeks a sound understanding of the characteristics, needs, and aspirations of potentially affected communities and populations. It facilitates the integration of local knowledge. It concentrates on distributive and poverty effects. It strives to identify and meet social development needs. It applies social performance standards, ensures effective follow-up, and seeks to contribute to social sustainability.

Substantive HIA practice effectively engages and integrates contributions from the health disciplines and professions. It assesses population health, addresses an extensive range of health alternatives (e.g., alternative goals, forms of knowledge, institutional choices), and focuses on health inequities. It is gender sensitive. It broadly defines health determinants and cumulative effects. It facilitates capacity building, ensures sufficient resources, and seeks to establish and raise industry-wide HIA standards. It addresses the need for better health information systems, enhanced knowledge of health impacts, and access to previous HIAs.

Substantive SA practice represents a means of integrating substantive issues into all IA levels and types. It employs a strong sustainability conception, focuses on sustainability outcomes (not just sustainability-oriented processes), and approaches sustainability from a systems perspective. It concentrates on critical decision windows. It is appropriate to the decision context. It avoids a reductionist approach. It explicitly commits to and applies sustainability purposes, principles, issues, objectives, criteria, indicators, and trade-off rules (e.g., no net loss, positive sustainability contributions). It seeks to facilitate resilient socioecological systems. It emphasizes climate change issues. It monitors from a sustainability perspective (i.e., protect the future). It applies models, conceptual frameworks, indicator systems, criteria, and other methods to structure the integration of sustainability concerns to assess progress toward sustainability. It ensures that preferred alternatives are resilient and adaptable, avoid undesirable trade-offs, and provide the greatest overall benefit.

SEA and project-level EIA are, by definition, substantive in the sense that they seek to broaden decision making to encompass environmental concerns. Unfortunately, perhaps in part because of the vagueness of their substantive aspirations, they have tended to focus on procedures over outcomes. As a consequence, the contribution of SEA and EIA to the realization of substantive environmental ends has either been uncertain or less than hoped. The need to make SEA and EIA more environmentally substantive in terms of outcomes has become a recurrent theme in recent years in IA theory and practice.

Substantive SEA practice is framed by broader environmental policies and requirements. It emphasizes tangible, positive environmental outcomes. It employs a holistic understanding of ecosystems and society. It systematically integrates substantive IA types (e.g., SA, SIA, HIA, EcIA), issues, and methods from the outset and extending through follow-up (e.g., longitudinal surveillance). It formulates and applies specific objectives and indicators. It employs multiple scales and time horizons. It clearly explains methods and assumptions. It recognizes the irreducibility of risks and uncertainties. It focuses on biodiversity, social, health, and sustainability-friendly options. It thoroughly assesses micro- and macroecological and social change processes. It systematically assesses the determinants of effects, the distribution of effects, and the cumulative effects.

Substantive project-level EIA practice focuses on valued ecological and social components, features, resources, and processes. It ensures scoping is appropriate to the context. It makes effective use of geographic surveillance systems. It systematically identifies, mitigates/enhances, interprets the significance of, and manages a broad range of positive and negative, direct and indirect ecological, social, health and cumulative options and effects. It emphasizes positive substantive environmental outcomes and enhancements. It seeks a fair distribution of effects, risks, and lasting benefits. It employs participatory approaches to assess options and effects. It seeks the free informed and prior consent of the affected public. It strives to strengthen local project governance (including transitional planning). It seeks to balance the best available science, competing societal objectives, and local political considerations.

6.6 CONTEMPORARY CHALLENGE—HORIZONTAL IA INTEGRATION

Integration in IA is a "hot topic" and a recurrent theme in IA theory and practice. Integration can refer to the integration of IA and planning/decision making (addressed in Chapter 3), vertical integration (e.g., as in tiering among policies, plans, programs, and projects) (addressed in Chapter 8), and horizontal integration (among, e.g., ecological, social, health, and economic considerations) (Lee, 2006; Pope, 2006). This section focuses on horizontal integration.

Integration can occur at the regulatory (e.g., laws, regulations, guidelines, policies) and/or applied levels. It can encompass procedural arrangements, organizational/institutional arrangements, and methodology (Fischer, 2006; Milner et al., 2005; Pope, 2006). Horizontal integration is generally displayed as a continuum ranging from no integration (e.g., minimize biophysical only or separate IAs for each substantive area) to full integration (e.g., integrated IA, sustainability assessment). Examples of middle ground options include (a) minimizing biophysical, direct heritage, and indirect socioeconomic effects; (b) minimizing direct and indirect negative biophysical and socioeconomic effects; (c) positive and negative biophysical, social, and economic effects; (d) trading off biophysical, social, and economic effects; (e) triple bottom line—biophysical, social, and economic; (f) mutually reinforcing positive, net gains—weak sustainability (natural capital can be fully substituted by man-made capital); (g) net gains sustainability and social justice; (h) strong sustainability in the use of natural capital, permitted only if they are fully replaced; (i) strong sustainability and precautionary principle; and (j) combinations (Fischer, 2007b; Kirkpatrick and George, 2006; Morrison-Saunders and Thérivel, 2006; Pisani and Sandham, 2006). As demonstrated in Table 6.5, there are strong arguments for and against integration. There also is little consensus regarding how far along the continuum is appropriate and under what circumstances.

The case for no integration pertains largely to keeping true to the original intent of IA (e.g., focused on biophysical; higher profile for natural environment; more easily understood and coherent; greater decision making weight, profile, and resources; greater advocacy role). It also avoids various risks associated with integration (e.g., "watering down" of biophysical, being associated with processes with limited decision-making effectiveness, being force-fit into inappropriate frameworks, being dominated by other decision-making factors). The same case could be made for keeping SIA and HIA separate, although the likelihood of institutional arrangements that provide sufficient resources to support parallel "action-forcing" systems that somehow come together at the end in the political arena appears remote.

A more likely scenario, if no integration is the preferred system, is competing systems, where there are winners and losers (e.g., formal biophysical and selective indirect social effects, informal, ad hoc, and secondary status for social and health concerns assessment). This is largely the pattern in the four jurisdictions. Also, having a separate system, with, for example, a biophysical emphasis, does not assure the decision-making effectiveness of such factors or a more prominent political position. It could instead contribute to the marginalizing of natural environmental concerns with the "real" decision making occurring outside the IA process. Moreover, there is a considerable price to be paid for no integration (e.g., ignores interconnections among types of environmental and cumulative effects, lacks holistic perspective, prevents "win–win" solutions, impairs sustainability potential, excludes valid environmental impacts and public concerns and preferences, is politically polarizing).

The case for integration is largely based on potential, positive environmental and procedural outcomes from integrated IA and/or sustainability assessment. The benefits of

Table 6.5 The Challenge—Integration of Substantive IA Types

	Not Integrated (e.g., Separate IAs, Biophysical Only)	Full Integration (e.g., Integration IA, Sustainability Assessment)
Efficient use of resources	(+) Resources focused on biophysical (−) Wasted resources if HIA and SIA have to take on tasks more appropriate to EIA and SEA (−) Danger that individual fields will sink under their own weight if continue to operate separately (−) Resources dispersed among multiple IAs; duplication and overlap	(+) Reduced duplication of reports and double-counting (+) Can facilitate more efficient and effective resource allocation (focused on potentially significant impacts and interactions); benefits of shared information and of time and resource savings (+) Takes advantage of established procedures, practices, and decision-making recognition (+) Provides opportunities for streamlined appraisal process (−) Resources dispersed over multiple effect types (−) Integration of little value if minimal capacity to address (−) Danger that integrated approaches could become too unwieldy; could become a "catch-all" for every conceivable topic
Decision-making effectiveness	(+) Privileges environmental factors; environmental concerns not marginalized (+) SIA and HIA not well suited to EIA/SEA (e.g., litigious environment, emphasis on quantitative/technical) (+) More suited for informing decision making if standalone (+) Avoids the danger of working with process often criticized for low effectiveness and weak implementation (+) Avoids the issue of resistance by EIA and SEA professionals for consideration of health and social concerns (+) May raise the profile of certain issues and help ensure decision-making weight (−) Value of standalone IA types limited if not broad, system-wide support across government for each IA type (−) Separating makes it harder to integrate into decision making; competing with other IA types (−) Decisions could be made based on considerations separate from IA process (−) If separate, some IA types will lack "action-forcing" mandate (−) Partial appraisals, including EIA, have a strong regulatory base	(+) All environment related decision making addressed at once under a legal mandate; comprehensive decision-making basis; greater credibility (+) Lifts the consideration of health and social issues to policy and project level; might not occur without trigger; provides regulatory standing and helps promote awareness (+) Decision makers provided with full range of impacts and options (+) Will lead to greater awareness of social and health issues and impacts (+) Helps build constituency and ensures legitimacy of HIA, EcIA, and SIA (+) Political appeal of sustainability could extent IA influence (+) Growing receptiveness to integrating different forms of IA; growing acceptance of sustainability (+) Growing recognition of the role of integration in public and private decision making (−) Tendency for economic to dominate; biophysical, health, and social effects marginalized; undermines the goal of building environmental considerations into decision making (−) Social and economic already dominate decision making; simply reinforces (−) If poorly integrated, runs a risk of marginalizing all substantive concerns (−) If sustainability too vague, substance will be lost; quantity over quality (−) Won't work if agencies lack capacity and/or are not convinced of benefits (−) IA, if broadened, weak tradition of addressing social and health issues; runs the risk of lowering standing of health and social concerns in eyes of decision makers (−) Integrated appraisal does not enjoy strong regulatory support (−) Integrated approaches often used loosely and uncritically
Transparency and ease of understanding	(+) Focused on narrower scope of environment; more easily understood (+) If kept separate, no uncertainty regarding what is best for the environment (−) Connections to excluded impacts and benefits; decision-making role not transparent	(+) All effects and interconnections addressed at once (+) Can provide a comprehensive evidence base; illustrates planning and decision-making process (+) All human judgments explicit (−) When all effects addressed, IA process and documents can be very complex and difficult to understand (−) Danger environmental issues and role submerged and undervalued (−) Sometimes so inclusive, loses clarity
Completeness and coherence	(+) More in-depth analysis if biophysical only (+) More coherent if confined within well-established biophysical frameworks and models	(+) Complete and coherent analysis of fully array of direct and indirect, positive and negative effects; easier to integrate under the umbrella of contribution to sustainability (+) Raises standard—net environmental benefits, explicit thresholds and trade-off rules, enhancement opportunities, and preferred futures

Table 6.5 *(Continued)*

	Not Integrated (e.g., Separate IAs, Biophysical Only)	Full Integration (e.g., Integration IA, Sustainability Assessment)
	(+) EcIA, HIA, and SIA will never reach full potential if subsumed under EIA and SEA; on own can develop methods and demonstrate value; if combined distorted and confined (−) When separate, tendency to be limited to mitigating negative (−) Many potential direct and indirect impacts excluded from the process (−) Interconnections between biophysical and socioeconomic not addressed and unclear (−) Inhibits ultimate integration of substantive environmental concerns	(+) When combined resources and mandates; facilitates sub-fields reaching full potential (+) Easier to trace cause effect chains (+) Facilitates integration of core values of SIA and HIA; human well-being fully addressed (+) Better cooperation and coordination (+) Sufficient good practice examples that full integration possible and practical (−) With effort dispersed across such a wide array of issues and effects, danger of superficial analysis (if any) of effects that should be addressed in detail (−) Danger that some impacts (e.g., social) will be "force-fit" into unsuitable frameworks (−) When terminology and objectives vague and unclear, inhibits the understanding and determination of positive outcomes (−) Regulatory agencies and practitioners lack social and health expertise and counter to entrenched practices
Bias and balance	(+) Eliminates potential for dominance of biophysical by socioeconomic; environmental dimension kept central (+) Maintains biophysical emphasis; easier to maintain integrity of methods (+) When separate, can focus on IA type (e.g., social, health) issues and not compromise values, frameworks, models, and methods (−) If separate, social and health issues not addressed or minimal consideration	(+) If properly scoped and managed, should be possible for all potentially significant effects and interactions to be adequately addressed (−) Tendency for secondary status for biophysical and social concerns—could "water down" treatment of such concerns; the illusion of balance (−) Risk of integrated appraisals being captured by dominant interests (−) Argument that SA overly promotes economic agenda and undermines environmental gains (−) Could result in reduction in the integrity of frameworks, models, and methods of individual IA types (−) Inhibited by professional bias of EIA/SEA practitioners (−) Tendency of quantitative, "comprehensive" approaches to bury uncertainty and undervalue qualitative (−) Can allow trade-offs between individual issues to be hidden
Natural environmental protection and enhancement	(+) Consistent with core reason for SEA and EIA (i.e., ecological rationality) (+) Focused on natural environment (+) Recognizes the ultimate goal—environmental regulation rather than integration per se (−) Other aspects of environment not considered or given secondary status (−) Could ultimately be detrimental to natural environment because lacks holistic perspective of natural/human environment interactions	(+) Emphasis on mutually beneficial solutions conducive to greater emphasis on environmental enhancement; more opportunities to identify WWW solutions (+) Possible to retain emphasis on natural environment by giving preeminent position to ecological sustainability (+) Comprehensive approach more conducive to effective environmental management (−) Dilution of IA focus could undermine the future of natural environment; breadth over depth (−) Danger that environmental quality and capability eroded under the guise of integrated IA (−) Danger that shifts emphasis from substantive to procedural (−) Danger of substance lost in sustainability rhetoric; malleable concept; perpetuates status quo (−) Danger of legitimizing (if three pillar approach) trade-off of environmental loss for economic gain and social benefits (−) Danger that environmental concerns undervalued if objectives set by proponents and decision makers (−) Not in best interests of the environment
Synergistic potential, cumulative effects, and sustainability	(+) Potential for synergies with individual IAs (+) Avoids the problem of environmental–socio–economic trade-off being submerged in documentation	(+) Provides a framework for systematic consideration of interactions and cumulative effects (everything connected, only way of ensuring all trade-offs addressed); all under the umbrella of sustainability assessment (+) Provides opportunity for multiple, mutually enforcing gains—greatest overall benefit and avoidance of undesirable trade-offs (+) Added value (e.g., environmental and social determinants of health)

(continued)

Table 6.5 (*Continued*)

	Not Integrated (e.g., Separate IAs, Biophysical Only)	Full Integration (e.g., Integration IA, Sustainability Assessment)
	(+) Better as standalone because of lack of progress in integration where has been undertaken (−) Limited potential for addressing synergies among separate IAs or between included and excluded aspects of environment (−) Only partial treatment of interconnections among effects, cumulative effects, and sustainability (e.g., ecological sustainability); even ecological sustainability difficult to address systematically because of limited consideration of interconnections to human environment	(+) HIA, EcIA, and SIA can benefit from each other and from literature and experiences of EIA and SEA (+) Makes it possible to fully explore potential "win–win–win" synergies and sustainability opportunities (+) Integration (the essence of sustainability) essential if all IA types have to be brought together under umbrella of sustainability (−) Analysis may be so wide ranging, vague, and superficial that synergistic opportunities and cumulative effects are addressed only superficially (−) Allows trade-offs among individual issues to be hidden (−) Sustainability may be so weakly defined that net negative effects or outcomes unclear (−) Argument that sustainability anthropocentric
Stakeholder and public involvement	(+) Stakeholder involvement focused on a small number of effects and issues; not "watered down" (−) Many public issues, stakeholders, and interconnections excluded from process; no public forum for addressing (−) Ignores needs and preferences of planners and policy-makers who need as much balanced and impartial information as can obtain (−) Inhibits mainstreaming of health, social, and gender interests (−) Public fatigue—asked to participate in too many IAs	(+) Possible to fully involve all interested and potentially affected parties; could facilitate stakeholder involvement (+) Broad definition of environment more conducive to public participation and collaboration; easier to identify and explore interrelationships among issues (−) With resources so widely dispersed, opportunities for "in-depth" involvement may be limited (−) Danger that overly complex IA analysis or overly superficial analysis (e.g., checklists) will inhibit involvement (−) Danger that public involvement procedures may be dominated by particular issues or stakeholders (−) Impossible to have public debates when everyone's opinion is equally considered ("smoothie model")
Democratic accountability	(+) Standalone IA can operate more effectively in terms of political advocacy (+) Political, value-based questions continue to be addressed in political arena (−) Danger that narrowly defined IA will be marginalized; reinforces role of decision-making factors outside the IA process (−) Excludes many valid public concerns from agenda (−) Danger that separate IAs will have to compete in political arena (−) Counter to interests of politicians who need more analytical and knowledge support (−) Separate treatment of substantive issues can be politically polarizing	(+) All factors affecting decisions within IA process; decision making accountable within IA process (+) Reflects breadth of public concerns (+) Helps institutionalize HIA and SIA (+) Reflects trend in central government departments and elsewhere (+) Helps raise awareness by decision makers of full array of potential consequences (+) Integrated assessment can help secure regulatory and social approval/acceptance (−) Decision makers and others may bypass the process regardless of how comprehensive the process is (−) Removes political/value-based questions (e.g., social vs. ecological) from democratically accountable decision making and instead addresses with technical/rational methods (−) Danger that expert-driven systems will undermine democracy (−) Weakens environmental advocacy role

Sources: Ahmed (2004), Azcarate and Balfors (2009), Bhatia (2007), Bina et al. (2011), Birley (2003), Bond et al. (2011), Burdett (2008a), Cole and Fielding (2007), Craik (2008), Dannenberg et al. (2006), Duncan (2008), Duncan and Hay (2007), Eales et al. (2005), Farley and Smith (2012), Fischer (2006, 2007b), Gibson (2006a), Gershberg (2002), Gibson (2010), Hacking and Guthrie (2011), Hacking and Guthrie (2008), Harris and Spickett (2011), Jackson and Illsley (2007), João et al. (2011), Kirkpatrick and George (2006), Lawrence (2009), Milner et al. (2005), Morgan (2011), Morrison-Saunders and Fischer (2010), Morrison-Saunders and Thérivel (2006), McCaig (2005), McCluskey and João (2011), Nilsson (2009), Orenstein et al. (2010), Ortolano (2008), Pisani and Sandham (2006), Pope (2006), Pope and Dalal-Clayton (2011), Pope et al. (2005), Scrase and Sheate (2002), Sadler (2011a), Sinclair et al. (2009), Thérivel (2010), Vanclay (2010), Wernham (2007), Wright et al. (2005).

integration have yet to be consistently demonstrated. Integrated forms of assessment may integrate all potentially significant environmental concerns (including their interrelationships) into decision making in a manner that facilitates and influences decision making and contributes to the realization of positive environmental outcomes. They may reflect a broad range of public concerns. They may effectively allocate resources. They may ensure transparency. They may avoid bias and imbalance, especially the domination of ecological, health, and social concerns by economic imperatives. They may protect and enhance the natural environment. They may systematically consider cumulative effects. They may facilitate stakeholder and public involvement. They may enhance the potential for sustainability (i.e., multiple mutually reinforcing gains, bad trade-off being avoided). They may contribute to democratic accountability. However, it is just as possible that integration, if improperly framed and conducted, can be wasteful, burdensome, unfocused, superficial, excessively procedural, vacuous, "captured" by narrow interests, incoherent, unsupportable in terms of institutional and resource capacity, excessively anthropocentric, reinforcing of the "status quo," of little value in protecting and enhancing the environment, and inhibiting of democracy.

In view of the above, the question then becomes—can such an integration (or some form of partial integration) be instituted, which can retain the positive features of no integration and realize the potential of integration, without suffering its' potential pitfalls—all within appropriate (designed to context) integrative regulatory and applied frameworks? The answer would seem to be a conditional yes. That is, integration is desirable but only if certain regulatory and applied preconditions are first satisfied (Azcarate and Balfors, 2009). There is no shortage of examples of such measures. They pertain to institutional, political, regulatory, and professional reforms, resources, and capacity building—measures that challenge the status quo (Biermann et al., 2012; Eales et al., 2005; Harris and Spickett, 2011; McCaig, 2005; Nilsson, 2009; Pisani and Sandham, 2006; Pope, 2006). They concern methodological integration research and testing, coordination and cooperation initiatives, and political awareness raising efforts (Ali et al., 2008; Bhatia, 2007; Birley, 2003; Dannenberg et al., 2006; Morgan, 2003b; Wright et al., 2005). They include environmental and risk/uncertainty thresholds and mitigation/enhancement objectives, measures, and positive outcomes that options and proposed actions must clearly satisfy (João et al., 2011; Mishra, 2009; Thérivel, 2010; Thérivel et al., 2009). They encompass transparency, public access, and public involvement requirements (Lawrence, 2009; Orenstein et al., 2010). They include clear, consistent, and broadly supported definitions for key terms and requirements, especially with regard to sustainability (Farley and Smith, 2012).

Particular thought needs to be given to questions such as (a) who integrates (e.g., not just proponents and their political supporters), (b) how to avoid turning political–social–ecological value-based choices into technical–rational choices, not always explicitly, by experts and expert methods, and (c) how integration is to take place (not just procedures but also roles and responsibilities) (Kirkpatrick and George, 2006; Lawrence, 2009; Morrison-Saunders and Thérivel, 2006; Pope et al., 2005)? Cumulative effects assessment and management is a further key concern that must be fully addressed in any exploratory integration initiatives (Fischer, 2011; Thérivel et al., 2009). It is essential that the measures adopted are mutually supportive, and designed, framed, and adapted to take full advantage of the potential of integration without succumbing to the pitfalls. Suites of good practice integration principles and criteria need to be identified and tested in varying settings, and from the perspectives of various stakeholders (Fischer, 2006; Orenstein et al., 2010). The product of these effectiveness reviews should be a clear set of priorities, leading to better regulation, guidance, and practice, and directly resulting in enhanced procedural and substantive outcomes (Morgan, 2011). In this way, the "added value" of integration can be demonstrated (Quigley and Taylor, 2003). With such initiatives, it is hoped that the debate can gradually shift from the relative merits of integration versus no integration to specifying conditions and perhaps milestones along the journey to satisfying those conditions.

6.7 SUMMING UP

This chapter is concerned with managing regulatory and applied IA processes to better integrate environmental perspectives, values, and knowledge. The two stories describe ways in which the IA process can become more substantive: (1) through the integration of substantive concerns into significance criteria and (2) through the systematic application of a sustainability test.

The general problem addressed is a shortfall between IA environmental aspirations and achievements. The more specific problem is the role that the IA process assumes in widening or narrowing that gap. Numerous ecological, social, and sustainability concepts and methods are presented. These concepts and methods provide the basis for characterizing how a substantive IA process might be applied at the regulatory and applied levels.

The relationship between IA process and substance has been approached from several perspectives. Some say that IA is an unnecessary diversion of resources. They argue that it reinforces the status quo or that it cannot be properly applied or that there are better ways to bring about environmental improvements. Others argue that the substantive benefits of the IA process are unknown because of knowledge gaps concerning decision-making effectiveness, outcome effectiveness, and environmental characteristics. Still others submit that the IA process can be conducive to environmental advancement but that refinements or modifications or major reforms are needed. Many acknowledge that issues and solutions vary depending on whether ecological, social, or sustainability concerns are being considered. This chapter explores how IA process management can better address this constellation of interrelated problems and solutions.

An overview of ecological, social, and sustainability concepts and methods provides the basis for the substantive IA process. The ecological concepts explored include—applied ecology, ecological impact assessment, environmental indicators, biodiversity—inclusive IA, the ecosystem approach, environmental planning and management, integrated environmental and resource management and assessment, adaptive environmental assessment and management, and traditional knowledge. Examples of recurrent themes displayed by these concepts include the need for an ecological systems perspective, to adopt a place-based approach, to employ sound ecological knowledge, to transcend disciplinary boundaries, to recognize ecological stresses and limits, to acknowledge knowledge and control limits, to continuously and adaptively manage impacts (both pre and post-approval), and for the process to be open, adaptive, creative, collaborative, iterative, selective, and action-oriented.

The social concepts considered include technical SIA, political SIA, participatory/community-based SIA, positivistic social science, functional, ecological, and systems theory, interpretative social science, critical social science, exchange theory, symbolic meaning, social learning, and phenomenological sociology. The overview analysis recognizes that people react in anticipation of and adapt to change and that social phenomena are very difficult to predict and influence in predictable ways. It demonstrates that there are multiple potentially applicable, but partially overlapping and conflicting social models, theories, perspectives, and frameworks. It points to the often-peripheral position of SIA in decision making. It describes the gulf between social sciences and applied fields such as SIA. It underscores the importance of being cautious regarding preconceptions and to design and adapt the process to fit the context. It emphasizes the need to see the world through the eyes of potentially affected parties. It acknowledges the value of a socially constructed IA process. It stresses the importance of exploring social impacts at multiple levels and from multiple perspectives. It demonstrates that the SIA/IA process can be beneficial or can exacerbate negative impacts. It illustrates how meaning and value are socially determined. It shows the central role of dialogue and social interactions in the process. It stresses that frameworks and methods should systematically address social interactions and choices, from multiple perspectives.

The overview of sustainability concepts describes the varying perceptions of the nature and purpose of sustainability. It illustrates how sustainability is refined through sustainability forms and ethical perspectives, directed by needs, aspirations, and principles; applied through instruments, procedures, and processes; integrated by strategies, visions, and frameworks; and adapted to contexts. It demonstrates that IA and sustainability are applied to varying environments and activities and can be integrated at conceptual, regulatory, and applied levels. It describes how sustainability extends and completes IA. Sustainability adds to the IA process a sound basis for interpreting significance and means for determining environmental limits; integrating measures of environmental change; interpreting present conditions; determining plausible, planned, and desirable conditions; integrating diverse perspectives and methods; adapting to context; and linking to other environmental management and sustainability instruments. Most importantly, sustainability helps make the IA process more effective in advancing substantive IA aspirations.

Several methods, potentially conducive to a more substantive IA process, are briefly described. The methods described include network analysis, systems diagrams, modeling, projection, forecasting, backcasting, visioning, scenario writing, storytelling, ecological footprint analysis, life cycle analysis, rapid rural appraisal, and participatory rural appraisal. Collectively, these methods effectively address interrelationships, interpret past and present conditions, identify ecological and social limits, portray plausible and desirable future conditions, determine how the gaps between plausible and desirable future conditions can be narrowed, manage uncertainties, and facilitate the involvement of interested and potentially affected parties.

Substantive IA requirements and guidelines in the four jurisdictions are briefly described. Each jurisdiction integrates process and substance in different ways, although there are many parallels. There are many positive and negative features and examples associated with how substantive environmental concerns are addressed in each jurisdiction. The appropriate mix of approaches will vary by jurisdiction. Additional effectiveness analyses are required. It seems advantageous for environmental substance to be directly integrated into IA requirements at the project, strategic, and regulatory levels. The selective merging of IA with other substantive environmental requirements can sometimes be beneficial. Further consideration should also be given to IA process adaptations that enhance the effectiveness of links to other environmental requirements and the potential for substantive environmental enhancements. Examples of regulatory approaches to integrating IA and substance are presented.

An example of substantive IA process is described. A context is established. Sustainability plans and strategies, sustainability constraints and opportunities, the need for action and stakeholder perspectives, and issues and positions are reviewed. The process is scoped. Potentially appropriate methods and frameworks are identified and refined. Procedural fairness standards and principles are determined.

Proposed actions and alternatives are identified. Base case models, scenarios, and stories are constructed and adapted to encompass alternative worldviews and value positions. Models, scenarios, visions, and stories are used to construct plausible and desired futures. Gaps between plausible and desired futures are determined. The gaps provide the basis for identifying sustainability principles, goals, objectives, and priorities. More specific sustainability targets, criteria, and thresholds are then formulated. Alternatives for closing the gaps are identified. Proposed actions and alternatives are refined and treated as potential sustainability catalysts.

The alternatives are screened and compared using sustainability thresholds, criteria, and decision rules. A sustainability assessment of the preferred alternatives is undertaken, extending from such conventional IA activities as baseline analysis, impact analysis, cumulative effects assessment, mitigation and enhancement analysis, and significance interpretations. Appropriate roles and responsibilities are determined. Residual limits, uncertainties, and implications are identified. Overall impact management strategies are prepared. Links to other sustainability and environmental management instruments are specified. These analyses provide the basis for proposal review and approval or disapproval.

Outcomes from the process are monitored and linked to sustainability and environmental indicators. The IA process is audited. The auditing results are widely circulated to help improve IA practice. The IA process is supported by such ongoing activities as public and agency involvement; comparable proposal review; data collection and analysis; applied research; the accommodation of traditional knowledge; and the preparation of interim, draft, and final documents.

A substantive IA process varies by IA level and type. Good practice examples for integrating substantive environmental concerns by IA level (SEA, project EIA) and by substantive IA type (EcIA, SIA, HIA, SA) are presented.

The contemporary challenge of horizontal integration (among, e.g., ecological, social, health, and economic considerations) is explored. Arguments for and against integration are summarized. The conclusion is reached that integration is possible and desirable subject to a range of conditions intended to retain the positive features of no integration, realize the potential of integration, and avoid potential integration pitfalls. Regulatory framing and contextual adaptation are both critical. These conditions need to be drawn together into complementary suites of measures and tested through effectiveness reviews assessed from multiple perspectives.

Chapter 7

How to Make IAs More Practical

7.1 HIGHLIGHTS

This chapter portrays a streamlined, efficient, and effective IA process—a practical process based on realistic expectations and competent practice.

- The analysis begins with three applied anecdotes (Section 7.2). The stories describe applied experiences associated with efforts to make IA practice more practical.
- We identify the problem in Section 7.3, which is the tendency for IA processes to be unfocused, disconnected from reality, weak on implementation, of variable quality, and slow to learn from experience and practice. The direction we identify is ways of making the IA process more focused, relevant, feasible, competent, and effective.
- In Section 7.4 we introduce a diversity of concepts bearing on how the IA process can become more focused (on what matters), realistic (in terms of how management and decision making take place), feasible (in terms of decision making and implementation follow-through), competent (in process execution), and effective (in facilitating IA process management learning).
- In Section 7.5 we draw together the insights and lessons presented in Section 7.4. We describe the properties of a practical IA process at both the regulatory and the applied levels. In Section 7.5.1 we explore how IA requirements could be more practical. In Section 7.5.2 we illustrate how a practical IA process could be expressed at the applied level. In Section 7.5.3, we present an overview of practicality measures for various IA types (SA, SEA, EIA, EcIA, SIA, HIA).
- In Section 7.6 we address the contemporary challenge of CEA good practice. We describe CEA characteristics, analyze its current regulatory and applied status, and provide examples of good practices.
- In Section 7.8 we highlight the major insights and lessons derived from the analysis.

7.2 INSIGHTS FROM PRACTICE

7.2.1 The Great Whale Scoping Process: Confronting the Challenges of Northern Mega-Project EIA

The wild rivers of northern Québec have great potential for hydroelectric development. The Province's energy company, Hydro Québec, has exploited some of these resources, resulting in megaprojects that transmit electricity to southern regions but also involve a range of negative social and ecological impacts, experienced primarily by Cree and Inuit communities in their traditional territories. The hydroelectric development proposals in the James and Hudson Bay regions—some already operational, others only in the planning stages—fit into a larger pattern of megadevelopment in northern Canada in the form of mining, oil and gas extraction, pipelines, and other infrastructure. This tension was aptly described, in the context of the proposed Mackenzie Valley Pipeline in the Northwest Territories, as hinterland versus homeland. It is, in many ways, the narrative of the Canadian north in the last century, as industrialization made remote megadevelopment possible.

Many proposed northern Canadian megaprojects remain unrealized. In northern Québec, one of the largest proposals, the Great Whale River hydroelectric project along with its associated infrastructure, was withdrawn or canceled twice, once in the 1980s and again in 1994. With a $12 billion budget (in 1990 dollars), the Great Whale project would have involved river diversion, flooded areas for reservoirs, dam construction, and transmission lines. Moreover, it would have involved a new highway connecting, for the first time, the remote Inuit and Cree community of Kuujjuarapik/Whapmagoostui with southern Québec, a specter that raised great concern about potential social impacts. While it has not been built yet, and may never be built, the proposed Great Whale project and its considerable controversy set the stage for an innovative environmental assessment process, with enduring lessons.

In Canada, much of the formal EIA that takes place is of the "routine" or low-profile variety; however, in rare cases, an alignment of factors results in a panel review, the highest

Impact Assessment: Practical Solutions to Recurrent Problems and Contemporary Challenges, Second Edition. David P. Lawrence.

level of EIA, with increased opportunities for public involvement and more elaborate requirements for proponents. Factors may include the scale of the proposed undertaking, risks and uncertainties, public concern, and court decisions resulting from legal actions on the part of stakeholders. Governments have tended to be reluctant to convene review panels, but in some cases they are forced to do so, and these high-profile cases, in effect, remove all participants—proponents, agencies, responsible authorities, scientists, interveners—from their respective comfort zones. Some of the most notable review panel cases have unfolded, not surprisingly, in the north, where proponents believe that the financial costs and risks of development tend to make it viable only on very large scales, hence megaprojects.

When the EIA process for the Great Whale project finally resumed in 1991, it followed in the footsteps of notable forerunners such as the Berger Inquiry, which examined the proposed Mackenzie Valley Pipeline in a famous case that predated EIA; and the Beaufort Sea review, an elaborate "concept assessment" examining the general prospect of hydrocarbon development, in other words a kind of strategic EA exercise before its time. The Great Whale EIA process also unfolded at a time of heightened environmental concern, gathering momentum for sustainability as an overarching concept, increasing recognition of the rights of First Nations, and the internationalization of environmental issues, in which it became more common for American and European environmental groups to intervene directly in Canadian resource development cases. These factors, along with a court judgment that required a combined (rather than a split) review of the Great Whale project along with its infrastructure, heightened expectations for a state-of-the-art EIA process—one that would do justice to its scale, risks, uncertainties, and issues.

These pressures resulted in a joint federal/provincial review panel that included Cree and Inuit representation. Public scoping hearings were conducted in nine communities over a period of 23 days. A total of 94 briefs were received and approximately 250 people made oral presentations. The process received national news coverage. The panel received technical support from a team of professionals, and commissioned a series of reports aimed at documenting state-of-the-art practices. The intercultural context of the process challenged the panel to make EIA accessible, relevant, and meaningful to the Cree and Inuit communities. This, in turn, required that the process be open to diverse knowledge systems and patterns of expression, and then address all of the public input from scoping meetings in the drafting of the EIS Guidelines that would be submitted to the proponent. The Great Whale EIS guidelines reflect these imperatives with explicit, and probably unprecedented, requirements, such as "a multicultural definition of the environment," and the need for the proponent to pay close attention to the conceptual and symbolic systems of local people. Likewise, in addressing cumulative effects, social cohesion, safe access to resources, valued or sacred sites, and many other issues, the guidelines departed substantially from standard practice and sought to define and promote new and expanded approaches to EIA in intercultural settings.

The process later reached an anticlimax. The proponent issued an EIS, but by that time, the Québec government had decided to withdraw or at least postpone the Great Whale project. The controversy is memorable for its protests, high-profile media coverage, and court decision, while the joint panel review itself was truncated, with only its ambitious scoping process completed. But it remains a highlight in the history of Canadian EIA, with important lessons for the challenge of conducting meaningful and effective environmental assessment in remote regions characterized by diverse knowledge systems, limited baseline data, irreducible uncertainties, and intense pressures for mega-development and rapid change. EIA evolves and improves, not only through the cumulative lessons of many routine cases but also periodically through experimentation with new approaches in unusual cases.

PETER MULVIHILL

Faculty of Environmental Studies, York University, Toronto, Canada

7.2.2 Intermediate Reports—A Practical Approach to Strengthening SEA Effectiveness and Consistency

During the strategic environmental assessment (SEA) of the urban plan of Ponteranica—a medium-size municipality in northern Italy—we experimented with the application of an approach based on the generation of a set of intermediate short reports. The purpose was to increase the interaction between SEA experts, planners, and decision makers, and, in turn, to enhance the overall effectiveness and consistency of the process. Existing regulations provide only for two reports to be produced during the SEA: a scoping report and a final environmental report. This was considered insufficient to guarantee a proper integration of SEA's outcomes into planning decisions. For this reason, immediately after the preliminary planning/SEA stages, a report detailing the criteria and indicators to be used to assess the sustainability of planning choices was produced. This report summarized the outcome of the first meetings with all actors involved, including formal public hearings. The indicators described were also used to generate a baseline reference study, which was then incorporated into a broader scoping report, made available on the web. As soon as major planning decisions emerged (e.g., concerning new infrastructures and facilities), an additional report was produced containing the reaction of the SEA team to those decisions, including comments for improvement and revision. This document was mainly used to facilitate interaction between technical experts and decision makers.

Later on, the plan started taking shape and its main features were sketched. A subsequent report was issued by the SEA team, this time containing an assessment of the alternatives concerning the main elements and decisions contained in those plan's features. The assessment was largely based on the indicators and guidance produced in the earlier reports. This increased the relevance of its arguments and conclusions. Finally, when a first draft of the plan was produced, a further report critically revised it. It focused on the elements that appeared to be inconsistent with the criteria and objectives set earlier in the process (including the objectives and expectations that emerged during the official public hearing) and captured in the previous reports. This led to the final revision of the plan. It was followed by a SEA short report suggesting appropriate compensation measures to offset the expected land requirements and impacts on natural ecosystems. Finally, the plan was submitted for consultation to the planning authorities. As required by SEA regulations, an environmental report was also issued and made public at this stage.

The approach was successful in strengthening the interaction between SEA experts, planners, and decision makers. It increased the consistency of the planning process, especially during critical moments (e.g., a change in the administration, following local elections that occurred midway through the process). The intermediate reports also served the purpose of recording the "storyline" of the process, keeping track of the evolution of the main planning decisions. This made it possible to document the role played by SEA in steering decision making, as well as some of its tangible results, which included:

- Reduction in the number of new urban expansion areas.
- Reduction of about 30% of the overall land take expected after full implementation of the plan.
- Improved design of the boundaries of land use conversion zones to minimize interference with the ecological network.
- Detailed guidance for ecological compensation interventions included in plan's regulations.

DAVIDE GENELETTI

Department of Civil and Environmental Engineering, University of Trento, Trento, Italy

7.2.3 Ohio River Mainstream Study—A Cumulative Effects Assessment Good Practice Example

Waterway navigation facilitated by locks and dams has existed on the main stem of the Ohio River for over 100 years. The 981 mile main stem, located in the Midwest region of the United States, stretches from Pittsburgh, Pennsylvania, to Cairo, Illinois. In Pittsburgh, the Ohio River is formed at the confluence of the Allegheny and Monongahela Rivers; and at Cairo, the Ohio River flows into the Mississippi River. Mined natural resources such as coal and stone and manufactured products such as petrochemicals and metals are moved up and down the main stem via towboats and various barge configurations. As such, waterway navigation has been and continues to be a vital link in the economic structure of the six contiguous states along the river (Pennsylvania, Ohio, West Virginia, Indiana, Kentucky, and Illinois). Further, such navigation is expected to remain a central influencing factor regarding commerce and economic growth in the region for the foreseeable future.

The design features, sizes, and locations of locks and dams on the Ohio River have evolved since the "wicket dams era" of a century ago. The U.S. Army Corps of Engineers (USACE) has been the lead federal agency since the inception of the navigation system. At the current time, there are 19 "high-lift" locks and dams either on the river, under construction, or authorized. Most locations are characterized by the presence of a main lock and a smaller auxiliary lock that is used during maintenance or rehabilitation periods for the main lock. Both locks may be used for commercial and/or recreational vessels.

Economic, engineering, and environmental issues were addressed in an Integrated Report including a System Investment Plan and Programmatic Environmental Impact Statement for navigation infrastructure to the year 2070 (US ACE, 2011). Economic issues were derived from a range of projections of navigation traffic increases, and cost inefficiencies that occur due to barge queuing when main locks are subjected to either scheduled or unscheduled maintenance or repair. Engineering issues encompassed the possibility of constructing increased sizes of auxiliary locks at several locations and development of risk functions and consequences of component failures that were used to proactively schedule major repairs, rehabilitations, and replacements at existing facilities. Environmental issues were addressed in a cumulative effects assessment and management (CEAM) study of the entire main stem navigation system (US ACE, 2011).

Due to the unprecedented nature of a CEAM study for the 981-mile main stem of the Ohio River, numerous challenges had to be addressed. These challenges provided the basis for delineating several good practice principles; such principles could be applied to other river-related and large-scale river basin studies. Five such principles are summarized below:

Principle 1. Because of multiple valued ecosystem components (VECs) that could be cumulatively affected over the long planning horizon, as well as multiple contributing actions to such effects, it was determined that *a systematic and flexible planning framework for the CEAM study should be utilized.* Further, after extensive discussions, it was also determined that *the framework should be applied on a VEC-specific basis.* The selected planning framework was the 11-step CEAM process promulgated in 1997 (Council on Environmental Quality, 1997a). The steps were adjusted to fit the study spatial boundaries and types of cumulative effects. Further, when the study team focused its thinking such that they represented a particular VEC, for example, freshwater mussels, it was feasible to apply all 11-steps, or subsets thereof, to freshwater mussels and other relevant VECs.

Principle 2. Because of the newness of large-scale CEAM studies, it was determined that traditional project-focused public scoping would not be sufficient. Accordingly, in addition to the central planning team (CPT) for the study, it would be beneficial *to establish an interagency working group (IWG) to provide advice and continuing participation throughout the study.* The IWG consisted of approximately 25 members representing federal and state agencies with responsibilities for environmental management, as well as three NGOs. Six members were from the U.S. Fish and Wildlife Service, with one serving as the coordinator for the USFWS group; two were from the U.S. Geological Survey; and one was from the U.S. Environmental Protection Agency. In addition, the Ohio River Valley Sanitation Commission (ORSANCO), a federally chartered compact among several states in the Ohio River drainage, had two representatives. The remaining members were from natural resources or environmental management agencies in the six states bordering the Ohio River. The IWG had periodic meetings with the CPT. These one-day meetings included information dissemination and updates related to the study, status reports on specific research projects, and working sessions on integrative topics such as impact matrices and indicators for environmental sustainability. In addition, numerous public scoping meetings and interchanges with specific agencies were held during the study. These activities provided an informal continuous scoping process.

Principle 3. Determining the incremental contributions of multiple past, present, and reasonably foreseeable future actions (RFFAs) in relation to both individual and cumulative effects were also a challenge. Again, after considerable discussion, it was determined that *RFFA matrices could be used to delineate cause-and-effect relationships between the multiple actions and the selected VECs and their indicators.* More specifically, a total of 22 RFFA matrices were initially developed for 12 VECs and their associated indicators. The RFFAs, which also included continuing past and present actions, were defined as:

Actions identified by analysis of formal plans and proposals by public and private entities that have primary (direct) or secondary (indirect) impacts on VECs associated with the Ohio River. RFFAs also include potential actions that are beyond mere speculation when incorporated in plans or documents by credible private or public entities. RFFAs may also include events forecasted by trends, probable occurrences, policies, regulations, or other credible data that may have bearing on the VECs.

A total of 87 types of RFFAs were identified and considered in the analyses; the types were divided into six categories: navigation investment actions, other USACE actions, "but for" actions (actions that would not occur "but for" the existence of the navigation system), actions by others, natural disasters, and regulatory environment. Each listed RFFA was characterized in terms of its anticipated time period of occurrence, probability of occurrence, and location on the River. The anticipated effects of each RFFA on each VEC or subcomponent were described in "smart cells" using Microsoft® Excel spreadsheets. Finally, the importance (high, medium, or low) of each RFFA relative to cumulative effects on each VEC or subcomponent was also described in "smart cells."

Principle 4. The ORMSS was conducted at a programmatic level; that is, the focus was on cumulative effects from a system-wide modernization plan for replacement and possibly new locks and dams. In this context, cumulative effects represented the integrative component of the effects of multiple actions on key VECs. Further, *the appropriate measure of such cumulative effects was assumed to be related to reaching or maintaining the environment sustainability (ES) of the key VECs.* Accordingly, a methodology for analyzing the historic, current, and future ES of selected VECs was developed. The methodology comprised four parts: (1) identification of "common effects" on the VEC or subcomponent thereof from the High and Medium importance RFFAs as delineated in the pertinent RFFA matrix; (2) selection of indicators of ES for the VEC or subcomponent thereof, and their tiered grouping, as appropriate; (3) description of the "connections" between the common effects (and related High and Medium importance RFFAs) and the indicator groups; and (4) assignment of a "bottom line" category to the ES of the VEC or subcomponent, based on considering the past, present, and future conditions. The ES categories included "not sustainable," "marginally sustainable," and "sustainable." Specific ES definitions were developed for each VEC or subcomponent. Further, it can be noted that the four-part approach represents the development of a specific conceptual model for each VEC.

Selected indicators were identified for the following key VECs—water and sediment quality, fish, mussels, riparian/floodplain resources, health and safety, and water-based recreation. The indicators were identified by the ORMSS study team in conjunction with the IWG. Using the selected indicators for each VEC, an overall determination of sustainability was made for each VEC at three time intervals (past, present, and future). Applying this methodology to each VEC in the CEAM resulted in past, present, and future characterizations of ES that were included in bar graphs accompanied by summary descriptions of conditions.

Principle 5. Unique CEAM studies will typically require special studies and considerations related to mitigation and management. Examples within ORMSS included research studies and planning for cumulative effects mitigation and management via usage of the expert elicitation process (EEP) (Swor and Canter, 2011). They included

- A study of various engineering and hydraulic factors related to Ohio River locks and dams, along with fish swimming velocities of 44 target species. The objective of this study was to assess upstream fish-passage opportunities through Ohio River main stem dams by relating historical hydraulic conditions at the dams to swimming capabilities of select native and nonnative fishes to determine if the dams are restricting upstream movements of fishes. Further, four additional field studies related to documenting actual fish movement and passage in the Ohio River.
- A study of winter habitat types used by fishes in two navigation pools, and comparison of the results between these downstream and upstream pools. The objective of this study was to examine how abiotic characteristics (e.g., channel morphology, latitude, and depth) regulate winter habitat use of critical fish species in the Ohio River.

A special planning study was focused on possible actions to restore aquatic and riparian ecological resources to a higher state of sustainability. The actions were identified via an EEP involving aquatic and riparian/terrestrial experts knowledgeable of Ohio River resources. The received information was synthesized into goals for the two selected VECs, actions or measures to attain the goals, and necessary monitoring to evaluate conditions. Finally, 26 types of ES actions were identified and classified into three broad groups of ES alternatives (Swor and Canter, 2011). They were then included in the decision-making stage along with four navigation improvement alternatives.

Final Observations. Planning and conducting regional-scale CEAM studies can be both challenging and complex. However, they can be accomplished by applying or modifying practices from EIA studies along with systematic approaches for addressing new themes such as analyses of environmental sustainability, and the development of strategic mitigation measures for the incremental effects from proposed actions along with regional management of cumulative effects resulting from multiple contributors. The above principles are illustrations of both modifications of existing EIA practices and the creation of new strategies for emerging CEAM challenges.

LARRY CANTER

School of Civil Engineering and Environmental Science, University of Oklahoma, Norman, OK, USA

TOM SWOR

U.S. Army Corps of Engineers, Nashville District, Nashville, TN, USA

7.3 DEFINING THE PROBLEM AND DECIDING ON A DIRECTION

The three stories address practicality in different ways. The first story describes a formal and ambitious scoping process that effectively integrated a diversity of stakeholder perspectives. The second story describes how the use of intermediate reports provided a practical approach to strengthening SEA effectiveness and consistency. The third story provides a good practice example of cumulative effects assessment. Key principles, with potentially broader application, are highlighted.

Practicality, as illustrated in Figure 7.1, has many dimensions. Each dimension encompasses numerous elements relevant to practicality in the IA process. The IA process, as expressed in IA theory and practice, can be much more practical.

Despite the widespread advocacy of screening and scoping too often, IA documents remain excessively descriptive, lengthy, and unfocused (Barrow, 1997; Ensminger and McLean, 1993; EC, 2009c; Ross et al., 2006; Tzoumis, 2007; Weston, 2011). IA processes continue to take too much time and to consume too many resources (Macintosh, 2010; Sadler, 1996; Wolfe, 1987). Alternatively, they operate within such severe time and budget restrictions that the potential for good practice is seriously inhibited (Clark, 1997; Offringa, 1997). Scoping, although demonstrably beneficial, is often either not applied or poorly applied (Morgan, 1998; Pinho et al., 2010; Sadler, 1996; Wood et al., 1996, 2006). Shorter IA reports, rather than being focused and streamlined, tend to be of poorer quality (Wood et al., 2006, 1996). Practice continues to lag well behind theory (Snell and Cowell, 2006). Practitioners continue to struggle with identifying major problems, root causes, priorities and impacts of real concern (Lee and Kirkpatrick, 2006; Sadler, 1996). They have difficulty in dealing with concepts like significance and sustainability (Retief et al., 2008). Key stakeholders and the public often are excluded from the scoping process (Snell and Cowell, 2006; Tsuji et al., 2011). Effectiveness ratings for study design activities such as problem definition, objectives determination, and terms of reference formulation leave considerable room for improvement (Lee and Kirkpatrick, 2006; Sadler, 1996). The dominance of efficiency arguments and ill-defined or excessive requirements, often in combination with gaps, overlaps, and coordination problems among government levels and across agencies, are still problems in many jurisdictions (Anderson, 2001; Ensminger and McLean, 1993; Offringa, 1997; Snell and Cowell, 2006). In short, IA documents, processes, and institutional arrangements could be more efficient and *focused.*

IA is plagued by such *reality*-related problems as (1) a gulf between how policy making and project planning take place and how IA processes and practices assume they take place, (2) IA theory not well grounded in or derived from IA practices and experiences, (3) IA practices poorly suited to

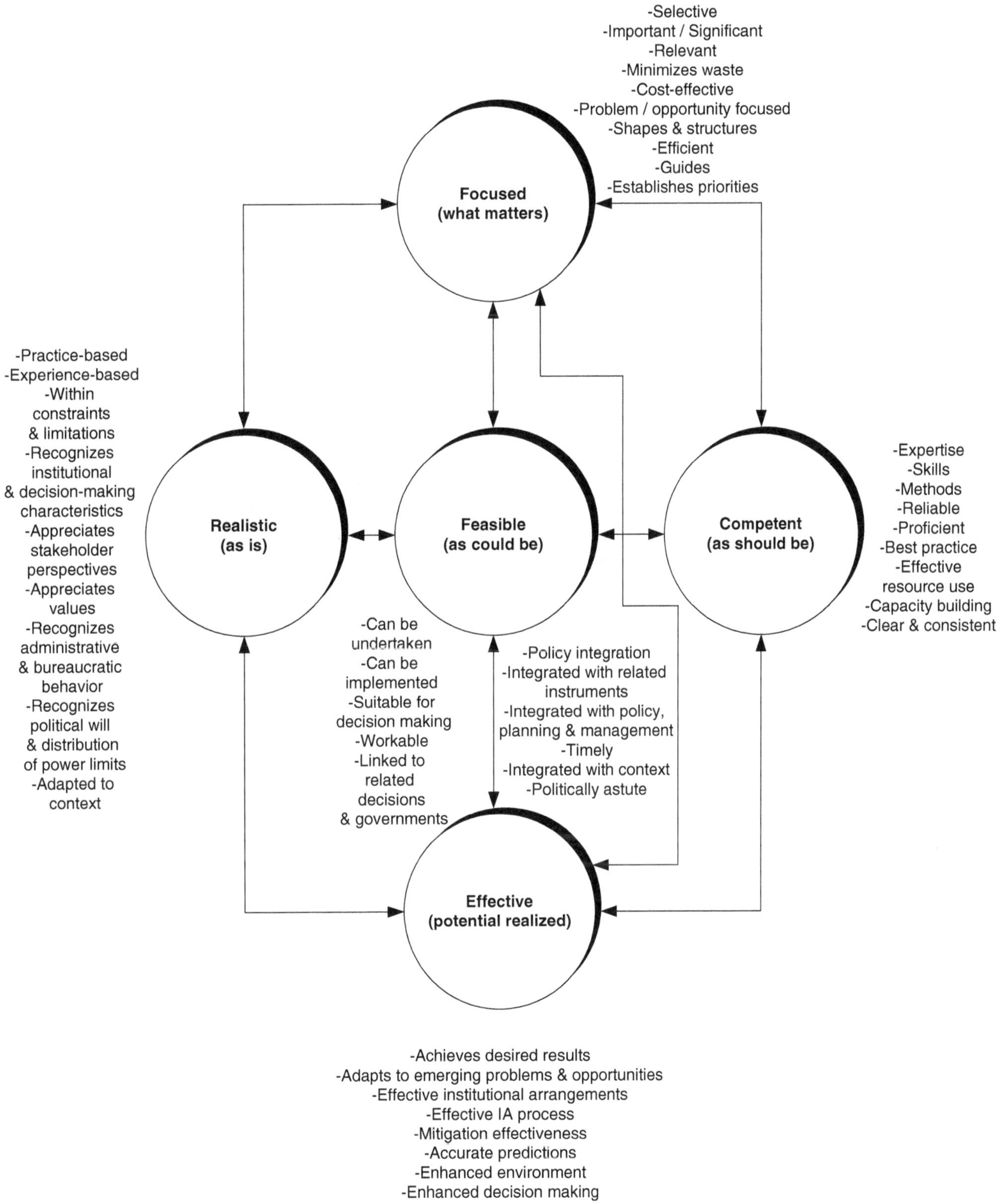

Figure 7.1 Examples of practicality dimensions.

the contexts in which they are applied, and (4) IA processes and practices that fail to appreciate and to integrate the perspectives, contributions, and implications of varying stakeholder values and perspectives. These problems suggest that IA theory and practice need a "reality check." Rational assumptions and fantasies (see Chapter 5) embedded in most IA process characterizations are rarely realistic (Snell and Cowell, 2006). Also unrealistic is the naïve assumption that better information leads directly to better decisions (Fuggle, 2006). Human, institutional, and political characteristics, perspectives, constraints, and behavioral patterns need to be better understood, especially regarding how decisions are made and implemented (Sandham and Pretorious, 2008; Weiner, 1997; Zhu et al., 2010).

More consideration could be given to policy and decision-making models that mimic planning and decision making as they are (such as bounded rationality), in contrast to how rationality advocates would like them to be (Nilsson and Dalkmann, 2001; Nitz and Brown, 2001). An enhanced understanding of the role of politics in IA-related decision making is critical (Fuggle, 2006). Also essential is a better appreciation of how and why authorities resist and disown IA requirements (Weston, 2011). Further consideration needs to be given to the barriers to effective IA and the reasons why IA practice largely ignores and falls well short of good practice standard (Noble, 2009a; Polido and Ramos, 2011). Stakeholder perspectives need to be better understood (Au, 2006). More attention needs to be devoted to how to more effectively integrate indigenous people and vulnerable and poorly represented segments of the public into decision making (Baker and McClelland, 2003). Additional emphasis needs to be placed on effectively characterizing and addressing the complex problems encountered in IA practice (Morgan, 2006). Decision-making and implementation constraints are not insurmountable. But first they must be understood. The naïve expectation that rational and/or scientific IA documents and processes lead inevitably to environmentally sound decision making and implementation is questionable at best and at worst can reduce the relevance of IA outputs to major project and policy decisions (Nitz and Brown, 2001). The late (in the decision-making process) initiation of IA requirements and the large number of major decisions not subjected to IA requirements remain recurrent problems in IA practice (US CEQ, 1997a). More systematic consideration needs to be given to how IA systems affect, are integrated into, and positively influence (i.e., added value) policy making and project planning (Bartlett, 1989; Noble, 2009a; Retief et al., 2008). IA practitioners need to be better informed about the nature of policy-making and project planning processes (Nitz and Brown, 2001).

There is an urgent need to learn from experience and good practice (Glasson et al., 1999; Retief, 2007b; Sadler, 1996). Often IA requirements are not being satisfied (Weston, 2011). Even when mandatory requirements are met, best practice is largely dismissed (Polido and Ramos, 2011). Lessons and insights, derived from good practice, need to be better integrated into IA regulatory requirements, guidelines, and practices (ERM, 2000; Spooner, 1998; Tang, 2010). IA regulators and theorists should strive for an enhanced understanding of stakeholders (e.g., bureaucrats, politicians, proponents, practitioners, nongovernment organizations, and members of the public), perspectives, interests, and needs (Morgan, 2006; Rowson, 1997). The IA process (or more exactly, multiple IA processes) needs to be designed to more closely match contextual characteristics (Greer-Wooten, 1997; Nilsson and Dalkmann, 2001; Nitz and Brown, 2001; Whitelaw et al., 2009).

A practical IA process also must be prescriptive. The shift to the prescriptive does not mean an abandonment of the "real." A practical IA process remains realistic but also seeks out *feasible* actions that can be undertaken and implemented in varying contexts. IA practice needs to balance practicality and prescription. Too often IA practice (1) neglects the needs of decision makers, (2) has minimal effect on planning and decision making, (3) fails to influence or facilitate implementation, and (4) is not effectively integrated with other environmental management instruments, project planning, and public policy making (Tinker et al., 2005; Thérivel, 2010).

Providing relevant and sound environmental information and advice to decision makers, although necessary and improving, is far from the whole picture. A less passive approach to decision-making integration is needed (Cashmore et al., 2004). The needs, values, and perspectives of decision makers and of other stakeholders are often neither identified nor addressed (Polido and Ramos, 2011; Spooner, 1998; Wood, 1995). Provisions for impartial and democratically accountable decision making tend to be weak (Hinte et al., 2007). Postapproval management and follow-up remain more the exception than the rule (AGC, 2004, 2008; Lundberg et al., 2010; Sadler, 1996). Strategies are required to ameliorate implementation obstacles such as overlapping mandates, delays, late triggers, unclear, incomplete, or contradictory requirements and guidelines, coordination and consistency difficulties, and the propensity of agencies to treat IA requirements as a rigid paperwork exercise and bureaucratic hurdle (Anderson, 2001; Clark, 1997; Evaluation Partnership, 2007; Sadler, 1996; US CEQ, 1997a; Weiner, 1997). The potential for IA as a strategic decision-making tool is not fully realized (Clark, 1997). The relationship between SEA and project-level EIA is poorly defined (EC, 2009d; Fischer et al., 2009; Noble, 2009a). IA requirements are still rarely applied to major government decisions or to nongovernmental actions with potentially significant environmental consequences (Andrews, 1997). IA documents, when prepared, are commonly treated as a bureaucratic and administrative exercise and as decision implementation rather than decision-making documents (Ensminger and McLean, 1993; Noble, 2009a). Too little attention is devoted to establishing complementary links between EIA and SEA and project planning, policy making, and other environmental management and sustainability instruments (EC, 2009d; Nitz and Brown, 2001). By neglecting these practical decision-making and implementation considerations, the IA process often falls short of its potential.

Sound execution of the IA process requires *competence*. Much advice regarding methods and procedures is offered in IA texts and literature. But IA practice, as reflected in requirements, guidelines, and documents, too frequently fails to meet even minimum good practice performance standards or to improve over time (Androulidakis and Kanakassis, 2006; Glasson et al., 1999; Hinte et al., 2007; Peterson, 2004; Sadler, 1996; Spooner, 1998; Tzoumis, 2007). Tested methodological frameworks are not broadly available or applied when they are available (Noble, 2009a). Public participation criteria are rarely met (Polido and Ramos, 2011). Poor quality and difficult (for the public and decision makers) to understand IA documents are still all too common (Alton and Underwood, 2003; Page, 2006; Ross et al., 2006). Often available methods are poorly applied (Ross et al., 2006). Activities such as scoping, the analysis of alternatives, significance determination, the consideration of substantive concerns such as health, the implementation of mitigation measures, the assessment of cumulative effects, and the design and application of follow-up continue to be poorly performed (Fischer, 2010; Jalava et al., 2010; Peterson, 2004; Ross et al., 2006; Sandham and Pretorious, 2008). As a consequence, methodological weaknesses all too often undermine the substantiation of findings (Lee and Kirkpatrick, 2006). The extreme variability in the quality of IA documents over time, from project to project and from region to region, is difficult to reconcile with the image of a "maturing" field of theory and practice (Ecologic et al., 2007; Renda, 2006). These shortfalls could simply be the result of a failure to apply available knowledge and insight. Perhaps the variability in quality could be largely explained by differences in the experience and expertise of practitioners (Barker and Wood, 1999).

Alternatively, IA literature could be "missing the mark" in meeting practitioner needs. Perhaps the guidance provided is too superficial, too scattered across numerous sources, too difficult to access, and too difficult to understand or apply (ERM, 2000). There may be insufficient time or money to apply "state-of-the-art" practice. Possibly the management skills and expertise required to "tie the pieces together" are insufficiently developed or inconsistently applied (Glasson et al., 1999). More capacity building could be required before IA practitioners achieve the necessary proficiency levels (Offringa, 1997). "Good practice" standards could be too general, contradictory, unrealistic, or poorly adapted to context (Põder and Lukki, 2011). Methods may need to become more cost-effective and more conducive to applying new technologies (Offringa, 1997). The skills and expertise required of practitioners could be more complex than the methods purveyors realize (Webster, 1997). Simply assuming that the necessary knowledge is available and that the problem will resolve itself as more experience is acquired is a dubious strategy given the continuing quality disparities after more than 30 years of IA practice. A more prudent strategy is to assume that IA competence deficiencies require an array of responses.

IA practicality "problems" are little more than impressions and the "solutions" offered no more than speculations if IA *effectiveness* is not systematically addressed. Considerable progress has been made in formulating and applying IA quality and effectiveness criteria and performance standards to documents, procedures, methods, and institutional arrangements (ERM, 2000; Sadler, 1996; Wood et al., 1996). These efforts, although laudable, barely "scratch the surface" in terms of what is required to "close the loop" from experience to learning (Glasson et al., 1999; Wood, 1995). The monitoring and auditing of actual environmental impacts (as compared with predicted effects), mitigation effectiveness, and decision-making effectiveness are still more the exception than the rule (AGC, 2004, 2008; Clark, 1997; Culhane, 1993; Lundberg et al., 2010; Morgan, 1998; Noble, 2009a; Sadler, 1996). The effectiveness of mitigation measures is rarely determined (Clark, 1997; Fischer, 2010). A much greater effort could be made to assess and compare IA methods, process designs, management strategies, and institutional arrangements, in varying contexts (Glasson et al., 1999; Zhu et al., 2010).

The contribution of IA to more environmentally sound planning and decision making and to a more sustainable environment is more often an assumption than a demonstrated outcome (Andrews, 1997; Retief, 2007b; Welles, 1997). SEA and EIA, although advanced as a solution to many environmental problems, have been far from effective in practice (Retief et al., 2008). The tendency has been for IA to be viewed as a procedure to be followed rather than as a proactive environmental management tool (Weston, 2006). Perception of results, in terms of influencing decision making and avoiding significant environmental effects, among stakeholders and among IA systems varies greatly (Heinma and Põder, 2010; Heinma and Põder, 2010; Macintosh, 2010; Pölönen et al., 2011). Evaluations of the quality of IA documents in terms of the treatment of substantive environmental concerns (e.g., biodiversity, health, and social) suggest little improvement (Bhatia et al., 2010; Kemm, 2005; Mandelik et al., 2005; Vanclay, 2010). Outcomes, in terms of contributions to sustainability, have tended to range from mildly positive to mildly negative (Thérivel et al., 2009). Application by the private sector of SA and SEA has been limited (Jay, 2005). The magnitude and nature of the contribution and which strategies and tactics effectively operate within constraints and overcome implementation obstacles are even less clear. In the absence of demonstrated contributions, it is difficult to argue for the continued allocation of resources to IA and for a well-defined role for IA within environmental management and sustainability strategies. What is required is a substantiated case that IA achieves desired results and adapts to emerging problems and opportunities.

The problem then is five clusters of interrelated problems, all bearing on the issue of practicality in the IA process. The direction is concepts and approaches for making the IA process more focused, realistic, feasible, competent, and effective. These approaches establish a foundation for practical IA regulatory requirements and practical IA processes.

7.4 SELECTING THE MOST APPROPRIATE ROUTE

7.4.1 Focused

The IA process, in its fullest expression, is, by definition, impossible. More environmental components and interactions, alternatives, and direct and indirect effects can always be suggested. The level of detail can always be increased. More parties can be involved. More participation can occur. More research can be undertaken of uncertainties. In short, there is no "stopping rule." As a result, the IA process can never be comprehensive. The real issue is how to focus or scope (used interchangeably here) the IA process to balance environmental objectives, available resources, and decision-making requirements. Figure 7.2 provides an overview of the major elements associated with focusing regulatory and applied IA processes.

Scoping focuses IA institutional arrangements, the IA process, and IA documents (Morgan, 1998). It determines what will and will not be examined (Wolfe, 1987). It establishes appropriate levels of detail for various analyses (Wolfe, 1987). It directs and structures the IA process, IA operational procedures, and institutional reforms. Scoping establishes priorities (Eccleston, 1999a). The benefits ascribed to scoping are considerable. Scoping, when effective, reduces the duration of IA planning and review processes, abbreviates IA documents, ensures efficient resource use, and identifies key issues, priorities, and problems early enough in the process to take appropriate action (Glasson et al., 1999; Sadler, 1996; US EPA, 1998b; Wolfe, 1987). It is conducive to early stakeholder involvement and can reduce the likelihood and severity of conflict among stakeholders (ERM, 2000; Sadler, 1996; US EPA, 1998b). Scoping focuses the process on potentially significant issues (for decision makers and the public), options, and impacts (Bond and Stewart, 2002; US CEQ, 2005a). It reduces the likelihood that resources will be wasted on insignificant concerns (ERM, 2000; US EPA, 1998a). Scoping contributes to higher quality IA documents and to more environmentally sound decisions (ERM, 2000; Sadler, 1996).

Narrowing the scope of the IA process, based on overview analyses and preliminary stakeholder discussions, however, can lead to the premature rejection of alternatives and to unanticipated effects (Erickson, 1994). The level of detail may not be sufficient to justify a clear distinction between significant and insignificant issues and impacts. Once an IA process is "scoped," the resulting study designs could be treated as "blueprints" to be followed regardless of changing circumstances. Sometimes a scoping process is unduly influenced by vested interests. These constraints can be ameliorated if alternatives and potential impacts are retained for further consideration if there is doubt regarding their suitability and significance. An open and staged process, with a high level of stakeholder participation, in combination with conservative assumptions, sensitivity analyses, peer review, and a careful scrutiny of the results of monitoring and effectiveness analyses, can reduce the likelihood of inadequately supported decisions (Morgan, 1998). Scoping works best, when there is early and ample stakeholder involvement (US EPA, 1998b; Morgan, 1998). Multiple perspectives should be brought to bear on scoping interpretations and decisions. Scoping need not be only a stage near the outset of the IA process. Instead it can be a scanning–focusing phase preceding more detailed analyses and prior to each IA process decision (Brown, 1998; Kennedy and Ross, 1992). Caution is essential, given the "broad brush" nature of the analyses. Sometimes this means scanning ahead. Sometimes previous decisions need to be reconsidered (Brown, 1998). Flexibility to adjust to changing circumstances is critical. Adequate consideration should be given to both biophysical and socioeconomic concerns (Erickson, 1994; Morgan, 1998).

Scoping can be applied at the outset to define the problem, establish the terms of reference, design the overall IA process, and set the study boundaries (Barrow, 1997; Sadler, 1996; Wood, 2000). It also can focus and structure each IA process activity and document leading up to (e.g., alternatives formulation and evaluation, baseline analyses, impact identification and prediction, public and agency consultation, document preparation) and subsequent to (e.g., monitoring, mitigation, auditing) proposal acceptance or rejection (Kennedy and Ross, 1992).

Scoping helps reform IA institutional arrangements. Screening distinguishes between actions subject and not subject to IA requirements. It also determines applicable approval streams. IA screening and scoping requirements are streamlined and focused by class or categorical IA requirements and by significance thresholds and criteria. Documents can be simplified by page limits, incorporation by reference, report format requirements, and page limits (Kreske, 1996). Review and approval can be expedited by timing and circulation limits, by merged and cooperative interagency and intergovernmental requirements, and by agency review guidelines. The latter directly link IA requirements to agency mandates, policies, programs, and priorities (Kreske, 1996). Meetings, workshops, study groups, task forces, expeditors, facilitators, mediators, and participant funding can constructively bring together interested and affected parties, both within and external to the government review process.

7.4.2 Realistic

A practical IA process is necessarily grounded in practice and experience. It is realistic. Figure 7.3 illustrates examples of distinctions potentially relevant to making the IA process more realistic. Table 7.1 presents a summary overview of the characteristics of several potentially relevant realism concepts. Collectively, these concepts suggest that knowledge is often subjective, pluralistic, experience-based, and socially constructed. Practice is concrete, action-oriented, critical, and experimental. Planning, policy making, and decision

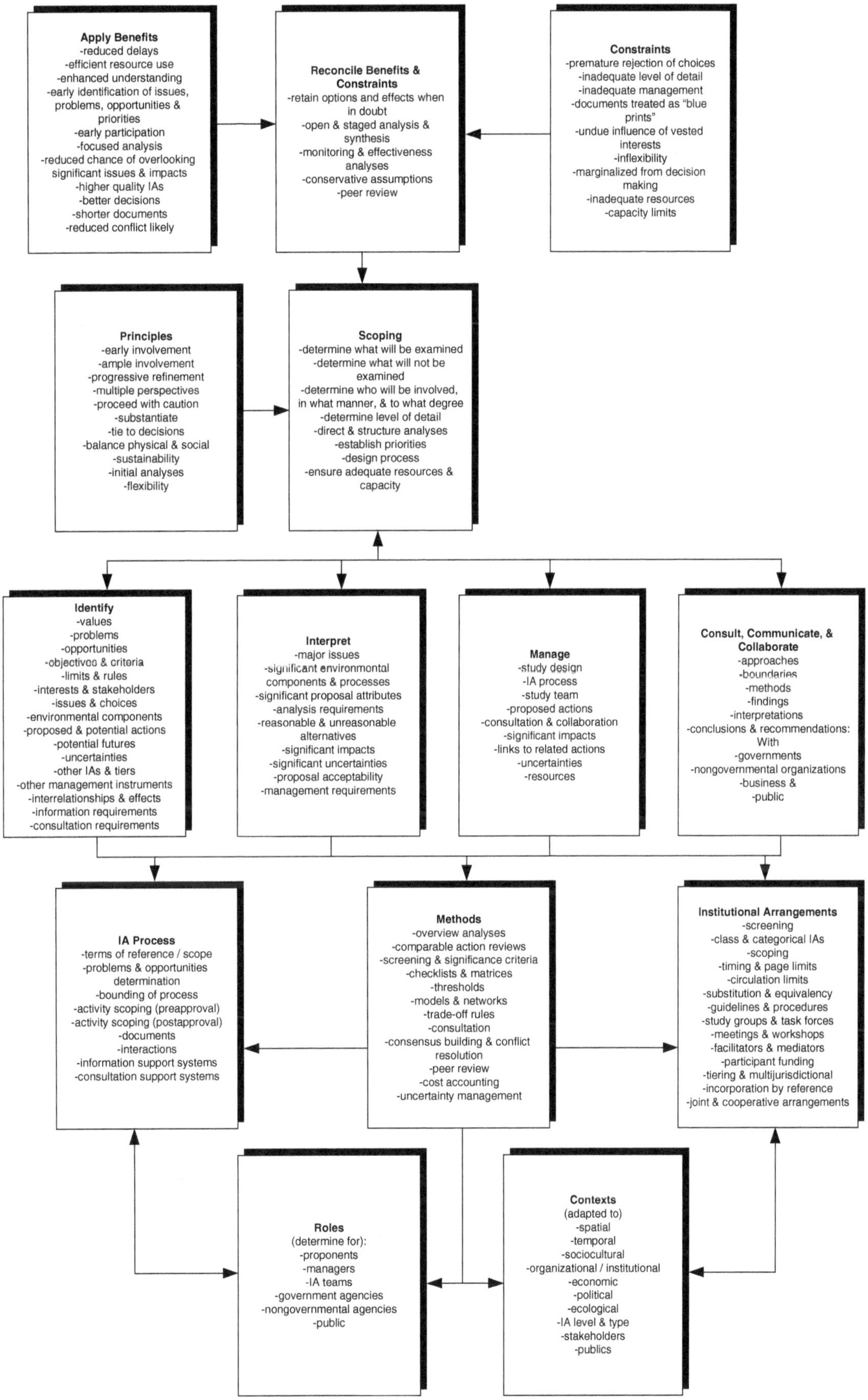

Figure 7.2 Focusing the IA process.

Figure 7.3 Realism in the IA process.

Table 7.1 Examples of Potentially Relevant Realism Concepts

Pragmatism	• A philosophy of everyday life; antifoundational; importance of dissent and irreverence • A plurality of shifting truths grounded in concrete experiences and language • Actions structured by subjective interpretations of the world; interpretations evaluated in terms of their practical implications; intent is to solve human problems • Concepts are socially constructed; truth not understandable outside the social and psychological processes and community that makes truth possible; justification from prior experience • Concepts, terms, and assumptions tentative and provisional—always open to further interpretation and criticism—fallibilism; prediction possible but limited • Focuses on concreteness, action, adequacy, facts, and power; turns from abstractions, verbal solutions, a priori reasons, closed systems, origins, fixed principles, and absolutes • Pluralistic—a plurality of traditions, perspectives, and philosophical orientations • No definitive formulation to problems and no clear solutions; all knowledge contingent • Learning by doing (learning and doing indivisible); learning a collaborative experience • Problems solved by common sense and experimentation; guided by changing experience
Empiricism	• Reliance on experience and observation alone • Founded on belief that all knowledge originates in experience or in the practice of relying on observation and experiment • Limited consideration of system or theory • Focuses on collecting facts and observation • Emphasis on information derived from human senses • Exemplified in studies of practice
Existentialism	• Point of departure experience rather than generalized concepts; all concepts derived from human perceiving, pattern forming, symbolizing, comparing, and conceptualizing • Emphasis on immediate experiences and individuals as autonomous moral agents • Terms described not defined; consistency only possible through repetition of experience; not possible to state assumptions and conclusions only reached based on implications • Different people reach different conclusions based on same information • Propositions have multiple meanings; communications failures expected • Reality only partially conveyed by symbols • Existence contingent (not independent of situation); only here and now meaningful and present experiences—complex, unique, correlated, uninterpretable, and uncommunicable
Phenomenology	• All knowledge is subjective • Analyzes and identifies basic features of subjective knowledge to understand individual and to make life more significant • Belief that people should be studied free from any preconceived theories and suppositions about how they act • Search for understanding of nature of act rather than explanation • Belief that for people world exists only as a mental construction; created in acts of intentionality
Incrementalism (also bounded rationality)	• Margin-dependent choices; successive limited comparisons; a process of gradual change (muddling through) • Restricted number of values, alternatives, and consequences; available means and solutions • Objectives adjusted to policies (means and ends overlap and reciprocal); no coherent set of goals • Analysis and evaluation—serial, remedial, socially fragmented, and unpredictable • Assumes ambiguous and poorly defined problems, incomplete information (baseline conditions, values, alternatives, consequences); thinking inseparable from context and experience • Decision making fragmented and largely reactive to external circumstances; not value-free; decision makers avoid uncertainty and adverse consequences • Appreciates human (especially expert) knowledge and control limits; political, social, and economic environments complex, uncertain and stable; planning incomplete, partial, collective, and episodic • Atomistic society and decentralized decision-making structures and procedures; policy making a negotiation and bargaining process involving a plurality of competing interests and values • Test of a good policy—agreement; driven by political circumstances; focus on political negotiations and coalitions • No to limited reliance on theory; adapted to limited cognitive capacities; influenced by free competition model of economics • Rationality bounded by cognitive limits, social differentiation, pluralistic conflict, and structural distortion

(*continued*)

Table 7.1 (*Continued*)

Knowledge and reflection in practice	• Each individual develops own way of framing (taken-for-granted assumptions) role (e.g., writer, organizer, advocate); may choose from profession's repertoire or fashion own interpersonal theory of action • Knowing in practice (common sense) directs and limits reflection in practice (thinking about what we are doing); self-reinforcing system in which role frames, action strategies, relevant facts, and interpersonal theories are bound together • Behavior understood in terms of problems set for self • Roles evolve in conversation with situation; practice as exploratory experiment (probing, playful) • Policies sometimes reframed in action; often as a result of reflecting on frame conflicts • Double visioning: awareness of own perspective and that of others • Rhetorical frames: underlies persuasive use of stories and arguments; action frames inform policy practice • Study of strategies to resolve frame conflicts (e.g., resistance, appealing to consensus, mapping one frame over another) • Policy design inevitably social, political, pragmatic, and communicative
Professional episodes	• A schematic framework for analyzing professional practice episodes • Theories are socially constructed; knowledge derived from action • Episodes analogous to dramas; practitioners construct performances with constituent others • Distinctions: institutional professional espoused theory vs. practitioner espoused theory, practitioner theory-in-use vs. practitioner espoused theory • Communications (talk is action) at core of professional episode • Practitioner strives for one or more of enhanced self-esteem, mastery of professional domain, cognitive and value consistency, self-actualization, or significant impact on world of contemporaries • Practitioner in performing concrete professional tasks is the ultimate theorist for each episode • Importance of concrete situation, language and communications, ambiguity, and evaluation
Critical theory	• Rationality types: cognitive–instrumental, moral–practical, and aesthetic–expressive • Reason anchored normatively (what individuals can mutually agree upon) and intersubjectively • Context-dependent • Forms of action: teleological/strategic and cognitive/instrumental • Action types: norm-regulated and moral–practical • Efficiency and truthfulness validity types • Distinction between deontological (means and ends chosen freely) and teleological (given ends, means to end) • IA must be free from control from any one party and from orientation toward a particular result • Open reflexive process (communicative rationality); criteria: truth, rightness, and truthfulness
Mangle of practice	• Human and nonhuman material agents (e.g., tools) intertwined and coevolve • Simultaneously objective, relative, and historical • Dialectic of resistance and accommodation • Favors antidisciplinary synthesis and multidisciplinary eclecticism • Multiple rather than monolithic conceptualizations, models, and approximation techniques • Data and theory not necessarily connected; approximations toward the truth • Practice aims to make associations (translations, alignments) between diverse elements

Sources: Blanco (1994), Bolan (1980), Braybrooke and Lindblom (1963), Elling (2007), Etzioni (1967, 1986), Forester (1989), Friedmann (1987), Hainer (1968), Lindblom (1965), Menard (1997), Nilsson and Dalkmann (2001), Pickering (1995), Schön (1983), Schön and Rein (1994), Simon (1976), Smith (1976), Verma (1998).

making are frequently constrained, decentralized, incremental, collaborative, communicative, political, and pluralistic. Society and environment (the context that circumscribes policy and decision making) are commonly fragmented, uncertain, complex, ambiguous, and unpredictable. Distinctions between theory and practice, ends and means, facts and values, and objectivity and subjectivity are artificial. These portrayals of "reality" challenge the value and validity of preconceived theories and suppositions, abstractions, fixed principles, absolutes, and symbolizing.

These realism concepts are largely a reaction against the rational assumptions (see Chapter 5) inherent in most policy, planning, and IA theories. Although they overstate decision-making constraints, they are a closer approximation of the environment within which most IA practitioners operate than the antiseptic versions of the IA process presented in most IA texts. These realism concepts, however, provide only impressions rather than a firm foundation for a practical IA process. What is required is a more detailed characterization of the "reality" of IA practice. IA literature and the literature of related fields such as planning offer a sense of many aspects of IA practice. A more complete picture would draw heavily upon related fields such as political science and would integrate relevant distinctions and concepts from decision making, public policy, organizational, and administration theory.

A realistic IA process would recognize how decisions are made, how organizations are structured, how people behave in organizations, and how organizational structures and procedures vary depending on contextual characteristics. How IA practitioners operate (effectively and ineffectively) and how knowledge is generated through practice would be understood. The mechanisms by which IA-related actions are integrated within decision making would be appreciated. Factors that promote and impede the integration of IA and organizational planning and decision making would be evident (Keysar and Steinemann, 2002). How power is exercised, how resources are allocated, and how actions are bounded and constrained would be acknowledged. The various ways in which actions are implemented, facilitated, inhibited, or prevented would be understood.

A realistic foundation has been partially constructed within IA and even more so in related fields such as planning and public policy. The relevant analyses are widely scattered. Cognitive limits and the inherent knowledge uncertainties associated with decision making have been considered (Beanlands and Duinker, 1983; Nilsson and Dalkmann, 2001). The constraints and opportunities posed by institutional arrangements, the exercise of political power, bureaucratic behavioral patterns, and the implications of ecological, economic, social, and cultural conditions are sometimes noted.

Institutional arrangements concern the government structures and procedures pertaining directly (e.g., legislation, regulations, policies, guidelines, staff, and budgets) and indirectly (e.g., related policies, programs and activities, departmental and agency jurisdictions and responsibilities, interactions among agencies and with other government levels, controls, resources, coordination and information transmission mechanisms, antagonisms, procedures for mediating conflicts and for representing interests, general efficiency, accountability, and flexibility) to IA (Nilsson and Dalkmann, 2001; Pressman and Wildavsky, 1973; Rickson et al., 1990a; Smith, 1993; Shoemaker, 1994).

IA practitioners often see themselves as objective, independent, and apolitical advisors. But the IA process is inherently *political.* Politics is a major determinant of if and how IA requirements are applied. The lack of political will is a major impediment to achieving environmental objectives (Caldwell, 1997). Realistic IA practice appreciates how political power is used and misused (Smith, 1993). If political power is highly dispersed, evasion and the dilution of reforms is the usual result (Pressman and Wildavsky, 1973). Highly concentrated political power can be authoritarian, coercive, and narrowly focused. Inequities in the distribution of prestige, power, and equity can make it difficult to realize social objectives (Rickson et al., 1990a). A centralized and hierarchical style of governance can prevent or severely inhibit the negotiation and consensus building needed to facilitate public understanding and possibly support (Conçalves, 2002). Stakeholder roles, formal and informal procedures for forming alliances, intervention rules and political structures, and decision processes need to be considered (Smith, 1993). Control (e.g., procedural, judicial, evaluative, development aid agency, professional, and direct public and agency) mechanisms need to be taken into account (Ortolano, 1993).

Bureaucratic structures, procedures, and patterns of behavior strongly influence IA effectiveness. Bureaucracies exhibit such characteristics as fixed official jurisdictional areas, rationalistic division of labor, official duties, hierarchical structure, management by written rules, and expert management (Hummel, 1977). The role of bureaucracies can be negative or positive. When negative, there is a gap between what is important to the bureaucracy (e.g., precision, stability, formal rationality, formalistic impersonality) and what is important in society (e.g., justice, freedom, poverty, illness) (Hummel, 1977). A failure to institutionalize new forms of public participation into IA requirements and practices, for example, can result in a gap between public aspirations and expectations and available participation forms (Conçalves, 2002). Such gaps often reinforce public distrust and contribute to community opposition.

Negative bureaucratic behaviors that can impede environmental initiatives include, for example, rigidity, classification (oversimplifying the world), interagency and intergovernmental antagonisms, favoring routine and prescribed rules over policy, displacing ends with means, preferential treatment of some client groups over others, "empire building," overcommitment, and ill-defined criteria (Pressman and Wildavsky, 1973; Rickson et al., 1990a,b; Sorenson and Auster, 1989).

Often government participants will agree with substantive environmental ends but will still oppose or fail to facilitate a proposed action. Pressman and Wildavsky (1973) identify several reasons for this behavior including (1) direct incompatibility with other commitments, (2) no direct incompatibility but a preference for other programs, (3) simultaneous commitments to other projects, (4) dependence on others who lack a sense of urgency in the program, (5) differences of opinion on leadership and proper organization, (6) legal and procedural differences, and (7) agreement coupled with lack of power. Bureaucracies do not always exhibit such tendencies. Frequently, government officials assume a positive and proactive role in advancing environmental objectives. Still, it is prudent to be aware of general bureaucratic tendencies (positive and negative) and of the specific constraints and opportunities posed by the structures, procedures, and behavioral patterns of each government department and agency involved in the IA process.

A realistic IA process considers interrelationships between process and ecological, social, economic, cultural, and political *contexts*. The dangers associated with uniformly applying standardized definitions of good IA practice and appropriate institutional arrangements are increasingly acknowledged. Particular attention has been devoted to the adaptations required to meet the needs of developing countries (Barrow, 1997; Lee, 2000; Rickson et al., 1990b; Smith, 1993). Comparable adaptations have been suggested for transitional economics, for northern environments, and for numerous other setting types. It is also necessary to make adjustments to suit the unique circumstances associated with a proposed action in a particular setting. The IA process is not simply designed to "fit" contextual realities. IA is an instrument for change. How contextual characteristics are likely to change, both positively and negatively, in response to changing IA requirements and practices should be considered.

Institutional, political, bureaucratic, and contextual constraints and opportunities are strongly influenced by the quality and effectiveness of IA practice. As documented in the "defining the problem and deciding on a direction" sections of Chapters 2–12, there remains a considerable shortfall between IA aspirations and achievements. Good IA practice can alleviate IA practice deficiencies. Part of good IA practice includes accounting for, and ameliorating where practical, institutional, political, bureaucratic, and contextual constraints. It includes taking advantage of opportunities. The full incorporation of realism into the IA process requires integrative frameworks, concepts, and distinctions, presented in a user-friendly format. It also necessitates a thorough canvassing of sources bearing on distinctions such as those presented in Figure 7.3. Most importantly, it requires empirical studies of IA practice—studies that systematically draw out positive and negative experience-based lessons and insights, with potential for broader application.

7.4.3 Feasible

A feasible IA process, as illustrated in Figure 7.4, is workable. It can be undertaken. It provides a decision-making basis. It can be implemented. It can be managed. It is appropriate to the context. It is built on a realistic foundation (i.e., it is experience and practice based). Consistent with realism, it is social, political, subjective, uncertain, and constrained. It is guided by strategies, informed by concepts, and aided by tactics—all of which are realistic and practical. It overlaps and is merged with decision making, implementation, management, and context. It contributes and adapts to IA reforms. Appropriate links are made to related tools and methods. It is integrated with related policies, programs, and plans. It is blended with organizational operations. It is linked to other decision-making levels, related environmental management instruments, and the relevant actions of other governments, the private sector, and nongovernmental organizations. It crosses disciplinary and professional boundaries. It is integrated, where practical, within synthesis (e.g., sustainability) frameworks.

Table 7.2 briefly describes a cross section of relevant feasibility strategies, concepts, and tactics. The IA process depicted in these sources is selective, cyclical, open, fluid, decentralized, and evolving. It is incremental but progressive. It learns through experimentation, reflection, and dialogue. It scans ahead and reconsiders past decisions. It continuously explores uncertainties, interconnections, complexities, and conflicts. It is reasonable rather than rational. It operates at multiple levels of detail. Preferred choices are not systematically and comprehensively compared. Rather, they are tested for agreement and feasibility. Then they are refined, adapted, and embellished to better meet agreed-upon needs, consistent with stakeholder perspectives. The process draws heavily upon knowledge derived from experience and practice. It operates within resource and other constraints. It transcends such false dichotomies as ends and means, technical and political, and objective and subjective. It freely crosses disciplinary and professional boundaries in the search for practical solutions to real problems. The process unifies planning, management, decision making, and implementation. It is connected, as needed, to related decisions, methods, and instruments. It is carefully matched to context—a context that is uncertain, complex, ambiguous, and subject to rapid and erratic change.

The unification of the IA process and *decision making* means that the process is built around decisions and the information, analysis, and interpretative needs of all parties involved in decision making. IA and organizational planning are ideally merged and concurrent (Keysar and Steinemann, 2002). At a minimum the IA process should strongly influence agency planning and decision making (see Chapter 3). Decision-making needs are anticipated, refined in consultation with stakeholders, and adapted as positions evolve and as new concerns emerge. IA documents cross-reference all requirements and comments. Reviewers and other interested and affected parties can, from the IA documents, readily determine how and where their concerns and requirements are considered. Reasons are provided for concerns and suggestions not addressed. The treatment of each requirement and suggestion is discussed with each party before documents are finalized.

A feasible IA process anticipates *implementation* requirements from the outset. Requirements are refined jointly with all parties likely to directly or indirectly influence implementation. Implementation commitments are clearly specified prior to approval. Close contact is maintained with agencies and departments likely to be involved in approvals and likely to impose conditions of approval. Conditions of approval are integrated into environmental management plans and strategies. Implementation includes such technical tasks as preparing monitoring reports, quality assurance, and assessing mitigation effectiveness. It also includes building coalitions of support, identifying and offsetting implementation obstacles, ensuring adequate resources to facilitate effective implementation, and making an effective, merit-based case, adapted to the needs, and

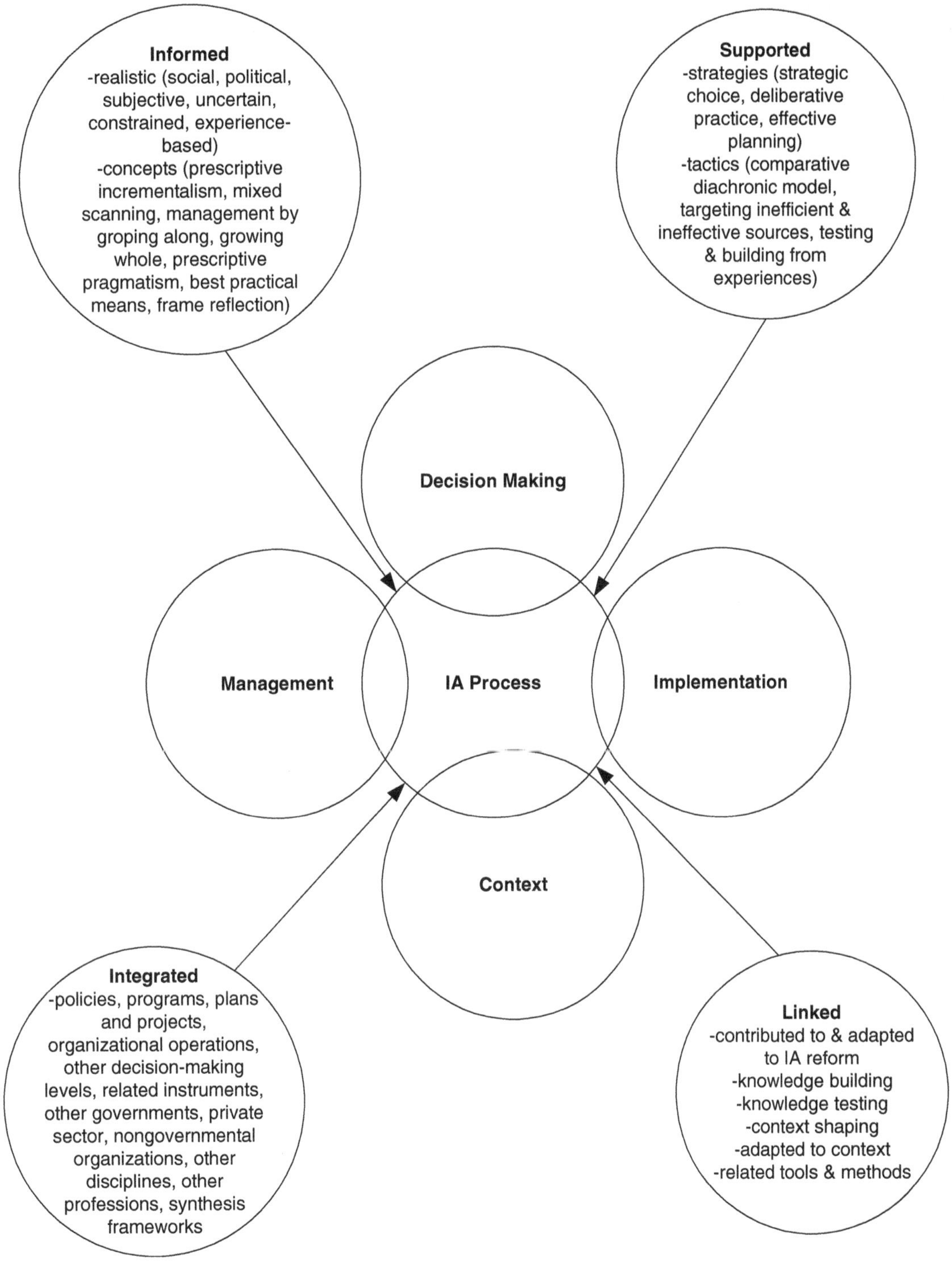

Figure 7.4 Feasibility in the IA process.

perspectives of each party associated with implementation (Wandesforde-Smith, 1989). Sometimes institutional capacity building is necessary prior to implementation.

The IA process is not designed and then adapted to the *context*. Process design commences only after the nature and potential implications of contextual characteristics are considered. The views of all interested and affected parties are actively solicited to ensure that the perspectives of each are reflected in the process (Rowson, 1997). The process is carefully designed and managed to account for IA regulatory requirements and institutional, ecological, social, cultural, political, and economic conditions, constraints, and opportunities (Barrow, 1997; Lee, 2000). Alternative IA process design types (e.g., rational, adaptive, conflict management) are appropriate for different contextual categories (e.g., varying levels of certainty and social conflict) (Nilsson

Table 7.2 Examples of Potentially Relevant Feasibility Strategies, Concepts, and Tactics

	Strategies
Strategic choice	• Strategic choice: choosing in a strategic way • Complementary aspects of any planning approach: technology, organization, process, and product • Four decision-making modes: shaping, designing, comparing, and choosing • Explores uncertainties about the working environment, guiding values, and related decisions • Distinguishes among: decision areas, links, schemes, and options • A sequence of structured workshops (can be supplemented by software) • Detailed guidance and practical advice provided (based on extensive experience from a variety of planning and development decisions) • Oriented toward interactive participation; a learning process • Issue oriented, cyclical, selective, and subjective • Systematically addresses lateral connections; addresses web of relationships between technical and sociopolitical streams • Addresses skill requirements and practicalities for each decision-making mode
Deliberative practice	• Process must be simultaneously interpretative, practical, political, and ethical (need to integrate theory, practice, pragmatism, and ethics) • Ethics not as standards to follow but as pragmatic action (the allocation and recognition of values) • Rationality is an interactive and argumentative process of marshalling evidence and giving reasons • Consensus building created on existing political stages (but also addresses power imbalances) • Need to improvise in complex and novel situations • Necessary to empathize with other parties and remain politically neutral at the same time • Critical listening, reflection-in-action, and constructive argumentation all interact • Importance of practical storytelling (letting stories supplement our limited rationality) • Challenge not to avoid, transcend, or displace conflict but to deal with practical differences in and through conflictual settings • A fluid process (issues formulated and reformulated) • Streams of choices, problems, solutions meet in unpredictable ways to shape ongoing, complex, and messy organizational outcomes • Argues for activist mediation (concern with process, efficiency, stability, and well-informed character of outcome)
Effective planning	• Planning as organizational learning • Importance of networking, learning from errors, experimentation, research, and pilot projects • Links strategic thought to implementation; a political and management activity • Need to build coalitions and networks to support proposal (building on shared values, importance of negotiations) • Technical and political considerations married • Stresses need to pay attention to logistics, reduce derailment potential, build trust, demystify, make contextual adjustments, and democratize • Process depends on transactions; continually evolving and engaged • Experience from the field the point of departure • Planning intertwined with management; argues for greater use of incentives and risk taking • Stresses value of making good on promises, correcting errors, removing mask of expertise, and delegating authority
	Concepts
Prescriptive incrementalism	• Various adaptations to incrementalism to make more prescriptive • Dialogical incrementalism (a dialogical process aimed at mutual understanding and agreement) • Purposive incrementalism (directed toward a purpose or vision and learning based)
Mixed scanning	• Comprehensive broad-angle analysis • Focusing on areas revealed by broad-angle analysis for more detailed scrutiny • Fundamental decisions set context for incremental decisions that lead to new fundamental decisions • Can be at several levels of detail and coverage
Management by groping along	• Experiment: determine what works and does not work • Progressively moves toward objectives; objectives well defined but means not; successes create new capabilities and motivate (strategy of small wins—facilitates learning and adaptation) • Test ideas before different audiences and gauge results; try different permutations and combinations

Table 7.2 (*Continued*)

	Strategies
Growing whole	• Goodness of fit between proposed action and context; sensitivity to stress and misfit • View of proposed action as a part of a growing whole (ecological, built, social, economic, cultural) • Assess how well preserves and enhances wholeness at many levels and in many ways • Move iteratively between ends and means, analysis and synthesis, rational and irrational, constraints and ideas; adds refinements, adjustments, and embellishments—all within a complex, ambiguous, organic, and evolving process • Involves multiple process designers; communications critical
Prescriptive pragmatism	• A philosophy of action; encompasses both doing good (moral and political) and being right (coherent and accurate technical analysis) • Includes human experience, practical activity, and democratic experience; importance of achieving and maintaining trust • A practical endeavor that links satisfying human needs with application experience
Best practical means	• Common approach to pollution control requirements • Practical taken to mean "reasonably practical" having regard to the state of technology, local circumstances, and financial implications
Frame reflection	• Scrutinize day-to-day tasks of practitioners; lessons from best practice and practice failure (practice wisdom, craft knowledge, experiential knowledge) • Transmit and exchange practical knowledge; policy evolves dialectically; policy discourse • Critically examine underlying assumptions, ideas, and beliefs; act from one perspective but be aware of others • Frame criteria: true, beautiful, just, coherent, utility, or fruitfulness
	Tactics
Comparative diachronic model	• A series of snapshots over time as development progresses; attempt to fill in what happens in between • Use of comparative and control studies to provide basis for impact study • Use of impact study (supported by comparative and control studies) as decision-making basis • Use of control study and postapproval impact analyses to manage impacts
Targeting inefficiency and ineffectiveness sources	• Counter negative bureaucratic tendencies by opening up systems, going outside the bureaucracies and with feedback and accountability loops • Undertake implementation analysis (consider implementation feasibility at early stages, anticipate implementation, backward mapping)
Testing and building from experience	• Use of pilot studies • Staged approvals • Identification and documentation of best practices (experimental knowledge, craft knowledge, true statements) • Use of empirical studies to provide insightful knowledge and accounts of practical constraints

Sources: Alexander et al. (1987), Behn (1988), Benveniste (1989), Burdge (1994), Etzioni (1967), Forester (1999), Friend and Hickling (1997), Gilpin (1995), Hoch (1984), Hummel (1977), Kørnøv (1998), Patton and Sawicki (1993), Sager (1994), Schön and Rein (1994), Sorenson and Auster (1989).

and Dalkmann, 2001). The role of the IA process as an instrument for changing the context is considered. Changing contextual characteristics are monitored up to and through implementation. The process is adjusted and refined as the context evolves. Multiple scenarios and sensitivity analyses ensure that the IA process and process outcomes are sufficiently "robust" to rapidly respond to changing conditions.

A feasible IA process treats impact *management* (project management is addressed in Section 7.4.4) as an ongoing function rather than as a stage at or near the end of the process. From the outset, consideration is given to how to avoid or minimize adverse effects, how to enhance benefits, how to offset inequities, and how to manage uncertainties. Mitigation is integrated into the alternatives analyses and into the proposal characteristics. Proponent and proposal-related impact management, compensation, and monitoring policies and strategies are formulated near the beginning of the process. They are refined jointly with stakeholders. Baseline analyses set up the environmental monitoring. Comparable action reviews, comparable environmental analyses, and pilot projects establish a foundation for impact analysis and management extending through the action life cycle.

Individual impact management measures (e.g., mitigation, compensation, local benefits, monitoring, contingency measures, financial security, funding, environmental liability) are consolidated within an impact management strategy. The strategy specifies management objectives and principles, variables, spatial and temporal boundaries, resources, responsibilities, testing protocols, methods, contingency measures, reporting requirements and stakeholder involvement, and conflict resolution procedures (Canter, 1996; Glasson et al., 1999; UNEP, 1997). It is linked to proponent policies, programs, and environmental management systems, to government requirements (e.g., compliance monitoring), and to environmental monitoring systems (Canter, 1996). Commitments to communities are formalized, where warranted, in impact management agreements. The impact management strategy is refined and adapted prior to approvals and throughout the implementation period. Impact management outcomes are documented in a form suitable for controlling impacts, for assessing mitigation effectiveness, and for validating and refining methods (Canter, 1996). Results are shared with stakeholders.

IA is an evaluation tool. Evaluation methods also are used within the IA process. *Additional relevant methods* sometimes are applied outside or partially overlap with the IA process. Examples include feasibility studies, needs assessments, life-cycle analyses, risk assessments, technology assessments, futures research, total quality management procedures, economic and social cost–benefit analyses, policy and program evaluations, environmental management systems, and conflict management procedures (Gilpin, 1995; Mayda, 1996; UNEP, 1997; Ridgway, 1999). A feasible IA process addresses links to methods used outside the IA process. Cross-referencing can minimize duplication. Inconsistencies are identified. Integration potential may be considered, at least to the point that a clear and consistent basis is provided for decision making and implementation.

A feasible IA process has a reciprocal relationship with *IA reform*. The IA system does not stand still while the IA process for a single proposal unfolds—especially for a process that takes years to complete. The "rules of the game" change. New requirements are instituted. Additional guidelines are issued. Perspectives and positions change. Sometimes a proposal, when caught in mid review, is "grandfathered." More frequently, the changes are subtle, particularly in terms of evolving agency perspectives and positions. A feasible IA process strikes a balance between consistent review positions over time and adaptations to changing circumstances. This generally means constructive discussions, some reconsideration of previous decisions, and some refinements to analyses and documentation. The potential for major changes can be greatly ameliorated if close contact is maintained with review agencies and if the IA process is undertaken in accordance with good practice standards in addition to meeting regulatory requirements. There is usually a lag between IA requirements and good practice. Sometimes an IA process (especially for a large, complex proposal, one involving new technologies or where there are major environmental uncertainties) cannot be adequately reviewed or managed without IA system changes (Lee, 2000). Auditing the experiences associated with individual IA proposal reviews contributes to IA system reforms.

IA requirements and the IA process are interwoven with the actions of others. A feasible IA process is necessarily boundary spanning. Project-level EIA requirements and procedures tend to be more effective when defined within the context of SEAs and SAs and in relationship to national, regional, and local sustainability, environmental, resource management, social and economic policies, strategies, programs, and plans (i.e., tiering) (Gilpin, 1995; Lee, 2000; Sadler, 1996). The auditing of project-level EIA experiences can contribute to SEA policy-, program-, and planning-level reforms. IA works best when IA roles (e.g., IA preparation, IA review) are a natural extension of agency objectives, policies, and operating procedures. Ideally, agencies consistently apply explicit environmental and resource quality performance criteria and standards. The concurrent application of environmental approvals and permitting requirements can expedite the IA process (Sadler, 1996). The IA process is further facilitated if the proponent has an environmental management system (EMS) in place (Barrow, 1997). An IA is often the impetus for instituting an EMS (Glasson et al., 1999).

Government IA responsibilities are often subdivided between head office and regional offices. There tends to be a greater concentration of specialists at the head office. Occasionally, both head office and regional office specialists comment on an IA, not always consistently. Such divisions of responsibility need to be closely scrutinized. The process is more complex when there are multiple government levels (see Chapter 8). Intergovernmental agreements, informal coordination, area-wide planning and management, and procedures to ensure a single process, a single IA document, and consistent timing requirements have all contributed to improved coordination and a clearer division of responsibility. The increased application of regional sustainability strategies and regional environmental and resource management tools (e.g, the ecosystem approach, integrated environmental management, and adaptive management) have further facilitated joint planning and management among government levels (Margerum, 1997). Some coordination difficulties will always remain. A feasible IA process ensures that such coordination difficulties do not unnecessarily "bog down" IA preparation and review. Interconnections among the disciplines and professions involved in IA preparation and review are identified and explored (UNEP, 1997).

7.4.4 Competence

Competence is a key aspect of practicality. When things go wrong with an IA process, the tendency is to blame unforeseeable circumstances. More often than the participants care to admit the problems that arise are foreseeable

Table 7.3 Competency: Examples Roles and Responsibilities

Project Management (Project Manager, Project Coordinator(s), Technical Writer, Editor, Administrator)

- Formulate (with proponent and study team) overall approach, study design, IA process (activities, events, inputs, outputs), and general methods (identification, prediction, evaluation, interpretation, cumulative effects assessment—CEA, participation, mitigation, compensation, monitoring, management); clear rationale for each
- Establish management structure and determine appropriate level of detail for each activity
- Assemble study team and related resources; establish team roles, norms, and environment for joint action; arrange, with proponent, contracts
- Control and manage team organization, activities, budgets, timing, and schedule; set work standards
- Monitor and update project plan continually; keep project and progress (task completion, budget completion, schedule) records; communicate progress
- Establish priorities, objectives, and milestones, solve problems, manage conflicts, negotiate trade-offs, and remove roadblocks
- Manage core team and support staff; identify stakeholders
- Determine report formats, hardware and software requirements, mapping scales and database management and GIS requirements; establish tracking and sign-off procedures
- Identify, with team, analysis gaps, and research and training requirements
- Coordinate analysis (proposal, environment, proposal–environment interactions), synthesis (data, criteria, significance, CEA, conclusions, recommendations), and interactions (internal—technical and management, external—agencies, elected representatives, groups, and individuals)
- Guide and challenge study team—scope, level of detail, database, assumptions, interpretations, judgments, conclusions, recommendations
- Document and present (with input and review by study team)—study design, study team organization, general frameworks and methods, general conclusions and recommendations, overall interim documents, draft and final overall IA documents, and summary documents
- Guide—documentation and presentation by individual study team members; review and edit each input for consistency, quality, and substantiation
- Coordinate documentation consistent with study schedule and decision-making requirements
- Organize, with proponent and with public consultation specialists, public involvement program
- Participate in interactions with management, agencies, elected representatives, and public; act as spokesperson for team
- Ensure overall efficiency, relevance, and adequacy
- Identify uncertainties and risks and develop a management strategy to address; decide, with proponent how to address unforeseen circumstances; prepare change orders

Specialists (Design and Engineering, Disciplinary, Professional, Methodology, Public Consultation, Mediation, Conflict Management, Legal)

- Formulate own methods and assumptions
- Manage internal organization, time, budget, and tasks
- Undertake data collection, analysis, and interpretation
- Document and present methods, analysis, role in synthesis, conclusions, and recommendations
- Participate in formulation of overall approach, synthesis, and interactions
- Specialist advisors address discrete problems, methodology, applied research, comparable proposals, and environments
- Peer reviews of interim and draft documents (for proponent, reviewers, or for other participants)

Proponent (Co-Proponents, Lead and Secondary Agencies, Sponsor Agencies)

- Overall schedule and budgets
- Corporate policies, programs, and operations
- Characteristics of existing operations
- Priorities and requirements
- Proposal characteristics
- Terms of reference
- Commitments
- Higher level agency interactions
- Participation in public involvement process

Agencies and Governments (IA Agencies, Specialist Review Agencies, Other Government Levels, Indigenous Peoples' Governments)

- IA requirements and guidelines interpretations
- Technical requirements and guidelines interpretations
- Data provision, analysis, and interpretation
- Technical expertise
- Policy, program, and priority interpretation
- Experience with comparable proposals and environments
- Output review

(*continued*)

Table 7.3 *(Continued)*

Public (Nongovernmental Organizations, Directly Affected Groups and Individuals, Indirectly Affected and Interested Groups and Individuals)

- Provision of data
- Participation in scoping
- Data review and interpretation
- Participation in determining criteria importance, alternatives preference, impact management measures, conclusions, and recommendations

Interactions

- Within study team
- Between project managers and proponent
- Between project management, study team members, and specialist advisors
- Between project management, study team members, and agency representatives
- Between project managers, proponents and study team members, and elected representatives and public
- Mechanisms: management committees, steering committees, advisory committees, task force, workshops built around frameworks, or models and meetings

Sources: Coe (2012), Erickson (1994), Glasson et al. (1999), Greenall (1985), Harrop and Nixon (1999), Holling (1978), Kreske (1996), UNEP (1997).

and are resolvable through competent IA practice. Competence is much more than the appropriate application of the methods and models presented in most IA texts. The knowledge of specialists extends well beyond the overviews of disciplinary analyses presented in such references. If they are properly selected and coordinated, the specialists do not tend to be the problem. More frequently, problems arise with the ways in which individual analyses are guided, integrated, and applied. Sometimes, roles and responsibilities are poorly defined. Table 7.3 lists examples of roles and responsibilities for various participants in the IA process. Good practices are not always applied for activities that transcend individual specialties. Table 7.4 provides good practice examples for study team management, study team participation, database management, the application of geographic information systems (GIS), report writing and documentation, financial control and budgeting, and the preparation of work programs and schedules.

Competence-related problems continue to occur in the IA process, notwithstanding the ample, readily available advice and guidance, which should minimize such problems. Perhaps this shortfall between knowledge and execution can be partially explained by a failure to focus on recurrent, avoidable, competence-related problems. Twenty examples of such problems, together with suggested solutions, are provided in Table 7.5. These competence-related pitfalls are largely avoidable. They are not always obvious. Care must be taken to minimize the likelihood and severity of their occurrence.

7.4.5 Effectiveness

The final aspect of practicality is effectiveness. Feasibility addresses the workability of the IA process (i.e., can it be undertaken, and can it be implemented?). Effectiveness considers how well it was undertaken. Competence deals with adequate practice levels. Effectiveness "raises the bar." As illustrated in Figure 7.5, it addresses the quality of the inputs (e.g., institutional arrangements, processes, methods, participant performance, documents) and the effectiveness of the direct and indirect outputs (e.g., goals achievement, environmental changes, methodological performance, management performance, contribution to practice).

Reviews of *institutional arrangements* evaluate the adequacy of EIA and SEA policies, laws, regulations, and guidelines (Halstead et al., 1984). Such reviews consider such matters as application to significant actions, environmental and effects definitions, scoping provisions, requirements to address alternatives and cumulative effects, public consultation requirements, transparent decision making, provisions for follow-up, enforcement and auditing, appeal and dispute settlement provisions, and methodological guidance (Gibson, 1993; Sadler, 1996; Spooner, 1998). The suitability of organizational structures and procedures to undertake IA-related responsibilities can be assessed (Kreske, 1996). The capability and capacity of organizational systems to conduct good practice IA regulation can be evaluated based on such considerations as IA and environmental staff qualifications, workload and the human, financial, and other resources devoted to IA administration and enforcement.

The quality of *individual IA processes* can be assessed overall and for individual IA activities and components (Lee, 2000). The analysis of the overall IA process can address consistency with good practice and appropriateness to context. The extent to which the IA process supports transparent and accountable decision making can be evaluated. The appropriateness and effectiveness of the political, public, and government agency involvement procedures can be considered. The choice and manner of application of all methods (e.g., data collection, compilation and analysis, prediction, interpretation, CEA, management, involvement)

Table 7.4 Competency: Good Practice Examples

Study Team Management

- Involve team in designing and scoping; clearly define objectives, approach, and anticipated inputs and outputs
- Ensure that roles and responsibilities are well defined; set priorities and maintain momentum
- Provide guidelines for text and table formats to ensure consistent inputs
- Ensure a coordinated approach to external contacts
- Provide for reciprocity of influence (manager and specialists); facilitate dialogue and integration; recognize different "mind sets" of specialist types
- Emphasize early drafts and initial outputs; scan ahead and "test water"; early opportunity for internal and external review
- Sketch out alternative approaches for dealing with problems and conflicts
- Test and challenge basis for all interpretations and conclusions; be aware of own limitations and those of others; take corrective actions
- Focus on and manage a collaborative, constructive, and creative response to all problems and disputes
- Ensure project manager is not the "bottleneck"; employ core team on larger projects
- Allow for regular meetings and workshops at key decisions
- Use subgroups to address problems and to address interconnections among specialists
- Often management functions shared between internal (proponent) and external (consultants, secondments, term contracts); tends to be more effective if ongoing proponent involvement
- Keep a record of findings, events, directives, changes in direction, comments, concerns, agreements, and decisions
- Leadership skills: analysis, integration, management, communications, presentation, negotiation, problem solving, general knowledge of each specialty, ability to ensure quality of work, detailed IA knowledge and experience, ability to delegate
- Leadership style (e.g., command and control, empowerment, learning) must match situation
- Leadership qualities: action and results oriented, self-confident, self-starter, visionary, enthusiastic, energetic, reliable, mature, even-tempered, adaptive, politically astute, tolerant of uncertainty, sense of humor, and patience

Study Team Participation

- Study team selection: availability, expertise, proposal-type experience, IA experience, local environmental knowledge, study team experience, personality and attitude, receptivity to viewpoints of others, work traits, range of interests (broader better), writing and communications skills, listening skills, adaptability, ability to interact with public and politicians, ability and experience with hearings, oriented to work to schedule, willingness to travel and make site visits, professional credibility, adaptability; often prudent to make process competitive
- Study team style: interdisciplinary (coordination at higher level) and transdisciplinary (coordination at all levels) rather than disciplinary (specialization in isolation), multidisciplinary (no cooperation), or cross-disciplinary (rigid polarization)
- Often core team, each member of which spans a few disciplines; a useful middle ground between project management and full team of specialists when large project
- Prompt and ongoing attention to small group problems (e.g., leadership—authoritarian or leadership struggles, blocks in group development, poor decision making, interpersonal conflicts, communications difficulties, goal ambiguity)
- Importance of clear purpose, expectations, and accountability; clear terms of reference for each team member
- Participate and contribute to overall team activities (e.g., team discussions, agency consultations, alternatives analysis, significance interpretations, public involvement, synthesis and summary document preparation and review, presentations at events, links to related disciplines)
- Undertake specialist analyses in accordance with good practice standards of field, guidance from project management, and expectations of regulators
- Adhere to scope of work, budgets, and timing requirements; address implications of limitations and uncertainties
- Respond promptly and fully to all questions and concerns raised about analyses from study team, regulators, public, and peer reviewers
- Work with related specialties in addressing interconnections across disciplines and in formulating and applying integrative frameworks (e.g., modeling, CEA, impact management strategy)

Database Management

- Tie information to decision-making requirements
- Database management involves determining what data are to be collected, when, by whom, at what level of detail, and how to be collected, compiled, analyzed, interpreted, integrated, applied, supplemented, refined, presented, and monitored
- Data collected throughout process; dependent on requirements of activities and decisions
- Data management: continuous, evolving, and dependent on context
- Ensure that all data is complete, accurate, and properly referenced
- Should reflect priorities, should be guided by data management strategy, and should systematically identify and explore implications of errors (correct), gaps (fill when necessary), inconsistencies (resolve), and uncertainties (allow for)
- Interpret data reliability (sources, methods for collecting, methods for compiling)
- Involve stakeholders in data collection, analysis, and interpretation; make effective use of local and traditional knowledge
- Important that data can be retrievable, cross-checked, and updated; important that dated and referenced
- Consider environmental data available in computerized information and retrieval systems (e.g., government agency information systems, environmental databases, and electronic bulletin boards)

(continued)

Table 7.4 (*Continued*)

Geographic Information Systems (GIS) Application

- Can store, retrieve, analyze, and display spatial data
- Importance of availability and quality of spatial data
- Useful for mapping, overlays, baseline analyses, modeling, monitoring (regular updating), visual displays, video imaging, testing of alternatives, route and site selection, CEA (incremental impacts, biodiversity), and public consultation
- Takes time to set up; high training and technical requirements; data often not available in digital format; potential data and user-related errors; weak analytical capabilities
- Can be combined with GPS (global position systems), imagery from satellites, aircraft, and internet
- Assumes importance of environmental impacts dependent on spatial distribution of impacts
- Pitfalls: not taking into account purpose of map, zooming in to improve accuracy, neglecting map projections and coordinate systems, failing to document and evaluate map sources, not including necessary map elements, presenting too much information, inappropriate type faces, misrepresenting qualitative and quantitative data, mapping absolute values, and neglecting data collection effects

Report Writing and Documentation

- Design to suit audience
- Ensure documents are scientifically sound, easily understood, feasible, legally defensible, and timely
- Build around preliminary and then detailed outline
- Focus on what is important (issues and needs of readers); space devoted to topic should be consistent with importance for decision making
- Ensure that audience can readily determine how major issues were addressed
- Engage interested parties in meaningful dialogue
- Ensure that regulators can readily determine that all requirements satisfied
- Use simple and familiar language; be succinct and clear; minimize generalities
- Be concise, consistent, and defensible
- Ensure well-structured and visually attractive presentation (ample use of visual displays)
- Check for technical errors and mistakes; ensure factually accurate; avoid plagiarism and bias (use neutral language)
- Ensure a consistent writing and presentation style; review and edit for consistency
- Identify limitations and uncertainties; identify implications and strategies for addressing
- Allow for planning, organizational, and editing mistakes
- Use consistent referencing and numbering system, indentations, titles, headings, and margins
- Avoid clichés and jargon; avoid defensive language
- Be honest, objective, frank, complete, fair, transparent, and vivid
- List acronyms and sources; define technical terms and explain technical concepts
- Use accessible graphics
- Utilize peer review to reduce documents to a brief summary designed to match needs and perspectives of each audience
- Provide reasons for data, methods, assumptions, findings, interpretations, conclusions, and recommendations
- Provide summaries and use appendices, cross-references, and tiering to streamline text

Financial Control and Budgeting

- Match staff to available budgets over time
- Track project expenditures regularly
- Recognize that expenditures build to a peak
- Conduct a postmortem of budgeting experience
- Tie each expenditure to decision-making priorities
- Team leader to monitor resource use
- Apply, as needed, graphs, charts, and computerized techniques to track expenditures and to compare task completion with budget expended

Work Programs and Scheduling

- Work program addresses goals, issues, and problems
- Includes activities, tasks, events, inputs, and outputs; purpose, sequence, duration, personal, hours, disbursements, and budgets for each
- Need to address interactions among activities
- Allow sufficient time for unforeseen circumstances (float time)
- Maintain flexibility; provide additional time for agency and public involvement activities; dangerous to cut short
- Use of graphs, charts, and computerized techniques to chart actual progress against scheduled progress
- Critical paths methods can help determine overall structure; often helpful to provide a range of time estimates (worst, most likely, quickest)
- Allow sufficient time for internal and external review, editing, and consideration of interconnections

Sources: Alton and Underwood (2003), Antunes et al. (2001), Barrow (1997), Bendix (1984), Buckley (1998), Canter (1996), Greenall (1985), Harrop and Nixon (1999), Hodgson and White (2001), Jantsch (1971), João (1998), Kent and Klosterman (2000), Kreske (1996), Moreno and Catchpole (2012), Page (2006), Ross et al. (2006), Thérivel et al. (1992), Verma (1995), UNEP (1997), US EPA (1998b), Webster (1997).

Table 7.5 IA Competence-Related Problems and Solutions

Problem	Nature of Problem	Possible Solutions
IA managers as "bottleneck"	A project manager, on a large IA, can be overwhelmed if she or he attempts to take on all the project management responsibilities	A core team approach is more appropriate for a large project. The same problem can occur on even an intermediate-sized project if the project manager "micro-manages" every aspect of the IA process. A good team and effective delegation are essential. Effective delegation means strategic management not the absence of control and guidance
Project managers as "autocrats"	Some project managers have a tendency to equate project management as giving orders without reasons and not asking for suggestions or even tolerating feedback	Close and ongoing communications and consultation should be maintained with proponents, with other study team members, and with stakeholders. The project manager should provide a clear rationale for all instructions. Often others have useful advice to offer. The project manager should be a good listener and should actively seek constructive advice and criticism. Openmindedness, flexibility, and an even temperament are all part of leadership
Project managers as "doormats"	Sometimes, a project manager will offer limited direction or boundaries and then accept inputs from other team members without questioning assumptions, methodology, or the basis for conclusions or recommendations	Project managers need to have a clear vision of where the IA process is to go and how objectives are to be achieved. The IA process cannot be allowed simply to drift. The project manager has to have sufficient self-confidence, experience, and general knowledge to challenge specialists when inputs are unsubstantiated, incomplete, inconsistent with requirements, misdirected, badly written, poorly structured, or of dubious quality. She or he also has to ensure adherence to budget, scope, format, and timing requirements. The project manager should exercise such responsibilities firmly and calmly
Team members who aren't team players	Sometimes, specialist team members see their role as no more than undertaking and documenting their analyses. They see team interactions, compliance with document format requirements, and other general project activities as unnecessary distractions to be avoided where possible or, if necessary, reluctantly tolerated	IA is a highly interdisciplinary, often transdisciplinary, activity. This necessitates the full participation of specialists in such joint IA activities as scoping, alternatives analysis, significance interpretation, CEA, agency and public involvement, impact management, and document preparation and review. A unified and consistent documentation approach also is essential
Not up to the task	Sometimes, specialists are involved in an IA process, who do not have sufficient relevant expertise and experience in their field, in IA, in applied knowledge situations, concerning the local environment, regarding the proposal type, or in working on a team	This type of problem can generally be minimized with careful team selection and effective project management. The competency problem is more problematic at the project management level. Having extensive project management experience is not the same as having extensive IA project management experience and expertise. Sometimes, specialists in other fields are competent IA project managers. However, an in-depth understanding of IA as a field of theory and practice coupled with extensive IA project management experience is essential
A failure to focus	Sometimes, there is the belief in IA practice that all topics are equally important and the more documentation the better	A practical IA process is necessarily focused. Without focus important concerns receive too little attention and unimportant concerns receive too much attention. The net result is a protracted and costly IA process and IA documents of dubious quality. Unfocused documents tend to be highly descriptive and very lengthy. Decision makers and stakeholders have difficulty determining if and how their concerns and priorities are addressed. IA is a decision-making tool. As such the IA process should concentrate on providing a sound basis for making and implementing environmentally sound decisions

(*continued*)

Table 7.5 (*Continued*)

Problem	Nature of Problem	Possible Solutions
Gaps and blind spots	IA practice is sometimes subject to "tunnel vision." Occasionally, for example, the analysis of alternatives is too narrow, superficial, and abbreviated. Social, cultural, cumulative, and sustainability effects, together with public concerns, also tend to receive insufficient attention	Reasonable alternatives need to be systematically generated and evaluated. Social and cultural concerns need to be fully addressed. More attention also needs to be devoted to indirect, cumulative, and sustainability effects, although the situation is improving. IA practice sometimes concentrates exclusively on meeting IA regulatory requirements. The appropriate treatment of stakeholder concerns and perspectives is frequently just as important in determining whether an IA will be approved and effectively implemented. Care needs to be taken to ensure that all potentially significant effects are fully assessed
A failure to integrate	IA documents, which represent little more than a compilation of specialist inputs, are of limited decision-making value	Competent IA processes and documents trace through the interactions among disciplinary inputs. They systematically undertake such integrative activities as alternatives assessment, model building, assessing cumulative effects, and formulating impact management strategies. Integration also entails creatively accommodating multiple study team, proponent, regulator, and stakeholder perspectives and interests
A failure to substantiate	Sometimes, IA documents are full of unsupported assertions, claims, interpretations, and conclusions. Professional judgment is not enough	Assumptions, methodology, interpretations, and conclusions should always be supported by evidence and explicit reasons. In this way, judgments can be independently tested and evaluated
Artificial timelines and false economies	Sometimes, artificial time and budget constraints are imposed either at the outset of a process or when a process is taking longer than expected	These constraints can result in superficial, error-prone, and inadequate analyses and truncated agency and public consultation procedures. The most common outcome from artificial limits is a much more time-consuming, controversial, and costly review and approval process and a much greater likelihood of process failure. A focused and well-structured IA process can be expeditious and economically executed. Occasionally, there are "hard deadlines," emergency situations, and severe resource constraints, which necessitate an abbreviated, selective, broad-level, and "streamlined" IA process. But there are limits
Quantify everything	Forcing the quantification of qualitative data can distort the analysis of impacts and inhibit the reasoned comparison of alternatives	The desire for precise, verifiable predictions and consistent comparisons is laudable. However, the database must be capable of supporting such efforts. The inappropriate application of quantitative methods can imply a greater level of precision and control than can be supported and can make it more difficult for decision makers and stakeholders to understand or participate in the IA process
A failure to quantify	It can be extremely exasperating to read an IA document full of vague generalities and ambiguous statements. Appreciating the limits of quantification does not mean abandoning all efforts to quantify	Quantified predictions should be provided wherever practical, with due allowance for uncertainties. In this way, predicted impacts can be monitored, the accuracy of predictions determined, and the suitability of predictive methods evaluated. Precision in specifying mitigation measures is necessary for the measures to be implemented and for mitigation effectiveness to be determined
Bias and advocacy	It is not an appropriate role for IA practitioners or documents to "win" approval or "make a case" through the selective and biased use of evidence	The standard of IA success should not be approval. Instead, it should be an environmentally sound decision-making basis and an enhanced environment. IA professionals cannot be objective or value free. However, consistent with professional codes of practice, they can work toward IA objectives in a manner consistent with good

Table 7.5 (*Continued*)

Problem	Nature of Problem	Possible Solutions
		practice standards. It is essential to the credibility of the IA process and documents for the professional integrity of the study team to be maintained. IA documents should be scrupulously checked to ensure that there is no bias
A failure to adjust	Except on the simplest IA projects, a "carved in stone" approach to IA process management is rarely effective. Modifications occur in activity characteristics, environmental conditions, available alternatives, stakeholder positions, and regulatory requirements. Unanticipated events occur	An IA process also must evolve and adjust in response to changing circumstances. A gulf between what is needed of a process and what it can provide will emerge and progressively widen with an inflexible IA process, usually to the point that a major crisis occurs. The outcome from the crisis will tend to be either the termination of the process or major, costly, and time-consuming modifications. Such crises can be avoided or greatly ameliorated with an adaptive IA process
A failure to anticipate	IA practitioners sometimes complain when things go wrong that they were "blindsided" by unanticipated events and changing circumstances. Sometimes the complaints are valid. Often, however, there are ample early warning signs	Early warning signs can frequently be detected by scanning ahead, by frequent consultations with other parties, through pilot projects, with systematic assessments of comparable situations and by "pretesting" interpretations, options, and conclusions. A flexible IA process also makes it easier to anticipate and rapidly respond to change
A failure to communicate	An IA process can be greatly hampered by poorly structured, badly presented, and awkwardly written IA documents, even if those documents are technically sound	Competent IA documents and presentations should be clear, succinct, and tailored to the audience. Effective communications channels into the IA process from regulators and from other interested and affected parties are also essential
Participation without involvement	A sure sign of a questionable IA process is the tendency to count the number of meetings, attendees, and submissions (i.e., inputs) without detailing the changes to the process and documents resulting from stakeholder comments and suggestions (i.e., outputs). Involvement also is inhibited if participation largely consists of presentations (i.e., one-way communications)	Events conducive to two-way communications (e.g., workshops and open houses) and continuous involvement procedures (e.g., advisory committees) are less likely to result in an IA process characterized by participation without involvement
A lack of perspective	Environmental specialists, proponents, regulators, nongovernmental organizations and Indigenous people will often interpret the significance and acceptability of impacts and proposed actions very differently. Sometimes, IA documents ignore or gloss over these differences or assume that the professional judgment should be the sole basis for interpretations	The IA process and documents should reflect and accommodate this multiplicity of perspectives. There are many ways of looking at the world and how it should be. It is especially important that judgmental activities such as scoping, significance interpretation, the evaluation of alternatives, proposal acceptability and the determination of appropriate mitigation, compensation, and monitoring be interpreted from the perspective of each interested and affected party in the process. Consultation programs should also be tailored to a variety of needs and perspectives
One size does not fit all	An IA process that operates effectively in one setting can be entirely inappropriate in another. Context matters	The IA process should be designed to suit activity type and setting type characteristics. Further adjustments to suit unique activity and environmental characteristics are also essential. The goal should be an IA process that (1) fits the context (e.g., ecological, social, political, institutional, economic) and (2) selectively and positively influences the context (i.e., IA as an instrument for environmental enhancement and sustainability)
Neglect of follow-up	A well-designed and executed process and sound IA documents are necessary. They are not sufficient	Adequate attention must be devoted to follow-up issues, procedures, and requirements. Such concerns need to be addressed both prior to and throughout implementation

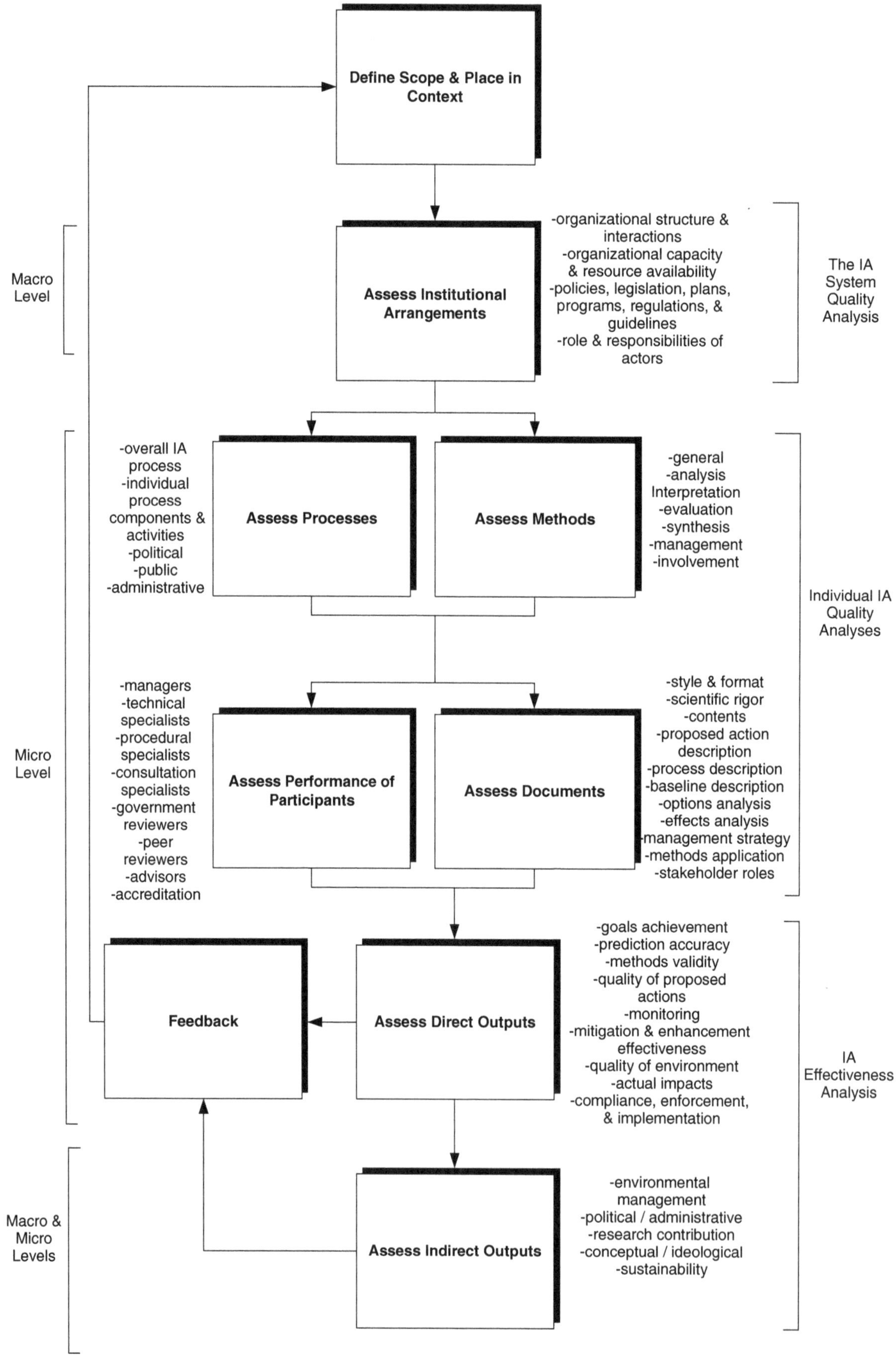

Figure 7.5 IA quality and effectiveness analyses. Adapted from Lawrence (1997a).

can be assessed (Ortolano, 1993). The qualifications, roles, and role performance of process participants (e.g., managers, technical and procedural specialists, government reviewers, peer reviewers, and advisors) can be analyzed. IA documents can be evaluated for style, format, content, and the treatment of individual IA activities, methods, and events (Barker and Wood, 1999; Wood et al., 1996). How well the documents focus on major concerns, comply with regulatory requirements, reflect stakeholder perspectives, and integrate public and agency concerns and contributions can be evaluated.

Direct and indirect outputs from IA processes can be assessed. Output analyses interpret results, both intended (relative to expectations) and unintended (positive and negative). Direct output analyses provide the basis for follow-up actions and practice refinements. Indirect output analyses are the means by which IA processes make substantive contributions to enhanced IA practice. Direct output effectiveness analyses address whether IA purposes, goals, and objectives have been achieved, the accuracy of environmental change and impact predictions, and the validity of methods (Barrow, 1997; Culhane, 1993; Tomlinson and Atkinson, 1987; UNEP, 1997). They can determine project modifications and quality, the suitability of monitoring measures, the effectiveness of mitigation and compensation measures, the quality of impact management, and the extent to which commitments are implemented, requirements are complied with, and adequate enforcement occurs (Glasson et al., 1999; Harrop and Nixon, 1999; UNEP, 1997; Wende, 2002). IA effectiveness reviews also can isolate factors that result in or impede effectiveness gains (Wende, 2002).

Indirect output analyses address the role of institutional arrangements and individual IA processes in furthering environmental management, environmental administration, and decision making, the IA knowledge base, and societal goals such as sustainability (Barrow, 1997; Glasson et al., 1999). Evaluations are undertaken of the contribution by IA to environmental objectives as compared to the costs and negative impacts incurred and relative to the achievements of other environmental management instruments. Such analyses can facilitate institutional arrangements reforms.

Numerous *methods* can be applied in effectiveness reviews (e.g., ad hoc procedures, checklists, applying principles, criteria or performance standards, the use of scaling levels) (Sadler, 1996; US EPA, 1998b). Effectiveness reviews can be undertaken by individual experts (internal or external, accredited or not accredited), panels of experts, public reviews, independent commissions, official inquiries, public reviews, or through legal proceedings such as court actions (Barrow, 1997; Tomlinson and Atkinson, 1987; UNEP, 1997). Various approaches can be adopted for conducting effectiveness reviews. A scientific-analytic, a management-efficiency, an interactive-interpretative, or an adaptive-evolving approach could be appropriate depending on such considerations as project complexity, data availability, degree of uncertainty, degree of controversy, and the rate and predictability of changing conditions (Culhane, 1993; Lee, 2000; Serafin et al., 1992). The interpretative nature of IA quality and effectiveness analyses underscores the importance of stakeholder involvement and perspectives (Spooner, 1998; UNEP, 1997; US EPA, 1998b).

Ideally an effectiveness review should include (1) a screening step (to reject an unacceptable action or document), (2) a performance analysis step (to evaluate actions or documents considered adequate but not necessarily consistent with good practice standards), (3) supplementary analyses (to overcome deficiencies), (4) clarifications (to resolve misunderstandings), (5) the documentation of findings at each decision (to ensure decision-making transparency), (6) provisions for agency and public involvement at each decision (to ensure full public and agency involvement in each step in the process), (7) monitoring or auditing analyses (to assess outputs), (8) an approval step (to provide a decision-making basis and to determine conditions), and (9) a modifications step (to adapt implementation to changing conditions).

7.5 INSTITUTING A PRACTICAL IA PROCESS

7.5.1 Management at the Regulatory Level

A practical IA regulatory system should (1) harmonize IA requirements among government levels, (2) ensure that IA roles among government departments and agencies, and among governments are well coordinated, (3) focus on what is important and minimize unnecessary costs and delays, and (4) ensure a minimum level of IA competence and contribute to an enhanced level of IA practice. The IA systems in the four jurisdictions (the United States, Canada, Europe, and Australia) all seek to achieve these objectives, albeit in different ways.

Table 7.6 identifies a range of approaches applied in the four jurisdictions for making IAs more practical. Practicality in IA processes, for example, can be facilitated by clearly defined environmental priorities at each government level, by measures to streamline, link, tier, harmonize, and integrate IA types and levels, and by efforts to enhance inter- and intragovernment coordination. It can be facilitated by review and dispute resolution mechanisms. It can be aided by measures to simplify and focus IA documents. It can be advanced by efforts to structure screening and scoping procedures. It can be furthered by initiatives to adapt IA requirements, guidelines, and procedures to different proposal types and settings. It can be advanced by efficiency and effectiveness guidance and by applied research. It can be facilitated by proactive efforts to involve agencies and the public in suggesting reforms.

Good practice examples, independent effectiveness reviews, and the auditing of IA requirements, processes, and documents can all further the cause of practicality. Examples of other potentially valuable practicality measures

Table 7.6 Positive and Negative IA Practicality Examples at the Regulatory Level

United States	Canada	Europe	Australia
(+) Provision to eliminate duplication with other governments (e.g., joint processes, documents, and public hearings), for addressing conflicts between federal and state or local planning, and for including state or local governments as joint or cooperating agencies (±) Programmic EISs for broad federal actions such as new programs and legislation; not widely applied (+) Emphasis on plain language (+) Numerous partnerships, shared databases and joint environmental planning and management among government levels and governments (+) Provides for and encourages tiering of IA documents (+) IA streaming requirements (e.g., separate requirements for categorical exclusions, environmental assessments, environmental impact statements) (+) Formalized and detailed scoping requirements and guidance (+) Considerable guidance on making process more efficient and timely (e. g., time lines, adoption of all or part by reference, concurrent circulation, early planning and scoping, dispute resolution, accelerated procedures, document length and focus, circulation limits, checklists) (US DOE, 1997)) (+) Interagency rapid response teams for transmission projects (+) Mitigation and monitoring requirements and guidance	(+) Purpose of Act includes ensuring EAs completed in a timely manner (±) Mitigation measures a requirement; limited to technically and economically feasible—issue of criteria for determining (±) Timelines for EA decisions and stages (+) Encourages the study of cumulative effects in regions (+) Offshore drilling boundary guidance (CEAA, 2003b) (+) Extensive guidance materials (CEAA, 2007a,b,c, 2009a,c,d, 2010a) (+) Provides for assessment of designated project in combination with other physical activities (±) Minister can establish a committee when activities entirely within federal lands (±) Focused on biophysical environment and effects within federal jurisdiction; issue of how selective approach can adequately address cumulative effects and sustainability (+) Screening provisions (+) Class EA provisions (CEAA, 2008a, 2009a) (+) Scoping provisions (+) Quality assurance and effectiveness analyses (Government of Canada, 2010b) (+) Inter-agency coordination and CEA guidance (CEAA, 2003a; CEAWG and AXYS, 1999)	(+) Proposed Project Directive (PPD)—clarifies screening procedure; modifies criteria and specifies content and justification (±) PPD—focused on projects with significant environmental effects; issue of criteria and cumulative effects with multiple smaller projects (±) PPD—decision and documentation time frames; uncertainty regarding IA quality implications (+) PPD—Detailed scoping requirements (e.g., decisions and opinions to be obtained, concerned authorities and publics, stages, duration, reasonable alternatives, significantly affected environmental features, information requirements and availability, methods) (+) PPD—required to address cumulation of impacts (+) PPD—detailed requirements regarding project characteristics, aspects of environment likely to be affected and likely significant adverse effects (+) PPD—efforts to harmonize IA and other environmental requirements (+) Screening and scoping required under SEA Directive (+) SEA Directive—requirement to address cumulative and transboundary effects (+) Extensive array of guidelines, applied research, effectiveness analyses, and data sharing (+) Efforts to facilitate tiering of SEAs and project-level EIAs	(+) Extensive array of multilateral and bilateral agreements and transboundary regulatory provisions (+) Accreditation procedure for equivalent state–territorial assessments and based on bilateral agreements; recognition that need explicit criteria and standards and transparent and inclusive procedure, including performance auditing (+) Clear division of responsibility—national level focuses on identified matters of national environmental significance(Australian Government, 1999, 2007a) (+) Recently accepted reforms should lead to greater use of SEA and enhanced coordinated tiering arrangements (+) Includes initial screening step (+) Streaming provisions allow for decisions based on referral information and preliminary documentation; recently accepted reforms should further simplify (+) Timing requirements for each decision (+) CEA guidance (Court et. al., 1994) (+) Class assessment provisions (+) Specific document content requirements (±) General references to practical and feasible alternatives; depends on how defined (+) Cost recovery consultation paper (Australian Government, 2011b)

(+) Effectiveness reviews and surveys (+) Pilot project soliciting agencies and public to nominate projects employing innovative approaches to completing environmental reviews more efficiently and effectively; involves nomination and selection process (+) Various measures to modernize and reinvigorate NEPA (US COR, 2005; US NTF, 2003) (+) Documentation of NEPA success stories (Environmental Law Institute, 2010) (+) CEA a requirement (+) CEA and mitigation guidance (US CEQ, 1997b, 2005b, 2011) (−) Lengthy documents and protracted process still a major concern	(+) Extensive array of guidance material (e.g., adaptive management); issue of application under new requirements (Government of Canada, 2010a,b; ESSA, 1982) (±) Auditing role of Commissioner of the Environment and Sustainable Development and of government committees (Office of the Auditor General of Canada (AGC), 2004, 2008, 2009; Warawa, 2012) (±) Focus on major projects; cumulative effects of small projects excluded (±) Substitution/equivalency provisions could reduce potential for duplication and overlap; issue of criteria to be applied (+) Cost recovery provisions (+) Coordinative roles of Agency identified (+) Agency leads a quality assurance program (±) Exemption of infrastructure projects; issue of impacts (−) Discretionary application of EA requirements leaves open opportunities for inconsistencies and quality issues (+) SEA cabinet directive includes a preliminary scan step (−) Minimal tiering	(+) Transboundary coordination efforts (e.g., SEA and EIA Conventions) (+) Detailed screening, scoping, and document review and good practice guidance at multiple levels (Department of Communities and Local Government, 2006) (+) Considerable internal and external advice and auditing (e.g., Group of EIA/SEA National Experts, European Court of Auditors, IA Board) (ERM, 2001a,b,c; EUCOTR, 2010; GHK, 2010; IEPP, 2004; Office of Management and Budget, 2008) (+) CEA a requirement of the SEA Directive and the Habitats Directive (+) CEA guidance (Walker and Johnston, 1999)	(+) IA legislation recently subject to major independent reviews (Hawke, 2009) (Australian Government, 2011e) (+) Annual reviews of systems performance and operational reviews (Environment Australia, 2001) (+) Performance audits by auditor general (Auditor General, 2002–2003) (+) Detailed mitigation and monitoring requirements and guidelines (+) Minister has authority to undertake compliance or performance audits (+) Consideration of proponent's environmental record (+) Extensive array of guidance materials including best practice guidance; provides both standard and tailored guidelines (+) Required to address cumulative and transboundary effects

include experimental institutional arrangements for expediting IA review procedures, systematic and explicit mitigation, and follow-up requirements, transboundary IA coordination provisions, the class or categorical assessment option, cost-recovery initiatives, and the consideration of the proponent's environmental record.

Efforts to make IA more practical need to be approached and applied with sensitivity to potential repercussions. Measures (e.g., timelines) to reduce the duration of various IA stages can be helpful provided IA quality and stakeholder involvement are not inhibited. Focusing IA requirements on "major" projects or impacts might lead to the more efficient and effective allocation of resources. But such measures can result in major cumulative effects from multiple small projects. They also can result in major adverse individual or cumulative effects when "major projects or impacts" are ill-defined, and when highly sensitive or significant environmental receptors are adversely affected. Focusing IA requirements on narrow and selective definitions of the environment, effects and alternatives can result in process efficiencies. However, they also can result in disjointed analyses and unnecessary and more severe environmental impacts. This is especially the case in terms of sustainability and cumulative effects. Effective sustainability-related and CEA requirements and guidance are essential for enhanced IA practicality, consistent with the argument that they are simply good SEA/EIA practice.

Measures to screen out alternatives, which are not "reasonable or feasible," can be helpful, or they can lead to inconsistencies and inhibit effectiveness if requirements do not include explicit definitions, criteria, and procedures. Measures allowing lower government level IA requirements to substitute for senior government level IA requirements can reduce potential duplication and overlap. However, such measures can result in unnecessary environmental impacts and reduced stakeholder participation if they are not structured, justified, and supported by explicit criteria, and if they do not include independent auditing and appeal provisions. For example, exemptions of infrastructure and "green" projects can result in economic/environmental benefits. But they can lead to unnecessary adverse environmental impacts. Such adverse effects often can be ameliorated or avoided with expedited and streamlined IA requirements for such projects. Considerable discretionary authority can facilitate the introduction of effectiveness/efficiency initiatives but can result in unnecessary inconsistencies in the application of IA requirements.

The desire to expedite and focus IA requirements is a recurrent issue in IA regulatory practice. The more efficient use of available resources and the desirability of focusing on major potentially significant actions and effects seem to be obvious and sensible goals for IA regulatory reform. Oftentimes the pressure to make IA requirements more focused and efficient stems from IA processes associated with a small number of highly controversial proposed actions—IA processes that generated a huge array of documents, cost vast sums of money, and dragged on for many years. Before proceeding too rapidly with introducing measures to prevent the reoccurrence of such processes, it would be worthwhile to independently review the major "problematic" processes. Although regulatory inefficiencies may have contributed to the costs and delays, it is also possible that other factors (e.g., poor IA practice, external political conflicts, unique proposal, and/or setting characteristics) may also have played a prominent role. With a better "handle" on what went right and wrong with the high-profile examples, broader questions of the appropriate mix of measures for making IA requirements more efficient and focused can be more systematically addressed.

As illustrated in the four jurisdictions, there are a host of measures available, at the regulatory level, for expediting and focusing IA document preparation and review. All jurisdictions have many already in place. In addition to reviewing the major "bad examples," systematic reviews of the relative effectiveness of the current suite of methods would be a useful departure point. For example, there is already considerable experience with screening and scoping. Lessons and insights from other jurisdictions, allowing for appropriate contextual adaptations, also can be helpful. It is important to bear in mind that expediting and focusing measures inherently mean that some matters will not be addressed at all, some concerns will be addressed much more rapidly and much more superficially, and some parties will not be involved or involved much less fully. This is well and good if a clear distinction can be drawn between significant and insignificant effects and proposals, if cumulative effects are not an issue, and if there is minimal uncertainty regarding who represent the "major" stakeholders. But this is often not the case. Such reforms also mean that the purpose of IA is shifted from the comprehensive integration of environmental concerns into decision making at all levels to the integration of selective environmental concerns (tied to specified areas of jurisdiction) into selective proposals in selective settings. Whether this is a desirable direction for IA regulatory practice should be the subject of open debate.

What the preceding distinctions suggest, at the very least, is the need to systematically explore the procedural (e.g., less open and inclusive decision making) and substantive implications (e.g., cumulative and sustainability effects addressed to a much more limited extent) of IA regulatory expediting and focusing approaches. The implications of some efficiency measures may be almost entirely beneficial. However, others could be highly problematic in both a procedural and a substantive sense. The responsible and practical course of action is to ensure a sound understanding of likely outcomes before proceeding, and to independently evaluate the effectiveness of such measures on a regular basis.

Measures to enhance IA competence and effectiveness are essential if IA is to improve over time at the regulatory level. As pointed out, all the jurisdictions provide guidance, support applied research, and undertake IA system effectiveness reviews and IA quality analyses. Independent reviews tend to have somewhat more credibility. Insights from other jurisdictions and from IA literature in general also can be helpful. Care needs to be taken to ensure that guidance neither restricts innovation nor "sets the bar too low" in terms of good practice. Many of the guidance documents in the four jurisdictions are dated. The impression is left that the suite of guidance materials and sponsored research that exists in the four jurisdictions is more the result of an "ad hoc" evolution of concerns and issues than the product of a systematic evaluation of what is needed to make the IA system operate more efficiently and effectively. Perhaps, a broader perspective is needed regarding the appropriate role of IA institutions in raising the level of IA practice at the regulatory and applied levels.

7.5.2 Management at the Applied Level

Figure 7.6 illustrates an example practical IA process. The figure and the process description that follows depict a focused, realistic, feasible, competent, and effective IA process. IA process managers can "pick and choose" the relevant and appropriate elements.

Start-Up Planning, decision making, and implementation are assumed to be integrated, constrained, decentralized, incremental, and partisan. It is recognized that many, often conflicting, parties and interests will need to be involved in the process. The context within which the IA process operates is expected to be uncertain and unstable.

The IA process is focused (through scoping) on what is relevant and important to regulators and other stakeholders. The initial scoping is supported by an overview of regulatory requirements and priorities, a scanning of key environmental and activity characteristics, and the identification of primary stakeholder issues and concerns. The problems to be solved, the needs to be met, and the opportunities to be taken advantage of are clearly identified. Priorities, boundaries, roles and responsibilities, major choices, sensitive and significant environmental components, major anticipated impacts, key activity characteristics, primary stakeholders, and critical issues (from the perspective of each stakeholder group) are determined. Scoping provides the basis for the IA process approach. The approach identifies major activities, events, inputs, and outputs. A study team, appropriate for addressing the identified issues, is assembled. An initial study design is prepared. The study design determines study organization, tasks, roles and responsibilities, budgets, and schedule. The approach, the study design, and political, agency, and stakeholder participation involvement approaches are formulated jointly with interested and affected parties.

The approach and study design are defined at a broad level of detail. They are expected to evolve and change through the process. There are major uncertainties regarding both ends and means. The "operating room" within which the process unfolds is highly constrained. Unforeseen and unforeseeable circumstances are anticipated to emerge through the process that will require approach and study design modifications and refinements. Ample "float time" is provided. Contingency funds are set aside for changing conditions.

Planning, Decision Making, and Implementation A practical IA process is nonlinear. It is iterative, cyclical, and incremental. It involves multiple stakeholders debating, discussing, negotiating, reviewing, analyzing, comparing, and bargaining about choices and constraints. The process is built around a series of decisions. It cycles back and forth among process elements. It is characterized by continuous learning. It provides for multiple interactions, for scanning ahead, and for feedback. It merges and transcends such conventional dichotomies as ends and means, objective and subjective, technical and political, analysis and synthesis, planning and implementation, and process and context. It crosses disciplinary and professional boundaries.

The process reflects bureaucratic and political requirements, preferences, and priorities. It operates within boundaries, acknowledges constraints and seeks out opportunities. It is focused, experimental, and action-oriented. Roles are negotiated. Ends are a general direction rather than precise objectives. Means are reasonable, available, and practical choices. The process is built upon a solid foundation of experience and practice-based knowledge, methods, insights, skills, and wisdom. Stakeholder perspectives, concerns, and preferences are integrated into the process. The process is designed to suit and refined to better match the context. Risks and uncertainties are freely acknowledged. The process proceeds cautiously and incrementally. Short-term time horizons largely predominate.

Choices that depart appreciably from current conditions, are highly uncertain, are potentially contrary to regulatory requirements, are controversial, are likely to be difficult or costly to implement, and are unlikely to be accepted by key stakeholders are quickly screened out. The key tests of a good option are regulatory compliance, stakeholder acceptance, ease of implementation, acceptable costs, and cost-effectiveness. Option comparison involves a reasoned exploration, from multiple perspectives, of implications and consequences rather than the formal application of evaluation methods. Once agreement is reached preferred choices are adapted, refined, and tested.

Outcomes from the process are formalized and documented in draft and final IA documents, consistent with regulatory requirements and agency expectations. Uncertainties regarding impact magnitude, impact significance, and mitigation effectiveness are incorporated into impact management strategies and tactics. Strategies are formulated, refined, and applied to facilitate approvals and implementation. Review and implementation tend to be incremental (e.g., phased approval), adaptive (e.g., continued focus on managing uncertainties), and conditional

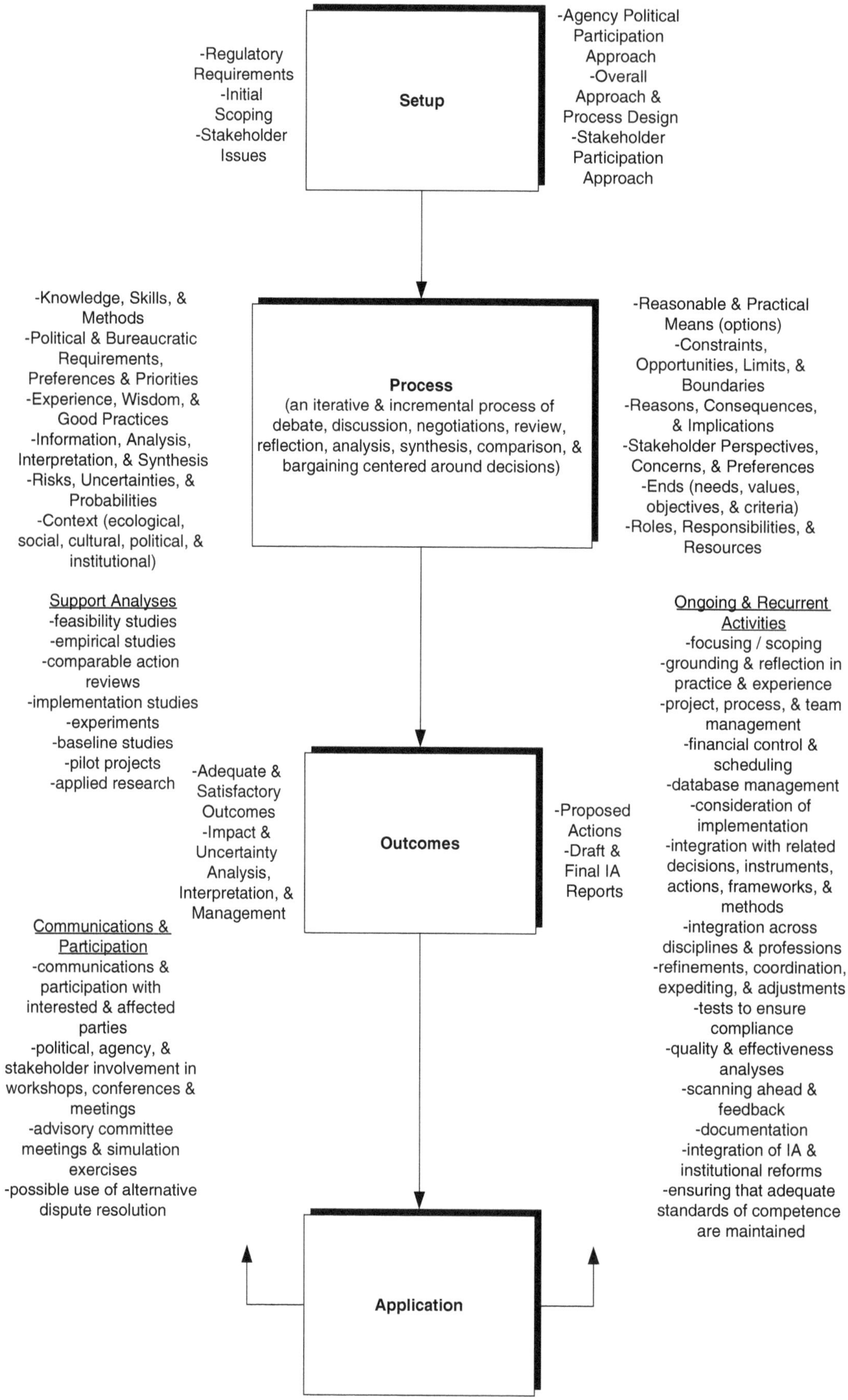

Figure 7.6 Example of a practical IA process. Adapted from Lawrence (2005a).

(e.g., ample provisional for monitoring and contingency measures). "Surprises" are expected to emerge during post-approvals. Surprises require both anticipation (to the extent practical) and prompt remedial action. The IA process extends through implementation. Direct and indirect outputs are assessed through effectiveness analyses. Knowledge, practice, and experience are considered tentative, contingent, partial, ambiguous, and uncertain.

Support Analyses Selective baseline analyses are periodically undertaken. These analyses focus on the requirements and expectations of regulators and of other stakeholders. Applied experience and practice-based knowledge are emphasized. Comparable actions, empirical studies, and implementation studies are reviewed to determine how power is exercised, how decisions are made, and how implementation obstacles and opportunities could be addressed.

Experiments and pilot projects are used to "test the water." Feasibility studies are undertaken to ensure that choices are cost-effectiveness and capable of implementation. Applied research is undertaken, where essential, to fill data gaps, which might impede decision making and implementation. Residual uncertainties are highlighted. Decision making and implementation implications of uncertainties are explored.

Ongoing and Recurrent Activities Focusing occurs with each cycle in the process. Documents, events, and interactions also are scoped. Analyses, interpretations, and conclusions draw heavily upon experience and reflection of practice. Action, process, and study team management, and financial and schedule control are maintained throughout the process. The database management system is continually updated and refined. Political feasibility and implementation requirements and implications are addressed, both prior and subsequent to approvals.

The IA process is integrated with decision making. It is linked to related decisions, actions, and environmental management instruments. Tiering and cross-referencing reduce paperwork, simplify review, and place the process and documents within a policy and strategic planning context. Related decision-making methods, such as cost–benefit analysis, feasibility studies, risk assessment, quality assurance, and technology assessment, are summarized and referenced, as appropriate. Critical links across disciplines and professions are identified. Individual analyses are integrated into methodological frameworks (e.g., sustainability assessment, integrated impact assessment) where needed to address cumulative effects and to ascertain progress toward broader environmental objectives.

Succinct and readily understandable interim, working, background, applied research, and consultation papers are prepared. They provide a clear decision-making basis, record decision-making process, and establish the basis for draft and final IA documents. Further refinements are introduced based on IA quality analyses of procedures, methods, documents, and participant performance and IA effectiveness analyses of interim outputs and comparable projects. IA documents incorporate stakeholder perspectives, demonstrate regulatory compliance, substantiate all assumptions, interpretations, and conclusions, and respond to the comments and suggestions of process participants. Interactions among process participants are coordinated. IA review, approval, and implementation are expedited. A minimum level of practice competence is maintained. Good practice is actively encouraged and facilitated. The IA process evolves in conjunction with IA and institutional reform.

Communications and Participation A practical IA process is open and interactive. Communications and participation with interested and affected parties are recurrent activities. Interactions are especially intensive leading up to and immediately following major decisions. Major perspectives are reflected in the analysis. All parties have an opportunity to review and respond to interim and draft IA documents. The concerns and priorities of all parties who could assume a significant role in approvals and implementation, in overcoming obstacles, and in building coalitions of support are solicited and documented.

Consultation methods conducive to identifying and accommodating differences (e.g., workshops, conferences, advisory committees) are applied as appropriate. Alternative dispute resolution methods, such as mediators and facilitators, are used when perspective and interest differences threaten the process. Close contact is maintained with regulators to minimize uncertainties regarding regulatory compliance. Consultation activities are both formal and informal.

7.5.3 Adaptations by IA Type

A practical IA process will vary by IA level and type. Table 7.7 provides examples of suggested SEA, EIA, EcIA, SIA, HIA, and SA practicality measures.

Crosscutting Themes Practical IA practices are not uniform among IA types. However, there are some themes that cut across IA types. All emphasize efficiency and effectiveness, the need for clear and understanding documents oriented to stakeholder interests and mandates, and the importance of ensuring adequate resources and IA capacity. They all recognize the importance of contextual variations and the central role of dialogue, persuasion, collaboration, and negotiations. Each seeks to accredit professionals in their field. They all appreciate the need to clearly understand the nature and most effective means of strategically influencing the bureaucratic/institutional planning/decision-making culture.

Practical SA Practice Practical SA practice focuses on the test of contribution to sustainability and the desire for multiple, reinforcing gains. Every aspect of the process must be geared to formulating and applying such tests.

Table 7.7 Practical IA Practice Characteristics by IA Type

Practical SA Practice	Practical SEA Practice	Practical EIA Practice
Uses thresholds, principles, and objectives to focus SA	Emphasizes role of collaboration, negotiation, and persuasion	Focuses on environmental regulatory requirements, issues, likely affected environmental components, and likely significant impacts
Focuses on multiple, mutually reinforcing gains rather than balancing; avoids undesirable trade-offs	Focuses on conflicts and policy and plan-making weaknesses and opportunities	Sets priorities and establishes temporal and spatial boundaries
Tests contribution to sustainability	Explores nature and validity of dominant arguments—clarifies SEA's purpose	Prioritizes participatory/open over technical/closed scoping
Looks for bridges to more lasting opportunities	Focuses on planning (plan-shaping activities) and decision-making links; focuses on implementation	Uses clear significance criteria
Screens for global warming impacts	Emphasizes organizational learning effectiveness	Emphasizes scoping of alternatives and mitigation
Recognizes interdependencies	Identifies concerns and information pertinent to subsequent tiers	Seeks consistency with local community environmental and social aspirations
and irreducibility of uncertainties and risks, accepts unpredictability and incomplete control; favors caution, resilience, and adaptability	Seeks enhanced understanding of stakeholder motives	Seeks enhanced understanding of context and decision-making culture
Identifies stakeholders and sources of resistance	Formulates, refines, and adapts conceptual models explaining how SEA works	Seeks enhanced understanding of links between EIA and community and regional planning
Seeks territorial/regional sustainability understanding; determines SA gaps	Seeks enhanced understanding of planning systems, cultures, and institutions to determine if effective SEA present	Determines why EIA documents poorly understood by decision makers and public
Uses frameworks for analyzing whether and how institutional changes for how sustainability can be achieved	Mediates between positivistic and postpositivistic approaches	Identifies lessons from court decisions
Uses frameworks for characterizing and grouping SA approaches and methods	Applies early enough to affect deliberation on purpose and options	States and justifies all assumptions
Identifies SA/SEA/EIA differences and good practice implications	Identifies how to break bureaucratic autonomy and ensure consideration of environment	Provides additional resources for scoping
Refines separate tools for high-level (international/national) SA	Seeks to enhance SEA ownership by planners	Institutes robust follow-up management and monitoring
Raises awareness of sustainability	Treats SEA as a social learning catalyst	Implements cost-effective mitigation, compensation, and management measures
Explores extent facilitates individual/organizational learning	Is both rational and collaborative	Seeks a better understanding of best management practices
Undertakes and facilitates sustainability capacity building	Enhances spatial data infrastructure	Identifies clear decision-making criteria
Tailors methods to SA context	Documents good practice SEA case studies and success stories	Provides more comprehensive and consistent guidance
Uses sustainability advisors	Develops and adapts SEA quality assurance checklists—identifies necessary and facilitating factors for SEA effectiveness	Institutes more training of practitioners and participants
Uses conceptual frameworks for assessing progress toward sustainability and for testing alternative SA approaches	Emphasizes SEA institutional capacity building	Makes EIA documents more understandable to decision makers and public
Assesses SA effectiveness for weak and vulnerable groups	Reinforces incentives to monitor	Seeks best practical approach (reviews and best practice)
Assesses procedural, substantive, transactive (worth time and cost), and normative effectiveness	Incorporates environmental limits	Assesses effectiveness in meeting community goals and aspirations
Shifts from procedural to substantive effectiveness emphasis (promoting SD)	Identifies effectiveness roles	Evaluates mitigation/compensation effectiveness
	Considers democratic, environmental, contextual and methodological effectiveness	Assesses consistency with best practices; recognize weaknesses of effectiveness tools (e.g., checklists)
	Identifies minimal SEA effectiveness requirements	Assesses effectiveness of requirements
	Assesses effectiveness—contribution to policy, planning, process quality, and quality of stakeholder involvement	Assesses effectiveness of impact and benefits agreements
	Seeks to demonstrate SEA added value, including indirect effectiveness and learning	

- Focuses biodiversity screening and scoping on composition, structure, and processes key to biodiversity creation and maintenance and on biodiversity threats
- Employs multiple spatial levels and long time horizons
- Prioritizes ecosystem services
- Focuses on protected areas and species
- Promotes more holistic databases (regarding ecological functions, connections, and networks) and approaches (e.g., ecological service types)
- Seeks an enhanced understanding of direct and indirect drivers of ecosystem change and of well-being of ecosystem service beneficiaries
- Promotes most environmentally friendly alternative
- Links IA and biodiversity-related treaties
- Seeks to make biodiversity part of corporate responsibility
- Considers use of use biodiversity screening maps and checklists
- Engages biodiversity experts, encourages biodiversity partnerships, and information networks, and institutes more training and good practice guidance
- Ensures sufficient resources
- Provides for technical guidance
- Assesses if and extent to which meets biodiversity and conservation objectives
- Monitors cumulative ecological effects
- Shares best practices and lessons

- Focuses on key social and cultural issues, rights, and purposes
- Emphasizes vulnerability of underrepresented and disadvantaged populations; stresses social distribution, environmental justice, and gender equality
- Seeks free, prior, and informed consent
- Seeks to understand social change and capacities to respond to change (resilience)
- Seeks an enhanced understanding of institutional financial and professional constraints to SIA
- Seeks an enhanced understanding of difficulties in applying social sciences in SIA
- Recognizes the contested and political nature of SIA
- Seeks to build social capital, social capacity, good governance, community engagement, and social inclusion
- Integrates community and traditional knowledge
- Formulates strategies for translating findings and recommendations into action
- Ensures skilled and locally experienced SIA professionals
- Facilitates SIA capacity building
- Ensures practitioner involvement consistent with ethical guidelines
- Accredits practitioners
- Assesses effectiveness in terms of achievement of social purposes
- Recognizes socioeconomic follow-up critical to effectiveness
- Assesses SIA effectiveness within or independent from SEA/EIA
- Assesses effectiveness in terms of realization of net social benefits

- Focuses on protecting and promoting public health, health determinants, health outcomes, affected populations, health equity, and vulnerable subpopulations
- Broadly defines health—environmental and social
- Places particular emphasis on avoidable, involuntary, adverse, irreversible, and catastrophic health effects
- In scoping assesses the quality of the evidence and designs monitoring surveillance systems
- Seeks an enhanced understanding of why health outcomes so rarely considered in SEA/EIA and
- varying levels of capacity and experience
- Develops and tests HIA practices
- Uses demonstration projects to test, refine, and demonstrate HIA methods and models
- Seeks intersectoral dialogue and action for health promotion and to address health disparities
- Seeks to demonstrate HIA decision-making benefits
- Undertakes critical appraisals to determine what is a good quality HIA and good quality health integration
- Seeks to raise decision makers' awareness of health and well-being consequences of choices
- Meets HIA practice standards
- Accredits and facilitates training of HIA practitioners
- Evaluates completed HIAs; enhances practice guidance and monitoring and evaluation guidance
- Disseminates review findings, case studies, and lessons
- Audits HIA utility of HIA, decision-making influence, predictive accuracy, and mitigation effectiveness
- Evaluates health outcomes (health improvements, reduced inequities)
- Assesses effectiveness within or separate from EIA/SEA; refines and creates alternatives

---

Sources: Adelle and Weiland (2012), Ali et al. (2008), Athanas (2005), Alton and Underwood (2003), Atkinson and Cooke (2005), Baker and McClelland (2003), Baker and Rappaport (2009), Barnes et al. (2010), Becker et al. (2005), Bhatia et al. (2010), Bina (2007), Birley (2003), Bond (2004), Bond et al. (2012), Bond et al. (2003), Buuren and Nooteboom (2009), Cashmore et al. (2004), Chaker et al. (2006), Cole (2004), Cole and Fielding (2007), Connor and Dovers (2004), Craik (2008), Donnelly et al. (2008), Égré and Senécal (2003), Esteves et al., (2012), Fischer (2002, 2005), Fischer and Gazzola (2006), Forsyth et al. (2010), Galbraith et al. (2007), García-Montero et al. (2010), Gazzola (2008), Genter et al. (2008), Gibson (2006a, 2011) Govender et al. (2006), Hacking and Guthrie (2006, 2008), Harris-Roxas and Harris (2011), Harris-Roxas et al. (2012), Hayes and Morrison-Saunders (2007), Hildén et al. (2004), Hinte et al. (2007), IAIA (2002a, 2005, undated b), ICPGSIA (2003), Jha-Thakur et al. (2009), Kemm (2005), Kemm and Parry (2004a,b), Kobus (2005), Krieger et al. (2003), Landsberg et al. (2011), Lundberg et al. (2010), Mandelik et al. (2005), McCluskey and João (2011), Moles et al. (2008), Morgan (2012), Morrison-Saunders and Hodgson (2009), Mulvihill (2003), Noble (2009a,b), O'Faircheallaigh (2009), Page (2006), Petäjäjärvi (2005), Péti (2012), Pisani and Sandham (2006), Põder and Lukki (2011), Pope et al. (2004), Pritchard (2005), Quigley and Taylor (2003), Rajvanshi et al. (2011), Retief (2007a,b), Retief et al. (2008), Ross et al. (2006), Rotmans (2006), Sánchez and Gallardo (2005), Sánchez and Silva-Sánchez (2008), Scanlon and Davis (2011), Singh et al. (2009), Slootweg (2005), Slootweg et al. (2006), Slotterback (2008), Slotterback et al. (2011), Snell and Cowell (2006), Söderman and Saarela (2010), Stoeglehner et al. (2009, 2010), Tang (2010), Tetlow and Hanusch (2012), Thérivel and Ross (2007), Theophilou et al. (2010), Thérivel et al. (2009), Tsuji et al. (2011), Utzinger et al. (2005), Vanderhaegen and Muro (2005), Vanclay (2006, 2010), Waldeck et al. (2003), Wende et al. (2005), Weston (2011), Whitelaw et al. (2009), Winkler et al. (2011), Wood et al. (2006), Zhu et al. (2010).

The IA process and outcomes are viewed as a bridge to lasting sustainability opportunities. Practical SA practice readily screens out unsustainable choices and makes effective use of available sources of sustainability-based knowledge. It focuses on institutional barriers and required changes to ensure sustainability. It identifies and remedies capacity limitations that inhibit sustainability. It stresses the substantive over the procedural. It is adaptive, proceeds with caution (appreciating the implications of risks and uncertainties), and fully addresses contextual implications. It fosters organizational learning and public awareness, focuses on the implications for the most vulnerable, and tests for normative/substantive effectiveness.

Practical SEA Practice Practical SEA practice focuses on dominant arguments, on potential conflicts, and on policy and plan-making weaknesses and opportunities. It is founded upon a sound understanding of the policy and decision-making process. It emphasizes effective organizational learning and SEA ownership. It seeks an enhanced understanding of stakeholder motives, institutional characteristics, and planning systems and cultures. It effectively draws upon conceptual models, effectiveness analyses, and case study, and auditing insights. It strives to overcome bureaucratic autonomy and resistance, ensure that effective SEA conditions are present, and demonstrates the added value of SEA.

Practical EIA Practice Practical EIA practice fully integrates regulatory requirements, issues, and guidance, clearly identifies and substantiates priorities, boundaries, assumptions, and methods, and draws heavily upon insights from court decisions, best practices, case studies, and effectiveness analyses (appreciating the strengths and limitations of such tools). It emphasizes the value of scoping and the need to understand the perspectives of each stakeholder. It employs clear significance and decision-making criteria. It seeks consistency between EIA outcomes and community aspirations. It effectively utilizes community and regional planning to frame EIA-related decision making. It provides for appropriate training and capacity building for IA practitioners and other participants, and ensures cost-effective and robust follow-up measures.

Practical Ecological Impact Assessment (EcIA) Key biodiversity attributes, major ecosystem services, and protected areas and species are priorities with practical EcIA practice. Practical EcIA practice employs multiple temporal and spatial levels. It seeks an enhanced understanding of the direct and indirect drivers of ecosystem change, and of the well-being of ecosystem service beneficiaries. It operates within the context of international ecological and biodiversity treaties. It is guided by clear biodiversity and conservation objectives. It promotes the most environmentally friendly alternative. It monitors cumulative ecological impacts. Practical EcIA practice seeks to make ecological concerns part of institutional and corporate responsibility. Adequate resources are sought. Biodiversity experts are engaged. Biodiversity partnerships and information networks are encouraged. Training, technical guidance, and best practice lessons are provided.

Practical SIA Practice Practical SIA practice focuses on key social and cultural issues. It emphasizes the vulnerability of underrepresented and disadvantaged populations. It stresses social distribution, environmental justice, and gender equality concerns. Free, prior, and informed consent is sought. Practical SIA practice seeks to understand social change and response capacities, institutional and professional constraints to SIA, and the difficulties in applying the social sciences in SIA. It strives to build social capital, social capacity, good governance, community engagement, and social inclusion. It recognizes the inherently political and contested nature of SIA. It integrates traditional and community knowledge. It facilitates SIA capacity building. It appreciates the critical role of follow-up. It assesses effectiveness in terms of the achievement of social purposes and the realization of net social benefits.

Practical HIA Practice Practical HIA practice is focused on protecting and promoting public health, health determinants, health outcomes, affected populations, health equity, and vulnerable subpopulations. Health is defined broadly, with a particular emphasis on avoidable, involuntary, adverse, irreversible, and catastrophic health effects. Care is taken to assess the quality of the evidence and the adequacy of proposed methods. Practical HIA practice seeks to understand and overcome the barriers to HIA and to demonstrate the utility of HIA. It strives to raise awareness by decision makers of the health and well-being implications of choices. It facilitates HIA capacity building. A particular effort is made to raise the level of HIA practice, both within and separate from SEA/EIA, by drawing upon, contributing to, and broadly disseminating demonstration projects, effectiveness analyses, and critical appraisals.

7.6 CONTEMPORY CHALLENGE—CEA GOOD PRACTICE

7.6.1 Definitions and Distinctions

Cumulative effects assessment (CEA) systematically analyzes and assesses cumulative environmental change (Cooper and Sheate, 2004). It focuses on the receiving environment, and on whether individually minor effects will be collectively significant (Canter and Ross, 2010; Dixon and Thérivel, 2011; Noble, 2008). CEA simultaneously assesses the positive and negative effects (additive, interactive, synergistic, irregular) on given receptor(s) from existing, planned, proposed, and potential human activities (Greig et al., 2004; Hanna, 2009a; Noble, 2009a). It is especially concerned with whether the environment's assimilative capacity will be exceeded.

CEA assumes many forms. A basic distinction can be drawn between regional cumulative effects studies (effects-based, measures environmental responses or valued ecosystem components, analytic) and project CEA (stressor-based, cumulative effects associated with a particular change agent) (Creasey and Ross, 2009; Noble, 2009a). Further distinctions can be drawn for EIA-driven CEAs between single project CEA and multiple projects/multiple component activities. For SEA-driven CEAs distinctions can be drawn between CEAs for plans or programs for a particular resource or industrial sector and for CEAs of multiple plans or programs across sectors (Harriman and Noble, 2008). CEA can be performed on its own or as an integral part of an SEA or project-level EIA process (Harriman and Noble, 2008).

One variation of an SEA-driven CEA (or perhaps more appropriately a CEA-driven SEA), which has received particular attention is regional SEA (RSEA). RSEA is a tool that can support regional planning by assessing the cumulative effects associated with alternative development scenarios (Johnson et al., 2011). It helps identify land use strategies and management approaches consistent with desired environmental, social, and economic outcomes, including regional sustainability (Gunn and Noble, 2009a; Johnson et al., 2011). Defining characteristics of RSEA include—strategic, region-based, futures-oriented, alternatives-based, VEC-based, multiscale, multitier, multisector, adaptive and opportunistic, structured and systematic, cumulative effects driven, learning-oriented (based on feedback from follow-up and regional monitoring), and integrated with planning and decision making (Gunn and Noble, 2009a; Noble, 2008; Sadler, 2011b).

7.6.2 State of Practice

Notwithstanding the semblance of CEA regulatory requirements and a range of guidance material and applied research in most jurisdictions, the CEA state of practice is consistently characterized as overwhelmingly weak (Devlin and Yap, 2008; Dixon and Thérivel, 2011). To date, SEA has not emerged as an effective tool for identifying and managing cumulative effects (Sadler and Jurkeviciute, 2011). CEA is also rarely an integral part of project-level EIA processes (Noble, 2008).

Commonly cited limitations associated with CEA practice (when it is undertaken at all) include (1) weak conceptualization and scoping, (2) a lack of understanding of the causes of cumulative effects, (3) undefined or ill-defined thresholds, (4) the poor consideration of past or likely future activities, (5) the weak treatment of cumulative ecological, socio-economic, and sustainability effects, (6) a failure to appreciate the implications of scale, boundary, and IA-type differences, (7) poorly defined methods, (8) limited cumulative effects monitoring and management, (9) limited multistakeholder CEA involvement, (10) difficulty dealing with uncertainties and issues such as biodiversity and climate change, (11) minimal or poorly defined tiering arrangements, (12) weak integration into IA analyses (i.e., an afterthought), (13) poorly integrated into policy and planning, (14) and a lack of commitment by decision makers to the assessment and management of cumulative effects (Baxter et al., 2001; Canter and Ross, 2010; Connelly, 2011; Crooks and de Witt, 2009; Duinker and Greig, 2006, 2007; Gunn and Noble, 2011; Law et al., 2005; Lien et al., 2011; Senner, 2008; Thérivel and Ross, 2007).

These limitations are partially attributable to the regulatory level. It is not sufficient to require the consideration of cumulative effects. Assessing cumulative effects is severely hampered if (1) the environment and effects are narrowly and selectively defined, (2) if specific CEA-related triggers are not identified in IA requirements, (3) if IA tiering arrangements are not instituted, (4) if IA requirements do not outline minimal requirements for the conduct of CEA, (5) if IA terms of reference and related CEA guidance materials are dated and overly vague, (6) if there is insufficient interagency collaboration in the provision of baseline data and in the analysis and management of cumulative effects, (7) if the necessary financial resources and expertise are not in place, (8) if a clear commitment to regional planning, informed by CEA, is lacking, (9) if there is no to minimal CEA monitoring and follow-up, and (10) if the central role of uncertainties and the need for a precautionary approach to impact and uncertainty management is not broadly acknowledged (Canter and Ross, 2010). However, the regulatory level is not entirely responsible for the weak treatment of cumulative effects. Clearly, there is a considerable gap in practice between CEA aspirations and the reality of CEA practice.

7.6.3 Aspirations

CEA should operate at both the strategic and the project level, and at the regulatory and applied levels. CEA is a tool for influencing the pace and scale of development, and for predicting sustainability. Arguably, the assessment of cumulative effects is IA as it should be. CEA and IA both seek (or should) to protect and improve the receiving environment, and to further the cause of sustainability (Dixon and Thérivel, 2011; Senner, 2008). CEA, if practiced well, can facilitate cumulative effects significance interpretations, cumulative biodiversity effects interpretations, potential and desired futures characterizations, the determination of thresholds and appropriate activities, environmental justices analyses, and government resource management and allocation (Canter and Ross, 2010; Harriman and Noble, 2008; Hegmann and Yarranton, 2011; Kreig and Faber, 2004; Treweek et al., 2011).

7.6.4 Process

Characterizations of the CEA process closely parallel SEA and EIA process depictions, at least in terms of major stages/activities. For example, reference is made to scoping, baseline descriptions, impact predictions, alternatives assessment, significance determination, mitigation and enhancement, government and stakeholder engagement, and impact management (Canter and Ross, 2010; Connelly, 2011; Dixon and Thérivel, 2011; Gunn and Noble, 2011; Johnson et al., 2011; João et al., 2011; Noble, 2008; Thérivel and Ross, 2007). However, CEA process depictions go on to stress the need to identify past, present, and likely future activities (cumulative change processes) that have or will affect receptors and lead to activities (Dixon and Thérivel, 2011; Noble, 2008; Thérivel and Ross, 2007). They construct and apply CEA frameworks (customized to the region and cumulative effects types). They systematically connect proposed and potential actions to selected VECs and their indicators. They formulate a regional vision. They generate and analyze alternative scenarios. They compare scenarios against significance thresholds. They build desirable and resilient futures (Atkinson and Canter, 2011; Canter and Ross, 2010; Connelly, 2011; Gibson, 2011; Gunn and Noble, 2009b, 2011; Johnson et al., 2011).

7.6.5 Methods

CEA methods have received considerable attention in IA literature in recent years. Commonly cited CEA methods include scenario development, causal chain and network analyses, conceptual frameworks, models, input–output analysis, adaptive management, overlay mapping and GIS, carrying capacity analysis, life-cycle analysis, and indicators and indices (Atkinson and Canter, 2011; Canter, 2008; Canter and Atkinson, 2010; Greig et al., 2004; Gunn and Noble, 2009a,b; Ross and McGee, 2006; Smit and Spaling, 1995; Thérivel and Ross, 2007). Given the dynamic, complex, and integrative nature of CEA, particular stress tends to be placed on quantification and on technical-/data driven methods (Dutta et al., 2004; Gunn and Noble, 2009b; Thérivel and Ross, 2007). More judgment-driven techniques, such as case study analyses, participatory appraisal, Delphi and multicriteria analysis, can assist in identifying critical linkages, in significance interpretations, and in facilitating stakeholder involvement (Crookes and de Witt, 2009; Gunn and Noble, 2009b).

The methods selected need to be adapted to the region and to cumulative effects types (Gunn and Noble, 2011). Care needs to be taken to identify and apply factors (e.g., salience, credibility, legitimacy) that facilitate or represent barriers to effective CEA (Kim, 2012). Flexibility and a VEC-based perspective are essential (Canter and Ross, 2010; Creasey and Ross, 2009; Duinker and Greig, 2006). Integrating the principles of the ecosystem approach, adaptive management, and a risk-based precautionary approach can be helpful (Canter, 2008; Canter and Atkinson, 2010; Sadler, 2011b; Treweek et al., 2011). Uncertainties, and related implications, should be acknowledged (Lien et al., 2011). Assumptions, key features, and a clear rationale for the methods selected should be provided (Canter, 2008; Thérivel and Ross, 2007). Resultant conclusions should be fully justified (Lien et al., 2011).

7.6.6 IA Levels

It has been consistently argued that CEA can, or at least should be, most effective at strategic levels (Cooper and Sheate, 2004; Duinker and Greig, 2006). A strategic level CEA complements and supports SEA, regional planning, and protected area management. It focuses on the resource and activity totality rather than only on assessing a proposed plan or program (Dixon and Thérivel, 2011; Greig and Duinker, 2007). It can help determine and apply environmental objectives, sensitive areas, and ecological thresholds. It can guide regional development (especially in relatively undeveloped areas). It can proactively identify and minimize cumulative environmental effects. It can provide an integrated appraisal of the ecological, social, and economic dimensions of sustainability. It can identify institutional and jurisdictional gaps. It can establish intergovernmental collaboration priorities (Connelly, 2011; Cooper and Sheate, 2004; Duinker and Grieg, 2006). It can help ensure that multiple activities, including the plan, do not have cumulatively significant effects (Dixon and Thérivel, 2011). CEA, at a strategic level, operates at different scales (e.g., subregional, regional, transboundary) (Cooper and Sheate, 2004). It can provide a pragmatic and balanced (potential activities vs. key receptors) approach whereby planners, stakeholders, and decision makers can understand and explore the potential cumulative effects of policies, plans, and programs (Cooper, 2011). Strategic level CEA can help frame, guide, and simplify project-level EIA (Connelly, 2011; Cooper, 2011).

Although cumulative effects can be more systematically addressed at the SEA level, there are also roles for CEA at the EIA project level. Project-level CEA can more systematically address indirect effects, reduce incremental contributions to cumulative effects, and place project impacts within the context of all potential impacts on receptors (Cooper and Sheate, 2004; Hinte et al., 2007). At the EIA level, CEA extends the IA process for projects. It emphasizes local project-based stressors and broadens the temporal and spatial scope of analysis to encompass other past, present, and likely future actions, (Dutta et al., 2004). There is some debate whether cumulative effects should be fully integrated into the EIA (assume all effects cumulative) or addressed in a separate section (own methodology, broader temporal and spatial boundaries, broader level of detail but still useful) (Bérubé, 2007; Duinker and Greig, 2006).

Regardless of the approach adopted, systematically including cumulative effects in project-level EIA enhances EIA practice (Connelly, 2011). However, CEA, at a project level, does not achieve a regional level of analysis and cannot effectively address regional issues such as environmental quality and biodiversity loss (Connelly, 2011; Duinker and Greig, 2006). SEA-level CEA can address cumulative effects at the source (policies and plans) (Johnson et al., 2011). It also has the temporal and spatial breadth necessary for encompassing cumulative effects that extend over long time periods and entail multiple sources and pathways (Johnson et al., 2011). But CEA at a regional level, because it is retrospective, has limited predictive capabilities. It also often lacks an institutional "home base" and an "action-forcing" mechanism for ensuring that decision making is actually influenced (Gunn and Noble, 2009b).

The issue, therefore, is not so much whether CEA is best applied at the strategic or project level but how cumulative effects can be integrated into both levels in a holistic and complementary manner (Hacking and Guthrie, 2006; Thérivel and Ross, 2007). Multilevel CEA involves more than tiering whereby SEA-level CEA scopes and frames project-level CEA. Foresight and dialogue between levels are essential (Gunn and Noble, 2011). Oftentimes a multilevel, multiscale analysis is more consistent with regional and project planning as it actually occurs in practice (João, 2007).

7.6.7 Decision making, Implementation, and Management

CEA can facilitate acceptability decisions by placing development within an historical context and within possible future trends (Hegmann and Yarranton, 2011). It can provide decision makers with reliable predictions and help ensure that the pace and scale of development remain within regional capacities (Gibson, 2011; Hegmann and Yarranton, 2011). CEA-related decisions are facilitated when there is a shared regional vision regarding the future state of the environment and development (Gunn and Noble, 2009b). It tends to be more effective when it is sensitive to key decision windows and when it is broadly recognized as more than "adding up" environmental effects (Gunn and Noble, 2009b). CEA should contribute to decisions directed toward mutually reinforcing and lasting environmental gains (Gibson, 2011).

CEA impact management should encompass both incremental effects mitigation and local and regional effects management (Canter and Ross, 2010). For CEA impact management to be effective, there should be adequate institutional support for regional monitoring and follow-up and a high degree of collaboration among agencies and stakeholders (Noble, 2008). A CEA impact management system should be independent and tiered (Gibson, 2011; Noble, 2008). It should address both effects and compliance management, and the anticipation and recognition of emerging concerns (Gibson, 2011).

7.6.8 Future Actions

The gap between CEA theory and practice will be narrowed when and to the extent that there is additional institutional support, applied research, good practice guidance, tiering mechanisms, agency-stakeholder collaboration, CEA training and education, and regional monitoring and follow-up (Burdge, 2004; Canter and Ross, 2010; Gunn and Noble, 2011; Noble, 2008). The consciousness and skills of consultants, proponents, and regulators regarding CEA need to be raised (Duinker and Greig, 2006). Potential priorities for targeted research and pilot projects include such matters as the bridging of SEA and EIA in CEA, the establishment of cumulative effects thresholds and boundaries, the management of the uncertainties associated with cumulative effects, efficient and effective cumulative effects management measures including offsets, and the testing and refinement of CEA methods (Connelly, 2011; Dowlatabadi et al., 2003; Lien et al., 2011).

7.7 SUMMING UP

This chapter portrays a streamlined, efficient, and effective IA process—a practical process based on realistic expectations and competent practice. The three stories address practicality in different ways. The first story describes a formal and ambitious scoping process that effectively integrated a diversity of stakeholder perspectives. The second story describes how the use of intermediate reports provided a practical approach to strengthening SEA effectiveness and consistency. The third story provides a good practice example of cumulative effects assessment. All three stories underscore the importance of matching process to context. The stories provide only a partial and preliminary impression of how an IA process can become more practical.

The problem is the tendency for IA processes to be unfocused, disconnected from reality, weak on implementation, of variable quality, and slow to learn from experience and practice. Several concepts are introduced to make the IA process more focused, relevant, feasible, competent, and effective. The concepts provide the basis for practical IA requirements and practical IA processes. A practical IA process is assessed against ideal IA process characteristics.

IA documents too often are unfocused and excessively descriptive. IA processes frequently take too much (or too little) time and consume too many (or too few) resources. Planning and IA processes and theory and practice are still widely separated. IA processes sometimes are poorly adapted to context. They do not always adequately integrate stakeholder values and perspectives. IA processes could make better use of experience and good practice. They neglect the needs of decision makers. They can fail to facilitate implementation. Sometimes, they are poorly integrated with other environmental management instruments and with public policy making. There is too much variability in IA competence levels. They do not adequately maintain and enhance IA quality and effectiveness.

A practical IA process *focuses* on what is relevant and important. Focusing or scoping can be applied to IA institutional arrangements, to IA documents, and to IA process activities, inputs, and outputs. Scoping is based on clearly defined and consistently applied principles. It identifies the possibilities, decides what is important, shapes and structures the process, and involves interested and affected parties. Numerous scoping methods are available. Scoping roles are clearly defined. The conduct of scoping varies depending on context.

A practical IA process is grounded in practice and experience. It is *realistic*. A realistic IA process understands how decisions are made, how organizations operate, how organizations are structured, how people behave in organizations, and how organizational structures and procedures vary in different settings. It accounts for how practitioners operate and how knowledge is generated through practice. It is aware of how actions are integrated into decision making, how power is exercised, and how resources are allocated. It considers how actions are implemented, facilitated, inhibited, and prevented.

A practical IA process is *feasible*. It is workable. It can be undertaken and implemented. It is informed by concepts, guided by strategies, and aided by tactics. It merges the IA process with decision making, implementation, management, and context. It is linked to related tools and methods. It contributes to and is adapted to IA reforms. It is integrated with policies, plans, and programs, with organizational operations, with other decision-making levels, and with related instruments. It is harmonized with other governmental requirements and is coordinated with private sector and nongovernment organizational activities. It addresses interconnections among disciplines and professions. It is embedded within synthesis frameworks.

Competence is essential in a practical IA process. It pertains to the qualifications of participants and to the conduct of analyses. It includes the choice and execution of roles and responsibilities by project managers, specialists, proponents, governments, and the public, both individually and collectively. It includes such joint IA process activities as study team management, study team participation, database management, the application of geographic information systems, report writing, documentation, financial control, budgeting, and the preparation of work programs and schedules. It actively avoids and minimizes recurrent, avoidable, competence-related problems.

A practical IA process is *effective*. Effectiveness addresses how well the process worked. It concerns the quality of inputs (e.g., institutional arrangements, processes, methods, participant performance, documents) and the effectiveness of direct and indirect outputs (e.g., goals achievement, environmental changes, methodological performance, management performance, contribution to practice). IA quality and effectiveness analyses raise the level of IA practice. They distinguish between acceptability and performance levels. They provide an opportunity to correct deficiencies. They involve interested and affected parties. They provide a clear rationale for all interpretations and conclusions. They document results in a form suitable for enhanced practice.

A practical IA regulatory system harmonizes IA among government levels, ensures that the IA roles of government departments and agencies are well coordinated, focuses on what is important, minimizes unnecessary costs and delays, ensures an adequate level of IA competence, and contributes to enhanced IA practice. The four jurisdictions apply multiple methods to achieve these objectives. The diversity of approaches points to the potential benefits of knowledge sharing. Uncertainties regarding performance suggest the need for additional effectiveness analyses.

A practical IA process is focused, realistic, feasible, competent, and effective. It focuses from the outset on regulatory requirements and stakeholder issues. It is based on a well designed but flexible overall approach and study design. The approach includes procedures for involving governments, politicians, the public, and other stakeholders. The process is built around decisions. It is iterative and incremental. It involves debate, discussion, negotiation, review, reflection, analysis, synthesis, comparison, and bargaining among interested and affected parties. The process integrates ends, means, constraints, analyses, knowledge, methods, skills, experience, uncertainties, requirements, perspectives, and concerns. It operates within available resources, clearly defines roles and responsibilities, explores consequences and implications, provides a rationale for interpretations and conclusions, and is adapted to the context. It applies impact and uncertainty analyses, interpretation, and management to identify adequate and satisfactory options and to select and refine proposed actions. It prepares draft and final IA documents. It formulates impact management and implementation strategies. It is merged with approvals and implementation. It evaluates the effectiveness of direct and indirect outputs.

A practical IA process is supported by baseline analyses, by applied experience, and by practice-based knowledge. It applies experiments, pilot projects, feasibility analyses, and targeted research to refine and test interpretations and conclusions. It identifies residual uncertainties and their implications. Many activities in the process are recurrent or continuous (e.g., focusing, management, considering implementation, competence tests, integration, quality and effectiveness analyses, the incorporation of knowledge and experience, documentation). Close contact is maintained with interested and affected parties. Communications and participation focus on issues and perspectives directly bearing on decision making and implementation. A practical IA process effectively integrates practicality measures associated with different IA types.

The effective analysis and management of cumulative effects is essential in a practical IA process. Effective CEA begins at the regulatory level. The assessment of cumulative effects should be a requirement at the SEA and project EIA levels. Mechanisms need to be in place to harmonize (e.g., tiering) and coordinate the analysis and management of cumulative effects among government levels and across IA types. There should be sufficient requirements and guidance in place to ensure an adequate quality of CEA practice. The treatment of cumulative effects in IA practice should facilitate the realization of CEA objectives, be consistent with CEA good practice standards, and be appropriate to the context. Good practice CEA includes effective follow-up. It also should contribute to the CEA knowledge base.

Chapter 8

How to Make IAs More Democratic

8.1 HIGHLIGHTS

This chapter describes a process where IA is "closer to the people" and people take the lead in IA process management. The decision-making role of the people and communities most directly affected by the proposed action is enhanced. Groups, segments of society, and perspectives, commonly excluded from or underrepresented in IA processes, are more prominently featured.

- The analysis begins in Section 8.2 with three applied anecdotes. The stories describe applied experiences associated with efforts to make IA practice more democratic.
- The analysis in Section 8.3 then defines the problem. The problem, from the perspective of many public "participants" in the IA process, is that the IA process is too far removed and the public's decision-making role is too often nonexistent to negligible. Frequently, members of the public are, or believe themselves to be, powerless in decisions that greatly influence their lives. The direction is bringing IA closer to the people and enhancing the decision-making role of the public, especially those most directly affected and most vulnerable to change.
- In Section 8.4 we provide an overview of democratic concepts and methods. We explore how (1) the decision-making role of the public can be enhanced, (2) the decision-making role of people affected by proposed actions can be enlarged, and (3) power imbalances in the IA process can be rectified.
- In Section 8.5 we apply the insights, distinctions, and lessons identified in Section 8.4. We describe the properties of a democratic IA process at both the regulatory and applied levels. In Section 8.5.1 we explore how IA requirements and guidelines could be more democratic. In Section 8.5.2 we demonstrate how a democratic IA process could be expressed at the applied level. In Section 8.5.3 we provide an overview of how democratic IA practice is expressed for various IA types (SA, SEA, EIA, EcIA, SIA, and HIA).
- In Section 8.6 we address the contemporary challenge of multijurisdictional IA. We identify key conceptual distinctions and suggest a range of good practice measures.
- In Section 8.7 we highlight the major insights and lessons derived from the analysis.

8.2 INSIGHTS FROM PRACTICE

8.2.1 The A4 Highway Extension and the "Dent in Democracy"

In the 1990s and 2000s, a series of EIAs were carried out to investigate the environmental and social impacts associated with a second traffic connection between The Hague and Rotterdam in the Netherlands. To be completed by 2015, the connection comprises an extension of roughly 7 km of the A4 highway to link the existing infrastructure around these two cities. It was premised that a second highway would significantly mitigate traffic congestion and road safety on the other highway, the A13, thereby making the Port of Rotterdam more accessible for transport over land and improving the Amsterdam/Antwerp connection. The proposed project, however, entailed probable adverse impacts related to soil, water, and air quality, biodiversity, archaeological value potential, nature recreation, and something as intangible as the "cultural–historic character" of the impact area. Economic, social, and ecological interests were pitted against one another.

The A4 highway extension has provoked a high level of public contestation ever since its first appearance on the political agenda in 1965. Stakeholders that opposed the project in recent years deployed two protest strategies: contra-expertise and environmental advocacy. Consistent with the legal requirement to scope the impacts for traffic flow associated with large infrastructural projects, the necessity of the extension in terms of congestion mitigation has been challenged by prominent leaders in the oppositional movement. A salient point for discussion has revolved around the state-induced notion of an "autonomous growth of traffic" in the impact area in decades to come. On the basis of probabilistic evidence of traffic growth, decision-makers

Impact Assessment: Practical Solutions to Recurrent Problems and Contemporary Challenges, Second Edition. David P. Lawrence.

estimated that a second connection was necessary and therefore justified. Public stakeholders, however, contradicted this scenario, arguing that a multitude of measures in the realms of public transport and congestion pricing are well equipped to harness traffic capacity. The second strategy pertained to emphasizing the critical socioecological function of the impact area. It was stated that the area is host to meadow birds, bats, and other small mammals; that it preserves large stocks of flora and fauna; and that it provides refuge for people that live and recreate there. The highway extension would provoke an irreversible loss of natural capital as it cuts through wetlands, fens, and ditches.

In the Netherlands, highways are planned, constructed, and maintained by the state. The practical consequences of decision outcomes in these domains are nonetheless borne by entities other than the central administration. The extension of the A4 has been planned in the province of South Holland and connects a number of municipalities. Highways, therefore, not only dissect the landscape, but also travel through different layers of democratic decision making. This could offer increased opportunities for public stakeholders to find institutional access and exert influence. In the case of the A4 highway extension, however, the Dutch government had enacted the Crisis and Recovery Law in March 2010 to speed up large infrastructural projects. The new legislation is meant to clear procedural barriers that hinder the progress of a project. It also accommodates procedures that emanate from certain legal domains, including EIA. As a consequence of this, subnational bodies have been prevented from engaging with the democratic decision making of projects that are planned in their jurisdictions. Public stakeholders have been furthermore curtailed in finding opportunities to collaborate with decision makers.

At the national level, the executive agency of the Ministry of Infrastructure and Environment, *Rijkswaterstaat*, has been responsible for the highway extension. Rijkswaterstaat is a technocratic agency that implements rather than deliberates a decision once it is mandated to do so. While the Netherlands has been repeatedly praised for its approach to concerted decision making with the civil society (*polderen*), public stakeholders involved in the extension have felt dismayed about the lack of democratic virtues in its procedure. They have been referred to Rijkswaterstaat as a "state within a state," a fourth branch of government that turns a deaf ear to citizen preferences and societal demands. It has been asserted that the agency embellished its own estimations, ignored or obstructed counter-impact assessment, and presented assumptions as facts, among others. Worst of all, Rijkswaterstaat has been accused of not reciprocating norms of good democratic practice. This has propelled the more radical public stakeholders to conclude that they were only involved to legitimate a decision already made.

The construction of the A4 highway extension was approved in July 2011. Albeit partly tunneled to mitigate its impact, a concession made to the opposition, the interest to invest in the mobility of the impact area has prevailed over ecological considerations. Some public stakeholders that coordinated action to stop this from happening have lamented the role that Rijkswaterstaat, under the aegis of the Dutch state, has played. Others have furthermore argued that EIA in its current form is too weak a decision-support tool to protect the environment. It may be safe to argue that the process has proved a Pyrrhic victory for democracy writ large. One particularly disgruntled stakeholder has reported a low level of confidence in extant opportunities for political participation and argued that poor decision making in the highway extension has led to a "dent in democracy."

JAAP G. ROZEMA

Science, Society and Sustainability Research Group, School of Environmental Science, University of East Anglia, Norwich, UK

8.2.2 Setting the Stage for Project EA Through Higher Tier Planning

In 2005, the province of Nova Scotia began to be approached by a number of developers of tidal energy about the possibility of testing and commercially developing a number of new tidal technologies in the Bay of Fundy, a marine bay along the western coast of Nova Scotia that boasts of the highest tides in the world. In response, the province of Nova Scotia first reviewed the existing regulatory system at the federal and provincial levels to determine how it would apply to tidal energy in the Bay of Fundy, where the current regulatory system was effective and efficient, and whether there were any gaps in the system in place.

One of the outcomes of this regulatory review was a recommendation that decisions about potential tidal pilots or commercial scale projects should not be made in a vacuum. Rather, ideally, they should take place in the context of an integrated regional plan, integrated regional management, and a strategic environmental assessment to consider how tidal energy might interact with other existing and potential future uses of the Bay of Fundy.

In 2007, the province of Nova Scotia decided to proceed with a strategic environmental assessment of offshore renewable energy with a focus on tidal energy in the Bay of Fundy. The SEA was carried out in cooperation with the province of New Brunswick, and in consultation with the federal government, as both jurisdictions clearly have some role to play in regulating activities such as tidal energy projects in the Bay of Fundy. In this sense, it is an example of interjurisdictional cooperation, particularly between the provinces.

Nova Scotia and New Brunswick cooperated in the gathering of background information to inform their respective processes. Each jurisdiction then conducted its own process and made its own separate decisions. The role of the federal government in the SEA could have been much stronger, particularly given its important regulatory role with respect to fisheries and transportation in the Bay of Fundy. It essentially played an observer role, even though it was requested by the provinces to become more actively involved.

The Nova Scotia process was placed in the hands of an independent organization, the Ocean Energy Environmental Research Association. The Association designed a process that consisted of the following key components:

- An interactive web site to provide information and seek input throughout the SEA (www.bayoffundysea.ca).
- A newsletter published regularly throughout the SEA process and posted on the SEA web site.
- Regular meetings of the SEA steering committee to guide the process.
- A consultant hired to serve as the "process-lead" for the SEA process, including chairing public meetings and writing the final report under the direction of the OEER subcommittee.
- Six community forums held in August 2007 in affected communities in Nova Scotia.
- Two rounds of participant support funding for community-based research and to provide opportunities for community groups to meet and discuss their perspectives about the potential arrival of this new industry in the Bay of Fundy.
- A background report prepared by an environmental consulting firm on the current state of knowledge of the various proposed technologies and the receiving environment, and the potential socioeconomic impacts of renewable energy development in the Bay of Fundy.
- A round table of about 25 interested stakeholders that met with members of the subcommittee approximately once a month between October 2007 and April 2008.

Early efforts to engage Nova Scotians were designed primarily to identify key issues to be addressed through the SEA process. The SEA SC decided that the scope of the SEA would be limited geographically to the Bay of Fundy and substantively to ocean renewable energy. The process remained open throughout to any issue relevant to informing decisions about whether, where, and under what conditions offshore renewable energy should be permitted and encouraged in the Bay of Fundy. The main purpose of the various efforts early in the process to engage affected communities and key stakeholders was to identify what issues the SEA should focus on, while leaving it open to participants throughout the process to raise new issues and to bring up new concerns.

The outcome of the SEA process was a consensus report of the stakeholder round table. The consensus was possible in spite of the short time frame because all of the participants shared full control over the scope of the SEA, the process, and the outcome. The process was assisted by the fact that there was a general recognition by all participants that while tidal energy posed risks to existing uses and natural systems in the Bay of Fundy, it offered the potential to provide a long-term sustainable supply of energy to Nova Scotians.

The focus of the SEA process and of the final report was on how to better understand and minimize negative impacts, how to determine whether it would offer net long-term benefits to the province, and how to ensure a fair distribution of impacts, benefits, risks, and uncertainties. The report concluded that the development of a tidal industry in the Bay of Fundy should be guided by a number of sustainability principles developed specifically for the Bay of Fundy context. A key principle was that the development of the tidal energy industry should only proceed incrementally, and that key issues had to be addressed at each step before deciding whether to proceed to the next level of development.

As a starting point, the SEA concluded that pilot projects could be tested in the Bay of Fundy under specified conditions. Key among the conditions was that pilots be used to fill information gaps in understanding the interaction between tidal turbines and the receiving environment in the Bay of Fundy. A second key condition was that turbines tested at this stage could not remain in the water for more than two years.

For more information on the process, see www.bayoffundysea.ca. For an assessment, see Doelle (2009).

MEINHARD DOELLE

Schulich School of Law, Dalhousie University, Halifax, Nova Scotia, Canada

8.2.3 Integrating SEA and Planning

When I was a doctoral student working on my dissertation, my research focused on integrating the principles of Strategic Environmental Assessment (SEA) into local land use comprehensive plans. The study area for my research was California, USA, which has a well-developed planning system and very strict environmental regulations at the state level. I, therefore, thought that I might receive very positive results for my hypotheses. I had hypothesized that the principles of SEA had been thoroughly integrated into local land use plans. However, after I reviewed the data, I found that this was not the case! Many jurisdictions' general plans had not been updated for many years. A few general plans were even older than me! I was shocked by these findings. I then sought to determine the factors influencing the limited integration of SEA and local plans.

The extent to which SEA and local comprehensive plans are integrated tends to depend on three factors: (1) *Motivation*: local decision makers have varying levels of motivation regarding sustainable development, economic development, environmental quality, and social equity. Most of the time immediate economic development and related job increases are a higher priority for local land use planning decision making than long-term, strategic, cross-boundary environmental decisions. (2) *Power and authority*: do local planners possess the power and authority to incorporate strategic environmental considerations into their local planning frameworks? The answer is yes! They have the power and authority to address uncertain, long term, regional environmental concerns. They have leadership, regulatory, spending–taxation, acquisition, and coordination powers, which enable them to fully participate in decisions relating to strategic environmental issues. (3) *Barriers and constraints*: local planners operate within a real and complex political decision system, characterized by numerous decision-making barriers and constraints. Local planners may not be adequately aware of or sufficiently knowledgeable about some strategic environmental issues, such as climate change. In some cases, the interests of planners may clash with those of other agency decision makers. In addition, planning for uncertainty is a major challenge for local planners when thinking about strategic environmental problems. Institutional constraints derive from spatial, temporal, and organizational challenges. Decision makers may deny planners' recommendations for such reasons as "not in my yard," "not in my term," or "not my business." Lastly planners are faced with technical, personnel, and budgetary challenges. With limited numbers, planners tend to focus on the most urgent, necessary, and immediate environmental challenges rather than uncertain, remote, and strategic environmental issues.

What kinds of strategies could be used to improve the integration of SEA and local land use planning decisions? A good departure point would be to: (1) improve planners' awareness of strategic environmental issues (biodiversity, ecosystem, climate change, etc.); (2) update local comprehensive plans regularly; (3) provide more technical assistance (e.g., GIS); (4) collaborate with other stakeholders; and (4) promote citizen involvement. This book may provide additional solutions.

ZHENGHONG TANG

Community and Regional Planning Program, University of Nebraska-Lincoln, Lincoln, NE, USA

8.3 DEFINING THE PROBLEM AND DECIDING ON A DIRECTION

The three stories presented in the preceding section, approach the question of democracy in contrasting ways. The first story describes how a project-level EIA was democratically ineffective in influencing decision making in favor of more environmentally sound choices. The second story provides an example of how collaborative, consensus-based, and democratic higher level planning can help direct and bound planning and decision making at the project EIA level. The third story illustrates how the integration of SEA and planning is a more complex, and sometimes constrained, endeavor than it is often assumed or understood. The stories provide an initial and partial sense of how democratic values should and should not be approached in IA process design and management.

The value of public participation in the IA process is widely acknowledged. Public participation leads to better-designed projects, an enhanced understanding of environmental conditions, more accurate predictions, a more focused IA process (on stakeholder issues and reasonable alternatives), and more effective impact management (Barrow, 1997; Bisset, 1996; Hughes, 1998; Lee, 2000). Public participation clears up misunderstandings and resolves conflicts (Barrow, 1997; Lee, 2000). The public is more likely to accept and support projects when there is high degree of public involvement in the IA process (Barrow, 1997). Approval is expedited. Proponents are more likely to achieve their objectives (Lee, 2000). Decisions are viewed as more legitimate, both by regulators and by the public (Barrow, 1997; Hughes, 1998). The public is better informed about the project. Public issues and concerns are more likely to be addressed (Hughes, 1998). Public participation skills are enhanced (Barrow, 1997). Decision making becomes more open, transparent, and less susceptible to lobbying by vested interests (Hyman et al., 1988). Everybody benefits.

This apparent consensus is more apparent than real. Participation can be defined in many ways. Perspectives vary greatly regarding the types and degrees of participation, which are necessary and appropriate. Suspicions concerning motivations abound. A seemingly shared interest in participation often masks deep ceded value and interest differences. Absolute positions, not conducive to consensus building or conflict resolution, are commonplace. The positions and perspectives of other parties are often misunderstood and overstated. The major parties frequently differ dramatically in their assessments of the effectiveness of the participation measures applied in IA practice.

Proponents, Politicians, and Public Officials have a tendency to be wary of public participation. They often see it as costly and time consuming (Kreske, 1996). They commonly are concerned that it will raise their profile, impede their relationships with authorities, and exacerbate rather than resolve conflicts (Bisset, 1996; Morgan, 1998; Thomas and Elliott, 2005). Many engage in public participation when it seems necessary for regulatory compliance and when it facilitates proposal approval and implementation. They are often adamant that they cannot share decision-making authority (UNEP, 1997). They shy away from conflict (Lee, 2000). They tend to be more comfortable with closed processes, or at least processes that are not opened up until major decisions have been made (Glasson et al., 1999). They generally give greater weight to technical analyses (Morgan, 1998). They sometimes characterize public inputs as subjective and inconclusive (Morgan, 1998). Some are quick to label concerned groups and individuals as ill informed, unrepresentative, and selfish (i.e., "Not In-My-Backyard," NIMBY) adversaries (Canter, 1996). Such proponents see the role of public participation as educating the public, neutralizing the opponents, and legitimating the process (Gerrard, 1995). They frequently are skeptical of collaboration. They tend to favor lobbying (of both regulators and politicians) and coercive measures (such as expropriation) for obtaining approvals (Gerrard, 1995).

Regulators often encourage participation, albeit within tightly circumscribed limits. Government agencies in many jurisdictions have a tradition of administrative discretion, secrecy, and limited public involvement (Glasson et al., 1999). Political interest groups often lobby government agencies and sometimes have considerable influence (Hyman et al., 1988). Public decision making is becoming more open and transparent, in part, because of IA requirements and procedures (US CEQ, 1997a). But many regulators remain reluctant to more fully open up policy and decision-making procedures. They also tend to have a limited understanding of the roles that SEA, in particular, can and should play in decision making (Noble, 2009b). Government agencies are commonly segmented and mission oriented. IA responsibilities are often secondary functions. Government officials, with IA review responsibilities, rarely have the time to address matters not immediately relevant to administrating regulatory requirements. Personal advancement is much more closely tied to meeting the desires and requirements of senior government officials and politicians than to satisfying the concerns of interested and affected members of the public. Sometimes societal goals are displaced by internal organizational goals such as organizational survival or expansion (Hyman et al., 1988). Government agencies often display a propensity toward centralized control, resulting in an inherent tension with the decentralizing tendencies of public involvement (Canter, 1996). Regulators tend to be comfortable with technical knowledge and less comfortable with conflict and confrontation (Canter, 1996). Regulators, in common with proponents, often preclude any possibility of shared or delegated decision making (Kreske, 1996).

IA Practitioners are often placed in the unenviable position of attempting to meet regulatory requirements, operating within time and budget constraints, and not offending proponents (March, 1998). At the same time, they attempt to adhere to good practice standards, including those related to the role of public participation in the IA process. In the ongoing effort to "juggle" these competing demands, they have a tendency to "control" the process, at least to the extent necessary to adequately perform their responsibilities (Maynes, 1989). They may be reluctant to apply "unproven" methods or to engage in open-ended processes where outcomes are difficult to predict or manage (Solomon et al., 1997). They have a tendency to view public participation as the proficient application of the appropriate suite of methods. They are generally in favor of public participation. They often support the use of or perform the role of mediators, facilitators, and similar procedural specialists to address conflicts as they emerge (Bisset, 1996; Fischer, 2007b). They tend to view local control as unrealistic, except in a very limited range of circumstances (e.g., local community infrastructure) (Morgan, 1998). As technical specialists, they usually favor technical over non-technical knowledge (Canter, 1996). They continue to embrace technical–scientific rationality as the ideal rather than acknowledging that they operate in a world of contested rationality (Richardson, 2005). They tend to assume and favor a rational planning IA model. IA practitioners tend to neglect the political aspects of IA, and often are naïve regarding the exercise of power in IA practice (Cashmore et al., 2010; Kørnøv and Dalkmann, 2011). They tend to give too little a consideration to their potential role in facilitating democratic effectiveness (Stoeglehner et al., 2009). Instead, they depoliticize political contests, which can reinforce existing power relationships (Cashmore et al., 2010).

The *public* generally acknowledges that public participation in the IA process can be beneficial. But they often are highly skeptical about whether and how often those benefits are realized in practice. They seldom see major differences among proponents, regulators, and practitioners. All are frequently seen as arrogant, patronizing, manipulative, and unworthy of trust or support (Arnstein, 1969; Gerrard, 1995; Parenteau, 1988). Proponents often are seen as relentlessly pursuing the implementation of predetermined proposals, regardless of environmental and local community consequences. Regulators are viewed as consumed with administrative matters and as highly susceptible to lobbying by vested interests (Maynes, 1989). Practitioners are seen as expeditors who work on behalf of proponents to meet minimal regulatory requirements. None of these parties is expected to seriously consider public concerns.

The public often sees itself as powerless or, at best, fulfilling a marginal role in the IA process (US CEQ, 1997a). They note that they are commonly treated as adversaries (US CEQ, 1997a). Decisions have generally been made before they are involved. They often have only a vague notion of the nature and role of SEA (Fischer, 2007b). They see communications as one way (US CEQ, 1997a). They often see their role in the process as tokenism and the IA process as a diversion away from where the real action is (Thomas and Elliott, 2005). They argue that the time allocated for their participation and the resources provided (if any) to facilitate their participation are so limited as to preclude even the semblance of "meaningful participation" (Thomas and Elliott, 2005). Frequently, they believe that they have no control over either the process or its outcomes. Occasionally, they hope that the IA process can be a tool for defending themselves against aggression and a means of redistributing power (Torgerson, 1980). But power is so concentrated in existing structures that the changes wrought through IA are usually seen as minimal (Torgerson, 1980). Sometimes they insist that they are being manipulated or co-opted. Often, they expect that, by the end of the process, power will be even more centralized in distant authorities. They often assume that agencies collude with proponents in subverting the intention of IA requirements (Burdge, 2004). The IA process, they commonly assert, does not prevent the approval of environmentally unsound projects (Mittelstaedt et al., 1997). Often the outcome, they point out, is the unfair distribution of facilities and services. Frequently, they conclude, it is the most vulnerable who bear the greatest burden and receive the least benefits (Gerrard, 1995).

Commentators on IA practice echo many of the concerns raised by the public. Some suggest that IA practice is increasingly open and collaborative. Others argue that IA requirements and practices are becoming less democratic (Ayre and Calloway, 2005). They acknowledge the many benefits that have accrued from increased participation by interested and affected parties in the IA process. But they also suggest that participation typically occurs too late in the process, is too infrequent, and is too narrowly defined (Devlin and Yap, 2008; Morgan, 1998). They suggest that IA processes, all too frequently, are shaped and sometimes corrupted by narrow political interests (Rickson et al., 1990a; Smith, 1993). Public participation in IA tends to reflect rather than alter the social distribution of power (Devlin and Yap, 2008). They argue that IA is too loosely connected to decision making. Consequently, all too often, the information and knowledge provided by IA is not used by those with real power who make decisions (Deelstra et al., 2003). They point to the disconnect between environmental and economic governance (Ayre and Calloway, 2005). IA practitioners are characterized as, at worst, compromised and at best, naïve regarding IA politics (Craig, 1990). They suggest that IA practitioners can contribute to making IA less transparent and less democratic by emphasizing the primacy of technical–scientific knowledge and methods (Binder et al., 2010; Persson, 2006). Social, cultural, and procedural issues, they argue, receive too little attention. Community power implications, they maintain, are neglected (Devlin and Yap, 2008; Thompson and Williams, 1992). Technical and quantitative concerns and methods are, they assert, overstressed (Solomon et al., 1997). Stakeholders' interests and conflicts are, they suggest, obscured (Persson, 2006). As a consequence of the neglect of politics and decision-making influence in IA practice, they conclude, IA findings tend to be marginalized in favor of nonenvironmental objectives and political considerations (Morrison-Saunders and Sadler, 2010; Jay et al., 2007).

Excessive reliance, they point out, is placed on a narrow range of public consultation techniques (Solomon et al., 1997). Insufficient attention, they argue, is paid to interactions among different forms of participation (O'Faircheallaigh, 2010). Too little consideration is devoted, they assert, to the role of IA (especially SEA) in questioning alternative value systems (Bonifazi et al., 2011). Public involvement guidance is, they conclude, highly variable, too general, and often unclear (Hughes, 1998). IA regulators, as reflected in IA requirements and guidelines, are admonished for not offering sufficient practical and up-to-date public involvement guidance and for not consistently and proactively supporting public participation (Hughes, 1998). IA practice commentators tend to be especially critical of the persistent use of coercive and reactionary (i.e., decide, announce, defend) involvement and implementation methods (Armour, 1990b; Halstead et al. 1984; Rabe, 1994). They point to the repeated failure of such approaches (Solomon et al., 1997). They describe NIMBY as a natural, reasonable, valid, and often constructive reaction to (1) a major threat to individuals and communities, (2) autocratic decision making, (3) a fundamental imbalance in the distribution of costs and benefits, (4) a legacy of poor communications, biased analysis, and inept management, (5) a failure to adequately address social impact, uncertainty, dread, stigma, and perceived risk concerns, and (6) belated, partial, and unduly restricted efforts to involve interested and affected parties (Mazmanion and Morell, 1994).

The commentators note (or, at least, hope) that, on a broader political front, the trend is toward more transparent, democratic, decentralized, and accountable decision making; greater access to information; more explicit and understandable documentation; earlier, more continuous, more learning oriented, and more collaborative public involvement, and the more direct involvement of nongovernmental organizations (NGOs) (Ayre and Calloway, 2005; Bisset, 1996; Glasson et al., 1999; Lee, 2000; Wood, 1995). IA, they argue, should be at the forefront rather than lagging behind such trends (IAIA and IEA, undated; Interorganizational Committee, 1994). Less clear are the unresolved issues of the appropriate roles of power and influence in IA practice (Hanna, 2005) (see Chapter 3).

IA practice commentators have devoted particular attention to the positive and negative experiences associated with siting locally unwanted land uses (LULUs) (see Chapter 4). Although the issues are complex and the experiences mixed, they point out that there have been several notable "successes" when siting processes have been built around a blending of local control, social equity, and shared management (Gerrard, 1995; Rabe, 1994; Seley, 1983). They contrast these experiences with the more uniformly negative experiences associated with autocratic, technical, top-down, and coercive siting approaches (Binder et al., 2010; Halstead et al., 1984; Solomon et al., 1997; Wolsink, 2010).

The conclusions of IA commentators are reinforced by the results of IA effectiveness reviews. These reviews consistently conclude that public involvement occurs too late and too infrequently in the IA process (Sadler, 1996; US CEQ, 1997a). Institutional provisions for public participation are highly variable. They are rated as excellent to good less than half the time (Sadler, 1996). More creative outreach is needed (US CEQ, 1997a). The performance of public involvement is as bad as often as it is good (Sadler, 1996). Public involvement has a significant influence on decision making only about one quarter of the time (Sadler, 1996). Clearly, the treatment of public involvement in the IA process, from a practitioners' perspective, is falling well short of its potential.

How best to determine the direction in light of these differing perspectives on the problem? One approach is to decide that there is no problem (i.e., the IA process is already sufficiently democratic). Many proponents would likely favor this approach on the grounds that they are already "bending over backwards" to facilitate public consultation. They also might argue that the appropriate form of public involvement is highly dependent on the situation. Flexibility is, therefore, essential. Practitioners and regulators, as illustrated by the IA effectiveness reviews, would suggest that there remains considerable room for improvement. The public and IA commentators would likely maintain that the status quo is more than inadequate—it is unacceptable. Overall, the case for no change in current practices seems highly dubious.

A second approach is to decide that all that is required is "polishing" and refining current practice. There should be more and earlier opportunities for public involvement. Requirements and guidelines should be clarified and should offer more practical advice. A wider range of involvement methods should be applied. Regulators should proactively support public participation. Public involvement methods should be used more effectively. Most regulators and practitioners would likely favor this approach. The public and commentators would probably see this approach as necessary but far from sufficient. They would argue that such changes would not instill trust or alter the peripheral decision-making role of the public. The public would continue to have limited influence over the process or its outcomes. It would not be able to adequately defend its interests or to ensure that its concerns are properly addressed. Others would decide its fate in processes where democratic decision making is more illusionary than real. This approach is not likely to appreciably narrow the gulf between the public and other participants in the IA process.

Assuming that the first two approaches are, respectively, unacceptable and insufficient, only two possibilities remain. One is to identify the problem as an undemocratic IA process. The only appropriate response is to delegate or share decision-making authority with the public. Delegating or sharing decision-making authority with the public is the heart of the IA process presented in this chapter. The second possibility is to accept that refining current public involvement practices is not enough but also to expect that there will be few situations where decision-making authority can be delegated to or shared with the public. The appropriate response is, therefore, a collaborative IA process. The public is an active and ongoing participant in a collaborative IA process. But final decision-making authority continues to reside with proponents and regulators. A collaborative IA process is presented in Chapter 9.

8.4 SELECTING THE MOST APPROPRIATE ROUTE

8.4.1 Definitions and Distinctions

Democracy is rule by the people, either directly and/or indirectly through periodically elected representatives. It is both a system of government and a political theory. It is a collective, social, and political (i.e., concerned with the exercise of power) endeavor (Nagel, 1987). It presumes or at least aspires toward such values and ethical principles as freedom of speech and assembly, equality of opportunity, minority rights, and majority rule. Democratic theory presumes that all citizens are or can participate equally in decision making and can be equally influential in the political system. Where this is not the case, democracy has a responsibility to correct power imbalances. Power is the ability of a person or group to know about, mobilize, and influence decision making (Burdge, 2004). Direct democracy is the ideal but is not always practical. Representative democracy is a compromise where there is a large population base and specialized roles. This does not mean that direct democratic principles, such as involving those affected by decisions, continuity of involvement, consensus building, discussion, action, and community need be abandoned (Nagel, 1987; Pateman, 1970). Instead, IA practitioners should actively seek to ensure that direct democratic principles are expressed and, wherever practical, fulfilled. The onus should be on those seeking to circumscribe the application of direct democratic principles to demonstrate why such limitations are essential to good and effective governance.

Figure 8.1 is a highly simplified illustration of key democratic participants and interactions. The ultimate source of authority, for the courts, politicians, and government officials, is the people. The courts, politicians, and government officials should all be responsive and responsible to the people. Ideally, they should share and delegate authority back to the people. Practical opportunities for direct democracy should be considered. The public should be fully informed of and involved in all decisions likely to affect them.

These ideals are commonly transgressed in practice. Politicians and government officials are often reluctant to inform, much less involve, share or delegate decision-making authority back to the people. Frequently, a tension develops between democracy and bureaucracy (the world of government officials) and between politicians (who believe that they are in charge once elected) and the public (who wish a continuing say in matters that affect them between elections). Increasingly, democracy is now seen as a continuous and dynamic process where even if governments retain the final authority, close public scrutiny is essential (Gilpin, 1995). Arguably, the trend is toward (or should be) a more participatory form of democracy. In a participatory democracy particular attention is devoted to the relationships between individuals and authority structures (Pateman, 1970). Public involvement is a right (Lee, 2000). The public fully understands the problems being addressed and the means proposed for addressing the problems (Canter, 1996). Public involvement occurs earlier. It is more frequent. It is sometimes continuous. The public has a direct and acknowledged role in decision making (Lee, 2000).

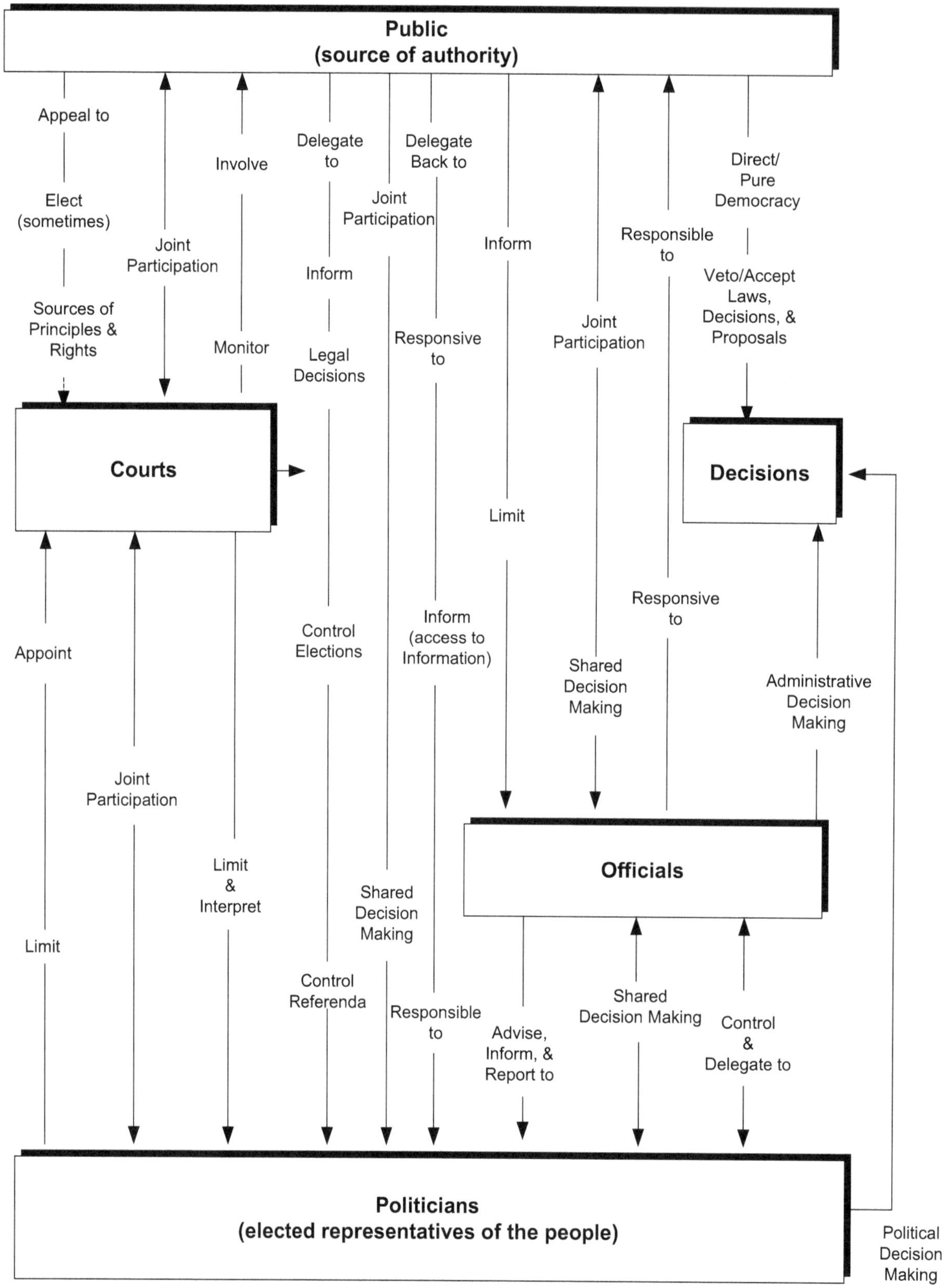

Figure 8.1 Examples of democratic participants and interactions.

The public has influence and, where practical, control over the decisions that affect them (Nagel, 1987).

Democratic IA practice should be viewed as a microcosm of participatory democracy (i.e., IA as a relational and pluralist construct) (Bonifazi et al., 2011). It should emphasize the right of people to participate in formulating, implementing, and evaluating actions that might affect their lives (Kemm and Parry, 2004b). There should be frequent, preferably continuous, public involvement provisions, full access to information, the right of appeal to an independent third party, the full involvement of interested and affected parties, and an explicit decision-making role for the public. IA is inherently political, involving value judgments about the relative importance of competing environmental receptors (Nilsson and Dalkmann, 2010). Arguably, IA practice has a responsibility to expand and extend democratic participatory tenets. It should contribute to political awareness, competence, and the capacity to influence decision making (Bonifazi et al., 2011). The protection of the environment is a critical democratic responsibility. Proposals, subject to IA requirements, invariably directly affect the environment and peoples' day-to-day lives. Thus, there is a democratic responsibility to ensure that people have a major say in the decisions that affect them. Impacts tend to occur at a site-specific and community-scale level. A community scale is often conducive to applying direct democratic principles and methods. Power and influence are seldom equally distributed among IA process participants. Therefore, power inequities must be offset. Environmental costs and benefits are rarely equally distributed. A proactive effort is required to ameliorate environmental inequities.

The IA process should be an expression of participatory democracy. It should emphasize issues of power and dominance (McCluskey and João, 2011). It should draw upon, adapt, and contribute to democratic theories (Bonifazi et al., 2011). It should be explicitly engaged with decision making and ensure influential decision-oriented IA (Nilsson et al., 2009). As illustrated in Figure 8.2, it should be built on a foundation of participatory democratic concepts. It should integrate democratic principles, methods, and decision rules. The public should have a major say over process activities, inputs, outputs, and events. The people should shape and guide the process. Particular attention should be devoted to the decision-making role of affected people and communities. Steps should be taken to prevent and ameliorate power imbalances. Every aspect of the process should be conducted within the boundaries of participatory democratic limits and requirements.

8.4.2 The People Decide

Letting the people decide is a simple concept. Implementing the concept is more complex. Do the people decide directly or indirectly through their representatives? Which people decide? Are the people's representatives members of community associations and other nongovernmental organizations or are they elected representatives? What are the roles of different levels of elected representatives? Where do nonelected government officials fit in? If the goal is to retain decision-making control as close to local people as practical in an IA process, as illustrated in Figure 8.3, members of the public most directly affected by the proposed action take the lead with the active participation of locally elected politicians. Alternatively, a team of ordinary citizens could take the lead. If the plan or agreement reached by the team receives broad public support, it is likely that political support will follow (Todd, 2002). As one moves down the ladder, decision-making control becomes progressively less direct and increasingly less subject to the control and influence of the local public and their representatives. The argument will be made that there are broader constituency interests represented by senior-level politicians and specialized areas of expertise only possessed by senior government officials. However, if the ideal is "rule by the people," it must be demonstrated why it is necessary to move down the ladder. Senior governments are also obliged to contribute to local capacity building (to facilitate local control), to ensure that the actions of regulatory officials are subject to local scrutiny and appeal, and to decentralize government operations wherever practical (to maximize local contact).

Letting the people decide also entails determining how decisions are to be made. Should they be made directly through referenda (e.g., local veto or acceptance)? Who sits on committees? Should decisions be made based on majority rule or is a consensus preferable? Should decision-making authority be delegated to an independent third party (e.g., an arbitrator, an environmental court, a public inquiry) (Westman, 1985; Halstead et al., 1984)? Should an advocate, a public defender, or peer reviewers assist the public (Kasperson et al., 1984)? Should alternative dispute resolution (e.g., mediator, facilitator) aid decision making? Should the local people decide by themselves or should decisions be made jointly with other parties?

Decisions must also be made regarding which IA process choices will be made by or shared with the public. Figure 8.4 identifies examples of potential choices. The potential choices encompass the major decisions leading up to approval and postapproval choices. It is necessary to determine for each decision when a community has decided (e.g., majority rule, more than majority rule, majority rule for more than one geographic sub-area, consensus among representatives). It also is necessary to decide if the range of choices available to the public should be bounded. The public, for example, could be permitted to choose among a set of alternatives all considered to be environmentally, technically, and financially acceptable. Alternatively, choices by the public could be subject to acceptability confirmation. Sometimes public choices are limited to certain project types (e.g., local community infrastructure) or to predefined geographic areas. The goal should be a consensus among the major parties concerning the public control "rules and boundaries."

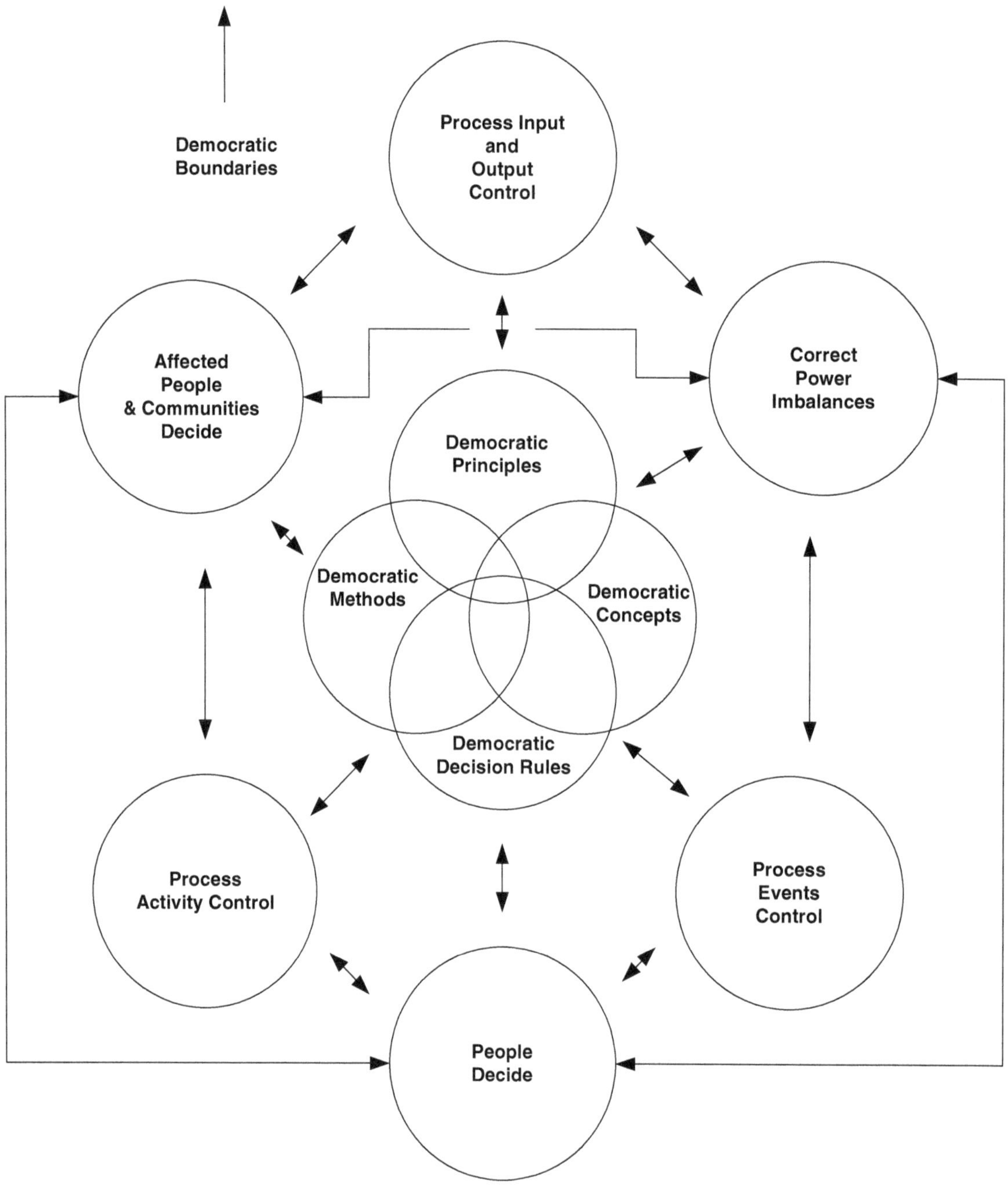

Figure 8.2 Democracy and the IA process.

The distinction between local and shared public control has been addressed in natural resource management. Community-based natural resource management occurs when the community is allocated ownership or authority (a form of delegated decision making) for natural resource management. Local populations arguably have a greater interest in sustainable resource use than do distant government officials. Co-management runs the spectrum from almost complete self-management to almost total state management. It has been widely applied in forest, fisheries, and wildlife management, especially with indigenous peoples. The parties make trade-offs among themselves and may adopt a legal agreement, which shares the legal authority for resource management (Harvey and Usher, 1996). Provision is made for funding, training, and staffing to support the partnership (Mittelstaedt et al., 1997). Co-management tends to work best when there is community consensus, a credible lead agency, a clear mission for the partnership, meaningful delegation, meaningful tenure, revenue autonomy, meaningful inclusion of interests, and reliance on existing structures (Harvey and Usher, 1996). Lessons and insights from both community-based resource management and resource co-management could greatly assist the design and management of democratic IA processes (Todd, 2002).

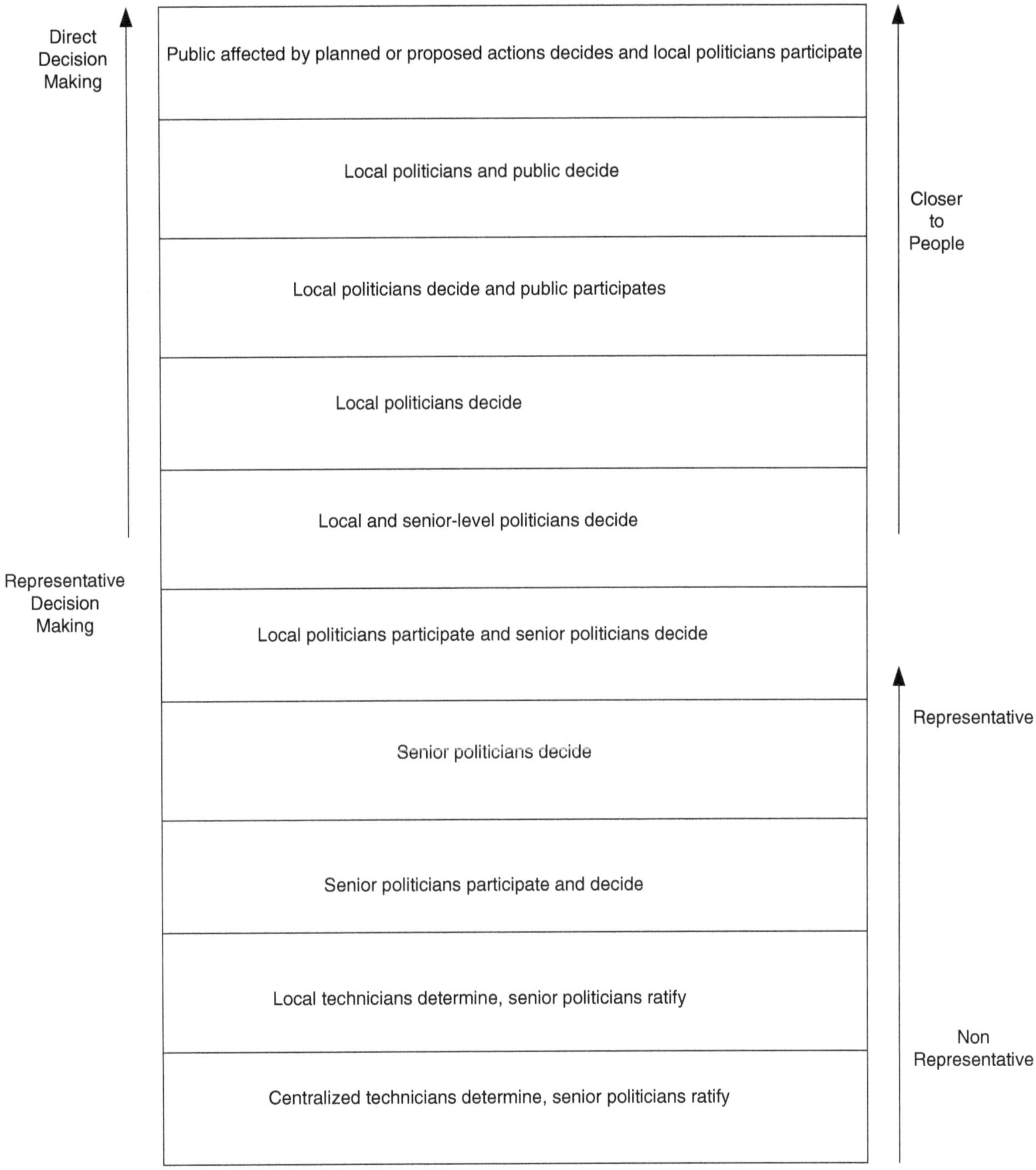

Figure 8.3 Ladder of democratic control.

Democratic, community-based IA processes are sometimes applied (Sinclair et al., 2009). Citizens are centrally located in the process (Armour, 1990b). The public is a partner with other parties (Lee, 2000). IA practitioners act as facilitators and collaborators (Armour 1990b). The parties jointly solve problems, control the process, conduct analyses, and reach decisions (preferably a consensus) (Lee, 2000; UNEP, 1997; Westman, 1985). The collaboration often extends through design and implementation (Lee, 2000). Continuous (e.g., committees) and interactive (e.g., workshops, conferences) techniques are applied in an ongoing effort to reach and retain consensus and to maintain contact with and support from broader constituencies.

Democratic IA processes have been fully developed and tested in the voluntary siting of LULUs (see Chapter 5, Section 5.6).

8.4.3 The People and Communities Affected by the Proposed Action Decide

A democratic IA process assumes that people and communities affected by proposed actions are willing and able to

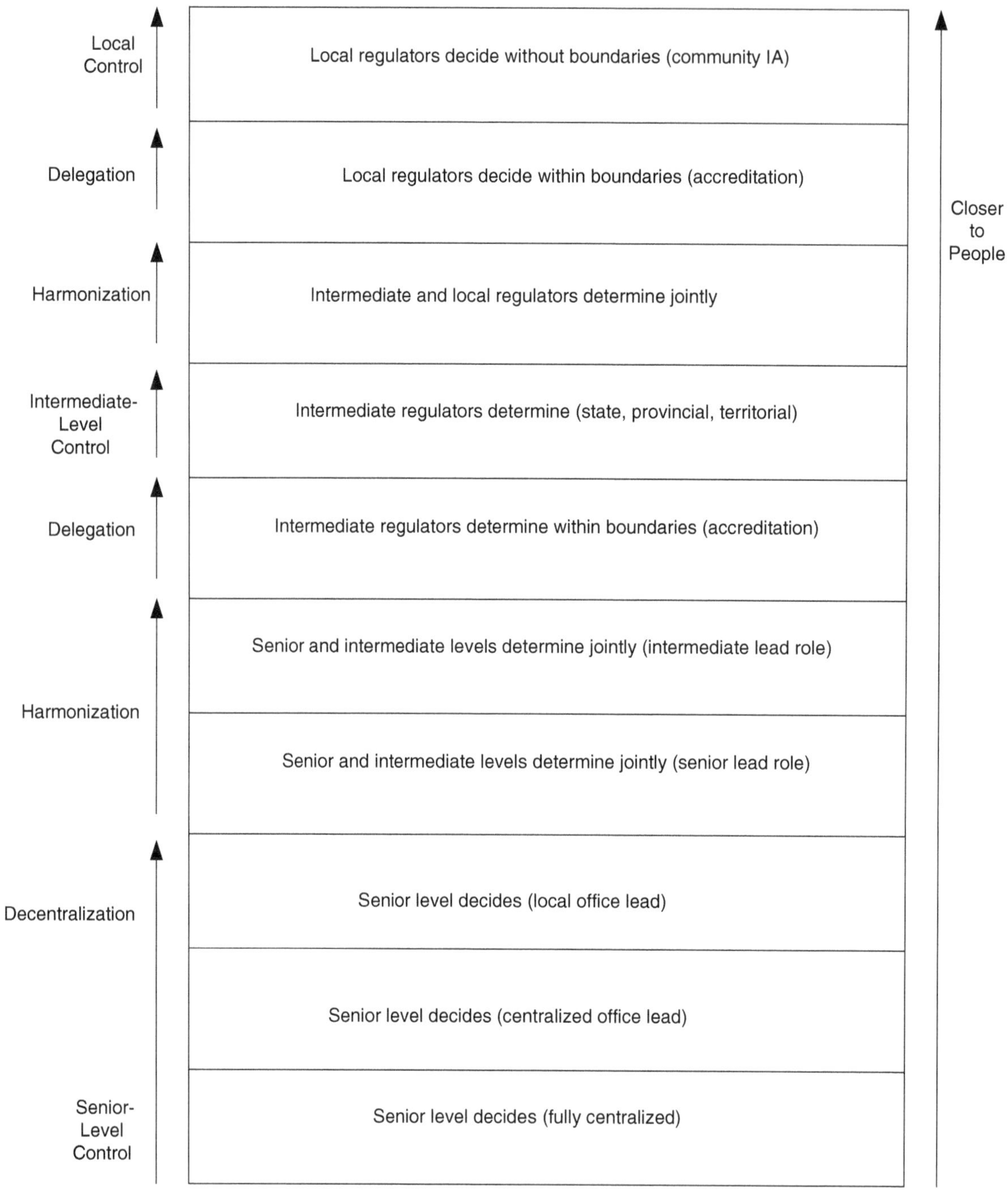

Figure 8.4 Ladder of regulatory control.

make the key decisions, alone or in a partnership arrangement. Some might argue that this is not a realistic or appropriate assumption. Table 8.1 highlights key characteristics of several concepts related to how people and communities can become more autonomous and thereby better able to make decisions.

Bioregionalism and communitarian approaches provide visions, principles, and strategies for more autonomous communities. IA practitioners (at both the regulatory and applied levels) can work with the public and communities in developing, refining, and applying the lessons and insights from such concepts. "Bottom-up" approaches to community autonomy, such as empowerment, community development, and mobilization, can be instructive to community and environmental activists and organizers. IA practitioners can encourage, support, and facilitate such efforts. Community-based IA illustrates how key parties in the community can collectively collaborate, in consensus building forums, to critically reflect on complex issues and reach consensus-based decisions. Concepts such as traditional knowledge and lay science, contribute to a community-oriented knowledge base—a knowledge base conducive to community

Table 8.1 Example of Potentially Relevant Local Autonomy Concepts

Bioregionalism	Focuses on developing self-reliant economic, social, and political systems Seeks to develop a territory's fullest potential through reliance on systems of production that draw on local resources, do not degrade the ecosphere, and require consideration of long-term implications as compared with short-term gains Requires basic changes in beliefs, attitudes, and values concerning human–natural environmental interactions Political power most effective at local level; leads to personal and community empowerment Stresses value of interdisciplinary analysis, experiential knowledge, and social learning Proposals viewed as catalysts for positive community and regional change Built around ecological principles; creates sustainable systems of production
Empowerment	People planning for themselves (self-respect, reliance, and determination); builds on community knowledge and capacity to organize and act; planning as a social struggle A multidirectional effort by grassroots groups to secure influence and control over procedures that affect their lives; challenges conventional power relationships; involves appropriating, extending, exercising, negotiating, and mobilizing popular support Participation leads to community empowerment and community improvement Ladder of empowerment (bottom to top): atomistic (individual unit), embedded individual (individual within larger settings or structures), mediated empowerment (in context of relationship between expert and client), sociopolitical empowerment (links individual to community through collective social action and challenging of oppressive institutional arrangements), and political empowerment (community or group the locus of change, operationalized through policy changes and access to community resources) Appropriates IA (especially SIA) to community priorities; facilitates community participation and control; extends into less formal settings where community influence is greater; negotiated participation in territorial campaigns for more acceptable local outcomes and mobilization of popular support Role of practitioner to establish mechanisms that allow for true participation and influence; responsibility shifted to affected people and community; planning collaborative and political; guided by ethic of empowerment
Community-based IA	Participatory and structured approach to EIA and SEA Use of workshops to reduce power differentials among participants Example techniques within workshops—visioning, brainstorming, and critical reflection Adapts approach to context Approach contributes to critical reflection and helps build democratic capacity in community and national institutions Benefits—meaningful engagement of community members in decision making, individual and social learning, and facilitates transition to sustainability
Community development	Process in which people in a community define their wants and devise means to satisfy them Concerned with process and outcomes Community building carried out by activists and community-based organizations Builds on local distinctiveness, escapes constrictions of local traditions, draws in outside opportunities, and limits potential for domination and exploitation Opposes dominating forces, reflects on experiences, overcomes barriers, and builds coalitions of support Reshapes boundaries and prevailing models; value-laden and political decision-making process Emphasizes social utility
Mobilization	Argues that people can be mobilized for political activities or special interest associations People plan and develop own projects People learn from own and other experiences, develop contacts with external institutions for resources and technical advice, but control how resources are used Counters efforts by industry and government to use IA to achieve scientific and technical legitimacy; raises questions about scientific legitimacy and environmental rationality IA as a catalyst for political mobilization; at the core of political debate and decision making IA a microcosm of a more democratic, culturally diverse and litigious society; IA a forum for democratic debate
Communitarian	Focuses on self-management practices of small communities; stresses the social side of human nature Belief that the human community is the best form of human organization for respecting human dignity, for safeguarding human decency, and for facilitating a way of life open to self-revision and shared deliberation Stresses individual human dignity and increased social responsibility Each community develops own agenda; not majoritarian but strongly democratic Seeks to make government more representative, participative, and responsive to all members of communities Supracommunity—a community of communities (the community of humankind) Actively maintains institutions of a civil society; communities and polities have a duty to be responsive to their members and to foster participation and deliberation in social and political life

(*continued*)

Table 8.1 (*Continued*)

Traditional knowledge	Generally refers to environmental knowledge possessed by indigenous peoples; more a way of life than a knowledge base; knowledge inseparable from people
	Integrates social, ethical, cultural, technical, scientific, and historical in a single process (within own context)
	IA should accommodate traditional knowledge based on trust, respect, equity, and empowerment
	Holistic; spiritual at the core
	Goal is co-existence rather than integration
	Different for each situation; flexible and fluid
	Often misunderstood and misapplied when viewed outside of indigenous contexts
Lay science	Interpretation considered a social act; cannot be separated from social context
	Lay research and discovery as an antidote to industry-dominated science
	Environmental professionals realize that there must be interaction between scientist and community; professionals help community volunteers participate in data acquisition and interpretation
	Lay participation in science (e.g., monitoring); basic training provided; residents and workers able to test and defend their own common sense
	Empowers community groups to take part in science and technology policy decisions
	Helps reduce public fears, increases public understanding, and contributes to better science
Analysis of power structure	Part of a political approach to IA and SIA
	Recognizes that proposed actions affect community power structures and stratification systems
	Analyzes effects on local autonomy, power of various factions, vertical linkages, local leadership structure, and processes of distributive politics
	Emphasizes issues of power and dominance
	Presumes political rationality (nonsequential, multiple interdependencies, extensive negotiations, and compromise)
	Identifies how power is exercised (e.g., social order, systems bias, systems of thought, tacit knowledge, reification, discipline, coercion)
	Considers degree roles, institutions, and values transformed and continuing success in providing structure and meaning to community life
	Outcomes complex and dependent on history and context
	Analysis can support management actions
	If community vitality and cohesion undermined could inhibit ability to cope with and manage change; if enhanced, communities in a better position to bargain with proponents and regulators and to control own fate
Conflict theory	Oriented toward the existence and identification of interest groups
	Concerned with how conflict affects decision making; a key test the ability to control the agenda
	Focuses on the exercise of power—the ability of a person or group to know about, mobilize, and influence decision making
	Identifies and analyzes conflict causes (root, proximate, triggers), stages, and management opportunities
	Draws upon concepts and methods developed from the study of conflict, controversies, contentious politics, and dispute intractability
	Addresses the distributional effects on community
	Helps identify strategies for alleviating concerns and tensions among community groups or across communities
	Also concerned with the capacity of locals to mobilize and respond effectively to outside pressure from government or industry
	Considers the glue holding groups and communities together
Capacity building	Long-term voluntary process of increasing ability of communities (and countries) to identify and solve own problems and risks and to maximize opportunities; aim is self-sufficiency
	Can involve developing and implementing educational and research programs and strengthening educational and research institutions
	Community level involves building up place-based institutional capacity; importance of openness to new relationships, knowledge flow from a wide range of sources, and enhancement of adaptive capability
	Role of IA and SIA as a learning process that contributes to the ability of communities and societies to change
	With IA can include developing a library of IA reports, maintaining databases, establishing practitioner networks, collecting examples of good practice, undertaking demonstration projects, producing newsletters, and inviting guest speakers
	Importance of linking IA to development planning, programming, and licensing

Sources: Barrow (2010), Bonifazi et al. (2011), Bredariol and Marini (2003), Burdge (2004), Campbell (2003), Cashmore et al. (2010), Craig (1990), Devlin and Yap (2008), Diffenderfer and Birch (1997), Etzioni (1993, 1995), Gagnon et al. (1993), Harris-Roxas and Harris (2011), Healey (1997), Heiman (1997), Kolkman et al. (2007), Lee (2000), McClendon (1993), McCluskey and João (2011), Novek (1995), Peltonen and Sairinen (2010), Rickson et al. (1990a), Rocha (1997), Thompson and Williams (1992), UNEP (1997), Weaver and Cunningham (1985), Welles (1997).

participation and empowerment. IA practitioners can support the development and accommodation of community knowledge. They can conduct, with public guidance and support, community power structure analyses. They can obtain relevant insights through the application of conflict theory. Community autonomy can be maintained and enhanced by such analyses. IA capacity building provides skills, knowledge, resources, and institutional reforms needed to further community autonomy objectives.

Each concept starts from the premise that communities should increasingly control their own affairs. Knowledge and control flows upward from individuals and communities rather than downward from experts and government. The local community takes on a larger share of the roles and responsibilities conventionally assigned to technical specialists, politicians, and government officials. Specialist, politicians, and government officials actively assist and support this reorientation. The existing distribution of power is challenged. Assumptions regarding scientific and political legitimacy are tested. The primacy of technical, scientific, and rational knowledge and methods is replaced by a far greater emphasis on the knowledge, experiences, and perspectives of local individuals, groups, and organizations. Disciplinary and other conventional categories are crossed and transcended.

Planning and decision-making processes become more social, political, adaptive, informal, subjective, participative, and context-dependent. The public is at the center of rather than peripheral to the process. The process is guided by community values and aspirations. Proposed actions are evaluated as catalysts for or against the satisfaction of community needs and aspirations. The process is broadened to encompass the political roles and activities of community activists and nongovernmental organizations. Governments and practitioners actively foster participation, local empowerment, community self-reliance, and the realization of community goals. Both the IA process and IA capacity building are guided by the ethic of community empowerment. IA becomes a tool for furthering community development. The potential community power and structure implications of proposed actions are analyzed. Communities are assisted, when requested, in their efforts to cope with and manage change, consistent with community ends.

8.4.4 Correcting Power Imbalances

Direct and participative democratic theory presumes that people have equal access to power. Such, of course, is rarely the case. Invariably, there are inequities in access to power among groups and segments of society. There also are perspectives conventionally underrepresented in the IA process. Table 8.2 highlights key characteristics of several concepts aimed at correcting power imbalances.

IA practitioners can help offset power imbalances. They can prepare alternative IA documents, interpretations, and analyses, consistent with advocacy theory. They can integrate social and environmental justice concerns into IA requirements, processes, and documents. They can, working within government, undertake and support social and environmental equity and progressive planning. They can advance concepts such as accountability, legitimacy, and governance, and tools such as discourse analysis, which

Table 8.2 Examples of Potentially Relevant Concepts for Correcting Power Imbalances

Accountability	About holding public bodies to account and/or making them responsible for their actions Requires justification of activities relative to other people's values (recognizes other potentially valid perspectives on decisions) Forms include democratic (political and bureaucratic), corporate, and professional Important role of formal procedures in delivering democratic governance Operates at legislative (provides accountability requirements) and implementation levels (where individuals, organizations, and authorities held to account for actions) Potential role for IA as a lever for increasing accountability of proponents and decision makers Could provide focus around which NGOs and public could coalesce and hold decision makers and public officials accountable Accountability perspective could help reframe IA from purely information and procedural instrument to one that proactively promotes accountability and sustainability at all levels
Advocacy	Argues that planning documents (such as IAs) reflect the existing distribution of power in society Promotes the preparation of alternative plans that reflect the interests and perspectives of underrepresented and vulnerable groups; assumes a pluralistic society and adversarial relationships; envisioned as a means of promoting democratic pluralism Practitioner an advocate; advocacy a bridge between political and professional Could result in multiple IAs, shared data but separate interpretation documents, single IA but client not proponent, or single IAs but separate peer reviews for each interest group Decisions might have to be made by an arbitrator or by means of a quasijudicial procedure Many variations (e.g., multiple advocacy: encompasses all major perspectives, ideological advocacy: advocate represents a perspective rather than a client, environmental advocacy: focuses on the development of a consciousness of a shared destiny)

(continued)

Table 8.2 (*Continued*)

Equity and progressive planning	Government-sponsored progressive planning; information a source of power
	Seeks to redistribute power, resources, and participation away from elites and toward poor and racial and ethnic minorities
	Planning based on substantive, political redistribution goals; aimed at reducing negative social conditions caused by disparities; aims to create justice, fairness, and equity
	Seeks to constrain and modify dominant power sources in process; promotes collaborative governance
	Faces challenge of growing ethnic and racial diversity; seeks coexistent viability of ethnic and racial groups
	Emphasis on citizen participation, power sharing, decentralization, and social self-determination
Legitimacy	Linked to justification of power and authority; creates political obligations; linked to moral justification
	Encompasses both procedural (e.g., public reason, democratic approval) and substantive (e.g., sustainability, distributive justice)
	Sources: tradition, faith in rulers, trust in rule of law, consent, democratic decision making, and beneficial consequences
	Sometimes defined as a criterion of minimal justice
	Public participation a critical variable in terms of civil legitimacy
	Potential role of IA (positively or negatively) in contributing to civil legitimacy of decisions
Governance	Shift from government to governance
	Network of governance: government only first among equals in designing and implementing public policy
	Social movements and NGOs represented more prominently through network settings in governance; can more easily oppose or rebut official government policy
	Encompasses policy regime, public and private agency collaboration, democratization of environmental governance and equitable allocation of funds, resources, and values
	Involves coalition of public and private actors, without necessary involvement of state
	Uses a nonstructured agenda to identify areas and issues for actor involvement
	IA a subset of environmental governance; a decision support tool for environmentally better decision outcomes
	Strengthened by democratic principles and by concepts of procedural and distributive justice
	Crucial role of participation in IA governance
	Necessitates being sensitive to IA bias and proactive effort to overcome institutional, technical, and cultural barriers to participation
Political–economic mobilization	Structural (class-based) analysis of inequities in the distribution of resources and power
	Practitioners said to legitimize the existing distribution of power and resources and to perpetuate inattention to incompatibility of democratic political processes with capitalist political economies
	Alternative role for practitioners to identify how inequities created and to define means of facilitating social control
	Equity considerations substituted for efficiency; requires state backing; linked to politics of empowerment, redistribution, and community
	Mobilization is all forms of political action that fall short of revolution; informed by political–economic structural (e.g., class) analysis
Social and environmental movements	Emergent action groups that seek social and environmental transformation; depends on successfully mobilizing social collectivities; relies heavily on the concept of power in numbers
	A source of political energy; a major role in encouraging citizen participation and in keeping issues on political agenda
	Underpinned by perceptions of common purpose and shared grievances; action-oriented
	May opt out of process and resort to protests, civil disobedience, demonstrations, boycotts, and so on
	Can assume an effective role in monitoring decisions and in acting as an ethical watchdog
Social and environmental justice	Refers to the fair treatment and meaningful involvement of all people, regardless of race, color, national origin, or income concerning the development, implementation, and enforcement of laws
	Minority, low income, and indigenous people should not bear a disproportionate share of negative impacts of government actions
	Analysis determines whether and to what extent injustices occur; may require measures to prevent, reduce, and offset
	Requires a demographic analysis, an impact assessment, and community involvement
	Can pertain to both procedural and substantive injustices; draws on theories of justice
	Seeks equal opportunity, affirmative action, and cultural inclusion; looks to examples of democratic citizenship in resisting injustices and in making the boundaries of community life more inclusive; inspiration from citizen activism
Critical social science and theory	Communications seen as political; process reflects systematic patterning of communications—influences community organization, citizen participation, and autonomous citizen action
	Critique of distribution of power in society and how reflected in practice; research is action-oriented; informs practical action; more complex than class—ecological, alienation, and interrelationships crises
	Practice as communicative action; importance of communications in the search for consensus
	Practice should seek to correct communications distortions and to institute communications enabling rules (e.g., speak comprehensively, sincerely, legitimately, the truth); ideal speech situation
	Seeks to reveal underlying sources of social relations and to empower people, especially the less powerful

Table 8.2 (*Continued*)

Deep ecology and ecological politics	Asserts that humans are but one species among many; no less valuable in own right than humans Biocentrism and self-realization at core; identification beyond humanity; favors noninterference with nonhuman parts of biosphere; concerned with quality of human life and health Committed to a decentralized and democratic ideology; generally allied with left (only way to give full expression to ecological demands); long-term perspective; driven by a sense of urgency Stresses harmony with nature, intrinsic worth of nature, simple material needs, global perspective, and natural limits Sometimes a political party; more often direct political action (e.g., protests, boycotts); linked to environmental movements; long-term goal of reestablishment of human society on a sustainable basis Some success in transforming traditional decision-making systems, in altering agenda, and in stimulating debate; expanded society's conscious; generally faces major obstacles Environmentalists: many differences but share core values; many political approaches; usually campaigns involve a coalition of groups; political action based on belief that it is necessary to sway public opinion to move politicians Environmental lobby groups: large multiissue groups, smaller, more focused groups, education, research, and policy development centers, law and science groups, and land conservation groups
Political SIA	Community conflict the focal point of practice; IAs engender conflict and are undertaken where there is conflict Major social process is one of conflict over resources; interests are the basic elements of social life Social order based on manipulation and control by dominant groups; seeks a more humane society Central role of conflict analysis and conflict management; stresses openness, participation, empowerment, community development; tendency to be holistic and subjective Consistent with critical theory; based on understanding of processes and structures of change
Feminism and eco-feminism	Assumes that women are exploited, oppressed, and devalued by society Interest in changing the conditions of women's lives; seeks extension of women's rights Importance of gender analysis in impact assessment (different roles, knowledge, and values) Eco-feminism: addresses links between androcentrism and environmental destruction; based on similar attitudes Seeks ethics-based environmental planning (grounded in responsibility and carrying reciprocity) Requires an interactive, face-to-face, democratic decision-making process needed to counteract the problem of internalized patriarchal structures and values
Institutional advisory groups	Formed on a permanent or ad hoc basis; offers guidance on a range of issues and activities including policy making Legitimacy depends on degree represents and is supported by interest community; since appointed no mandate from interest community; legitimacy determined by source and quality of advice Roles: means of testing public reaction, forum for expressing public opinion, places controversial issues into objective opinion arena, involves expert critics in decision process Role in IA includes panels for reviewing IA documents and providing advice to government Can bring valuable insight and criticism into decision-making process; offers analysis and alternative viewpoints at modest expense; in recent years broadened to extend representation (e.g., gender, language, region, sector) Dangers: used to placate opposition, co-opt opponents, offer symbolic response to problem, delay action, persuade opponents and provide publicity and patronage instrument Increasingly environmental nongovernmental organizations have developed capability of offering equally high caliber advice; sometimes with financial assistance
Discourse analysis	Discourses: the frames through which groups of actors give meaning to aspects of issues and decisions Goal to stimulate decision makers and stakeholders to critically reflect on and reconsider their discourse Environmental information inconsistent with dominant discourse runs the risk of being ignored IA a potential means of impacting the dominant discourse Discourse analysis allows enhanced understanding of arguments considered or not considered legitimate Takes into account stakeholders and context dimensions Facilitates strategies to enhance decision-making contribution and influence Promotes discourse reflection among stakeholders and on the part of IA practitioners
Participant and intervener funding	Can be provided for participation in review and approvals (e.g., mediation, panel reviews) Sometimes provided for participation in earlier stages of IA process Usually based on criteria Applied to such costs as peer review, legal services, administration, coordination, and travel
Impact management and benefits agreements	Legally binding agreement between a proponent and community Serves to ensure that communities have the capacity and resources to maximize project benefits Initially focused on employment opportunities and investment; more recently includes environmental restrictions, social and cultural programs, dispute resolution, and revenue sharing provisions

Sources: Bass (1998), Barrow (2010), Benveniste (1989), Birkeland (1995), Burdge et al. (1994), Checkoway (1994), Craig (1990), Davidoff (1965, 1978), Day (1997), Devall (1985), Fainstein and Fainstein (1985), Filyk and Côté (1992), Forester (1989), Friedmann (1987), Gorz (1980), Greer-Wooten (1997), Halstead et al. (1984), Healey (1997), Hoch (1993), Maynes (1989), Metzger (1996), Morrone (1992), Nevman, 1997Neuman (1997), Noble and Birk (2011), Peter (2010), Pickvance (1985), Rawls (1971, 2001), Ritzdorf (1996), Rothblatt (1978), Rozema et al. (2012), Runhaar (2009), Runhaar et al. (2010), Sheate (2012), Smith (1993), Vanclay (2003), Wilson (1992).

demonstrate the deficiencies of nondemocratic decision-making practices and illustrate the operation of democratic decision making when power imbalances have been corrected. They can advocate and support decentralization and deconcentration. They can operate outside government as advocates, activists, and community organizers. They can encourage and support ecological political activities and environmental and social movements. They can undertake and accommodate critiques of power imbalances. These critiques could draw, for example, upon political–economic mobilization, critical social science and theory, deep ecology and eco-feminist theories and analyses. They can be open to perspectives and critiques (e.g., deep ecology, eco-feminism) that fall outside the "mainstream" of IA practice.

IA practice, consistent with political SIA, can focus on the analysis and management of community conflict. The IA decision-making process can be broadened to establish, support, and give greater weight to the analyses and suggestions of institutional advisory bodies and nongovernmental organizations. Intervener and participant funding can be provided so that unrepresented and underrepresented segments of society can participate more fully in the IA process. IA capacity building efforts (see Chapter 9) can identify and ameliorate power imbalances. Impact management and benefits agreements can formalize the redistribution of power.

The net result of these "power corrective" actions should be an IA process consistent with direct and participatory democratic principles. The IA process and outcomes also are more likely to be fair, just, and equitable (see Chapter 10).

8.5 INSTITUTING A DEMOCRATIC IA PROCESS

8.5.1 Management at the Regulatory Level

Table 8.3 presents a range of positive and negative examples, derived from the four jurisdictions, pertaining to the role of the regulatory level in making IA practice more democratic. The measures employed encompass a host of initiatives to clarify, harmonize, integrate, delegate, coordinate, share, simplify, and enhance IA requirements among multiple government levels. They also address efforts to enhance the decision-making roles of the public and indigenous populations.

As a general proposition, democratic IA requirements should seek to bring government closer to the people. Power relationships within and among governments, IA levels, and related laws and requirements should be streamlined, structured, clarified, harmonized, substantiated, and coordinated. Senior governments should, wherever practical and appropriate, decentralize and deconcentrate their IA operations. Senior level IA requirements should be delegated to intermediate level governments wherever appropriate and practical. They should be harmonized where delegation is inappropriate. Intermediate and local level IA systems should be encouraged and supported. Intermediate governments should explore, with local governments, delegation potential and harmonization opportunities. Close and frequent consultation should be maintained with governments closer to the people. Senior levels should justify when delegation is not practical and centralized control is essential. Harmonization and coordination with lower government levels should not be a matter of senior level discretion. All governments should promote and support the early and ongoing involvement and influence of interested and affected parties and communities. Support and encouragement includes measures to ameliorate and rectify imbalances in the distribution of power through, for example, public participation requirements, intervener and participant funding, IA capacity building, and environmental justice requirements.

Considerable care needs to be taken when selecting, combining, and adapting specific measures intended to operationalize democratic IA principles. It should not, for example, be assumed that IA harmonization means modeling the intermediate IA level after the senior IA level, regardless of contextual differences and regardless of the flaws in the senior level. Tiering approaches may be of limited value if they do not explicitly address such matters as triggers, decision procedures, procedures for addressing gaps and overlaps, procedures for addressing cumulative effects, guidance, and applied research. A range of coordination measures does not ensure effective coordination. Such measures should be individually and collectively assessed for effectiveness, from multiple perspectives.

There also has been a heated debate in IA literature and practice regarding whether, and if so, how IA should be centralized at senior governmental levels and head offices or decentralized to intermediate level governments and regional/local offices. On the surface, consistent with democratic IA principles such as subsidiarity and proportionality, there would seem to be a strong argument in favor of relegating IA to the lowest level practical (i.e., closer to the people). However, perspectives vary, and how decentralization is undertaken is crucial in terms of if it is on balance, beneficial, or detrimental to the environment.

Transparency and collaboration are essential to any efforts to facilitate democratic IA practice. Decisions should be substantiated based on explicit, substantiated, and consistently applied standards and criteria. Procedural and substantive outcomes need to be independently audited. The sharing of approaches and experiences among jurisdictions and independent, applied comparative research would be a good place to start.

8.5.2 Management at the Applied Level

Figure 8.5 illustrates an example democratic IA process. The figure and description that follow depict an IA process that fosters the decentralization of power and local autonomy, and maximizes public influence and control, especially by

Table 8.3 Positive and Negative Regulatory Level Democratic Examples

United States	Canada	Europe	Australia
(±) Tiering encouraged as a means of framing and focusing the analysis; limited use of programmatic EISs (+) Numerous references to eliminating duplication and resolving conflicts (+) Provisions for state agencies, local agencies (with similar qualifications), and tribal governments to act as cooperating agencies, and for joint planning processes, joint environmental research and studies, joint documents, and joint public hearings (+) Several examples of federal agencies delegating authority to states as a means of expediting the process (in special circumstances) for federally funded projects (+) Instances where state agency grantees have served as co-lead, have been given the authority to determine NEPA categorical exclusions, and where (in a pilot program) states have assumed federal authority under NEPA (LeBoeuf et al., 2010) (+) References to notifying, involving, and making documents available to states, local agencies, Indian tribes, and the public (+) Federal agencies expected to cooperate with state and local agencies to reduce duplication (+) Tribal/agency agreements have been used to define IA responsibilities	(±) Consultation and coordination provisions with other jurisdictions; mentions provinces but not territories (+) Federal authority must make information or knowledge available within a specified period (+) Environmental effects include effects outside Canada (+) If appropriate, Minister must, on request from province, approve the substitution of EA process; only if satisfied that listed factors considered and public access and opportunity to participate (+) Equivalent assessment provisions: minister must approve substitution or exempt designated project, taking into account implementation of appropriate mitigation measures, whether significant adverse environmental effects, follow-up program, and any other conditions (−) Criteria and who determines basis for substitution or equivalency still very general and no provision for independent audit or review (+) Description of designated project includes EA and regulatory requirements of other jurisdictions and references to federal areas of jurisdiction (±) Narrowing definition of effects to areas of federal jurisdiction will reduce the potential for duplication and overlap with other governments but could result in inconsistencies and inhibit the consideration of sustainability and cumulative effects (+) Panel must consider opportunities for cooperation with other jurisdictions (powers, duties, functions) (+) Provides for agreements with other jurisdictions (+) Emphasis on importance of coordination and cooperation with aboriginal people, the value of public and aboriginal participation, the respect of aboriginal and treaty rights, the importance of public and aboriginal access, and the critical	(+) Proposed Project Directive (PPD): provides for coordination and integration of EIA assessment procedures with other EU legislation; seeks to enhance coherence and synergies with other Union legislation and policies (+) PPD provides links to Europe 2020 strategy; priority: sustainable growth (+) PPD respects subsidiarity principle: limits to minimum requirements; aims to comply with international conventions; all Member States must take measures to comply with minimum requirements (+) PPD provides for coordinated or joint procedures when subject to other Union legislation (e.g., SEA, conservation of wild birds, water policy, industrial emissions, conservation of natural habitat, and of wild flora and fauna); 1 EIA (+) PPD: provisions do not apply if objectives of Directive met by national legislation; Member States inform application of this provision every 2 years (+) PPD scoping provisions include information and knowledge available and obtained at other decision-making levels (+) PPD: provision for consultation when transboundary effects (+) Applied research undertaken of links between SEA and EIA Directives (Imperial College London Consultants, 2005); limited experience in operating together (±) Applies the proportionality (only what is necessary to achieve objectives) and subsidiarity (as close to citizens as possible) principles; aims to minimize intrusion on Member	(+) IA harmonization facilitated by clearly identified national environmental priorities (i.e., matters of national environmental significance) (+) Cross references to obligations under various international conventions and agreements (+) Provides for the accreditation of state/territorial IA processes, in whole or in part, if they adequately address relevant impacts; in response to the recommendations of an independent review of the Australian requirements, the government agreed to the publication of accreditation standards and criteria, an effort to reach agreement with states and territories regarding standards, and the performance auditing of the accreditation process to determine if claimed outcomes are being achieved (+) Government also agreed to improved intergovernmental cooperation measures, investigating national EIA standards, and considering joint assessment panel opportunities (+) Refers to involving governments, the community, landowners, and indigenous peoples in a cooperative approach to environmental protection and conservation (+) Notes the role of indigenous people in conservation and in the sustainable use of Australia's biodiversity; promotes the use of indigenous peoples' knowledge of biodiversity and the involvement and cooperation of the owners of the knowledge (+) Significance to indigenous tradition is a significance criterion for species and ecological communities

(continued)

Table 8.3 (*Continued*)

United States	Canada	Europe	Australia
(+) NEPA guidance refers to government to government relationships with Indian tribes (+) Extensive array of environmental justice requirements and guidance (−) Progress in transboundary IA, as part of North American Agreement on Environmental Cooperation, limited (Garver and Podhora, 2008) (+) Guidance (US CEQ, 2012) to make process more efficient and timely; addresses such matters as early NEPA integration in planning, scoping, intergovernmental (state, local, tribal) coordination, coordinated reviews, and documents under other applicable laws, adoption, incorporation by reference, expediting responses to comments and clear time lines for NEPA reviews)	role of community and traditional knowledge (+) Federally funded projects on reserves subject to EA; potential for variation or exclusion where land claims (+) Considerable applied research regarding aboriginal and public participation (+) Use in practice of environmental agreements to help secure aboriginal participation (+) Agency objects including promoting EA uniformity and harmonization (with potential role in negotiating agreements) and facilitating aboriginal participation (−) Minimal use of tiering (+) Long history of IA harmonization including multilateral and bilateral agreements (+) Some instances of regional studies to address cumulative effects (−) Considerable political/administrative discretion regarding such matters as exemptions, designated projects, project definition, scope, treatment of alternatives and approvals (even if significant adverse environmental effects); a form of centralization (±) Broadened public access/participation once process instituted (including participant funding) but access/participation constrained by a narrow range of effects, options and projects, time restrictions, involvement after planning, no mediation option, and NEB hearing restriction to "interested parties" (±) Canada ratified, in 1998, the ESPOO Convention; still considerable room for refinement and clarification (−) Progress in transboundary IA, as part of North American Agreement on Environmental Cooperation, limited (Garver and Podhora, 2008)	authority; can inhibit the maintenance of good practice standards (+) Required to address cumulative and transboundary effects under both the SEA and the proposed EIA Directives (+) Transboundary impacts addressed through UN ECE Convention on EIA in a Transboundary Context (+) Citizen participation in decision making facilitated by cross references to Aarhus Convention on access to information, participation in decision-making and access to justice on environmental matters (+) Emphasis on consultation with environmental authorities and early involvement of public (+) Balance between local autonomy and good practice facilitated by an extensive array of guidance documents and sponsored applied research	(+) Provides for public notification, public access to assessment documents, and opportunities for public comments; refers to describing the public consultation that occurs, identifying affected parties, and including a statement mentioning any communities that may be affected and describing their views (+) Provides a framework for informing and involving the public; planned, consultation-related initiatives include more generous timing provisions, the greater use of public inquiries and joint assessment panels for major projects, and the development of principles and guidelines for best practice public consultation (+)Seeks to involve indigenous people in biodiversity conservation and enhancement (+) Factor considered when deciding whether or not to approve an action is a person's environmental history (+) Prepared IA training resource manual for developing countries (−) Could devote more attention to community autonomy, enhanced public influence, and shared decision making

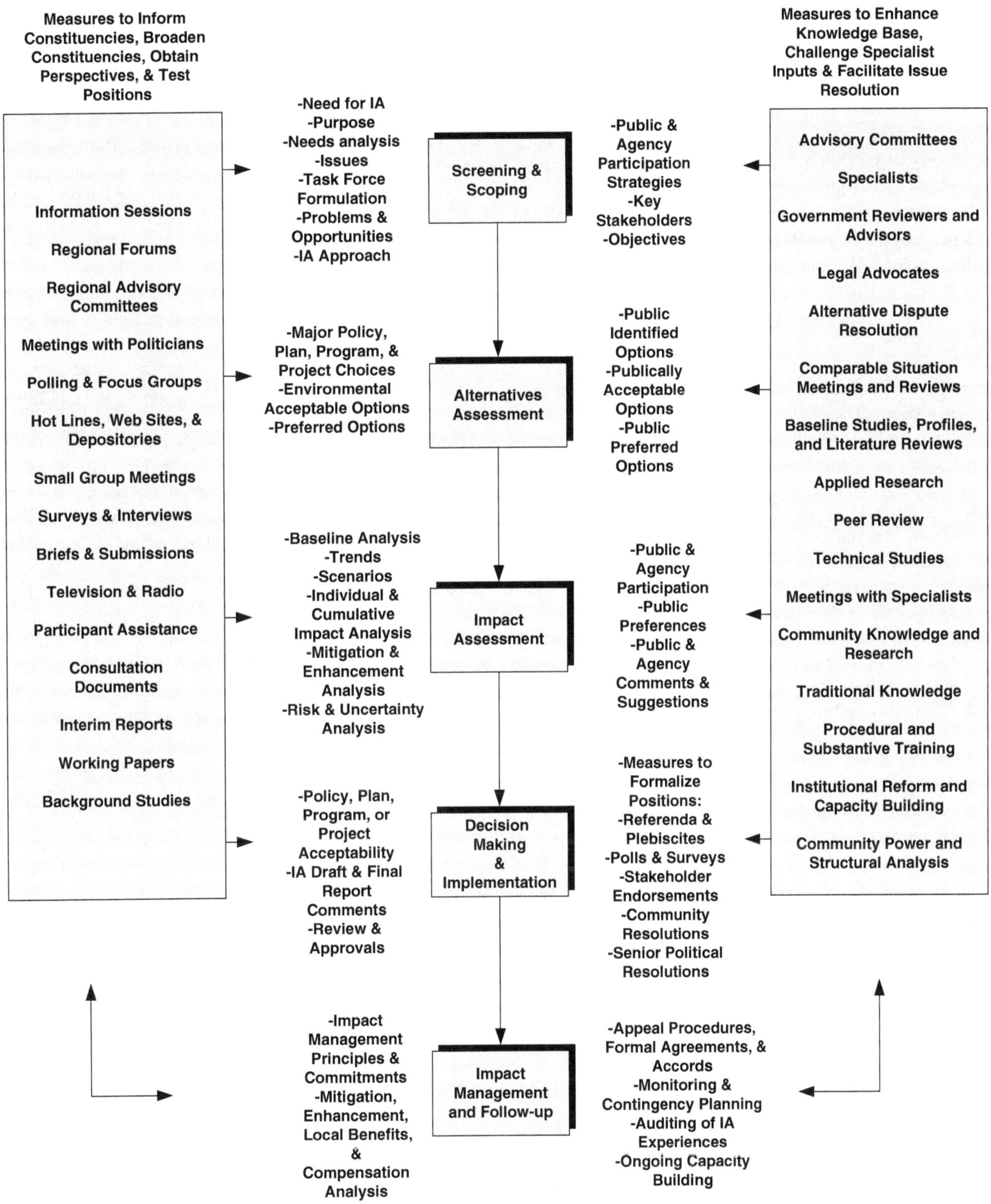

Figure 8.5 Example of a democratic IA process. Adapted from Lawrence (2005b).

individuals, organizations, and communities most directly and severely affected by proposed actions. The democratic IA process is also concerned with identifying and ameliorating power imbalances.

The process is built around the concept of delegated decision making. It is equally well suited to a co-management or shared decision-making approach. The process is an expression and fulfillment of public perspectives. It seeks to build public confidence, trust, and acceptance. Equal and independent parties, with well-defined rights and responsibilities, voluntarily come together to make mutually beneficial decisions. Participants can opt in or out of the process at any time. IA practitioners act as facilitators and collaborators.

Start-Up One or more public task forces guide and manage the IA process. The process begins with meetings or workshops among individuals, groups, and organizations interested in or concerned about a problem, an opportunity, and/or a proposed action(s). These initial meetings identify major issues and perspectives. They also help determine participant expectations. Participants jointly identify potential candidates who are willing and able to be task force members. The members collectively represent a cross section of values, perspectives, and interests relevant to the situation. The parties may include respected and credible independent individuals. Two or more task forces are established if there are major opinion differences regarding task force membership. These task forces operate in parallel. They meet periodically through the process to reach, where possible, common positions.

The major principles, to guide the process and structure the task force operations, are identified. Measures (e.g., participant assistance) are instituted to ameliorate power inequities and to ensure the full and fair participation of all parties. General "ground rules" for operating the task force are established. Major process decisions are identified. Measures, analyses, and procedures to provide a decision-making basis are determined. Procedures for involving and testing the support of the public at large and for involving politicians and government officials are established. These decisions are integrated into an overall IA approach. The approach also addresses resource requirements and schedule. The approach is progressively refined through the IA process. The problem, need, or opportunity to be addressed is thoroughly explored, both within the committee and with potentially affected communities. Key issues and potentially relevant choices are identified. The IA process is scoped to focus on major issues, choices, perspectives, stakeholders, and sensitive and significant environmental components.

Preapproval Decisions The IA process leading up to review and approvals is, like most IA processes, built around a series of decisions. Unlike most IA processes, the public (operating through the task force) assumes the lead role. Specialists, politicians, and government officials are informed, consulted, and involved, as needed. The public controls the process, inputs to the process, and outputs from the process. The task force guides the analyses and undertakes the interpretations and evaluations associated with each decision.

A proposed action may be "on the table" from the outset. Alternatively, there may simply be a problem (or opportunity) and a general sense of ways for solving the problem or taking advantage of the opportunity. In either case, the major available and reasonable choices are identified, screened, and compared by the task force. The task force seeks a consensus regarding a preferred course of action. General impact management, including compensation and local benefits, principles, and commitments are formulated.

A voluntary siting approach is applied if location is a choice. A broadly acceptable (e.g., appropriate physical, social, and economic conditions) region or area is identified. Mitigation, local benefits, and compensation policies are formulated. Voluntary communities and, in turn, sites are solicited within the acceptable region and area. Communities volunteer based on municipal council resolutions. They can withdraw from the process at any point. Environmental and social justice considerations are integrated into the analysis. The task force is modified to add members from the voluntary communities. Extensive consultation is undertaken within the voluntary communities. Local advisory committees are established in each area. Referenda are undertaken in each voluntary community to determine the level of public support. The preferred community is selected, taking into account the degree of public support, environmental suitability, social equity, and economic constraints. Further analyses are undertaken within the volunteer community to identify potentially environmentally suitable areas and sites. The selected site(s) also are voluntary (e.g., public land or private property from a willing vendor) within the acceptable areas. Environmental suitability and the degree of support from neighbors surrounding the site and along access routes to the site are considered when making the final determination.

An analysis of baseline environmental conditions is undertaken. The community helps identify sensitive and significant environmental components and processes. Practical follow-up alternatives are identified. The task force screens and compares the alternatives. Community preferences and concerns are central when selecting the preferred alternatives. Consideration also is given to environmental effect, uncertainty, and technical and economic differences.

The task force, supported by technical analyses and ongoing community consultation identifies, predicts, and interprets individual and cumulative environmental effects. Ways of preventing and offsetting negative effects and of enhancing benefits are determined. The impact significance interpretations take into account mitigation and enhancement potential. Compensation and local benefits policies and measures are refined, based on local conditions. Calculated and perceived risks and uncertainties and their potential implications are explored. Uncertainties are reduced by supplementary analyses, where practical. Community preferences regarding risk and uncertainty acceptability and management play a key role. Individual impact management measures are consolidated into an overall impact management strategy. The impact management strategy includes monitoring, contingency measures, compensation, local benefits, and postapproval consultation.

Periodic workshops and meetings are held if there is more than one task force. The task forces compare analyses and attempt to come to common positions. Where practical, sensitivity analyses explore differences. Conciliation, facilitation, and mediation help identify and expedite discussions. An arbitrator, an independent commission, or a review panel may be used when a consensus position cannot be achieved. Referenda are used to determine community positions on the acceptability of the proposed action(s). Polls or surveys in directly affected areas are undertaken. Resolutions, endorsing or opposing the proposed action(s), are sought from elected representatives (at each relevant government level) and from each nongovernmental organization involved in the process.

Knowledge Base The task force draws upon a diverse knowledge base: baseline studies, literature reviews, and applied research. It determines its own research and advice requirements. It has the final say regarding terms of reference and in selecting specialist advisors. Consistent with advocacy theory, it supports analyses from multiple perspectives. Analyses are tested by peer review. Legal advice is provided where appropriate. Support analyses are structured around the knowledge requirements associated with each decision in the IA process. The support analyses include an assessment of community power structure implications.

Heavy reliance is placed on community involvement in baseline analyses, on community knowledge and, where pertinent, on traditional knowledge. The community actively participates in data collection, analysis, and interpretation. Technical training, institutional improvements, and capacity building measures are instituted where needed to facilitate community understanding, involvement, and control.

Technical advisory and government agency review committees are constituted where needed. These committees provide advice regarding the technical soundness of analyses and the likelihood of regulatory compliance. Training in effective group methods is provided to task force members, as needed. A concerted effort is made to minimize communications misinformation, distortions, and barriers.

Validating the Decisions The activities of the task force are only "legitimate" and "accountable" in a democratic IA process when they reflect the perspectives, interests, and preferences of the overall public. The process links task force activities to broader constituencies. Close contact is maintained with elected officials at all pertinent government levels. Ongoing political participation and support is seen as essential to community understanding and acceptance.

The participation activities extend well beyond the conventional meetings, open houses, and other measures commonly employed in IA to inform and to obtain feedback from the public. Techniques such as small group meetings, workshops, forums, and conferences provide a more interactive, personal, and collaborative approach to addressing process related issues and choices. Polling, surveys, interviews, and focus groups are employed to obtain more in-depth and structured feedback. Advisory committees provide continuity of involvement. The extent of involvement is broadened through hotlines, web sites, television, and radio. Referenda are used at key decision points (e.g., whether a community should volunteer, proposal acceptability) to obtain a comprehensive community response. Inputs from groups, organizations, and segments of society, traditionally underrepresented in the IA process, are actively solicited. Participant assistance is provided to facilitate involvement.

Numerous consultation (adjusted to the needs of varying constituencies) documents are prepared and broadly circulated. Interim, background, and working reports are prepared to provide a sound basis for each decision. The public is provided with ample opportunities to comment on draft and final documents. Several community document repositories are established to ensure that documents are widely and readily available. Documents are also available at a proposal web site. Briefs and submissions from individuals and nongovernmental organizations are actively solicited. Responses are provided to all comments and suggestions received.

Approvals and Postapprovals The proposed action only proceeds to review and approve when there is clear community acceptance and preferably support. An appeal is available to an independent review body for participants dissatisfied with the process or its outcomes. Analyses and consultations are detailed in draft and final IA documents. The ongoing involvement of government agencies minimizes the likelihood of unforeseen regulatory concerns. The public, through the task force and an array of consultation procedures, is a full participant in determining approval conditions. If the proposed action is approved, commitments to the community and to neighbors are formalized first in draft accords and then in final agreements. The agreements address such matters as measures to reduce need, local benefits, monitoring, contingency measures, and compensation.

The agreements provide the foundation for continued public influence and control right through implementation. Community representatives have a major say in follow-up planning and management. The experiences associated with the process are documented and made available to others wishing to apply a similarly democratic IA approach.

8.5.3 Variations by IA Type

The treatment of democratic concepts and principles varies by IA type, as illustrated in Table 8.4. This will result in

Table 8.4 Democratic IA Practice by IA Type

Democratic SA Practice	Democratic SEA Practice	Democratic EIA Practice
Uses SA to frame deliberation about policy controversies Clarifies roles of and decision-making rules for each decision-making level Identifies sources of unsustainability (often deeply embedded in policy and institutional context) Requires strong political commitment to SA Role for SA in directing decision making toward sustainability Focuses on how to design and undertake SA so that it is more influential Builds on close link between governance and participatory and representative forms of governance Redefines agency mandates to address sources of unsustainable (e.g., pace and scale of development) Is consistent with decentralized decision making Uses community-based IA approaches to facilitate transition to sustainability Requires close connection between environmental and economic governance Emphasizes inter- and intragenerational equity Is open, inclusive, and participatory Central role of early discourse and dialogue with stakeholders and affected citizens Recognizes danger that SA can contribute to dominance of socioeconomic considerations Is consistent with bottom-up participation approach and methods; demands greater public involvement in decision making and greater accountability to public Manages development scale and pace to be within regional capacities	Clarifies roles of experts, decision makers, and citizens Identifies and clarifies planning issues, constraints, sources of resistance to good governance and SEA, and strategies for overcoming Uses SEA to contribute to problem definition, helps break down the self-sufficiency of institutions and interest groups, and forces planning and decision making to recognize environmental aspects brought out by SEA SEA is decision oriented and focuses on clarifying planning–decision-making links and role of planners in influencing decisions Emphasizes SEA roles in improving planning process and contributing to sustainability rather than only for approval/permitting purposes Power and interest brokerage central to SEA; effective SEA requires political support and ownership Requires up-front consideration of politics SEA can help create cross-sectoral governance networks Role in breaking autonomy of planning bureaucracy Emphasizes planning and decision-making integration (making a difference) Considers democratic effectiveness Role of SEA practitioner as mediator for negotiations Broadens application to more policies Extends to private sector—a form of environmental advocacy Identifies and addresses sources of tension associated with efforts to apply SEA Potential role for SEA in fostering democratization and promoting public participation Seeks enhanced public control over policy making and participants' empowerment Considers implications of ethics from outset Enhances competence and capacity in influencing decision making Role of SEA in contributing to political pluralism (democratic SEA) Applies community-based SEA approaches Views SEA as a social struggle over problem definition and future choices Recognizes that SEA effectiveness depends on how well fits into planning context; requires analysis of political situation and of context variables such as institutional organization Facilitates SEA capacity building Assesses intended and unintended consequences of SEA	Seeks to raise awareness about the political character of EIA (EIA as a political arena) Learns about the beliefs, values, interests, and aspirations of stakeholders Works with an understanding of power and contested rationality Focuses on conflicts as communications starting point Seeks to involve and empower marginalized groups Seeks to facilitate power sharing and changes to decision-making structures and power relationships Seeks to reframe decision making Recognizes that the legitimacy of the EIA is a key issue Seeks to project international environmental norms into domestic EIA processes Favors institutions that enhance collaborative planning Seeks to help grassroots movements create and participate in a more democratic way Reinforces decentralization Focuses EIA on making a positive contribution to democratic decision making Focuses on interpreting the meaning and implications of plural interpretations of effectiveness Accepts the political nature of EIA and views EIA role as facilitating bargaining Recognizes that social equity cannot be dealt with, without addressing political inequality Focuses on procedural and distributional justice Explores potential role of EIA as an arena for democratic deliberation; experiments with dialogue-based tools Assesses effectiveness in terms of opening up opportunities to involve local people in decision making Seeks to shift the balance of power

Democratic EcIA Practice	Democratic SIA Practice	Democratic HIA Practice
Facilitates inclusive approach; involving all societal sectors Ensures intersectoral cooperation Seeks to resolve inconsistent government processes	Recognizes that SIA inherently and unavoidably political; should demonstrably influence decision making Includes all the values and interests of parties in goals	Seeks to broaden SEA/EIA to include health effects Treats citizen concern with a health issue as a starting point Ensures adequate provision for external accountability

Table 8.4 *(Continued)*

Democratic EcIA Practice	Democratic SIA Practice	Democratic HIA Practice
Draws upon deep ecology perspectives (biocentrism, identification beyond humanity, favors noninterference with nonhuman parts of biosphere, concern with quality of human life and health, commitment to a decentralized and democratic ideology) Seeks to transform traditional decision-making systems, alter agenda, stimulate debate, and expand society's conscious Fosters alliances Identifies negative effects of strong centralized control over management Enhances enforcement Integrates deep ecology and ecological politics—some success in transforming traditional decision-making systems, in altering agenda, in stimulating debate, and in expanding society's conscious Integrates eco-feminism perspectives Seeks ethics-based environmental planning (grounded in responsibility and reciprocity) Treats EcIA as a form of environmental advocacy Requires capacity building Enhances stakeholder involvement in decision making and co-management Fosters greater use of societal tools and less reliance on regulatory tools Involves environmental lobby groups: large multi-issue groups, smaller, more focused groups, education, research and policy development centers, law and science groups, and land conservation groups Potential role for direct political action (e.g., protests, boycotts)	Starts from the principle that people have the right to be involved in the decision making about planned interventions that affect their lives Acknowledges that sometimes values incommensurable and compromise impossible Focuses on understanding, mediating, and managing conflict Is closely linked to "rights" and principles Is consistent with multisectoral integration (tiering) Mitigation and enhancement measures are culturally appropriate Analyzes social development needs Overcomes tension between the political and the technical Serves as a mediator or a forum through which competing knowledge claims, values, and interests can be discussed and linked to options and interventions Promotes inclusiveness, community development, and empowerment Community-based IA is structured and interactive and empowers community-based forums Is consistent with subsidiarity principle (decision-making powers decentralized close to individual citizens) Utilizes community outreach Seeks local community benefits; local discretion essential—empowerment of beneficiaries Seeks shared decision making (through agreements) Seeks to facilitate social consensus and empowerment, enhances marginalized groups, reduces dependency, fosters capacity building, develops social capital, and contributes to community cohesion and greater equity Argues that interventions should be broadly acceptable to the members of the community likely to benefit from or be affected by intervention (public acceptance a decisive consideration) (free, prior, and informed consent) Fully describes and analyzes environmental justice; emphasizes vulnerability of underrepresented and disadvantaged populations Respects community and traditional knowledge Aboriginal control is an essential prerequisite to effective SIA; necessary to end marginalization Stresses equitable sharing of benefits Emphasizes the creation of participatory processes, social learning, and local content	Views HIA as a means of advancing a political agenda; assumes political rationality (extensive negotiations and compromise, multiple interdependencies) Role for HIA in engaging other sectors to consider health effects Seeks evidence that effective in changing decisions and policies, programs and projects Views HIA as a means of empowerment at the community level Can employ community-led HIA approach; building capacity of people to becoming active participants in decisions that affect community well being Fosters indigenous community engagement in HIA research Can use HIA as a means of advocacy Identifies facilitators and barriers to health integration (e.g., partnerships, management and resources, appraisal processes) Views the HIA as a means of empowerment at the individual level Treats full and active stakeholder involvement as an underlying value of HIA Treats community development as HIA building block Fosters community partnerships Facilitates HIA capacity building Is based on explicit concern for health and equity Blends indigenous perspectives with public health data

Sources: Adelle and Weiland (2012), Ayre and Calloway (2005), Banken (2004), Barrow (2010), Bhatia (2007), Binder et al. (2010), Bond and Morrison-Saunders (2011), Bond et al. (2012), Bonifazi et al. (2011), Brown et al. (2003), Burdge (2004), Cameron et al. (2011), Carmichael et al. (2012), Cashmore et al. (2010), Bredariol and Marini (2003), Craik (2008), Esteves and Vanclay (2009), Esteves and Barclay (2011), Esteves et al. (2012), Fischer (2007a), Gibson (2006a, 2011), Harris-Roxas and Harris (2011), Harris-Roxas et al. (2012), Harris et al. (2003), IAIA (undated b), ICPGSIA (2003), Jay (2007), Karjalainen and Järvikoski (2010), Kontić and Kontić (2012), Kørnøv and Dalkmann (2011), Kemm and Parry (2004b), Kwiatkowski (2011), Kwiatkowski et al. (2009), Lane et al. (2003), Lee (2006), Manou and Papathanasiou (2009), McCarthy and Utley (2004), McLauchlan and João (2012), McCluskey and João (2011), Mindell et al. (2004), Mittelmark et al. (2004), Morgan (2012), Morrison-Saunders and Fischer (2010), Nilsson et al. (2009), O'Faircheallaigh, (2009, 2010), Peltonen and Sairinen (2010), Persson (2006), Pope and Grace (2006), Putters (2005), Richardson (2005), Rowan and Streather (2011), Scott (2011), Shepherd et al. (2008), Stoeglehner et al., (2009), Runhaar (2009), Sheate (2012), Sinclair et al. (2009), Taylor et al. (2003), Tetlow and Hanusch (2012), Thomas and Elliott (2005), Youngkin et al. (2003), Vanclay (2003, 2006), Walker (2003), Wernham (2007), Wikland (2005), Wlodarczyk and Tennyson (2003), Wolsink (2010).

some variations in process characteristics between, for example, the SEA and project EIA levels and among substantive IA types (e.g., SA, EcIA, SIA, and HIA).

Democratic SA Practice Democratic SA practice uses sustainability to frame, direct, and represent the critical measure of outcome effectiveness for all decisions. SA, by its nature, requires strong political support, cuts across all decision making and planning levels, and is closely allied to participatory, bottom-up forms of governance. It is generally compatible with and conducive to decentralized decision making, community-based IA approaches, and the systematic consideration of intra- and intergenerational equity. Democratic SA practice challenges the existing institutional arrangements and practices. It is effective only when there is a close connection between economic and environmental forms of governance. Politically ineffective SA can contribute to the dominance of socioeconomic considerations.

Democratic SEA Practice Democratic SEA practice is inherently political and closely linked to (arguably intertwined with) decision making. It is decision oriented. It goes well beyond informing and clarifying decision making. It actively seeks to demonstrably influence decision making toward the realization of more beneficial environmental outcomes. To be effective, it invariably must challenge and reform the prevailing institutional and bureaucratic structures and practices. Power and interest brokerage are central features of democratic SEA practice. SEA is ineffective without strong political support and ownership. The effectiveness of democratic SEA practice is contingent on being able to demonstrate that decision making is more democratic (e.g., more open, transparent, and collaborative, closer to the people). SEA practitioners help identify and resolve tensions and facilitate/mediate negotiations among stakeholders. Democratic SEA practice views strategic planning and decision making as a social struggle. It is community-based, appropriate to the context, and sensitive to procedural and substantive ethical concerns. It fosters democratization, promotes public participation, enhances the potential for public empowerment and control, and facilitates the capacity of the public to influence and reshape decision making at all levels.

Democratic EIA Practice Democratic EIA practice is inherently political and community driven. It focuses on conflicts (contested rationality). It brings to the fore the beliefs, aspirations, values, and interests of stakeholders. It especially emphasizes involving and empowering marginalized groups, and facilitating involvement, bargaining, and dialogue at the community level. While still driven by environmental norms and principles (adapted to the regional and community context), democratic EIA practice actively seeks to decentralize decision making and to make decision making more collaborative, inclusive, equitable, accountable, and legitimate (from multiple perspectives).

Democratic EcIA Practice Democratic EcIA practice draws heavily upon deep ecology, biocentric, and eco-feminist perspectives. It is driven and bounded by environmental ethics and principles. It seeks to transform the value systems, interests, and institutional structures and practices that dominate decision making. It actively encourages the networking of environmental alliances and stakeholder participation (especially among environmental NGOs). It represents a form of environmental advocacy. Where warranted, it supports direct action.

Democratic SIA Practice The need to make decision making and IA practice more overtly political, accessible, collaborative, and democratic is a major theme in SIA theory and practice. Democratic SIA practice is inherently political and, consistent with the subsidiarity principle, decentralized. It favors delegated (i.e., community control) and shared decision making. It insists that decisions be broadly acceptable to the community (i.e., free, prior, and informed decisions). It focuses on understanding, mediating, and managing conflict. It explicitly integrates values, interests, rights, social needs, environmental justice, and ethical principles. It promotes inclusiveness, active participation, community development, empowerment, reduced dependency, social learning, local content, social capacity building, and the enhancement of marginalized groups. It respects community and traditional knowledge. It is community based. It actively seeks local (equitably shared) community benefits, the development of social capital, greater equity, and enhanced community cohesion. It is especially concerned with the implications of potential choices upon underrepresented and disadvantaged populations.

Democratic HIA Practice Democratic HIA practice starts from citizen health issues. HIA is viewed as a means of advancing a political agenda—an agenda focused on building the capacity of people to actively participate in and influence decisions affecting community well-being. Democratic HIA is an advocacy tool for broadened the scope of decision making to more effectively address, from a community perspective, health concerns and effects. It seeks to demonstrably influence health-related decisions and decision making. It helps identify political and institutional facilitators and barriers to health integration. It fosters community-based health partnerships and public and indigenous engagement in HIA health research and health service delivery. It facilitates health-related capacity building, active stakeholder participation, and community empowerment. It is especially concerned with health-related issues pertaining to disadvantaged and marginalized groups, populations, and communities.

8.6 CONTEMPORARY CHALLENGE—MULTIJURISDICTIONAL IA

8.6.1 The Roles of Power in Multijurisdictional IA

The roles of power within a multijurisdictional IA situation, as illustrated in Figure 8.6, are many and varied. It encompasses

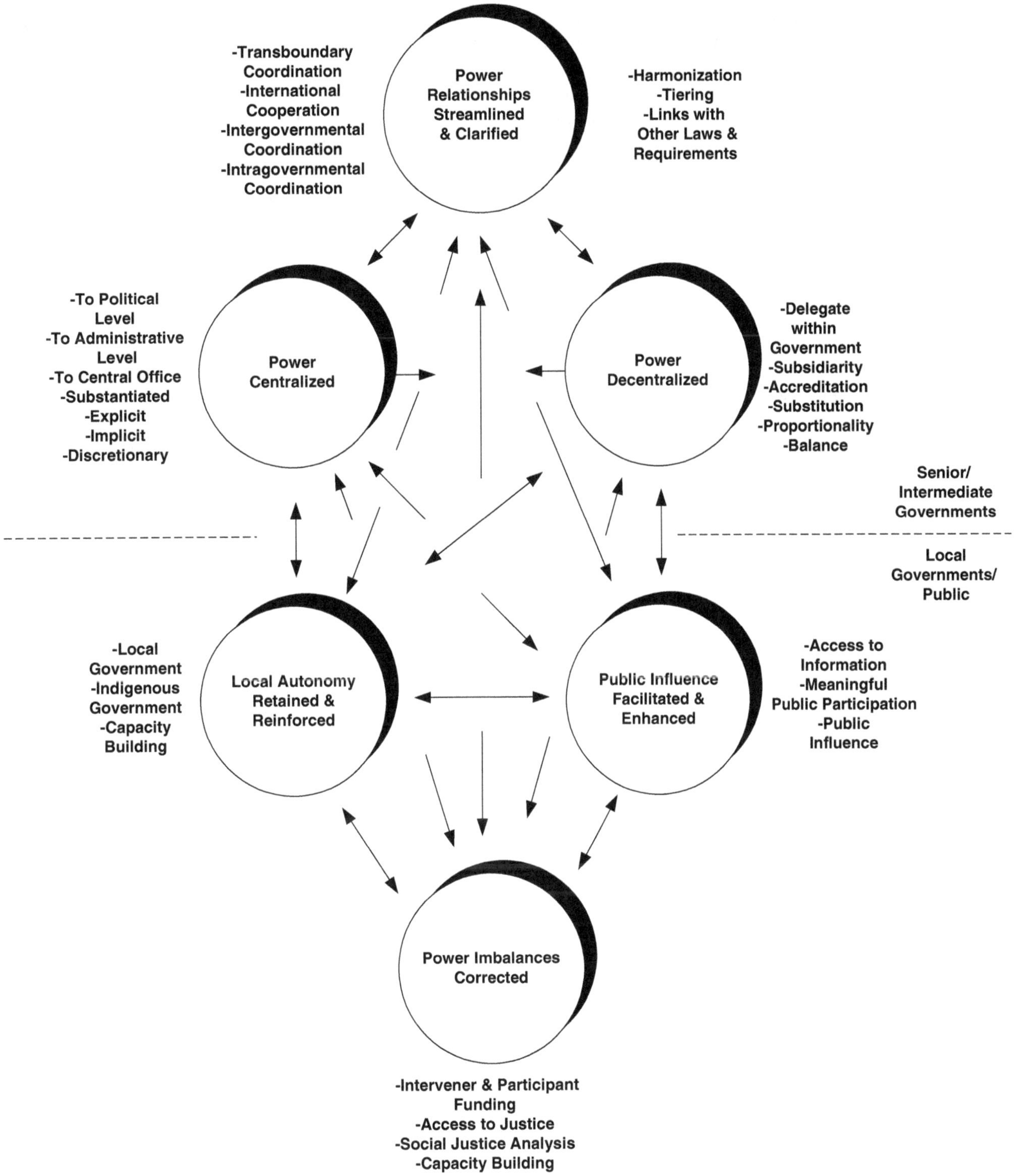

Figure 8.6 The roles of power in multijurisdictional IA.

the streamlining and clarification of power relationships, the upward, downward, and horizontal flow of power among governments (e.g., senior, intermediate, local community, indigenous), power interconnections between government and the public, and initiatives to correct power imbalances.

Power Relationship Streamlined and Clarified Multiple jurisdictions can be involved in IA in a variety of ways. Proposed actions and effects from proposed actions can cross jurisdictional borders (transboundary IA). There can be vertical and horizontal interconnections among planning (policy, plan, program, project) levels (tiering). Measures can be taken to progressively modify, laws, regulations, and administrative procedures, when more than one set of IA requirements apply to the same action, to achieve a degree of homogeneity and enhanced functioning (harmonization). Harmonization does not necessarily, and rarely does, lead to standardization. IA requirements and procedures can be explicitly linked by, for example, cross-referencing and circulation procedures to the administration of other laws policies, plans, and programs. Steps can be taken to facilitate cooperation among and across government types and levels (e.g., international, national/subnational intergovernmental, subnational/regional/community intergovernmental, intragovernmental head office/regional field office).

Power Centralized and Decentralized IA-related power can be centralized at the political level and/or at the administrative level. Centralization can involve the upward flow of power to more senior governments or centralized offices within governments. Centralization of IA-related decisions often takes the form of a greater level of implicit political-/administrative discretion, sometimes as part of the IA process, and sometimes at the end of and external to the IA process. Centralization tendencies can be directed and bounded by such mechanisms as requirements to justify decisions, explicit criteria that must be considered, and appeal procedures.

Decentralization in IA can entail delegation within government (as when decision-making power is devolved to a regional office), and the shifting of IA-related responsibilities in whole or in part to, for example, intermediate governments, indigenous governments, or community/municipal governments, through such mechanisms as accreditation, substitution, and exemption. The exercise of power through such decentralization initiatives can be directed and circumscribed by such measures as explicit criteria and thresholds, adherence to explicit good practice standards, external advisory groups, and the independent auditing of procedural and substantive effectiveness.

The flow of power in IA practice is sufficiently complex that the decentralization and centralization of power, in different forms and in different ways, often occurs simultaneously, and varies among sectors, geographically, and over time. The somewhat naïve view of IA simply informing decision making fails to reflect the complex and evolving patterns in the exercise of power and influence in multijurisdictional settings. A failure to understand these flows of power and influence, and then to act on this understanding, can result in the marginalization of IA requirements and procedures and, in turn, the marginalization of the role of IA practitioners in influencing the exercise of power by decision-makers.

Local Autonomy, Public Influence, and the Correction of Power Imbalances Multijurisdictional IA literature and practice tend to focus on the role of power as it is exercised within and between senior and intermediate levels of government. The potential flow of IA-related power to and from local and indigenous governments tends to receive much less attention. However, consistent with the principle of subsidiarity (decisions made by the smallest, lowest, or least centralized authority), an argument can be made for an enhanced decision-making role for community-based IA and IA by indigenous governments—on the grounds that they are closer to the people, more accountable, and more sensitive to local conditions and circumstances. Such a downward shift of power, or perhaps more legitimately, such a retention and reinforcement of local autonomy, might, for example, necessitate capacity-building measures, the use of formal agreements, contextual adaptations, and variations in the scope of responsibilities entailed.

The tendency in multijurisdictional IA institutional arrangements is to limit the role of the public to notification requirements and involvement opportunities. These matters need to be addressed explicitly and comprehensively. However, a democratic IA process, regardless of the number of governmental and nongovernmental levels and participants, should include the roles of the public in actually influencing decision making. Such a reorientation shift, would, for example, necessitate demonstrating if and how decisions were shaped and influenced by the public. It could, even in a multijurisdictional setting, suggest such procedural performance standards as the free, prior, and informed consent of potentially affected publics. It also suggests a predisposition toward and embracement of collaborative and shared decision-making models.

Multijurisdictional IA practice is prone to power imbalances. This suggests that multijurisdictional IA institutional arrangements need to encompass a range of measures to ameliorate such imbalances, especially for disadvantaged and marginalized groups and populations. Measures such as participant funding, access to justice provisions, the analysis of environmental justice implications, and various capacity building initiatives are likely to be necessary in order to allow all potentially interested and affected parties to fully participate and demonstrably influence multijurisdictional IA requirements, procedures, and practices.

8.6.2 Multijurisdictional IA: Good Practices

Table 8.5 lists a range of possible multijurisdictional IA good practices. Collectively, the measures seek to make multijurisdictional IA institutional arrangements and practices more transparent, comprehensive, consistent, collaborative, substantiated, efficient, effective, and democratic. The categories contained in the table are far from mutually exclusive. An effective range of harmonization/coordination measures might, for example, represent a viable alternative to decentralization. Or, partial decentralization could work effectively in tandem with various harmonization/coordination measures. Measures, to enhance public influence, reinforce local autonomy, and correct power imbalances,

Table 8.5 Multijurisdictional IA—Good Practices

Power Relationships Streamlined and Clarified Tiering

- Clarify tiering meaning, types, and dimensions
- Encourage agencies to tier SEAs and EIAs
- Recognize that tiering is not always hierarchical and linear
- Identify tiering objectives, principles, and frameworks
- Clearly identify roles and issues appropriate to each level
- Explore tiering options
- Make tiering process more explicit
- Identify gaps, potential conflicts, and overlaps
- Integrate sustainability into each level
- Provide tiering guidance
- Sponsor tiering pilot projects and applied research
- Assess the effectiveness of tiering initiatives from multiple perspectives

Harmonization

- Define harmonization
- Apply when actions subject to more than one assessment process
- Identify environmental mandate, interest, objectives, priorities, and policies for each level
- Identify harmonization objectives and principles
- Address consistencies and inconsistencies among SEAs and EIAs and seek to harmonize procedures and significance criteria
- Institute an explicit, substantiated, and collaborative harmonization procedure(s)
- Provide for IA-related agreements and joint or coordinated processes and documents
- Include provisions for enhanced coordination and disputes resolution
- Encourage national IA performance standards
- Openly address unresolved harmonization issues and place within the context of broader environmental quality enhancement initiatives

Other laws, policies, plans, and programs

- Make explicit cross-references to related laws
- Seek to resolve conflicts and inconsistencies with other laws, policies, plans, programs, and projects
- Explore linkages between EIA/SEA and substantive environmental and procedural requirements
- Seek mutually supportive approaches and requirements between EIA/SEA and land use and environmental planning and management
- Assess effectiveness of requirements, individually and collectively, from multiple perspectives

Transboundary IA

- Require consideration of transboundary effects at scoping stage
- Institute interjurisdictional transboundary IA agreements
- Explicitly identify and apply transboundary IA principles
- Provide opportunities for other, potentially affected jurisdictions to cooperate in IA processes and in hearings/court cases
- Require information on IA and regulatory requirements of other jurisdictions
- Clarify such matters as contacts, languages, terminology, division of responsibilities, document contents, notification procedures, public access and consultation procedures, dispute resolution and appeal procedures, competence of authorities, cost allocation, follow-up, and contextual adjustments
- Assess effectiveness of transboundary provisions from multiple perspectives

International cooperation

- Identify shared and country-specific objectives and priorities, and clarify geographic areas over which have jurisdiction
- Explicitly link to international treaties, accords, conventions, goals, and environmental norms
- Seek to strengthen international environmental treaties and manage conflicts among international treaties (e.g., independent inquiry)
- Seek to increase accountability and legitimacy
- Ensure binding rules and compliance with IA good practices in international areas and adapted to context and indigenous populations
- Build on and contribute to good practices of international aid organizations
- Participate in international IA good practice initiatives
- Assess effectiveness of cooperative measures

Intergovernmental cooperation

- Identify shared and government-specific objectives and priorities
- Clarify coordination and substantive roles and responsibilities

(continued)

Table 8.5 (*Continued*)

- Institute coordinated approach to IA applied research, IA harmonization, and joint or concurrent but linked IA processes
- Ensure community-based regional planning in place prior to IA project review
- Require information on IA and regulatory requirements of other jurisdictions
- Facilitate intermediate level IA capacity building
- Assess effectiveness of cooperative measures

Intragovernmental cooperation

- Identify shared and level-specific objectives and priorities
- Clarify coordination roles and responsibilities
- Define substantive and procedural roles and responsibilities of each party
- Institute coordinated approach to IA applied research and document review
- Assess effectiveness of cooperative measures

Power Centralized

- Clearly define centralization objectives and principles (e.g., transparent, substantiated, and collaborative)
- Only centralize when essential to environmental mandate, necessary expertise, and independence
- Apply subsidiarity and proportionality principles to any power centralization proposals
- Recognize advantages and disadvantages of centralization
- Only reallocate IA responsibilities among government levels with the involvement and consent of those most directly affected
- Approach centralization with caution, fully substantiate, and test for effectiveness and consequences

Power Decentralized (Delegation, Subsidiarity, Accreditation, Substitution, and Exemption)

- Decentralize to regional/local government offices where practical and appropriate
- Clearly define decentralization types under consideration and clearly define and substantiate decentralization objectives and principles
- Independently assess effectiveness of current IA institutional arrangements
- Collaboratively and transparently develop, with interested and affected parties, any decentralization initiatives
- Compare selective and comprehensive decentralization options with harmonization and mixed harmonization/decentralization/tiering options
- Clearly identify redefined roles and responsibilities in formal agreements, where practical
- Explore opportunities for selective delegation
- Provide for lower level substitution if consistent with explicit criteria and if auditing and appeal provisions
- Apply subsidiarity and proportionality principles to any power decentralization proposals
- Only undertake decentralization if consistent with level's mandate and capacity, if demonstrable substantive environmental and procedural benefits, if no reduction in public access or participation, and if independence of review not compromised
- Undertake in conjunction with IA harmonization and tiering measures, and initiation of joint IA good practice standards
- Undertake applied research of decentralization measures
- Assess effectiveness of decentralization measures

Local Autonomy Retained and Reinforced

Local government

- Proactively seek to enhance local government decision-making influence and autonomy
- Seek to obtain community acceptance/approval
- Explicitly and collaboratively identify community involvement, participation, and influence objectives and principles
- Make provisions for full participation of local jurisdictions in IA processes, including the possibility of formal agreements
- Encourage and support community level IA
- Explore opportunities for delegating IA responsibilities to community level
- Ensure IA process designed and managed to take into account local community characteristics and preferences
- Consider delegating selective IA responsibilities to communities but ensure sufficient capacity
- Consider use of formal agreements in facilitating participation and influence of local governments
- Assess consistency of all proposed actions with community objectives, policies, plans, programs, and laws; seek to resolve all conflicts
- Assess implications of all proposed actions for local decision-making authority and autonomy
- Fully integrate community knowledge
- Require explicit responses to all community proposals and positions, and require demonstration of local decision-making role
- Assess effectiveness from community perspective

Indigenous government

- Institute government to government IA arrangements with indigenous peoples
- Explicitly and collaboratively identify indigenous community involvement, participation, and influence objectives and principles
- Consider use of formal agreements in facilitating participation and influence of indigenous peoples

Table 8.5 *(Continued)*

- Explore opportunities for selective or complete delegation of senior level IA responsibilities to indigenous groups or communities
- Proactively seek to enhance decision-making influence and autonomy of indigenous peoples
- Seek to obtain indigenous community acceptance/approval
- Ensure early and ample time for indigenous peoples' input, review, and comment at each decision point
- Ensure appropriate structures and procedures and adequate financial and other resources for "meaningful participation"; capacity build as needed
- Encourage and support indigenous government IA
- Collaboratively adapt IA process to characteristics, needs, and aspirations of indigenous peoples
- Respect treaty and other rights of indigenous communities
- Consider and apply, where practical and appropriate, indigenous community-based IA model
- Assess consistency of all proposed actions with indigenous community objectives, policies, plans, programs, and laws; seek to resolve all conflicts and consistencies
- Assess implications of all proposed actions for indigenous peoples' decision-making authority and autonomy
- Respect and fully integrate traditional knowledge
- Require explicit responses to all indigenous community proposals and positions
- Provide for indigenous community to assess effectiveness from own perspective

Public Influence Facilitated and Enhanced

Access to information

- Identify explicit access to information objectives, principles, and procedures (jointly determine with stakeholders)
- Provide timely public access to all IA documents and files
- Independently assess effectiveness of access to information provisions

Meaningful public participation

- Identify explicit public participation objectives, principles, and procedures (jointly determine with stakeholders)
- Ensure early and ample time for public input, review, and comment at each decision point
- Broadly define public and interested parties
- Proactively seek more open and transparent IA processes; draw upon and contribute to IA public participation good practices
- Require consideration of community knowledge and indigenous traditional knowledge
- Provide for public review of draft guidelines, codes of practice, agreements, arrangements, and criteria
- Independently assess public participation implications of changes to IA institutional arrangements and effectiveness of public participation measures

Public influence

- Require demonstration of public's role in decision making
- Require justification of IA-related decisions, consistent with democratic principles and public values
- Explicitly respond to all public comments and proposals
- Seek to enhance public decision-making influence and effectiveness
- Apply collaborative, shared decision making and delegated participatory IA approaches
- Seek effectiveness of measures to enhance public's role in planning and decision making, from multiple perspectives

Power imbalances corrected by participant funding

- Provide funding for full participation in IA process; additional funding for disadvantaged groups
- Provide funding for full participation in hearings and legal proceedings
- Provide funding for independent peer review
- Independently assess effectiveness of participant funding mechanisms

Access to justice

- Provide standing to public in hearings
- Provide ability to appeal IA decisions
- Independently assess effectiveness of access to justice provisions

Environment justice analysis

- Integrate environmental justice analysis requirements
- Provide environmental justice analysis guidance
- Analyze distribution of effects; seek to minimize adverse impacts on and to enhance benefits to less advantaged
- Independently assess effectiveness of environmental justice analysis requirements

(continued)

Table 8.5 (*Continued*)

Capacity building
• Provide additional resources for disadvantaged communities, groups, and segments of society • Ensure that all interests are represented at the table, power imbalances are ameliorated, and community accountability and influence is enhanced • Independently assess effectiveness of capacity building measures

Sources: Albergaria and Fidelas (2006), Albrecht (2008), Arts et al. (2011), Biermann et al. (2012), Bina et al. (2011), Bonvoisin (2011), Burdge (2004), Carter and Howe (2006), Craik (2008), Donnelly and Mahoney (2011), Eccleston (2008), EC (2008b, 2009b, 2012c), Fischer (2007b), Fitzpatrick and Sinclair (2009a,b), Galbraith et al. (2007), Glasson and Bellanger (2003), Hacking and Guthrie (2011), Hemmings and Roura (2003), Herring (2009), Imperial College London Consultants (2005), Jiricka and Pröbstl (2009), Koivurova (2008), Lawrence (1999), Lyhne (2009), Mas (2003), de Mulder (2008), Noble (2009a), Noble and Gunn (2009), O'Faircheallaigh (2006), Pinho et al. (2010), Pope and Dalal-Clayton (2011), Pope and Grace (2006), Purnama (2004), Renda (2006), Ritsatakis (2004), Risse et al. (2003), Ruddy and Hilty (2008), Sánchez and Silva-Sánchez (2008), Sadler and Jurkeviciute (2011), Sinclair et al. (2012), Taylor et al. (2004), Thérivel (2010), Tomlinson (2004), Whitelaw et al. (2009).

could complement streamlining and power clarification measures at senior government levels.

Many of the measures listed in Table 8.5 are reflected, to varying degrees, in existing IA multijurisdictional IA institutional arrangements. What is much less clear is whether or to what extent these measures (either individually or collectively) are effective, complementary, publicly supported, and appropriate to the context. The state-of-practice of multijurisdictional IA also is severely hampered by a "rush to judgment" based on preconceived notions regarding what is right or wrong with the current system, and what reforms should be introduced to remedy perceived deficiencies and to enhance perceived strengths. Perspectives tend to vary dramatically among stakeholders regarding the state-of-practice of multijurisdictional IA. Based on this scant and impressionistic knowledge base, positions rapidly become entrenched. Depending on who has the "balance of power," reforms are made to the existing IA system or the entire system is replaced, without a clear sense that the problems that they are intended to remedy are valid, if other problems might be more serious, whether the proposed remedies will have the desired effect, any unintended "side effects," and what contribution other interested and affected parties might have made to the reform process.

Ideally, what should emerge from ongoing debates and discussions surrounding multijurisdictional IA is, first and foremost, a consensus concerning multijurisdictional IA core principles, goals, unacceptable practices, and minimally acceptable practices. The existing multijurisdictional IA system should then be evaluated for effectiveness from multiple perspectives. Armed with a clear, and preferably consensual understanding of what is right and wrong with the current system, a baseline can be established for moving forward. Individual "good practice" candidate measures can then be knit together, through collaborative forums, into a coherent strategy.

The strategy would be guided and bounded by a clear set of objectives and principles. It would encompass, for example, tiering procedures, harmonization measures, transboundary consultation, and analysis provisions, links to related requirements, inter- and intragovernmental coordination mechanisms, selective delegation and centralization mechanisms, and a range of proactive measures to facilitate local autonomy, public influence, and the correction of power imbalances. All proposed measures would be fully explained and justified. In the event that there are major divisions in perspectives regarding proposed measures, these measures could be deferred, subject to intense discussions and negotiations, or alternative strategies or strategy components could be formulated. These strategies or strategy components could then be compared and evaluated (again in collaborative forums) against the multijurisdictional IA goals and principles.

The preferred strategy would extend from the minimally acceptable practices and selectively integrate complementary elements from other strategies. It also would be adapted to the institutional, environmental, and cultural context. Elements of the preferred strategy, with which there is ample experience and broad support, could first be progressively integrated into the existing system. Other elements, where there is a greater level of uncertainty and/or support, could be tested through pilot programs associated with individual IAs. The multijurisdictional IA system would progressively evolve over time and subject, periodically, to independent review and stakeholder review and discussion. The whole enterprise could be informed by experience elsewhere, and IA-related literature and good practice standards. The effectiveness reviews of the multijurisdictional IA reforms also should contribute to the wider IA knowledge base.

8.7 SUMMING UP

This chapter describes a democratic IA process—a process that shifts power from specialists and politicians to the public. The people most directly affected by proposed actions have a major say in whether and how actions proceed. Groups, segments of society, and perspectives, commonly excluded or underrepresented, assume a more prominent position in the IA process.

The three stories presented in the preceding section approach the question of democracy in contrasting ways. The first story describes how a project-level EIA was democratically ineffective in influencing decision making in favor of more environmentally sound choices. The second story provides an example of how collaborative, consensus-based, and democratic higher level planning can help direct and bound planning and decision making at the project EIA level. The third story illustrates how the integration of SEA and planning is a more complex, and sometimes constrained, endeavor than is often assumed or understood. The stories provide an initial and partial sense of how democratic values should and should not be approached in the IA process design and management.

Although the value of public participation in widely acknowledged, the public too often has a minor role in the IA process. Members of the public frequently are or believe themselves to be powerless in major decisions that affect their lives. This problem is exacerbated by imbalances in the distribution of power. The most vulnerable segments of society tend to be the least influential. The solution is an IA process that delegates or shares decision-making authority with the public.

Democracy is rule by the people. The IA process should be an expression and fulfillment of direct and participatory democratic concepts and principles. The courts, politicians, and government officials should involve, delegate power to, share power with, and be responsive to the public.

The IA process should be designed to facilitate a high degree of direct public influence and control. The public, working closely with local politicians, should assume the lead role in shaping and guiding the process. There should be an influential public role for each decision within the process. There should be a clear rationale for how the public is defined, what represents a public choice, and any bounding of the public's role. The IA process should draw upon the principles, insights, and experiences of community resource co-management and voluntary siting approaches.

The IA process should help make people and communities more autonomous and better able to make decisions about matters that affect their lives. Lessons should be derived from community-based IA and from visions, principles, and strategies for more autonomous communities. The IA process should facilitate community empowerment, development, and mobilization. It should promote and accommodate community and traditional knowledge. It should assess conflict and community power structure implications. It should facilitate community IA capacity building and institutional reform. Roles, conventionally assumed by politicians, government officials, and technical specialists, should be shifted to the public, wherever practical. The public should be at the center of the process. Community activists and nongovernmental organizations should be prominently featured. The IA process should be guided by community values. Proposed actions should be assessed as catalysts for or against the realization of community needs and aspirations.

The IA process should be designed and managed to minimize power imbalances. This may necessitate preparing alternative IA documents, interpretations, and analyses. Social and environmental justice concerns should be integrated into the process. Decision making and the IA process should be accountable and legitimate. Imbalances should be ameliorated by decentralizing and deconcentrating power. Collaborative and inclusive governance networks should be facilitated. Dominant discourses should be identified, critically evaluated, and collaboratively reformed. Environmental equity and progressive planning concepts and principles can be instructive. Power relationships should be formalized and redefined, when appropriate, by using tools such as impact management and benefits agreements. The IA process should encourage and support community advocates, activists, and organizers. Social and ecological political activities and movements also should be encouraged and supported. Identifying and characterizing power imbalances should be aided by drawing upon critiques from pertinent political, social, ecological, and feminist theories and analyses. The process should be opened up to perspectives and critiques outside mainstream IA practice. The analysis and management of community conflict should receive particular attention. The IA process should draw upon the analyses and suggestions of institutional advisory bodies and nongovernmental organizations. Financial and other assistance should be provided to underrepresented and unrepresented segments of society so that they can more effectively participate in the process. IA capacity-building efforts should devote greater attention to identifying and rectifying power imbalances.

Senior governments should, wherever practical, decentralize and deconcentrate their IA operations. IA responsibilities should be delegated to intermediate government levels where appropriate and practical. A clear rationale as to why central control is essential should be provided whenever delegation does not occur. The autonomy of local IA systems should be promoted and supported. IA systems should promote and encourage early and ongoing public participation and influence. Opportunities for delegated or shared decision making with the public should be provided where practical and appropriate. IA systems should ameliorate power imbalances. They also should resist the propensity to centralize.

A democratic IA process seeks to maximize public influence and control, foster local autonomy, and correct power imbalances. Decision-making authority is delegated to or shared with the public. The process is designed and managed to build public confidence, trust, and acceptance.

One or more public task forces guide and manage the example democratic IA process. IA practitioners act as facilitators and collaborators. The process begins with the task force formulation, the identification of major values and principles, the formulation of public and agency participation strategies, and the development of a preliminary IA approach. Major issues and choices are identified. The process is scoped. The need for action is determined. The balance of the process, leading up to review and approvals, is built around decisions. The task force, drawing upon an extensive knowledge base and maintaining close contact with the broader public, makes each decision. Major available alternatives to the proposed action are identified, screened, and compared. Impact management principles and commitments are determined. Proposed facilities are located using a voluntary siting approach. Proposal characteristics are formulated. Community and traditional knowledge are prominently featured in the baseline analyses. Community concerns and preferences provide the basis for mitigation, local benefits, and compensation policies and measures. Facility design, operations, closure, and postclosure options are evaluated. Individual and cumulative impact analyses, taking into account mitigation potential, compensation measures, risks, and uncertainties, are undertaken. An overall impact management strategy is formulated. Differences among task forces are reconciled, where practical. Measures such as referenda, council resolutions, and organizational endorsements are used to gauge public and stakeholder acceptability of proposed actions.

The public task force draws upon an extensive knowledge base. Reference is made to technical studies, baseline studies, community profiles, literature reviews, applied research, peer reviews, and visits to comparable facilities. Community and traditional knowledge is supported and accommodated. Community power structure implications are explored. The task force works closely with technical specialists, government officials, legal advocates, and procedural specialists. It receives technical and procedural training, as needed. Institutional reforms and capacity building occurs where needed to support the process. The task force makes an ongoing effort to reflect the perspectives, interests, and preferences of the overall public. Close contact is maintained with politicians. A variety of methods are applied to communicate with, involve, and collaborate with the public. Interim, background, and working documents provide a sound decision-making basis. The documents are readily available, are adapted to the needs of different publics, and fully integrate public concerns and preferences. The IA process proceeds to review and approval only when there is clear community acceptance and preferably, support. There is ample public and agency involvement in preparing draft and final IA documents. An appeal procedure to an independent review body is available. The public participates in determining approval conditions. Commitments to individuals and communities are formalized in accords and agreements. Proposed facilities are comanaged. The public has a major influence in closure and postclosure planning. Community experiences with the process are documented. The process is designed and adapted to suit the IA level (e.g., SEA, project EIA) and type (e.g., SA, EcIA, SIA, HIA).

A democratic IA process proactively seeks to influence decision making by bringing it "closer to the people." It focuses on the exercise of power and seeks to broaden the range of participants in IA-related decision making, enhance the role of community-level decision making, decentralize decision making to the lowest practical level, make IA public participation more "meaningful" (in the sense of being more influential), and correct power imbalances. It effectively draws upon community and indigenous knowledge and perspectives. A democratic IA process is not always effective in overcoming inertia, resistance to change, and political maneuvering outside or even within the IA process.

Multijurisdictional IA is a major challenge in contemporary IA practice. It encompasses the consideration of the streamlining and clarification of power relationships, the centralization and decentralization of decision making, the retention and reinforcement of local autonomy, the facilitating and enhancement of public influence, and the correction of power imbalances. A range of possible multijurisdictional IA good practices is identified. Collectively, the measures seek to make multijurisdictional IA institutional arrangements and practices more transparent, comprehensive, consistent, collaborative, substantiated, efficient, effective, and democratic.

Many of the available measures are reflected in existing IA multijurisdictional IA institutional arrangements. What is much less clear is whether or to what extent these measures (either individually or collectively) are effective, complementary, publicly supported, and appropriate to the context. The state-of-practice of multijurisdictional IA also is severely hampered by a "rush to judgment" of behalf of many IA stakeholders.

What is needed is a consensus concerning multijurisdictional IA core principles, goals, unacceptable practices, and minimally acceptable practices. What should then occur is the systematic and collaborative assessment of the effectiveness of the existing array of measures. This would be followed by the knitting together, through collaborative forums, of individual potential improvements to the system into a coherent strategy. This strategy should be progressively introduced and tested for effectiveness. This multijurisdictional IA system would progressively evolve over time and be subject, periodically, to independent review and stakeholder review and discussion. The whole enterprise could be informed by experience elsewhere, and IA-related literature and good practice standards. The effectiveness reviews of the multijurisdictional IA reforms also should contribute to the wider IA knowledge base.

Chapter 9

How to Make IAs More Collaborative

9.1 HIGHLIGHTS

In Chapters 8 and 9 we pose two choices concerning how IAs might become more democratic. The first choice, delegating or sharing decision-making authority with the public, is addressed in Chapter 8. This chapter is concerned with the second choice, taking steps to ensure that the public is an active and ongoing participant in a collaborative IA process. Final decision-making authority, however, continues to reside with the proponent and/or with the regulators.

Collaboration is defined broadly to encompass all forms of public participation short of delegation or shared decision making. Collaboration implies a joint endeavor of the public and of other "stakeholders." Therefore, forms of public participation, not fully collaborative, are included in the analysis but as prerequisites to or subsets of a collaborative IA process. Nonparticipation, either warranted or not warranted, is not considered. Also not addressed, except in the sense of dangers to guard against, is illegitimate participation (e.g., deliberately incomprehensible, insincere, untruthful) (Forester, 1989). The differences between collaborative and democratic IA processes are largely a question of degree. Both processes seek to enhance the role and influence of the public in decision making.

- The analysis begins in Section 9.2 with three applied anecdotes. The stories describe applied experiences associated with efforts to make IA practice more collaborative.
- The analysis in Section 9.3 then defines the problem, which is the gulf between the potential benefits of collaborative IA processes and the more modest benefits achieved by public participation approaches commonly evident in IA requirements and practices. The direction is exploring the potential for and means of making IA processes more collaborative.
- In Section 9.4 we first consider the possibility that valid and significant disadvantages and constraints largely preclude a collaborative IA approach to public participation. This explanation, although partially valid, is found wanting. The second possibility is that there is an extensive foundation of sound analysis and good practice, which could provide the basis for collaborative IA processes. But the relevant source materials are immense, of varying quality, and scattered across numerous related fields. What is required is a succinct presentation and analysis of the major "building blocks" of a collaborative IA process. Major distinctions drawn in the analysis include (1) principles and practices, (2) consultation, (3) communications, (4) mutual education, (5) negotiations, and (6) collaboration.
- In Section 9.5 we apply the insights, distinctions, and lessons identified in Section 9.4. We describe the properties of a collaborative IA process at both the regulatory and applied levels. In Section 9.5.1 we explore how IA requirements could facilitate collaborative IA practice. In Section 9.5.2 we demonstrate how a collaborative IA process could be expressed at the applied level. In Section 9.5.3 we illustrate how a collaborative IA process could be expressed for various IA types.
- In Section 9.6 we address the contemporary challenge of IA capacity building. We identify key conceptual distinctions and offer good practice guidance.
- In Section 9.7 we highlight the major insights and lessons derived from the analysis.

9.2 INSIGHTS FROM PRACTICE

9.2.1 Collaboration and Technical Expertise Do Not Always Achieve Everyone's Idea of Sustainability

Sustainability appraisal was introduced by the government in 2004 to apply to regional and local spatial plans in England. It is a process, whereby emerging plans are compared against a sustainability framework comprising criteria selected, based on locally relevant issues, and tested through indicators. As such, it attempts to predict the consequences of various draft policies, leading to the selection of preferred policies based on the consideration of sustainability outcomes. Its effectiveness, however, is highly contested because of the value-based nature of the assumed goal (sustainable development), because effectiveness can be viewed through different theoretical framings, and because good governance does not guarantee sustainable outcomes

Impact Assessment: Practical Solutions to Recurrent Problems and Contemporary Challenges, Second Edition. David P. Lawrence.

(a more detailed explanation can be found in Bond et al., 2011). In practice, sustainability appraisal is typically conducted by consultants (paid by authorities) who develop the framework, the authorities consult on it, and then the consultants apply it behind closed doors.

A research study in England examined the implications of large-scale land-use change through focusing on the potential for large increases in land area for biomass crops such as *Miscanthus* and Willow. As part of the research process, sustainability appraisal was undertaken in the South West and East Midlands regions of England to examine these implications; the results highlighted potential solutions for the three issues highlighted above, and these are considered, in turn, below.

The normative nature of sustainable development has been recognized as a confounding factor when attempting to achieve consensus on future planning. It is argued that the many interpretations of sustainability make legitimization of any decisions difficult. Research undertaken on sustainability appraisal of land-use plans in England has suggested that it favors socioeconomic benefits at the expense of the environment. It is further suggested that this is inevitable, given that the plan-making process inherently involves trade-offs that are likely to lead to interpretations of sustainability in which the environment loses out (weak sustainability). In the research project, affected stakeholders were brought together in order to develop the sustainability appraisal framework, and they acknowledged the inevitability of detrimental trade-offs at the decision stage. This possibility was mitigated through the use of constraint mapping, whereby areas of critical natural capital (as defined by the stakeholders) were excluded from the appraisal process (and from the possibility of future planning), rendering subsequent trade-offs environmentally benign.

Historically, assessment processes were based on the theory of rational choice, whereby decision makers would make better decisions, given better information. However, the effectiveness of decision making can be viewed through different theoretical framings and few people argue that decision making is entirely rational, with arguments made that assessment processes have different roles in decision making, ranging from imposing institutional values on the decisions, through to being purely symbolic to give the impression of rationality where none exists. Recently, more arguments are being made that powerful actors in the process manage to subvert assessment processes to reach particular conclusions. The research acknowledged these different theories associated with effectiveness of sustainability appraisal and addressed them through embedding greater pluralism in the process, both in terms of framing the sustainability appraisal itself (through its objectives and indicators), and through the interpretation of the results. This demanded considerably more consultation than is the norm for such appraisals, with at least two workshops to define and refine the sustainability framework.

Based on the Commission of the European Communities, definition of "good governance" as being underpinned by five principles—openness, participation, accountability, effectiveness, and coherence—it can be argued that uninformed democracy is likely to lead to more unsustainable outcomes, whereas expert-led decision making is likely to lack legitimacy. The argument is simply that a lack of comprehensive understanding of natural systems by laypersons makes it somewhat difficult to agree on solutions that benefit, rather than negatively impact, the environment. A solution proposed, coming out of the research study, was to undertake more analytic-deliberative sustainability appraisal, whereby affected stakeholders had the opportunity to work with experts to set the sustainability appraisal framework (in the defining and refining workshops mentioned above). However, an interesting finding was that national stakeholders, who were involved because of their legal responsibilities, covering areas like water quality, heritage, and nature conservation, typically failed to apply a local understanding (or if they had local understanding, it was blanked out by the institutional line), which made it more difficult to generate an appraisal framework specific to a geographical area. A particularly revealing feature of the work was that the affected local stakeholders were unwilling to walk away from the process after the framework was agreed, arguing that only people with a local understanding and a stake in the sustainability of an area could properly interpret the results of a sustainability appraisal and, most importantly, agree on trade-offs. This led to follow-up meetings to allow affected stakeholders to be involved in the application of the framework rather than have it take place behind closed doors. Such engagement with locally affected stakeholders helps to embed ecosystem services in the appraisal, given that those people who use and value those services are involved with the appraisal.

ALAN BOND

School of Environmental Science, University of East Anglia, Norwich, UK

9.2.2 Active Stakeholder Support for an Effective EIA Process

There is an old adage in EIA circles, especially held by regulators, to the effect that "a good EIA process is one in which all of the stakeholders are little unhappy." If the proponent is a little unhappy, it likely means that while they have received an approval for their proposal, they are being required to implement more rigorous or demanding mitigation and follow-up measures than they would prefer. And if an environmental group is a little unhappy, it likely means that their concerns have been listened to and at least partially met, although the approved mitigation measures have not gone as far as they would like them to (or, of course, the proposal has not been rejected outright as they might wish were the case). Where EIA stakeholders are extremely happy or extremely unhappy, respectively, it probably spells trouble for regulators and/or the environment. Thus the aim is to keep everyone equally engaged, albeit in a slightly disgruntled state. This has seemingly long been the ambition of EIA regulators in Western Australia, as it was from a senior staffer within the Office of the Environmental Protection Authority (EPA) that we first heard this adage many years ago. We have since heard EIA regulators from other countries express a similar sentiment on this matter.

Perhaps an even better test of a good EIA process though is one that all stakeholders actively defend. We have witnessed this in recent years during meetings of the Stakeholder Reference Group (SRG) that was established to provide advice to the Western Australian EPA on a review of EIA that commenced in 2009. Membership of the SRG includes various government agencies (e.g., health, planning, mines and petroleum, local government), peak industry group representatives (e.g., mining, petroleum, land development, environmental consultants association), and environmental groups. While it might be expected that there could be friction or even hostility between some of these groups, not to mention each having its own agenda for seeking changes to EIA that might especially benefit their own sector, we have witnessed a high level of cooperation overall and a sense of working for the greater common good. Furthermore, at times, the SRG members have explicitly sought to strengthen or improve the EIA process or the EPA's capacity to implement EIA in Western Australia. An example of the latter was an offer for industry and consultants to assist with the preparation of EIA guidance materials in light of serious staff and capacity shortages experienced by the EPA. The industry and conservation groups alike realized that a weak EPA, and by extension a weakened EIA process, would actually be counter to their interests even coming as they often are from opposite ends of the "green spectrum."

A good EIA system appears to be highly valued by industry, especially because of the financial ramifications that ensue. Getting a recommendation of approval from the EPA is good for business with company share prices often going up in immediate response (and the converse is also the case). An EPA nod of approval (with subsequent endorsement by the Ministry of Environment) also seems to be imperative for ventures requiring capital from banks or other lenders. Thus, the environmental credentials extended to a company through a favorable EIA process leading to an approval decision have considerable value. The EPA of Western Australia have always grounded their assessments in the best available science and with a high degree of consistency between individual proposals but also have set out incrementally higher standards and expectations for environmental protection over time. This appears to have earned them high respect from all stakeholders and is important for making EIA more influential than otherwise might be the case.

While we are heartened by the cooperation of diverse stakeholders in working toward an improved EIA process in Western Australia, we should note that there is one stakeholder group that we have never known to be actively and explicitly represented: future generations. Despite the addition in 2003 of "intergenerational equity" as an object of the *Environmental Protection Act 1986* and as a guiding principle for EIA in Western Australia, this matter seemingly gets little attention. Instead, all stakeholders appear to be mainly driven by immediate interests. For the EPA itself, it appears to be intent on weathering the storms of short-term government cycles with the changes to departmental structures, efficiency, promises (i.e., chiefly for faster development approvals), and budget cuts, either actual or threatened. For industry stakeholders, time horizons appear to be based around 5–10 years resource development plans. Finally, for environmental groups, there is seemingly always a current campaign for conservation to be fought for in response to industry and government activities and the media profiling of current events in society. Sadly, the long-term view of sustainability and the interests of future generations do appear to get drowned out by the immediate concerns of EIA stakeholders.

In summary, a strong EIA system is in the best interest of all stakeholders in the process and in my mind a sign of a robust or effective EIA system is one that the stakeholders actively support or defend. However, an ongoing challenge for all practitioners is to enable the rights of future generations to be represented in current EIA processes and decision making.

ANGUS MORRISON-SAUNDERS

School of Veterinary and Life Sciences, Murduch University, Murdoch, Western Australia, Australia

9.2.3 Making Policy-Level Participation Effective: Territorial Impact Assessment (TIA) of European Union Policy Proposals

Anticipated and unanticipated impacts of public policy can occur at different geographical scales. In Europe, various such effects have been observed in relation to numerous EU policies, for example, the so-called Seveso II Directive on the control of major-accident hazards (Directive 96/82/EC). In one case, a new stadium in the Netherlands could not be built adjoining a railway station given the potential for trains carrying dangerous chemicals to also use the railway and pass in close proximity to the site. This created a conflict with existing Dutch transport and land-use policy, which strongly supports good public transport access to stadia.

Since 2003, the European Commission has undertaken Impact Assessment (IA) of its policy proposals to detect and evaluate potential positive and negative economic, social, and environmental impacts. These IAs are essential documents, prepared by experts (usually consultants). While they are a laudable effort, making reliable predictions at this level of decision making is notoriously difficult, particularly as impacts normally vary quite substantially across 27 European Union member states with over 500 million inhabitants on

4.3 million km^2 of land and may depend in particular on the unique characteristics of a specific region or locality. As a consequence, IAs have usually remained rather vague and are unlikely to detect conflicts of the kind described above.

There have been various research projects looking into how regional and local effects may be assessed better and more reliably, funded by ESPON (European Observation Network, Territorial Development and Cohesion; www.espon.eu). These have mostly revolved around spatial (computer) modeling of impacts, focusing on ex post assessment. In addition, modeling tools for ex ante assessment have also been developed. To date, these have been used to carry out some exploratory assessments. However, overall, the perceived accuracy has remained contested. Also, data requirements for models, aiming at assessing impacts on several hundreds of regions in 27 countries in a consistent manner, are enormous. As a consequence, exercises so far have taken up to several years to complete. They have thus remained impracticable for use in "real" policy-making processes that only allow up to a few weeks for establishing evidence on potential impacts.

Against this backdrop, between 2010 and 2012, ESPON funded a stakeholder-based and participatory project on developing a Territorial Impact Assessment (TIA) methodology of European policy proposals, to be applied at the level of EU member states, with input from regional and local levels, called ESPON and TIA (EATIA). In this context, territorial impacts are seen both as impacts on spatial usage (e.g., sprawl or new infrastructure) and also as broader socioeconomic and environmental impacts that may differ in different regions or localities across the EU territory.

The EATIA project was initiated by stakeholders from the ministries responsible for spatial planning in the United Kingdom, Slovenia, and Portugal and was delivered by a project team from the Universities of Liverpool, Porto, Ljubljana, and Delft. Key priorities of the project were that TIA should provide a supportive tool in the policy-making cycle to enable European, national, and regional/local policy makers and practitioners to anticipate or adapt to emerging sectoral policies and that it should be able to be undertaken ex ante and without being delayed by heavy data requirements.

A TIA methodology was developed, covering governance, methods, and procedural aspects. In the development of the methodology, interactive learning networks played a key role. These consisted of 15–25 public and private sector stakeholders with an interest in spatial planning, territorial policy, and impact assessment in each of the three stakeholder countries. These provided for critical feedback and suggestions throughout the lifetime of the project. The work of the interactive learning network was key to devising a methodology that, overall, is perceived to be practicable and simple and which has the support of those who are supposed to use it (i.e., national, regional, and local administrations). Furthermore, the methodology was tested by other national, regional, and local practitioners in the three countries and their feedback was used to refine the approach.

The emerging TIA framework is based on a highly efficient communicative process, which is to be coordinated with other existing assessment instruments (e.g., regulatory impact assessment or strategic environmental assessment), whenever possible. The process consists of four main stages: screening, scoping, assessment, and evaluation. Key to a successful TIA is skilled interdisciplinary teams, representing various public administrations that come together with a cooperative spirit, and that reflect a high level of familiarity with the assessed policy area and wider territorial and sustainability expertise. The process is thus practitioners' driven and qualitative.

Screening (whether or not to conduct TIA) is led by central government departments responsible for a particular policy area, supported by the department responsible for spatial planning and/or impact assessment. Once a decision is made to conduct TIA, the scope of assessment is to be established by the same bodies. Testing has shown that with some experience, screening and scoping can be completed in half a day.

Impact Assessment is done by regional- or local-level planning bodies. In this context, existing planning processes and teams are used that should be able to complete the exercise within a half-day workshop. Evaluation of assessment results is done in terms of territorial policy objectives at national levels. This is based on amalgamating and analyzing information generated at regional/local levels, which can be achieved efficiently by using web-based reporting sites. The results are to be fed into the policy negotiation process at the European level. Methods used in the TIA methodology include logical chains and matrix evaluations, based on sets of between 15 and 20 economic, social, environmental, and governance-related indicators. Furthermore, collaborative workshops and web-based feedback mechanisms play important roles.

There are good chances that the developed TIA framework will be promoted in all EU member states. In addition, efforts to develop more effective impact modeling techniques at the European level will also continue. Ideally, both, participatory national/regional/local-level assessment exercises and European-level quantitative modeling tools would be applied in a complementary manner, providing a strong evidence base for making informed European policy decisions.

For further information on the approach, go to http://www.espon.eu/main/Menu_Projects/Menu_TargetedAnalyses/EATIA.html

THOMAS B. FISCHER

4imPiAct Research Team, School of Environmental Sciences, University of Liverpool, Liverpool, UK

9.3 DEFINING THE PROBLEM AND DECIDING ON A DIRECTION

The preceding stories illustrate some of the complexities and subtleties associated with collaborative IA processes. The first story demonstrates that collaboration among stakeholders does not always lead to consistently understood and supported substantive environmental enhancements. Sustainability is not the inevitable outcome of a process that generates a consensus among participants. Interactions between process and substance, and between specialist and local knowledge, need to be thoughtfully structured such that sustainability principles and substantive knowledge bound and direct IA procedures, mutual education is facilitated, and the many benefits of open engagement and collaboration with locally affected stakeholders are retained and enhanced. The second story makes the point that an effective IA process or system does not mean that all stakeholders should be uniformly satisfied with the outcome. Either extremely happy or unhappy, IA stakeholders can be problematic. The aim instead should be an IA system and process that all stakeholders can actively defend, where there is a high degree of cooperation, and where all parties have a sense of working for the "common good." Time horizons, however, need to be extended, especially regarding the interests of future generations. The third story addresses the difficult task of facilitating stakeholder engagement at the policy level. It describes a stakeholder-based and participatory approach to TIA. The TIA methodology for assessing European policy proposals integrates both technical methods (e.g., logical chains and matrix evaluation, quantitative modeling) and collaborative procedures (e.g., interactive learning networks, collaborative workshops, web-based feedback). The story illustrates that stakeholder collaboration at the policy level is practical, and can be undertaken in a manner that is highly communicative, technically sound, and effectively coordinated with existing assessment instruments.

Numerous benefits, as highlighted in Figure 9.1, have been ascribed to effective public participation in public and private decision making. Public participation is intrinsically beneficial to participants and to society (Day, 1997; Nagel, 1987; Pateman, 1970). It is consistent with human nature. It is ethically just to involve the public in decisions that could affect their lives (Praxis, 1988).

Effective public participation has considerable developmental value (Nagel, 1987). It can facilitate a greater level of interest and involvement in public life (Morgan, 1998; Pateman, 1970). It can enhance the confidence of and political skills of participants (Day, 1997; Dunning, 1998). Citizens can more ably articulate their preferences and demands (Burdge, 2004; Day, 1997). They can be better informed about and more ably participate in planning and decision-making processes (Burdge, 2004). The application of these skills can empower people, further community identity and development, foster environmental sensitivity, and contribute to a more democratic and responsive political system (Barrow, 1997; Day, 1997). Also, social learning is promoted, skills transfers are facilitated, local service capital is created, and community buy-in is fostered (Buchan, 2003; Noble, 2009b).

Public goals can be advanced by effective public participation. Decision making can become more transparent, balanced, and accountable (Alshuwaikhat, 2005; Sinclair and Diduck, 2009). Democracy is facilitated (Sinclair and Diduck, 2009). "Regulatory-capture" by proponents can be avoided and the public can have more direct access to decision making, prior to final determinations (Hyman et al., 1988; Shepherd and Bowler, 1997; Sinclair and Diduck, 2009). Public values, perspectives, and preferences are incorporated into decision making (Morgan, 1998). Relations among diverse stakeholders may be improved (Tamburrini et al., 2011). Unrepresented people are able to present their views. The public is able to examine expert knowledge and to weigh and test the positions and decisions of elected representatives (Pateman, 1970). Public alienation and marginalization are less likely to occur. The public has a better understanding of environmental conditions, project characteristics, local issues, and potential impacts (Lee, 2000; O'Faircheallaigh, 2010). Decision making is more likely to reflect and be responsive to stakeholder and community needs (Shepherd and Bowler, 1997; US EPA, 1988b). Decision making is fairer and more accountable to the public (Barrow, 1997). A check on government and private action is provided. The balance of power (i.e., empowering marginalized groups) can be shifted and decentralized (O'Faircheallaigh, 2010). There is a greater potential for individuals, groups, and communities to use IA and other decision-making tools to help solve their own problems and to better influence their own futures (SERM, undated). The ability of the public and communities to cope with change can be enhanced (Burdge, 2004).

Better decisions and more effective, balanced, and efficient decision making can result from effective public participation (Burdge, 2004; Elling, 2011; Howell et al., 1987; Lee, 2000; Noble, 2009b). A means is provided for obtaining local and traditional knowledge and for determining local issues, perspectives, and values (Bisset, 1996; Glasson et al., 1999; Morgan, 1998; Noble, 2009b). The public can help diagnose and clarify problems, opportunities, and solutions, illuminate goals and objectives, formulate and evaluate options, identify, predict, and integrate impacts and socially acceptable solutions, interpret impact significance, determine appropriate mitigation, compensation, and monitoring measures, and decide on proposal acceptability (Barrow, 1997; Greer-Wooten, 1997; Hughes, 1998; Noble, 2009b; Sinclair and Diduck, 2009; US EPA, 1988b). Public contributions to decision making can make it easier to establish priorities. Access is provided to local and traditional knowledge (Sinclair and Diduck, 2009). Management expertise can be enhanced and new financial, human, and in-kind resources accessed (Praxis, 1988; Sinclair and Diduck, 2009). Decision makers are better able to plan for and adapt to change (Day, 1997; Lee, 2000). Costs and delays associated with public opposition are less likely (Glasson et al., 1999; Noble, 2009b). Project management can focus on key public issues (Praxis, 1988). Approvals and

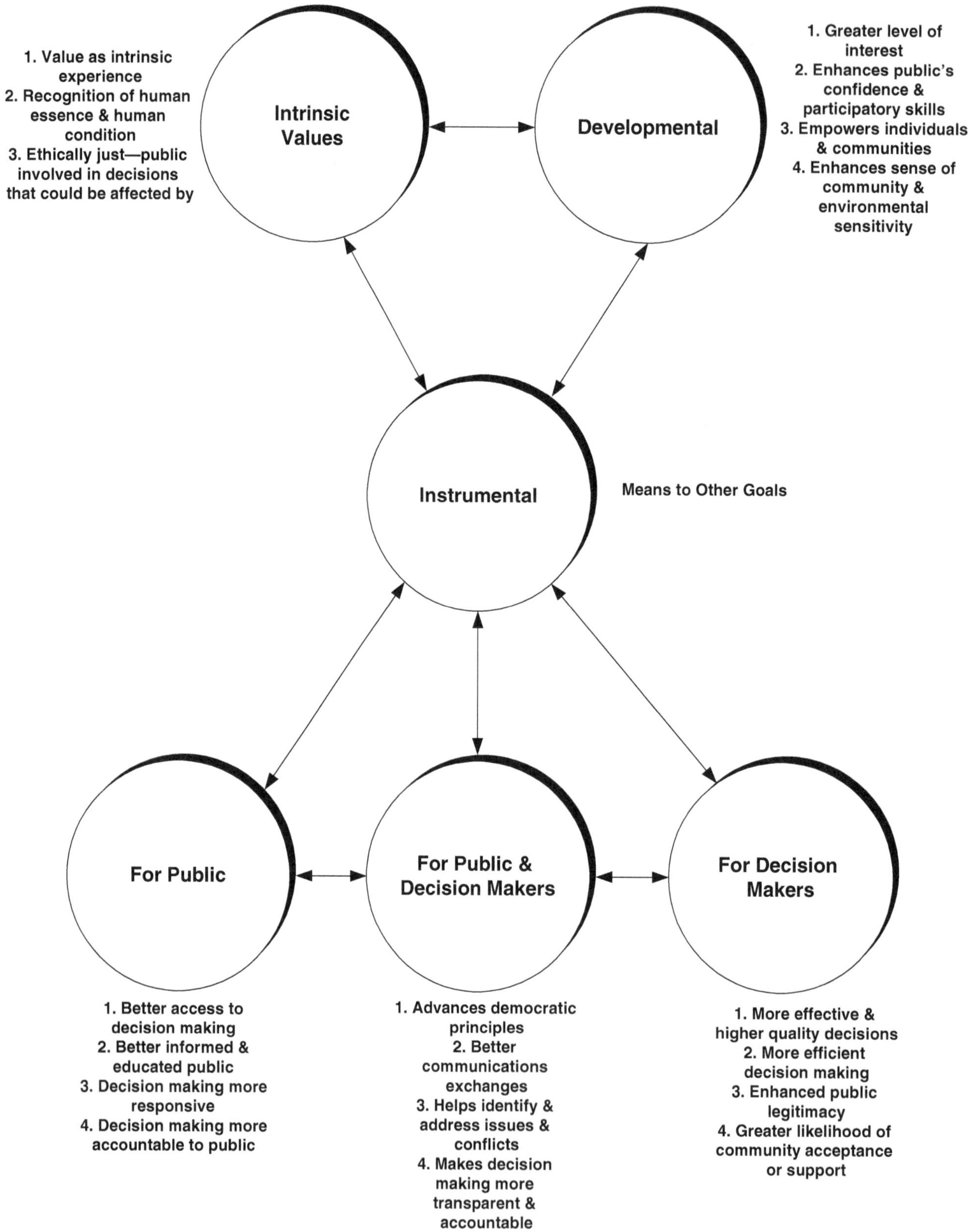

Figure 9.1 Examples of potential benefits of effective public participation.

implementation tend to be less complex and confrontational (Noble, 2009b; SERM, undated). Decision makers are viewed as more credible and the decision-making process is perceived as more legitimate (Barrow, 1997; Creighton et al., 1983; Smith, 1993). The possibility of legal challenges is reduced (Noble, 2009b). There is likely to be less hostility and a greater level of community trust, acceptance, and sometimes support for both the process and for the decisions resulting from the process (Hyman et al., 1988; Lee, 2000; Shepherd and Bowler, 1997; Tamburrini et al., 2011).

The public and decision makers can benefit jointly from a decision-making process based on popular sovereignty and political equality principles. All parties benefit when misunderstandings are clarified and when information and knowledge are effectively shared (Hughes, 1998). Effective public participation provides a means of identifying and a forum for resolving issues (Bisset, 1996; Hughes, 1998). It offers a mechanism for building consensus and for avoiding, reducing, and resolving conflicts (Greer-Wooten, 1997; Praxis, 1988; Tamburrini et al., 2011). It can contribute to more open, transparent, and democratic planning and decision making (CCMS,1995; Shepherd and Bowler, 1997). Narrow technical biases can be ameliorated. A countervailing force is established to offset administrative and political power concentrations. Broad public involvement and support also can facilitate sustainability initiatives (Barrow, 1997; SERM, undated). Effective community-based public participation approaches can facilitate learning, build IA capacity, and contribute to better practice (Sinclair et al., 2009).

It would be difficult to realize most of these ascribed benefits in an IA process characterized by late public involvement and/or through periodic public involvement events intended largely to inform the public. Instead, early public involvement and a more continuous and interactive IA process seem more in order. Effective two-way communications and mutual education seem essential. Mechanisms to anticipate, avoid, and resolve disputes, to build consensus, to collaboratively plan, to solve problems, and to take advantage of opportunities all appear necessary. These elements of effective public participation, as illustrated in Figure 9.2, need to be guided by general principles, goals, and good practices, structured by integrative frameworks and bounded by limits of acceptable practice.

It is far from clear if or how well these elements of effective IA public participation are being or are likely to be satisfied. There are both positive and negative patterns and trends. There is also considerable variability in the quality of IA practice. IA requirements and practices have significantly increased public information and input into agency decision making (US CEQ,1997a). Earlier and more continuous public consultation is being emphasized to a greater extent. There are numerous examples of sincere, creative, and effective approaches for involving the public, for resolving disputes, and for collaboratively solving problems (Carpenter, 1991; Creighton et al., 1983, 1999; DeSario and Langton, 1987; Gray, 1989; Susskind and Cruikshank, 1987; Susskind et al., 1999). A concerted effort has been made to provide detailed guidance and to identify and interpret the lessons of public involvement, alternative dispute resolution, and collaborative planning practice (Creighton et al., 1983, 1999; CSA, 1996; PCSD, 1997; Praxis, 1988; US DOE, 1998; US EPA, 2001a,b). In recent years, there has been a greater emphasis on two-way communications, on transparent and accountable decision making, on outreach to traditionally underrepresented groups and NGOs, on facilitating procedural and distributive justice, and on community empowerment. These trends, analyses, and guidance materials, although pertaining to many forms of public and private environmental decision making, also are largely applicable to IA practice.

Concurrent with these positive trends and developments, there has been a tendency for public agencies to opt for forms of IA that preclude or severely restrict public involvement (Shepherd and Bowler, 1997; Solomon et al., 1997). The trends toward deregulation, the application of business principles and concepts to public administration, and privatization could further inhibit public involvement in IA practice (Bisset, 1996; Sinclair and Diduck, 2001). The treatment of public participation in IA guidelines is highly variable. Too frequently, guidelines are confusing, lacking in practical guidance, and weak in proactively advocating public involvement (Hughes, 1998). The public role in IA practice is often poorly defined (Harrop and Nixon, 1999). Much of the time, public involvement begins after major decisions have been made and only occurs at two or three key decision points in the IA process (Freudenburg, 1983; Shepherd and Bowler, 1997; Solomon et al., 1997). Oftentimes, public consultation is limited to disseminating information and gathering public comments, frequently in poorly structured processes. The range of public consultation methods employed in practice remains narrow and there is a tendency to overemphasize quantitative methods and biological and physical impacts (Solomon et al., 1997). Public participation rarely extends into the postapproval stage (Harrop and Nixon, 1999).

The gulf in perspectives between proponents and the public regarding the need for and role of public participation is still considerable (Fell and Sadler, 1999). The level of public distrust and cynicism remain high concerning the motives of decision makers and the weight they attach to public comments and suggestions (Mittelstaedt et al., 1997). Citizens sometimes are frustrated and feel that they are being treated as adversaries rather than as welcome participants (US CEQ, 1997a). In some cases the role of citizens is substantial and influential. But in others it is largely symbolic. There is a particular need for reforms to enhance the involvement and influence of indigenous peoples and governments (Mittelstaedt et al., 1997; Paci et al., 2002; US CEQ, 1997a). Despite the many benefits of effective public participation and the array of instructive analyses and guidance materials, the reality of public participation still falls well short of the promise (Lawrence et al., 1997). This

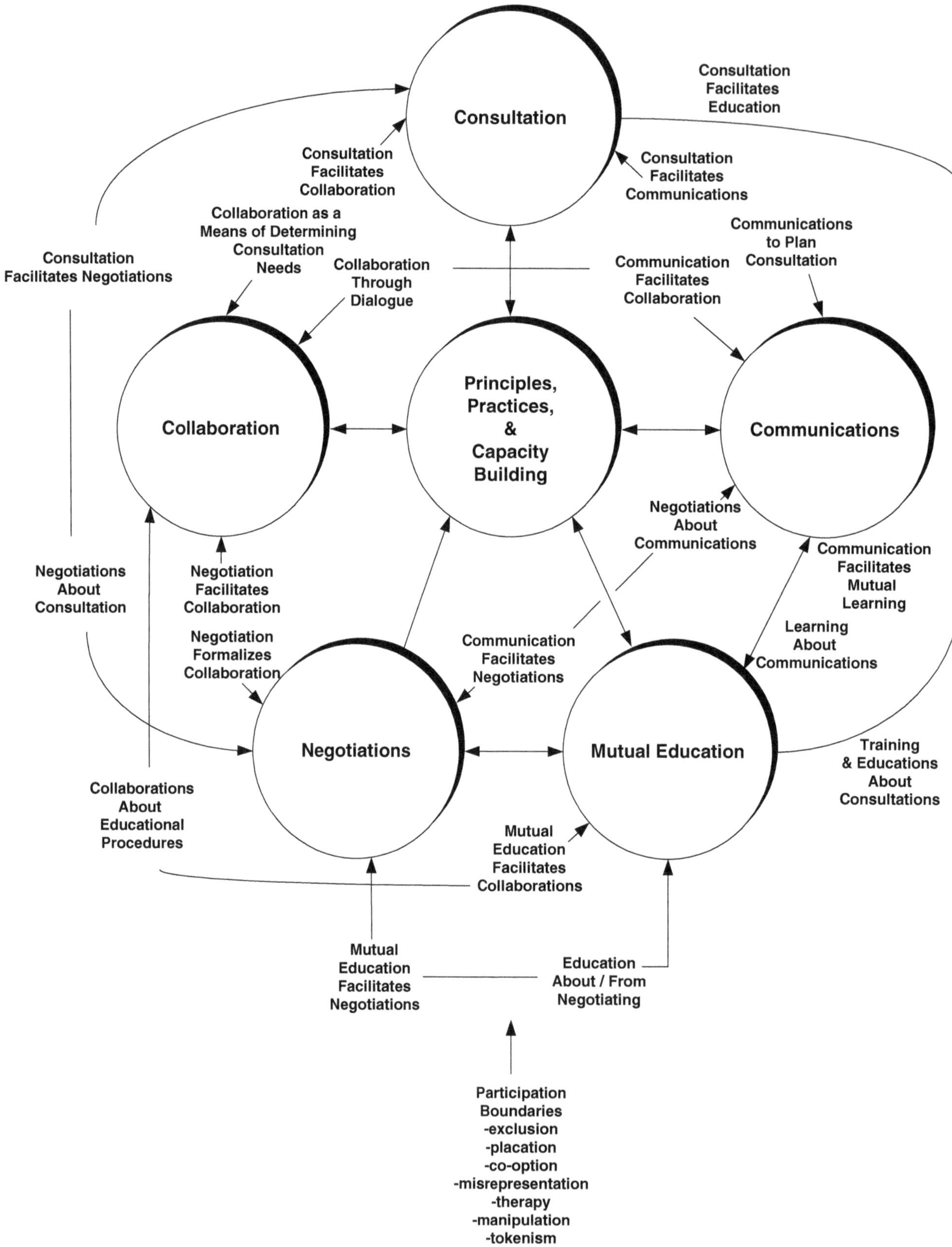

Figure 9.2 Examples of interactions among IA public participation types.

discrepancy between potential and performance raises several questions. (1) Are there major disadvantages, which largely offset the benefits of effective public participation? (2) Are there unresolved, and perhaps irresolvable issues, which generally preclude an enhanced level of public participation in IA practice? (3) Is the problem simply one of applying insights and lessons from the available literature and guidance materials? (4) Are the relevant prescriptive materials too scattered and in need of consolidation and succinct presentation? (5) Is it necessary to integrate additional concepts and frameworks? In Section 9.4 we answer each of these questions. We also seek to maximize the benefits of effective public participation (as highlighted in Figure 9.1) and to address and integrate the elements of effective IA public participation (as displayed in Figure 9.2). The Section 9.4 analyses provide the foundation for the collaborative IA process presented in Section 9.5.

9.4 SELECTING THE MOST APPROPRIATE ROUTE

9.4.1 Disadvantages and Issues

Public participation is a generic term for all types of activities designed to include the public in the decision-making process, prior to and after a decision (Ross and Thompson, 2002). IA public participation entails the active involvement of the public in the IA process (Sinclair and Diduck, 2009). The role of the public is direct and acknowledged (Lee, 2000). Members of the public can influence or attempt to influence decision-making outcomes (Nagel, 1987). Meaningful public participation (i.e., more than a one-way flow of information) encompasses such essential participation elements as integrity and accountability, influence, fair notice and time, inclusiveness and adequate representation, fair and open dialogue, multiple and appropriate methods, two-way communications and education, adequate and accessible information, participant support, and informed participation regarding all environmental aspects of the proposed action (Law et al., 2005; Sinclair and Diduck, 2009; Sinclair et al., 2012; Stewart and Sinclair, 2007). Most parties generally accept that public participation in IA practice is desirable. However, qualifications and exceptions are rapidly identified. There also are varying definitions of what represents public participation (e.g., involvement, influence) and when it is most appropriate for it to occur. "When" concerns both the circumstances under which public participation is appropriate and when in the IA process. Table 9.1 lists examples of reasons commonly advanced for why public participation is undesirable or why it should be severely limited. Comments regarding the validity of the reasons also are provided. Public participation may be viewed as undesirable by proponents and regulators or by members of the public, albeit for different reasons.

The case against public participation, as detailed in Table 9.1, is usually either dubious or overstated. If public participation is accepted as necessary (i.e., a right) and generally desirable, then arguably the burden of proof should be on those seeking to prevent or curtail public participation. Moreover, to realize the benefits of public participation (as highlighted in Figure 9.1), it is generally more desirable for public participation to start early in the IA process and to occur either continuously or at frequent intervals throughout the IA process. Circumstances do vary, however. Occasionally, there may be valid reasons for precluding or limiting public participation activities. At a minimum, consistent with the goal of decision-making transparency, a clear rationale should be provided for such limitations.

The concerns listed in Table 9.1 can be valid when a public participation program is poorly designed or executed. Care should be taken to ensure that avoidable public participation "disadvantages" do not occur. The public participation program should be appropriate (i.e., suited to the situation), efficient (i.e., time and other resources are not wasted), and effective (i.e., achieves the shared objectives of the participants). The scope and limits of the program should be jointly determined and should be clearly specified from the outset. The program should evolve in conjunction with the IA process. It should include an appropriate blending of consultation (e.g., information exchange, continuous involvement), communications (e.g., publicity, dialogue), education (e.g., stakeholder, proponent, mutual), negotiations (e.g., to identify, avoid, and resolve disputes), and collaboration (e.g., to build consensus, to jointly solve problems, to create "win–win" solutions) elements.

Except in very special circumstances, the "disadvantages" of public participation in the IA process do not appear to be valid or can be avoided or minimized through good public participation practice. Therefore, public participation disadvantages do not provide an adequate explanation for the discrepancy between the potential and the performance of public participation in the IA process.

Public participation issues can be more of a challenge. Figure 9.3 highlights some issues often encountered in IA public participation practice. Clearly, the management of public participation activities in the IA process requires numerous, complex, difficult, and subjective decisions. These public participation issues do not provide a rationale for not undertaking or severely restricting public participation activities. Quite the opposite! The issues are not resolvable through technical analysis or political expediency. Instead, they should be addressed jointly with interested and affected parties. A breadth of perspectives can make it easier to identify and explore issues. A sounder foundation also can be established for reaching and substantiating interpretations and decisions.

Public participation has been a component of IA practice for more than 30 years. The quality and effectiveness of practice (both in IA and in related fields) has advanced rapidly, especially over the past 10–15 years. Thus, there are ample good practice examples and guidance materials to help in designing and executing public participation and dispute resolution programs, for anticipating and managing

Table 9.1 Analyses of "Disadvantages" of Public Participation

Proponents and Regulators	
It costs too much	• The public participation program can be designed to stay within available resources • Public participation is likely to be a tiny fraction of overall project costs. It is a cost of good practice • The public often contributes insights that lead to cost savings • Public opposition is more likely with a closed process. The additional costs associated with approvals, litigation, and implementation are likely to be far greater than the costs of public participation
It will lead to delays	• Public participation can extend timelines but is an essential facet of democratic decision making • A properly designed and managed public participation program should not result in significant delays • The time associated with preparing for and participating in lengthy, adversarial hearings can be considerable. Hearings can often be avoided or greatly abbreviated with effective public participation • The implementation timetable, even if approval occurs, will likely be extended because the "stage has not been set" through effective public participation
It will make decision making less efficient	• Short-term efficiency gains are likely to be more than offset by the costs and delays associated with not consulting the public • Public knowledge and experience can provide insights that lead to greater efficiencies • Public issues can help focus the IA process (i.e., scoping) • The public has a right to be involved in decisions that affect them • IA leads to more open and balanced decision making • One of purposes of IA is to broaden the decision making to encompass more than efficiency concerns
It is divisive. It will result in a partisan process	• Not involving the public can be even more divisive. Conflict is deferred and usually exacerbated • Effective public participation can help build consensus and avoid, ameliorate, and resolve conflicts • Public participation is an effective check against partisan "technical" analyses and interpretations
The public lacks the knowledge and skills to contribute to the process	• Part of effective IA participation involves enhancing public understanding • The public usually possesses valuable local knowledge and experiences, which can enhance the process • Much of IA practice involves interpretations of significance and acceptability. The public can and should contribute to such interpretations
The public will adopt a "not in my backyard" position	• NIMBY is a natural, reasonable, and appropriate response to a potential intrusion into the community • Early public involvement can mean that the public is a partner in identifying and comparing choices • There are generally multiple publics with multiple perspectives • Perspectives often change through the course of the public participation program • NIMBY is less likely with a voluntary siting approach
It will raise the project profile and empower opponents	• The public has a right to be informed about decisions that might affect them • An effective public participation program can clarify misunderstandings that are often the basis for conflict. A worthwhile project should be able to stand up well to public scrutiny • Effective public participation can reduce opposition and lead to a greater level of community acceptance and support. It could help scope or even avoid legal action • A demonstrably inadequate public participation can be a source of even greater power for opponents
The public only has one point of view (e.g., environmentalists), which we know	• There are multiple publics • The public has multiple values, perspectives, and interests • Public perceptions, attitudes, and positions often change through the process • Even if positions are understood, they should still be expressed. It is unlikely that the bases for positions will be fully understood without effective public participation
Short-term local interests may have to be overruled by regional, long-term needs	• Agencies and regulators can retain final decision-making authority • Local participation is essential to fully understand the trade-offs involved • Local participation can help in addressing mitigation and compensation options and measures • The dichotomy can be a false one. Sometimes local participation makes it possible to identify "win–win" solutions
We will lose control of the process	• Proponents and regulators can always retain final decision-making authority • The delegation or sharing of control is a choice. It can increase decision-making credibility and legitimacy • There are many forms of public participation that do not involve power sharing or delegation
We lack the necessary participation skills	• There are numerous specialists in public participation and alternative dispute resolution • There are many "user-friendly" public participation resource materials, which are readily available • Public participation training programs are widely available
The process will be "hijacked" by activists	• There is a danger that a few individuals can dominate a public meeting • There are many other public participation methods that are less prone to such problems • Activists usually have something to contribute and should be consulted • Approaching multiple publics using a variety of methods can largely offset such problems

Table 9.1 (*Continued*)

Proponents and Regulators	
The participants are not "representative" of the public	• Public participation is a voluntary process. Some element of self-selection is inevitable • Ensuring that the full range of relevant values and interests are integrated into the process is usually more important than how representative an individual is of a larger constituency • A valid concern or suggestion is of value regardless of the level of support • The use of multiple consultation methods and ample opportunities for stakeholder representatives to consult with their constituencies can further ameliorate the issue of representation
It will lead to land speculation	• Land speculation is an issue only during site selection • Land can be purchased or expropriated at values prior to site announcements. Land can be optioned • The validity of the choice can be subject to public scrutiny before final decisions are made
It will result in the release of confidential materials	• This is rarely a valid issue • It is usually possible to protect proprietary information in an IA process • Confidentiality issues tend to arise more frequently as part of hearings and court cases. Effective public participation can reduce the likelihood and scope of such proceedings
It will confuse the process. It will be less rational. There will be multiple perspectives and possibly errors	• There should be minimal confusion with a well-designed and well-managed public participation program • Data can be checked for accuracy. Facts and values are mixed in practice • Multiple perspectives simply reflect the value context. Different perspectives can facilitate interpretations, address uncertainties, and help resolve problems • Complex projects require more, not less, participation to ensure that potential impacts and uncertainties are adequately identified and interpreted. Effective public participation planning is essential
It will create expectations that we cannot fulfill. The results could be inconclusive	• The scope and limits of the public participation program can be clearly defined from the outset • Public participation provides a means of determining and of transmitting commitments to the community • Inconclusive public participation results could simply reflect multiple perspectives and divisions within the community. This does not preclude taking the comments and suggestions provided into consideration • Sometimes public participation can help build consensus and identify and narrow differences
Public issues will dominate the process. The environment will suffer	• Public issues generally closely parallel environmental issues and uncertainties • The public is often a useful source for identifying and interpreting potential environmental impacts and uncertainties. Perceptions of impacts and uncertainties are "real" social impacts • The public participation program provides a means of addressing public misconceptions • The assessment of public issues does not preclude considering other issues and concerns. The public is usually supportive of addressing impacts, uncertainties, and concerns identified by others
The public will lose interest or will not be interested in policies or programs	• Participation is voluntary • A focused and well-planned participation program is more likely to maintain interest • Participation for plans, policies, and programs requires alternative approaches (e.g., the involvement of national and regional NGOs, surveys) rather than no participation
The project is too urgent	• If the need is genuine, the public is often supportive of an accelerated project schedule. "Urgent" timelines are frequently artificial • Public participation can usually be designed to meet a project schedule • Even when the need is urgent, it is still essential to minimize adverse impacts and uncertainties. The public can contribute to such efforts
It is too early in the process. We don't yet have a proposal	• Involving the public when there is a need or opportunity and no clear proposal is the best time to commence public participation. In this way the public can fully participate in the decisions leading up to the proposal determination • One of the major public objections tends to be that the decisions have already been made
The project is too small. There are no or negligible impacts	• Small projects with no to negligible impacts are not (or should not be) subject to IA requirements • Such projects can be addressed with categorical or class assessments or by means of a streamlined IA process. There should be a "bump-up provision" for significant impacts. Some level of public participation can still occur • There is still an argument for decision-making openness and transparency regardless of impact scale • Cumulative effects can be an issue with multiple small projects or projects in sensitive settings. The public can assume a valuable role in addressing such issues • Negligible is a matter of interpretation. The public can contribute to such interpretations

(*Continued*)

Table 9.1 *(Continued)*

Nongovernmental Organizations and Individual Members of the Public	
The decisions have already been made. What's the point?	• There may be cases of "tokenism" and "placation" where decisions have already been made. Where this is clearly the case, it may be appropriate not to participate • Sometimes participating can lead to a reconsideration of or a reversal of a decision • Often, decisions are tentative. Public scrutiny can lead to reversals
The public will be co-opted or manipulated	• The public can participate while making it clear that they do not endorse either the process or the outcome • As with proponents and regulators, the public should not make up their mind from the outset • Changing positions based on new knowledge and thoughtful deliberations is not co-option or manipulation • It is the proponent and regulator's responsibility to avoid bias and misrepresentation. The public, with adequate support, can test data and interpretations (e.g., independent peer reviews)
We don't have the time	• The public participation program should be designed so that the time requirements for individuals and groups to participate are no more than absolutely necessary for effective participation • The use of a range of methods to involve various publics should reduce the time burden on any one group or individual • The timing and duration of public participation activities should respect the other demands on the time of participants (e.g., planting season). Outreach methods (e.g., kitchen table meetings) can sometimes be helpful
We don't have the resources	• Sometimes participant or intervener funding is warranted. The criteria for funding should be clearly specified • Payments for expenses can often be helpful, as well as the provision and sharing of resources • Public participation programs can be designed to ensure that resource constraints do not preclude or seriously inhibit public participation, especially for traditionally underrepresented groups

Sources: Barrow (1997), Canter (1996), Day (1997), Glasson et al. (1999), Nagel (1987), Petts (2003), Priscoli (1982), SERM (undated), Shepherd and Bowler (1997), UNECE (2006), UNEP (1997).

problems and dilemmas, for making difficult judgments, for reconciling or accommodating conflicting perspectives, for ameliorating obstacles, and for managing uncertainties. Dilemmas, obstacles, and problems, which appear "impossible" in the abstract, can generally be worked through in practice, especially by effectively drawing upon the knowledge, experience, and judgment of the public. The difficult issues that emerge in a public participation program reflect the complexities of decision making in a pluralistic society. There are no "quick fixes." A sufficient record of public participation "successes" and "failures" has been amassed to suggest that practical resolutions or accommodations to issues such as those listed in Figure 9.3 can often be reached.

At the same time, much public participation literature is not derived from theory nor does it provide a coherent basis for deriving theory. It is difficult to measure public participation effectiveness. Codes of good conduct are largely very general. Analyses of public participation effectiveness tend to be qualitative and anecdotal. Given the difficult issues often encountered and the uncertain conceptual foundation, it is not surprising that the quality and effectiveness of IA public participation efforts are highly variable. The limits of and difficulties sometimes encountered in practice reinforce the need for more effective public participation. They may occasionally explain why public participation efforts fall short of aspirations. They do not justify the status quo. Also, they do not imply that potential public participation benefits are either inappropriate or unattainable. They do underscore the need to thoughtfully and jointly plan and execute public participation programs, to draw upon the lessons of public participation practice, and to anticipate and effectively address the many types of issues that often emerge in practice.

9.4.2 Principles and Practices

Public participation in IA practice has advanced to the point that a core body of prescriptive knowledge is emerging. Table 9.2 identifies general, consultation, communications, mutual education, negotiations, and collaboration examples of public participation goals, principles, and good practices. The table demonstrates that there is a considerable knowledge base potentially relevant to IA process design and management.

It is apparent from Table 9.2 that the various elements of IA public participation are highly interdependent. Goals and principles guide good practice. Good practices extend from and contribute to goals and principles. The general goals, principles, and practices provide a framework for the

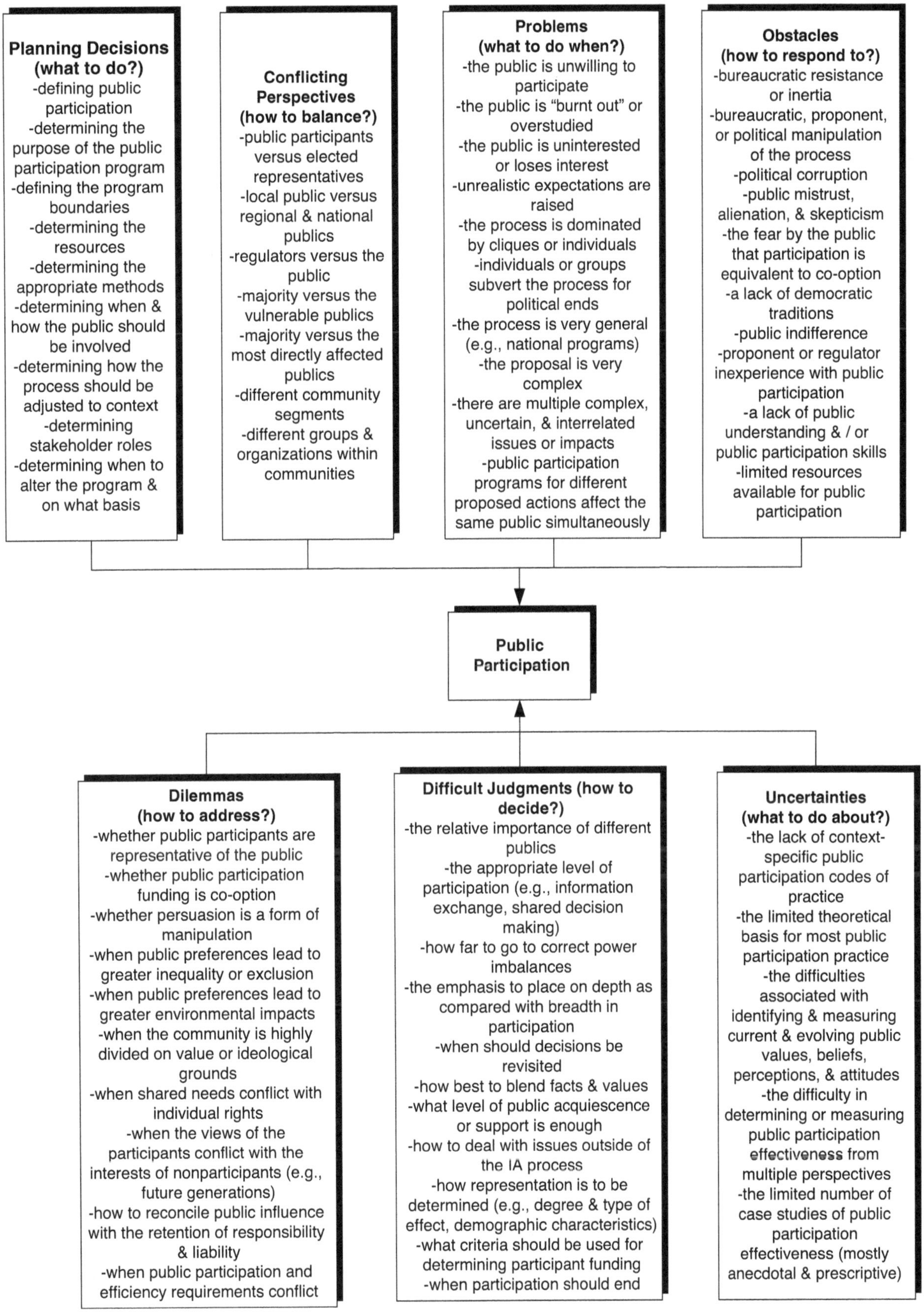

Figure 9.3 Examples of public participation issues.

Table 9.2 Examples of IA Public Participation Goals, Principles, and Good Practices

Goals	Principles	Good Practices
	General	
To produce better decisions To earn the trust of public participants To increase decision-making transparency and accountability To promote fairness To build credibility with those who might be affected To integrate public participation and decision making To contribute to more environmentally responsible decisions To make it possible for participation to cascade from policies to plans to projects	Be honest, open, inclusive, and responsive Explicitly identify public participation objectives; clearly identify decisions to be made Explicitly identify public participation limits Design public participation efforts to match situation Ensure that all interests are represented; define the public broadly; inclusive and equitable Ensure adequate resources, including time, for effective public participation Recognize that the people have a right and a responsibility to manage their own affairs Design process to be responsive to community needs Recognize public contribution to process Recognize "nestedness" of IAs, policies, and projects Be sensitive to the role of professional cultures in limiting the effectiveness of the IA process and related participation Adapted to context	Undertake community or social profiling Select methods to match objectives, context, issues, publics, and stage in IA process Focus on issues as identified by the public Design the process to accommodate stakeholder values Clearly define roles and responsibilities Ensure sufficient time and flexibility for adequate public participation Seek to make interactions informal and personal Recognize and ameliorate barriers to participation Design the process to motivate public to participate constructively Clearly and frequently inform public of process progress Evaluate, with public, by stage, effectiveness of public participation measures; adjust and supplement as needed Address the multiplicity of stakeholders and types of knowledge, data, and interests in a comprehensive way Employ qualified and unbiased participation specialists
	Consultation	
To involve interested and affected parties early and throughout the IA process To provide effective notice To identify public values and concerns To provide to the public relevant information regarding the proposal, possible options, and potential impacts To make decision making more transparent and accountable To respect indigenous world views and rights and take indigenous concerns seriously To obtain feedback from the public concerning values, perspectives, preferences, and suggestions To ensure that public concerns are taken seriously To provide in-depth involvement opportunities To achieve consensus around desired outcomes To involve traditionally unrepresented and underrepresented groups and segments of society To contribute to institutional reform	Involve members of the public in decisions that might affect them Work for broad participation Ensure that decision making facilitates public scrutiny The public has a right to information relevant to potential decisions that might affect them Provide an opportunity for those otherwise unrepresented to express their views (outreach); provide the resources necessary to ensure their effective participation Ensure awareness of range of stakeholder views Obtain and accommodate local and traditional knowledge Seek to better understand public perceptions of change including perceived causes and effects Balance the needs of current and future generations Consultation most effective when strong stakeholder networks and high expertise level	Support and institutionalize public participation Utilize SEA as a bridge to better public participation Identify relevant interest groups; actively seek to involve NGOs Provide participant funding Involve before IA begins and after it is completed Prepare a public involvement plan Involve the interested and affected publics in formulating the public involvement plan Select and adapt involvement methods to stakeholder characteristics Design consultation for the convenience of the public Interview representatives of each group to identify potential concerns Share information openly Adopt inclusive scoping approach; restrictive scoping can exclude potentially affected stakeholders Clearly explain how public input will be used; provide explanations if input rejected; provide prompt responses

Table 9.2 (*Continued*)

Goals	Principles	Good Practices
	The public should have an opportunity to comment prior to each decision in the process Responses should be provided to all public comments and suggestions Involve the public early in the process (e.g., problem definition, alternatives identification, criteria identification, public identification) Involve regulators from the outset Fully document record of public involvement	Take into account public attitude toward proposal and process (e.g., acceptance, rejection, ambivalence, support) Maintain the visibility of the public consultation program Identify and ameliorate barriers to information flow (e.g., lack of awareness, legal, financial, technical) Place greater emphasis on interactive formats, such as workshops or coffee klatches, in preference to public hearings or large public meetings Involve the public in approvals and implementation Critically evaluate effectiveness of public participation methods Use social media to magnify public participation
	Communications	
To enhance public, proponent, and regulator understanding and perceptions of reality To facilitate the interchange of ideas among citizens To encourage respectful speaking and listening To establish and maintain a dialogue between those responsible and those affected by possible actions To minimize communications distortions To be informative, proactive, and communicative To enhance the social relevance of scientific and technical knowledge To ensure that information is accurate, relevant, and unbiased To provide an opportunity for those otherwise unrepresented to express their views (outreach) To employ effective communications skills To stimulate constructive collaboration and produce common meanings	View IA as a form of communicative action Minimize inaccuracies Communications materials should be adapted to the needs of each participant group Provide the public with accurate, timely, pertinent, and understandable information Interpretations, ideas, options, and management measures should be substantiated and open to reasoned criticism Recognize that feelings equal facts Facilitate interagency communications and cooperation Promote reciprocal communications between competent authority and public at all stages Recognize that listening is a critical element of participation Recognize that communication is two way Provide channels for receiving, evaluating, and responding to individual, group, and societal public concerns and suggestions Establish a working rapport with all stakeholders Ensure that documents clearly communicate local sentiments to decision makers Facilitate public access to expert information Ensure more effective information dissemination Provide a grievance mechanism for affected populations Use social media as a effective communications and engagement tool	Learn to speak the public's language Allow plenty of time for public to understand and assimilate information; provide opportunities to cross-examine Simplify technical and scientific language Recognize that process communicates content Use professional expertise to create opinions not to kill them off Recognize the importance of early discourse; design communicative strategy to suit context Seek to enhance the capacity to listen and hear the stories of others Look at the range of values, not just the numbers Ensure that documents are well planned, organized, edited, and presented Guard against a public relations approach that seeks to minimize or diffuse conflict in a manner that legitimizes the discourses, power, and positions of the dominant coalition Minimize or explain technical language where must include; avoid jargon and excessive detail Design documents to suit audience Use third-party mechanisms when there are arguments over facts Independently facilitate dialogue where appropriate Be proactive in communicating with the public Inform public of communications channels to IA team, to regulators, and to decision makers Ensure that communications is clear, concise, and noncondescending

(*continued*)

Table 9.2 (*Continued*)

Goals	Principles	Good Practices
		Make effective use of visual techniques and multimedia packages (e.g., photo-simulation) Ensure follow-up results are understandable and broadly available
	Mutual Education	
To enhance public knowledge about possible actions, environmental conditions, and possible impacts To enhance proponent and regulatory knowledge about local conditions, values, needs, and concerns To promote mutual, social, organizational, collaborative, and transformative learning To foster cognitive enhancement (the acquisition of knowledge) and moral development (growth in the ability to make judgments about right and wrong) To promote critical IA education (education about and through IA)	Seek out and make use of public knowledge Treat traditional knowledge as a valid form of knowledge Distinguish among technical, conceptual, social, and sustainability learning Recognize learning as a step to conflict avoidance or resolution Treats decision making as a process of learning and negotiations among multiple actors Successful stakeholder involvement requires agency staff training or expert assistance Facilitate learning about facts, values, and social identities Ensure learning is free from coercion and distortion Be open to alternative perspectives Ensure the free expression of attitudes, feelings, and intentions Seek to improve the intelligence capacity of government agencies and of communities	Provide for local capacity building (to participate more effectively) where needed Provide for participant training Plan educational programs/activities in partnership with stakeholders Seek to integrate personal/experiential/contextual knowledge with processed knowledge Distinguish between cognitive (knowledge dominant) and social learning (responsive communications leading to policy reframing) Reflect critically about presuppositions Pay close attention to fairness and competence Use dialogic and argumentative processes to promote learning Foster and recognize interactions among critical listening, reflection-in-action, and constructive argumentation Integrate learning from practice stories Treat the IA process as a learning process (e.g., contributes to the ability of communities and societies to learn and change)
	Negotiations	
To avoid and reduce conflict To develop decisions that are mutually acceptable to interested and affected citizens To search for new conceptions of values To meet a mix of people's substantive, procedural, and psychological interests To reduce the risk of subsequent misunderstanding To ensure a just and equitable process To ensure just and equitable outcomes To further advancement toward social, environmental, and sustainability ends	Ensure information to support process is complete and accurate Ensure that the full range of interests are represented and that all are free to negotiate with other stakeholders Correct power imbalances Ensure that third parties (e.g., mediator) have adequate training and experience and are acceptable to participants Ensure that all parties have sufficient resources and authority Provide for early and ample opportunities for conflict resolution Negotiate over interests not positions Consider a wide range of alternatives that reconcile differences Agree on principles or criteria to evaluate alternatives Seek to realize sustainability visions through a process of dialogue and	Identify potentially controversial issues and seek resolution with the appropriate parties Plan conflict resolution process, especially prenegotiations; ensure agreement on rules and procedures Start with joint fact-finding Highlight underlying assumptions Seek to identify low cost trades Design the process to suit the type of conflict Understand the role of interpersonal dynamics and help people to move on Define measures of success (e.g., products, acceptance, interests protected, responsibilities defined, relationships established and maintained) Stave off angry confrontation Seek points of mutual agreements; focus on options for mutual gain

Table 9.2 (*Continued*)

Goals	Principles	Good Practices
	negotiations in which several possible discourses on each topic are discussed Document the agreement Agree on the process by which the agreements are to be revised Ensure commitments are observed Ensure outcomes are monitored and enforced	Seek to provide discussion platforms for multiple negotiations between stakeholders and decision makers Provide sound technical data and support to process and stakeholders Employ practical approaches for dealing with disruptive behavior Clarify the presumed liability of participants, confidentiality agreements, legal agreements, and extent to which precedents are or are not being set Visibly isolate extremes Keep public informed of progress Conduct post hoc evaluations of effectiveness
	Collaboration	
To build consensus To build and sustain trust To build support for and acceptance of decisions To make the IA process and decision making more accountable to interested and affected parties To ensure procedural and outcome fairness To foster collaborative and creative explorations of problems and opportunities To obtain tangible environmental and sustainability outcomes (i.e., goodness of decision) Ensure that principles of free, prior, and informed consent observed and human rights are respected	Ensure information to support process is complete and accurate Involve the public in idea generation and problem solving Facilitate interagency collaboration View IA as a partnership of proponent, state, and community Do not substitute compromise for good problem solving Treat analysis as a joint effort rather than a battle over facts Seek to enhance the role of IA in making decision making more accountable to public Define the problem rather than propose solutions or take positions View the situation as an opportunity for collaboration, not competition Recognize the interdependence of process and substance Seek to define common goals and shared visions of the future (community and environment) Recognize that consensus not always possible	Be clear regarding boundaries, who are invited to participate, expectations of contributions by participants, how facilitators are chosen, how information generated will be used, and who owns Separate people and their personalities from the problem Ensure that process is flexible and where appropriate, experimental Provide sound technical data and support to process and stakeholders Keep public informed of progress Undertake documentation in partnership with community leaders Favor participatory (e.g., workshop-based approach) and transdisciplinary methods over top–down assessment methods, wherever practical Provide deliberative rather than adversarial fora Be attentive to the distribution of power by stakeholders and facilitators Adopt activist mediation model (process and outcome) Make effective use of methods for creatively redefining problems and for generating, selecting, and evaluating ideas Conduct post hoc evaluations of effectiveness

Sources: Alshuwaikhat (2005), Bauer and Randolph (2000), Binder et al. (2010), Bisset (2000), Bond and Morrison-Saunders (2011), Booth and Skelton (2011a), CIER (2009), Clark (1994), Cline and Lamb (2005), Creighton et al. (1999), Cuppen et al. (2012), Daniels and Walker (1996), Devlin and Yap (2008), Diduck and Sinclair (1997), Doelle and Sinclair (2006), Donnelly et al. (2007), Ebrahim (2008), Égré and Senécal (2003), Elling (2011), Esteves et al. (2012), Fell and Sadler (1999), Fischer et al. (2009), Forester (1999), Gibson (2010), Glasson et al. (1999), Healey (1997), Howell et al. (1987), Hodge (2004), IAIA (2006b), Interorganizational Committee (1994), Jiliberto (2011), Kirkpatrick and George (2006), Kørnøv and Dalkmann (2011), Kørnøv and Thissen (2000), Lawrence (2009), Lavallée and André (2005), Lemon et al. (2004), Manring et al. (1990), March (1998), Maser (1996), Maynes (1989), Moore (1986), Morgan (1988), Morgan et al. (2012), Morrison-Saunders and Sadler (2010); Motion (2005), Negev (2012), Noble and Gunn (2009), Okello et al. (2009), Praxis (1988), Peterlin et al. (2006), Petts (2003), Peirson-Smith (2012), Priscoli and Homenuck (1986), Pope and Grace (2006), Priscoli (1982), Rickson et al. (1990a), Robinson and Bond (2003), Sadler (2011b), SERM (undated), Sheate (2012), Sinclair et al. (2007), Smith (1993), Smith et al. (1997), Smith and Schin (2004), Stewart and Sinclair (2007), Susskind and Cruikshank (1987), Susskind (1999), Täbara and Pahl-Wostl (2007), Tuinstra et al. (2008), UNEP (1997), US EPA (2001b), Vicente and Partidário (2006), Webler et al. (1995), Weiss (1989), Whitelaw et al. (2009), Wirutskulshai et al. (2011), Wolsink (2010).

consultation, communications, mutual education, negotiations, and collaboration elements. Effective consultation is conducive to effective communication. Mutual education is more effective when built on a base of effective consultation and communications. Negotiations (to address differences) and collaboration (to build shared visions) are complementary. Both negotiations and collaboration are enhanced when they extend from effective consultation, communications, and mutual education. Negotiations and collaboration can foster more effective consultation, communications, and mutual education. Sensitivity to these interdependencies is essential to effective public participation in the IA process.

Public consultation and negotiations (especially alternative dispute resolution) have received the most attention in the IA literature and in the literature of related forms of environmental management. Communications is commonly characterized as communicating to the public. Education has tended to be defined as public education. Public education is sometimes equated with persuading or even manipulating the public. Collaboration is often seen as an extension of negotiations (i.e., building on "win–win" solutions to conflicts). More attention should be devoted to principles, concepts, and methods of two-way communications, mutual education, and creative and substantive collaboration (see subsequent subsections of this chapter).

Additional effectiveness reviews could help derive, refine, and test public participation principles and practices. Such analyses could demonstrate which practices contribute the most and the least to achieving public participation goals. They could illustrate critical interdependencies. They could identify when principles or practices are complementary and when they operate at cross-purposes. They could contribute to more effective public participation planning and management and to more effective integration of public participation into the IA process. They could also demonstrate how the IA process could be reformed and adapted to foster more effective public participation.

Public participation concepts and categories of methods often are displayed as continua, as illustrated in Table 9.3. Continua are useful for grouping methods. They can illustrate which categories of methods are best suited to achieving alternative citizen participation goals (e.g., citizen control, citizen autonomy, citizen influence, citizen involvement). They can indicate which groupings are more appropriate to situations characterized by varying mixes of cooperation and conflict. They can assist in role definition for public agencies, the public, and third parties. They can provide a general sense of major methods' characteristics (e.g., degrees of formality, continuity, and intensity). There are some inconsistencies in the placement of various categories along the continua. These differences reflect varying definitions of categories (e.g., one- versus two-way communications and education) and varying role interpretations.

A continuum clearly and succinctly displays major differences. But only differences for a single criterion can be displayed at a time. The impression can be created that only one category can be used (it is possible to use several in an IA process) and that categories further along the continuum are somehow better (it is more often a case of matching the methods to the context). Public participation methods can be classified in ways that do not involve continua. They can, for example, be categorized by function (e.g., information dissemination, information collection, initiative planning, reactive planning, decision making, participation process support) or by operational characteristics (e.g., large group meetings, small group meetings, organizational approaches, media, community interaction, legal mechanisms). Matrices can display differences along more than one dimension. Table 9.4, for example, clusters methods by public participation element (e.g., consultation) and by role (e.g., information exchange, continuous involvement, formal involvement).

Public participation texts and manuals generally describe the characteristics, advantages, and disadvantages of numerous individual methods. Sometimes, connections are drawn between the methods and public participation goals. Ideally, a consistent set of criteria (based on public participation goals and principles) would be applied to each method grouping and/or each method. The application of scaling procedures could help ensure that differences are consistently addressed. Such analyses would be further strengthened if linked to IA process activities (e.g., scoping), if assessed for varying contexts (e.g., third-world countries), and if supported by systematic reviews of IA effectiveness analyses.

9.4.3 Consultation

With public consultation or involvement the public is informed about proposals. They also express their views about proposals. These interactions occur prior to decision making (Parenteau, 1988). The decision-making process should facilitate public scrutiny and encourage effective public participation (Gibson, 2010). Public concerns and suggestions are taken into account by decision makers (US EPA, 1988b). Public inputs inform but do not dictate decisions. Proponents and regulators retain final decision-making authority (Smith, 1993).

Figure 9.4 illustrates how a public involvement process might unfold. Early consultation provides a general sense of such matters as issues, levels of interest, key people and organizations, organizational mandates and decision-making procedures, and barriers to participation. Early consultation activities provide a basis for initial consultation planning. Initial consultation planning establishes preliminary goals, determines the special characteristics of the situation, indicates study team requirements, highlights institutional constraints and stakeholder vulnerabilities, and identifies decision makers, the various publics, and other stakeholders.

Stakeholder identification is a critical element of public involvement planning. Individuals, groups, organizations,

Table 9.3 Examples of Public Participation Continua

Manipulation	Therapy	Education	Information Feedback	Engagement	Joint Planning	Comanagement	Designated Authority	Self-Determination
Low Citizen Control								**High Citizen Control**
Information Out	Persuasion	Tokenism	Consultation	Dialogue	Extended Involve-ment	Collaboration	Partnership	Empowerment Local Control
Low Level of Citizen Participation & Influence	Public Information			Consultation		Joint Planning		**High Level of Citizen Participation & Influence**
		Information Feedback			Enhanced Involvement			
Low Degree of Citizen Autonomy	Passive Participation	Participation by Consultation				Self-Mobilization		**High Degree of Citizen Autonomy**
		Participation in Information Giving		Functional Participation				
Low Degree of Third-Party Involvement/ Control	Consultation	Policy Dialogue	Negotiations	Mediation	Active Mediation	Administrative Hearing	Litigation	**High Degree of Third-Party Involvement/ Control**
	Collaborative Problem Solving	Conciliation	Facilitation	Nonbinding Arbitration		Binding Arbitration	Adjudication	
	Informal Procedures	Cooperative Decision Making	Relationship Building Assistance	Procedural Assistance	Substantive Assistance	Advisory Nonbinding Assistance	Binding Assistance	
Low Intensity of Continuity of Involvement	Education & Information		Information Feedback	Involvement & Consultation		Extended Involvement		**High Intensity of Continuity of Involvement**
	Public & Government Education	Two-way Communications	Advisory Committees & Structures		Public Consultation & Dialogue	Partnerships	Joint Decision Making	
	Share Information	Obtain Feedback			Provide Forum for Dialogue & Interaction		Joint Planning or Shared Decision Making	

(*continued*)

Table 9.3 *(Continued)*

Low end	Position relative to arrow	Points along the continuum (left to right, low → high)	High end
Low Degree of Administrative Formality	Above arrow	Fact-Finding; Mediation; Nonbinding Arbitration; Administrative Hearing	**High Degree of Administrative Formality**
	Below arrow	Unassisted Procedures; Facilitation; Dispute Review Board; Binding Arbitration; Litigation	
High Degree of Cooperation	Above arrow	Unassisted Procedures (Informal Discussions, Negotiations); Third-Party Decision Making (Administrative Hearing, Binding Arbitration, Litigation)	**High Degree of Conflict**
	Below arrow	Third-Party Assistance (Conciliation, Facilitation, Mediation, Nonbinding Arbitration, Settlement Judge, Dispute Review Board, Partnering); Nonviolent Coercion (Civil Disobedience)	
Low Degree of Decision-making Influence	Above arrow	Public Information; Procedural Public Participation; Consensus Seeking Public; Participation; Negotiations/Alternative Dispute Resolution	**High Degree of Decision-Making Influence**
	Below arrow	Be Informed of the Decision; Be Heard Before the Decision; Influence the Decision; Agree to the Decision	
Low Degree of Agency Control	Above arrow	Decision Maker Exchanges Information; Decision Maker Develops Agreements; Partners Develop Recomme-ndations; Capacity Building Information Exchanges; Capacity Building Develops Agreements	**High Degree of Agency Control**
	Below arrow	Decision Maker Develops Recommendations; Partners Exchange Information; Partners Develop Agreements; Capacity Building Develops Recommendations	
Low Degree of Agreement	Below arrow	Outreach; Information Exchange; Collaboration & Recommendations; Agreement	**High Degree of Agreement**

Sources: Arnstein, (1969), Bisset (2000), Creighton et al. (1999), Fell and Sadler (1990), Health Canada (2000), Hughes (1998), Parenteau (1988), Praxis (1988), Smith (1993), Susskind and Madigan (1984), Westman (1985).

Table 9.4 Examples of Methods

Consultation		
Information Exchange	*Continuous Involvement*	*Formal Involvement*
Information in (e.g., interviews, surveys, polling, focus groups, public comments, community profiling, call in television, direct e-mail, hot lines, mail-in response forms, door-to-door canvassing, responsiveness summaries, briefs, submissions, content analysis, cumulative brochures, letters to the editor, electronic chat room) Information out (e.g., briefings, exhibits, displays, contact person, telephone network/phone tree, computer bulletin boards, community liaison officer, political preview, demonstration projects, document circulation, feature articles) Town meetings, open houses, and workshops Conferences, roundtables Contests Stakeholder meetings Ombudsperson or representatives Televoting, 1–800 numbers, media-based issue voting, web polls Workbooks and community mapping	Advisory committees, councils, groups, and boards Task forces and groups People's panels Citizens' review board Breakfast meetings Community impact committee Community planning council Citizen action committees	Hearings Litigation/adjudication Referenda and plebiscites Commissions Inquiries

Communications		
Publicity	*Dialogue*	*Enhanced Dialogue*
Traditional publications (e.g., newspaper inserts, information kits, brochures, newsletters, fact sheets, mail-outs, paid advertisements, plain language communications) Audio/visual (e.g., film presentations, video, slide presentations, tape) Media (e.g., radio and television interviews, web sites, media releases, public service announcements, press kits, newspaper inserts, news conferences) Information fairs/exhibits Translations Group presentations	Coffee klatches Kitchen table meetings Search and consensus conference Constituent assembly Roundtables Retreats Computer-assisted participation and interactive www/e-conferencing Online discussion groups Participatory television/cable television Community-sponsored meetings Field offices Advice and argumentation	Relationship building assistance (e.g., counseling/therapy, conciliation, team building, informal social activities) Search and consensus conference Issue conference Capacity building and outreach Technical assistance and participant funding Structured workshops Citizen employment Citizen honoraria Coordinator or coordinator catalyst Discourse ethics Combating misinformation (critical theory) Procedural justice

Mutual Education		
Community Education	*Proponent, Regulatory, and Specialist Education*	*Mutual Education*
Technical and financial assistance Citizen training Lectures and workshops Computer-based programs Publications and translation Site visits, depositories, and resource materials	Procedural training Substantive training Networking Comparable proposal and environment review Citizens' juries and panels Community profiling	Storytelling Relationship building assistance (e.g., counseling/therapy, conciliation, team building, informal social activities) Coaching/process consultation Participatory research Study circles and study groups

(*Continued*)

Table 9.4 *(Continued)*

Mutual Education		
Community Education	*Proponent, Regulatory, and Specialist Education*	*Mutual Education*
Formal education, integration into existing curricula Simulation exercises and photo-simulation Citizen training programs Seminars, discussions, and position papers Media campaigns Speaker's bureau and panels of experts Technical advisors and peer reviewers Demonstrations and demonstration projects	Traditional knowledge Citizen surveys	Participatory drama Social and collaborative learning Transformative learning Deliberative learning
Negotiations		
Unassisted	*Third-Party Assistance*	*Third-Party Decision making*
Informal discussion Negotiation Conciliation Information exchange meetings Interest-based negotiation Policy dialogue	Fact-finding Conciliation and facilitation Mediation Conflict anticipation Conflict assessment Technical advisory board Minitrial and nonbinding arbitration Settlement judge and dispute review board Settlement conference Negotiated rule making Community dispute resolution centers	Dispute prevention Advisory nonbinding assistance (e.g., nonbinding arbitration, summary jury trial) Administrative hearing Binding arbitration Med-arb Mediation, then arbitration Dispute panels (binding) Private courts/judging Litigation/adjudication
Collaboration		
Joint and Collaborative Planning	*Joint Management*	*Creative Collaboration*
Roundtables, conferences, and working groups Cooperative/collaborative problem solving Role-planning Joint planning Coalition building Strategic choice Large-group response technique Consensus building Collaborative planning Charrette Niagara process Trade-off games The Samoan process Multicriteria group decision-making models Constructive engagement	Comanagement boards and councils Partnering and partnership agreements Cojurisdiction Steering committees Public authorities Community representatives on boards Citizen assemblies Community forums	Community visioning and shared vision planning Brainstorming, brainsketching, and brainwriting Delphi process Nominal group process Lateral thinking methods Think tanks Active mediation Simulation, modeling, and scenario writing Creative problem solving

Sources: Canter (1996), Creighton et al. (1983, 1999), Daniels and Walker (1996), De Bono (1992), Forester (1989, 1999), Friend and Hickling (1997), Glasson et al. (1999), Health Canada (2000b), Howell et al. (1987), Morgan (1998), Praxis (1988), SERM (undated), Sinclair and Diduck (2001), Smith (1993), Susskind et al. (1999), US DOE (1998), US EPA (2001a,b).

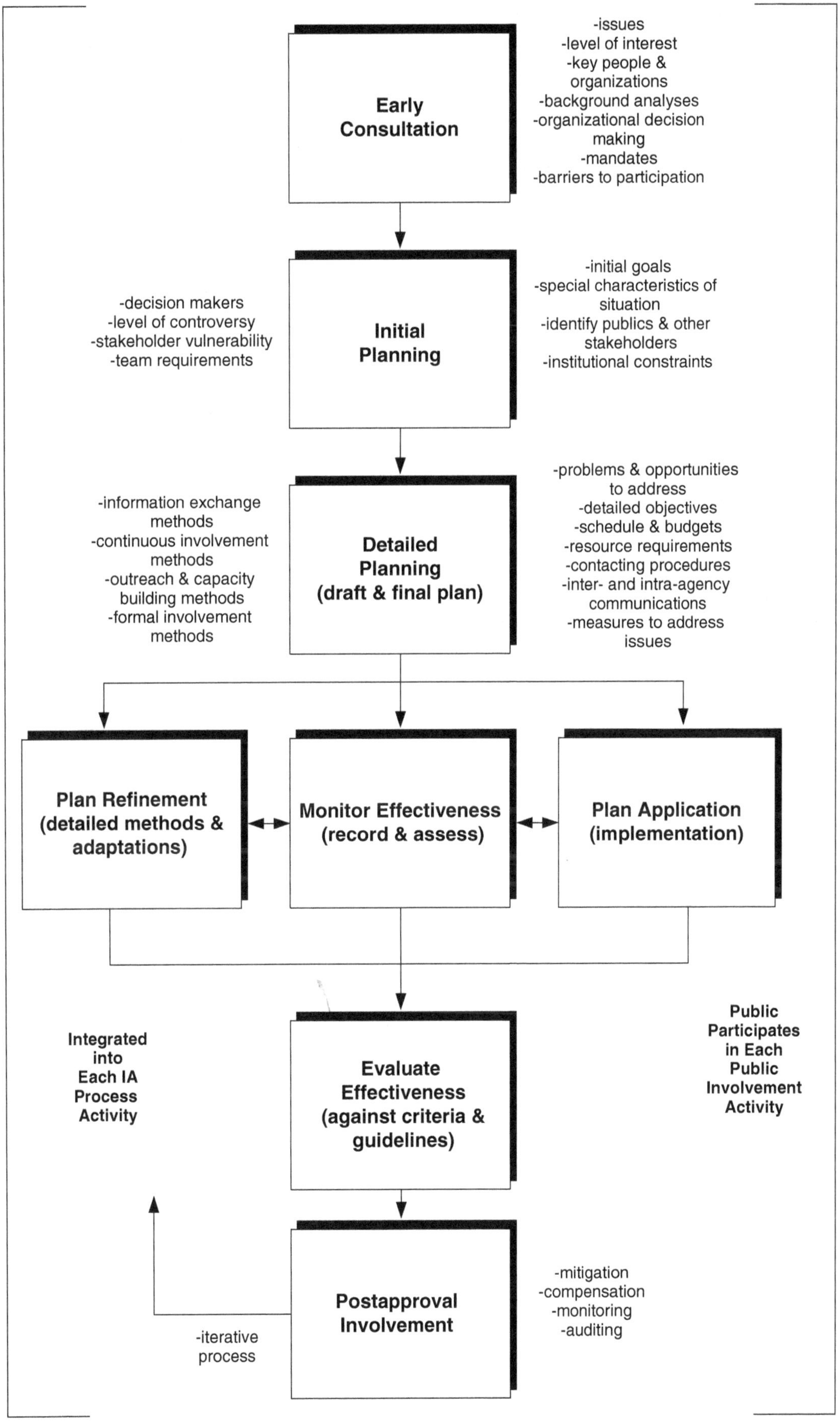

Figure 9.4 Examples of an IA public consultation process.

and segments of society can be differentiated based on, for example, location (e.g., local, regional, national), interests (e.g., industry, environment, community service), and characteristics (e.g., social, cultural, economic, political) (Bisset, 2000; Hughes, 1998; Praxis, 1988). Further distinctions can be drawn among types of individuals (e.g., community leaders, local informants, local experts, politicians, practitioners), groups (e.g., professional, environmental, cultural, recreational, service), and organizations (e.g., government agencies, businesses, institutions, media, labor unions) (Canter, 1996; Priscoli and Homenuck, 1986). There will be differences among stakeholders in the extent to which they are involved or not involved, informed or not informed, organized or not organized, united or divided and supportive, opposed to or apathetic to the proposed action (Priscoli and Homenuck, 1986; Praxis, 1988). Stakeholders, third parties, and staff can identify interested and affected parties. Geographic, demographic, historical, and comparative analyses also can help in stakeholder identification. The characteristics, perceptions, and positions of each stakeholder can be determined, appreciating that positions and alliances change, sometimes rapidly.

Detailed consultation planning generally involves preparing a draft and then a final public involvement plan. The plan is likely to characterize problems, determine objectives, establish schedules and budgets, allocate resources, establish contacting procedures, identify communications channels, and determine specific procedures for addressing identified issues (Praxis, 1988; UNEP, 1997; US EPA, 1998b). A public involvement plan can include information exchange, continuous involvement, outreach and capacity building, and formal involvement methods.

Information exchange methods provide a useful means of transmitting information to the public (e.g., newsletters, briefings, displays, background papers), of receiving information, comments, and suggestions from the public (e.g., surveys, public comments, response forms, direct e-mail), and for agencies and the public to exchange information and viewpoints (e.g., open houses, meetings, workshops). Information exchange methods can be geared to large or small audiences. If undertaken effectively, they can reach a major proportion of the population. They generally occur periodically in an IA process.

Continuous consultation methods (e.g., an advisory committee) involve a small number of stakeholder representatives. They facilitate bottom-up participation and transcend individual disciplines (Binder et al., 2010). They are inclusive and discourse-based (Bond and Morrison-Saunders, 2011). The committee meets frequently throughout the IA process. The committee can address issues, analyses, options, and trade-offs in much greater depth than is possible in information-exchange sessions (Maynes, 1989; Praxis, 1988; US EPA, 1988b). Continuous involvement is more effective when aided by facilitation and when adequately supported by influential agencies. It is ineffective if there is a lack of trust or if the viewpoints expressed through the committee are not taken seriously (Priscoli, 1982). Continuous involvement and information exchange methods can be complementary. Continuous involvement methods can explore issues and concerns identified in information-exchange sessions. Information exchange stresses breadth of involvement. It provides a mechanism for committee "representatives" to test positions and concerns with constituency groups. Continuous involvement emphasizes depth of involvement.

Outreach and capacity-building methods help bring into the IA process traditionally unrepresented or underrepresented groups and segments of the population. Voices previously outside become central to the process (Lawrence, 2009). Outreach and capacity building can take the form of technical or financial assistance. It can entail the supply of technical resources such as phone conferences and e-mail support and the provision of translation and facilitation services (US EPA, 2001b). Such methods can enhance the capacity of organizations and groups to participate effectively in the IA process. Communities also can be empowered to leverage additional resources and to capitalize on existing civic assets (US EPA, 2001b). Outreach and capacity building can support and supplement both information exchange and continuous involvement methods.

Formal involvement methods, such as hearings, commissions, and inquiries, tend to assume the characteristics of judicial procedures. Such methods can provide a useful way of presenting and testing evidence. They are often adversarial, however, and can be intimidating to the public, especially if technical and financial resources are not made available to public groups and organizations (Maynes, 1989). Referenda and plebiscites provide a formal mechanism for testing agreements obtained through continuous involvement procedures or for obtaining public feedback on major proposals and options. Such procedures can be costly, can oversimplify complex choices, and are occasionally divisive.

Public involvement plans are not simply implemented. Refinements and adjustments occur throughout the IA process, based on an ongoing assessment of changing circumstances and of methods effectiveness (Howell et al., 1987). Public involvement procedures and the public involvement plan are merged with rather than distinct from the IA process (Burdge and Robertson, 1994). Separate consultation objectives and methods are selected and adapted to meet the requirements of each IA process activity (e.g., scoping, alternatives formulation, impact assessment) (Burdge and Robertson, 1994). The overall effectiveness of the public involvement procedures is assessed prior to approvals. This helps identify supplementary involvement measures, which should be instituted to address identified gaps and weaknesses (Wolfe et al., 2001). Public involvement extends into the postapproval period to ensure that public concerns associated with project implementation, mitigation, compensation, and monitoring are fully considered. Public involvement, in common with the overall IA process, is highly iterative (Praxis, 1988).

Public involvement procedures can establish a strong foundation for a collaborative IA process. Although necessary, they tend not to be sufficient. Although partially addressed through good practice guidance, insufficient attention tends to be devoted to the specific mechanisms by which effective two-way communications and mutual education can occur. Public involvement methods, by themselves, tend to be ineffective in avoiding, managing, ameliorating, and resolving conflicts. They also tend to neglect the development and application of specific techniques and procedures for building consensus, for creative problem solving, and for collaboratively contributing to the realization of substantive environmental objectives.

9.4.4 Communications

Communications involves interactions among people. It is the bridge between environmental analysis and decision making (Holling, 1978). The communications act includes the parties involved (who), the message (what), the form by which the message is encoded (how), the audience (to whom), and a result (with what effect) (Bishop, 1975, 1983). Effective communications can facilitate understanding, conflict resolution, consensus building, and decision making. Ineffective communications can lead to a lack of understanding or to misunderstandings. It also can undermine consensus, exacerbate conflict, and inhibit decision making.

References to communications in IA literature tend to revolve around communications in IA documents and communications of good science and information to decision makers (Ayre and Calloway, 2005). Stress is placed on facilitating understanding through clear, focused, and consistent document presentation and organization (Morgan, 1998). Documents should focus on the needs and concerns of and be readily understandable to the target audience. General references also are made to developing and refining the verbal and written communications skills of IA practitioners (Daniels and Walker, 1996). Two-way communications can be furthered by process facilitators and through strong stakeholder networks (Dusik and Sadler, 2004; Kain and Söderberg, 2008; Kirkpatrick and George, 2006). Increasingly, emphasis is placed on communicating sustainability issues (Devuyst, 2000).

Table 9.5 highlights the characteristics of several communications concepts relevant to IA process management. These concepts underscore the central role of dialogue in IA practice. IA, in common with planning, is a collective, dialogic, practical, and moral activity. It is, therefore, essential to provide conditions conducive to open, unencumbered, undistorted, and noncoercive dialogue. This may require the formulation and application of mutually acceptable communicative ethical principles. IA discourses may need to be systematically analyzed. Steps may need to be taken to correct power imbalances. Ideal speech characteristics could be explicitly recognized. A concerted effort could be made to identify, avoid, and minimize communications misinformation and distortion. Communications networks may need to be established and reinforced.

The characteristics of effective and ineffective advice giving could be considered. IA practitioners (and related procedural specialists such as facilitators) could help establish and maintain the conditions required for effective and appropriate argumentation, persuasion, and storytelling. They could provide and derive important insights from IA practice stories. They could help participants explore the characteristics of and potential for accommodating diverse arguments and stories within the IA process.

The institution of effective IA communications measures can contribute to enhanced public consultation. Such measures also are conducive to a collaborative IA process. Additional building blocks, concerned with mutual education, conflict resolution, and consensus building, however, are still required.

9.4.5 Mutual Education

Education in IA practice is conventionally depicted as using information dissemination and general instruction to create public awareness of proposed actions and issues, to encourage more responsible environmental stewardship, and to facilitate informed decision making through enhanced understanding (Morgan, 1998; Praxis, 1988; SERM, undated). It is generally recognized that educational activities and programs should be jointly planned with stakeholders.

Education is sometimes acknowledged as a necessity and a precondition for advanced levels of public involvement, conflict resolution, and collaboration (Diduck and Sinclair, 1997; Maser, 1996). The IA process has been characterized as a technical, a conceptual, and a social, sustainability learning process—a process that can help communities and societies change, improve the intelligence capacity of government agencies, and further the cause of sustainability (Ebrahim, 2008; Rickson et al., 1990a; Täbara and Pahl-Wostl, 2007; Tuinstra et al., 2008).

The assumption tends to be made that learning is one way (i.e., to the public). It is sometimes recognized that proponents, regulators, and practitioners also need to receive training (e.g., in consensus building and conflict resolution techniques) and can learn much from the public. Increasingly, reference is made to the critical role of feedback, to the mutual learning and education that occurs through dialogue and debate among stakeholders, to the educational support role of computer modeling and sustainability frameworks, and to more encompassing and substantive forms of social and sustainability learning (Ayre and Calloway, 2005; Daniels and Walker, 1996; Diduck and Sinclair, 1997; Siebenhüner and Barth, 2005; Täbara and Pahl-Wostl, 2007; Tuinstra et al., 2008; Webler et al., 1995; Wiek and Binder, 2005).

Table 9.6 highlights the characteristics of various mutual education concepts relevant to IA process management. These concepts demonstrate that there are many forms of mutual learning (e.g., cognitive, communicative, social,

Table 9.5 Examples of Potentially Relevant Communications Concepts

Dialogue	• A form of communications in which understanding and respect are goals; intentions include maintaining social contacts and affiliation, eliciting and gaining information, promoting relationships, and changing the environment and others • In dialogue, participants present their own perspectives, listen carefully to the perspective of others, remain open to change, speak for selves and from personal experience, allow others to express their perspectives safely, learn significant new things about selves and others, find shared concerns with people holding different perspectives, explore doubts and uncertainties, ask questions based on true curiosity, explore the complexity of issues without polarization, and collaborate to create better futures • Communications process models include diffusion processes (to public), collection processes (from public), and diffusion–collection processes (information disseminated with intent of obtaining response) • Messages can be received at the perceptual, the cognitive, and the judgmental levels • Roles in dialogue: sender of message (determine own believes, motives, and beliefs); receiver of message (listening for decisions, listening with empathy, nonverbal communications)
Discourse analysis	• The frames through which groups of actors give meaning to aspects of policy issues and decisions supported by IA • Reflects conflicts of interests, power plays, and ways in which actors perceive and understand the world • Decision making conceptualized as a system of competing discourse coalitions and their struggles to control shared meanings and gain acceptance of how issues are framed • IA can impact upon the dominant discourses • Discourse analysis a useful tool for analyzing the assessment purpose and the larger sociopolitical climate; may help unravel how democracy and science guide the discussions • Discourse analysis allows for understanding of arguments considered legitimate or not • Role of IA practice in contributing to decision making by promoting stakeholder discourse reflection • Potential role for public participation and participatory appraisal in improving the civil legitimacy of decisions
Communicative planning and action	• Sees planning as an exercise in collective, participatory action • Argues that the best window onto planning practice is planning discourse; discursive interaction the most important element of planning practice • Recognizes that planning process may manipulate citizen action and lead to systemic distortions; systemic distortions are avoidable • Seeks to facilitate informed, open, unforced, and unmanipulated citizen action • Seeks sincere, comprehensive, and appropriate communications; self-consciously chooses actions to overcome institutional barriers and to become more egalitarian • Seeks a deliberative style of debating issues and interests; an open dialogue among equals • Principles: all important interests (identified and articulated) at table; all stakeholders fully and equally informed and able to represent their interests; all equally empowered in discussion; power differences from other contexts must not influence who can speak or who is listened to or not; allows all claims and assumptions to be tested and all constraints tested; all participants able to assess the speaker's claims; all must speak sincerely, honestly, and comprehensively; groups should seek consensus • Can be obstructed by social inequities and tensions and conflicts among groups; criticized as politically naïve
Discourse or communicative ethics	• A procedural approach to moral justification; procedural morality • Assumes that the basic unit of meaning is the speech act, that meaning is inseparable from the role of language in structuring practices and social interactions, and that truth and normative rightness are essentially discursive matters • Seeks to engender ideal speech situation: freedom of access, equal rights to participate, truthfulness on the part of participants, and absence of coercion in taking positions • Endeavors to ensure that all relevant voices get a hearing, the best arguments given the present state of knowledge are brought to bear, and only the unforced force of better arguments determine the yes or no of participants • Requires that all members be prepared to listen for differences not only in interests but also in values and cultural references • Sets itself the tasks of deriving argumentation rules for discourse in which moral norms can be justified • Role of discourse ethics: to examine the normative validity of public action-guiding norms, to examine not just whether all affected participants might accept a norm but whether the norm deserves to be accepted by them, given the process in which they might consider them
Misinformation	• Various forms of misinformation impede and distort communications • Managing comprehension (e.g., deliberate ambiguity, jargon, ideological language, obscure messages) • Managing trust (e.g., false assurances, symbolic decisions, marshalling respectable personage to gain trust, ritualistic appearance of openness)

Table 9.5 (*Continued*)

	• Managing consent (e.g., decisions reached without legitimate representation of public interest, arguing technically when acting politically, appeals to adequacy of participation, not addressing systemic failures) • Managing knowledge (e.g., decisions that misrepresent actual possibilities to the public before a decision is made, misrepresenting costs, benefits or options, ideological or deceptive presentation of needs) • Managing control (e.g., withholding information, misleading information or judgment, inconsistencies in what is being said, gaps in argumentative chain, undue persuasion, professionalization of debate) • Need to address and combat the effects of unequal power relations and misinformation; most misinformation avoidable even when systemic
Advice, argumentation, and persuasion	• Requirements of advice: relationship of persons of trust and truth, a basis in the world through knowledge and experience, expressed in reasonable and justifiable stories, and a public understanding of who we are as a community • Planning as a dialogic and argumentative process; involves marshalling evidence and giving reasons, minimizing the exclusion of relevant information, encouraging the testing of conjectures, and welcoming rather than punishing value inquiry • Planning as action in a flow of persuasive argumentation; expressed in awareness of differing or opposing views • Need to meander skillfully: arguing for visions, constructing inclusive processes, negotiating the meaning of key concepts, responding to unexpected events, taking existing rules and prior decisions into account (while seeking to change problematic ones), relying on own substantive knowledge (while being open to other forms of knowledge and expertise), configuring arguments (in the face of contestable configurations), and arguing persuasively in diverse media and forums • Argumentation affected by conflicts with others over meaning, media in which persuasive efforts occur, events that create new opportunities and constraints, institutional rules and previous decisions, legalistic procedures that inhibit understanding and innovation, social and institutional factors, and opposition to open and inclusive processes • Rhetorical frame (persuasive use of story and argument in policy debate) as distinct from action frame (frames that inform policy practice)
Story telling	• Planning arguments are characteristically expressed as stories • Stories describe events, provide explanations, warn of dangers, identify benefits, report relevant details, search for others' meanings, confess mistakes, justify recommendations, and prepare others • Need to understand the significance of the very messiness, complexity, detail, and moral entanglement of living stories • Stories are accounts of value and identity, of abiding concern, and of complexities; ignored at practical risk • Stories are morally thick, politically engaged, and practical • Seeks to enhance the capacity of participants to both listen and hear the stories of others • The discursive process needs to be designed to explore different storylines about possible actions • Suggested convergence strategy when varying stories: a pluralistic strategy; embrace rather than seeking to resolve or ignore controversy; consistent with an open moral community
Networks	• IA support role in creating and fostering cross-sectoral governance networks • Involves citizens and civil society organizations • Emphasizes constructive contributions to political pluralism • Enhances empowerment of participants • Role of self-help networks in IA capacity building • All participants should consider themselves equally responsible within IA networks • Social network properties include greater awareness, competence, and capacity to increase decision making through IA

Sources: Bishop (1983), Bonifazi et al. (2011), Dusik and Sadler (2004), Fischler (2000), Forester (1989, 1999), Habermas (1993), Healey (1997), Hodge (2004), Huxley and Yiftachel (2000), Innes (1998), Krieger (1981), Lauria and Soll (1996), Mandelbaum (1991), Patton et al. (1989), Rozema et al. (2012), Runhaar (2009), Schön and Rein (1994), Sager (1994), Taylor (1998), Throgmorton (2000).

practical, collaborative, transformative, critical, emancipatory, organizational, deliberative, sustainability, traditional knowledge) possible in IA practice. They illustrate how mutual learning integrates the cognitive, the moral, and the practical. They demonstrate how knowledge is brokered. They indicate how it can be facilitated (e.g., by the expert elicitation process). They show how learning can be approached from multiple perspectives, how it varies depending on the historical, social, and cultural context, and how it integrates and transcends such distinctions as personal and processed knowledge, facts and values, and people and the environment.

Mutual learning is interactive, social, reflective, critical, practical, affective, holistic, collaborative, deliberative, and democratic. It facilitates learning about facts, values, issues,

Table 9.6 Examples of Potentially Relevant Educational Concepts

Mutual education	• No single party, organization, or discipline holds the key to understanding; therefore, mutual learning critical • Types of learning: about what is (facts and explanations), about what should be (values), and about participants • In mutual learning, personal experiential and processed knowledge are integrated • Ideal conditions for learning: accurate and complete information, freedom from coercion, openness to alternative perspectives, ability to reflect critically upon presuppositions, equal opportunity to participate, and ability to assess arguments in a systematic manner and to accept rational consensus as valid • Learning involves various thinking modes (e.g., concrete experience, reflective observation, abstract conceptualization, active experimentation); combined to form learning dialectics
Communicative and social learning	• Social learning: the process of framing issues, analyzing options, and debating choices in the inclusive deliberation • Occurs when citizens involved in working out mutually acceptable solutions mature into responsible democratic citizens and reaffirm democracy • Views IA as a vehicle for social and individual learning • Distinction between cognitive learning (where knowledge is a dominant variable) and social learning (based on responsive communications leading to the reframing of a policy issue) • Two general component of social learning: cognitive enhancement (i.e., the acquisition of knowledge) and moral development (i.e., growth in the ability to make judgments about right and wrong) • For social learning to occur there must be a free expression of attitudes, feelings, and intentions • No predetermined outcomes; supported by information from multiple perspectives, citizens add value, there are serious and substantive discussions, and discussions are supported by neutral facilitators
Organizational learning	• IA can provide opportunities for organizational learning • Involves examination of information sharing, information interpretation, organizational memory, and learning outcomes • Organizations have a variety of structures that facilitate learning • Can be single-loop (improving performance within the existing system) or double-loop (changes the system) • Seeks to enhance organizational structures and procedures to foster learning • IA process can foster organizational learning (e.g., accurate information, freedom to participate, freedom from coercion) • Need to consider how organizational structures and dynamics facilitate or inhibit learning
Practical and deliberative learning	• Practitioners learn and reflect as they act with others in practical situations • Practitioners reflect in action, make moves, evaluate results of moves, and reconsider working theories; practice can lead theory; theory and practice integrated • Practitioners learn alone or from or with others; can learn from systematic studies and by listening to practice stories from thoughtful practitioners; deals explicitly in the everyday language of practical life • Double visioning: ability to act from one perspective while holding awareness of other possible perspectives • Reflective transfer: the process by which patterns detected in one situation are carried over as projective models to other situations where used to generate new causal inferences and are subjected to new, situation-specific validity tests
Collaborative learning	• Sees IA as a learning and civic discovery process where people act together and find new solutions • Designed to address complex and controversial issues; combines elements of systems methods, mediation/dispute management with experiential learning theory • Process: introduction to process, identify situation to be improved, share situation perceptions and description, dialogue about interests and concerns, develop transformative models, compare models with reality, and collaborative arguments about desirable and feasible change • Emphasizes learning and negotiation interaction as the means through which learning and progress occurs • Attributes: stresses improvement (rather than solution), situation (rather than problem or conflict), concerns and interests (rather than positions), systems thinking (rather than linear thinking); recognizes that considerable learning about science, issues, and values will have to occur before implementable improvements are possible
Critical, emancipatory, and transformative learning/critical IA education	• Critical pedagogy: accepts the transformative possibilities of willed human action on an individual and social level; student centered with emphasis on democratic dialogue • Major descriptors of critical pedagogy: participatory, situated (in student thought and language), critical, dialogical, desocializational (students desocialized from passivity in classroom), multicultural, research oriented, activist (classroom is active and interactive), and affective (interest in broad development of human feelings) • Transformative theory of social learning: explores not only how our arguments change in dialogue and negotiations but how we change as well; transforming ends, ideas, and ourselves • Transformative learning: a comprehensive theory of how adults learn; focuses on learning process and accommodates social context; describes how individuals improve instrumental (how to control and manipulate

Table 9.6 (*Continued*)

	the environment) and communicative competence (trying to understand what someone means when they communicate) • Learning not just through arguments, reframing of ideas, and critiques of expert knowledge; also through transformations of relationships, responsibilities, networks, competence, and collective memory and memberships • Critical IA education encompasses both education about IA and education through IA; includes education about project, environment, how decision-making processes and project decisions can be challenged, and how members can work together to pursue their own goals • Critical IA education: contributes to human democratic liberation, to assessment activities, and to fostering of critical consciousness (enables public to evaluate dominant discourse and to present forceful counterarguments)
Sustainability learning	• Focuses on generating and applying a specific type of content of what is learned • Learning to develop the capacity to manage options for the adaptation of human societies to the limits and changing conditions that are imposed on socioecological systems • Seeks to overcome such dualisms as individual and collective, human and natural systems, structure and change, internal and external system properties, human agency and natural conditions • Advocates a more hybrid, relational, and coevolutionary holistic understanding of human–natural interactions • Citizenship encompasses transcendence of human spirit for a self-sustaining humanity on a life-supported planet
Knowledge brokerage	• Involves designing opportunities to facilitate knowledge exchange and transfer as part of IA process • IA can serve role as platform for knowledge brokerage, identifying and bringing together people interested in an issue and helping develop evidence-based solutions • Needs an appropriate range of stakeholders, resources, time and space for open dialogue, and nonjudgmental environment in which exchanges can take place • Role of IA to facilitate processes and mutual learning and sharing • Can help build stakeholder capacity
Expert elicitation process	• Method promotes collaboration between key scientists and policy makers • Requires careful planning, implementation, and documentation • Utilizes professional knowledge, experience, and judgment • Involves meetings of environmental experts to identify VEC goals and measures to achieve • Identifies need, selects study leader, identifies issues to address, selects experts and peer reviewers, prepares premeeting materials and disseminate, conducts meetings and document outputs • Can be used for promoting environmental sustainability and for impact management planning and implementation
Traditional knowledge	• A cumulative, dynamic body of knowledge, practices, and beliefs about the relationship of living beings with one another and their environment handed down through generations by cultural transmissions; biophysical, cultural, and cosmological; represents a cognitive spiritual awareness based on the relationship of indigenous people and their environment • Acquired from experience, culture, or interactions with land or resources over time • Built up over time and continuing into the present, by people living in close contact with the natural environment • An attribute of societies with historical continuity of resource use practices (generally indigenous or tribal); is unique to each tradition and is closely associated with a given territory; varies among different indigenous societies • Usually linked to a belief system that stresses respect for the natural world; takes a holistic perspective which stresses the place of humans with the natural system; four perspectives: taxonomic, spatial, temporal, and social • Oral communications; taught through observation and experience; explained based on spiritual and social values • Can assist with building relationships between proponents and indigenous peoples; barriers—perceptual, skepticism of scientific community, and political obstacles; needs to be controlled at the community level • Fosters meaningful indigenous participation in IA; an IA process where scientific and traditional knowledge is balanced

Sources: Berkes (1993), BC EAO (2001), Brascoupé and Mann (2001), Daniels and Walker (1996), Diduck and Mitchell (2003), Fitzpatrick (2006), Forester (1999), Gadgil et al. (1993), Healey (1997), Johannes (1999), Landry et al. (2009), Mezirow (1994), Noble (2009b), Paci et al. (2002), Saarikowski (2000), Schön and Rein (1994), Sheate and Partidário (2010), Sinclair and Diduck (2001), Sköllerhorn (1998), Swor and Canter (2011), Täbara and Pahl-Wostl (2007), Webler et al. (1995).

decision-making processes, and the participants in the process. It can further democratic values. In IA it encompasses environmental, organizational, political, and societal perspectives (Faber et al., 2010). Participants in mutual learning are transformed by the experience. Mutual learning is conducive to learning about and through the IA process. It is more likely to occur when supported by accurate information and a noncoercive environment. Third parties, such as facilitators, can help participants adapt and apply mutual learning.

Mutual learning concepts, coupled with more conventional educational methods, such as the training of participants, can contribute to more collaborative IA processes. Education in and through IA broadens and reinforces the base established through public involvement and communications measures. Additional measures, however, are needed to address conflicts, to build consensus, and to advance substantive environmental goals.

9.4.6 Negotiations

Negotiation in IA practice is based on a conflict and interest-oriented view of society. Negotiations can be aided or unaided. Aided negotiations can follow the route of litigation through the courts or can employ alternative dispute resolution (ADR) mechanisms. This analysis focuses largely on the potential roles of various forms of ADR in the IA process. It does not preclude unaided negotiations. It also recognizes that ADR tools can be applied for purposes other than avoiding, managing, and resolving conflict.

ADR is based on the theory that the people involved in a controversy, because they know their own needs and interests, are best able to develop reasonable and lasting solutions (US EPA, 2000a). ADR is voluntary and flexible (Bingham and Langstaff, 1997). It involves stakeholders discussing differences and working together as a group to solve problems or to address issues (SERM, undated). Neutral third parties (e.g., a facilitator, a mediator) often assist the parties in reaching mutually acceptable accommodations. Third parties (e.g., active mediation) are not always neutral. They can help ensure equitable procedures and fair, enduring, and environmentally sound outcomes (Susskind and Madigan, 1984). Authorities retain final decision-making authority with some forms of ADR (e.g., facilitation, mediation) but not with others (e.g., binding arbitration) (Susskind, 1999). Parties to the process are not contractually liable for their actions during negotiations (McGlennon and Susskind, undated). ADR seeks to avoid, mitigate, and resolve conflict, without resorting to litigation and where existing administrative procedures are ineffective (US EPA, 2001a).

ADR has been applied in many situations (e.g., adjudication, rulemaking, policy development, enforcement actions, permit issuance, contract administration, IA) (US EPA, 2001a). The types of conflicts, which can be addressed through ADR, are many and diverse. They can, for example, concern resource allocation, policy priorities, jurisdiction, environmental quality standards, data, values, interests, and relationships (Campbell and Floyd, 1996; Moore, 1986; Priscoli, 1999). Conflicts can be perceived or latent (whether parties conscious of), manifest or potential (whether taking place), real or displaced (whether actors correctly conceive), system dependent or independent (whether generated internally or externally), zero or variable sum (win lose or win–win potential), cooperative or noncooperative (information exchange and coalition building potential), means or ends oriented (objectives or instruments), formal or informal, institutionalized or ad hoc (Sager, 1994). Conflicts can also be characterized based on the number of parties involved, the types of parties involved, and the extent to which there is agreement regarding goals and problem definition (Priscoli, 1999). Outcomes from ADR can include pacification, settlement (i.e., procedural accommodation), or solutions (i.e., substantive improvements) (Sager, 1994).

ADR is not always appropriate. It should be possible to identify and include all relevant parties. All parties must be willing and able to voluntarily come to the table, to negotiate in good faith, and to reach a settlement on behalf of their constituents. Each party should formally accept ADR as potentially preferable to either not participating or to litigation (i.e., they gain some value) (Amy, 1987; Bingham, 2001; Susskind et al., 1999). The agency with final decision-making responsibility should support the process. The participants should be prepared to accept the negotiating ground rules and the negotiating structure (Bingham, 2001). The issues should have crystallized or "ripened" to the point that a common purpose can be agreed to, alternative courses of action can be determined, trade-offs and compromises can be identified, and solutions capable of joint acceptance are possible (Amy, 1987). Agreements reached through negotiations should be reasonable and capable of implementation (Moore, 1999). Legal challenges should be unlikely (Rodwin, 1982). ADR is not a good idea if policy precedents are likely to be set or if unacceptable environmental conditions could result (Bingham, 2001; Moore, 1999). ADR is very difficult, but not impossible, when there are fundamental clashes of values or principles.

It should be possible to ensure a relative balance of power among the parties (Amy, 1987; Armour and Sadler, 1990). Adequate resources and relevant data should support the process. There should be a deadline and some urgency for a decision (Bingham, 2001; Moore, 1999). There must be sufficient time for consensus building (Susskind et al., 1999). Third-party support (e.g., facilitation, mediation) and training for participants (if needed) should be available (Emond, 1990; Susskind et al., 1999). It should be possible to address both technical and nontechnical issues. Information should be freely shared among the parties (Bingham, 2001). It should be possible to create a clear map outlining how consensus is to be built (Susskind et al., 1999). Communications with broader interests should be maintained throughout the process (Bingham, 2001). Final agreements should be written and signed by each participating party representative. The resulting document should be legally binding and enforceable. Opinions vary as whether the process should be transparent or confidential (Emond, 1990; Bingham, 2001).

Figure 9.5 highlights some potential characteristics of an ADR process. The process starts with an overview of the factors (e.g., issues, range of parties) that determine whether ADR is possible, appropriate, and timely compared with the available alternatives (such as litigation or conventional administrative procedures). It may for desirable to formalize this review in a conflict assessment (Susskind, 1999). Parties associated with each interest are identified. Credible representatives for each party are determined and recruited (Susskind and Madigan, 1984). Funding commitments are obtained (Susskind, 1999). Assurances are sought that decision makers will take the process outcomes seriously and will allow sufficient time to ensure a sincere consensus-building effort. Procedural rules for the process are drafted and refined in consultation with parties. An overall strategy or plan is formulated and refined (Moore, 1999). The strategy addresses such matters as schedule, timing, resource requirements, training needs, communications methods, third-party assistance needs, roles and responsibilities, contacting procedures, decision-making links, and procedures for maintaining communications with constituents (Susskind, 1999). A mediator or facilitator is identified, together with a recorder (Susskind and Madigan, 1984). An agenda for the initial negotiations session is prepared (Moore, 1999).

The negotiations process is highly iterative but appears to coalesce into four overlapping steps: (1) initial deliberations, (2) focusing, (3) detailed deliberations, and (4) final refinements. During initial deliberations, the underlying interests of each party are identified, background data are obtained and exchanged, the negotiations skills of participants are enhanced (where necessary), the committee structure is determined, supplementary data collection, analysis, and review (i.e., joint fact-finding) takes places, the concerns and priorities of each party are identified, initial concept statements are formulated, and general efforts are made to build trust, rapport, and cooperation among the parties (Moore, 1999; Susskind, 1999). Focusing involves identifying key issues, determining points of agreement and disagreement, establishing the scope and boundaries for the negotiations' packages, identifying possible negotiations packages, generating texts to focus discussions, and excluding clearly unacceptable packages (Moore, 1999; Susskind, 1999). In detailed negotiations, packages are presented, concessions and commitments are advanced and traded, the consequences of the packages are determined, the packages are evaluated and possibly combined, solutions are sought which are mutually acceptable and which will maximize joint gains, measures are identified to prevent and offset negative features, a preferred package or package combination is identified, and commitments are obtained from each party to the preferred package (Susskind and Madigan, 1984). Final refinements elaborate on the preferred package to add implementation, monitoring, compensation, and postnegotiations evaluation provisions. A reopener or dispute resolution mechanism is often included (Susskind, 1999). The final package is prepared in a form suitable for signing by the parties. The parties are held responsible for signing the agreement.

During negotiations, the parties are expected to be responsible (act in good faith), open (all concerns explicit), respectful (all heard and taken seriously), trustworthy (nothing held back, no hidden agendas), fair (power inequities offset), flexible (discussion based on interests rather than predetermined positions), and constructive (search for outcomes that meet and further the interests of all) (Innes, 1996; Innes and Booher, 1999; Praxis, 1988; Susskind, 1999). The parties should strive for but not insist on consensus (Nagel, 1987). Consensus can be either unanimity (a good idea but not always practical) or overwhelming agreement (in contrast to a bare majority) (Susskind, 1999). During negotiations, authority does not have to be given up (any party can walk away at any time) and principles and interests need not be abandoned (the process seeks solutions which respect and further the interests of each participant) (Susskind et al., 1999). A third party, such as a facilitator or a mediator, can assist in consensus building, conflict resolution, and joint problem solving (Smith, 1993). A record of the process should be kept (Susskind, 1999). As the process unfolds it may be necessary to revisit earlier stages.

Implementation can involve ratification by constituents. The parties are held to their agreements (Susskind and Madigan, 1984). It may be necessary to ensure that informal agreements are incorporated into formal mechanisms. The provisions of the agreement need to be implemented. Implementation should be monitored and evaluated. Monitoring results may necessitate renegotiations or modifications to elements of the agreements. The lessons and insights obtained through negotiations and monitored should be incorporated into the organizational learning mechanisms of the process participants (Susskind, 1999).

Conflicts or potential conflicts, which might warrant ADR, can arise at any stage in the IA process where interests, values, or perspectives might clash (e.g., scoping, significance interpretation, impact management). They can concern the design and execution of public participation activities. They can occur both prior and subsequent to approvals. They can pertain to procedure or to substance. ADR procedures can help avoid conflict and scope issues when instigated near the outset of the IA process. They can help keep the process "on track" when dealing with difficult interpretative issues such as alternatives evaluation and impact significance interpretations. They can help bring the process to a successful conclusion when dealing with troublesome issues surrounding mitigation, compensation, monitoring, and implementation. ADR can represent an alternative to or can scope a hearing or court action.

Various ADR methods may be appropriate at different stages in the IA process. Table 9.7 lists examples of

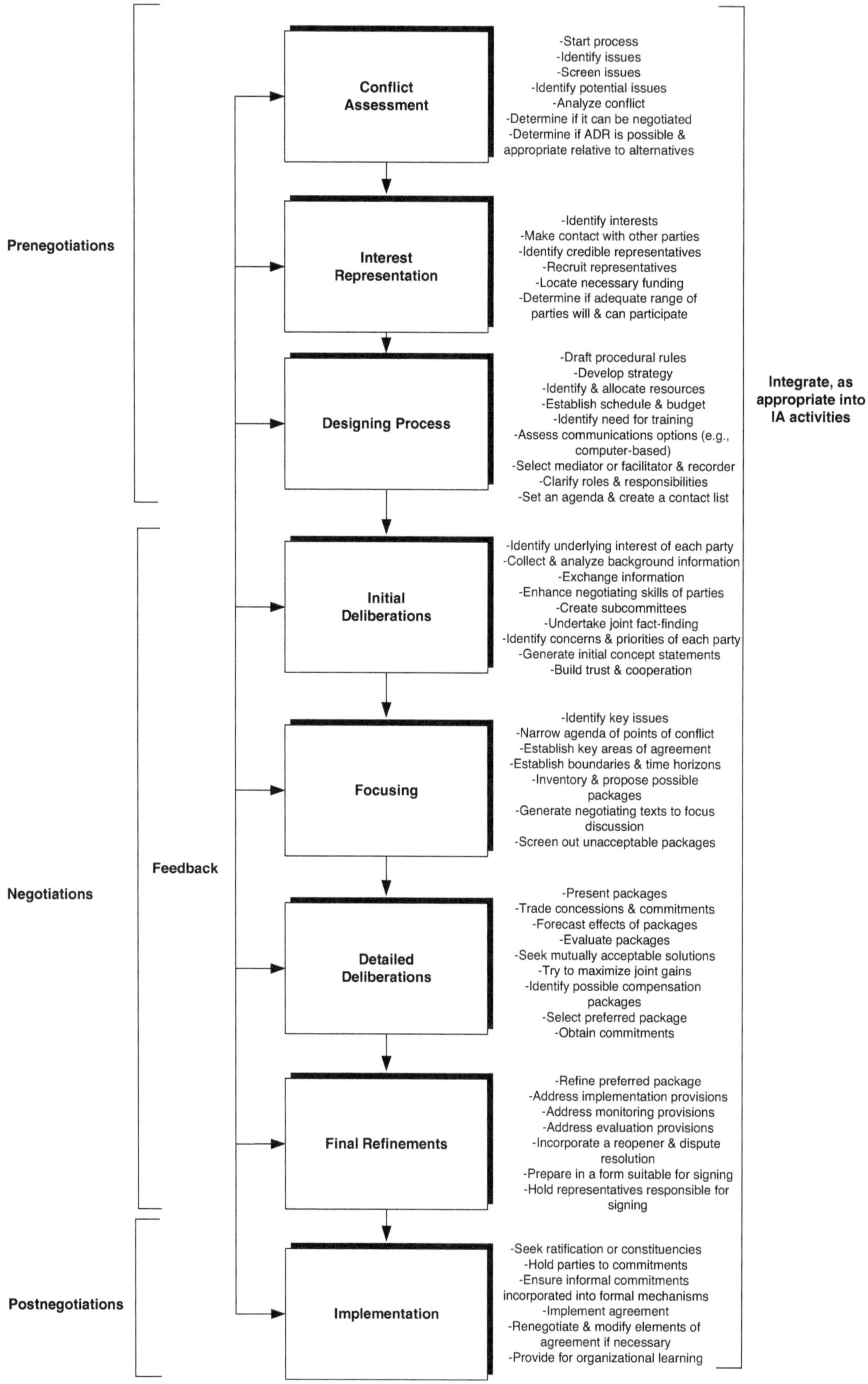

Figure 9.5 Example of an ADR process.

Table 9.7 Examples of Negotiations, Methods, and Mechanisms

Unaided negotiations	• Contending parties work out differences without help • Distinction between position (each side argues from positions) and interest (alternative solutions that meet interests or needs) negotiations • Differences resolved based on compromise or interest-based principles • Informal: serves to soften hard positions, explore underlying interests, develop options, and reach a mutually acceptable resolution
Conflict theory anticipation, assessment, and management	• Conflict theory: identifies different interests, distributional effects, and possible strategies to alleviate • Dispute prevention: third party identifies potential disputes before opposing positions fully identified • Involves communications, building personal relationships, and establishing procedures for addressing issues before they become disputes • Conflict assessment: a document that spells out what the issues are, who are the stakeholders, where they disagree and where they find common ground; usually prepared by a neutral outsider based on confidential interviews with key stakeholders • Potential role for SIA in managing conflicts, utilizing conflict mediation methods • Allows the assessor to explore the parties' incentives and willingness to negotiate in good faith; creates an opportunity for the assessor to educate stakeholders about what it takes to bring consensus process to successful conclusion • Produces recommendations regarding who has a stake, what issues are important to stakeholders, and whether it makes sense to proceed given constraints and circumstances • Phases: introductions, information gathering, analysis, process design, report writing, and report distribution
Information exchange and joint fact-finding	• Information exchange meetings: parties share data and check out perceptions of each other's issues, interests, positions, and motivations in an effort to minimize unnecessary conflicts over facts • Used to establish a shared framework for analyzing a dispute, resolving disputes on matters of fact, and clarifying disagreements of fact • Fact-finding can be used in scientific, technical, or business disputes in which knowledge is highly specialized • A third-party subject matter expert is chosen by the parties to act as a fact-finder or independent investigator • Can identify facts and areas of agreement and disagreement; expert then submits a report or presents the findings
Conciliation	• Neutral party, generally with no stake in the dispute, to informally communicate separately with disputing parties for the purposes of reducing tensions, build trust, and agreeing on a process for resolving the issue • Attempts to assist negotiators in searching for accommodations • Helps establish a basis for direct negotiations
Facilitation	• Facilitator: a nonpartisan or neutral trained specialist who helps people design effective meetings and problem-solving sessions and then acts as the meeting leader on behalf of the group; does not have the authority to make substantive decisions on behalf of the group • Focuses on process and uses available tools to create and foster an environment conducive to joint problem solving • Assists parties in coming together to exchange views, share information, and clarify differences; helps promote meetings that are purposive, efficient, productive, and civil • Examples of responsibilities: taking care of meeting logistics, reminding parties of ground rules, intervening when someone violates the ground rules, keeping discussion on track and schedule, summarizing and focusing discussion, identifying key points, clarifying issues and interests, orienting the group to objectives, promoting effective communications, eliciting creative options, and maintaining a nonthreatening environment that encourages people to participate
Mediation	• Involves the intervention of an acceptable, impartial, and neutral third party, who has no decision-making power, to assist contending parties in voluntarily reaching their own mutually acceptable settlement of issues • Mediator is only concerned with process issues, works hard to ensure that process is fair and unbiased in the eyes of the parties at the table; process is voluntary, informal, and confidential • Assists in clarifying issues, facilitating information collection, exchange and evaluation, identifying key issues, uncovering hidden interests, designing an effective negotiations process, developing options for dispute resolution, and helping to identify and formulate areas of agreement

(continued)

Table 9.7 (*Continued*)

	• Mediation process: decision to start, mediator contacts parties and outlines process and logistics, mediation meetings held, mediator's report is prepared and report signed by parties and submitted to approving authority • Conventional mediator: no stake in outcome; active mediator: works hard to ensure that process is fair, unbiased, and open to all parties affected by the outcome, whether they sit or do not sit at negotiating table, seeks to ensure that the outcome is viewed as fair by the community at large, is reached efficiently, and remains stable after bargaining
Minitrial, dispute review board, or disputes panel	• Minitrial: expedited presentation of positions and evidence to a panel composed of senior decision makers representing each participant and a third party; authorized representatives hear case and negotiate agreement; parties can present summary proofs and arguments; third parties can advise, mediate, or make advisory opinion • A voluntary, expedited, nonjudicial, informal, and confidential procedure; used to address complex technical issues where litigation costs would be high and senior decision makers want maximum control of terms of settlement • Dispute review board or dispute panel: provides the parties with an objective evaluation of the dispute by fully qualified experts; opinion of the board is advisory, with the parties negotiating a final resolution
Inquiries, public hearings, and adjudication	• Formal judicial or quasijudicial proceeding; parties meet in an adversarial setting before an impartial judge or panel • Can ensure thorough presentation and testing of evidence; can address issues not suitable for or remaining after ADR • Can be intimidating, time-consuming, and expensive • Can exaggerate conflicts; encourages people to take positions rather than share fundamental interests or to engage in problem solving; sets up a "win–lose" situation
Arbitration	• Parties select a neutral individual or panel with expertise on issues at dispute and set rules or norms to apply • Parties (or counsel for each party) present their case (facts, positions, and formal arguments); arbitrator recommends a basis for settlement • Nonbinding arbitration: parties are not bound to submit to arbitrator's decision; but advise normally carries a great deal of weight • Binding arbitration: parties agree to live by arbitrator's decision even if they are unable to reach a voluntary decision themselves • Each party seeks to design the most reasonable outcomes for presentation to arbitration; emphasis on design rather than argument • Results in an unbiased judgment, avoid problems of litigation, prevents loss of face, and encourages a decision
Negotiated rule making	• A process of bringing together representatives of various interest groups and a government agency to negotiate the text of a proposed rule; goal is for the committee to reach consensus on the text of a proposed rule • Parties need to perceive that transaction costs of developing, implementing, and enforcing regulations in the usual fashion are high and that significant costs savings are possible from a different approach • Involves convening the appropriate parties, clarifying roles and responsibilities regarding consensus-based decision-making process, reaching and testing the scope of the agreement (joint problem solving and fact-finding, option development and evaluation, selection of a preferred option), and binding parties to their commitment (draft rule published and subject to comment period)
Negotiated environmental agreements	• Involves negotiated agreements between proponents and communities/indigenous peoples • Commits monetary and nonmonetary return back to community (e.g., direct and indirect employment, local procurement, financial payments) • Facilitates communications and community–industry partnership • Formalizes public role and pre and postapproval decision making • See enhancement as core means converting project risks into sustainability opportunity • Culturally appropriate enhancements • Addresses such matters as community benefits and impact management • Help engage stakeholders and build local capacity and empowerment of beneficiaries

Sources: Bingham (1986, 2001), Bingham and Langstaff (1997), Burdge (2004), Campbell and Floyd (1996), Creighton et al. (1999), De Bono (1992), Esteves and Barclay (2011), Esteves et al. (2012), Forester (1999), Lowry et al. (1997), Morgan (1998), Moore (1999), Noble and Birk (2011), O'Faircheallaigh (2010), Peltonen and Sairinen (2010), Rowan and Streather (2011), Sager (1994), Smith (1993), Susskind and Madigan (1984), Susskind et al. (1999), Weber (1998), Westman (1985).

negotiations and ADR methods and mechanisms. The methods vary in the roles they can perform (e.g., conflict avoidance, fact-finding, determining if ADR is practical, expediting meetings, problem solving, dispute resolution, the siting of "locally unwanted land uses"). They also vary in the role of participants, in the formality of proceedings, in the degree of confidentiality, and in the types of situations into which they are applied. The application of ADR methods in IA should take into account these differences, general ADR characteristics, strengths and limitations, the specific characteristics, strengths and limitations of individual ADR methods, and the match between methods and context.

ADR methods have had considerable success in contributing to settlements (Creighton, 1999; Sipe and Stiftel, 1995). Participants tend to be very satisfied with the process (Sipe and Stiftel, 1995). ADR has been effective in identifying the rationale for settlements (Creighton, 1999). It is often faster and less costly than litigation (Campbell and Floyd, 1996; Harashina, 1995). It is credited with building and enhancing relationships, facilitating higher quality decisions, identifying and solving conflicts and problems, furthering procedural and substantive equity, enhancing the likelihood of approvals, implementation, and compliance, and contributing to stakeholder empowerment (Campbell and Floyd, 1996; Creighton et al., 1999; Harashina, 1995; Innes and Booher, 1999; Harrop and Nixon, 1999; Smith, 1993; Smith et al., 1997). Managers often favor ADR because it is voluntary, nonjudicial, confidential, and does not necessitate control delegation or sharing (Creighton et al., 1999). Stakeholders may prefer ADR because it is informal, nonintimidating, and conducive to joint problem solving.

There are, however, many situations, as noted earlier, where ADR is inappropriate or impractical. Some suggest that agreements reached through ADR tend to be vague and general (Neuman, 2000). It is feared that, by operating on the fringes of institutional structures, ADR will either be ineffective or will undermine representative democracy and state intervention (Fischler, 2000; Neuman, 2000). ADR, it is argued, could lead to ethical, democratic, or environmental sacrifices if objectives, principles, or interests are compromised in the quest for consensus (Smith, 1993). Citizens may find themselves at a disadvantage because of imbalances in training and expertise (Smith, 1993; Smith et al., 1997). They may find the time commitments too onerous (Canter, 1996; Smith, 1993). They may resist the need for coalitions with other interests. They may fear co-option (Canter, 1996). Agencies could be reluctant to participate because ADR is inconsistent with their conventional operating procedures (Manring et al., 1990; Smith, 1993). They may hesitate to involve others in decision making on the grounds that they will lose control or that they will be opening up their decision-making processes to scrutiny and legal challenges (Bingham and Langstaff, 1997). They too may lack negotiating skills and experience. ADR could run counter to the financial interests of outside counsel. Some argue that it is difficult to enforce implementation and monitoring requirements obtained through ADR (Smith, 1993).

The many ascribed benefits of ADR are, according to some, overstated (Bingham, 2001). While acknowledging that ADR has a high settlement rate, they suggest that it does not necessarily result in a higher compliance rate (Sipe and Stiftel, 1995; Sipe, 1998). Each ascribed ADR benefit, it is suggested, should be treated as a success measure and should be systematically tested against experience (Bingham, 2001). ADR brings parties together to resolve disputes and to solve problems. Disputes are frequently resolved. Most ADR disadvantages and constraints can, with judicious application and a heightened awareness of potential pitfalls and limitations, be avoided or ameliorated. It is less apparent whether and how ADR generates creative solutions to problems or proactively generates "win–win" options and opportunities, which move beyond the reconciliation of interests (De Bono, 1992). The specific techniques for creative collaboration are not well developed in ADR. It is possible that groups, by relying excessively on the skills of facilitators and mediators, will be slow to develop or to apply their own collaborative skills (De Bono, 1992). ADR, applied appropriately and in the right circumstances, can assume a vital role in IA public participation. But alternative approaches and a wider range of creative methods are likely needed before an IA process can be said to be fully collaborative.

9.4.7 Collaboration

Collaboration is about people cooperatively working together in a joint endeavor with substantive aspirations. The orientation, with collaboration as compared with negotiations, shifts from interests and positions to perspectives, from problems to visions and opportunities, from conflict management to creative exploration, and from negotiations to collaboration. Negotiations may (or may not) be an element of or precede collaboration. Collaboration can build on a base of but is more than the sum of effective consultation, communications, mutual education, and negotiations. Effective collaboration transcends the other elements of IA participation.

Table 9.8 highlights the characteristics of several collaborative concepts relevant to IA process management. Some concepts, such as joint fact-finding, active mediation, and consensus building, also are used in negotiations. Others, such as procedural justice, closely parallel such related communications and educational concepts as discourse ethics. These commonalities underscore the many interactions among IA participation elements. Collaboration encompasses, integrates, and transcends involvement, communications, mutual education, and negotiations.

The concepts exhibit numerous ideal collaborative process characteristics. Collaboration, for example, should be inclusive. It should include multiple interests, issues, values, and perspectives within a multistakeholder planning system

Table 9.8 Examples of Potentially Relevant Collaboration Concepts

Joint fact-finding	• Information exchange: parties share data and compare perceptions, perspectives, and motivations • A fact-finder is an independent expert chosen by the parties to conduct investigations • Fact-finders have technical expertise; they investigate and analyze issues • Joint fact-finding can provide a shared factual and analysis basis for collaboration
Joint planning	• Representatives of the interested parties participate in a committee with the power to make a binding decision • Involves dialogue, shared responsibility, multiple perspectives, and in-depth deliberations • Tends to work best when serious environmental impacts and trade-offs, a wide range of complex issues, many concerned public and agency groups, sufficient time for planning, public strongly desires formal involvement program, and agency has capacity to support • Works best when mutual respect and trust, recognition of knowledge, experience and respect of each participant, recognition of individual and joint rights, and responsibilities and agreement on meeting procedures
Partnering	• A formal or informal means to improve and build the relationship between government and another party and/or to work with one or more parties to achieve a common goal • Used primarily during contract performance; goal is a more cooperative, team-based approach • Partners share some level of responsibility, planning, and decision making and ownership of process and product; resources, expertise, energy, and risks shared • Partnerships can be formal or informal • Built on dialogue, trust, and alignment of purpose and effort • Partnering includes key scientists, policy makers, proponents, and the public • Parties jointly define a clear vision, goals, and action items and then work together to achieve; process is working when sharing, clear expectations, trust and confidence, commitment, responsibility, courage, understanding, and respect, synergy (outcome more than sum of partners), and excellence
Group problem solving and opportunity seeking	• Problem solving: group identifies problem, analyzes problem, identifies and evaluates possible solutions, and develops a plan for implementing the problem solution agreed to by the group • Opportunity seeking: begins with search for positive possibilities rather than reaction to something going wrong • Seeks to adhere to certain virtues of group inquiry such as clarity, honesty, open-mindedness, and attention to detail • Various characterizations of group development process (e.g., forming, storming, norming, performing) and of group maintenance behavior (e.g., harmonizing, gate keeping, encouraging, compromising, standard setting and testing and relieving tension) • Numerous individual and small-group methods for redefining and analyzing problems, for generating ideas, for evaluating and selecting ideas, and for implementing ideas (e.g., brainstorming, nominal group process, forced relationships or free associational, related or unrelated stimuli, lateral thinking techniques, charrettes, simulation games)
Coalition building/ networking/ working groups	• Networking refers to linking stakeholders through formal or informal channels so as to bring about plan formulation and implementation • Networks exist over time, are invitational, are numerous, have a limited capacity, are only as good as their members, depend on exchanges and incentives, tend to focus on selected actions, are channels of action, are a source of power, and take place in a symbolic context • Network tasks: map the terrain, gain information and identify actors; identify the relevant leverage points; select the tentative coalitions of support; float the initial image to symbolize the possibility of action; adapt the technical argument to the requirements of support and opposition; organize the coalition to trigger the multiplier and maintain and feed the coalition moves toward implementation; can be extended to collaboration in institutional design • Potential network roles: information provider, watchdogs, and pressure groups • Potential formation of networks of national, local, and international NGOs; can enable more efficient use of resources and expert knowledge
Active mediation	• Challenge to move beyond role as process people; mediators are nonpartisans but concerned with representation of affected parties in the mediation process and the efficiency, stability, and well-informed character of potential mediation outcomes • Moves beyond search for acceptable agreements within a given space of interests; the space is altered and the participants are transformed and empowered • Searches for new possibilities and agreements; concern with the decision's goodness and the quality of agreements • Involves a broader conception of political life, the public interests, deliberation and debate, mutual recognition and discussion, learning, and civic discovery
Procedural and environmental justice	• Assumes that the procedures used to arrive at decisions are significant determinants of satisfaction separate from the effect of outcomes; procedures perceived as unfair might reduce satisfaction with what are otherwise judged as objectively fair decisions • Practice of fairness: talk to trust, to outside selves, to mutual respect, and to joint search for mutually acceptable reasons and measures

Table 9.8 *(Continued)*

	• Fairness rules: decision consistency, suppression of bias, accuracy, error correction, representativeness of groups of affected individuals, ethically compatible with fundamental moral and ethical values and correction of power imbalances • Also necessary to address environmental justice issues
Collaborative planning	• A process through which parties who see different aspects of a problem can constructively explore their differences and search for solutions that go beyond their own limited visions of what is possible • Embraces sustained dialogue, stresses common ground, and promotes shared vision of the future • Involves shared power, open discussions, and shared values • Characterized by face-to-face dialogue, mutual learning and voluntary participation; critical listening, reflection-in-action, and constructive argument all interact • Fosters the inclusion of all members of political communities while acknowledging their cultural diversity
Cooperative modeling	• Group of people work together to develop a model to better understand a complex system and explore consequences of various 'what if' scenarios • Modeling can help identify subsystems in which impacts are significant and structure deliberations • Can help create a sense of community • A healthy level of conflict can be beneficial • Importance of ensuring interest representation and local knowledge • Necessary to explicitly decide level of participation individuals want, ground rules for interaction and communications methods
Collaborative public participation process	• Characteristics: shared vision and objectives and measurable outcomes; process is equally managed by stakeholders; involves up-front planning, conflict resolution and open communications among participants; balanced and inclusive stakeholder participation; strong leadership; capacity created for stakeholders to understand information; possible help from facilitators • Sound informational base, rests on sound democratic participation principles, honors a full spectrum of values, and holds everyone responsible for success • Begins with mutual education; no one leader; no one excluded from table; works together with community to define shared visions that sustain the community and the environment • Requires advance planning, management support, funding for a facilitator and other expenses, and sufficient time to reach useful results; crucial to maintain a balance of power among participants
Collaborative governance	• Starts from proposition that large and growing fraction of capacity to create public value exists outside government • Brings public and private stakeholders together in collective forums with public agencies to engage in consensus-oriented decision making • Process is formal, consensus-oriented, and deliberative • Process involves face-to-face dialogue, trust-building, commitment to the process, shared understanding, and intermediate outcomes • Effectiveness influenced by such variables as prior history of conflict or cooperation, incentives for stakeholders to participate, power and resource imbalances, leadership and institutional design • Potential role for IA in contributing to collaborative governance; can assume important roles in joint fact-finding, in realizing meaningful stakeholder involvement, in interactions between people and experts, and in identifying problems and alternative solutions
Comanagement	• Government and stakeholders work cooperatively to undertake integrated management of the environment and natural resources in a sustainable way consistent with goals of parties; primarily applicable to indigenous communities; allows parties with an interest in the ownership and management of natural resources to power share; various mechanisms (e.g., a cooperative, a comanagement council) • Makes it possible to integrate local community interests with third party and government interests • Principles: public ownership and government responsibility, cooperate as partners, stewardship of natural resources and environment, integration of environment/natural resources, economic development and social well-being and inclusive process • Lessons: government commitment, open debate about long-term direction, meaningful third-party agreement, a reflective and evolving process, real teeth and shared decision making, direct community input to resource inventory and planning, community and local government staff, formal agreements, a broad range of interests, coalitions of interests, multiresource framework, interim measures, comanagement plan and implementation, action linked to information and understandability
Consensus building	• A process in which people agree to work together to resolve common problems in a relatively informal, cooperative manner; two meanings of consensus: unanimity and positive support from large proportion of participants

(continued)

Table 9.8 (*Continued*)

	• A good consensus-building process: includes all relevant and significantly different interests, is driven by a purpose and tasks that are real, practical, and shared by the group, is self-organizing, allows participants to decide on ground rules, objectives, tasks, working groups, and discussion topics, engages participants, keeps participants at the table, interested and learning through in-depth discussion, drama, humor, and informal interaction, encourages challenges to the status quo and fosters creative thinking, incorporates high-quality information of many types and assures agreement on its meaning, seeks consensus only after discussions have fully explored the issues and interests and significant effort has been made to find creative responses to differences • Represents a way to search for feasible strategies to deal with uncertain, complex, and controversial planning and policy issues of common concern; properly designed can produce results that approximate the public interest • Creative efforts can be enhanced by techniques such as scenario formulation and role-playing simulation
Shared visions planning	• A way to use computers to help stakeholders to participate in rigorous planning analyses; developed by Institute for Water Resources, U.S. Army Corps of Engineers • Shared vision models are built using user-friendly graphical simulation models; built with decision maker and stakeholder involvement; models used to evaluate alternative plans according to decision criteria • Marries systems engineering, public policy, and public involvement; similar to adaptive environmental assessment and management • Because experts and stakeholders build models together, conducive to developing a consensus view of how system works as a whole and how it affects stakeholders and the environment • Model flexibility makes it easy to analyze sensitivity of conclusions to errors in data, changed forecasts, or conflicting assumptions • Other more general visioning approaches combine team building with an alternative futures planning process to produce shared visions; factors provide a basis for themes, which are, in turn, built into scenarios and strategies which are compared, which then form the basis for short- and medium-terms goals and action plans; process completed with assignments and target dates
Constructive engagement	• An approach that brings groups together to establish and monitor a facility's environmental activities through a cooperative, nonadversarial partnership • Takes many forms (e.g., citizen advisory groups, stakeholder negotiations, formal mediations, "Good Neighbor Agreement" processes, oversight committees, independent organizations) • Have dealt with issues such as site location, facility operations, emissions and waste controls, worker health and safety, regulatory relief, site cleanup, and pollution prevention • Offers an approach to improving communications among stakeholders and for finding creative solutions about facility activities
Democratic deliberation	• Build upon the concepts and ideas of Habermas • Applies Habermasian principles; a useful ideal type for investigating deliberation in IA • Sees IA as having "hidden" deliberative potential • Treats IA as an arena for democratic deliberation • Seeks to clarify whether IA public participation fulfills the deliberative idea notion of discourse • Applies inclusive and dialogue-based participatory tools • Seeks to offset resource distribution inequities • Considers role of institutional arrangements in facilitating or inhibiting discourse • Assesses if and in what ways views of participants change as a result of deliberation
Community engagement and community-based IA	• Seeks to engage the broader community in developing broader visions and higher level policies and in addressing key issues • Involves a partnership of the state, the community, and the proponent • The public is meaningfully engaging, both formally and informally • Use of dialogue-based methods (e.g., interactive workshops) for visioning, brainstorming, and critical reflection • Helps reduce power differentials among participants • Conducive to individual and social learning and can facilitate a transition to sustainability

Sources: Ansell and Gash (2007), Bauer and Randolph (2000), Benveniste (1989), Blake (2010), Buuren and Nooteboom (2010), Cockerill et al. (2010), Creighton et al. (1999), De Bono (1992), Donahue (2004), Forester (1999), Gauthier et al. (2011), Gray (1989), Healey (1997), Innes (1996), Innes and Booher (1999), Koivurova (2008), Krieger (1981), Lawrence et al. (1997), Laws (1996), Maynes (1989), Moore and Woodrow (1997), Mosley et al. (1999), Nagel (1987), Patton et al. (1989), Pope and Grace (2006), Praxis (1988), PCSD (1997), Ryu et al. (2004), SERM (undated), SIFC (1996), Sinclair et al. (2007), Sinclair et al. (2009), Susskind et al. (1999), Susskind and Madigan (1984), Susskind and Cruikshank (1987), Swor and Canter (2011), US EPA (2000a,b, 2001a,b), Westman (1985), Wikland (2005), Witty (1994).

(Kain and Söderberg, 2008). It should decentralize decision making (Gibson, 2006a). It should include interagency and stakeholder collaboration (Noble and Gunn, 2009). The participants should bring to the table a diversity of relevant knowledge and experience. It should be substantive—systematically addressing intergenerational equity and the limits of life support systems (Govender et al., 2006). A collaborative process should be jointly undertaken and owned by the participants. Links among participants should be facilitated and strengthened by such mechanisms and methods as joint planning, partnering, networking, coalition building, the use of working groups, mediation, cooperative modeling, comanagement, group problem-solving opportunity seeking, consensus building, shared vision planning, community engagement, and community-based IA. Power and responsibility should be shared. It should enhance the capacity of participants to both listen and hear the stories of others (Hodge, 2004). Technical specialists (e.g., fact-finders) should assume a support rather than a lead role. Collaboration also can be facilitated by technical support methods such as modeling, environmental systems analysis, and life-cycle analysis (Cockerill et al., 2010; Sinclair et al., 2007). A collaborative IA process should be directed toward and guided by substantive environmental management, environmental justice, and sustainability ends. The process should be positive and purposive. It should strive to develop a complex and collective vision of what the parties wish to sustain (Hanna, 2009b). Third parties (e.g., active mediators) should support the realization of stable, efficient, and "good" outcomes. It should extend beyond IA approvals to include citizen-centered monitoring (Hunsberger et al., 2005).

The participants should generate and then pursue visions, goals, objectives, and opportunities. They should rectify and ameliorate problems. They should identify priorities and establish limits (Hermans and Knipperberg, 2006). The process should be open, voluntary, informal, flexible, cooperative, and consensus seeking. It should be guided by procedural, democratic principles such as clarity, honesty, commitment, mutual recognition, mutual respect, trust, open-mindedness, confidence, fairness, and attention to detail. The process should be procedural just. Power imbalances should be offset. The process should be creative and synergistic. Participants should jointly discover and explore new ideas and solutions. They should make effective use of methods (e.g., models, scenarios, role-playing simulation, brainstorming, lateral thinking procedures, nominal group process) and group development and maintenance techniques, conducive to fostering and applying creativity. Collaborative processes can be enhanced by drawing upon insights and distinctions identified by such concepts as procedural and environmental justice, collaborative governance, and democratic deliberation.

Participants in an effective collaborative process are altered and empowered by the experience. The process should not be confined to the group. Ongoing contacts should be maintained with constituents. The process, to be more than an interesting experience, must be practical and real. Outcomes should be formalized (e.g., a plan, a strategy, an agreement, a contract, a rule, an IA, facility operations) and capable of implementation. The collaborative process, to be effective, needs to be supported by management (e.g., resources, time) and by the public. Implementation may necessitate networking and coalition building and maintenance. Sometimes, institutional design and reform may be required.

9.5 INSTITUTING A COLLABORATIVE IA PROCESS

9.5.1 Management at the Regulatory Level

Table 9.9 highlights examples, from the four jurisdictions, of regulatory level measures for facilitating agency and public collaboration in the IA process.

The four jurisdictions generally include methods to ensure that information is provided to the public and that comments and concerns are obtained from the public. Some progress has been made to further the involvement of traditionally unrepresented and underrepresented segments of society. There are very limited provisions for continuous, in-depth public involvement. There remains room for improvement in ensuring early and frequent public involvement, in involving the public in IA screening, scoping, and postapproval decisions, and in demonstrating how public concerns and preferences have influenced decision making.

All four jurisdictions have addressed IA communications in terms of communicating with the public either directly or through documents. Considerable progress has been made in ensuring electronic access to IA documents and comments. General references are made to effective two-way communications. Some scattered references are made to adapting communications strategies and documents to meet the needs of communities and of various segments of society. More specific advice could be provided concerning communications principles, communications skills, offsetting communications distortions, and applying measures to foster and enhance dialogue among IA process participants.

The four countries have devoted considerable attention to educating IA practitioners and the public about IA, both in general and with specific reference to public participation principles and practices. There also are numerous, albeit scattered, efforts to integrate community and traditional knowledge into IA requirements and practices. Very little attention, except for general references, has been devoted to mutual learning concepts, methods, and practices. ADR, as an option for resolving IA disputes, does not appear to have received much attention in either Europe or Australia. The logical departure point for assessing the potential role of ADR in IA is to draw upon the extensive US and Canadian knowledge base. Additional attention could be devoted to the specific adjustments needed to both IA and ADR requirements and practices to more effectively link and integrate these two related environmental management fields.

Table 9.9 Positive and Negative Collaborative Examples at the Regulatory Level

United States	Canada	Europe	Australia
(+) Numerous references to open decision making, public notification, early public involvement, timing requirements, and public involvement at key decision points; criteria for when public meetings or hearings appropriate (+) Agencies and individuals consulted must be listed, and documents related to the public's participation must be included (+) Guidelines address document dissemination to other governments (states, Indian tribes, local agencies) and public; comments must be invited (+) Continuous contact with nonagency stakeholders recommended (+) Opportunities for public involvement during scoping and during draft and final IA document review (+) Potential for judicial review of the final EIS; an EIS can be rejected if it fails to foster informed decision making and public participation (+) Extensive range of public involvement guidelines; some refer to the participation of indigenous groups and tribal citizens (US DOE, 2010) (+) Citizen's guide to NEPA (US CEQ, 2007b) (+) Stakeholder directories (US DOE, 2002a) (+) U.S. environmental justice requirements conducive to broadening the range of publics (+) Presidential proclamation has sought to further promote public transparency and involvement in implementing NEPA (U.S. Administration of Barack H. Obama, 2009) (+) Updated NEPA web page (+) Efforts to facilitate enhanced communications are evident from the preparation of stakeholder directories, from the soliciting of advice from stakeholder groups and	(+) Purposes of Act and EA Agency objects include promoting coordination and cooperation with aboriginal peoples, and promoting opportunities for meaningful public participation (+) Comments from the public are identified among the factors to be considered; community and aboriginal traditional knowledge also may be taken into account (+) The definition of effects on aboriginal peoples is broader, encompassing health, heritage, cultural, and resource and land-use effects (+) Process requires a federal coordinator (+) Provisions for public access to screening report and draft EA (+) Decision posted online; Internet notification procedures for panels; panel must consider public concerns (+) Reference is made to cooperation and coordination between federal and provincial governments, among federal departments and agencies, and with aboriginal peoples (+) Reasonable public notice and opportunity to comment on draft guidelines (+) Canada provides considerable IA guidance, some oriented to IA practitioners and some geared to more general audiences; IA training oriented to overall requirements and to specific types of IA documents is provided (+) Public opportunity to comment on initial steps in determining if EA required, during conduct of EA, on draft EA, and during panels (+) Participant funding provisions (CEAA, 2008b, 2009b) (+) Public participation guidance (CEAA, 2010b) (+) Detailed provisions: CEA Registry (government-wide online registry); web site, project files	(+) Proposed Project Directive: EIA includes consultation with public concerned and environmental authorities (+) PPD: highly collaborative approach to development of Directive (EC, 2012a,b,c,d) (+) PPD: provision for reasonable time frames for different phases; time frame for consulting public not less than 30 days and not more than 60 days (+) PPD: scoping includes determining the public and authorities likely to be concerned (+) PPD: information to public—decision, conditions, basis for decision including public participation process, main mitigation measures, and monitoring measures (+) PPD: opportunity to comment while options still open to competent authority (+) EIA and SEA directives include general references to notifying and informing the public, making screening reasons and documents available, providing early and effective public participation opportunities, enabling the public to express opinions on draft documents, the preparation of nontechnical summaries, allowing sufficient time for consultation and transparent decision making (+) SEA directive requirement to make draft plan or program available to authorities and public (+) The SEA directive refers to the public availability of monitoring results and showing how consultations influenced the final plan or program; includes consultations with environmental and health authorities (+) Public notification and involvement guidance (DETR, 2000; EC, 2009e; European Union, 2003) (+) Links to Aarhus Convention (public access to information,	(+) Includes provisions regarding notification procedures, document availability, the soliciting of public comments in response to draft IA documents, the description of public consultation procedures and of the views of affected communities, and procedures for conducting public inquiries (Australian Government, 2007a) (+) Refers to a cooperative approach involving governments, communities, landowners, and indigenous peoples (+) IA documents required to identify affected parties, to indicate how the communities may be affected, to describe the views of the public regarding the proposed action, and to take into account public comments (+) Several provisions aimed at facilitating involvement of indigenous people and for promoting the use of indigenous peoples' knowledge (−) No opportunities for public involvement in minister's decisions regarding IA type (+) Accepted reforms include measures to make IA documents, submissions, and public comments readily available, to make the basis for significant decisions more transparent, and to incorporate minimum timelines (+) Commitment to prepare principles and guidelines for best practice public consultation, and to perform auditing of public awareness and effective engagement (Australian Government, 2011d) (+) References to clear and understandable documents and to effective two-way communications with the public (+) References to special arrangements for affected groups with particular

Table 9.9 *(Continued)*

United States	Canada	Europe	Australia
advisory committees, and from the provision of status reports to local parties (±) Two-way communications, dialogue, and improved communications are recurrent themes in IA guidance documents; some still argue that communications are often one way (+) Numerous basic and advanced NEPA courses (+) Community outreach is a major focus of environmental justice requirements and guidelines (+) Highly controversial, a significance factor; conflicts with the plans, policies, and controls of other governments must be identified (+) ADR has been used extensively in U.S. IA practice (+) Handbook on collaboration in NEPA (US CEQ, 2007a) (+) Tribal capacity building guidance (US CEQ, 2004) (+) NEPA regulations refer to compliance capability; guidance on enhanced tribal capacity	(+) There has been considerable experience in Canada in the negotiations of impact management and benefits agreements and land claims agreements, in the establishment of aboriginal IA regimes, and in resource comanagement (O'Faircheallaigh, 2006) (+) The tri-party IA approach (federal, territorial, indigenous people) adopted in the north is a form of joint planning (Armitage, 2005) (−) Mediation option remove (−) Public involvement provisions under the new Act have been criticized as belated (i.e., after registration document filed by proponent), and inhibited by tight time restrictions, a partial definition of the environment, a narrow range of projects subject to the Act, discretionary project definitions, an ambiguous public role during scoping, a restricted range of types of alternatives, a narrow definition of interested parties for National Energy Board hearings, uncertainties regarding IA substitution criteria, and the possibility of projects being approved notwithstanding significant adverse effects, if justifiable (−) Federal IA requirements also have been characterized as being limited to notification and consultation (rather than collaboration and shared decision making), and as being weak regarding public involvement at the strategic level (Noble, 2009b; Sinclair and Diduck, 2009) (−) Canadian IA systems have been criticized for emphasizing process over outcomes and for favoring bureaucratic elites and experts over the general public and local forms of knowledge (Galbraith et al., 2007)	public participation, access to environmental justice) (+) Requirement to prepare a nontechnical summary (+) Devotes particular attention to transboundary IA communications procedures (+) European EIA and SEA guidelines describe the potential roles of public participation in screening, scoping, and IA document review (+) Europe launched a major initiative to facilitate SEA capacity building in Eastern Europe (UNECE, 2007); networking among government officials has proven to be an especially effective IA capacity-building strategy (UNDP and REC, 2006) (±) IA requirements and guidelines in Europe largely pertain to collaboration among European states; public participation provisions, at both the EIA and SEA levels, have tended to be weak (Benson, 2003); strengthened somewhat in Proposed EIA Project Directive	communications needs having adequate opportunity to comment on proposed actions (+) Reference to taking into account the proponent's environmental history (+) Requirement that community and stakeholder comments must be taken into account (+) Draws upon a broader knowledge base with the use of a series of independent advisory committees (e.g., Threatened Species Scientific Committee, Biological Diversity Advisory Committee, Indigenous Advisory Committee) (+) Prepared an IA training resource manual for use in developing countries (+) Commitment to strengthen the involvement of indigenous peoples (Australian Government, 2011d) (+) Emphasis on a cooperative approach among governments, communities, landowners, and indigenous peoples; addresses cooperative arrangements between the Commonwealth and states and territories in considerable detail (+) Accepted reforms allow for more time for public comments and the greater use of SEA

Collaboration appears to be well developed in the IA requirements of the four jurisdictions regarding interactions among governments. Some attention has been devoted to mechanisms for facilitating stakeholder involvement and collaboration (e.g., participant funding, environmental justice). Considerable experience has been acquired in applying collaborative approaches in such related fields as resource, environmental, and facility planning and management, especially in the United States, Some attention, has been devoted to general collaborative planning approaches. There are scattered examples of good practice collaborative IA approaches in all four jurisdictions. But these examples are more the exception than the rule. Practical collaborative planning methods and frameworks and the adaptations to IA requirements needed to facilitate collaborative IA practice could receive more attention. Experiences with collaborative IA approaches, and the role of the regulatory level, should be evaluated from multiple perspectives. Comparisons among jurisdictions would be worthwhile, appreciating the need for contextual adjustments.

All four jurisdictions engage in IA capacity building, to varying degrees. Although the four jurisdictions offer some worthwhile examples of IA capacity-building initiatives, there remains ample room for a more systematic, comprehensive, and inclusive approach. The effectiveness of capacity-building efforts should be evaluated from multiple perspectives. Initiatives to substitute IA requirements with those of lower decision-making levels should be preceded by independent evaluations of IA capacity and, where needed, capacity-building measures. Contextual differences and differences in stakeholder interests and preferences should be taken into account. IA capacity building is addressed in greater detail in Section 9.6.

9.5.2 Management at the Applied Level

Figure 9.6 illustrates an example of collaborative IA process. The figure and the description that follows depict a process consistent with IA public participation goals, principles, and practices. Specific consultation, communications, mutual education, negotiations, and collaboration concepts, methods, and processes are incorporated into the process. The process seeks to enhance collaborative planning and decision making by IA stakeholders. Communications, mutual education, negotiations, and collaborative elements are grafted onto and integrated into the IA process. The process is broadened to encompass numerous publics. It is supported by a sound knowledge base. The process is informal, open, inclusive, interactive, and people-centered.

Start-Up and Planning The collaborative IA process begins with initial consultation. Issues are identified. Level of interest is determined. An overview of environmental characteristics is undertaken. Historical grievances are noted. Remedial actions are taken where practical. Key people and organizations are identified. Pertinent organizational mandates, characteristics, and constraints are described. This initial context scanning provides the basis for a conflict assessment. The conflict assessment considers issues, potential parties, and potential conflicts. It then decides whether alternative dispute resolution is possible and appropriate. Key parties, who might be interested in or potentially affected by the proposed action, are identified. The parties are contacted. Appropriate representatives are identified and recruited as members of an advisory committee. The committee includes a diversity of proponent, government, and public stakeholder representatives. The committee does not have final decision-making authority. However, because of the membership breadth and the in-depth deliberations anticipated, the findings and recommendations that emerge from the committee are expected to have considerable decision-making "weight."

The committee establishes and agrees to procedural rules and principles. It identifies an appropriate range of subcommittees. IA and public participation planning are integrated. The overall IA/public participation plan addresses such matters as general principles and goals, issues, and problems to address, activities, and tasks, schedule, resource requirements, technical specialist input requirements, budgets, roles and responsibilities, and public involvement procedures. The plan also includes communications (e.g., information exchange procedures, communications goals, principles and methods, measures to overcome communications barriers, plans for groups and organizations with special communications needs), mutual education (e.g., community, proponent and regulator training and education requirements, mutual education goals, principles and methods, plans for groups and organizations with special educational needs), negotiations (e.g., third-party assistance, measures to offset power imbalances, negotiations goals, principles and methods, conflict identification, and management procedures), and collaboration (e.g., core values and preliminary visions, collaboration goals, principles and methods, outreach and capacity-building methods, participant assistance requirements, procedures to foster creative collaboration) elements. A draft plan is first prepared. The draft plan is modified based on comments from committee members and from a broader audience. Further refinements and adjustments to the plan occur as the IA process unfolds. An evaluation of the effectiveness of the start-up and planning activities is undertaken.

Application The core application activities largely mirror those commonly associated with IA processes. The overall process is scoped. The need for action is assessed. Alternatives, of varying types and at different levels of detail, are systematically generated, screened, and compared. Baseline and proposal characteristics are determined. Individual and cumulative impacts, risks, and uncertainties are identified and predicted. Impact significance, with and without mitigation, is interpreted. Compensation, monitoring, and contingency measures are determined. An overall impact

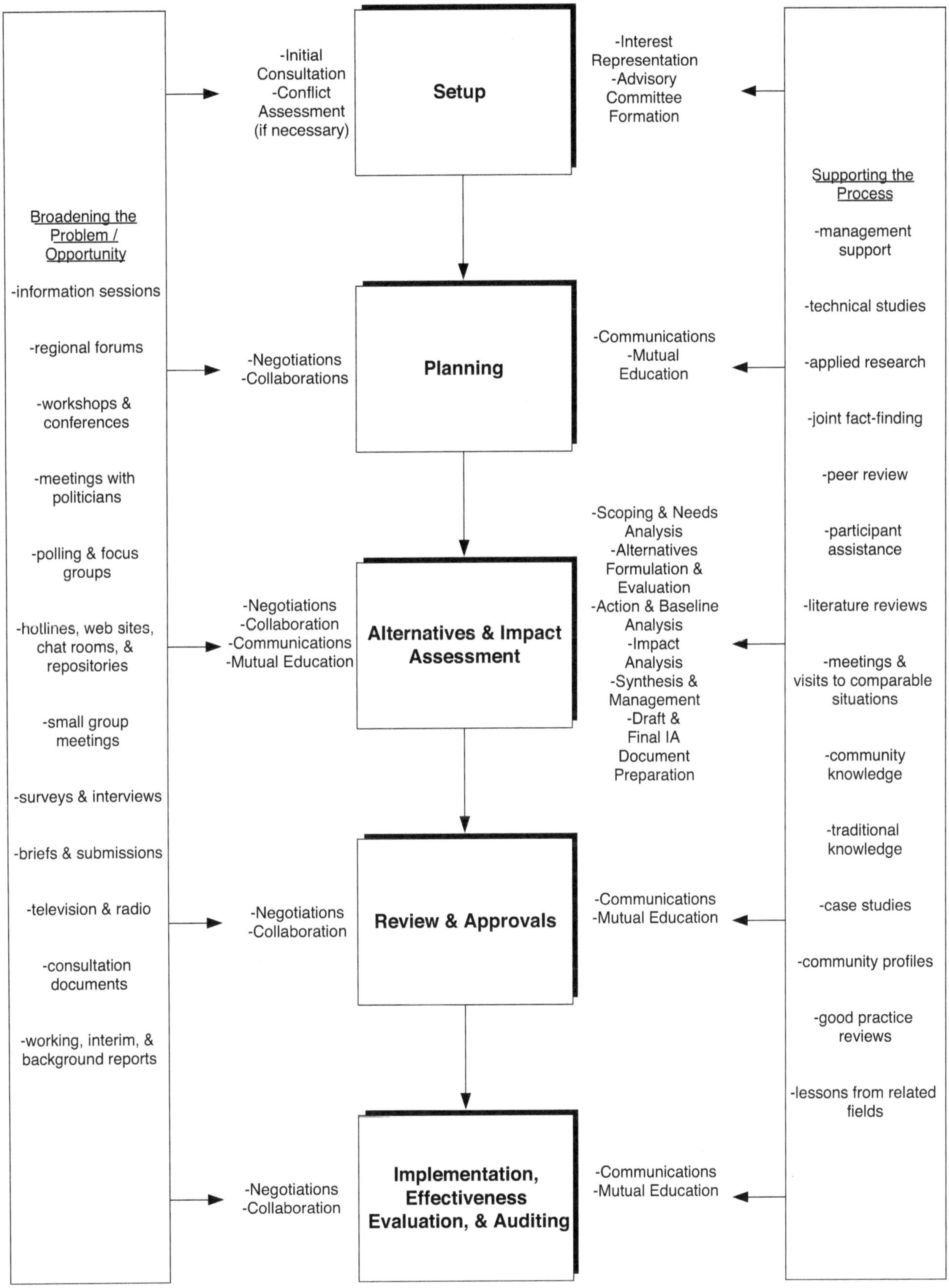

Figure 9.6 Example of a collaborative IA process. Adapted from Lawrence (2005b).

management strategy is formulated. Techniques such as sensitivity analyses test the robustness of conclusions. Draft and final IA documents are prepared. The modifications to the draft IA documents reflect both regulatory and public concerns and preferences. The process is highly iterative. Unlike most conventional IA processes, it is the advisory committee that takes the lead, and it is the IA and other specialist practitioners who assume a support role.

The committee, with the assistance of specialists, guides both the technical and the procedural (i.e., public involvement, communications, mutual education, negotiations, collaboration) activities. Public involvement includes numerous procedures to provide information out to the public, to obtain information, comments, and concerns from the public and to exchange information and perspectives with the public. Communications involves interactions both within the committee and between the committee and constituents. It includes activities such as preparing clear, succinct, accurate, and readily understandable documents, identifying and offsetting misinformation, distortions, and communications barriers, facilitating and maintaining dialogue, ensuring that appropriate adjustments are made to reflect the varying communications requirements of different groups and segments of society (e.g., based on language and cultural differences), establishing and keeping communications channels open, ensuring that IA documents are known about and are readily available, and ensuring that local perspectives and concerns are accurately reported and effectively integrated into IA documents.

Mutual education extends well beyond community education. Public knowledge of proposed actions, options, environmental conditions, potential impacts, and management measures is enhanced. Training and education opportunities are provided to public committee members and, where appropriate, to other members of the public. Mutual education also entails educating the proponent, the regulators, and the specialists, ensuring that the process fully accommodates community and traditional knowledge, facilitating learning about facts, values, issues, decision-making processes and participants, integrating learning from practice stories, ensuring that learning is free from distortion or coercion, and promoting mutual, social, transformative, and critical learning.

Negotiations are tailored to the conflicts that emerge and to the characteristics of individual IA activities. Concerns, priorities, issues, and points of conflict are identified. Conflicts are, where practical, avoided, "staved off," and ameliorated. Remaining conflicts are, wherever possible, resolved. Third parties (e.g., facilitators, mediators) assist the negotiations. Possible conflict resolution packages are identified, screened, and compared. Commitments are obtained. Where warranted, the selected packages are incorporated into agreements, suitable for signing by stakeholder representatives. Provisions are included to ensure the monitoring of agreement implementation and to permit mutually acceptable adjustments to suit changing circumstances. Collaboration builds rapport, trust, consensus, and support. Common goals and shared visions of the future are formulated. Procedural inequities are offset. The creative and collaborative exploration of problems and opportunity is fostered. Effective use is made of synergistic methods for generating novel approaches to complex issues and concerns. The process is focused on achieving environmental and sustainability improvements without sacrificing outcome equity, especially for disadvantaged groups and segments of society.

The customary IA activities are all undertaken in accordance with regulatory requirements and good practice, but in a manner that integrates all elements of effective collaboration. The effectiveness of the IA and public participation process and outcomes are evaluated both during and subsequent to the completion of the application activities. The evaluation leads to both procedural and substantive adjustments and refinements.

Review, Approvals, and Postapprovals A sound decision-making basis is provided for all parties that should have a say in whether and how the proposed action proceeds. Communications and mutual education ensure that all parties are fully appraised, in a form suited to their needs, of all matters relevant to their deliberations. Care is taken to avoid and offset communications breakdowns and knowledge deficiencies. Negotiations and collaboration enable the parties to avoid, reduce, and resolve disputes and to generate creative approaches and solutions that serve both the interests of the parties and the broader public interest. Effective negotiations and collaboration eliminate the need for or dramatically scope formal public review proceedings (e.g., hearing, litigation before the courts).

Collaboration activities extend into the postapprovals stage. Communications and mutual education ensure that all parties are fully informed of implementation activities. Proponents and regulators are made aware of community concerns and preferences as they emerge. Communications among the parties is facilitated. Prompt action is taken to correct misinformation and communications distortions. Knowledge deficiencies are corrected. Conflicts are avoided where practical. Negotiations serve to ameliorate and resolve residual conflicts. Collaboration ensures that impact management measures are efficient, focused, and effective. Local benefits, rapport, trust, and cooperation are maintained and enhanced. The effectiveness of all elements of the IA/public participation process is evaluated immediately following approvals. Additional effectiveness evaluations are undertaken periodically through implementation and whenever major unforeseen circumstances and remedial actions occur.

Broadening and Supporting the Process Confining participation to an advisory committee composed of stakeholder "representatives" places an impossible burden on committee members. Committee members can only be assured that they are representing the views, interests, and concerns of their constituents if there are ample opportunities for the broader public to be involved in the process. The process provides for such opportunities prior to each major decision. A variety of involvement procedures inform and obtain input from the public. The public is provided with an ample range of different types of involvement opportunities (e.g., information sessions, small group meetings, television, radio, web sites). Involvement procedures are tailored to meet the needs of various groups, organizations, and segments of society. A particular effort is made to involve those groups and segments of the community likely to be the most directly affected and which are especially vulnerable to change. Positions adopted by the committee or by subcommittees are tested with such techniques as surveys, polling, and interviews. Committee members carefully compare their perspectives and positions against those contained in briefs and submissions. The broader public is provided with an opportunity to respond to background, interim, and draft reports. Expanding the base of public understanding and involvement contributes to an enhanced level of comfort for both committee members (that they are effectively representing the views of their constituents) and the public (that their concerns, interests, and preferences are being adequately represented).

A collaborative IA process is highly dependent on an adequate level of support. There must be a strong management commitment to the approach—a commitment reflected in adequate resources and sufficient time for the process to "proceed at its own pace." The activities of the committee are supported by sound technical studies, applied research to address areas of uncertainty, community knowledge, and where applicable, traditional knowledge. Community involvement in establishing a sound knowledge base is supported, where needed, by participant funding and by procedural and substantive training. Specialists assist joint fact-finding. Committee members draw upon the insights and lessons obtained through visits to comparable situations, community profiles, case studies, literature reviews, reviews of experiences in related fields, and good practice reviews. Peer reviewers test technical analyses. Good communications practice is reflected in how documents are structured, presented, and edited. The support activities reflect the needs, preferences, priorities, and expectations of the committee members, of regulators, and of the broader public.

9.5.3 Collaboration IA Process Characteristics by IA Type

Crosscutting Characteristics As highlighted in Table 9.10, each IA type provides for participant collaboration, albeit to varying degrees and in different forms. All IA types presuppose a multistakeholder system. They all provide for discourse, networking, mutual learning, and collaboration among participants. Each emphasizes the need for transparent, open, and inclusive decision making. They all stress the need to integrate the concerns and suggestions of interested and affected publics early in the IA, prior to key decision points throughout the process and as part of follow-up. They all encourage capacity building to facilitate a more open, fair, and effective planning and decision-making process. The differences among the collaborative versions of the various IA types pertain more to orientation and emphasis.

Collaborative SA Practice With democratic SA practice the collaboration of interested and affected parties is strongly encouraged but always with a purpose—sustainability. Process and substance are fully integrated. Collaborative multistakeholder planning systems are guided and bounded by sustainability ends and limits.

Collaborative SEA Practice Collaborative SEA practice emphasizes the importance of dialogue, mutual learning, feedback, and participation. However, because SEA is so closely connected to planning, policy making, and decision making, and so interdependent with its organizational context, great stress is placed on understanding and reforming institutions (e.g., institutional learning, ameliorating bureaucratic inertia and resistance), on the political dimension of decision making (i.e., enhancing IA decision-making influence), on dialogue within and among public agencies, and on opening up an often closed planning and decision-making process to greater levels of public involvement and participation.

Collaborative EIA Practice Historically, project-level EIA has tended to involve a top–down, technically-driven, and often adversarial EIA process. Public participation, in such cases, tends to be limited to public education and tightly circumscribed forms of public involvement. With collaborative EIA practice, interested and affected parties (especially local community and potentially affected groups and individuals) operate in a partnership with regulators, proponents, and other stakeholders to derive and implement mutually beneficial solutions. Such procedures are open, transparent, inclusive, and interactive. They are characterized by communicative learning, dialogue, and negotiations. Resource inequities are ameliorated through such measures as participant funding. Secondary benefits of collaborative EIA processes include individual and collective learning, community empowerment, community capacity building, and the advancement of community objectives.

Collaborative EcIA Practice Collaborative EcIA practice tends to operate at two levels. At one level, it is concerned that the knowledge and insights of ecological and biodiversity specialists are fully integrated into the IA process and, in turn, into planning and decision-making procedures. Collaboration at this level involves narrowing the gap between the scientific community and proponents and government officials, and strengthening alliances among environmental specialists. At another level, collaborative EcIA practice entails integrating community and public knowledge and perspectives into each process step. In common

Table 9.10 Collaborative IA Process Characteristics by IA Type

Collaborative SA Practice	Collaborative SEA Practice	Collaborative EIA Practice
Guided by sustainability decision-making protocols Stresses communications of sustainability issues Seeks transformative outcomes; social learning an essential element Redefines sustainability assessment as collaborative sustainability assessment Founded on principles of civility, democratic governance, and decision-making decentralization Draws upon a multistakeholder planning system; involves partnership of proponents, state, and communities Different perspectives and definitions gradually combined, through a sustainability discourse, to form complex and collective vision of what wish to sustain Framed discourse refines sustainability perspectives, debates trade-offs, identifies priorities, and establishes limits Encompasses environmental, organizational, and societal perspectives Facilitated by sustainability advisors Seeks stories of various interests as a means of engagement; links perceptions, history, and values; formally and informally meaningfully engages broader community in developing broader visions and addressing key issues Balances the needs of present and future generations Utilizes iterative approach; continual reflection back to original vision Favors integrated participatory and transdisciplinary methods, methodological pluralism, and stakeholder involvement as basis for more informed decision making Integrates SA training at local level Utilizes citizen-centered monitoring in the interests of sustainability livelihoods Emphasizes capacity building (e.g., encouraging learning, knowledge brokerage)	Entails strong institutional and vertical coordination and interagency collaboration SEA viewed as a form of communicative action, a mechanism for promoting public participation and social dialogue, and a means of enhancing communications among stakeholders and democratizing decisions SEA roles: knowledge brokerage, facilitator of strategic decision making, and catalyst for organizational learning Involves early discussion of SEA objectives Employs a participatory prestudy process to aid in data collection and objective formulation Improves planning/decision-making transparency, provides space for dialogue and for individual/organizational learning Utilizes participatory-dialogue-based methods (e.g., workshops, mediation, mediated modeling, consensus conferences, citizen juries, cooperative discourse) Emphasizes transparency, feedback, learning, and the promotion of public participation Seeks to foster learning and institutional reform and to facilitate consensus building, joint analysis, capacity building, and social mobilization Supported by stakeholder analysis, institutional analysis, IA, beneficiary analysis, and participatory poverty assessment Bridges a political dialogical approach and systemic disciplined inquiry Integrates multiple problem visions, establish communications links, guides communications strategies, and stimulates constructive collaboration and the production of common meanings Promotes stakeholder discourse reflection and provides for continuing dialogue, training, and education Seeks to broaden the SEA mindset (e.g., to systematically address cumulative effects)	A community-based approach to EIA, with a heavy emphasis on capacity building, dialogue, and empowerment Authority accepts value of participation Seeks an open and fair process Engages stakeholders and affected citizens in early dialogue Identifies the values and interests of interested and affected publics Promotes transparency, participation, discursiveness, and the active and critical Incorporates public input prior to each decision in IA process; all participation activities integrated into EIA Involves exchange of ideas among proponents, regulators, and public Seeks to combine different world visions while remaining flexible and inclusive Views the IA process as an arena for democratic deliberation Links design of public participation and EIA processes with nonformal education, individual and collective learning, social action, and sustainability Brings to bear a wealth of knowledge and diverse perspectives; links education, participation, and learning outcomes Emphasizes communicative learning EIA involves cooperation and even a partnership among regulators, proponents, and other stakeholders; builds on dialogue, trust, and alignment of purpose and effort Provides a grievance mechanism for people affected by project Ensures sufficient capacity to support process Local people participate in workforce and supply chain (i.e., local content requirements) Seeks substantive positive outcomes (e.g., participant learning, resource provision, participant influence, sustainability)
Collaborative EcIA Practice	**Collaborative SIA Practice**	**Collaborative HIA Practice**
A learning by doing and people-centered approach to integrating biodiversity considerations into broader livelihoods sustainability context Places within the context of international biodiversity treaties and conventions Integrates stakeholder biodiversity use, needs, and objectives Roles of different parties clarified, especially during scoping Treats resource management as a matter involving societal choice, involving all sectors and disciplines	Founded on principle that people have a right to be involved in decision making about planned interventions that will affect their lives SIA creates participatory process and deliberative spaces to facilitate community discussions regarding desired futures, impact acceptability, proposed benefits, and IA process inputs Identifies and describes interested and affected stakeholders and communities (profiles); assumes that individuals respond and adapt to change in different ways	Broadly defines health (community, social, mental, spiritual) Determines attitudes of affected communities toward issues of disease, risk, and health Founded on the right of people to participate in formulation and decisions that may affect their lives (including their health) Integrates local knowledge into health determinants Seeks an enhanced understanding of people's complex responses to change and potential health implications

Table 9.10 *(Continued)*

Collaborative EcIA Practice	Collaborative SIA Practice	Collaborative HIA Practice
Seeks to enable decision-making process to be gradually permeated by an ecological rationale	Provides meaningful participation and reassurance; SIA role in facilitating and coordinating stakeholder participation	Comprehensively addresses the multiplicity of stakeholders and types of knowledge, data and interests
SA frameworks guide stakeholders on biodiversity interventions	Entails proactive public involvement (e.g., community outreach); occurs continuously through the planning and decision process (e.g., interactive community forums)	Community involvement as full and active stakeholders a value underlying HIA; public should be engaged, informed, and influence decision making
Emphasis on sensitization and biodiversity capacity building	Builds on local knowledge and uses participatory processes to analyze concerns of interested and affected parties	Ensures meaningful and inclusive public participation; treated as a tool to engage and build relationships
Seeks to extend expert-based and technical biodiversity IA to citizens	Community participation fully integrated into SIA process	Recognizes value of and integrates local knowledge
Stakeholders select indicators of ecological health, including biodiversity	Founded on principle of free, prior, and informed consent; public acceptance a priority	Sometimes representatives from key stakeholder organizations and from affected communities overview the HIA process and outcomes (community partnerships)
Ensures equitable sharing	Includes interested and affected parties in identifying issues and goals, in deciding environmental and social value and impacts, in evaluating their importance, in analyzing alternatives, in making consensus-based choices, and in monitoring the planned intervention	HIA process is community led; proactively involves individuals and group and organizational representatives who have an interest or are affected by the proposal
Seeks to strengthen biodiversity partnerships and information networks	Community values and perceptions (e.g., risks) considered	Involves and engages health professionals; close communications between health professionals and decision makers
Biodiversity specialists engaged with planners and decision makers	Recognizes and builds on local and indigenous knowledge; recognizes indigenous rights (e.g., consensus model to address traditional environmental knowledge process)	Intensive stakeholder involvement throughout HIA process, especially during scoping; active stakeholder involvement in identifying health issues and determinants; emphasis on collaboration
Collaboratively involves ecologists and other specialists	Emphasizes vulnerability of underrepresented and disadvantaged populations; seeks to enhance marginal groups and facilitate greater equity; respects human rights	Focuses on accentuating positive health outcomes, avoiding adverse health impacts; distinguishes between voluntary/involuntary risks, incorporates equity issues (health equity IA), and emphasizes vulnerable subpopulations and distribution of health impacts among groups
Provides for communications and negotiations with stakeholders	Uses the knowledge and experiences of individuals most affected by proposed changes for projecting impacts	Inclusive communications of principal findings; distributed to stakeholders with meaningful review; publicly accessible
Assesses negative impacts on priority ecosystem services in terms of changes in beneficiaries' well-being	Fosters community engagement, social inclusion, building of social capital, capacity building, gender equity, community empowerment, and good governance	Public engaged in monitoring, implementation, and effectiveness evaluation
Should lead to understanding and support by nonspecialists	Involves an open and transparent IA process	
Management actions at appropriate scale and decentralized to lowest level	Engenders social learning by proponent, planners, regulators, and the public; role for SIA in managing conflicts	
Strengthens cooperation between planning and environmental authorities and stakeholders		
Monitoring undertaken jointly by government, research institutes, and capacity-building organizations		
Requires capacity building and the fostering of alliances		

Sources: Adelle and Weiland (2012), Ardern (2004), Ayr and Calloway (2005), Azcarate and Balfors (2009), Baker and McClelland (2003), Becker et al. (2003, 2004), Bhatia (2007), Bhatia et al. (2010), Bina et al. (2011), Binder et al. (2010), Blake (2010), Bond (2004), Bond and Morrison-Saunders (2011), Bond et al. (2004), Bonifazi et al. (2011), Booth and Skelton (2011b), Buchan (2003), Burdge (2003b, 2004), Craik (2008), Devuyst (2000), Diduck and Mitchell (2003), Donnelly et al. (2007), Égré and Senécal (2003), Elling (2011), Elliott et al. (2004), Esteves et al. (2012), Faber et al. (2010), Gasparatos et al. (2007), Gibson (2006a), Govender et al. (2006), Gunn and Noble (2011), Gunning et al. (2011), Hanna (2009a,b), Hansen and Kørnøv (2010), Harris-Roxas and Harris (2011), Harris et al. (2003), Hermans and Knippenberg (2006), Hodge (2004), Hunsberger et al. (2005), IAIA (2003, 2005, 2006a, undated b), ICPGSIA (2003), IEEM (2006), Jiliberto (2011), Kemm (2005), Kende-Robb and Van Wicklin (2008), Khera and Kumar (2010), Kørnøv and Dalkmann (2011), Landry et al. (2009), Lane et al. (2003), Landsberg et al. (2011), Law et al.(2005), Lawrence (2009), LaPierre (2012); Lavallée and André (2005), Letsela et al. (2010), Lobos and Partidário (2010), Lockie et al. (2008), McCarthy and Utley (2004), Mindel et al. (2004), Morgan (2012), Negev (2012), Noble and Gunn (2009), Partidário and Coutinho (2011), Peltonen and Sairinen (2010), Peterson (2004), Pope and Grace (2006), Quigley and Taylor (2003), Raphael (2012), Rauschmayer and Risse (2005), Ross and Thompson (2002), Runhaar (2009), Scanlon and Davis (2011), Sheate and Partidário (2010), Shepherd (2008), Sherrington (2005), Simpson et al. (2005), Sinclair and Diduck (2009), Sinclair et al. (2009), Sinclair et al. (2008), Smith and Schin (2004), Söderman and Saarela (2010), Tamburrini et al. (2011), Tetlow and Hanusch (2012), Thérivel et al. (2010), Treweek et al. (2011), Tugwell and Johnson (2011), Tuinstra, Jäger and Weaver (2008), Vanclay (2003, 2006), Vicente and Partidário (2006), Walker (2003), Wernham (2007), Whitelaw et al. (2009), Wlodarczyk and Tennyson (2003), Youngkin et al. (2003).

with collaborative SA, substantive ends and limits guide and bound the process. While emphasizing the value of public knowledge and the desirability of public participation, collaborative EcIA practice tends to stop short of advocating shared or delegated decision making.

Collaborative SIA Practice SIA has arguably gone the furthest in fully articulating the characteristics of highly collaborative IA practice. Collaborative SIA practice is founded on clearly defined procedural principles (e.g., the right of people to be involved in decisions that will affect their lives, free, prior, and informed consent, respect for human and aboriginal rights). Local and traditional knowledge is recognized and built on. It is also viewed as equally or more important than the substantive and procedural knowledge of social scientists, and of SIA and public participation specialists. Collaborative SIA practice is not limited to insisting on open and transparent decision making and public involvement prior to each IA decision. It also seeks to foster mutual learning, achieve consensus-based choices, apply fair and inclusive procedures, and ensure equitable outcomes. Particular emphasis is placed on the concerns and interests of underrepresented and disadvantaged groups and populations. Public participation is proactive (e.g., community outreach) and continuous. It seeks to rectify power imbalances (e.g., by applying capacity-building measures). IA-related public participation is expected to be conducive to such social ends as community engagement, social inclusion, social capital building, gender equity, community empowerment, and good governance.

Collaborative HIA Practice Collaborative HIA practice defines health broadly, with a particular emphasis on public perspectives, rights, values, attitudes, knowledge, and interests. In common with other substantive IA types, collaborative HIA practice entails effectively engaging specialists (e.g., health professionals), integrating local knowledge (e.g., into health determinants), and building health-related partnerships with communities. Collaborative HIA practice is community led—focusing on community health concerns and seeking to achieve positive health outcomes for the community in general, and for vulnerable subpopulations in particular. Collaborative HIA practice is transparent and inclusive. The public is fully and actively engaged in identifying health issues and determinants, in comparing options and selected preferred actions, in reviewing findings, and in monitoring, implementation, and effectiveness evaluation.

9.6 CONTEMPORARY CHALLENGE—IA CAPACITY BUILDING

9.6.1 Definition and Distinctions

Capacity building is a term increasingly evident in IA literature and good practice guidance. In general terms, it entails a knowledge-intensive process that creates, utilizes, improves, transfers, and sustains a range of experiences, abilities, relationships, and values for addressing specific challenges or opportunities (UNEP, 2006; Virji et al., 2012). Capacity building tends to focus on understanding and on progressively and gradually eliminating obstacles to development objectives (e.g., sustainability). In so doing, conditions are created for efficiently and effectively undertaking specific tasks or responsibilities (Partidário and Wilson, 2011).

Capacity building can operate at various levels (e.g., individual, organizational, institutional, country) and scales (e.g., national, regional, local) (Virji et al., 2012). Capacity-building components (e.g., values, structures, skills, knowledge, procedures, resources, incentives, technology, training, regulatory reform, and networking) vary depending on the objectives (Dixon and Thérivel, 2011; Tamas, 2008). This suggests that capacity building entails a process (e.g., stages and activities) that integrates individual elements and activities into a coherent strategy, directed toward tangible ends. Capacity-building procedures are, or at least should be, founded on and bounded by ethical principles and limits (e.g., democracy, participation, development, continuous improvement, shared learning opportunities, equal access to opportunities) (Partidário and Wilson, 2011).

Capacity building can be applied to the field of IA as a whole and/or more specifically to various IA types, activities, settings, population subgroups, proposal types, impact types, environmental components, issues, time horizons, and organizational types and levels. As highlighted in Figure 9.7, IA capacity building necessarily entails collaboratively designing and managing the capacity-building process (definitions and distinctions), determining the area of application (applied to what, where, and how), selecting the constituency and capacity-building facilitators (for whom and by whom), deciding on the capacity-building objectives (for what purpose), and identifying, adapting, and integrating the capacity-building tools (by what means). Capacity building, in common with other IA aspects, builds on good practices at both the regulatory and applied levels (see Table 9.11).

9.6.2 Applied to What, Where, and When

IA has far from realized its full potential. There remains a considerable gulf between theory and practice. The quality and effectiveness of IA practice varies greatly from proposed action to proposed action, among jurisdictions, among IA types, among IA activities, and in relation to specific environmental components, issues, organizational types, and population groups. Improvements over time have been far from uniform. Arguably, in some instances, the quality of IA practice, as a result of both internal and external factors, has either declined or remained static. Also, good practice IA standards are highly context-dependent. The role of IA capacity building is to narrow the gulf between the aspirations of IA and its tangible achievements, appreciating that perceptions of the desired roles for and objectives of IA will

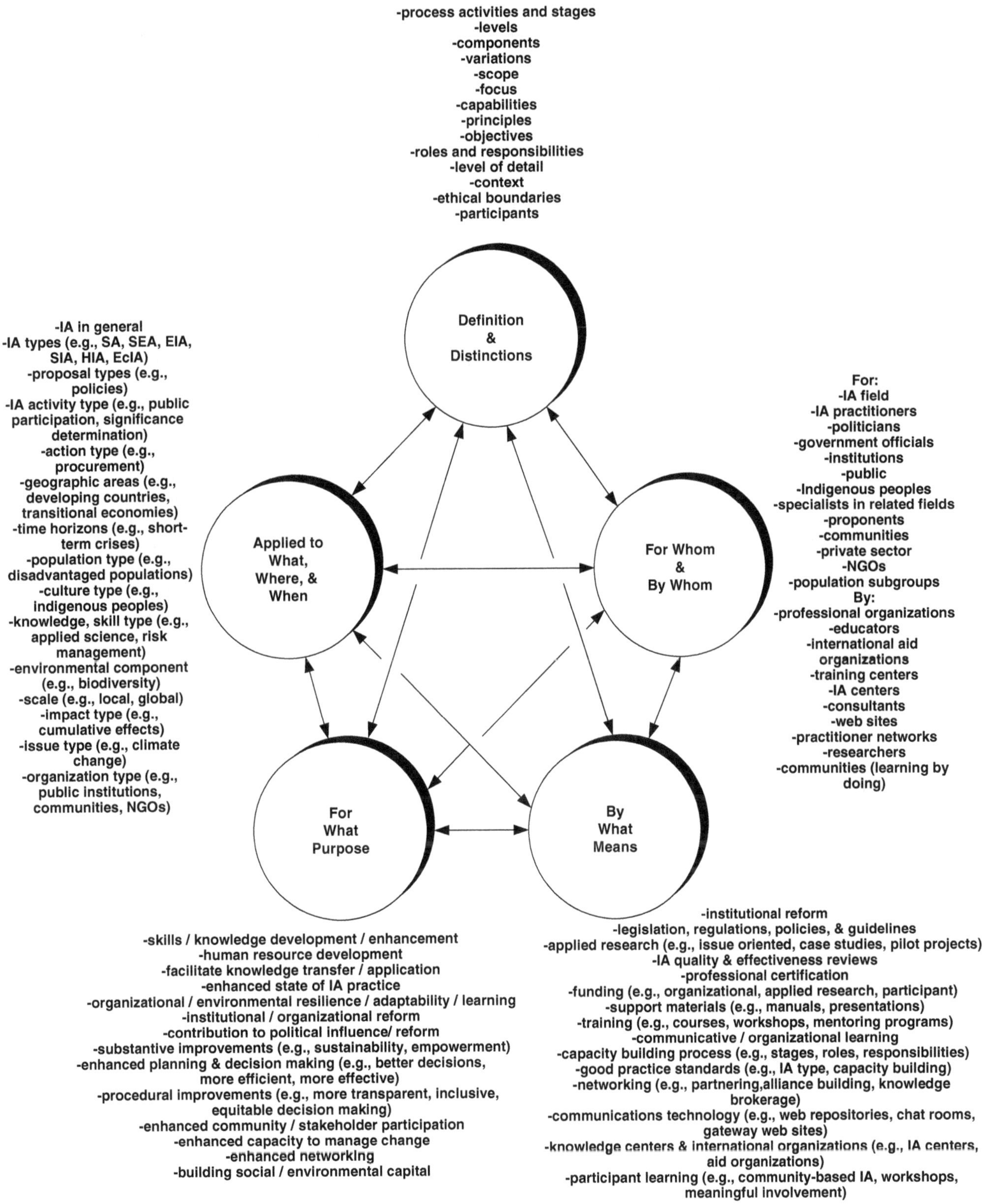

Figure 9.7 IA capacity building.

Table 9.11 Examples of IA Capacity Building Good Practices

Regulatory Level—Facilitating Capacity Building	Applied Level—Undertaking Capacity Building
Fund IA capacity building and applied research Institutionalize capacity building at multiple levels (e.g., international, national, regional, local); regional/local levels often especially in need of additional capacity building Focus on institutional capacity; identify roots of institutional weaknesses; foster political and community support for IA and for capacity building Encourage active networking of individuals, groups, organizations, and institutions; foster alliance building and coordination Institute and support IA knowledge centers Develop coherent institutional frameworks and governance approaches Develop broad system (political and government) support for capacity building by IA type across government Seek government–academic partnerships (e.g., courses, applied research, training) Foster a participatory, adaptive, continuous learning, and integrative planning model across government Foster greater IA awareness; make enhanced IA guidance a priority Certify IA professionals Link IA capacity building to other planning levels and sectors Recognize that education, mutual learning, and capacity building a continuing need and process Facilitate communications between agencies and IA practitioners Evaluate capacities of IA authorities (leadership, structure, human resources, financial resources, interorganizational links), proponents, knowledge organizations (consultants, universities), and other stakeholders to effectively perform responsibilities and to participate effectively in process; target capacity-building initiatives to fill gaps Ensure feedback to government structures regarding capacity building effectiveness; upward information flow critical Provide publicly available repositories of IA-type experiences Pilot test new capacity-building approaches Seek to harmonize IA capacity-building efforts with broader governance and environmental visions Commit to meaningful participation of public and of indigenous peoples Strengthen monitoring/auditing functions Link capacity building to actual assessment and reporting Links capacity building insights to IA legislative and regulatory reform and IA guidelines	Be context specific Don't rush Target the right people to build a critical mass Emphasize mutual respect, accessibility, inclusiveness, openness, transparency, timeliness, trust building, and efficiency Identify needs and build on existing capacity Clearly define capacity-building objectives (e.g., sustainability, greater decision-making influence, enhanced skill sets, strengthened organizations, ability to commission, prepare or review IAs, more effective communications of IA) Focus on needs and priorities of participants; respect value system of participants and foster self-esteem Promote innovation and diversity; introduce innovative tools and methods Focus on problem assessment and remediation and major policy/planning choices; occurs within a framework of integrated/interdisciplinary problem solving Challenge mindsets and power differentials Test capacity building with pilot programs Use a wide range of capacity-building approaches Emphasize active participation; establish positive incentives Make the training of trainers approach work Ensure that capacity is built to acquire knowledge, improve skills, and apply outputs from skills and knowledge Emphasize participant learning (i.e., learning by doing); foster participant "ownership" Emphasize organizational learning Ground capacity building on empirical evidence Integrate case studies and examples Promote best practice Promote more equity focused decisions and greater gender parity Outreach to vulnerable groups Stay engaged under difficult circumstances Ensure process accountable; seek to demonstrate added value of capacity building; remain accountable to ultimate beneficiaries Apply specific criteria when assessing effectiveness of capacity building Document experience gained and lessons learned Share experiences (e.g., regional exchanges, self-help networks)

Sources: Birley (2004, 2007), Bond (2004), CIER (2009), Dannenberg et al. (2006), Devuyst (2000), Diduck and Mitchell (2003), Dimento and Ingram (2005), Dixon and Thérivel (2011), Doberstein (2003, 2004), Dora (2004), Dusik and Sadler (2004), Greig and Duinker (2011), George et al. (2001), Harris and Spickett (2011), IISD (2008), Kakonge (2006), Kolhoff et al. (2009), Kwiatkowski et al. (2009), Marshall et al. (2005), Morrison-Saunders and Bailey (2009), OECD (2006), Partidário (2011), Partidário and Wilson (2011), Sheate (2011), Shippey (2004), Simpson et al. (2005), Sinclair et al. (2009), Stewart and Sinclair (2007), Tang (2010), UNEP (2006), Van Gent (2011), Varghese and Nadeau (2004), Virji et al. (2012), Waldeck et al. (2003), Zhu et al. (2010).

vary among stakeholders and over time, and that context-specific adjustments will be critical. The resources available for capacity building often will be limited, especially in the current economic and political climate. Thus, difficult choices and trade-offs will be required.

The theme of capacity building is a recurrent one in IA literature (Bina et al., 2011). It tends to come up more often when the quality of IA practice is viewed as especially weak (e.g., cumulative effects assessment, follow-up), when institutional arrangements tend to be lacking or very limited in terms of supporting IA practice (e.g., SEA, SA, HIA, SIA, policy IA, international treaties, transborder projects and effects), when particular environmental issues are considered especially critical (e.g., climate change, biodiversity), when certain population groups tend to be marginalized from IA practice (e.g., indigenous people, local communities, gender inequities), and in geographic settings characterized by severe environmental concerns and limited resources (e.g., developing countries, transitional economies, "commons" areas such as the oceans, the Arctic, and the Antarctic) (Alshuwaikhat, 2005; Birley, 2004, 2007; Bond, 2004; Bond et al., 2011; Canter, 2008; Connelly, 2011; Craik, 2008; Dannenberg et al., 2006; Dixon and Thérivel, 2011; Doberstein, 2004; Dora, 2004; Égré and Senécal, 2003; Erlanger et al., 2008; Harris-Roxas and Harris, 2011; Kakonge, 2006; Larsen et al., 2012; Marshall et al., 2005; OECD, 2006; Parry and Kemm, 2004; Partidário, 2011; Partidário and Wilson, 2011; Simpson et al., 2005; Sinclair et al., 2009; Taylor et al., 2004; Vanclay, 2003, 2006; Varghese and Nadeau, 2004).

The range of situations where IA capacity-building initiatives would be desirable is, as suggested above, very wide. Compounding the problem is the propensity in many jurisdictions to curtail the resources available to government agencies to actively participate in the environmental review of proposed actions, thereby undermining existing IA capacity. These realities suggest that a general call for more IA capacity building, in a host of areas, is not likely to be effective. Priorities need to be collaboratively established, based on the consistent application of transparent and fully substantiated criteria. The effectiveness of existing capacity-building measures should be systematically evaluated. A particular effort needs to be made to develop and apply suites of capacity-building measures (e.g., networking strategies) that consume limited resources, are appropriate to the context, and have proven to be effective in a range of settings.

9.6.3 For Whom and By Whom

The target constituencies for IA capacity-building initiatives include, for example, IA practitioners, politicians, government officials, proponents specialists in related fields, the general public, local communities, the private sector, NGOs, indigenous people, and population subgroups (Booth and Skelton, 2011b; Chaker et al., 2006; IAIA, 2006a; Harris-Roxas and Harris, 2011; UNEP, 2004). An IA capacity-building initiative should necessarily be designed for and with the active participation of the target group or population.

A great many organizations and individuals are involved in facilitating and supporting IA capacity-building initiatives. International aid organizations (e.g., World Bank, UNEP, Asian Development Bank), IA and environmental professional organizations (e.g., IAIA, NAEP), and various knowledge centers (e.g., IA centers) have considerable experience in furthering the cause of IA capacity building. Also, numerous universities, consultants, individuals, educators, researchers, and training centers have participated in IA capacity building (Dimento and Ingram, 2005; Greig and Duinker, 2011). Increasingly, networks of IA professionals, NGOs, government agencies, and concerned and interested individuals have shared IA knowledge and experiences (often through gateway web sites and chat rooms) in a manner conducive to IA capacity building.

Considerable care is needed to ensure an appropriate match is provided between IA capacity-building providers/facilitators and the IA capacity-building constituency. Familiarity with and appropriate adjustments to the context is essential. Again, the creative use of limited resources is likely to be essential. IA capacity-building initiatives, whereby the target populations "learn by doing" (e.g., community-based IA, community-based research), tend to "blur" the boundary between constituents and providers/facilitators (Cameron et al., 2011; Kwiatkowski, 2011; Sinclair et al., 2009). Such initiatives tend to consume limited resources and have secondary benefits such as community empowerment.

9.6.4 For What Purpose

A host of primary and secondary objectives have been identified for IA capacity-building initiatives. At the broadest level, there is the desire to raise the level of IA practice, both in general and in relation to particular IA types, settings, and impacts. Intermediary objectives tend to pertain to such matters as the transfer of skills and knowledge, enhanced networking, and human resource development. An enhanced level of IA practice (aided by capacity building) can contribute to more environmentally sound projects and decision making, to organizational/institutional reform, to greater political influence, and to more informed, open, transparent, adaptive, inclusive, efficient, and effective planning and decision making (Dimento and Ingram, 2005; Greig and Duinker, 2011; UNEP, 2006; Partidário, 2011). Enhanced IA practice, in combination with reformed institutions, planning, and decision making, can potentially contribute to such substantive environmental objectives as biodiversity enhancement, sustainability, growth management, the enhancement of social capital, and community empowerment (Bina et al., 2011; Gibson, 2011; IAIA, 2003; Partidário, 2011; Varghese and Nadeau, 2004; Wirutskulshai et al., 2011).

This chain of connections and consequences portrays IA capacity building as a catalyst for the realization of a host of organizational, procedural, and substantive objectives. For capacity building to reach its full potential, this suggests the need to systematically identify the interconnections between IA capacity-building initiatives (e.g., training, funding, participant learning) and broader institutional, procedural, and substantive objectives, coupled with the systematic evaluation of the effectiveness of IA capacity building. Such analyses should take into account other factors that could influence, positively or negatively, the expediting role of IA capacity building.

9.6.5 By What Means

IA capacity building can be facilitated by and/or can contribute to institutional reforms and changes to IA legislation, regulations, and guidelines (Waldeck et al., 2003). Applied research, IA quality and effectiveness reviews, good practice standards, and IA knowledge centers and international organizations can all contribute knowledge, skills, and experience conducive to IA capacity building (Canelas et al., 2005; Greig and Duinker, 2011; Van Gent, 2011). Care must be taken to make appropriate contextual adjustments.

Training (e.g., courses, workshops, seminars, briefings, the training of trainers), support materials (e.g., manuals), the certification of IA professionals, and the provision of funding can augment institutional capacity, enhance practitioner and public authority skills and knowledge, and enable stakeholders to more effectively participate in IA (Birley, 2004; Dannenberg et al., 2006; Égré and Senécal, 2003; IAIA, 2006a; IISD, 2008; McCabe and Sadler, undated; Peterson, 2004; Shippey, 2004; Sinclair et al., 2012; UNEP, 2004). The capacity to conduct and meaningfully participate in IA can be augmented by networking (e.g., dialogue, mentoring, information exchanges, partnering, alliance building, coordination, regional exchanges) and communications technology (e.g., web sites, chat rooms, repositories) (Bond, 2004; Dannenberg et al., 2006; Dusik and Sadler, 2004; Harris and Spickett, 2011; Morrison-Saunders and Bailey, 2009; OECD, 2006; Shepherd, 2008; UNEP, 2004; Virji et al., 2012). The direct participation of interested and affected parties in IA (e.g., through community-based IA, meaningful public participation, and meaningful involvement of indigenous people) is both a potentially effective form of capacity building and a possible contributor to mutual learning, greater public influence in decision making, and community empowerment (CIER, 2009; Diduck and Mitchell, 2003; Fitzpatrick, 2006; Sheate and Partidário, 2010; Stewart and Sinclair, 2007).

IA capacity building is a process. It starts from a clear understanding of the factors influencing the capacity gap (Tang, 2010). It involves a series of stages and activities, and entails identifying the roles and responsibilities of participants (e.g., conceptual framework, stakeholder engagement, needs assessment, organizational structure, acquiring skills and resources, implementation, monitoring, and assessment) (OECD, 2006; Virji et al., 2012). The process should be designed and managed to meet the objectives, to be consistent with explicit ethical principles, to suit the context, and to be consistent with the needs, expectations, and values of the participants (Kwiatkowski et al., 2009; OECD, 2006; Zhu et al., 2010). Given the wide array of available capacity-building measures, particular care should be taken to ensure that the measures are appropriate and complementary (Kolhoff et al., 2009; Partidário and Wilson, 2011). Reference should be made to capacity building good practices and experiences in IA and in related fields (Dora, 2004; Partidário and Wilson, 2011; UNEP, 2006). Adjustments should be made as the process unfolds based on feedback from the participants. The effectiveness of the process, and its individual components should be independently evaluated against the objectives and from multiple perspectives. IA capacity-building experiences should be shared among IA practitioners and with other participants, and should contribute to IA theory building. It is especially important, given contemporary resource constraints and political/bureaucratic resistance, to demonstrate the "added value" of IA capacity-building initiatives, and to obtain political and government support (Harris and Spickett, 2011).

9.7 SUMMING UP

This chapter describes a collaborative IA process—a process where the public is an active and ongoing participant. The three stories illustrate some of the complexities and subtleties associated with collaborative IA processes. The first story demonstrates that collaboration among stakeholders does not always lead to consistently understood and supported substantive environmental enhancements. The second story makes the point an effective IA process or system does not mean that all stakeholders should be uniformly satisfied with the outcome. The aim instead should be an IA system and process which all stakeholders can actively defend, where there is a high degree of cooperation, and where all parties have a sense of working for the "common good." Time horizons, however, need to be extended, especially regarding the interests of future generations. The third story addresses the difficult task of facilitating stakeholder engagement at the policy level. It describes a stakeholder-based and participatory approach to TIA. The story illustrates that stakeholder collaboration at the policy level is practical, and can be undertaken in a manner that is highly communicative, technically sound, and effectively coordinated with existing assessment instruments.

Collaboration encompasses all forms of public participation short of delegation or shared decision making. Stakeholders jointly undertake a collaborative IA process. Noncollaborative forms of public participation are prerequisites or subsets. The problem is the gulf between the potential benefits of collaborative IA processes and the more modest benefits achieved by public participation approaches commonly evident in IA requirements and practices. The direction is exploring the potential for and means of making IA processes more collaborative. Six complementary elements of effective public participation practice are identified: (1) core principles and practices, (2) consultation, (3) communications, (4) mutual learning, (5) negotiations, and (6) collaboration.

Effective public participation has intrinsic benefits. It furthers human potential and offers numerous individual and joint benefits for the public and for decision makers. IA public participation practice often fails to realize these benefits fully. The possibility is raised that the shortfall occurs because of disadvantages associated with public participation or because of largely irresolvable issues associated with public participation practice. The public participation disadvantages are generally either dubious or overstated. They can generally be avoided, overcome, or largely ameliorated. Restrictions to public participation in the IA process should always be justified. There are numerous issues associated with IA public participation practice. These issues pertain to conducting public participation activities, balancing conflicting perspectives, treating problems, obstacles, dilemmas, and uncertainties, and making difficult judgments. Rather than justifying restrictions to public participation, the issues underscore why issue identification, exploration, and resolution should be collaborative. They also point to the need to learn from public participation practice.

Examples of IA public participation goals, principles, and good practices are identified. Public participation methods are often placed along continua to demonstrate major methods' characteristics, their suitability for achieving different objectives, and their match to varying situations. Public participation methods can also be characterized by function or by operational characteristics. The characteristics, advantages, and disadvantages of categories of and individual methods should be considered before they are applied. Methods should be consistent with IA process goals and should be appropriate to the context.

Public consultation or involvement includes informing the public, integrating the views of the public, and interacting with the public, all prior to decision making. A public consultation process includes early consultation, initial and detailed planning, plan application and refinement, the monitoring of plan effectiveness during and subsequent to plan application, and postapproval involvement. Public involvement should begin early in the IA process and should be integrated into each IA process activity. Stakeholder identification is especially important. Information exchange, continuous involvement and outreach, and capacity-building methods can be applied. Formal involvement methods sometimes occur near the end of the process to present and test evidence. Public involvement plans are usually refined and adjusted through the process.

Communications is concerned with clear, focused, and understandable documents, the communications and advice-giving skills of IA practitioners, and undistorted and noncoercive dialogue and networking among IA participants. Applying procedural ethical principles, insights from IA practice stories, and effective argumentation could enhance communications in the IA process.

Education in the IA process is conventionally portrayed as educating the community. *Mutual education or learning* works both ways. It involves the parties learning together and potentially being transformed. The IA process should be treated as a learning process—an opportunity for all parties, individually and collectively, to enhance their knowledge and their intelligence capacity. The application and accommodation of traditional knowledge and concepts such as knowledge brokerage and organizational, social, collaborative, practical, critical, transformative, and sustainability learning could facilitate learning about and through the IA process.

Negotiations is concerned with avoiding, resolving, and ameliorating conflict in the IA process. ADR provides the IA process with a means of negotiations not limited to conventional administrative procedures or to litigation through the courts. Third parties such as mediators or facilitators should generally assist the process. ADR is not always appropriate. The appropriate conditions must be satisfied. ADR processes generally involve prenegotiations (e.g., conflict assessment, interest representation, designing the process), negotiations (e.g., initial deliberations, focusing, detailed deliberations, final refinements), and postnegotiations (i.e., implementation). ADR can potentially be integrated, in different forms, into several IA process activities. ADR methods range from unaided negotiations through procedural assistance (e.g., conciliation, facilitation, mediation) to quasijudicial mechanisms (e.g., minitrials, public hearings, arbitrations). ADR can culminate in formal agreements. Methods should be selected to suit the process objectives and the situation. ADR advantages and disadvantages should be considered when determining if, how, and when in the IA process ADR methods are to be applied.

Consultation, communications, mutual education, and negotiations, individually and collectively, can contribute to *collaboration* in the IA process. Collaboration is about people working together in a joint endeavor with substantive aspirations. Collaboration emphasizes perspectives, visions, opportunities, and creative joint exploration by IA process participants. Effective collaboration is inclusive and open, involves multiple perspectives and forms of knowledge, is procedurally fair, is jointly undertaken by stakeholders, seeks to creatively solve problems and take advantage of opportunities, and is directed toward and guided by substantive environmental management, environmental justice, and sustainability ends. Collaboration extends from a sound knowledge base and proactively involves all interested and affected parties, especially the public and local communities. It can be facilitated by networking, joint fact-finding, joint planning, consensus building, shared vision planning, and constructive engagement. It can be formalized through partnering and comanagement arrangements and agreements.

The IA administrative arrangements in the four jurisdictions are partially conducive to collaborative IA processes. They help ensure information exchanges with the public and outreach to less involved groups and segments of society. Additional emphasis could be placed on earlier, frequent, and continuous involvement, on involvement in screening documents, and on decision-making links. Communications to the public through IA documents is partially addressed. More advice could be provided regarding communications principles, communications skills, communications distortions, and facilitating dialogue. Much attention is given to educating the public about IA participation opportunities and practitioners about participation approaches. More attention could be devoted to integrating community and traditional knowledge and to applying mutual learning concepts, methods, and practices. Considerable guidance regarding ADR is provided in the United States. More consideration could be given to the potential roles of ADR in the IA process. Collaboration among governments is well developed but in need of further refinement. Measures to facilitate stakeholder collaboration have received some attention. More consideration could be given to collaboration planning methods and frameworks and to facilitating collaborative IA processes.

The example of a collaboration IA process begins with an initial round of consultation, a conflict assessment, the formation of an advisory committee (the focal point of the process), interest representation, and public participation and IA planning. The IA process and public participation planning and execution are fully integrated. The various elements of effective public participation (involvement, communications, mutual education, negotiations, and collaboration) are integrated into IA and public participation planning, into major IA activities (e.g., scoping and needs analysis, alternatives formulation and evaluation, proposal and baseline characteristics, impact analysis, synthesis, and management, draft and final IA document preparation), into review and approvals, and into implementation. The effectiveness of public participation actions is evaluated during and subsequent to each stage in the process. The process is built on a solid knowledge foundation and includes frequent and numerous links to the broader public.

Collaborative IA is expressed in different ways for various IA levels and types (e.g., SA, SEA, EIA, EcIA, SIA, HIA). A collaborative IA process needs to be designed and adapted to suit the IA types that it encompasses. Insights regarding collaborative approaches should be shared among IA type practitioners, appreciating the importance of contextual adjustments.

IA capacity building involves measures to enhance the quality and effectiveness of IA requirements and practices, as interpreted from multiple perspectives. It can be applied to IA as a whole or more specifically to various IA types, activities, settings, population subgroups, proposal types, impact types, environmental components, issues, time horizons, and organizational types and levels. IA capacity building necessarily entails collaboratively designing and managing the capacity-building process (definitions and distinctions), determining the area of application (applied to what, where, and how), selecting the constituency and capacity-building facilitators (for whom and by whom), deciding on the capacity-building objectives (for what purpose), and identifying, adapting, and integrating the capacity-building tools (by what means). Capacity building, in common with other IA aspects, builds on good practices at both the regulatory and applied levels.

Chapter 10

How to Make IAs More Ethical

10.1 HIGHLIGHTS

In this chapter we respond to the challenge to make IA processes and outcomes more fair, equitable, and just. We also seek to identify and advance rights (especially those of disadvantaged groups) to ensure that duties are fulfilled, and to address ethical issues (professional ethics and accreditation) and challenges (the determination of significance). These concerns are all ethical. More precisely, they fall under the umbrella of normative (what ought to be), applied (directed toward the resolution of practical problems), and practical (moral questions, the answers to which involve commitments to action) ethics. We also illustrate how ethical concerns can be integrated into IA requirements and processes.

- The analysis begins in Section 10.2 with three applied anecdotes. The stories describe applied experiences associated with efforts to make IA practice more ethical.
- The analysis then defines the problem (Section 10.3—Defining the Problem and Deciding on a Direction), which is an insufficient effort to explicitly and systematically integrate ethical considerations into IA processes and process outcomes. In this section we demonstrate the ubiquitous nature of ethical concerns in IA practice.
- In Section 10.4 we define key terms, describe relevant ethical concepts, and highlight major distinctions. We present examples of procedural fairness principles, distributional fairness principles, and ethical rights and duties. These concepts, principles, and distinctions provide the basis for defining an ethical IA process. We also address the subject of IA professional ethics and consider the issue of professional accreditation.
- In Section 10.5 (Instituting an Ethical IA Process) we detail how an ethical IA process could be implemented at the regulatory and applied levels. In Section 10.5.1 we infuse ethical concepts and perspectives into IA regulatory requirements and guidelines. In Section 10.5.2 we integrate an ethical perspective into an applied IA process. In Section 10.5.3 we highlight the characteristics of an ethical IA process for different IA types.
- In Section 10.6 we address the contemporary challenge of significance determination. We identify key conceptual distinctions and offer good practice guidance.
- In Section 10.7 we highlight the major insights and lessons derived from the analysis.

10.2 INSIGHTS FROM PRACTICE

10.2.1 Are Impact Assessment Practitioners Craftsmen, Tradesmen, or Professionals?

During the Middle Ages, those making their livings in towns mostly identified themselves as craftsmen, tradesmen, or professionals (gender sensitivity was not high in those days). Craftsmen underwent an apprenticeship and acquired through practice the skills necessary to produce specific goods from raw materials—for example, garments, tapestries, metal items, jewelry. They were skilled at making things, and to protect special interests those of similar occupations grouped together in guilds (e.g., tailors, blacksmiths, goldsmiths). These guilds regulated who would be deemed a master in a particular craft, and also established performance and quality standards. Buying and selling the goods made by crafters, as well as farm produce and forest products, was the prerogative of traders. Expertise in commercial transactions was developed through experience and success was closely associated with developing networks to match supply with demand. Those who did not make or sell things, but provided services based on knowledge and extended formal education, were known as the professionals—lawyers, physicians, clergy, engineers. Groupings of individual professionals styled themselves Institutes or Associations and decided on appropriate training and entry criteria to protect society from unqualified persons.

A professional seeks to make a living by exercising particular expertise, so there is obviously a personal interest in obtaining work both to support oneself and family as well as employed staff. A client employs a professional in order to meet a specific need of their own—this might be to cure an illness, design a house, or defend a legal action. The client expects the professional to earn the agreed fee by acting in their (the client's) best interests. The professional also

Impact Assessment: Practical Solutions to Recurrent Problems and Contemporary Challenges, Second Edition. David P. Lawrence.

expects to make a living by providing a service that meets the need of the client. The nature of the marketplace is such that a professional who is perceived by a client not to be acting in their best interests will forfeit the business. It has nevertheless become recognized that scientists, teachers, journalists (the so-called information professionals) must have a primary allegiance to the accuracy of the information they disseminate and that this must take precedence over a client's desires if a tension occurs.

For many professionals, two common ethical dilemmas arise (1) deciding between one's personal interest and the interest of one's client, and (2) deciding between serving the desires of a client and serving the needs of society. Most professional associations or institutes deal with the first problem by having professional codes that require the client's interest to be placed ahead of personal considerations. They require that conflicts of interest be declared and that the basis for professional remuneration should be formally agreed and accord with the norms of the profession. The second problem is not as easily resolved when a client's desires and societal needs are not aligned. The question then arises: Is the practitioner undertaking impact assessments (IAs) in a different position to other professionals with respect to meeting a client's desires?

The term "sweetheart report" has been used to categorize impact assessments that are deemed favorable to a developer's interest at the expense of societal interests. And since the inception of environmental impact statements—as mechanisms to inform decision makers as to the likely consequences of proposed actions, so that decisions can be taken in society's best interest, sweetheart reports have been a problem. Conventional wisdom is that "he who pays the piper calls the tune," so no matter how impartial an impact assessment practitioner attempts to be, this perception persists. The fact that very few impact assessments hostile to a development are made public feeds the "sweetheart report" perception. It is illogical for the proponent of a project to appoint a professional who will not add value to the activity. However, value can be added by "planning with nature" and ensuring that no ecosystem services are lost, by optimizing project design to ensure sustainable operation, by reducing public opposition to a project, or by reducing bureaucratic delays due to missing data. All impact assessment practitioners should seek to give their clients value in one or more of these ways—not by producing biased sweetheart reports.

Various mechanisms have been proposed to ensure that impact assessments are fair and unbiased. One that has not found favor is that the proponent of a project should not appoint the professional team undertaking the impact assessment. Project proponents reject this approach because this undermines the client–professional relationship and is not conducive to trust and openness. The generally favored mechanism is that of third-party review of an impact assessment before it is passed to the relevant decision maker. This, in turn, is not without its problems. Who appoints the reviewers? Who remunerates them? What professional competency should they have? The wide variation that is occurring around the world with respect to review of impact assessments is unfortunately not conducive to public trust. It is clearly in the interest of all impact assessment practitioners that mechanisms be sought to assure decision makers and the public that sweetheart reports are not acceptable to the profession and are not the norm. Unfortunately, many countries are failing to regulate the quality and ethics (as opposed to the content) of impact assessments.

All impact assessment professionals need to remember that the profession grew out of social demands that environmental parameters need to be considered alongside technical and economic factors when decisions that will affect communities are made. Accountability to society as much as to a client is thus implicit in our professional activities. Impact assessment professionals have a primary allegiance to the accuracy of the information they disseminate and must be regarded as information professionals rather than service professionals. They must also be committed to the promotion of sustainability, the freedom of access to information, and the right of citizens to have a voice in decisions that affect them.

Practicing personal integrity and ensuring professionally competent reviews that promote sustainability should be the norm for all environmental professionals. It needs to be clear to the uninitiated that impact assessment practitioners are not craftsmen intent on producing good-looking reports made to regulatory specifications. Neither are they traders making money through their knowledge of the network of supply and demand for impact assessments. Greater attention must be given to promoting the professionalism of impact assessment practitioners.

RICHARD FUGGLE

Environmental Evaluation Unit, University of Cape Town, Cape Town, South Africa

10.2.2 Making the Attribution of Significance to Social Impacts More Rigorous

Social impact assessments (SIAs) with no significance methodology, community impact assessments with insufficient analysis, and SIA identification lacking cohesion with approaches used by other environmental specialties were commonplace results when I became an SIA practitioner in the mid-2000s—after 10 years of rural development experience.

SIA follows the normal impact assessment methodology—establishing the baseline, predicting impacts, attributing significance to impacts, and identifying mitigation and enhancement measures. The first SIA chapters I reviewed were almost completely baseline, either scantily or verbosely. Impacts were identified as positive or negative with little commentary, analysis, or justification.

At that time, the main sources of readily available SIA guidance were the International Association for Impact

Assessment's (IAIA's) *International Principles for Social Impact Assessment* and the World Bank's *Social Analysis Sourcebook: Incorporating Social Dimensions into Bank-Supported Projects.*

The IAIA's 2003 *International Principles* identified why principles were needed and presented a useful definition of SIA. It conceptualized social impacts, identified SIA activities, and elaborated core values, principles, and guidelines. The WB's 2003 *Sourcebook* identified five entry points for social analysis at the project appraisal stage and for social assessment throughout the life of the project. Neither mentioned attribution of significance.

After having skimmed through other EIA chapters, it was staggeringly apparent that SIA was an afterthought, written at the end of a study, or by someone who happened to have been involved in some basic consultation. The science of SIA was not present. There was no cohesion relating significance of social impacts with significance of impacts in other chapters. The overall impression was that the socioeconomic chapter was the least objective and systematic of the various disciplines. Ways to address this impression became a priority.

Other EIA chapters used magnitude and value/sensitivity criteria for the attribution of significance. Ensuring a human element was essential. The number of people affected was an obvious criterion, but not the only one; other magnitude criteria including duration, spatial extent, likelihood/probability of the impact occurring and reversibility seemed worthwhile. Soon, the concept of reversibility was dropped, not only because it is very unlikely but also because many of the projects had poor socioeconomic conditions where reversibility would not be desirable.

Magnitude criteria recently used are: **Determination**

- ***Major adverse/beneficial.*** A probable impact that affects the well-being of groups of many people or business entities within a widespread area beyond the project life.
- ***Moderate adverse/beneficial.*** A possible impact that will likely affect either the well-being of a group of people or business entities beyond the local area of influence into the wider area of influence or continue beyond the project life.
- ***Minor adverse/beneficial.*** An impact that may affect the well-being of a small number of people and/or households or businesses, or occurs exceptionally, mostly within the project area of influence and does not extend beyond the life of the project.

For sensitivity criteria, looking at what could be used to describe a group as sensitive, it was questioned what are projects trying to prevent, preserve and contribute to? Vulnerability seemed a logical concept. The next question was: vulnerable to what?

For developing countries, and recognizing the universal acceptance of the Millennium Development Goals (MDGs) movement, vulnerability to impoverishment risks made sense. Impoverishment risks identified by Michael Cernea are landlessness, joblessness, homelessness, marginalization, increased morbidity and mortality, food insecurity, interruption of education, loss of access to common property, and social disarticulation. In the United Kingdom and other developed countries, the discourse on "social exclusion" prevention outcomes seemed parallel to poverty eradication goals. Social exclusion has been identified as what can happen when people or areas face a combination of linked problems such as unemployment, discrimination, poor skills, low incomes, poor housing, high crime, bad health and family breakdown.

Sensitivity criteria also need to include acknowledgement that the ability to deal with change differs temporally and contextually. The ability of people to absorb changes caused by projects depends on their demographic profile, current life load, and access to resources that support adaptation, among other factors.

Sensitivity criteria recently used are:

- ***High.*** An already vulnerable social receptor with very little capacity and means to absorb proposed changes or with very little access to alternative similar sites.
- ***Medium.*** An already vulnerable social receptor with some capacity and means to absorb proposed changes or with little access to alternative similar sites.
- ***Low.*** A nonvulnerable social receptor with limited capacity and means to absorb proposed changes and with some access to alternative similar sites.

Like other disciplines, the SIA magnitude and sensitivity criteria are combined using a matrix to determine the category of impact as "major," "moderate," or "minor" significance or "insignificant." Impacts are also identified as adverse or beneficial.

Clients have accepted this more objective approach to attributing significance for SIA impacts. Readers of the socioeconomic or community chapters understand its logic. Assumptions embedded in the various criteria can be referred to in discussions regarding SIA impact significance.

MARIELLE ROWAN

Social Sustainability Team, Mott MacDonald Group, Brighton, UK

10.2.3 Recognizing the Significance of Small Projects—Smaller Does Not Mean Insignificant

Often attempts to "expedite" IA processes focus on eliminating small project assessment. There are thousands of so-called small projects for which conducting an impact assessment is often portrayed as a bottleneck clogging up the "approval" process. If these projects were all actually small in scale and very likely not to cause significant impacts, alone or cumulatively, such as a window installation in a government building, or a new picnic table in a national park, one could make the case that there would be few, if any, adverse environmental effects. But the term "small projects" is often a misnomer, disguising the scope of many quite large projects that clearly have the potential to negatively impact the environment. For example, here are some projects with complicated environmental interactions that were considered "small" under the Canadian Environmental Assessment Act:

- A small mine project that requires many kilometers of new access road and sends industrial runoff into an aquatic ecosystem.
- The installation or replacement of a culvert that has the potential to alter water-level regimes and aquatic species movements between large areas of wetland.
- A single small bridge across a stream in a remote natural area that opens up several hundred kilometers to resource extraction and public activity.

Assessment of the significance of the environmental effects, including cumulative effects, of projects like these appears to be expendable when there are pressures for a more streamlined approach to IA approval processes in the interest of accelerated economic development and improved IA efficiency. The federal government in Canada recently rushed to abandon the environmental review process for small projects even though it had concluded that most small project reviews do not take an unacceptably long time to complete. In analyzing some 18,056 small project assessments, the Quality Assurance Program also found that 90% of the projects that underwent a screening level assessment appeared "unlikely to cause more than minor adverse effects." This type of data is often quoted to falsely attest to the fact that these projects do not need IA. However, it should be noted that these determinations of minor adverse effects are made *after* the assessment process has been completed and mitigation measures have been identified through that process. These figures do not address the question of whether the project would have had significant environmental impacts in the absence of an environmental assessment. The data do demonstrate that having a meaningful environmental planning and review process is an effective tool leading to thousands of better projects across Canada 90% of the time.

Knowing the environment is a very complex system with a myriad of links and interactions and that small project impacts can also be synergistic, it is surprising that there is often a lack of willingness to consider the significance of small disturbances, or the accumulation of seemingly unrelated disturbances, that small projects can have, and that can result in significant implications for the natural environment, human health, and in many cases the stewardship (or not) of financial and natural capital in our economy. To ask people to be less vigilant about the environmental effects of the majority of proposed projects because of their scale or that they do not seem to have major environmental effects is like asking accountants to cancel thousands of financial audits just because a number of them contained no discrepancies.

We feel there are ways to create efficiencies in the assessment of small projects, rather than eliminate them, to ensure that their impact significance is determined. One approach could be the expanded use of a class IA process so that truly small and routine projects can be grouped together in a single advance assessment. This approach has, for example, been applied to numerous municipal infrastructure and business license projects in national parks in Canada. In these cases, the assessment of routine projects is simplified to a questionnaire with strategic questions that can be completed quickly followed by the implementation of recognized environmental practices associated with the specific type of project. Another approach could be the use of standard environmental operating procedures to ensure that routine projects apply mitigation measures. Such procedures have been developed in Canada, for example, by the Department of Fisheries and Oceans for projects that may affect wetlands and, if used in appropriate circumstances identified through IA, they can ensure a quick and efficient environmental review.

JOHN SINCLAIR

Natural Resources Institute, University of Manitoba, Winnipeg, Manitoba, Canada

PETER DUCK

Banff, Alberta, Canada

GARY SCHNEIDER

Environmental Coalition of Prince Edward Island, Belfest, Prince Edward Island, Canada

10.3 DEFINING THE PROBLEM AND DECIDING ON A DIRECTION

The three stories point to the ethical nature of IA in different ways. The first story is concerned with IA professional ethical issues and dilemmas. The remaining two stories address the contemporary challenge of significance determination in two different ways. One is concerned with good practice approaches to interpreting the significance of social impacts. The other addresses the issue of the propensity to assume that smaller projects are, by definition, insignificant. The three stories provide a partial and preliminary sense of how IA processes and outcomes can become more ethical. A more detailed exploration of the potential role of values and ethics in IA requirements and practice, however, is required.

Some critics suggest that IA practitioners and potentially affected groups and individuals are often "talking a different language." The IA practitioners tend to take great pains to demonstrate how the procedures they employ are systematic and consistent; to highlight the many opportunities for public involvement; and to show how overall adverse impacts are minimized. The public stresses that the IA process is unfair. They argue that benefits and adverse impacts are unfairly distributed. They insist that their rights have been ignored or diminished. They suggest that proponents and regulators have not made clear, verifiable, and enforceable commitments to the public and to the environment. Both parties are frustrated because their "message is not getting through."

Both IA practitioners and the public in IA processes grapple with the values associated with human conduct (Jiliberto, 2002). These values provide the principles and standards applied by each party to assess whether the proposed action and the IA process are "good" or "bad," "right" or "wrong." Ethics is a branch of philosophy that addresses whether actions are moral (i.e., good, bad, right, or wrong). Neither party tends to view their conflicting positions, perspectives, and interests as elements of an ethical debate. However, by acknowledging the ethical nature of IA, the first step is taken toward establishing a framework for accommodating perspective and interest differences.

IA is an inherently ethical activity. It seeks to advance environmental values (Hettwer, 1991; Jiliberto, 2002). It is subjective, moral, and value-full (Mostert, 1996; Finsterbusch, 1995). Value-based interpretations and judgments are made in every IA activity (Enk and Hornick, 1983). Oftentimes IA is perceived as biased, sometimes with good reason (Beder, 1993). The ethical basis for interpretations and judgments is occasionally explicitly presented. Too often it is not. As detailed in Chapters 4 and 5, IA practice tends to be shrouded in the language and pretense of objectivity.

IA is prescriptive, predictive, and interpretative. Uncertainties, ambiguities, and alternative interpretations abound, especially when predicting future conditions (with and without a proposed action) and when determining impact significance. Notwithstanding the inevitable ethical uncertainties and dilemmas, IA seeks to provide a sound decision-making basis. This leaves IA practitioners, at both the regulatory and applied levels, with considerable administrative discretion. Consequently, they have an ethical obligation to justify their positions and actions. They also have a responsibility to seek out and respond to the values and ethical positions of other participants in the IA process.

Issues of procedural fairness are inherent to the IA process. The IA process must be perceived as fair, from multiple perspectives, if it is to be accepted as legitimate (Laws, 1996; Firth, 1998). Procedural fairness is both an end in itself (consistent with democratic decision-making values) and a means of reducing public dissatisfaction and of enhancing the potential for public acceptance (Lawrence et al., 1997; Kasperson et al., 1988).

The proposals assessed through IA result in temporal (e.g., exacerbating historical inequities, adverse effects on future generations), spatial (e.g., inequities in the distribution of costs and benefits and of services and facilities), and social group (e.g., disadvantaged groups bearing a greater share of the burden of adverse impacts) inequities (Interorganizational Committee, 1994). They also contribute to changes in the distribution of political power. The potential for distributional inequities tends to be a particular concern when siting LULUs (locally unwanted land uses) and when assessing the social and environmental justice implications of proposed actions (e.g., legacy issues, pace and scale options, community and environmental resilience) (Hermans and Knippenberg, 2006; Gibson, 2011; Liu, 1997; Morell, 1984).

IA is a form of applied research. Hence there is a need for IA practitioners to apply ethical research standards and to consider the ethical dimension of different forms of social inquiry (Chase, 1990; Fuggle, 2005b). Procedural ethical principles and standards come to the fore in public consultation and in joint efforts with stakeholders to negotiate mutually acceptable solutions. IA practitioners, as environmental professionals, also should comply with the ethical standards of professional organizations such as the National Association of Environmental Professionals (NAEP) and the IAIA (IAIA, 2010). The extent to which such codes of conduct should be formalized (i.e., accreditation) and independently audited is a subject that has engendered considerable debate and discussion within the IA practitioner community (Montague, 2004; Morgan et al., 2012; Pisani and Sandham, 2006; Reinstein, 2010).

IA is one among many instruments for advancing sustainability and for furthering the cause of environmental and social justice. Social equity has been identified as a key element (some would say a prerequisite) of social sustainability (Boyce, 1995; Gardner and Roseland, 1989; Leith, 1995; Weaver et al., 2008). The unequal distribution of environmental hazards has become a major public policy concern (Weinberg, 1998; Albrecht, 1995). The recognition that the proposals assessed through IA requirements can exacerbate such inequities has resulted in initiatives to integrate environmental justice concerns into U.S. IA requirements.

IA does not operate in a vacuum. It is inevitably influenced by "the rights revolution," by debates concerning the role of justice in public policy, and by alternative characterizations of human and natural environmental relationships (Ignatieff, 2000; Chase, 1990; Etzioni, 1995; Rawls, 2001). Often these debates are or could be framed in ethical terms. Applied fields, such as IA, environmental management and planning, increasingly draw upon ethical theory to more systematically and explicitly explore and apply ethics in public policy (Harper and Stein, 1992; Finsterbusch, 1995; Beatley, 1989).

It is evident from the above that ethics is and should be a central attribute of IA practice. The question then is how best to proceed from the recognition of the role of ethics in IA to its full integration into the IA process.

10.4 SELECTING THE MOST APPROPRIATE ROUTE

10.4.1 Definitions

Ethics is a branch of philosophy concerned with the moral rules, principles, and standards that govern conduct. Ethics depend on values. IA is a prescriptive field of practice. Therefore, *normative ethics* (which seeks to arrive at moral conduct standards) and *applied* or *practical ethics* (which study specific practical problems and involve a commitment to action) are especially relevant. Moral obligations represent the kind of "ought" statements, which tell us how to behave within a framework of valued social norms and conventions (Jasonoff, 2003). They guard against arbitrary and unreasonable decision making (Jasonoff, 2003). This chapter focuses on integrating ethical principles and standards into the IA process.

Judging from the criticisms of IA practice, the ethical concepts—equity, fairness, justice, rights, and duties—seem especially pertinent. *Equity* concerns treating people impartially (i.e., treating everyone in the same way). *Fairness* involves treating people reasonably, consistent with moral rules or standards. *Justice* is concerned with moral rightness (an end). Justice also involves determining rights and administering rewards and punishments (a means). *Rights* are the expression of values to which people have a moral and sometimes legal claim. *Duties* or responsibilities represent a moral and sometimes legal obligation from one person to another.

As illustrated in Figure 10.1, these five ethical concepts are highly interrelated. Each concerns moral principles and standards of human conduct. Each involves judgments regarding right or wrong behavior. Equity, justice, and fairness are commonly used interchangeably. Although their meanings clearly overlap, there also are distinct differences. Equity, for example, could be viewed as a subset of fairness (i.e., equity is not the only standard of fairness). Fairness, in turn, could be considered a subset of justice (i.e., fairness is not the only standard of justice). Justice determines and enforces rights and duties. It also represents a means to achieve equity and fairness. Rights can be a precondition to fairness, justice, and equity. Duties implement rights, fairness, justice, and equity. There are equity, justice, and fairness rights and duties.

Integrating ethical concerns into IA practice involves considering the potential role of equity, fairness, justice, rights, and duties in both the IA process (i.e., a procedural focus) and in outcomes from the process (i.e., a substantive or distributional focus). Ethical concerns can take many forms in IA practice. They can be issues, objectives, principles, criteria, standards, decision rules, or requirements. They can be integrated into methods, into planning and decision-making processes, and into organizational structures and procedures.

10.4.2 Distinctions

Ethics, even when limited to normative applied ethics, is a diverse field of theory and practice. Therefore, it is necessary to be selective regarding potentially relevant ethical distinctions, sub-fields, and concepts. Table 10.1 lists several examples of potentially relevant ethical concepts, sub-fields, and distinctions. Key characteristics and potential IA process implications are identified.

Table 10.1 demonstrates that the IA process is a forum within which practical ethics are expressed and applied. The IA process applies (or should apply) both professional and research ethics. Multiple ethical standards, principles, and decision rules are available for assessing proposed actions and for conducting IA processes. It will sometimes be helpful to apply a plurality of ethical principles, standards, and decision rules. The preferences for and the manner in which ethics are applied will vary depending on the value systems of process participants. Ethical perspectives and positions, as with values, change and evolve. Making the evolving ethical perspectives and principles of IA process participants explicit can reduce confusion and sometimes ameliorate conflict.

Ethical trade-offs and dilemmas are highly likely with multiple perspectives, values, participants, and potential ethical principles and standards. An ethical analysis should seek to identify and address ethical issues, trade-offs, and dilemmas.

Ethical principles and standards can be applied to both procedures and proposed actions. They also can be applied to individual process activities (e.g., research, significance interpretation, consultation, communications). Ethical principles and standards are likely to vary by discipline (e.g., social, political, ecological, sustainability) and by perspective (e.g., feminist, traditional knowledge). Substantive ethical principles and standards can only be determined after analyzing the potential distribution of effects over time, over space, and among social groups (e.g. by undertaking environmental justice analysis). Measures will often be necessary to prevent and offset procedural and substantive distributional inequities. Efforts to address substantive fairness and equity issues are likely to be inhibited if participants perceive the IA process to be unfair.

Part of an ethical analysis involves making the rights and duties of participants explicit. Rights extend beyond process participants to future generations and to the environment. Rights will often conflict. It is, therefore, necessary to identify and assess the implications of conflicting rights. Duties are not limited to proponents. They extend to regulators and to all process participants. Interpretations of "appropriate" duties will likely vary among participants. These varying interpretations also need to be explored.

An ethical analysis can build from ethical codes of practice, applied research ethical principles, natural justice principles, and efforts to integrate social and environmental ethics into corporate planning. Practice-based precedents, such as environmental justice initiatives and the application of substantive equity principles could be particularly relevant. Many useful concepts, principles, and distinctions can be culled from applied ethics literature, especially efforts in directly related fields of practice such as environmental management and planning. Varying conceptions of the role of ethics within broader integrative frameworks and

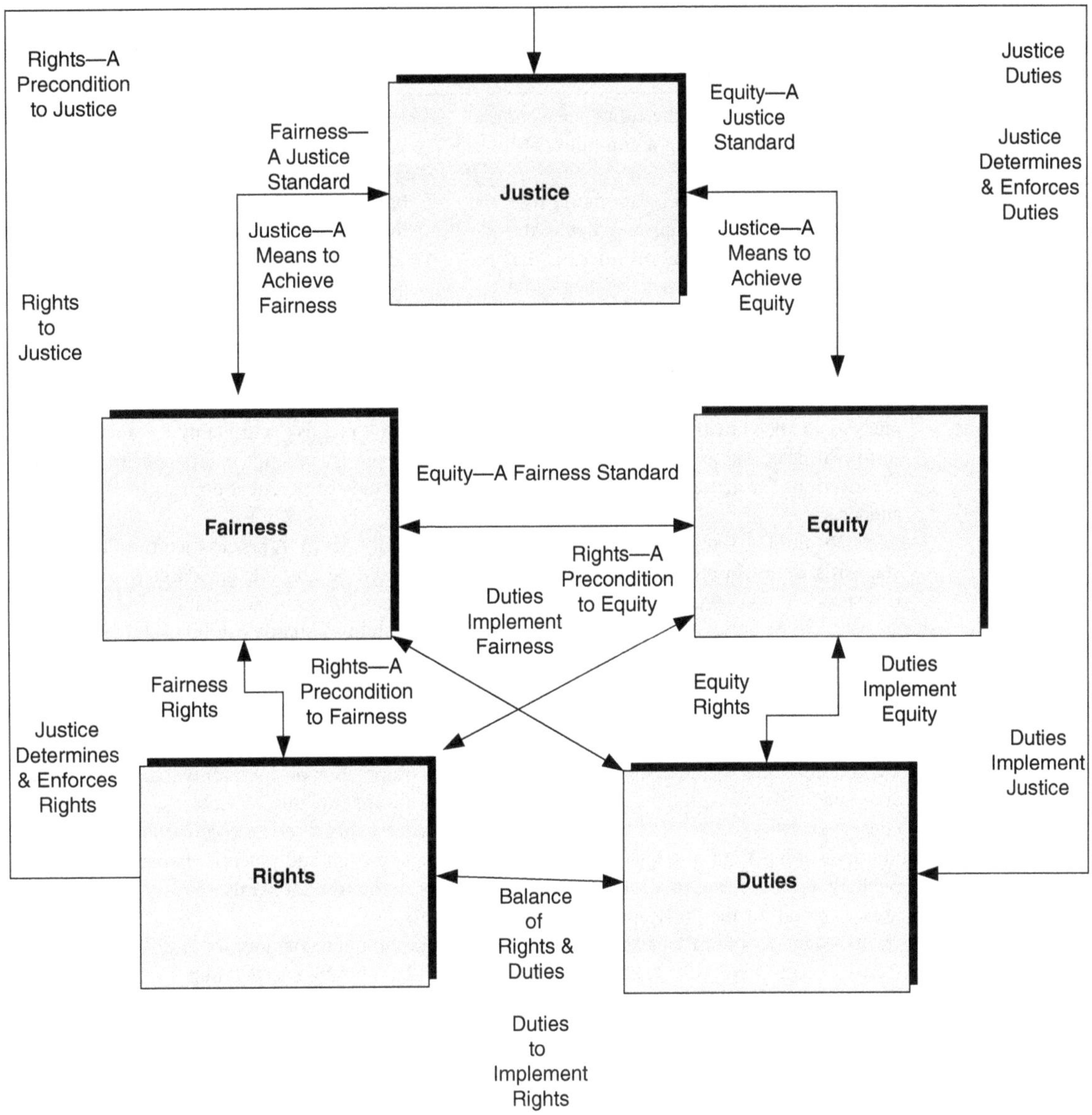

Figure 10.1 Examples of interactions—justice, fairness, equity, rights, and duties.

concepts, such as sustainability, could be especially instructive to IA process management.

10.4.3 Procedural Fairness

Procedural fairness is concerned with the fairness of the IA regulatory requirements and the IA process. It includes both how consultation with interested and affected parties is undertaken and how choices are made (Kasperson et al., 1988; Lawrence et al., 1997). Minimum standards for procedural fairness are addressed through natural justice principles (Morrison-Saunders and Early, 2008). Procedural fairness principles and standards can pertain to the rights of participants and to the duties and responsibilities of proponents, government regulators, review bodies, IA practitioners, IA team members, and process participants. Natural justice is owed to those likely to be adversely affected by a decision (Morrison-Saunders and Early, 2008). Procedural justice or fairness can contribute to the legitimacy of proposed actions (Karjalainen and Järvikoski, 2010).

IA regulatory requirements should, consistent with natural justice and procedural fairness principles, include provisions regarding such matters as adequate notice, reasonable opportunity to make representation, clear statements of actions, reasons for actions, adequate notice of right of review or appeal, and full disclosure (Morrison-Saunders and Early, 2008). All interested and affected parties have a right to effectively participate in the IA process. They also may see it as their right to be involved in designing and adapting the IA process. They are likely to be particularly concerned with timely access to all relevant information and analysis, and with timely (e.g., prior to major decisions) and adequate (e.g., sufficient time to formulate, review, and respond) involvement provisions.

Table 10.1 Potentially Relevant Ethical Concepts

Concepts	Key Characteristics	IA Implications
Practical Ethics	Addresses specific moral questions; answering a question involves making a commitment to action Can entail an appeal to a relevant moral rule, the questioning of the relevant moral rule, the justification of general moral rules, or the resolution of moral dilemmas (when same action falls under two different but acceptable rules) Moral reasoning normally involves clear definitions, the evaluation of arguments, the analysis of social institutions, the collection of historical data, and assent to a series of prescriptions for action that the reasoning supports	IA must grapple with moral questions and involves a commitment to action The decision rules that guide IA decision making, because they are value-based, are arguably relevant moral rules IA decision rules (moral rules) can be appealed to, questioned, and justified IA often involves conflicting values and moral positions; frequently, it is necessary to choose between or seek to reconcile conflicting moral rules (i.e., resolving moral dilemmas) The IA process is arguably a form of moral reasoning; hence IA practice can benefit from the insights and lessons of practical ethics
Deontological Ethics	Applies absolute or foundational normative standards, duties, or principles of moral conduct (irrespective of consequences) Duty based (e.g., duties applicable to every situation)	Participants in IA processes often judge proposed actions based on absolute standards (e.g., nuclear power or clear-cutting unacceptable) Absolute standards can be useful in screening Also helpful in understanding basis for stakeholder positions and in determining proponent duties
Teleological or Consequentialist Ethics	Normative principles of choice Rightness or wrongness of an action depends on the consequences of the action (total good consequences outweigh the total bad consequences) Utilitarianism is a form of teleological ethics (the public good is the sum of all preferences or the greatest good for the greatest number)	Implicit in much of IA practice Tendency in IA to focus on minimizing the negative rather than on comparing total good versus total bad or on maximizing benefits Once explicit, can recognize limitations with utilitarian approach and potential benefits of applying other normative standards
Rawlsian Ethics	Right to extensive system of basic liberties (restrictions to liberty for sake of liberty of others) Equality of opportunity Protect resources for future (just savings) Greatest benefit to the least advantaged Lowest cost to least advantaged	Explicit consideration of distribution of costs and benefits by social group Rationality contested; IA within a world of discursive and deliberative democracy A moral rule consistent with focus on environmental and social justice (maximize utility of worst off) Combines consideration of social justice, liberty, and resource protection
Libertarianism	Elevates individuals and their rights above all others	Tendency, especially for public proposals, to assume that greater "public interest" should always prevail over individual rights Points to need to consider, and, to the extent practical, minimize losses of individual rights and freedoms
Natural Justice	Legal principles that collectively constitute procedural fairness in administrative decision-making Applies to a person with rights at issue Legal examples—right to ruling free of bias, right to a fair hearing, right to due notice, right to be heard when might be adversely affected, right to know case against, right to judgment based on evidence, how evidence was used is communicated	Natural justice principles should never be compromised in IA IA should specifically address need for natural justice when new information arises that may be significant for decision-making IA-related examples—adequate notice, reasonable opportunity to make representation, clear statement of action, reasons for action, adequate notice of right of review or appeal, full disclosure, unbiased IA analysis Natural justice owed to those likely to be adversely affected by a decision Should include credible and effective grievance mechanisms

Table 10.1 (*Continued*)

Concepts	Key Characteristics	IA Implications
Environmental Justice Analysis	Occurs when communities or segments of communities bear a disproportionate burden of negative externalities Includes distributional analysis of winners and losers and of benefits and burdens Complex—can ameliorate, generate, or sustain conflicts; inevitably normative and political	Environmental justice should be fully described and analyzed Emphasizes well-being and vulnerability of underrepresented and disadvantaged populations Takes into account cumulative environmental injustices Examples of methodological issues—procedures for selecting populations, choice of variables, health and well-being indicators, spatial analysis (e.g., GIS) and comparison areas, statistical methods, interpretations
Discourse or Communicative Ethics	Seeks to counteract misinformation Seeks to ensure procedural fairness	Recognizes that IA is a dialogue among interested and affected parties Recognizes need to minimize communications distortion and to facilitate procedural fairness
Procedural Fairness or Equity	The fairness of consultation and choice procedures Seeks to enhance democratic decision-making processes	Provides a basis for determining when procedures are unfair and for formulating and applying rules and measures to prevent and offset May require additional measures to facilitate the involvement of traditionally underrepresented groups and organizations
Critical Ethics	Focuses on inequalities in the distribution of power and knowledge as a catalyst for action	Recognizes power inequities as component of an ethical analysis; inequities can be exacerbated by proposed action (e.g., centralization of authority) Provides a basis for efforts to reduce and offset political inequities
Communitarian Ethics	Focuses on the normative values of and control by local communities Normative values arise from the community	IA proposals can inhibit or enhance local empowerment Local control (e.g., voluntary communities) one approach to the siting of "locally unwanted land uses"
Egalitarian Ethics	Stresses the need to treat people equally and for those who receive the benefits to accept the burdens Merit of action dependent on whether the process distributes basic rights and duties justly and equitably	The unequal distribution of benefits and burdens is a recurrent issue Provides a basis for identifying and, where practical and appropriate, preventing or offsetting inequities
Distributional or Outcome Equity or Fairness	Focuses on the distribution of resources, benefits, and costs over time, over space, and among social groups	Distributional inequities is a recurrent issue Provides a basis for identifying and, where practical and appropriate, preventing or offsetting inequities Consistent with social and environmental justice Seeks just outcomes
Research Ethics	Concerned with the ethical standards applied in natural and social science research	IA is a form of applied research Many research guidelines available A particular concern when undertaking research involving indigenous peoples
Professional Ethics	Concerns conduct of professionals in practice Codes of conduct applied by specific professions such as planners, engineers, scientists, environmental managers, and IA specialists	Many professions involved in IA process Professional codes of conduct facilitate ethical behavior Codes of conduct for environmental professionals conducive to good environmental practice
Environmental Ethics	Concerned with the moral basis of environmental responsibility Extends ethical rights to other organisms and to ecological communities Ecocentric perspective	Ethical responsibilities to environment a central attribute of IA practice Helpful perspective in assessing environmental and impact significance
Feminist Ethics	Focuses on women's issues and women's moral reasoning Emphasizes responsibility, obligation, and care more than rights, rules, and justice	Consistent with a discursive, inclusive, relational, nonexploitive and nonmanipulative IA process Helpful model for integrating ethical with technical Useful perspective on balancing rights and duties in fair processes

(*continued*)

Table 10.1 (*Continued*)

Concepts	Key Characteristics	IA Implications
Sustainability Ethics	Responsibility of current generations to future generations; includes intragenerational ethics Ethics on a global scale Distributional equity—a key element of social sustainability Requires socioecological civility and democratic governance	Provides a framework for integrating ethical with other decision making considerations Extends distributional analysis to include rights of and responsibilities to current and future generations Integrates explicit limits, criteria and trade-off rules Addresses social equity and sustainability links Broadens IA temporal and spatial perspectives Encompasses human and nonhuman needs and wants
Ethical Pluralism	Addresses ethics from multiple perspectives using multiple methods and standards	A range of ethical perspectives, methods, and standards could be used to assess options Conflicting ethical standards and perspectives are possible and can be addressed through the IA process

Sources: Beatley (1989), Bond and Morrison-Saunders (2011), Burdge (2004), Connelly and Richardson (2005), Cronin (1993), Erikstad et al. (2008), Etzioni (1995), Finsterbusch (1995), Forester (1989), Gardner and Roseland (1989), Gibson (2006a, 2011), Harper and Stein (1992), Hendler (1994), Hermans and Knippenberg (2006), Howe (1990), IFC (2009), Jackson and Illsley (2007), Kasperson et al. (1984), Kreig and Faber (2004), Lawrence et al. (1997), MacNiven (1982), McCluskey and João (2011), Morrison-Saunders and Early (2008), Patton and Sawicki (1993), Rawls (1971, 2001), Richardson (2005), Rose et al. (2005), Rowan (2009), Taylor (1986), Walker (2010), Walker et al. (2005).

IA practitioners must avoid conflicts of interest and their analyses must be unbiased (Werling and Turner, 2010). Rights also concern the "ground rules" for participating in and withdrawing from the process. They can extend to how participants are treated, to how their knowledge is incorporated into the process, and to the provision of grievance mechanisms (Smith and Schin, 2004). Procedural rights can vary among social groups. Specific provisions may be necessary to meaningfully involve; to offset procedural inequities; and to respect the rights, knowledge, and perspectives of disadvantaged groups, traditionally underrepresented groups, and indigenous peoples (Booth and Skelton, 2011a; CIER, 2009; IAIA, undated b; Lajoie and Bouchard, 2006; O'Faircheallaigh, 2009; Whitelaw et al., 2009). Sometimes it will be necessary to address historical inequities at the outset before some parties will accept the process and its outcomes as potentially legitimate.

Proponents and IA team members have a duty to establish a clear and understandable IA process. Time frames should be reasonable. The need for the proposed action should be established. Reasonable alternatives should be considered. Assumptions, interpretations, conclusions, and recommendations should be explicit and substantiated. The IA process has a role in redressing procedural inequities (e.g., ensuring that the interests of disadvantaged groups are influential and reflected in outcomes) (Connelly and Richardson, 2005). The proponent and team members, together with the IA process, should be sufficiently flexible to adjust to changing circumstances and to adapt to language and cultural variations. They should respond to the concerns and suggestions of other process participants. They should respect the rights and values of other participants. They are obliged to prevent bias (e.g., burying uncertainty, recognizing role of professional culture), to provide accurate data and analysis, to correct errors, to identify uncertainties and their implications, to comply with regulatory requirements, and to record and fulfill commitments (Duncan, 2008; Morgan et al., 2012). They should seek to remove barriers to understanding and participation. The IA team members should comply with applicable ethical codes of conduct. Grievance mechanisms should be incorporated into the IA process (Smith and Schin, 2004). The IA process and methods should be consistent with good practice standards. The IA process should be monitored and efforts made to enhance its effectiveness.

In exchange for the fulfillment of rights and duties, such as those cited above, all parties are commonly expected to participate in "good faith." They also are accountable for their actions and should maintain contact with and be accountable to their constituents. They should not engage in rhetoric or make sensational charges. Depending on the process, all parties could attempt to reach a consensus or accommodate conflicts.

It may sometimes be advantageous to formalize procedural rights and duties through written agreements. These agreements will likely evolve in conjunction with the IA process. Procedures also may be necessary to address situations where rights conflict, or where there are conflicting interpretations of duties. Appeal procedures may be needed for such matters as the timely provision of all relevant information. A commitment could be made to institute an open, fair, impartial, and independent review at the end of the IA process.

The choice and application of rights and responsibilities will vary among IA processes. A suite of good practice ethical procedural principles could evolve over time. These principles could be adapted to suit individual proposal and environmental circumstances and to meet the needs and expectations of process participants.

10.4.4 Distributional Fairness

Distributional fairness pertains to the distribution of risks, costs, and benefits over space (community/regional/state/provincial distribution, fair/unfair locations), over time (historical inequities, current, future), and among social groups (income, ethnic, indigenous peoples, class, age, other susceptible populations). It can refer to the allocation of services and resources (inter and intrageneration/receipt, fair/unfair distribution, opportunities) (Bond and Morrison-Saunders, 2011). It can concern the extent to which individual liberties and local decision-making powers are reduced or enhanced. Distributional fairness takes into account the fairness of cumulative effects; the carrying capacity; the vulnerability to change of social, economic, and ecological systems; and the overall contribution to sustainability (Dalal-Clayton and Sadler, 2004). It can help legitimize proposed actions (Karjalainen and Järvikoski, 2010).

The aggregation of distributional fairness concerns can take many forms. Total impacts and costs can be compared or net benefits to society can be determined. Both these approaches are consistent with a utilitarian ethical approach. Alternatively, net benefits by social group (e.g., gender analysis, indigenous populations, low-income populations, communities of color) and by geographic area can be determined (Burdge, 2004; Kreig and Faber, 2004; Sharma, 2010; Walker, 2010; Walker et al., 2005). The latter approach is more conducive to identifying and addressing social and environmental injustices. Environmental justice analyses, especially for vulnerable and disadvantaged segments of society, can be undertaken (Kreig and Faber, 2004; Rose et al., 2005; Walker, 2010). Fairness and acceptability determinations can be influenced by the availability of reasonable alternatives, by the potential for avoiding and mitigating inequities, by historical injustices and cumulative burdens, and by applying equity compensation measures and local benefits (Connelly and Richardson, 2005).

Distributional fairness, as with procedural fairness, involves rights and duties. Distributional rights could pertain, for example, to avoiding unnecessary adverse effects, to providing net community benefits, to reducing and mitigating adverse effects, and to compensating for significant adverse effects, which cannot be mitigated to acceptable levels. The rights of potentially affected populations to receive benefits and to avoid impacts could vary depending on the degree of potential harm, on the extent of adverse impacts already incurred, and on the degree to which the potentially affected population is socially and economically disadvantaged (Kreig and Faber, 2004).

IA, at the regulatory level, could require environmental justice analyses and provide guidance regarding the conduct of such analyses (Eccleston, 2008; McCluskey and João, 2011). The proponent and the IA study team have a duty to determine the distribution of costs and benefits and to ascertain the vulnerability of various groups. They should strive for the greatest overall benefits by geographic area and over time (Gibson, 2006a, 2011). They should seek to identify and redress historical and current burdens and hazards, to protect the interests of current and future generations, to accept the burden of proof, to emphasize enhancement, and to bear the full costs of the IA process (Connelly and Richardson, 2005; João et al., 2011; Thérivel, 2010). Benefits and unavoidable risks should be equitably shared and there should be fair access to compensation based on clearly defined and consistently applied criteria. Meeting societal needs should be viewed as a shared responsibility. Some parties could suggest that risks should only be imposed voluntarily, that the greatest benefit should accrue to the least advantaged, and that a community once selected should be assured that it will not be selected for future facilities.

10.4.5 Rights and Duties

A *right* is a claim on others that a person or a group of persons has and is enforced by law, custom, or education (MacNiven, 1982). Rights express and often give a legal meaning to values. They tend to highlight some injustices (e.g., barriers to access to information) and to devote less attention to others (e.g., economic inequities). They protect our right to be equal (e.g., equal protection under the law) and to be different (e.g., minority rights). Privileges are possible within a rights system (e.g., affirmative action). Rights help determine what is right. Rights often conflict. However, rights systems tend to provide a means of adjudicating rival claims. Rights and duties have a reciprocal relationship. Each right entails an obligation (Ignatieff, 2000). The appropriate balance between rights and duties is often highly contentious, with competing claims centered on different notions of justice (Walker, 2010). The IA process is one among many forums within which rights and duties are expressed and applied. SIA has made a particular effort to integrate rights (e.g., fundamental human rights, right to equal justice, right to live and work in an environment conducive to good health) and duties (e.g., polluter pay, cost internalization, prevention, multisectoral integration) into its procedures and practices (Vanclay, 2003).

An ethical IA process determines and applies rights and duties. As highlighted in Table 10.2, there are various types of rights (rights about). Rights are possessed by different segments of the population (rights of) (e.g., rights of indigenous peoples) (IAIA undated a). Rights address a range of concerns (rights to). Different parties exercise duties (duties by). The duties concern specific subjects (duties about). There are those who benefit from the conscientious application of duties (duties to). Rights and duties are often expressed as principles (e.g., intergenerational equity, precautionary principle, enhancement of marginalized groups, recognition and preservation of diversity) (Vanclay, 2003).

Rights can be possessed by, for example, proponents, communities, indigenous people, consumers, workers, landowners, and governments. Rights can be extended to the

Table 10.2 Examples of Rights and Duties

Rights		
Rights to	**Rights About**	**Rights of**
Basic needs	Fundamental freedoms	Proponents
Fundamental human rights	Health and safety	Regions and communities
Information	Democratic principles	Indigenous peoples
Consultation	Community notification and involvement	Consumers
Legal protection	Legal requirements	Workers
Free, prior, and informed consent	Social concerns	Landowners
Liberty	Economic concerns	Interested and affected parties
Equal justice	Equality of access and treatment	Governments
Continued resource use	Minority populations	Future generations
Diversity	Traditional activities	Environment
Coexistence	Resource use	
Privacy	Ecological integrity	
Health and safety	Environmental quality	
Fair treatment	Mitigation and compensation	
Self-determination and consent		
Language and culture		

Duties		
Duties to	**Duties About**	**Duties of**
Other governments	Health and safety	Proponents
Indigenous peoples	Environmental stewardship	Governments
Regions and communities	Planning and decision making	IA practitioners
Workers	IA process design and management	Professionals
Knowledgeable individuals	Information access	Researchers
Nongovernmental organizations	Notification	Current generation
Constituents	Consultation	Nongovernmental organizations
Field of practice	Preventing and paying for pollution	Participants in the IA process
Environment	Internalizing costs	
Current generations	Legal liabilities	
Future generations	Research procedures	
	Institutional controls	
	Training and employment	
	Compensation	
	Local benefits	
	Risk and uncertainty management	
	Treatment of rights	
	Treatment of risks and uncertainties	
	Work and living environment	
	Treatment of historical grievances	
	Respect of culture and values	
	Sustainability	

environment and to future generations. There are basic or fundamental human rights and freedoms. Rights can, for example, concern such matters as health and safety protection, the application of democratic principles, compliance with legal requirements, social and economic concerns, the equality of access or treatment, responsibilities to minority populations, the continuation of traditional activities, the mitigation of and the provision of compensation for adverse effects, the protection of renewable and non-renewable resources and the maintenance and enhancement of environmental quality and ecological integrity. Rights can apply to such matters as how basic human needs are to be fulfilled, information is to be shared, consultation activities are to be conducted, decisions are to be made, personal freedoms and privacy are to be maintained, continued resource use is to occur, and safety is to be assured. Rights might concern how parties are to be treated fairly, how their languages and culture are to be protected, how they are to coexist, and how they are to continue to determine their own futures.

Many parties could have duties in an IA process including, for example, proponents, governments, professionals, researchers, nongovernmental organizations (NGOs), and individuals. The duties could be directed toward governments, communities, workers, NGOs, constituents, fields of practice, the environment, and future generations. The duties could concern how health and safety is to be determined and protected. They could relate to environmental objectives and performance standards, to responsibilities to prevent and pay for pollution, to internalizing costs, and to contributing to sustainability. They might pertain to the design and application of the IA process, including, for example, information generation and sharing, public consultation, research procedures, and respect for culture and values. Duties often extend to organizational obligations regarding such matters as legal liabilities, training and employment procedures, compensation and local benefits procedures and policies, social and environmental performance standards, and risk and uncertainty management standards and procedures.

Additional rights and duties are likely to be needed when indigenous peoples are involved in the IA process. These latter duties could involve such matters as legal duty to consult with indigenous people; respecting self-determination goals and aspirations (i.e., free, prior, and informed consent); respecting treaty rights and indigenous property and resource rights and sovereignty; seeking to preserve the culture, identity and way-of-life of indigenous people; recognizing the uniqueness and cultural heritage of each indigenous group; taking in account indigenous perspectives and worldviews; treating traditional knowledge and western knowledge with the same respect; adapting planning, decision making, and research procedures (e.g., community-led capacity building, land-use planning and resource comanagement); negotiating on a government-to-government basis; ensuring indigenous control of SIA; redressing procedural inequities (e.g., participant funding); and providing compensation and local benefits to help indigenous people advance their own goals (Booth and Skelton, 2011a; CIER, 2009; IAIA, undated a; Landry et al., 2009; Lajoie and Bouchard, 2006; Noble, 2009b; O'Faircheallaigh, 2009; Whitelaw et al., 2009).

10.4.6 Professional Ethics and Accreditation

Ethical issues and dilemmas frequently occur in IA professional practice. They often arise regarding the procurement of IA services (e.g., who appoints? who pays?), and concerning the nature and boundaries of professional competency (Birley, 2007; Fuggle, 2005b). IA professionals must decide whom they are accountable to (e.g., self, employed staff, clients, the profession, society, the environment) (Fuggle, 2005b). They must deal directly and explicitly with value and interest-related issues (Woodward, 2003). On occasion, for example, IA practitioners have been known, because of a higher obligation to their profession and the environment, to ally themselves with NGOs by leaking information (Craik, 2008).

IA practitioners need to stay within the limits of their professional competency (Birley, 2007; Taylor et al., 2004). They must retain their independence, avoid conflicts of interest, and provide analysis and advice based on good practice rather than just regulatory compliance (Morgan et al., 2012; Ross and Thompson, 2002; Werling and Turner, 2010). They need to resist the urge and pressure to "be team players," which, in effect, means the propensity to, for example, minimize direct impacts, ignore or obfuscate indirect and cumulative impacts, and dismiss alternatives, concerns, assertions, and complaints raised by groups and the public (King, 2012). They have an ethical responsibility to keep abreast of and contribute to their field (e.g., by networking with other environmental professions, by ongoing training, through applied research, by keeping up with the IA literature, by participating in the activities of and adapting the resources of professional organizations such as IAIA and the NAEP), and by ensuring that that they fully understand the institutional and environmental context (and associated implications) within which they operate (Reinstein, 2010). They should contribute to "raising the bar" of IA practice, such that the IA knowledge base is enhanced (i.e., theory building), the gap between IA theory and practice is narrowed, the IA approaches and methods utilized are appropriate to the setting, knowledge is effectively shared and actions coordinated (e.g., the use of partnering agreements), and the procedural (e.g., more inclusive and transparent decision making) and substantive ends (e.g., sustainability) of IA are advanced (Morrison-Saunders and Bailey, 2009; Weaver et al., 2008). Table 10.3 (prepared by Richard Fuggle on behalf of the IAIA) provides an overview of the role of ethical matters in IA practice, cites examples of ethical dilemmas and possible solutions, and outlines examples of good practices.

Professional accreditation is an often-suggested mechanism for enhancing the competency of IA professionals. Professional accreditation encompasses such matters as the mandatory registration or certification of IA professionals, the potential exclusive use of professional titles, the potential exclusive right to practice, standing as expert witnesses before courts and tribunals, a code of conduct, the accreditation of institutions of higher learning, accredited training and outreach opportunities, complaints review and professional misdemeanor procedures, formal links to related fields, good practice guidance, knowledge and experience sharing, and government oversight (Reinstein, 2010). The International Association for Impact Assessment (IAIA) and the National Association for Environmental Professionals (NAEP) are part way down this path (e.g., codes of conduct, membership qualifications, promotion of good practices, the promotion of knowledge sharing and outreach, training opportunities). Of particular note, the IAIA formulated and broadly distributed "Guideline Standards for IA Professionals" (IAIA, 2010). Adopted by the IAIA Board of Directors in October 2010, these qualitative performance standards address such themes as code of conduct, education and training, experience, understanding of IA methods, IA study management, sustainable development, IA administrative systems, and professional development and mentoring for varying levels of IA practitioners and IA administrators.

Professional accreditation is considered desirable because it clearly distinguishes between qualified and unqualified practitioners, and between acceptable and unacceptable conduct standards (Reinstein, 2010). It clarifies the ethical obligations and responsibilities of practitioners (Fuggle, 2005b). It enhances the likelihood that the IA will be conducted by qualified personnel (Taylor et al., 2004). It provides the opportunity for the third-party auditing of the professional system, and of the conduct of individual practitioners (Van Der Vorst et al., 2010). It builds on accreditation models from related fields (e.g., planning, EMS) (Van Der Vorst et al., 2010). It makes it possible for IA practitioners to draw upon the wisdom of an entire profession, as exhibited through conference proceedings, journal publications, newsletters, viewpoints, mentors, training opportunities, and other resources (Looney, 2011; Pisani and Sandham, 2006). It underscores the necessity of continuing education, provides an opportunity for independent oversight, and enhances the likelihood of a higher standard of professional competency (Birley, 2007). For IA practitioners it formally recognizes competency, facilitates career choices and mobility, provides easier access to indemnity insurance, enhances professional credibility, provides opportunities for interaction with other environmental professionals, and assists in providing a due diligence defense (Shippey, 2004). For the environment industry, it fosters professional recognition; contributes to public confidence in the advice provided; harmonizes professional standards;

Table 10.3 Professional Ethics in IA—FASTIPS # 2 (April 2012)

The Social Contract

The social contract between impact assessment professionals, civil society, and decision makers is that impact assessments will be conducted with integrity and will be free from misrepresentation or deliberate bias.

Ethics in IA

Ethical considerations are important in impact assessment, as pressure can be applied to professionals to disregard the tenets that underpin good impact assessment. Good impact assessments enhance the free flow of complete, unbiased, and accurate information to decision makers and affected parties. Impact assessments address broad social and health rights as well as issues of sustainability and biodiversity. Consideration of all pertinent matters and respect for human rights and human dignity must underpin all assessments. Nevertheless, stakeholders (proponents, clients, donors, employers) sometimes want impact assessments to emphasize their position, possibly underplaying or overplaying certain elements. Failure to comply with pressure to sway the conclusions of the assessment may result in losing a contract or future work. The party paying for the assessment may also refuse to meet the costs of work that is necessary for a full and balanced impact assessment, forcing the professional to make an ethical decision. The FASTIPS that follow are intended as a reminder to impact assessment professionals that they have a duty of care to both present and future generations and that the assessments they undertake are to serve the interest of society through facilitating decisions that are equitable, sustainable, and accurate.

Examples of Ethical Dilemmas in IA	Things You Can Do When Faced with Such Dilemmas	Five Important Things to Know	Five Important Things to Do
Impact assessment professionals are faced with an ethical dilemma when - The terms of reference for the assessment unreasonably constrain the study - Pressure is exerted to limit the scope of the assessment or to influence the results - Clients, authorities, or affected parties refuse to engage with a study - Budgetary limitations affect the ability to conduct an adequate analysis or adequately engage all appropriate stakeholders - The time allocated for the assessment is inadequate for a proper study	Do not suppress or hide your sense of unease; discuss the matter with your manager, or if you are the principal, with a trusted and respected colleague or mentor Show the party that is causing the unease a copy of the IAIA Code of Professional Conduct and point out to them how their request, expectation, or assumption is at odds with the ethical code of your profession	**1.** The social contract between impact assessment professionals and civil society and decision makers is that (a) impact assessments will be conducted with integrity and will be free from misrepresentation or deliberate bias, and (b) impact assessments will respect citizen rights to participate in decisions that affect them **2.** An impact assessment professional's beliefs and cultural preferences must not interfere with the fair representation of the potential impacts of policies, plans, programs, and projects. It is also improper to advance private interests to	**1.** Be open and honest with yourself, your clients, and the public. Conduct your professional activities with integrity and professionalism, free from any misrepresentation or deliberate bias **2.** Conduct your professional activities only in subject areas in which you have competence through education, training, or experience. If asked to undertake work outside your field of competence, you should subcontract to, or work together with, other professionals who do have the competence you lack **3.** Take care that your professional activities promote sustainable and equitable actions

Table 10.3 (*Continued*)

Examples of Ethical Dilemmas in IA	Things You Can Do When Faced with Such Dilemmas	Five Important Things to Know	Five Important Things to Do
- There is a request not to undertake certain specific specialist studies - It is suggested that they emphasize or exaggerate, omit, or not disclose certain topics - Requests are made to change words or the emphasis in draft reports that could be seen as a change in interpretation with which the assessor does not concur - They become aware of inaccurate reporting by clients, sub-consultants, government agencies, NGOs, donors, or the public - All or part of their remuneration is conditional on project approval - There is a request to issue either a favorable or critical review of a completed assessment - They are asked to assess a project in which they have a financial or personal interest or that will affect them directly	Do not be tempted to accept a compromise that remains in conflict with your code of professional conduct. Remember, a reputation for integrity and for making accurate honest assessments takes time to build but can be quickly lost	the detriment of the public, clients, or decision makers **3.** Impact assessment professionals should seek sustainable and equitable outcomes from human actions that affect ecosystem and social functions and have due regard to the rights and interests of future generations. As different groups in society experience benefits and harm in different ways, impact assessments should strive to promote equitable access to, and use of resources **4.** Impact assessments must be conducted and implemented in a manner that averts infringement of the human rights of any section of society and does not condone the use of violence, harassment, intimidation, or undue force **5.** Impact assessment professionals must strive for excellence by maintaining and enhancing their own knowledge and skills, by encouraging the professional development of coworkers, and by fostering the aspirations of potential members of the profession	**4.** Refuse to provide professional services whenever you are expected to exclude reasonable alternatives from assessment, favor specific alternatives, omit or distort facts, or bias your analysis to arrive at a predetermined result **5.** Disclose all personal or financial interests that could reasonably raise concerns that there may be a conflict between your private and your professional interests

Further Reading:

1. IAIA Professional Code of Conduct and Ethical Responsibilities (www.iaia.org/about/mission-vision-values.aspx).
2. Baines, J. T., and C. N. Taylor, in F. Vanclay and A. M. Esteves, eds., "Ethical Issues and Dilemmas," *New Directions in Social Impact Assessment: Conceptual and Methodological Advances*, Edward Elgar, Cheltenham, 2011, pp. 96–113.
3. Chadwick, R., ed., *Encyclopaedia of Applied Ethics*, Vols. 1–4, Academic Press, San Diego, 1998, especially chapters by T. Airaksinen (Professional Ethics), R. E. Spier (Science and Engineering Ethics, Overview), M. Jarvela et al. (Environmental Impact Assessment).
4. Fisher, R., "Anthropologists and Social Impact Assessment: Negotiating the Ethical Minefield," *The Asia Pacific Journal of Anthropology* **9** (3), 231–242 (2008) (http://dx.doi.org/10.1080/14442210802251670).
5. Howitt, R., "The Importance of Process in Social Impact Assessment: Ethics, Methods and Process for Cross-Cultural Engagement," *Ethics, Place & Environment* 8(2), 209–221 (2005) (http://dx.doi.org/10.1080/13668790500237336).
6. Vanclay, F., "International Principles for Social Impact Assessment," *Impact Assessment & Project Appraisal* 21(1), 5–11 (2003).

Prepared by Richard Fuggle, with significant input from Frank Vanclay, Rita Hamm, and Charlotte Bingham (www.iaia.org) for International Association for Impact Assessment.

promotes the export of professional practice; promotes environmental knowledge and awareness; advances ethical and competent environmental practice; helps define minimum education, experience, and core competency standards; and encourages members to meet those standards (Shippey, 2004).

IA professional accreditation, however, is not without potential drawbacks, uncertainties, and dilemmas. Professional accreditation may hamper IA quality and effectiveness if the necessary resources (e.g., training programs) and infrastructure are not in place. Valid questions can be raised regarding whether professional accreditation is in the public interest because it raises professional standards or it could be argued that it inhibits competition and unduly restricts practice. The nature of IA, as a field of theory and practice, also raises issues. For example, does an amorphous, interdisciplinary field that encompasses multiple IA types, varies greatly depending on context, is far from settled regarding good practice standards and effective outcomes, and is evolving rapidly lend itself to formal practice requirements and standards (Morgan et al., 2012)? What roles should proposal type and regional/local experience assume in accreditation? How should accreditation requirements be varied for different IA types, and from country-to-country (Pisani and Sandham, 2006)? Should IA accreditation be connected to other forms of environmental management or subsumed under the umbrella of broader environmental management organizations? These issues have been faced, and addressed in varying ways by other interdisciplinary fields of theory and practice—planning, for example. To hold back from scrutinizing and, where appropriate, applying accreditation procedures creates the potential to inhibit the credibility and professionalism of IA practitioners and challenges the relevance of IA as a potentially effective environmental management tool.

10.5 INSTITUTING AN ETHICAL IA PROCESS

10.5.1 Management at the Regulatory Level

The four jurisdictions (the United States, Canada, Europe, and Australia) all address ethical issues, as highlighted in Table 10.4, although rarely as explicit ethical principles and obligations. They tend to spell out (in varying levels of detail) the IA process-related responsibilities of the proponent and government. All four jurisdictions require the interpretation of the significance of impacts and offer guidance regarding significance determination criteria and procedures. They generally address public procedural rights to the extent of including minimum public notification and access to information and public involvement requirements. These requirements are not generally portrayed as rights. There has been a general move in each of the jurisdictions to facilitate the involvement of disadvantaged groups, and to take into account the rights, knowledge, culture, and traditional activities of indigenous people.

Overall, the regulatory approaches in the four jurisdictions fall well short of the measures described in the previous chapter sections. There remains some latitude for more specific procedural and distributional fairness provisions. Such provisions might have a "harder edge" if they were described in terms of rights and duties, perhaps along the lines of and extending from the Australian legislation. The approaches employed for consultation with indigenous peoples in Canada, the United States, and Australia could be compared. It could be worthwhile to adopt a more formalized approach to equity impact assessment and to access to information, decision making, and justice rights, as has occurred in Europe. It might be advantageous to formalize the requirement to undertake a fairness distributional analysis, as has occurred in the United States. Such distributional analyses could be more broadly defined. As IA requirements move toward a greater emphasis on sustainability, it could be necessary to introduce specific provisions concerning the rights of future generations. The stress placed on local benefits in northern Canada seems to have the potential for broader application.

Some potentially interesting ethically related measures, introduced in individual jurisdictions, are worthy of review for potential application elsewhere. Examples include environmental justice requirements (United States), scientific and information integrity requirements (United States), participation funding (Canada), public access rights to information, public participation and justice (Europe), the treatment of the rights of future generations (Australia), and an auditing of professional environmental practice role by government (Australia). The issues surrounding IA accreditation, codes of conduct and the auditing of quality of practice, as raised in Australia, Europe, and the United States, are clearly worth further exploration. The ethical implications of measures to make IA requirements more efficient and focused in all four jurisdictions should be carefully scrutinized for ethical implications.

The approaches being applied in the four jurisdictions offer some worthwhile insights regarding significance determination approaches. The Australian approach points to the value of each regulatory level explicitly and systematically identifying its environmental priorities. The European and American approaches point to the need to explicitly identify significance criteria and procedures, while maintaining sufficient flexibility for contextual adjustments. The Canadian and European approaches underscore the need to focus on those settings and types and scales of projects likely to induce significant adverse effects. At the same time, they point to the potential dangers of equating proposal scale and type with significance, and the potential inconsistencies and gaps in application associated with selective definition of effects and potential participants, the assumption of equivalency among IA levels, and a high degree of discretion in the application of IA requirements.

The broader application of ethically related measures, applied in individual jurisdictions, could, on first inspection, be very appealing. However, each measure should be fully evaluated for effectiveness. Care should be taken to ensure a match between approach and context. But there is a danger in too much precision at the regulatory level. The interested and affected parties vary among proposals and settings. IA processes frequently involve a negotiation of procedural and distributional rights and duties. These negotiations occur both between proponents and regulators and among interested and affected parties. It could be worthwhile, in many cases, to formalize such negotiations. In this way, confusion can be minimized and conflict contained. The establishment

Table 10.4 Positive and Negative Ethical Examples at the Regulatory Level

United States	Canada	Europe	Australia
(+) Detailed IA-related environmental justice requirements and guidelines; requirements pertain to the potential for disproportionately high and adverse impacts on minority populations, low-income populations and Indian tribes (+) Individual departmental environmental justice strategies (US DOE, 2007) (+) Required to investigate the direct, indirect, and cumulative environmental justice effects on communities (+) EPA—agency created mapping tool aimed at identifying "environmental justice" hot spots (McArdie, 2012) (+) Executive orders address notice, consultation, and coordination with Indian tribal governments, environmental justice, and the protection of children from health and safety risk (+) Required notice to Indian tribes when effects may occur on reserves (+) A recent Presidential memorandum and follow-up guidance stress the need for scientific integrity in government activities (+) Requirement—full and fair disclosure of significant direct and indirect environmental impacts (+) Significance determination on a case-by-case basis considering context and intensity; regulations define context and specify intensity factors (e.g., unique environmental characteristics, risk and uncertainty, quality of environment, cultural or historical significance, endangered species or habitat, cumulative effects, controversy, legal compliance)	(+) EA and panel must consider likelihood of significant adverse environmental effects (+) Participant funding for designated projects (+) Guidance and extensive sponsored research on significance in EA (+) Requirements, guidelines, and related sponsored research stress the need to promote coordination and communications with aboriginal people; to assess impacts on the traditional use of land and resources by aboriginal people; to fully consider heritage impacts; to take into account aboriginal traditional knowledge; to respect aboriginal rights, values, and worldviews; to facilitate the meaningful involvement of aboriginal people; and to consult with aboriginal people on policy issues (+) Provides for assessments by band councils (+) IA requirements in the northern territories of Canada stress the need for proposed actions to optimize benefits for northern residents and communities (+) Local benefits guidance (EISC, 1999) (+-) Decision making takes into account potential for significant adverse effects and if such effects justified (−) Potential for procedural and outcome fairness being undermined by such measures as limiting involvement in National Energy Board hearings to interested parties (directly affected or having relevant information or expertise), the elimination of the mediation option,	(+) Proposed Project Directive (PPD)—when determining whether significant environmental effects likely should identify relevant criteria and information; required to take into account Annex III criteria; state how taken into account; reasons for requiring or not requiring an EIA; specify mitigation measures and make available to public (+) PPD—more detailed criteria should be taken into account when determining significance of environmental effects (e.g., projects affecting valuable natural resources, proposals for environmentally sensitive locations, projects with potentially hazardous or irreversible effects) (+) PPD—requirement that developer shall ensure environmental report prepared by accredited and competent experts or verified by accredited and technically competent experts and/or committees of national experts; qualification determination by Member States (+) PPD—requirement that if it will have adverse environmental effects must consider with authorities whether report should be modified, and need for additional mitigation/ compensation (+) PPD—Annex IV provisions—description of likely significant effects should cover direct and indirect, secondary, cumulative, transboundary, short, medium, long-term, permanent and temporary, positive and negative effects; measures to prevent, reduce, and where possible, offset significant adverse effects and proposed monitoring measures, including postproject analysis of adverse environmental effects (+) SEA—required to assess likely significant adverse effects and reasonable alternatives; criteria for determining significance provided (Annex II)	(+) Several requirements to accommodate the traditions, needs, and knowledge of indigenous people; reference made to promoting a cooperative approach with aboriginal people (+) Stresses that native title rights will not be affected (+) Addresses the rights (including principles of ecologically sustainable development and the rights of future generations) and duties of various parties (+) Explicitly defines matters of national environmental significance; detailed requirements and guidance are provided for each matter of national environmental significance (Australian Government, 2009a) (+) Policy statement—significant impact guidelines (+) Act lists principles of ecologically sustainable development including explicit reference to intergenerational equity (+) Recent review of the Australian legislation recommended that the Australian government, in consultation with the environmental and planning consulting industry, develop an industry code of conduct for IA consultants and audit (Australian Government, 2011d) (±) Government did not see code of conduct as necessary; cross-referenced Certified Environmental Practitioner Program (Environmental Institute of Australia and New Zealand—a professional body of environmental

(*continued*)

Table 10.4 *(Continued)*

United States	Canada	Europe	Australia
(+) Desirability is stressed for IA training and the accreditation of environmental professionals, through the National Association of Environmental Professionals (NAEP) (Eccleston, 2008) (+) Guidelines for ensuring and maximizing the quality, objectivity, utility, and integrity of information disseminated by federal agencies (+) Final guidance clarifying appropriateness of findings of no significant impact	substitution/equivalency provisions, discretionary scoping provisions, the assumption that small projects will have no significant impacts, a selective definition of the effects (within areas of federal jurisdiction only), and the discretion to approve projects, even with significant adverse effects, if justified	(+) European Commission has instituted requirements to facilitate public access to information, public participation in decision making, and public access to justice (+) Guidance documents address the need to take into account fundamental rights and to consider the implications of rulings of the European Court of Justice (EC, 2010, 2011a; Stec, 2003; UNECE, 1998) (+) Some individual jurisdictions in Europe (e.g., the Greater London Authority) have instituted equality impact assessment requirements aimed at preventing discrimination against vulnerable and disadvantaged populations	practitioners); agreed with auditing recommendations—will develop guidelines for the publication of auditing reports (+) Review suggested that consideration be given to how the code is to be enforced and that the government assume an auditing role; government agreed to develop guidelines and to publish auditing reports (+) Environmental offsets draft policy and discussion paper (Australian Government, 2007b,c)

Sources: Stec (2003), UNECE (1998).

of general ethical "ground rules" at the regulatory level could expedite proposal-specific discussions and negotiations. Also, consistency with natural justice standards and principles should be a basic requirement. However, the parties also must have sufficient latitude to come to agreements and accommodations, which best match local circumstances, and are consistent with the needs and aspirations of the participants. The auditing of proposal-specific experiences in treating ethical concerns could help identify recurrent issues where direction and guidance from the regulatory level did or could facilitate the IA process.

10.5.2 Management at the Applied Level

Figure 10.2 is an example of an ethical IA process. Figure 10.2 and the process description that follows incorporate many ethical IA elements. IA process managers and participants can "pick and choose" the relevant and appropriate elements.

Start-Up The process begins with an overall study design. The study design incorporates a preliminary public and agency consultation plan. This step ensures that the IA process is structured and focused. Consideration is given to redressing past grievances and injustices (a historical equity issue).

A concerted effort is made to identify ethical issues and conflicting ethical perspectives and positions. The analysis is based on both secondary source reviews and discussions with interested and affected parties. An initial overview of applied ethical literature is undertaken to identify pertinent concepts, theories, and distinctions. These analyses contribute to study design refinements and to scoping the IA process.

Ethical Foundation Once the start-up activities are completed, the emphasis shifts to identifying procedural fairness principles (to guide and structure interactions with stakeholders) and distributional fairness principles (to guide the analysis of distributional effects). The procedural fairness principles address such concerns as timely and complete access to information, the fair treatment of participants (including assistance to disadvantaged groups), the right to fully participate in planning and decision making, the removal of participation barriers, and access to an open, fair, impartial, and independent review process. The distribution fairness principles concern such matters as undertaking a distributional analysis (with a particular emphasis on adverse effects on and benefits to minority, low income, indigenous and other susceptible populations), assessing the fairness of cumulative hazards (including the consideration of social and ecological carrying capacity), and instituting measures to manage equity-related impacts (mitigation, compensation, local benefits, monitoring).

Methods for determining distributional differences are formulated. The rights and duties of each major participant in the IA process are identified. The principles, methods, rights, and duties are refined and adjusted based on stakeholder discussions. Ethical research rules and professional standards are formulated for environmental and ethical specialists. These rules and standards refine and adapt professional and disciplinary codes of practice. Measures

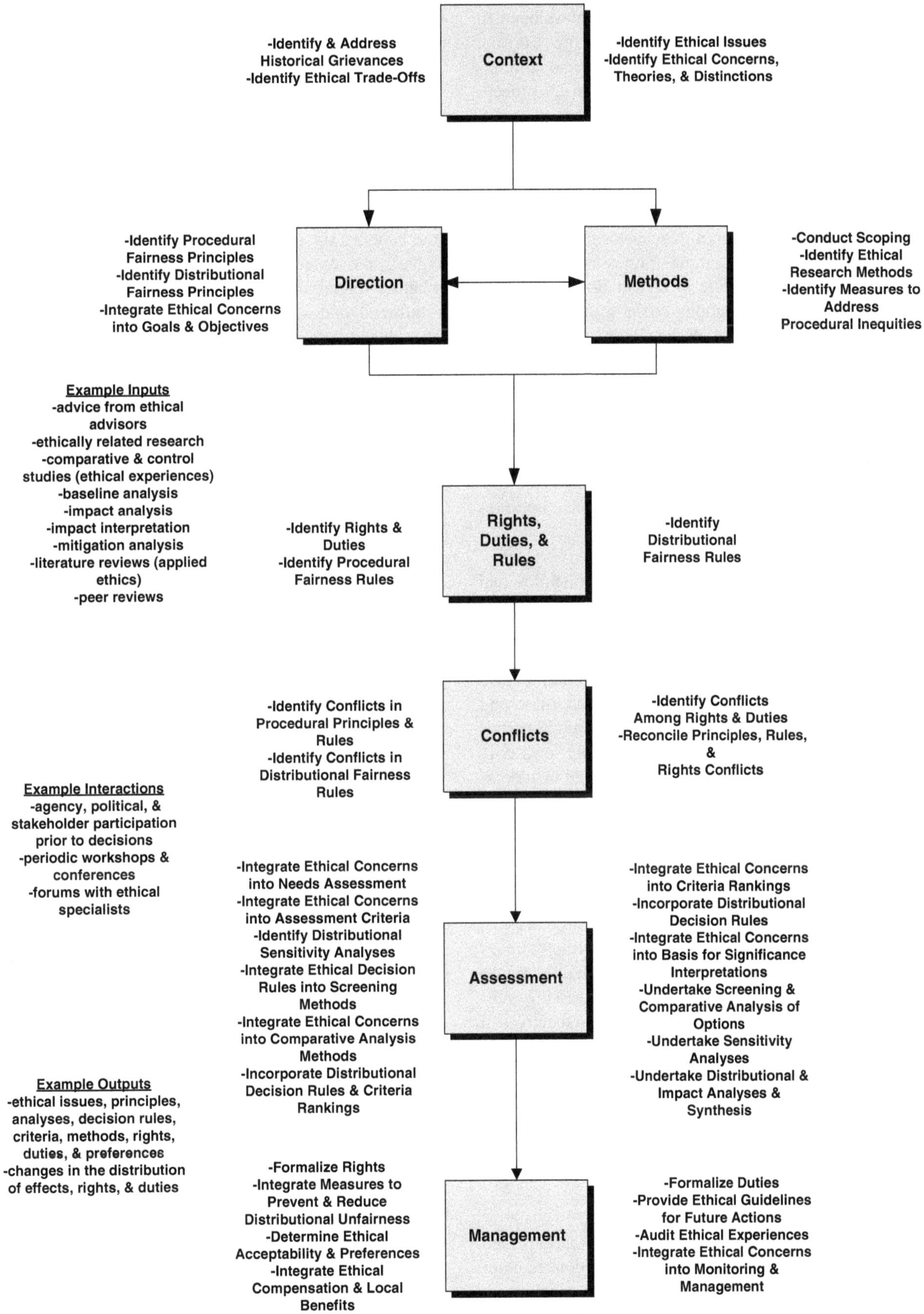

Figure 10.2 An example of an ethical IA process. Adapted from Lawrence (2005b).

(e.g., participant funding, expert advice) are developed to offset procedural inequities. These measures focus on the needs of disadvantaged groups, organizations, and communities. Ethical considerations are built into the IA process goals and objectives.

The procedural fairness principles, in combination with the measures to offset procedural inequities, lay the groundwork for procedural fairness rules. The procedural fairness rules are jointly determined with stakeholders. They ensure that the dialogue and debate minimize distortion and are fair to all participants. The distributional fairness methods determine the magnitude and nature of distributional inequities over time, over space, and among social groups. Distributional fairness decision rules link the distributional analysis to decision making. They address such matters as minimizing overall adverse impacts to society (especially to the least advantaged), maximizing benefits to society (especially to the least advantaged), and minimizing undue burdens on future generations. Ethical concerns are integrated into such impact analysis activities as needs determination, assessment criteria formulation, and impact significance interpretation factors.

Refinements Conflicts among procedural principles and rules generally occur. Rules and principles are adjusted to resolve or at least accommodate conflicts. A concerted effort is made to reach a consensus among parties regarding changes to procedural principles and rules. Conflicts also are likely among distributional principles and rules and among rights and duties. Again, a consensus among parties is sought. Residual differences are addressed by sensitivity analyses. Methods such as mediation are applied to address conflicting perspectives. A composite list of distributional sensitivity analyses is formulated. Ethical considerations are integrated into screening, option comparison, and criteria ranking methods.

Integrating ethical considerations in assessment methods includes formulating ways to explore ethical concerns, trade-offs, and dilemmas; procedures for assessing choices against ethical goals and consequences (including future generational implications); approaches for assessing individual and cumulative effects from multiple perspectives (a pluralistic approach); methods for combining or modifying options to enhance ethical benefits; measures to prevent and offset inequities and to recognize and reinforce rights; procedures for testing outcomes against varying principles, theories, and methods; and methods for testing choices against ethical regulatory policies, laws, standards, guidelines, positions, and preferences.

Application and Decision Making The options are screened using distributional decision rules. Options remaining after screening are compared taking into account the distributional analysis as well as the ethical inputs to criteria selection and rankings. The ethical distributional analysis extends and refines the impact analysis. Distributional principles and decision rules help determine the need for mitigation and the acceptability of the proposed action. Sensitivity analyses address uncertainties regarding distributional decision rules, the allocation of duties, and the likely effectiveness of ethically oriented mitigation, compensation, and local benefit measures.

Ethically preferred options are selected. Whether the proposed action is ethically acceptable is determined. Net benefits to society are determined. The distribution of costs and benefits among social groups (especially the disadvantaged), the extent to which rights are infringed upon, whether spatial and temporal inequities are ameliorated or reinforced, and whether political inequities are exacerbated or reduced are all considered. Whether the anticipated allocation of duties is appropriate and is likely to result in the adequate management of potential injustices and inequities also is taken into account.

The duties and rights associated with implementing the proposed action (if approved) are formalized and built into terms and conditions. Measures are instituted to monitor the actual distribution of effects, the effectiveness of measures to address inequities, the extent to which rights are maintained, and the extent to which duties are fulfilled. The effectiveness of the IA process in addressing ethical concerns is reviewed. Ethical guidelines are prepared for future related actions.

Inputs, Outputs, and Interactions The process is supported by advice from ethical advisors and peer reviewers and by ethically related research and literature reviews. The experiences of comparable and control communities in addressing ethical concerns are considered. Ethical analyses are combined with other planning and decision-making factors. Ethical considerations assume a pivotal role within broader planning and decision-making activities.

Agencies, elected representatives, and stakeholder groups are highly involved in the ethical IA process, consistent with procedural fairness principles. A variety of involvement approaches are applied. Ethical specialists could formulate the principles, rules, rights, and duties. Modifications could be made based on the comments received. Alternatively, interested and affected parties could take the lead, with ethical specialists providing a support function. A possible middle ground entails the proponent, ethical specialists, and stakeholders jointly integrating ethical concerns into the IA process. Forums and workshops are convened to broaden the range of ethical perspectives. Both specialists and nonspecialists participate in such forums. Depending on the location of proposed actions, it could be especially important to accommodate the ethical perspectives associated with traditional knowledge. Interpretations and conclusions are tested from multiple ethical perspectives.

There are numerous interim documentary outputs (e.g., ethical issues, principles, distribution analyses, decision rules, methods, rights, duties). A clear documentary trail is provided of how and why ethical concerns were addressed in the IA process. The ethical analyses are expected to contribute to positive changes in the distribution of effects, rights, and duties within affected communities and among affected populations (both human and nonhuman).

10.5.3 Ethical IA Practices by IA Type

Crosscutting Issues Ethical IA practices for different IA types, as summarized in Table 10.5, are all founded on ethical principles, explicitly identify rights and duties, and address issues pertaining to procedural and distributional fairness. As Table 10.5 indicates, there also are important differences in orientation and emphasis. Such differences partly reflect variations in the nature, orientation, and priorities of each IA type. There remains, however, considerable potential to jointly address equity issues for mutual learning and coordinated capacity building. Reforms to IA requirements and guidelines to facilitate more equitable IA processes also could be linked and, at least partially, integrated. The systematic and independent evaluation of the effectiveness of equity-related initiatives, from multiple perspectives, is an essential first step. Care should be taken to respect the valid differences among IA types, and the very different contexts within which ethically related IA measures and procedures have been applied.

Ethical SA Practice Ethical SA practice treats sustainability as the touchstone for each interpretation of appropriate rights, duties, and responsibilities. Procedural fairness, democratization, distributional fairness, and equality of opportunity, although important, are also viewed as means to further the cause of sustainability. Care is taken to maintain a broad and holistic perspective (e.g., intra and intergenerational equity, maximum net gains rather than minimizing adverse effects).

Ethical SEA Practice Ethical SEA practice has focused, as a matter of procedural fairness, on opening up traditionally closed policy and planning processes to broader public and community involvement, participation, and mutual learning. Increasingly, SEA practice has adopted a more substantive ethical orientation, with a greater emphasis on the distributional consequences of policy and planning options, on the promotion of the interests and perspectives of disadvantaged populations, on greater public decision-making influence, and on the role of SEA as a sustainability instrument.

Ethical EIA Practice Ethical EIA practice acknowledges the value-full, political, and distributional nature of project-level EIA. Initially, this meant ensuring procedural fairness by, for example, making decision making more open, transparent, and inclusive through public notification and public involvement opportunities, and by assessing distributional fairness through the explicit and systematic analysis of the distribution of risks and impacts over time, over space, and among population groups. Over time procedural fairness has been broadened to facilitating meaningful involvement, community empowerment, and local autonomy. Substantive fairness has evolved into a proactive effort to redress inequities, maximize local benefits, and facilitate the realization of community objectives. The formalization of rights and responsibilities, through impact and benefits agreements, represents a logical extension of this pattern.

Ethical EcIA Practice Ethical EcIA practice is founded on ethical ecological principles. Citizens are seen as having a right to be involved in environmental decision-making. Society is seen as having a duty to the nonhuman world. This duty is reflected in ethical ecological principles such as the promotion of conservation, the prevention of pollution and ecological degradation, and the securing of ecological sustainability. IA is envisioned as one among many sustainability instruments—hence the need for close interconnections with related forms of environmental management. The driving force behind ethical ecological IA practice is the ethical imperative to operate within ecological limits and to facilitate ecological sustainability.

Ethical SIA Practice SIA has gone the furthest in terms of explicitly integrating ethically based principles, perspectives, and methods into IA requirements and processes. Ethical SIA practice is founded on social ethical principles. Ensuring procedural fairness, especially for the disadvantaged and traditionally excluded populations and groups, is a priority. Community and traditional knowledge are expected to be fully considered and treated with equal respect to scientific and technical knowledge. Systematically assessing the distribution of effects and then redressing inequities, especially for the disadvantaged, is considered essential. In common with other IA types, the orientation of ethical SIA practice has moved from avoiding and minimizing the negative to community-based imperatives such as maximized local benefits, reduced dependence, enhanced quality of life, capacity building, community empowerment, the development of human potential, and the realization of community aspirations.

Ethical HIA Practice Ethical HIA practice integrates equity-related concerns into each step in the IA process. Ethical evidence standards are applied. The distribution of health-related effects is systematically assessed, with particular emphasis on implications for disadvantaged groups and populations. Health-related inequities are reduced or avoided when they are avoidable and unfair. A conservative approach is adopted for considering risks and uncertainties (e.g., application of the precautionary principle). Community-related health priorities, perspectives (e.g., perceived risks), and knowledge are fully integrated. Health-related duties and responsibilities are explicitly identified. Ethical HIA practice seeks to enhance community health, well-being, and resilience (broadly and largely locally defined),

Table 10.5 Ethical IA Practice by IA Type

Ethical SA Practice	Ethical SEA Practice	Ethical EIA Practice
Governed by procedural fairness principles (e.g., transparency, feedback, bottom-up participation, open) Emphasizes inter and intragenerational equity; seeks to manage legacy issues Presupposes equal right to most extensive system of basic liberties, equality of opportunity, and greatest benefit to least advantaged Adopts an holistic perspective Seeks maximum net gains, enhancement, and value to community Ensures fair distribution of benefits and risks Seeks to protect the future Contribution to sustainability test Treats sustainability as a higher order societal goal (analogous to democracy, equity or justice) Requires trade-off justification	Addresses inter and intragenerational equity Stresses need to give vulnerable a voice and participation at the policy and planning levels Seeks to ensure that disadvantaged groups' interests are influential and reflected in outcomes Integrates and supports poverty reduction strategies Includes distributional analysis Promotes public participation Seeks enhanced governance and development outcomes Seeks to enable agreement across different beliefs, values, roles, experiences, convictions, roles, experiences, and worldviews Seeks to stimulate constructive dialogue and produce common meanings Supports an open learning process Integrates an environmental justice perspective; SEA role in redressing injustices and achieving just outcomes	Recognizes values and value conflicts as central to process (rationality as a contested concept) Seeks meaningful involvement of interested and affected parties; emphasizes shared and decentralized decision making Provides for inclusive scoping Integrates pace and scale alternatives Seeks positive legacy and enhancements Seeks equitable distribution of risks and benefits Proactively seeks to redress procedural inequities (e.g., least advantaged) Respects indigenous values, rights, knowledge, positions, and worldviews Integrates procedural ethical principles (e.g., open, initiated early, sustained, transparent) Integrates mechanism for grievances for directly affected people Includes impact and benefits agreements; seeks to maximize local development benefits and opportunities

Ethical EcIA Practice	Ethical SIA Practice	Ethical HIA Practice
Founded on ethical principles (e.g., conservation, no net biodiversity loss, sustainable use of biodiversity resources, net conservation benefit, biodiversity conservation) Biodiversity values integral to process Seeks net benefits for biodiversity and ecological enhancement Consistent with strong sustainability Goes beyond no net loss and compensation (e.g., restored biodiversity, improved ecosystem services, increased biodiversity security, improved ecological system resilience) Aims to maximize societal benefits Based on right of citizens to be involved in environmental decision making (ecological citizenship); necessitates access to reliable information about ecological problems, their causes, and their consequences Aims to promote global and environmental justice IA seems as an instrument for environmental justice; reflects duties toward the nonhuman world (e.g., prevent pollution and ecological degradation, promote conservation, secure ecologically sustainable development, justify social and economic development)	Ethical principles at core of SIA (e.g., fundamental human rights, diversity, subsidiarity, present and future generations, health and safety, safe living and working environment, legal protection of rights, equal justice) Consistent with social performance standards Puts people first; proactively seeks to include least advantaged and historically excluded Ensures process procedurally fair Fully describes and analyzes environmental justice issues, including gender analysis Seeks to reduce disproportionate burdens of negative externalities; burden on vulnerable groups a prime concern Applies ethical principles to actions of government and proponents (e.g., polluter pay, cost internalization, prevention, precaution, multisectoral integration of social issues) Takes into account local and traditional knowledge Ensures community capacity to meaningfully participate; fosters participatory democracy and community empowerment	Seeks to be responsive to a range of population health concerns and purposes Equity concerns integrated into each IA step (e.g., screening, scoping, impact identification, assessment, recommendations, management) Assesses both effects on health of population and distribution of health effects within population (e.g., age, sex, ethnicity, socioeconomic status) Focuses on impacts (analysis and management) on excluded or vulnerable groups (e.g., age, disease, ethnicity, deprivation) HIA is gender sensitive; recognizes gender as a major health determinant Seeks to reduce health inequities from factors considered avoidable and unfair Integrates precautionary principle Promotes multisectoral responsibility for health and well being Distinguishes between voluntary and involuntary risks Seeks to foster good health and resilient communities Ensures ethical use of evidence (e.g., transparent and rigorous process, best available evidence, all evidence valued, impartial recommendations, evidence supports judgments and recommendations)

Table 10.5 *(Continued)*

Ethical EcIA Practice	Ethical SIA Practice	Ethical HIA Practice
	Maximizes positive effects and quality of life, reduced dependence, greater equity, development of human potential, empowerment, enhancement of marginalized groups and capacity building—all from community perspective Assesses social development needs, and seeks to reduce dependence and enhance, maximize, and equitably share community benefits Applies local content requirements	Clarifies management responsibilities for health mitigation and enhancement; seeks to build capacity of people to become active participants in decisions affecting community well-being

Sources: Adelle and Weiland (2012), Ahmadvand and Karami (2009), Ayre and Calloway (2005), Binder et al. (2010), Bond and Morrison-Saunders (2010), Bond et al. (2012), Booth and Skelton (2011a), Bredariol and Marini (2003), Brown (2003), Burdge (2004), Cameron et al. (2011), CIER (2009), Connelly and Richardson (2005), Dovers (2005), Égré and Senécal (2003), Esteves et al. (2012), Esteves and Vanclay (2009), Fischer (2011), Galbraith et al. (2007), Gasparatus et al., (2007), Geneletti (2002), Ghanimé et al. (2011), Gibson (2006a, 2011), Gunning et al. (2011), Harris-Roxas and Harris (2011), Harris-Roxas et al. (2012), Harris et al. (2003), Hermans and Knippenberg (2006), Herring (2009), IAIA (2005a,b, 2006a, undated a), IFC (2007, 2008), João et al. (2011), Jackson and Illsley (2007), Karjalainen and Järvikoski (2010), Kemm (2005), Kemm and Perry (2004a), Kende-Robb and Van Wicklin (2008), Khera and Kumar (2010), Kirk (2000), Kørnøv and Thissen (2000), Knaus et al. (2006); Kreig and Faber (2004); Kwiatkowski (2011), Lajoie and Bouchard (2006), Lane et al. (2003), Landsberg et al. (2011), Landry et al. (2009), Mackenbach et al. (2004), McCluskey and João (2011), Melo-Escrihuela (2008), Mindel et al. (2004), Morgan (2012), Morgan et al. (2012), Noble (2009b), O'Faircheallaigh (2009), Parry and Kemm (2004), Rajvanshi et al. (2011), Richardson (2005), Rowan and Streather (2011), Sharma (2010), Simpson et al. (2005), Smith and Schin (2004), Tamburrini et al. (2011), Tetlow and Hanusch (2012), Thérivel (2010), Treweek et al. (2011), Verloo and Roggeband (1996), Villani (2011), Vicente and Partidário (2006), Whitelaw et al. (2009), Utzinger et al. (2005), Wegner et al. (2005), Vanclay (2003), Winds and Voices Environmental Services Inc. (2000).

rather than simply attempting to avoid and minimize health-related risks and impacts (as narrowly and technically defined).

10.6 CONTEMPORARY CHALLENGE—SIGNIFICANCE DETERMINATION

10.6.1 Definition and Rationale

Definition Significance determination is a vital but often problematic IA activity for every IA type. Significance determination in IA practice makes judgments about what is important, desirable, or acceptable. It interprets degrees of importance. It focuses on relevance to decision making. It considers the interplay among impact characteristics and the characteristics of the receiving environment. It varies by context and perspective. It is structured and partially determined by institutional arrangements. It takes place at both the regulatory level and at the applied level. It applies procedures to determine impact significance. It can be defined narrowly or broadly.

Rationale IA practice can never be fully comprehensive. It is always possible to address more potential impacts, interactions, and alternatives over a wider area, for a longer time period, and to a greater level of detail. With no "stopping rule," value-laden judgments must be made and substantiated regarding what should and should not be examined, and to what level of detail. Systematic, explicit, open, and thoughtfully supported significance judgments help ensure that the value-basis for decisions is explicit. They can aid in ensuring that resources are allocated efficiently and effectively. They can help ensure that the many uncertainties associated with value judgments and the prediction of future conditions are effectively managed. They can treat comparable situations in a consistent manner. They can provide a sound technical/scientific basis for decision making. They can effectively integrate community knowledge, concerns, attitudes, values, perceptions, and preferences into decision making. They can ensure that proposed actions and IA processes and outcomes are consistent with and supportive of government policies, plans, standards, objectives, and priorities.

10.6.2 Objectives and Properties

Significance determination in IA practice, if properly undertaken, should identify and seek to achieve both procedural (how significance determinations are made) and substantive (outcomes from significance determinations) objectives or performance standards.

Procedural Performance Standards Examples of procedural significance determination performance standards include (1) focused, (2) efficient, (3) explicit and clear, (4) logical, (5) substantiated, (6) systematic and traceable, (7) appropriate, (8) consistent, (9) open, (10) inclusive, (11) collective, (12) collaborative, (13) effective, (14) adaptable, and (15) combinations.

Substantive Performance Standards Examples of substantive significance determination performance standards include (1) regulatory compliance and policy consistency, (2) the avoidance and reduction of potentially significant negative impacts, (3) the avoidance and reduction of all negative impacts, (3) the reduction of all adverse impacts considered potentially significant, as defined by significance thresholds, to acceptable levels, (4) net positive impacts (benefits outweigh negative impacts), (5) the public interest (public a net beneficiary), (6) the greatest good for the greatest number (utilitarianism), (7) the greatest good for the least advantaged (distributional equity), (8) local and regional benefits exceed adverse local and regional impacts, risks, and costs (local and regional communities and environment net beneficiaries), (9) issue resolution or management (major points of contention resolved or ameliorated to acceptable levels), (10) consensus among major parties (major parties or stakeholders can reach an accommodation on major points of disagreement), (11) net benefits to the environment, (12) sustainability (contributes to rather than inhibits sustainability), and (13) combinations.

Properties There are several inherent properties associated with impact significance judgments in IA practice. Each property has implications for how significance determination procedures can and should be conducted. Significance determinations, for example, are subjective, normative, and value dependent. They are imprecise. They vary among IA activities and for different types of effects and environments. They are context dependent. They are political and often controversial. They are not the same as the magnitude of change. They involve a process. They are collective. They are complex and difficult.

10.6.3 Significance Determination Processes, Approaches, and Roles at the Regulatory and Applied Levels

General Characteristics Impact significance determination processes vary considerably, depending on the approach and methods selected. However, four general characteristics commonly exhibited in most significance determination procedures include staged, iterative, internal and external involvement, and internal and external support.

Applications at the Regulatory Level Significance determinations occur at the regulatory level in determining triggers for the process; the types of IA requirements, scoping requirements; and the impact significance objectives, principles, thresholds, and criteria applied in legislation, regulations, guidelines (e.g., criteria, scaling levels, process, methods, sources), decision making (e.g., acceptance, rejection, conditions), and judicial review (e.g., legal interpretations).

Applications at the Applied Level Significance determinations occur as part of each IA activity. They take place, for example, during scoping (e.g., focusing on what is important); baseline analysis (valued ecological and socioeconomic components); alternatives analysis (e.g., reasonable, acceptable and preferred choices); the characteristics of proposed actions (e.g., most likely to induce significant impacts); impact prediction (e.g., the choice of impacts, boundaries, methods, criteria and criteria levels, cumulative effects thresholds); impact interpretation (e.g., impact acceptability, impact importance); consultation (e.g., major issues and stakeholders); impact management (e.g., when warranted and if effective); documentation (e.g., rationale for interpretations, assumptions, conclusions, and recommendations); and decision-making (e.g., basis for decisions, proposal acceptability, conditions).

Significance Determination Approaches As highlighted in Figure 10.3, technical, collaborative, and reasoned argumentation are three general significance determination approaches. Under the *technical approach*, significance is broken down into constituent parts and then combined using technical methods. This approach aims to provide a sound technical and scientific decision-making basis. It emphasizes consistency, transparency, and the ability to replicate. It employs thresholds (e.g., legal, environmental, impact), criteria, scaling levels, and decision rules. It relies on expert data, analysis, and knowledge. It uses qualitative and quantitative procedures (often favors the latter). It tends to be expert-centered with agency and public input.

With the *collaborative approach*, interested and affected people jointly determine what is important, why, and to what degree. This approach is undertaken in interactive forums. It involves close connections to broader constituencies. It substantiates interpretations and conclusions through joint reasoning. It is supported by technical analysis and by facilitators. It stresses bottom-up and inside-out decision making. It emphasizes communications, mutual learning, and negotiations. It entails numerous forums and methods. It provides a central role for the public, politicians, and local and regional perspectives, and a support role for technical and scientific analysis.

The *reasoned argumentation approach* has a long legal and academic tradition. It involves reasoned judgments supported by technical and nontechnical knowledge and evidence. It is usually qualitative (with some quantitative support). It integrates technical/community, facts/values, objective/subjective, multiple perspectives, and qualitative/quantitative analyses. It incorporates oral and written arguments. The effectiveness of this approach depends on how the analysis is structured (e.g., choices, disciplines, impact types, study areas, time horizons, project characteristics), the

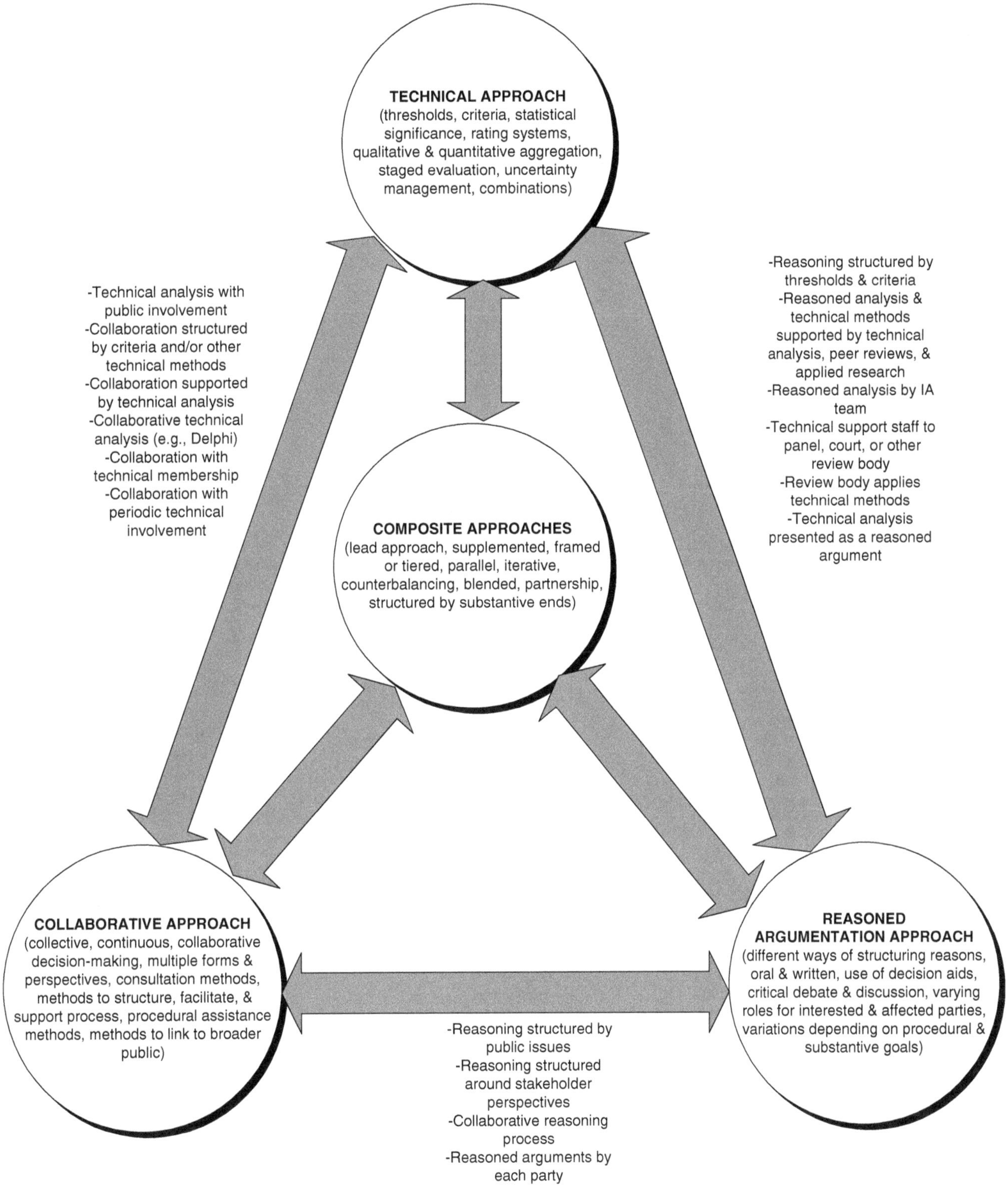

Figure 10.3 Composite significance determination approaches. Adapted from Lawrence (2007b).

adequacy and soundness of the knowledge base, and full and active stakeholder participation. This approach is most evident in summary IA documents and in judicial and quasi-judicial decisions.

Composite and general approaches, which combine elements of the three approaches, can reinforce the positive and offset the negative tendencies of each approach. They can vary approaches to suit contextual variations. Depending on how they are designed or applied, they can be more or less than the "sum of the parts." In addition to these three approaches (together with variations and combination), significance determination procedures can draw upon numerous support methods and a wealth of good general impact significance practices.

As highlighted in Figure 10.4, there are *significance determination roles* for IA specialists, technical and scientific specialists, decision makers, government agencies, and the public. Many significance determination roles involve two or more parties. Coordination among significance determination participants, therefore, is crucial.

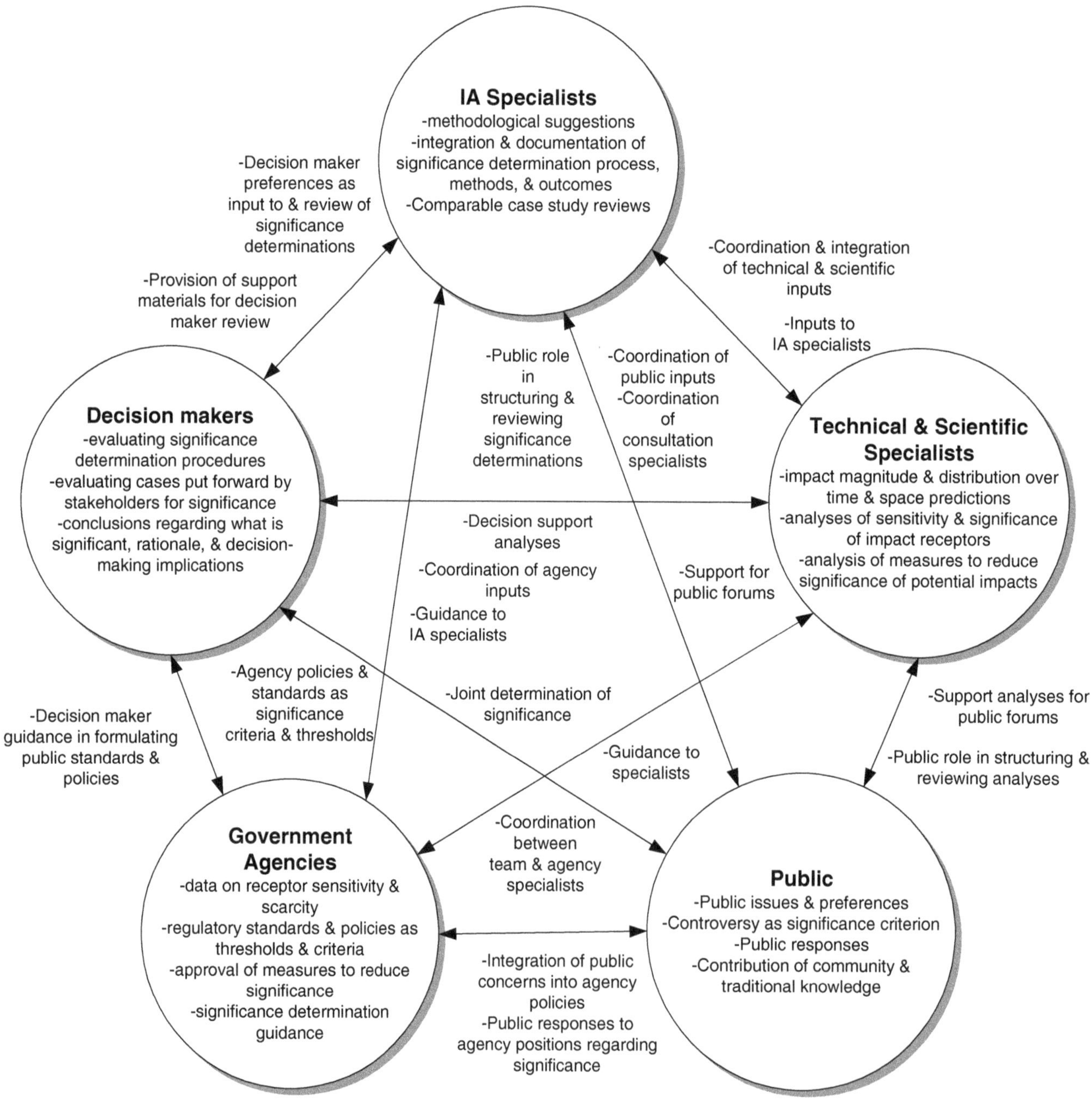

Figure 10.4 Examples of IA significance determination roles. Adapted from Lawrence (2007c).

Table 10.6 Significance Determination—Examples of Good Practices

Technical Approach	Collaborative Approach	Reasoned Argumentation Approach	General Practices	Selective Lessons
Integrate incrementally, explicitly, and consistently all decision-making factors Ensure methods clearly defined, appropriate to context, fully substantiated, easily applicable, and sensitive to discontinuities Adjust for differences in impact types (e.g., social), reliability of data and varying public and agency perspectives Allow for degrees of significance Apply to multiple spatial scales and time horizons Use ecological and social/economic evaluations as aid for determining significance of area (e.g., VECs) Consider implications of uncertainties and consequences of being wrong (e.g., precautionary, sensitivity analyses) Reconsider after mitigation, as part of cumulative effects analysis and in terms of extent supports or undermines sustainability	Focus on local and regional issues, tradeoffs and aspirations Seek to involve most directly affected and most vulnerable; offset procedural inequities Ensure membership in interactive forums representative Ensure effective links to broader publics Integrate community, technical, and traditional knowledge View significance from multiple perspectives Ensure open, transparent and inclusive process Fully document rationale for interpretations	Tie into procedural and substantive objectives Systematically integrate all forms of knowledge and data Incorporate relevant distinctions (e.g., choices, perspectives, study areas) Adapt to context View significance from multiple perspectives Guard against advocacy and bias Ensure rationale for significance determinations explicit, systematic and consistent	Recognize that inherently subjective and central to decision-making (what is important) Focus on community and regional issues, values, tradeoffs, and valued ecological and socio-economic components Ensure basis for judgments (e. g. criteria, methods) understandable, consistent, appropriate and fully substantiated; critical importance of how communicated to stakeholders Make effective use of and fully substantiate choice of support methods (e.g., qualitative and quantitative analysis and evaluation, public consultation and communications methods, group interaction methods and procedures)	Use of thresholds and criteria (properly applied) can enhance decision-making (consistent, explicit, informed) Sometimes helpful to define significance thresholds, explicit criteria, decision rules, and related guidance at regulatory level Fertile middle ground between standardized and case-by-case interpretations (e.g., classes of procedures for classes of situations with context-specific adaptations) Some types of impacts tend to be more significant across context types (e.g., health, displacement, impact triggers, community-level impacts, impacts that affect capacity and sustainability, impacts that increase vulnerability and that exacerbate inequities) Sometimes appropriate to acknowledge significant from outset and to focus on acceptability, alternatives, and mitigation Focusing on key issues, questions, themes and linkages can help scope significance determinations Technical approach tends to be more effective as support tool for a collaborative, culminating in a reasoned argumentation, approach

(continued)

Table 10.6 (*Continued*)

Technical Approach	Collaborative Approach	Reasoned Argumentation Approach	General Practices	Selective Lessons
			Apply methods to establish local and regional perspectives, values, and aspirations, identify distributional inequities, assess comparable actions and environments, characterize values, explore uncertainties and understand systems interactions Define broadly (e.g., positive, negative, direct, indirect, cumulative, biophysical, socioeconomic) Ensure procedures are consistent and explicit Ensure the direct, early, and ongoing involvement of interested and affected parties Offset procedural inequities Ensure that significance determination culminates in decisions regarding impact and action acceptability	Systematic consideration of uncertainty in significance determinations can be critical Interpretations aided when placed in context of broader socioeconomic, ecological, and sustainability initiatives, regional issues, planning and management, international standards, conventions and guidelines, corporate social and ecological policies, legal requirements and agreements, and research frameworks Traditional knowledge can make a valuable contribution to significance from a community perspective Importance of exploring tradeoffs among study areas in impact significance interpretations Support methods (e.g., sensitivity/vulnerability maps) can be helpful Focus on conflicting perspectives among stakeholders regarding which impacts are more significant and why Systematically consider nature of differences between public perceptions and technical and scientific interpretations regarding impact magnitude, likelihood, and significance Make the proponent responsible for demonstrating insignificance (shift the burden of proof) Adapt approach to problem and context Systematically integrate significance interpretations into impact management

Makes it possible to compare positive and negative effects to determine if in public interest (i.e., net benefits) Makes it possible to address inequities in the distribution of benefits and adverse effects over time and space (e.g., locally concentrated adverse effects and dispersed benefits, leakage of benefits, "boom and bust") Makes it possible to better address unacceptable impacts regardless of benefits Makes it possible for impact management measures to focus on vulnerability concerns (e.g., through community benefits agreements) Actions can be better designed and managed to further community and regional goals and aspirations Sets up significance–sustainability links	Impacts from announcement and planning Behavior altered in anticipation of impacts; process and public perceptions alter impacts Significance varies among groups, communities and sectors Socially determined meaning; dialogue central Affected by SIA and social science limits Assess significance from community/local perspective Recognize that IA subjective and value-full Importance of defining social groups and community resources; community well-being and vulnerability key Public reasoned judgment and collaboration, with technical support, generally favored; filtered through multiple values, beliefs, and perspectives	More holistic and inter-disciplinary methods More emphasis on goal setting and impact management, SEA, partnerships, and sustainability strategies Enhanced role for uncertainty and uncertainty management Collaboration among parties especially important Helps integrate project-level EIA, SEA, CEA, and SA	Uncertainty an explicit factor and/or can elevate a significance rating More tentative and cautious interpretations Uncertainty potentially a basis for action rejection or management requirements Greater weight to uncertainty and harm avoidance; can trigger need for mitigation and monitoring; shifts burden of proof to proponents to demonstrate safe and harm avoidance More stress on population vulnerability, harm avoidance and reduction and uncertainty management Significance determinations tend to be more open, adaptive, iterative, transparent, and democratic	Alternatives screened and compared for sustainability Shift from minimizing damage to maximizing gains and opportunities (e.g., proposed action as catalyst or impediment or as means for advancing community aspirations) Proposed actions assessed against likely and desired futures; net contribution to sustainability Use of ecological and sustainability thresholds and criteria (e.g., undermines or contributes substantially to socio-ecological integrity, contributes to serious or irreversible damage to sustainability foundations, contributes positively in a mutually supportive way to several or all aspects of sustainability) Link dependence of proposed actions on ecological services and biodiversity implications More emphasis on human and resource resilience, the most vulnerable and links to broader sustainability initiatives; maximizes long-term gains and opportunities for multiple parties

Sources: Azcona and Palmada (2011), Baxter et al. (2001), Cloquell-Ballester et al. (2007), Geneletti (2002), Gibson et al. (2005), IEEM (2006), Ijäs et al. (2010), Khera and Kumar (2010), Kirk (2000), Lawrence (2004, 2007a,b,c), Landsberg et al. (2011), Noble (2009b), Ross et al. (2006), Rowan (2009), Söderman and Saarela (2010), Tomlinson (2004), Wale and Yalew (2010), Wegner et al. (2005), Wilkins (2003), Winds and Voices Environmental Services Inc. (2000), Wood (2008).

10.6.4 Good Practices, Selective Lessons, and Future Directions

Objectives for significance determination procedures should be determined early in the IA process, should be broadly defined, should encompass both procedural and substantive aspects, and should be clearly and consistently applied. Care should be taken to avoid unacceptable significance determination approaches. Examples of such approaches include determining significance without substantiation (e.g., professional judgment only), demonstrable bias, serious factual inaccuracies as the basis for significance determination, major factors relevant to significance determination not considered, failure to consider perspectives of major parties, and an approach clearly inconsistent with decision-making requirements (e.g., regulatory requirements).

Significance determination good practices vary by significance determination approach (see Table 10.6). General significance determination good practices and selective good practice lessons are summarized in Table 10.6. Examples of new directions in significance determination include the interpretation of positive effects, the interpretation of social and economic effects, the interpretation of cumulative effects, links between significance determination and the Precautionary Principle, and significance interpretation for sustainability. Examples of significance determination good practices for each of these new directions are summarized in Table 10.6.

10.7 SUMMING UP

This chapter responds to the need to make IA processes and outcomes more fair, equitable, and just. It is concerned with identifying and advancing rights and ensuring that duties are fulfilled. It describes, with three stories, the ethical nature of IA. It provides the conceptual underpinning for an ethical IA process. It describes an ethical IA process as it might be applied at the regulatory and applied levels and allowing for variations by IA types. It addresses the contemporary challenge of significance determination.

Establishing a foundation for an ethical IA process begins with three stories. The first story is concerned with IA professional ethical issues and dilemmas. The remaining two stories address the contemporary challenge of significance determination in two different ways. One is concerned with good practice approaches to interpreting the significance of social impacts. The other addresses the issue of the propensity to assume that smaller projects are, by definition, insignificant.

An overview of major ethically related shortcomings in IA practice is presented. The critics emphasize that insufficient attention is being devoted to the fairness of the IA process; to the distributional consequences of proposals subject to IA requirements; to the rights of participants in the process; and to the duties of proponents, regulators, and other process participants.

The major concerns raised by the critics (fairness, equity, justice, rights, and duties) are all concerned with the moral rules, principles, and standards that govern human conduct (i.e., ethics or more specifically in this case, normative, applied, practical ethics). These terms are each defined. Interconnections are highlighted. Several key ethical concepts are briefly described, together with implications for IA process management. The concepts largely concern situations in which ethics might be applied and alternative ethical standards for judging behavior.

An overview of procedural fairness is presented. Procedural fairness is concerned with both how consultation takes place and how decisions are made. It includes principles and rules pertaining to the rights and duties of process participants. Several examples of procedural fairness rights and duties are presented. Examples of distributional fairness distinctions, principles, and duties are identified. Distributional fairness pertains to the distribution of risks, costs, and benefits over space, over time, and among social groups. It can refer to the allocation of services and resources and to impacts on individual liberties and on local decision-making powers. It considers the fairness of cumulative effects and the relationships to social, economic, and ecological carrying capacity and to vulnerability to change. A description is provided of possible rights and duties. Rights express and give legal meaning to values. Duties entail obligations. Rights and duties can be expressed and applied through the IA process. There are various types of rights and duties. They apply to different population segments and they concern a range of subjects. Some rights and duties are established through regulatory requirements. Others are determined, often through discussions and negotiations, during individual IA processes. The role of professional ethics in IA is described. Examples of ethical dilemmas and potential solutions are provided. Good practice advice is offered. The issue of the accreditation of IA professionals is explored.

IA regulatory requirements in the four jurisdictions identify some proponent and government review duties. They identify minimum public notification and involvement rights. More consideration is being given to measures to facilitate the involvement of disadvantaged groups and to accommodate indigenous rights, knowledge, culture, and traditional activities. Varying approaches are being taken regarding such matters as environmental justice requirements and guidance, access to information, measures to enhance procedural fairness, the provision of local benefits, the treatment of intergenerational equity, and the accreditation of IA professionals. More could be done to address ethical concerns at the regulatory level. The effectiveness of each measure should be fully evaluated. Natural justice standards should be adhered to. A balance should be maintained between greater structure and guidance and the need to make proposal and setting specific adaptations and refinements.

An example of an ethical IA process is described. The process begins by considering historical grievances and by identifying ethical issues and trade-offs. Relevant literature and experiences are canvassed. These overview analyses provide the basis for identifying procedural and distributional fairness principles and methods. Rights and duties, ethical research methods, and measures to address procedural inequities are addressed. Ethical concerns are integrated into IA process goals and objectives.

Procedural and distributional fairness principles are refined into decision rules. Ethical concerns are integrated into significance determination factors, assessment criteria, and the needs analysis. Conflicts among procedural principles and rules, among distributional principles and rules, and among rights and duties are identified. The conflicts are resolved or accommodated to the extent practical. Residual conflicts are addressed through sensitivity analyses. Ethical concerns are integrated into screening methods, comparative analysis methods, and criteria ranking.

Distributional analysis, distributional decision rules, and criteria rankings are incorporated, where applicable, into the screening of options, the comparison of options and the impact analysis. Ethical uncertainties are addressed by sensitivity analyses. Ethically acceptable and preferred options are selected. Ethical concerns are built into mitigation, compensation, and local benefit measures. Rights and duties are formalized. Ethical concerns are incorporated into monitoring and management. Ethical guidelines are prepared for future actions. The IA experience in addressing ethical concerns is audited.

The IA process is supported by advice from ethical advisors, applied research, and reviews of comparable situations. Ethical concerns are integrated with other planning and decision-making activities. Agencies, elected representatives, and stakeholders participate in identifying and applying ethical concerns. Efforts are made to broaden the basis of involvement and to ensure that multiple ethical perspectives test interpretations and conclusions. The role of ethical concerns in the IA process is fully documented. Lessons and insights regarding the treatment of ethical concerns, as expressed in the literature and practice of various IA types, are considered and adapted.

An ethical IA process systematically addresses the contemporary challenge of significance determination. Procedural and substantive significance determinations are explicitly defined and justified. Unacceptable significance determination approaches are avoided. The significance determination approach adopted is transparent, inclusive, adaptable, and appropriate to the context. Significance determination roles and responsibilities are clearly defined and justified. The significance determination approach is broadly defined, encompassing such matters as positive effects, social and economic effects, cumulative effects, uncertainty management, and sustainability. The significance determination process and methods are consistent with good practice standards.

Chapter 11

How to Make IAs More Adaptable

11.1 HIGHLIGHTS

In this chapter, we address how IA processes can adaptively anticipate and respond to the uncertainties associated with difficult problems in chaotic and complex environments. It is commonplace in IA literature and practice to emphasize that the IA process should be adaptive, flexible, and iterative. Specific means for accomplishing this aim are less evident. The major approaches advanced for managing uncertainties are controversial and only partially or indirectly connected to IA process management. This chapter provides a systematic, integrated approach to managing uncertainties in the IA process.

- The analysis begins with three applied anecdotes (Section 11.2). The stories describe applied experiences associated with efforts to make IA practice more adaptive.
- The analysis in Section 11.3 then defines the problem. The problem is a failure to adequately characterize and manage uncertainties in the IA process. In this section, we explain why it is necessary to formulate IA processes that adaptively manage uncertainties.
- In Section 11.4 we explore the potential roles in IA of difficult problems and complex and chaotic environments, uncertainty and adaptation, the analysis and management of risks, the precautionary principle, and human health.
- In Section 11.5 we detail how an adaptive IA process could be implemented at the regulatory and applied levels. In Section 11.5.1 we explain how regulatory requirements and guidelines can facilitate uncertainty management and adaptation. In Section 11.5.2 we describe how the uncertainty concepts, strategies, tactics, and approaches can be linked and combined in practice. In Section 11.5.3 we describe the characteristics of an adaptive IA process for different IA types.
- In Section 11.6 we address the contemporary challenge of climate change. We define key terms, draw major distinctions, describe a climate change impact assessment process, and summarize good institutional arrangements/capacity building and regulatory and applied level practices.
- In Section 11.7 we highlight the major insights and lessons derived from the analysis.

11.2 INSIGHTS FROM PRACTICE

11.2.1 Adapting Strategic CEA to the Needs of Institutional Partners

Initiated in 2001, the Transboundary Crown of the Continent Manager's Partnership (CMP) was formed to address cross-boundary cumulative environmental effects within the Crown of the Continent ecosystem, a shared region of the Rocky Mountains between Alberta and British Columbia, Canada, and Montana, United States of America. More than 20 government agencies responsible for land management in the region joined forces in a strategic partnership facilitated by the Mistakiis Institute at the University of Calgary. By pooling knowledge and resources, land use managers on both sides of the Canada–U.S. border planned to formulate strategic actions and partnerships to proactively influence developmental trends in their respective jurisdictions. The partnership was also motivated by a common need for strategic, landscape-level information to provide context for the assessment of individual project proposals.

The methodological approach to the strategic cumulative effects assessment initiative was straightforward: following the establishment of a collective regional vision for the future of the ecosystem, the partners contributed baseline and trend data to an ALCES (A Landscape Cumulative Effects Simulator) modeling initiative. ALCES is a "stock" and "flow" simulator of ecosystem dynamics that can assist in understanding how overlapping land uses and natural disturbance regimes can combine to alter terrestrial and aquatic landscapes over time. Using ALCES, the likely environmental effects of various regional development scenarios in the Crown of the Continent ecosystem were analyzed. Attempts to continuously improve the modeling exercise were made through ongoing regional collaboration.

This modeling-intensive approach to cumulative effects assessment was ultimately fraught with unanticipated

Impact Assessment: Practical Solutions to Recurrent Problems and Contemporary Challenges, Second Edition. David P. Lawrence.

roadblocks. When ecosystem-scale results were eventually produced, it proved difficult to translate them into field-level operational actions in the various jurisdictions. Further, the CMP found progress toward strategic goals hard to track due to the number of "puzzle pieces" involved. Unexpectedly, rapidly shifting regional baseline conditions, both environmentally and politically, rendered modeling results obsolete almost as fast as they were produced; perhaps most surprisingly, a curious form of political pushback occurred when the modeling exercise delivered results that were perceived as "pointing fingers" at certain interests and were ultimately unpalatable. This forced the CMP to abandon the original "effects prediction" approach in favor of a strategic initiative to establish regional environmental targets and thresholds for change. The switch ensured that partners retained a sense of autonomy in determining how to meet ecosystem management goals.

More than 10 years on, the CMP has shown remarkable adaptive propensity and persistence in the face of early challenges. The recently released 2011 Strategic Plan identifies five additional strategic initiatives including documenting ecological health trends and developing institutional capacity. This story demonstrates that significant thought must go into making impact assessments at the strategic level, not just practical, but adaptive. Strategic assessment is not simply assessment outside or "upstream" of a project context, as the CMP initiative was, but an assessment that adopts a strategic mindset with regard to institutional and political realities, which it eventually did. In this case, significant adaptations to the strategic assessment exercise were necessary so that, in effect, it was sensitized to the ability and willingness of partners to respond, not overreaching what was possible to do, given the current state of ecosystem management and institutional development.

JILL A.E. GUNN

Department of Geography and Planning and School of Environment and Sustainability, University of Saskatchewan, Saskatoon, SK, Canada

11.2.2 A Technically Driven Impact Assessment in an Evolving Social Environment

Environmental assessment of a new greenfield mining and industrial project in the small locality of Anitápolis, Southern Brazil is a story that spans over two decades, starting in 1989, when a first EIS was prepared. This initial EIA process was very conventional, as the EIS itself, and led to the project approval by the State environmental authorities. However, due to a U-turn in economic policy that made imported produce (fertilizer) much cheaper, the proponent didn't go ahead with construction. The project remained on the shelf for 15 years, when it was revived by a new proponent. In fact, the ownership of the parent company changed completely and the new management had no memory of the previous EIA process. More than that, they acknowledged that the most consistent information about the project was documented in the EIS, not in the company's files. The new proponent was a joint venture between two multinational corporations, each partner running their own operations in Brazil and abroad.

A project team was established and commissioned new engineering and environmental studies. A decision was made to hire the same environmental consultancy that had prepared the first EIS, largely due to the fact that individuals working for the firm held part of the project's memory.

The second EIS was prepared between 2005 and 2006, in parallel with engineering design. Environmental fieldwork updated and constructed a new baseline, capturing a few important changes. While the local economic basis remained virtually unchanged, a small municipality of ~5000 inhabitants featuring declining population (due to migration to the capital city) and small-scale agriculture, in the project's area the natural vegetation kept regenerating and parts of the property featured outstanding ecological value. Another change, whose implications were not adequately captured by the EIA team, was a slow shift toward a conversion of small rural properties into secondary or hobby farms and development of land into secondary housing estates.

Although the proponent's project team was prepared to make changes in project design to accommodate environmental mitigation, they were not attentive to initiating public involvement early enough. On the other hand, they were very keen on preparing a solid environmental assessment document and hired an experienced third party to review successive drafts of the EIS, aiming at filing an excellent document.

Similar care was not exerted toward communications with stakeholders. Many individuals in the host municipality featured positive expectations about the revival of the project, as they did in the late 1980s. By then, a new hotel had been built by a local investor to cater to the influx of people during construction, but this initiative was frustrated. Similarly, there was a hope that many new jobs could possibly be assumed by local people who had left for bigger cities. Hence, the rebirth of the project was positively received by part of the local community.

A different perspective was assumed by a group of residents in the neighboring municipality of Rancho Queimado. Situated in the mountain range near the capital city of Florianópolis, they were benefiting from a new rurality represented by the secondary residences and hobby farms and projecting a future of increased rural and ecotourism activities boosting the local economy. They saw a mine and an industrial plant, albeit situated 30 km downstream, as incompatible with such a bright future, especially because the project would induce increased truck traffic on the roads and possible correspondent truck drivers' services.

In between was the small rural locality of Rio dos Pinheiros, where most significant impacts would be felt. Locals voiced mixed feelings about the project. Acknowledging benefits in terms of jobs and increased access to services, they also felt that a few thousand construction workers would mean potential trouble for a very small community, as well as dozens of lorries circulating during the production phase.

On the other hand, a new law protecting the remnants or particular type of rainforest natural to the area came into force. In the 15+ years since the first EIS, the forest stand that would need to be felled, regenerated to higher conservation status, which made any approval for clearance more difficult.

The staff at the State EIA agency was well aware of the challenges and performed a careful review of the EIS. In the review process, they twice asked for supplementary information to be provided, but their standpoint was similar to the consultants: to perform technically sound tasks and to comply with relevant laws and regulations.

A public hearing was called for, heavily attended by citizens from every affected community. Only in preparation for this public hearing did the proponent seriously considered engaging with the public. But the proponent didn't look at the "public" as segmented into different groups of stakeholders, each with different perspectives, being differently affected by the project, and having different influence or leverage to influence the environmental licensing decision. This proved to be a fatal flaw.

Brazilian law entrusts public prosecutors to litigate in order to protect environmental and cultural values. Many systematically follow-up every EIA in their jurisdiction and it is not uncommon that lawsuits challenge the outcomes of the environmental approvals process, not only in terms of observance of legal procedures, but also in terms of contents of EISs. The best EIS will always be imperfect, baseline can always be more extensive and detailed, modeling more sophisticated, and mitigation more detailed (and expensive).

As opponents campaigned against the project, the prosecutor pressured for a second public hearing to be held and subsequently filed a lawsuit. After considering the case, the judge ruled that the license issued by the State should be dismissed. At the time of writing (mid-2012), one of the shareholders had been sold to a bigger mining company and no new developments are known.

The main message from this story is that it describes an EIA process that was very classical and technically led, but did not pay enough attention to stakeholder involvement and did not consider the failure risks (under the proponent's point of view) derived from judicial challenges. An early and thorough stakeholder mapping process would have provided the proponent with a more accurate picture of the potential beneficiaries, as well as of groups that could perceive themselves to be adversely affect by the project. In preparing the EIS, the consultant could have addressed the differentiated distribution of impacts over different social groups, looking for more focused mitigation.

LUIS E. SÁNCHEZ

Escola Politecnica – PMI, University of São Paulo, São Paulo, Brazil

11.2.3 Understanding Policy Chaos Through the National Environmental Policy Act

The Bonneville Power Administration (BPA), a federal Power Marketing Agency for the Pacific Northwest region under the U.S. Department of Energy, needed to make a decision on how to achieve a comprehensive and consistent policy to guide the implementation and funding of its fish and wildlife mitigation and recovery efforts. After decades of spending billions of dollars on previous attempts to address the decline of some populations of fish and wildlife in the Columbia River Basin, some due to hydro generation, BPA needed a new policy strategy. The agency was spending over $100 million annually and species were still being proposed for protection listing. The major difficulties were coordinating attempts in the region where nine federal and four state agencies, over 50 Indian tribes, and numerous other pertinent interests struggled with their different and conflicting values and priorities; lack of clear and agreed-upon scientific answers; and conflicting directives and jurisdictions of regional authorities. Some form of a unifying strategy to assess a comprehensive and consistent policy fit well into the National Environmental Policy Act (NEPA) process. The Fish and Wildlife Implementation Plan Environmental Impact Statement (FWIP EIS, DOE/EIS-0312, April 2003) became the instrument to assess impacts and alternatives at the policy level, rather than the previously typical practice of proposing an uncoordinated series of program and project actions.

One of the many unique tools within the FWIP EIS was the *Policy Finder*. It represented the process for reviewing relevant regional actions by all major decision makers and creating a structured assessment of the perfunctory functioning regional policy. This process permitted BPA to break down the proposed actions in the individual regional plans and strategies and place them into a set of five BPA basic and distinctly different policy direction alternatives developed during the public scoping efforts; Natural Focus, Weak Stock Focus, Sustainable Use Focus, Strong Stock Focus, and Commerce Focus. Each basic policy alternative had a subset of the same 40 key issues that denoted the must have buy-in issues drawn from public participation of vested regional parties to be satisfied with a policy plan. By placing the proposed regional actions using this process, BPA was able to discern where the different regional plans and strategies fell within the five basic policy direction alternatives. This also highlighted how the regional entities' plans overlapped multiple policy directions and lacked fulfillment of all the key issues.

The final step of the Policy Finder process for BPA was to merge all of the assessed regional plans into a workable consolidated regional policy direction. It honed in on the heart of what policy direction or blend of policy directions were at work. BPA was then able to reasonably infer from this operational regional policy direction what missing key issues were needed to fully complete a comprehensive and consistent strategic policy. The culmination of this process was the development of BPA's own Preferred Policy Direction alternative. Not only could the Policy Finder be used to discover the working regional policy and a Preferred Policy Direction but it had the ability to, at any time in the future, to be used to recast what regional policy is at work and assist in modifying an agency's policy direction to correct for the changes. It also could be reversed and used as a desired policy direction to determine what types of proposed actions would be necessary to implement it.

The BPA Fish and Wildlife Implementation Plan Environmental Impact Statement policy finding process accentuates the ease and importance of understanding public policy and the human environment. Many federal agencies continue to face the lack of an undefined or unrecognizable policy at work in their area of influence and responsibility. The policy doesn't only have to be made up of government agencies but can also include what the private sector is promoting. It is less important that all parties agree to how their proposed actions are sorted than to have an understanding that they have been systematically distributed into key issues and the different basic policy directions drawn from a public process. Without a way of gaining this level of understanding for agency decision makers and the public concerning the current public policy atmosphere, proposed regional program and project actions go unchecked with regard to a comprehensive, consistent implementation plan. Additionally, this process leaves the agency with a more adaptable decision-making tool fueled by the ability to quickly and efficiently assess public policy and potential implementing programs and projects while understanding the human environment consequences within the context of the spirit and letter of the law for NEPA. This is truly a practical strategic environmental assessment process for informing both the agency decision makers and the public.

CHARLES C. ALTON

Alton Strategic Environmental Advisor,
New Port Richey, FL, USA

11.3 DEFINING THE PROBLEM AND DECIDING ON A DIRECTION

The three stories address uncertainty in different ways. The first story describes a strategic cumulative environmental assessment that adaptively evolved in response to unanticipated methodological challenges and the changing needs and expectations of the major partners in the process. The process involved careful planning, institutional capacity building, a practical approach to problem solving, sensitivity to political resistance and database uncertainties, and an adaptive propensity and persistence in the face of rapidly changing conditions and major challenges. The second story illustrates the risks associated with a technically driven IA process that does not appreciate or mitigate potential sources of proposal failure, does not recognize changing institutional arrangements, is insensitive to varying stakeholder perspective and interests, and is unwilling or unable to adapt to an evolving regional context. The third story describes how a proponent learned from past mistakes and built on past successes. The process adopted made it possible to rapidly keep abreast of, integrate, and adjust to changing policies and requirements; to operate effectively in a complex, multistakeholder institutional environment; to fully integrate evolving public and private sector perspectives, concerns, and preferences; and to focus on key issues, obstacles, and opportunities. The three stories underscore the need for IA processes to be open and adaptive if they are to operate effectively in complex, rapidly changing decision-making environments fraught with uncertainty.

Uncertainty is about not knowing and about not being sure (Yoe, 1996). IA practice has been faulted both for being overly deterministic (i.e., unsupportable precision) and for being overly vague (i.e., a lack of precision). In the former case, uncertainty is not acknowledged or adequately addressed (Byer et al., 2009). In the latter case, vagueness stems more from a lack of effort to be precise than from an acknowledgement of uncertainty and its implications. IA practice tends to give limited consideration to uncertainties and errors (Byer and Yeomans, 2007; Tickner, 2003b). Instead, it tends to assume that a single number can represent the range of values potentially associated with a measured or predicted parameter (Carpenter, 1995). Such thinking fails to acknowledge or account for natural variation, knowledge gaps, or indeterminacy (Power et al., 1995). IA tends to operate under the illusion that present and future conditions can always be readily and precisely measured, predicted, and controlled (i.e., optimistic bias) (Gardner, 2010; Hodgson and White, 2001). The predictive models so prevalent in IA practice, moreover, tend to diminish the disclosure of uncertainty (Duncan, 2008). Often, decision makers and other stakeholders are not made aware of uncertainties or the nature of uncertainties and their implications are poorly communicated (Byer et al., 2009; Tennøy et al., 2006).

IA rarely understands the irreducibility of risks and uncertainties and generally fails to adopt unpredictability and incomplete control assumptions, even when addressing issues such as climate change, when effective uncertainty management is crucial (Byer et al., 2011; Govender et al., 2006). IA analyses and decision making, based upon such thinking, neither acknowledge nor explicitly consider uncertainty (Lobos and Partidário, 2010; Reckhow, 1994). Equally unacceptable are vague, unsupported, qualitative statements about current or future conditions (Culhane et al., 1987; Malik and Bartlett, 1993). Such statements provide the reader with minimal insight regarding what is known (or knowable) or unknown (or not known with precision). IA predictions tend to appear more certain than they are and the processes for deriving the predictions are often not transparent (Tennøy et al., 2006). False assurances of certainty are misleading. Vague and unsupported "musings" are similarly uninformative. IA approaches that employ unconnected indicators and that fail to address interconnections across disciplines inhibit and undermine the search for innovation, and adaptive and transdisciplinary solutions (Wiek and Binder, 2005). IA effectiveness reviews demonstrate that accurate forecasts, the use of confidence limits (as a means of acknowledging uncertainties), and monitoring (as a means of testing the accuracy of forecasts and the effectiveness of mitigation measures) are still more the exception than the rule (Culhane et al., 1987; Lobos and Partidário, 2010; Sadler, 1996). As IA practice is broadened to address new and critical issues such as climate change, the need to more effectively understand and manage uncertainties becomes much more critical (see Section 11.6).

Accepting the need to address uncertainties and for adaptive IA processes is only the beginning. Good practice uncertainty management principles need to be identified (Sadler, 1996). Dubious assumptions (e.g., equating vagueness with flexibility, assuming that reducing uncertainty will increase certainty) need to be avoided (Hodgson and White, 2001). Identifying uncertainty types and sources, characterizing uncertainty concepts, and formulating, adjusting, and applying adaptation approaches, methods, and concepts all require further attention. IA literature and practice often acknowledge the uncertainties associated with difficult issues and problems and with complex and chaotic environments. Reference is frequently made to matching the IA approach to the problem and the environment. Less attention is devoted to characterizing those problems and environmental conditions most prone to uncertainty and to exploring IA management implications.

Uncertainty is commonly coupled with risk as alternatives to certainty. Arguably, risk is a subset of uncertainty (i.e., a form of uncertainty to which probabilities can be attached). Risk, however, goes further in considering potential negative implications. Risk combines probabilities with harmful outcomes—to people, to property, and to ecological systems. The treatment of risk in IA guidelines and practice is often either superficial or highly variable (Eccleston, 1999b; Malik and Bartlett, 1993; Sadler, 1996). Increased attention has been given to if and how risk and more particularly, risk assessment and management could be linked and integrated with IA, although in practice IA and risk assessment are rarely used in a complementary manner (Barrow, 1997; Canter, 1993b; Carpenter, 1995; Demidova and Cherp, 2005; Erickson, 1994; Harrop and Nixon, 1999; Ugoretz, 1993; Westman, 1985). Although the need for IA to consider risk is broadly acknowledged, the merits of elements of risk assessment and management and whether and how the two fields might best be linked, integrated, or modified have been intensely debated. The debates extend to comparisons with alternatives to risk assessment and management (e.g., performance standards, semiquantitative hazards assessment). In recent years it has broadened to encompass alternative risk, uncertainty, and health effects management approaches (e.g., human health impact assessment, the precautionary principle, adaptive environmental assessment and management, emergency planning and management)—approaches that could provide a framework for, be subsumed within, or represent an alternative to risk assessment and management.

The precautionary principle has been identified as a sustainability principle (Sadler, 1996). It is integrated into IA requirements in some jurisdictions (Australia and Europe, for example), although implications have yet to be fully determined. There are, however, numerous definitions, a host of positions concerning potential IA and decision-making roles, and a lengthy list of ascribed advantages and disadvantages. Adaptive environmental assessment and management (AEAM) blends scientific, ecological model building with adaptive, heuristic group planning and decision making. It has been widely applied in environmental and resource management, although frequently only partially and sometimes with mixed results. AEAM has been identified as an effective approach to uncertainty management in IA (US CEQ, 1997a). As with the precautionary principle, there is an intense debate surrounding AEAM. Its potential IA practice roles and attributed strengths and deficiencies are often overstated.

Harm, in the sense of human health effects, is a component of risk. Although human health risk is invariably a major public concern, human health effects are often not or are only superficially, partially, inconsistently, and inadequately addressed at both the SEA and project EIA levels (Arquiaga et al., 1994; BMA, 1998; Bond et al., 2011; Canter, 1990; Davies and Sadler, 1997; Dora, 2004; Harris and Spickett, 2011; Noble and Bronson, 2006; Ortolano and Shepherd, 1995; Steinemann, 2000). IA guidelines tend to devote limited attention to health concerns or define health issues very narrowly (Sadler, 1996). The propensity to equate health-related concerns with environmental standards ignores the health implications of substances and processes not covered by standards (Arquiaga et al., 1994). HIA is rarely applied at the policy level (McCaig, 2005). The treatment of the nature, significance, and distribution of health effects, at both the EIA and SEA levels, generally falls well short of good practice standards (Kørnøv, 2009). When it is applied, the tendency is to focus on positive impacts and to ignore or superficially address negative impacts (Kauppinen and Nelimarkka, 2004). A major response to these types of deficiencies has been health impact assessment (HIA). Several jurisdictions have issued HIA guidelines in recent years (Enhealth Council, 2001b; IPHI, 2001; Health Canada, 2004; WHO, 1987, WHOROE, 2001a,b). The effectiveness of these guidelines is yet to be determined. HIA institutional relationships, the relationship between HIA and other forms of impact assessment, and the relationships between HIA and other risk and uncertainty management approaches require additional attention. HIA practice needs to more proactively influence the choice of preferred options rather than just mitigate the adverse effects of predetermined proposals (Fischer et al., 2010).

The problem then is a combination of confusion regarding the nature of uncertainty and the related concepts of risk and health effects and ambivalence concerning the most appropriate approach (or combination of approaches) for managing uncertainties in the IA process. The direction, as illustrated in Figure 11.1, is: (1) to classify the types of problems commonly associated with high levels of uncertainty; (2) to identify the relevant properties of chaotic and complex environments and systems; (3) to describe uncertainty sources, types, and concepts; (4) to provide an overview of general adaptation strategies and tactics; (5) to explore the role of uncertainty in adaptive IA processes; (6) to describe adaptive environmental assessment and management and potential IA process links; (7) to

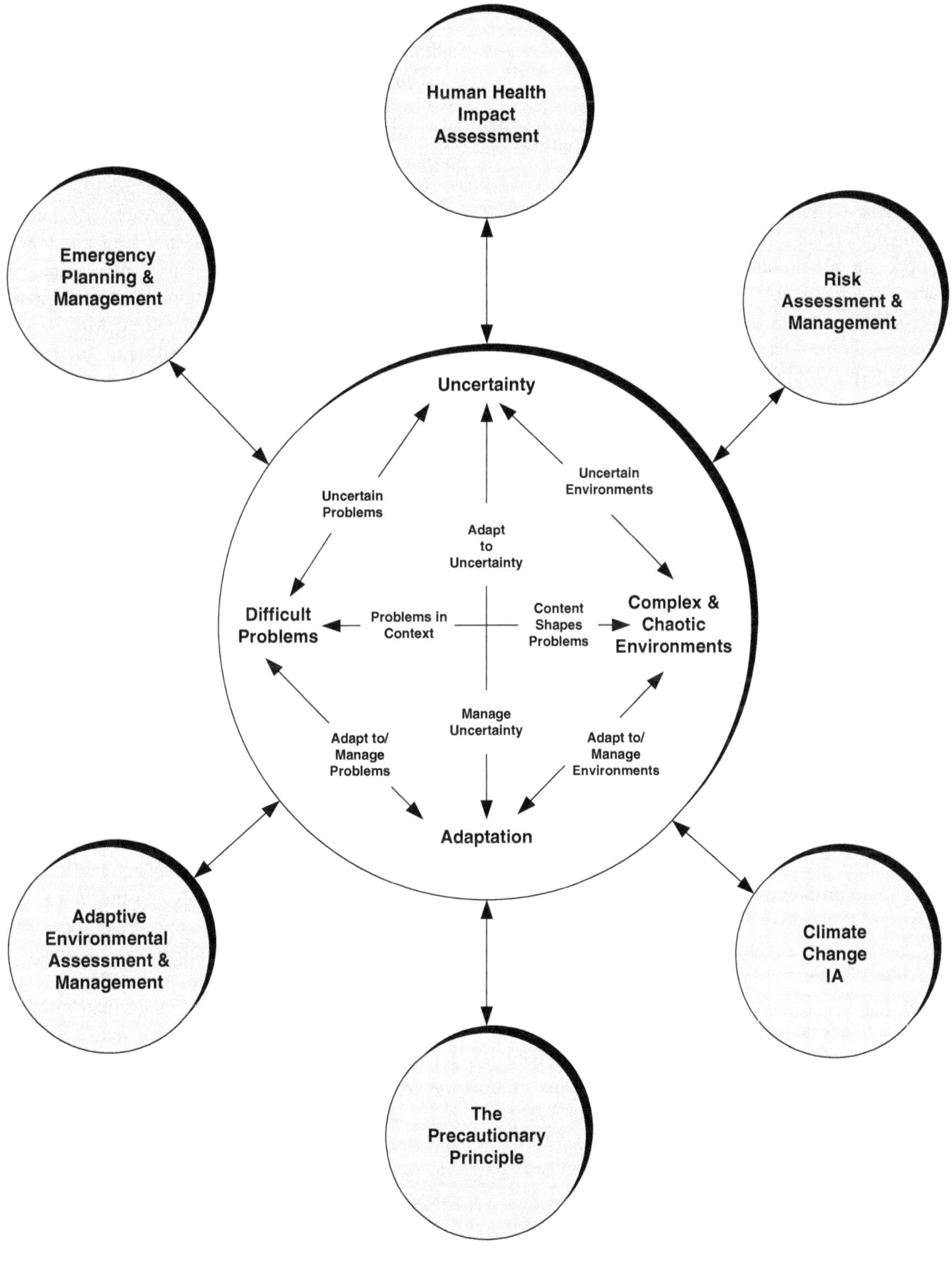

Figure 11.1 Examples of uncertainty management elements.

describe risk assessment and management and potential IA process relationships; (8) to address the treatment of emergencies in IA practice; (9) to describe the precautionary principle and potential IA process roles; (10) to describe human health impact assessment and potential IA process connections; and (11) to address the treatment of climate change in IA practice. These analyses collectively provide the basis for adaptive IA process management at the regulatory and applied levels.

11.4 SELECTING THE MOST APPROPRIATE ROUTE

11.4.1 Difficult Problems and Complex and Chaotic Environments

Difficult Problems The impetus for an IA process is often a desire to solve a problem or take advantage of an opportunity (the opposite of a problem). Problems and opportunities also arise during the IA process. A problem has the following general properties: (1) a question, an issue, or a situation triggers the problem; (2) it has negative connotations (e.g., it is perplexing, vexing, or distressing), and (3) it needs to be dealt with, solved, or addressed. Perceptions of the incidence and nature of problems vary among stakeholders (i.e., problems are subjective) (Cartwright, 1973). Problem-solving processes often begin by identifying, describing, defining, bounding, and stating the problem (Bardwell, 1991; VanGundy, 1988). The initial problem statement is progressively refined through the process. IA practice frequently assumes that the problem is "obvious," that the proposed action will "solve" the problem and that additional problems will not arise. More attention to problem delineation in IA could reduce such recurrent mistakes as solving the wrong problem, stating the problem so it cannot be solved, solving a solution, stating the problem too generally, and trying to obtain agreement on the solution before there is agreement on the problem (International Associates, 1986).

IA, planning, and environmental management problems tend to fall within four broad, overlapping categories: (1) simple or tame problems; (2) compound or semistructured problems; (3) complex or ill-structured problems; and (4) crises or metaproblems (Cartwright, 1973; Miller, 1993; VanGundy, 1988). Uncertainty progressively increases from Level 1 up to Level 4. The adverse consequences of uncertainty are particularly acute at Level 4. IA practice has tended to focus on or to assume Level 1 and 2 problems. Conventional IA approaches are poorly suited to addressing Level 3 and 4 problems, defined here as difficult problems. Table 11.1 outlines several concepts relevant to difficult problems. An adaptive IA process should be designed to cope with difficult problems.

Simple or tame problems are well defined. Ends can be readily established. Much is known about environmental conditions, technologies, methods, and available alternatives. Simple problems can generally be resolved with standardized, often quantitative, procedures and methods. Some but not all the parts are known with compound or semistructured problems. There may be varying perspectives regarding ends. There are likely to be a mix of calculable variables, uncertainties, knowledge gaps, and surprises. Routine procedures will not suffice. Additional analyses are required to fill data gaps. Experimentation, innovative approaches, and practical procedures are needed to deal with new, emerging, and unanticipated issues. Frequent and ongoing stakeholder consultation, mediation, and bargaining are required to cope with varying and conflicting perspectives, values, and interests. Well-defined and managed "good practice" IA procedures are generally adequate for compound or semistructured problems.

High levels of uncertainty and variability and low levels of understanding and control characterize complex or ill-structured problems. Complex problems are dynamic, interdependent, "messy," ambiguous, unique, and real. They involve multiple variables, interactions, and interdependencies. They often defy simplification. Models of complex systems and problems frequently fail to capture critical components and interrelationships. Analytic science and rational/synoptic planning tend to be poorly suited to complex problems. Complex problems are often less amenable to quantification. The past is of limited value either in understanding the present or as a basis for prediction. Ends are not agreed to and means are not known. There are usually multiple perspectives concerning which methods best suit complex problems. Many uncertainties cannot be understood with additional analysis nor effectively managed with "good practices." Complex problems transcend disciplinary boundaries. They often extend over broad geographic boundaries, involve multiple jurisdictions, and are long term. Traditional hierarchical institutional structures and analytical methods rarely cope well with complex problems. Complex problems can be ameliorated but not solved. They require creative approaches tailored to their unique and changing characteristics. Flexibility and adaptability are essential to anticipate and accommodate change and surprise. Scientific approaches that transcend individual disciplines (e.g., trans-scientific, postnormal science, complexity science) are more appropriate when addressing complex problems. Conventional IA approaches are not well suited to managing complex problems.

Metaproblems or crises are more than difficult or even intractable—they are deadly or "wicked." Crises take many forms and are often interdependent. Efforts to address metaproblems frequently encounter paradoxes, dilemmas, and contradictions. Metaproblems are impossible to fully understand or manage. Untended, they can rapidly become disasters or catastrophes. No experimental intervention is consequence free. Incremental adjustments and other adaptive behaviors can exacerbate the problem or even trigger an irreversible chain of deadly consequences. Crises often emerge or occur because of a widening gap between the

Table 11.1 Examples of Difficult Problems

Real Problems

- Problems that exist in the real world
- Principal characteristics—large size, high spatial and temporal variability, not conducive to experimental control, ambiguous and poorly defined

Complex Problems

- Occur in systems where there are multiple interactions among numerous variables, there are many unknown variables and relationships are hard to identify and understand
- Absence of deterministic and complete information about the options, impacts, and interest groups; also multiple interests and multiple and often conflicting objectives and perceptions of problem
- Implications—only superficial control over problems, character of problem often misunderstand, not possible to address problems through training, past nor a good guidepost for future and many social and economic hierarchies unworkable

Transboundary Problems

- Effects cross-jurisdictional boundaries (within or among countries) or affecting global commons
- Multiple jurisdictions, each with different priorities
- Need to create new institutional mechanisms to address; tension because of fears regarding loss of sovereignty

Trans-scientific, Postnormal Science, and Complexity Science Problems

- Crosses and transcends disciplinary boundaries
- Not amenable to analytical scientific methods
- Requires intelligent scanning; succession of judicious nudges
- Addresses situations where facts are uncertain

Paradoxes, Dilemmas, and Contradictions

- Paradox—variety of meanings—(1) something that appears contradictory but that is true; (2) something that appears true but that is contradictory; (3) a series of deductions from a self-evident starting point that leads to a contraction
- Both visual and linguistic paradoxes
- Example—arrow impossibility theory—demonstrates that no method of combining individual preferences to produce a social choice that meets all democratic choice conditions

Impossibilities/Insoluble/Intractable Problems

- Cannot discover all truths
- Types—incompleteness, undecidability, logical and practical impossibilities and technological, cosmological, human and deep limits

Wicked/Messy Problems

- No definitive formulation; no stopping rule; solution are not true or false or good or bad; no immediate and no ultimate test of a solution; every solution is a "one shot" opportunity, because there is no opportunity to learn by trial and error, every attempt counts significantly; do not have an enumerable or an exhaustible describable set of potential solutions, nor is there a well described set of permissible operations; every wicked problem is essentially unique; can be considered a symptom of another problem; the existence of a discrepancy representing a wicked problem can be explained in numerous ways—the choice of explanation determines the nature of the problem's resolution; the practitioner has no right to be wrong
- Involves complex and dynamic situations of changing and interdependent problems
- Analysis and solutions cannot be standardized into general laws or theories; cannot be managed through traditional analytical science

Latent Time Bombs/Catastrophes/Crises

- Latent time bombs—potentially major, sudden disasters such as earthquakes, droughts, floods, or financial collapse; can be interpreted spatially and temporally and can take many forms (e.g., physical, ecological, social, and economic)
- Concerns major events, predictions about them are credible, early intervention is understood to be possible and potentially beneficial, costs associated with advance preparation are significant and highly visible
- Catastrophes—as move through a family of functions a stable fixed point of the family loses its stability; this change of stability forces the system to move abruptly to the region of a new stable fixed point
- Tendency of governments to take action after, rather than before, threatening events occur because of the need to engage in cost distribution

Ingenuity Gaps

- A shortfall between the rapidly rising need for ingenuity and adequate supply
- Problems intrinsically harder to understand and knowledge slow to develop; result critical time lag between problem recognition and delivery of sufficient ingenuity; converging complexities and connections result in need for high-speed decision-making and associated management difficulties
- Human knowledge and ingenuity progress at different rates in different domains; impeded by human cognitive limits, intrinsic complexity of subject matter, nature of scientific institutions and slow and unwieldy economic, social, and political systems
- Roles for markets, science, and democracy but failures and constraints associated with each (e.g., market failures, cognitive limits, varied rates of scientific progress in different domains, rising costs, political gridlock and corruption, social turmoil)

Sources: Allen and Gould (1986), Barrow (1998), Casti (1994), Dery (1997), Gasparatos et al. (2007), Homer-Dixon (2000), Patterson and Williams (1998), Rittel and Webber (1973).

need for and supply of ingenuity. Early intervention is possible, potentially beneficial, but costly. Markets, science, and democratic institutions can all contribute to avoiding, ameliorating, and staving off crises but all have limitations, some profound. Crises require a unique mix of sustained ingenuity, commitment, institutional reform, capacity and network building, leadership, high-speed decision making, and precaution. The likelihood and severity of crises can be reduced, sometimes delayed, and occasionally reversed. Major uncertainties will remain, notwithstanding best efforts. Conventional and even "good practice" IA approaches tend to fare poorly when coping with metaproblems and crises.

Complex and Chaotic Environments It is often stated that IA processes should match the environment or the context. Environment or context generally encompasses ecological, political, social, economic, institutional, and technological components and systems. Some context types are more uncertain than others. Classification systems for environments, contexts, situations, systems, or futures generally involve a continuum from the simple to the highly complex (Barrow, 1998; Hodgson and White, 2001; Trist, 1980). Simple systems have a limited number of variables and interactions and a slow and usually, predictable pace of change. Such systems are not commonly constrained by human or natural limits. Command and control management approaches tend to work well in such situations. Intermediate levels have greater complexity, more interactions and interdependencies, and a higher level of uncertainty. Decision making is more constrained by environmental conditions. Operating in such environments requires effective planning, consultation, and coordination. Turbulent, complex, and interdependent systems are very difficult to understand, predict, or influence. Decision making in such environments is more effective when oriented toward social learning, judicious experimentation, and the proactive anticipation, review, and selective management of risk, error, and uncertainty.

Context classification systems closely parallel those for problem types. Not surprisingly, the most difficult situations encountered in IA practice often involve a combination of difficult problems in complex and chaotic environments. In such cases the problems and the context are poorly defined and overlapping. Practitioners and decision makers face the double dilemma of not knowing when to begin (i.e., separating the problem from the context) and not knowing when to end (i.e., no "stopping rule" for determining that the problem response has been adequate). The related concepts of chaos and complexity are highly relevant to IA practitioners seeking to operate in highly uncertain environments. Table 11.2 outlines some key properties of chaotic and complex environments. Simply put chaos involves lower level order (i.e., system apparently governed by a small number or rules) that evolves into higher level disorder or randomness (i.e., rules are transcended at the higher level). Chaotic systems are highly sensitive to initial conditions. Complexity begins with disorder or randomness but order emerges. Such self-organizing behavior results from feedback mechanisms. Complex systems can be organized or disorganized. They involve multiple variables, interactions, and interdependencies. Chaotic systems are not always complex. Complex systems are often chaotic. Both chaotic and complex systems evolve, often abruptly, in unpredictable ways. Both tend to be irreducible, incomputable, irreversible, incoherent, unstable, dynamic, and nonlinear. Errors and surprise are inevitable with complex and chaotic systems.

Table 11.2 Examples of Characteristics of Chaotic and Complex Environments and Systems

Chaos—General

- Order without predictability; deterministic randomness
- Characteristics—outputs transcend rules, local rather than system order, self-referential, sensitive to initial conditions (the Butterfly Effect), loss of information about initial conditions; basic cause–effect processes still operate among system components but interactions over time and large-scale behavior unpredictable; some systems flip back and forth between chaos and order
- Chaotic systems not always complex (chaos can be observed in simple systems)
- Analysis and interpretation implications—impossible to know a system's exact initial state (incomputable), prediction logically impossible, errors of measurement and calculation inevitable, amplifies uncertainties, impossible to infer from its present state how it got there, can never be fully understood, surprise inevitable, strategy of breaking down whole into components does not apply, value of introspection, humility and pluralism, recognizes that the world infinitely complex and the human mind fallible
- Planning and management implications—ensemble of forecasts and simple models, sensitivity analyses applied to initial state, look for patterns of system behavior, local and incremental predictions, planning as a succession of judicious nudges rather than a step-by-step recipe, unlike natural systems humans can learn and can change behavior to avoid chaos, underlying systems cannot be reduced to equations, future will be determined by an almost infinite array of interlocking contingencies

Complexity—General

- Intricate tangles of shifting and often opposing forces that unfold in unpredictable and frequently surprising ways
- Complex systems not always chaotic but common in complex systems
- As complexity rises, precise statements lose meaning and meaningful statements lose precision
- Slightest error in knowledge of initial state eventually grows
- Complex systems tend to be highly decentralized; end result—large number of local choices and bottom-up information flow

Table 11.2 (*Continued*)

- Characteristics—multiple variables, interactions and feedback, and feed-forward loops; absence of deterministic information, acausal, diffuse authority, new laws come into play when the level of complexity increases, sensitive to the smallest changes and perturbations, behavior can flip from one mode to another suddenly and dramatically, openness to outside environments, global behavior outlasts behavior of component parts and exhibits different characteristics at different scales
- Types—disorganized complexity (millions or billions of variables only approached by statistical mechanics or probability theory), organized complexity (moderate number of variables but all variables interrelated)
- Critical processes—coevolutionary diversity (competition and interdependence of system entities), structural deepening (individual entities become steadily more sophisticated to improve performance) and capturing software (systems take over or task simpler systems)
- Analysis and interpretation implications—counterintuitive behavior of system, as complexity increases not only limited but self-limiting (theory predicts cannot predict), current events heavily influence the probabilities of many kinds of later events, increasing model complexity does not necessarily lead to error reduction, objectivity and subjectivity must be in balance and inform one another, usual statistical approaches may be misleading, variance in outcomes may be enormous
- Planning and management implications—unable to predict or manage the behavior of complex systems, results in confusion and sometimes fear, importance of ability to switch between different modes of behavior as environmental conditions are varied, resulting flexibility and adaptability introduces notions of choice and of collective or social learning, still possible to make general warnings, can explore possible scenarios and expose fragilities but no mathematical model may exist to tell how system will behave, expected harm may be higher than medium, potential for nasty surprises, characteristics support a precautionary approach to risk management, as understand system dynamics can move to adaptive management

Self-Organizing

- A self-organizing system produces complex organization from randomness without external intervention; self-organizing systems use feedback to bootstrap themselves into a more orderly structure
- In self-organizing systems, orderly patterns emerge from lower level randomness; opposite of chaotic systems where unpredictable behavior emerges out of lower level deterministic rules
- If equilibrium of self-organizing systems destroyed pushes system to edge of criticality and perhaps chaos
- Complex systems have a tendency to organize themselves into critical states that are optimally sensitive
- Self-organizing/self-learning; intelligence builds from bottom-up; macrointelligence (system learning) and adaptability derive from local knowledge
- Elements renew and reorganize after change; adaptive capacity resides in memory, creativity, innovation, flexibility, and diversity
- Properties (e.g., the whole is greater than the sum of the parts, self-controlled within larger scale constraints, they evolve)
- Aim of management should be to enhance the capacity of the system for self-management, with active management being used to steer it away from large discontinuities

Emergent

- Properties of a system that the separate parts do not have; the idea that simple elements that are governed by a few simple rules and operate through trial and error with interaction and feedback can produce persistent and systematic patterns that are quite unlike the original elements
- Complex structures seem to display thresholds that, when crossed, give rise to sudden jumps in complexity
- Rarely a smooth, steady increase in the consequences of similar changes in complexity
- Emergent properties cannot be computed
- Self-organizing; dependent on feedback loops
- Emergent system-design principles—more is different (critical mass), ignorance is useful (better to build from simple elements), encourage random encounters, look for patterns and pay attention to neighbors (local information can lead to global wisdom)
- A top-down analytical approach (dissecting whole into parts) will miss emergent or synergistic properties
- Making an emergent system more adaptive generally entails tinkering with different kinds of feedback (positive and negative)
- Emergent organizational systems (e.g., a more cellular, distributed network of small units) tend to be more innovative and adaptable to change than hierarchical models
- Recognizes emergent issues and environmental threats, considers emergent strategies and prioritizes early warning

Turbulent/Unstable/Dynamic

- Turbulence—the apparently random eddying and twisting of the flow; a special case of chaotic behavior
- A dynamic system is in constant flux; the higher its variety, the greater the flux
- Instability and commotion are common

Nonlinear/Feedback

- A change in a system can produce an effect that is not proportional to its size
- Does not obey the laws of addition; generally produces complex and frequently unexpected results
- Small changes can produce large effects; large changes can produce small effects
- Feedback involves some element of the system looping back on itself and either driving the effect up or dampening it down
- Often a consequence of positive feedback, which tends to amplify small perturbations

(*continued*)

Table 11.2 *(Continued)*

- Characterized by multiplicative or synergistic relationships among components or variables
- Entails nonlinear knowledge generation and social learning

Irreducible/Synergistic/Irreversible/Antagonistic

- Nonreductionist; behavior cannot be decomposed into parts
- Synergy—whole more than sum of parts
- Antagonism—whole less than the sum of parts (offsetting)
- Irreversible—the one way time evolution of a real system
- Irreducible risks and uncertainties

Variable/Random/Heterogeneous

- The number of possible states of a system is called its variety; a measure of complexity in a system
- Inherent randomness or variability (stochasticity); difficult to reduce because an inherent characteristic of system being assessed
- A population's natural heterogeneity or diversity, particularly that which contributes to differences in exposure levels or in susceptibility to the effects of chemical exposures
- In risk assessment arises from differences in the nature and magnitude of a population's exposure to hazards and from variations in people's susceptibility to hazardous exposures; quantities vary from time-to-time and place to place
- The only thing that control variety is more variety; variety absorbs variety (Ashby's Law—The Law of Requisite Variety)
- Randomness—uncertainty that is impossible to reduce

Incoherent

- Aspirations and activities do not integrate with one another
- Do not cohere conceptually, operationally, linguistically, or socially
- Function of forecasts and planning—to enhance focus, direction—coherence—for whatever ends

Unpredictable/Surprise/Incomplete Control/Uncomputable

- Chaotic systems are unpredictable (lack of predictability inherent rather than situational); starting situation is never the same between two circumstances; outcome can never be predicted and solution to achieve the desired outcome will need to be created in each new situation
- Complex systems are deterministic but not predictable or manageable
- Uncomputable—systems that cannot be accurately modeled using equations
- Need to accept the fact that what we would like to predict will forever be unpredictable
- Interventions should accept unpredictability and design for surprise

Interdependent

- An increasingly interdependent world
- A dense web of causal connections among components
- Interdependencies of social and natural systems
- The density, intensity, and pace of interactions sharply increases with complex systems
- Positive (reinforces or amplifies initial change) and negative (counteracts the initial change) feedback among system components

Resilience and Stability

- Resilience determines the persistence of relationships within a system in the face of sharp and unexpected external pressures
- Stability is the ability of a system to renew, reorganize and return to an equilibrium state after a temporary disturbance; it is quantified in terms of return time
- Part of resilience is the potential to create innovation opportunities
- Other definitions emphasize conditions with more than one stable equilibrium, where instabilities can flip a system into another regime of behavior
- A concept that relates to a system's ability to absorb, cope with and benefit from change, without losing its basic integrity
- Originally developed in an ecological context but since applied to economic, social, and political systems
- Policy design criteria—maintain different distinct modes of behavior because of rather than despite variability; the more that variability in partially known systems is retained the more likely it is that both the natural and management parts of the system will be responsive to the unexpected (i.e., adaptive capacity)

Sources: Axelrod and Cohen (1999), Barrow (1998), Calow and Forbes (1997), Carpenter (1995), Cardinall and Day (1998), Cartwright (1991), Casti (1994), Cherp et al. (2007), Coveney and Highfield (1995), Croal et al. (2010), Dearden and Mitchell (1998), Dimento and Ingram (2005), Donnelly et al. (2007), Duncan (2008), Farber (2003–2004), Gardner (2010), Gibson (2011), Gleick (1988), Govender et al. (2006), Greene (1999), Grinde and Khare (2008), Hermans and Knippenberg (2006), Hodgson and White (2001), Hollick (1993), Homer-Dixon (2000), Jasonoff (undated), João et al. (2011), Johnson (2001), Innes and Booher (1999), Michael (1989), Nicolis and Prigogine (1989), Nooteboom (2007), Orwell (2007), PCCRARM (1997b), Radford (1988), Rothman and Sudarshan (1998), Rotmans (2006), Rowe (1994), Slootweg et al. (2010), Suter (1993), Stern and Fineberg (1996), Tickner (2003b), Treweek (1999), Trist (1980), Yoe (1996), US ACE (1992), US EPA (1998c).

There are no standardized approaches to operating in chaotic and complex environments. It is prudent to be sensitive to initial conditions, to behavioral patterns at the local level, and to interdependencies. Confusion, fear, and surprise should be expected. An ensemble of simple models, in combination with local, incremental predictions are likely to provide more insights than a single, grand model and long-term system-level forecasts. Variety should be matched to variety. Multiple sensitivity analyses (i.e., tinkering with positive and negative feedback mechanisms) can reveal critical interdependencies and potential thresholds. Adaptability and creativity are essential. A cellular network of small organizational units is usually more innovative and flexible than hierarchical models. Organizational and social learning, synthesis, and the capacity to respond quickly as conditions change are attributes to be fostered. Limits, errors, risks, and uncertainties should be priorities. Approaches that selectively intervene to enhance the self-management capabilities of systems, that maintain and reinforce resilience, and that steer systems away from large discontinuities are often more appropriate. Approaches that evolve and change in parallel with complex and chaotic systems are more likely to be effective in coping with uncertainty.

11.4.2 Uncertainty and Adaptation

Uncertainty Uncertainty, broadly defined, is any situation where we are not absolutely sure (i.e., the opposite of certainty) (Yoe, 1996). There is doubt, incertitude, or lack of clarity. There may be an absence of knowledge (something is not known or knowable), knowledge may be partial, or knowledge may be imprecise. Uncertainty, narrowly defined, focuses on situations where the direction or system characteristics are known but the nature of the outcome or its probability is unknown (Carpenter, 1995; Dearden and Mitchell, 1998; Hyman et al., 1988). Risk is included where probabilities can be ascertained. This analysis (i.e., risk as a subset of uncertainty) applies the broader definition. With both risk and uncertainty, it is largely a question of degree (of certainty or uncertainty), with overlapping or highly permeable boundaries between the two concepts. Interpretations of when there is uncertainty and how much uncertainty exists are subjective, social, and political (Gullett, 1999).

Uncertainty is ubiquitous and unavoidable in environmental decision making and in IA practice because humans operate in complex, unpredictable, and uncertain systems, and because IA is inherently concerned with a difficult to predict and manage future (Gibson, 1992; Tennøy et al., 2006; Tickner, 2003b; Tonn, 2000). Uncertainty is not well understood. As Tonn points out, there can be (1) too much uncertainty (inadequate effort to reduce), (2) too much certainty (a failure to consider the consequences of inaccurate predictions), (3) conflicting perspectives on certainty and uncertainty, (4) misrepresented certainty, (5) misunderstood uncertainty, (6) the confounding of uncertainty and values, and (7) a lack of foresight (Tonn, 2000). Uncertainties occur throughout the IA process (Gibson, 1992). Not acknowledging uncertainties or addressing uncertainties with simplistic and unsupported "safety factors" can impair decision making and contribute to inequities (e.g., increased uncertainties in the lives of the weak) (Cardinall and Day, 1998; Marris, 1996). Explicitly addressing uncertainties and, by extension, follow-up is essential to good IA practice (Government of Canada, 2001; Sadler, 1996; Yoe, 1996). Paralysis is not inevitable (Gibson, 1992). Uncertainty analyses help hedge away from large losses and aid in avoiding and reducing the potential for nasty and tragic surprises (Gibson, 1992; Reckhow, 1994). An uncertainty-oriented IA process is necessarily flexible, adaptive, and iterative. Flexibility (anticipating and rapidly adjusting to changing conditions) and iterative (linking IA activities and stages with feed forward and feedback loops) are commonly identified as good practice principles (IAIA, undated b, Sadler, 1996).

There are many uncertainty forms. Uncertainty can be quantitative or qualitative, objective or subjective, shallow or deep/extreme (CEC, 2000; Dearden and Mitchell, 1998; US ACE, 1992). There can be scientific or methodological uncertainties concerning the choice of parameters, the measurements made, the conditions of observation, the samples drawn, the models used, and the causal relationships employed (Carpenter, 1995; CEC, 2000; CRAM, 1993; Rowe, 1994; US EPA, 1998c, 1999). Perceptions of uncertainties often vary between scientific and technical specialists and lay observers (Grima et al., 1986). There can be substantive knowledge or epistemological uncertainties regarding organizational or environmental systems (Cardinall and Day, 1998; Friend and Hickling, 1997; Mostert, 1996; US EPA, 1998b,c). Knowledge uncertainties can sometimes be reduced through additional analysis but also can be inherent (i.e., fundamental limits to our knowledge of the world) (Tonn, 2000). Uncertainties can pertain to the past, to the present, or to the future (Rowe, 1994). There can be uncertainties concerning guiding values and desires, especially when values, perspectives, and interests conflict, interact, and change (Cardinall and Day, 1998; Friend and Hickling, 1997; Tonn, 2000). There can be contextual uncertainties (e.g., technology, infrastructure, politics, the perceptions and interests of actors) and operational uncertainties (e.g., related to IA process) (Lyhne, 2009).

Many sources can contribute to uncertainty. There can be a lack of data, knowledge, experience, or understanding (Carpenter, 1995; US ACE, 1992; US EPA, 1998c; Yoe, 1996). Theories, explanatory paradigms, methods, and models can be inadequate (Carpenter, 1995; US ACE, 1992; US EPA, 1998c; Yoe, 1996). Time, expertise, and other resources can be insufficient (Carpenter, 1995). Analyses can lack focus because of an absence of direction or poor management (US ACE, 1992). Institutional capacity constraints, or deficiencies in IA requirements and guidelines can contribute to uncertainty (Mostert, 1996). Uncertainties can be exacerbated by poor communications, errors, bias, conflict, and dubious judgments (Carpenter, 1995; Rowe, 1994; Treweek, 1999; US EPA, 1998c; Yoe, 1996).

Uncertainties can result from inherent variations, changing proposal characteristics, randomness, and the multiplicity of intervening variables associated with complex systems (Carpenter, 1995; US EPA, 1998c; Yoe, 1996).

Novel situations and new technologies, materials, and methods tend to be especially uncertain and prone to surprise (Carpenter, 1995). Uncertainty is generally heightened as analysis scales are increased, as time horizons are extended, and as study schedules are abbreviated. Uncertainties in IA can pertain to systems functioning and to cause–effect chains (Seidler and Bawa, 2003). Examples of uncertainty sources, often evident at the SEA level, include natural environmental variability, predicting future conditions, determining carrying capacity, understanding environmental behavior, data inadequacies, future technological changes, socioeconomic conditions, changes in political and economic priorities, how strategic actions will be integrated into projects, unanticipated changes during implementation, effects from other strategic actions and projects, model simplifications, errors in modeling application, analysis and interpretation uncertainties, how results are presented, and value judgments (João, 2007). Uncertainty is often introduced into IA by means of decision rules. Examples of uncertainty-related decision rules include minimizing the maximum regret, maximizing the minimum outcome, and providing for a mix of optimism and pessimism (e.g., Hurwicz alpha criterion) (Byer et al., 2011).

Table 11.3 briefly describes several key uncertainty-related concepts. Recognizing ignorance or lack of knowledge can be humbling. It acknowledges inevitable knowledge gaps. It can stimulate efforts to reduce knowledge deficiencies. Errors and bias will always occur in IA practice. It can be helpful to understand the different types of errors that can occur, to be sensitive to the conditions that contribute to errors and bias, to focus on those types of mistakes likely to have the most serious implications, and to proactively anticipate, minimize, and correct mistakes and bias. Indeterminism and inconclusiveness underscore the limits to uncertainty reduction and the need to ensure that conclusions are not more definite than supporting analyses.

Table 11.3 Examples of Uncertainty Concepts

Ignorance/Incomplete Knowledge

- Lack of knowledge; not all outcomes are known; also ignorance of own ignorance (don't know what don't know)
- Scientists surprised by the outcome; they do not know, but, with hindsight, can usually explain it
- Two faces—positively—a humble admission that we don't know what we don't know; negatively—the practice of making decisions without considering uncertainties
- Cure—obtain knowledge (e.g., education, training, talking to experts, acquiring experts through hiring, contract or coordination)
- Culpably ignorant—need for certain supplementary specific information or measures to avoid harm recognized but failure to obtain

Errors/Mistakes/Bias

- Type I errors (false positives—concluding that there is an effect when, in fact, there is none); Type II errors (false negatives—concluding that there is no effect when, in fact, there is); Type III errors (wrong problem)
- Errors of measurement, calculation and judgment
- Bias in data acceptance (can treat research too leniently or too harshly) and bias in data interpretation (e.g., overemphasis on avoiding Type 1 errors)
- Measures to address errors should provide new information, should not destroy the experimented and should not cause irreversible environmental change; when errors occur it should be possible to learn from error (a source of new information) and to start over

Uncertainty Principle (Heisenberg)

- Places an absolute, theoretical limit on the combined accuracy of certain pairs of simultaneous, related measurements
- Specifically gives a theoretical limit to which a particle's position and momentum can be measured simultaneously
- Recognizes that knowledge of social and natural world is incomplete and can never be complete given constant change and variations over space and time
- Has been elevated by some to the status of a philosophical principle, called the principle of indeterminacy, which has been taken to limit causality in general

Indeterminism/Inconclusiveness

- Means that the uncertainties are of such magnitude and variety that they may never be significantly reduced
- Scientific knowledge is inadequate; causal chains and networks are open and not understood
- Reflects the lack of direct causal linkages in open ended systems
- Potentially relevant concepts from new physics (e.g., new ideas of time, space and causation evident in theories of relativity and quantum indeterminism)
- Inconclusive means information that cannot lead to conclusive or definitive results

Fuzziness/Vagueness

- Fuzziness—vagueness; haze at the edges; degrees of truth; arguable that probability is a special case of fuzziness
- Fuzzy thinking is not precise; it reflects truths, not facts or statistics; a convenient way to approximate nonlinear systems

Table 11.3 (*Continued*)

- Asks if a particular conclusion, which is always tentative, is more true than untrue, or more untrue than true; by progressive steps, it backs and fills its way, merely reflecting the observed phenomena; a sliding scale
- Addresses intensity, extent, and persistence
- Admits the possibility of partial membership in a class, generalizing what might be otherwise crisp sets into ones where class boundaries are, or cannot, be defined clearly
- Reflects judgments that permeate all scientific inquiry and decision making; fuzzy set theory addresses nonspecificity and fuzziness
- A potential bridge between probabilistic risk assessments and qualitative assessments; quantifies the qualitative, while preserving imprecision; also can be integrated into IA simulation models (e.g., fuzzy cross-impact simulation)

Deep/Extreme Uncertainties

- Causal mechanisms not available
- No probability distribution
- Parties value outcomes of alternative decisions differently
- Extreme uncertainties resulting from both limitations of current scientific tools and nature of complex systems

Ambiguity/Nonspecificity

- Having more than one possible meaning; intentionally or unintentionally, obvious (patent) or hidden (latent)
- Also pertains to vague, uncertain or doubtful meaning or interpretation
- When faced with ambiguity about rules, obligations, promises, mandates and duties, practitioners tend to look for precedent, tradition, a source of legitimacy, a consensually based interpretation and an appropriate and fitting response

Approximations

- Simplifications of complex real systems
- Four types—(1) can be solved exactly but do not know correct equation; (2) to solve problem is impossible so resort to approximation; (3) simplify equations (a further abstraction from reality); (4) solution too complicated to understand (approximate to make result understandable)

Doubt

- Occurs in more complex decision-making contexts; issues and problems typically have no exact precedent or involve several parties with divergent or conflicting interests; also insufficient or unreliable data, disagreement over the importance of variables and the fact that some variables may not be quantifiable
- Generally handled through the rules and structures of the procedure within which the parties interact

Confusion/Linguistic Imprecision/Dissonance

- Dissonance—pure conflict (one statement is true and its rivals are false); addressed by probability theory
- Confusion—pure and potential conflict; addressed by possibility theory
- Procedural confusion—the complexity and uncertainty of the situation exceeds the problem-solving capacity of existing decision-making techniques, procedures and institutions
- Linguistic imprecision—imprecise communications

Surprise

- A manifestation of uncertainty that cannot be predicted; a qualitative disagreement between observations and expectations
- Typology—local (created by broader scale processes for which there is little or no previous local knowledge); cross-scale (similar to local surprise but larger scale fluctuation intersects with slowly changing internal variables to create an alternative stable, local system state) and novelty (something truly unique, in which new variables and processes transform the system into a new state)
- Revenge effects—ironic, unintended consequences of mechanical, chemical, biological, or medical ingenuity
- Surprise generating mechanisms and effects—logical tangles (paradoxical conclusions), catastrophes (discontinuities from smoothness), chaos (deterministic randomness), uncomputability (output transcends rules), irreducibility (behavior cannot be decomposed into parts), and emergence (self-organizing patterns)

Uncertainty Analysis

- Analysis of information about risks that are only party known or unknowable; describes the degree of confidence in the assessment
- Quantitative uncertainty analyses explicitly describes the magnitude and direction of uncertainties
- Qualitative descriptions of uncertainties avoid false sense that know precisely extent of risk, helps identify uncertainties with the largest impacts, explains differences in risk estimates generated by different stakeholders and suggests research opportunities

Sources: Benveniste (1989), Burdge et al. (1994, 2004), Calow and Forbes (1997), Carpenter (1995, 1997), Casti (1994), Cardinall and Day (1998), Cartwright (1991), CEC (2000), Coveney and Highfield (1995), Crossley (1996), Dearden and Mitchell (1998), Gunderson (1999), Hodgson and White (2001), Holling (1978), Homer-Dixon (2000), Jones and Greig (1988), Kaiser (2003), Lein (1992), McNeil and Freiberger (1994), Parashar et al. (1997), Peche and Rodríguez (2011), PCCRARM (1997b), Rothman and Sudarshan (1998), Rowe (1994), Stern and Fineberg (1996), Suter (1993), Thissen and Agusdinata (2008), Tickner (2003b), Tickner and Raffensperger (1998), Treweek (1999), Westman (1985), Yoe (1996), US ACE (1992), US EPA (1998c), WHOROE (2001c).

Science offers potential insights into the nature of indeterminism and into its implications for understanding and action. The uncertainty principle (also from science) illustrates the limits of measurement and causality. Fuzziness or vagueness demonstrates that boundaries often are permeable, blurred, and overlapping (i.e., degrees of truth). Methods (e.g., fuzzy set theory, fuzzy cross-impact simulation) can apply this conceptual insight to bridging the qualitative and the quantitative. Ambiguity and nonspecificity point to vague or multiple meanings and to the need to scrutinize the meanings associated with statements and observations. Deep or extreme uncertainty illustrates that there are degrees of uncertainty, up and including a complete inability to know or understand. Approximations illustrate how complex real systems are simplified. Doubt can be both a healthy attitude (consistent with good scientific practice) and an acknowledgment that stakeholders (often with good reason) tend to be skeptical of "experts," specialist analyses, and IA processes. Confusion can arise because of miscommunications, conflict, and poorly adapted problem-solving approaches. Perceptions of confusion frequently vary among IA process participants. Surprises cannot be predicted. The generating mechanisms for surprises can be characterized. Uncertainty analysis, a common stage in risk assessment, illustrates that uncertainties can be systematically assessed. There is a danger in confining the analysis of uncertainties to a single stage in the IA process.

Managing Uncertainty in the IA Process Addressing uncertainty in IA should begin with an attitude or perspective change. Uncertainty is a fundamental process attribute, rather than as "a distasteful transition to attainable certainty" (Holling, 1978). A dynamic perspective, oriented toward constant improvement, is essential (Faber et al., 2010). The irreducibility of risks and uncertainties is understood and unpredictability and incomplete control is accepted (Govender et al., 2006). Priorities shift from prediction and control to adaptability and responsiveness. It is necessary to learn from error, live with, and obtain benefits from uncertainty, avoid the unwarranted appearance of certainty, and address uncertainties throughout the IA process (Canter, 1993b; Dickman, 1991; Hollick, 1993; Mostert, 1996; US ACE, 1992). The IA process becomes an ongoing investigation rather than a one-time prediction of impacts (Holling, 1978). The process is iterative (anticipatory scanning and feedback loops), open, and adaptive (Gibson, 1992; Mulvihill and Keith, 1989). It evolves with and selectively and proactively influences both the problem and the context. It incorporates insights from postnormal and complexity science (Gasparatos et al., 2007). This perspective shift is necessary at both the regulatory (e.g., performance-oriented requirements and guidelines) and applied (e.g., IA process management) levels. It also should be present in each IA activity.

An adaptive IA process begins with a thoughtful, open (to divergent perspectives and interests), and systematic search of the problem space or situation. Uncertainties in the problem definition and in governing norms, values, and interests are explicitly identified (João, 2007; Mostert, 1996). Care is taken to use uncertainty language consistently (Tonn, 2000). The IA process is carefully bounded (US EPA, 1998c). It is highly conscious of and sensitive to the relevance of context (Hindling-Rydevik and Bjarnadóttar, 2007). Climate change concerns are fully integrated into the process (Larsen et al., 2012; Wilson, 2010). Constraints, ambiguities, and cognitive and resource limits are openly acknowledged (Cardinall and Day, 1998; Feldman and Khademian, 2008; Kørnøv and Thissen, 2000). Measurement limits are recognized. Vulnerabilities and blind spots associated with predictive models are identified, ramifications are explored, and appropriate adjustments are made (Duncan, 2008). The process brings together the best of the qualitative and the quantitative (Hodge, 2004). Risk and uncertainty issues are identified, objectives are formulated, and methods are determined (US ACE, 1992). A resilient mix of reliable solutions is identified—each treated as a case study. The proposed action(s) encompasses components intended both to prevent failure (i.e., fail-safe) and aimed at responding and surviving if failure occurs (i.e., safe-fail) (Holling, 1978). The proposed action (or more likely, suite of actions) is adapted and refined as circumstances change, both during and subsequent to approvals (Hollick, 1993). Ideally, the action is suited to staged approval (i.e., self-contained components) and implementation. In this way, the monitoring results can lead to modifications to and, where warranted, termination of the action (Hyman et al., 1988).

The process is open and involves multiple parties and perspectives in a creative and heuristic search for reversible, low magnitude, flexible, simple, error-friendly, proven reliable, safe-fail, and harm reducing options that hedge away from large losses or catastrophic effects; provide benefits even if problems are less serious than feared (i.e., no regrets); involve simple, known, and predictable environmental conditions; have minimum potential for synergistic effects; protect and enhance environmental integrity and sustainability; and can be harmonized with surrounding natural and social systems (Gibson, 1992, 2006a; Hollick, 1993). Baseline conditions are explored with scenario analyses (Hacking and Guthrie, 2008; João, 2007). Risks and uncertainties are recognized and managed (Govender et al., 2006). Complexity is carefully considered (Grinde and Khare, 2008). A proactive effort is made to identify and reduce interdisciplinary obstacles (Gee and Stirling, 2003). Assumptions and the basis for scale and data choices are clearly stated and fully substantiated (João, 2007). Multiple sources of information are utilized (Gardner, 2010). Radically different futures and options are considered (Wilson, 2010). Both possible and preferable futures are considered (João et al., 2011). Option evaluation criteria reflect these types of properties. Alternative criteria and criteria rankings and multiple sensitivity analyses test varying assumptions and perspectives. The evaluation narrows the list of options, but several, potentially acceptable, options and option combinations are carried forward into the process as far as practical. Retaining multiple options enhances action and process flexibility (Hollick, 1993, Gibson, 2006a). The preferred options are those best able to adapt to changing conditions, are lasting and resilient, will do well in most possible future circumstances, pose the least threat to the vulnerable environmental components and systems (assuming flawed predictions and ineffective mitigation), can be adapted as the future unfolds, and make the most positive contribution to sustainability (Gibson, 1992, 2006a, 2011; Homer-Dixon, 2000; Thissen and Agusdinata, 2008).

Uncertainty is a central consideration in baseline and impact analysis, interpretation, and management. Vulnerable (to impact, change, and surprise) environmental components, interactions, and systems are identified. The analysis focuses on change processes and identifies key variables and processes likely to amplify fluctuations (Gibson, 1993; Hollick, 1993). Uncertainties are systematically identified and assessed (Thissen and Agusdinata, 2008). Major uncertainties are identified once data are obtained (João, 2007; Yoe and Skaggs, 1997). Supplementary analysis and research reduce the uncertainties. Knowledge from multiple disciplines is integrated and connected (Dovers, 2005). Multiple models are developed, refined, and applied to characterize the system. Future conditions are predicted in ranges (João, 2007). Uncertainties are analyzed using a combination of scenario and probabilistic analyses (Byer and Yeomans, 2007). Sensitivity analyses, wide error margins, and confidence ranges test assumptions, assess the consistency of relationships, and bound uncertainties (Hacking and Guthrie, 2008; Hyman et al., 1988). Mitigation and adaptation measures are linked (Larsen et al., 2012). Considerable uncertainties remain despite such measures. Predictions are difficult, sometimes impossible. More emphasis is placed on understanding the system, on ensuring that real world conditions are accounted for, and on obtaining a rough sense of the possibilities and probabilities rather than on accurate prediction (Gardner, 2010; Gee and Stirling, 2003; Holling, 1978). The future is addressed by exploring planned and unplanned alternative futures, using such techniques as scenario analysis and sensitivity analysis (Hollick, 1993; João, 2007). Allowance is made for a wide range of errors and outcomes. More stress is placed on avoiding Type II errors (predicting no impacts when impacts occur) than on avoiding Type I errors (predicting impacts when no impacts occur) (Interorganizational Committee, 1994).

Impacts are assessed over the life cycle of the proposed action under normal and abnormal conditions (Tonn, 2000). Extreme and worst-case scenarios are formulated. Broad safety margins and conservative assumptions are employed. Predictions are expected to be inaccurate. Mitigation measures are assumed to be ineffective (Gibson, 1992). Experiments and pilot projects (both at the site and for comparable undertakings and settings) help refine and test the analysis (Treweek, 1999). Uncertainty, vulnerability to change, reversibility, resilience and adaptability, and consequences of error and failure are major impact significance factors. The known and the unknown are rationally connected to decision-making choices (Atkinson et al., 2006). A collective judgment is sought (Gardner, 2010). A good decision is one that delivers positive results in a wide range of futures (Gardner, 2010). Robust, diverse, and adaptable solutions, caution, and design for learning are favored (Gee and Stirling, 2003; Gibson, 2006a). Effective use is made of local knowledge as well as relevant expertise (Gee and Stirling, 2003). Learning is fostered and reinforced through organized evaluation, innovation, integration, and interaction (Tuinstra et al., 2008). Conclusions are questioned, challenged, and tested (i.e., metacognition) (Gardner, 2010). Adaptive management encompasses both ignorance and uncertainty (Gee and Stirling, 2003).

The mitigation analysis stresses emergency and contingency planning, early warning systems, reversibility, adaptability, and the availability of fallback positions and damage control systems (De Bono, 1992). Natural mitigation approaches, which require minimal intervention and which recognize and cultivate the self-organization capacity of systems, are favored over methods reliant on a high degree of intervention, control or "engineering" (Hollick, 1993). A risk and uncertainty analysis identifies, analyzes, interprets, and determines appropriate management measures for risk and uncertainty types and sources (Reckhow, 1994; Tonn, 2000; US ACE, 1992; Yoe and Skaggs, 1997). Uncertainty management includes targeted research and error reduction procedures. Uncertainties are presented in a form suitable for decision making and monitoring (Glasson et al., 1999; Holling, 1978). Monitoring begins early in the IA process by assessing comparable environments, comparable undertakings, and pilot projects. It continues during and following a staged review and approval process. Monitoring focuses on maintaining and enhancing the health of vulnerable environmental components, on detecting emerging discontinuities, and on contributing to environmental integrity and sustainability (Axelrod and Cohen, 1999). Monitoring uncertainties are explicitly acknowledged. Actions taken in response to monitoring err on the side of environmental protection.

Public attitudes toward uncertainty, including value differences regarding uncertainties, are identified (US ACE, 1992). All potentially affected parties are involved in addressing uncertainties (Mostert, 1996). The IA process is open and collaborative. The need for multiple perspectives is recognized (Govender et al., 2006). Reciprocity facilitates trust. Trust among social groups is acknowledged as essential for ameliorating complex, collective, action problems (Ostrom, 1998). Consultation is a social learning opportunity, where the knowledge limits of all parties are recognized. What is and is not certain, the magnitude and consequences of uncertainties, what is and is not being done about uncertainties, and the rationale for all actions taken or not taken in response to uncertainty are transparently communicated to all parties (Hance et al., 1990; Tennøy et al., 2006). Interpretations are open to challenge and comment and subject to independent peer review (Yoe, 1996).

Study teams are selected and managed with uncertainty in mind. Study team leaders identify productive areas of uncertainty and confusion and lead the team toward opportunities (Hodgson and White, 2001). The study team is willing and able to explore ambiguities, handle uncertainties, tackle difficult and unknown problems, readily adapt to changing situations, span boundaries, focus on essentials, scan ahead, communicate effectively, and accommodate conflict (Hodgson and White, 2001). IA documentation explicitly identifies risks, uncertainties, limitations, and constraints. Simplifying assumptions and subjective interpretations and choices are fully justified (Mostert, 1996). The limits of data, technologies, methods, and procedures are acknowledged (US EPA, 1998c). Key uncertainty issues and how they were addressed are described (US ACE, 1992). Aspects of uncertainty most likely to affect decision making are identified (Reckhow, 1994).

Decisions are supported by uncertainty analysis and management techniques (e.g., fuzzy set analysis, bounding analysis, subjective probability analysis, expert panels, scenario analysis, sensitivity analyses, life cycle analysis, simulations, comparative analysis, decision and event trees, exploratory modeling and analysis) (Bro-Rasmussen, 2003; Duinker and Grieg, 2007; Gibson, 1992; Thissen and Agusdinata, 2008; Tonn, 2000; Treweek, 1999; US ACE, 1992; Yoe, 1996). Proposed actions with high uncertainties and potentially grave and likely irreversible consequences are generally rejected (Gilpin, 1995; Hyman et al., 1988). Actions are deferred if short-term studies and pilot projects can reduce high uncertainties and can manage the remaining uncertainties (Gilpin, 1995; Wende et al., 2012). Actions are more likely to be approved if experience elsewhere suggests low magnitude impacts and uncertainties that can be adequately addressed by conditions and monitoring (Gilpin, 1995). Staged approval is applied, where practical, to maximize the opportunity to monitor, adjust, defer, or even terminate proposed actions. Decision making is risk aversive. It hedges decisions away from large losses (e.g., no or least regrets, minimize the maximum regret, maximize the minimum value) (Byer et al., 2011; Gibson, 1992). It recognizes that the need for follow-up is greatest when there is inherent uncertainty (Marshall, 2005). Adaptive management and governance approaches are utilized (Govender et al., 2006). Monitoring is connected to adaptive design and management (Hunsberger et al., 2005). Outcomes of similar actions are monitored (João, 2007). The effectiveness of adaptive methods and procedures is continually assessed. Early warning is prioritized (Donnelly et al., 2007). Feedback loops are closely monitored (Grinde and Khare, 2008). Institutional constraints and implications are identified. Adaptive organizations provide for rapid and continuous knowledge acquisition; have effective information flow and communications networks; have regenerative–restructuring capability; have a bias toward action, preventative planning, and monitoring; and are vertically and horizontal integrated. They tend to be collaborative, experimental, flexible, creative, reliable, and evolving. They have error detecting and error correcting mechanisms, encompass varying critical and systems perspectives, are open to scrutiny, and are responsive to interested parties and diverse interests (Homer-Dixon, 2000; Michael, 1989; Mulvihill and Keith, 1989). Adaptive organizations are much like self-organizing, emergent, complex systems.

Adaptation Concepts and Distinctions Table 11.4 identifies several potentially relevant adaptation concepts. Design confronts complexity with positive visions, pragmatic, tested concepts, and the creative use of analogies, analogs, and models. It progressively explores, evaluates, refines, and embellishes. It structures incremental adjustments that build toward a coherent whole, effectively fitted within larger systems. Ingenuity is concerned with innovative (both new ideas and novel applications of known ideas) and practical, technical, and social solutions to difficult problems in complex environments. Social ingenuity (both in an institutional reform and in a policy sense) is generally a prerequisite to technical ingenuity. Creativity is often mentioned but rarely understood or systematically applied as a means of coping with uncertainty. Sufficient advances have been made to provide a good general sense of the creative process and to offer numerous practical individual and group techniques for fostering creativity. Strategic choice offers a well-tested mix of frameworks, methods, and procedures for exploring and managing multiple uncertainties in high pressure planning situations. It employs an array of useful concepts pertaining to problems, comparisons, decisions, and interactions, all within a highly flexible and iterative group decision-making process. Consilience demonstrates how science has and can interlock facts and fact-based theory across disciplines and branches of learning. It offers a science-based explanatory model of the convergence and unity of knowledge—a model potentially capable of coping with complex system uncertainties. Resilience thinking focuses on interdependencies and on enhancing the ability of critical and irreplaceable system elements and vulnerable parties to adapt and recover from change. Holistic science provides an alternative perspective and framework for spanning boundaries, for escaping the constraints of analytical thought, and for understanding the interdependencies and interconnectedness of complex living and nonliving systems. Sustainability science is concerned with how the equilibrium of complex, interdependent, self-organizing natural, social, and economic systems can be maintained and made more adaptive and resilient. Integrated sustainability assessment applies an integrated systems approach to complex societal problems. It transcends individual disciplines is nonlinear and adaptive, draws upon complex systems theory, is reflective and learning oriented, and seeks to foster the potential for innovation and the realization of transformative outcomes.

Table 11.4 Examples of General Adaptation Concepts and Methods

Design

- Idea of a growing whole; operates at many levels in many different ways; object is to incrementally produce wholeness or coherence
- Establishes a context for future actions; visionary—experienced and then expressed as a vision; a vision that can be communicated to and felt by others; positive—aim to create a positive character based on urban, ecological, and sustainability visions and principles
- Begins from a few well tested concepts that pragmatically respond to prevailing conditions (initial ideas often from analogies, analogs, and models); consequences explored and re-evaluated; ongoing refinement, adjustment, and embellishment

Ingenuity

- Sound sets of instructions; minimum ingenuity requirement—shortest set of instructions to solve problem
- Ideas that can be applied to solve practical technical, social, and environmental problems
- Amount of ingenuity dependent on intrinsic difficulty of achieving goal and kinds and amounts of available resources
- Innovation (truly new ideas) and application of known ideas in different ways and in different contexts
- Technical (helps solve problems in physical world) and social (well functioning markets, institutions, social arrangements) ingenuity; social a prerequisite to technical ingenuity
- Within social ingenuity can distinguish between structural ingenuity (used to create or reform institutions) and policy ingenuity (for actions pursued within an existing institutional framework)
- Measure of ingenuity—quantity (number of instructions) and quality (how well works in practice)
- Can distinguish between ingenuity applied to short-term and to long-term problems; entangled with social and political processes (context)

Creativity

- Key characteristic of brain; the ability to improvise in novel situations; intelligence implies flexibility and creativity; capacity for analogy and metaphor especially important; value of affect or emotion in higher level integrative brain functions
- Concerned with the changing of concepts and perceptions and with the generation of new concepts and perceptions
- Creativity operates on more than one plane and is liberating (defeats habit by originality), visionary, nonlinear; involves both differentiation and integration; also transcends logic (thinking aside or laterally), rationality (i.e., the extrarational), language and science
- Creative thinking; can be fostered deliberately with approaches (e.g., lateral thinking, synectics) and by specific techniques (e.g., brainstorming); facilitated by training
- Individual and group creativity; better in combination; creativity involves preparation, incubation, illumination, and verification
- Potential role in improvement, in problem solving, in realizing value, and in taking advantage of opportunities; particular need for creativity in order to generate future possibilities and to devise ways of coping with multiple possibilities
- More than a way to make things better; approaches and techniques seek to break out of old structures, patterns, concepts, and perceptions
- Stresses value of employing analogs, metaphors, imagery, illusion, simulation, story-telling and games, of exploring apparent contradictions, of searching for patterns and interconnections, of transcending dualisms, of identifying hidden assumptions, of formulating theme variations, and of connecting concepts

Strategic Choice

- Choosing in a strategic way; stresses interconnections among decisions; focus on planning under pressure; seeks opportunities for managing uncertainties through time
- Links technology, organization, process and product in a process that involves shaping, designing, comparing, and choosing
- Creatively manages multiple uncertainties (about the working environment, guiding values, and related decisions)
- Employs various concepts concerning problems (e.g., current and modified decision problems, broader planning problems, problem focus or foci), options (e.g., option bars, option graphs, exploratory options, composite options), comparisons (e.g., comparison area, relative assessment, advantage comparison, working shortlist, evaluation framework), decisions (e.g., decision areas, decision links, decision schemes, immediate decisions, future decision space, action scheme, commitment packages), and interactions (e.g., lateral connections, switching, looping, coalescing decision areas)
- Systematically explores uncertainties (e.g., eliciting limits of surprise, identifying uncertainty areas, linking uncertainties to decision areas, reformulating composite uncertainty areas, comparing alternative responses to uncertainty, weighing uncertainty against urgent decision making, accommodating uncertainty in future decision spaces)
- Applies in a flexible and iterative group decision-making process; provides skill development and practical advice

Consilience

- Consilience—proof that everything in our world is organized in terms of a small number of fundamental natural laws that comprise the principles underlying every branch of learning
- Argues for fundamental unity of knowledge; entails the interlocking of facts and fact-based theory across disciplines and branches of learning (e.g., biology, social science, ethics, environmental policy) to create a common groundwork of explanation (i.e., a "jumping together of knowledge")
- Argues for extending the habits of thought (e.g., reductionism, integration, competing hypotheses, no claim accepted as final) that have worked so well in material world into social sciences and humanities; natural sciences already has constructed a webwork of causal explanations ranging from quantum physics to brain sciences and evolutionary biology—already converging
- Consilience (i.e., units and processes of a discipline that conform with solidly verified knowledge in other disciplines) a criterion for theoretical quality; others include parsimony, generality, and predictiveness

(*continued*)

Table 11.4 (*Continued*)

• The greatest challenge is the accurate and complete description of complex systems
Resilience Thinking
• Fully recognizes complex dynamics and interdependencies between human and natural systems • Looks at past behavior and historical systems patterns • Focuses on resilience of what and to what, distinguishing between what can and cannot be managed • Focuses on resilience of whom—most vulnerable parties • Identifies critical and irreplaceable systems elements • Emphasizes role of governance, institutions, and management in enhancing society's ability, commitment, and preparedness to manage systems for resilience; stresses adaptive comanagement
Holistic Science
• Interconnectedness and interdependence of all living and nonliving systems; social and ecological systems coevolve • Emphasizes—complexity, surprise, nonlinearity and emergence and on the need for creative, intuitive, adaptive, integrative, normative, trans-scientific and pluralistic approaches; suspends the constraints of analytical thought • Broadens the conception of the problem and of context; stresses ecological limits, equity, integration, and holistic perspective
Sustainability Science
• Economics, society, and nature—complex adaptive systems in a world of uncertainty • Self-organizing systems exist in equilibrium; depend on feedback loops • If destroy equilibrium push system to edge of criticality and perhaps chaos • Need to understand biosphere from holistic-systems perspective • Need to consider complexity and closely monitor feedback loops • Seeks to foster adaptive capacity and resilience
Integrated Sustainability Assessment
• Applies an integrated systems approach to complex societal problems embedded in a process-based context • Applies an interdisciplinary science perspective that utilizes complex systems theory, nonlinear knowledge generation, social learning, and systems innovation • Seeks to capture nonlinear dynamics and adaptive behavior • Seeks transformative outcomes • Requires a participatory process, adaptability, reflection, evaluation, and learning • Reinforces learning through organized evaluation, innovation, integration, and interaction

Sources: Alexander et al. (1987), Benveniste (1989), Clark et al. (2011), De Bono (1992), Dearden and Mitchell (1998), Friend and Hickling (1997), Gibson (1992), Gordon (1961), Grinde and Khare (2008), Haney and Power (1996), Hermans and Knippenberg (2006), Hodgson and White (2001), Hofstadter (1985), Homer-Dixon (2000), Iles (1996), Koestler (1964), Miller (1993), Mulvihill and Keith (1989), PCCRARM (1997b), Porritt (2000), Reckhow (1994), Rotmans (2006), Rowe (1991), Slootweg et al. (2010), Tuinstra et al. (2008), Wilson (1998, 2010).

Adaptive Management Adaptive management enables actions in the face of uncertainty (Feldman and Khademian, 2008). It allows decisions to be made notwithstanding imperfect information (Clark et al., 2011). Adaptive management can be active or passive (Kwasniak, 2010). It can serve as a follow-up tool within IA (Canter and Atkinson, 2010). It can facilitate cooperative resource management between indigenous peoples and other stakeholders (i.e., adaptive comanagement) (Landry et al., 2009). It can make it easier for environmental and IA policies to adjust in a rapidly changing institutional environment (Cherp and Antypas, 2003). Adaptive capacity refers to the ability to implement adaptation measures (Byer et al., 2011). It does not mean that those measures will be implemented. Examples of potential roles for adaptive management within IA include reducing cumulative effects uncertainties, informing decision makers of measures to reduce incremental effects, managing regional cumulative effects from multiple contributors, cooperative resource management, and adapting proposed actions to future climate change (Byer et al., 2011; Canter and Atkinson, 2010; Landry et al., 2009). IA good practice guidance has long called for IA to be adaptive (e.g., adjusted to changing realities, issues and circumstances, iterative, incorporating lessons as processes unfolds), while not compromising process integrity (IAIA, 1999).

AEAM is a form of adaptive management. It treats environmental management as a quasiexperiment (i.e., probing ecosystem responses to human actions) (Johnson, 1999; Lee, 1999). Managers learn while doing. Subsequent decisions are adjusted and enhanced from feedback (Reinke and Swartz, 1999; Wieringa and Morton, 1996). The AEAM process is an iterative cycle of planning, implementation, monitoring, research and reexamination (IEMTF, 1995). Each cycle facilitates the selection of more appropriate management actions, helps change stakeholder behavior, and provides a learning opportunity (Lal et al., 2001).

The process is structured around a series of workshops. An example AEAM process is depicted in Figure 11.2.

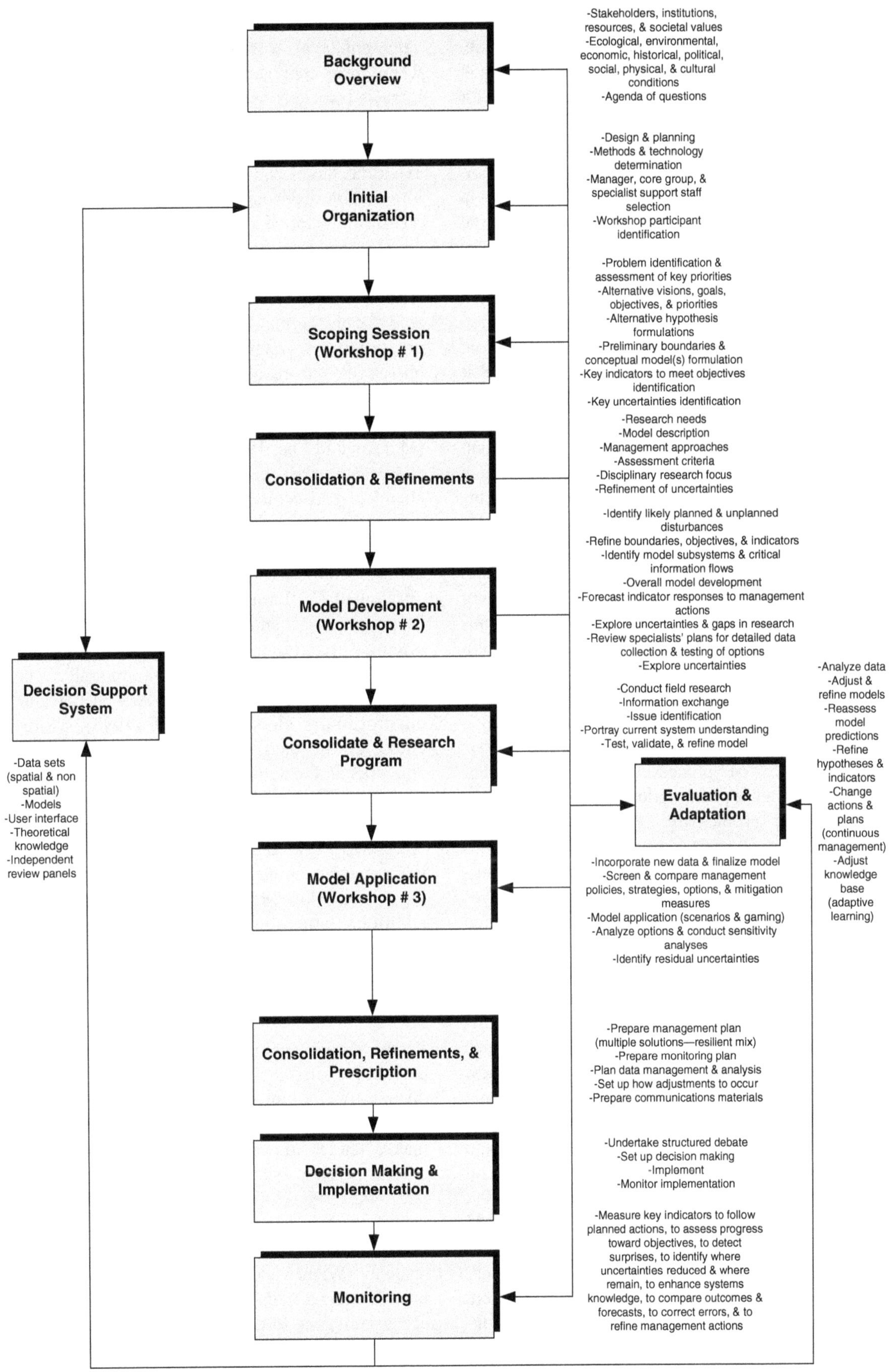

Figure 11.2 Example of an adaptive environmental assessment and management process.

AEAM has much to offer to IA, especially in actively adapting to and managing the uncertainties associated with complex problems and/or complex environments (Hyman et al., 1988; Johnson, 1999; Noble, 2000b). It is a useful tool for assessing mitigation and regional management measure effectiveness, and for assessing the likelihood of cumulative and large-scale effects (Canter and Atkinson, 2010; CEAA, 2009e). It recognizes the value of science in identifying and diagnosing surprise but acknowledges the limits of scientific methods, prediction, and control (Lee, 1999). It appreciates that the uncertain, the unexpected, and the unknown are normal facets of planning and management (Dearden and Mitchell, 1998). It facilitates rapid knowledge acquisition rates and rapid detection of changes (Canter and Atkinson, 2010; McLain and Lee, 1996; Smith, 1993). It recognizes that more information is not always desirable and can hinder decision making (Hyman et al., 1988). It permits learning by doing and underscores the value of monitoring (Morgan, 1998). It is able to handle indirect effects and cumulative effects (Hyman et al., 1988). It provides an integrative systems approach, which links and transcends disciplines and perspectives and helps alleviate the problems associated with fragmented research and coordination (Holling, 1978; Noble, 2000b). The workshop format, built around model building and testing, provides a potentially constructive and nonconfrontation approach for stakeholders to build a common understanding of the problem, to synthesize existing knowledge, to highlight key uncertainties, to clarify assumptions, to stimulate creativity, and to generate innovative options (GBC, 1999; McLain and Lee, 1996; Noble, 2000b; Smith, 1993).

The IA and AEAM processes are very similar, especially at the strategic planning level. The project level is potentially more problematic. Projects, particularly those involving large up-front costs, allow for contingencies but are often not amenable to midcourse corrections, which radically depart from original project objectives (Carpenter, 1997). Project-level EIA is largely oriented to obtaining data and to making specific predictions for decision-making purposes. AEAM is more focused on reaching a policy, resource, or environmental management strategy consensus (Morgan, 1998). Organizational resistance to AEAM will often occur because of a reluctance to admit uncertainty, to make mistakes or to try new solutions, a lack of interest in developing an organizational learning capacity, a perception that it will challenge bureaucratic self-interests, short-term perspectives, and an expectation that scientific research will be costly and of little administrative and political value (GBC, 1999; Gunderson, 1999; Lee, 1999; Walters, 1997). AEAM principles may be less applicable in unique situations where lessons are not transferable (i.e., no spatial replication), where impacts are curable rather than chronic, where uncertainties are limited and manageable, where reasonably accurate impact predictions can be formulated, where continuing surveillance of environmental systems is unwarranted, where ecological components and systems are not resilient, where stakeholders are inflexible (i.e., major value conflicts), where multiple systems are involved, and where there are potentially significant irreversible risks and impacts associated with experimentation (Gunderson, 1999; Johnson, 1999; Morgan, 1998; Noble, 2000b; Walters, 1997). The full application of AEAM modeling and field experimentation can be costly (Walters, 1997). AEAM is information dependent. The information needed to support AEAM may not be available and there may be no means to develop the information (IEMTF, 1995).

AEAM tries to cover multiple objectives by including representatives of various disciplinary backgrounds in the study team (Hyman et al., 1988). Since the process does not specify a systematic way of dealing with multiple objectives, the results are very sensitive to study team composition (Hyman et al., 1988). AEAM is largely based on applying ecological, often linear system models (McLain and Lee, 1996). Such models can have difficulty addressing cross-scale linkages (e.g., between physical–chemical and ecological processes), the nonadditivity of parameters and effects and difficult and emergent processes (Walters, 1997). Nonscientific and qualitative information, knowledge, and experiences are sometimes discounted (McLain and Lee, 1996). Such models may be unable to accommodate fundamental conflicts among scientists regarding facts and assumptions and among policy makers concerning community preferences (McLain and Lee, 1996). Most AEAM literature focuses on procedural elements (Smith, 1993). More attention could be devoted to social, cultural, and economic concerns and to substantial contributions to sustainability (Smith, 1993; UNEP, 1997). Methods of obtaining institutional support, the institutional structures required for AEAM to work, and the procedures for overcoming data inadequacies, model inadequacies, and misunderstandings about AEAM concepts and methods, all require additional attention (Jones and Greig, 1988; McLain and Lee, 1996).

11.4.3 Risks

Risks Risk is a combination of a frequency (in the past) or probability (in the future) and a usually harmful consequence for the human or natural environment (Eccleston, 1999b; Erickson, 1994; Whyte and Burton, 1980). Decision makers tend to know the alternatives but each alternative has several possible outcomes (i.e., outcomes are not certain) (US ACE, 1992). Adverse human outcomes or harm can include injury, disease (morbidity), death (mortality), impaired quality of life, financial loss, property damage or delay (Wiener and Rogers, 2002). Ecological harm can include damage to individual plants or animals, to species, to ecosystems, and to ecological diversity. There can be economic, health, and environmental risks (Grima et al., 1986). Risks can result from natural (e.g., natural disasters) and from human (e.g., human actions that result in exposures to chemical, microorganisms, radiation) sources. There can be high- or low-probability risks. There are best risk estimates and high value risk estimates (e.g., worst case, varying safety margins) (Kamrin, 1993). There are chronic (e.g., diseases resulting from persistent or repeated exposure) and acute (e.g., from abnormal events) risks. Levels of uncertainty and the magnitude of consequences can vary among risks. There are risks to the overall population and to sensitive or susceptible populations (e.g., asthmatics, fetuses, infants, young children, elderly). There can be objective (quantitative) or subjective (estimated or perceived) risks (US ACE, 1992). All risk estimates or calculations have an element of subjectivity. There are deterministic (exposure quantified as a point estimate) and probabilistic (probability distribution incorporated for each variable) risk estimates. Risk estimates can be based on scientific evidence (empirical), predictive models, or heuristic judgment and qualitative reasoning. Risks within a IA process can pertain to potentially significant adverse effects; risks to effective implementation associated with safety; natural hazard risks; security or political stability; risks for which data, process, or methods may be unreliable, invalid, or discredited; risks associated with information gaps, unknowns, major uncertainties; and risks connected to the ability to implement promptly and effectively (Catchpole and Moreno, 2012; Croal et al., 2010).

Table 11.5 outlines some major risk concepts. Risk assessment is concerned with how risks are characterized, described, and estimated. Perceived risks are subjective risk interpretations by individuals and groups. Risk communication involves the exchange of risk-related information and opinion between specialists and the public. Comparative risk assessment compares and ranks risk types. Risk evaluation determines the tolerance for, acceptability of, and desirability of options and proposals. Risk management is an umbrella term for all activities concerned with identifying, assessing, interpreting, communicating, and evaluating risks. It also includes measures to prevent and reduce risks, measures to take advantage of risk-related opportunities, decision-making, implementation, and monitoring. Integrated risk management includes organizational objectives and procedures. A hazard is an intrinsic property that can, under some circumstances, be harmful. A disaster is the realization of the hazard. Hazard identification determines

Table 11.5 Examples of Risk Concepts

Risk Estimation, Analysis, and Assessment

- Risk analysis—the systematic, scientific characterization of potential adverse effects of human or ecological exposures to hazardous agents or activities; performed by considering the types of hazards, the extent of exposure to the hazards, and information about the relationship between exposures and responses, including variation in susceptibility
- Adverse effects or responses could result from exposures to chemical, microorganisms, radiation, or natural events
- Distinction between chronic (diseases occurring as a result of repeated or persistent exposures) and acute (abnormal events) sources and effects; between human health and ecological risks
- Distinction between deterministic (quantifies exposures as point estimates) and probabilistic (incorporates probability distributions for each variable) risk assessment
- Distinctions among empirical risk assessment (based on scientific evidence), model-based assessment (uses predictive models in place of empirical evidence) and qualitative risk assessment (draws on heuristic judgment and qualitative reasoning)
- Need to consider risks to highly exposed populations (e.g., asthmatics, fetuses, infants, and young children, socioeconomic groups, elderly)
- Methods—relative risk indices, event trees and decision networks, environmental transport and fate models, dose response models
- Essential that limitations and negative tendencies of methods be acknowledged, avoided, and minimized

Perceived Risks

- The subjective perception of risk by members of society both individually and collectively; varies from individual to individual and group to group for the same risk and from one risk to another
- Recognizes that lay public knows something that the experts do not and have good reason not to be convinced of all expert evidence
- Public has deep emotional investment in beliefs (e.g., anomie, resentment, distrust, sabotage, stress)
- Public attitudes toward risks are real experiences that determine how people feed and act
- Perceived risk affected by person-related (age, sex, personality type, personal stake, sensitive populations), situation-related (e.g., beyond control of individual, involuntary, children at risk, scientific controversy, high media attention, victim identity), and risk-related characteristics (e.g., origin, immediate threat, consequences for health, dread hazard, catastrophic consequences, unfamiliar hazard, uncertainty, controllability, effects on future generations, reversibility, accident history)
- Also affected by public trust in institutions, fairness, media attention, benefits, and evidence
- Need to consider factors that contribute to outrage (e.g., involuntary exposures, lack of previous knowledge, dread of effects, severe consequences, inadequate or unclear benefits, outside personal control, artificial rather than natural risk, insidious dangers, unknown duration, associated with memorable events, unethical or unfair distribution of risk burden, managed by untrustworthy information sources, effects on children)
- Should be serious consideration in determining risk acceptability

(continued)

Table 11.5 *(Continued)*

Risk Communications

- An interactive process involving the exchange among individuals, groups and institutions of information and expert opinion about the nature, severity, and acceptability of risks and the decisions taken to combat them
- Involves providing citizens with scientific information about risk, making risk information genuinely meaningful, facilitating public involvement in processes where risk analyzed and managed and obtaining public conceptions of risk (risk perceptions); also serves to encourage risk reduction measures, to increase mutual trust and credibility, and to resolve conflicts and controversy
- Need for two-way interaction—learn about patterns of exposure, peoples' perceptions of risk acceptability, and peoples' concerns, values, and knowledge; should describe risks and uncertainties openly and understandably; must begin before important decisions made
- Risk communications strategies and methods adopted can either ameliorate or reinforce perceived risk concerns
- Distinction between informational (provides information necessary to understand characteristics and magnitudes of risk faced and methods for ameliorating) and persuasive (goal of changing people's behavior with respect to a particular risk)
- Need to explicitly consider uncertainty and public issues and to communicate with diverse ethnic and socioeconomic groups
- Elements of risk communications—objectives (why undertaken?), content (what is being conveyed?), form of communications (how should transmit?), feedback from audience (what is being received?)

Comparative Risk Assessment

- The process of comparing and ranking various types of risks to identify priorities and to influence resource allocations
- Useful for ranking the social impacts of various forms of environmental degradation
- Analyzes several different hazards or sources of harm to the same individual, or valuable ecosystem site, in terms of relative risk
- Relative differences in risks are significant and can assist in settling priorities among alternative environmental programs so as to get the most risk reduction per unit expenditure; CRA can lead to risk-based strategic planning
- Examples (progressively less acceptable)—first class risk comparison (same risk/different occasions, risks against existing standards, different estimates of the same risk), second class (with and without activity, risks of different alternatives, same risks in other sites), third class (average risks against most serious risks, risk by source against risks by all sources producing same effects), fourth class (risk/cost ratios, risks vs. benefits, risks vs. risks from same source, risks vs. other causes of same illness or trauma), fifth class (unrelated risks)

Risk Evaluation and Acceptability

- Concerned with the desirability of options or proposals; value-full; expertise dispersed throughout society
- Need to consider all feasible options (modify wants, modify technology, prevent initiating event, prevent release, prevent exposure, prevent consequences, mitigate consequences)
- Risk evaluation—the determination of the importance of risks; risk is context dependent; need to consider cultural, social, and psychological factors; evolves; represents a subjective/political, value-full decision requiring the involvement of all sectors of society and often necessitating alternative dispute resolution
- Need to consider all major consequences (e.g., economics, environment, societal resilience, equity); need to compare against background, alternative actions, other familiar risks, and benefits of continuing the project and taking the risk
- Takes into account such considerations as predicted effects, public perceptions, risk–benefits, background and comparative risks
- Risk evaluation methods (e.g., professional judgment, cost–benefit analysis, cost effectiveness analysis, weight-scoring, decision analysis such as event tree)
- Acceptable risk is a risk whose probability of occurrence is so small, whose consequences are so slight, or whose benefits (real or perceived) are so great that a person, group or society is willing to take that risk; risk tolerance would be a more accurate characterization of the concept
- Risk acceptability of technology dependent on information people exposed to, information choose to believe, values held, social experience, dynamics of stakeholder groups, political process, and historic moment

Risk Management

- Risk management is a systematic approach to setting the best course of action by identifying, assessing, understanding, acting on, and communicating risk issues; integrated risk management—process for building into organizational objectives and procedures
- The process of identifying, evaluating, selecting and implementing actions to reduce risk to human health and to ecosystems; answers the question—what shall we do about it?
- Goal is scientifically sound, cost effective, integrated actions that reduce or prevent risks while taking into account social, cultural, scientific, technological, economic, ethical, political, and legal considerations
- Risk management—an umbrella term that encompasses risk analysis or assessment, risk evaluation (the determination of the importance of risk), risk mitigation, and monitoring
- Some argue should aggressively seek alternatives to command and control (e.g., environmental accounting, education, market-based, incentives, consensual decision-making approaches)
- Stages—define the problem and put it in context, analyze the risks associated with the problem in context, examine options for addressing the risks, make decisions about which options to implement, take actions to implement the decisions, conduct an evaluation of the actions; conducted in collaboration with stakeholders; uses iterations if new information is developed that changes the need for or nature of risk management

Table 11.5 *(Continued)*

- Examples of methods—education/information, incentives, substitution, regulation/prohibition, monitoring, surveillance, research, and risk compensation (for the anxiety created as a result of the hazard potential and for the consequences when a hazardous event occurs)

Disasters and Hazards

- Hazard is an intrinsic property of a substance, which is activated upon an event; a factor or circumstance that may under some circumstances be harmful or injurious; can produce a particular type of adverse health or environmental effect
- A hazard is a perceived event or source of danger that threatens life or property or both; a disaster is the realization of a hazard
- Hazard assessment seeks to recognize things that give rise to concern
- Hazard identification—addresses what can go wrong
- Examples of hazard identification methods (e.g., literature review, plant visits, brainstorming, hazard and operability studies, failure modes, effects and criticality analysis, safety audit)
- Hazard accounting or analysis—establishes boundaries of analysis and determines likelihood of events
- Examples of failure/risk assessment methods—preliminary hazard analysis (identifies hazards as early as possible), event tree analysis, fault-tree analysis, failure modes and effects analysis (attempt every possible way each component or interface among components could fail, then consider effect of failure on system), human reliability analysis (identifies how people interacting might cause to fail)
- Disasters include natural disasters and human induced (e.g., accidents, terrorism); emergency planning and management can be integrated into IA; potential for rapid environmental assessments in disasters
- Climate change a long-term disaster; can be addressed through climate change IA or through integration into other IA types; includes ecological, health, social and economic impacts, risks, and mitigation/adaptation measures; addresses impacts on and from climate change; emphasizes prevention; pertains to policies, plans, programs, and projects; systematically assesses and communicates uncertainties; and provides for adaptive management

Human Health Risk Assessment

- Assesses risk of cancer and from noncancerous (e.g., reproductive, neurotoxic, developmental, immunologic) health effects (alone and in combination); also from abnormal events (acute)
- Carcinogen risk assessment includes hazard assessment (whether agent poses carcinogenic hazard to humans and how might be expressed), dose-response assessment (evaluates potential risks to humans at exposure levels of interest), exposure assessment (the qualitative and quantitative determination of magnitude, frequency and duration of exposure), risk characterization (integrates risk assessment results in nontechnical discussion)
- Examples of issues (e.g., animal testing of potential carcinogens, modeling of carcinogenesis, overly conservative exposure assumptions, risk communications, perceptions and acceptability)

Ecological Risk Assessment

- A process used to estimate the likelihood of adverse effects on plants and/or animals from exposure to stressors
- Evaluates the probability and resulting adverse effects from one or more environmental hazards or stressors (nonendemic events or chemicals), which when introduced have the potential to accumulate, biomagnify, and genetically mutate species, poison, or in any other way impact a species or ecological system in an area
- Examines the extent of damage from a stressor (e.g., defined toxic agents and pollutants) or possible effects to a system or species as a result of a stressor; can be used to predict the likelihood of future adverse effects (prospective) or evaluate the likelihood that effects are caused by past exposure to stressors (retrospective)
- Tiered approach can be helpful—(1) descriptive risk assessment (simple qualitative data and/or comparative methods using literature); (2) semiquantitative (models and data collected to analyze priority issues); (3) site-specific data and predictive models
- Numerous methodological issues (e.g., ranking environmental problems and ecosystem sites, defining endpoints, selecting indicator species, determining scale, managing and quantifying uncertainties, extrapolations across scales, validating predictive tools, valuation, elements of a uniform approach) and areas requiring additional research (e.g., effects of multiple chemical, physical, and biological stressors)
- Extensive debate surrounding appropriate ecological risk assessment paradigm and whether same decision process should be used for human health and ecological risk assessment (e.g., no equivalent to lifetime cancer risk estimate)
- Example elements—problem formulation, receptor identification (partitioning assessment, biological characterization, system organization), hazard identification, endpoint identification (the target species or system that is subject to an environmental hazard), relationship, exposure characterization, ecological effects characterization, risk characterization, and uncertainty analysis

Sources: Agrawala (2010), Arquiaga et al. (1992), Baker and Rappaport (2009), Barrow (1997), Brown (2003), Burdge et al. (1994, 2004), Byer and Yeomans (2007), Byer et al. (2009), Byer et al. (2011), Canter (1993b), Carpenter (1997), Covello et al. (1988), CRAM (1993), Demidova and Cherp (2005), Dooley (1985), Eccleston (2008), EnHealth Council (2001a), Fischhoff et al. (1981, 1982), FPTCCCEA (2003), Furlow (2010), Gardner (2010), Grima et al. (1986), Health Canada (2000), Hood and Nicholl (2002), Hyett (2010), Kamrin (1993), Larsen et al. (2012), Kelly (2005), Lein (1992), Morganstern et al. (2008), Montague (2004), Powell (1984), Power and Adams (1997), PCCRARM (1997a,b), Rahm-Crites (1998), Sandman (1992), Slovic (1987), Spickett et al. (2011), Stackelberg and Burmaster (1994), Treasury Board (2000), US EPA (1998c), US NRC (1983, 1997, 2011a), Yoe (1996), Watson (2010), Weisner (1995), Wende et al. (2012), Weston (2010), WHOROE (2001c), Wilson (2010), Whyte and Burton (1980).

what can go wrong. Hazard assessment bounds the analysis and determines the likelihood of events. There are both natural and human-induced disasters. Human-induced disasters can be both accidental (e.g., spills, collisions) and deliberate (e.g., terrorist acts). Not all disasters are short term (e.g., climate change). Human health risk assessment estimates individual or cumulative risks to people from abnormal events, from cancer, and from noncancerous health effects. Ecological risk assessment estimates the likelihood of adverse effects on plants or animals from exposure to one or more environmental hazards or stressors. There are numerous methods and methodological issues associated with each risk concept.

Emergencies and Disasters There is a tendency in IA practice at the regulatory level to assume that IA requirements are not appropriate in emergency situations. The assumption tends to be that IA requirements are simply too time consuming, and would impede the prompt responses necessary in the event, for example, of a major flood, accident, or terrorist attack. IA requirements, for that reason, generally provide for exemptions to IA requirements for emergency situations. To the extent that major disasters are addressed in IA practice, there seems to be a propensity to "reinvent the wheel," in terms of disaster planning, with each successive IA. Some jurisdictions have addressed, in part, the relationship between IA and emergency planning and management. In the United States, for example, NEPA-related guidance has been provided regarding the analysis of accidents and intentionally destructive incidents (Eccleston, 2003; Luther, 2007). In the wake of the Deepwater Horizon explosion in the Gulf of Mexico, a federal government report addressed policies, practices, and procedures related to outer continental shelf oil and gas exploration and development (US Administration of Barack H. Obama, 2010). The report addresses, among other matters, the role of programmatic and site specific NEPA analyses in assessing, mitigating, and managing potential impacts and risks. Europe requires specific reasons for exemptions (e.g., urgent and substantial need for project and inability to undertake the project earlier and inability to meet the full requirements of the Directive) (EC, 2006a). Guidelines for rapid environmental assessments in disasters also have been prepared in Europe (Kelly, 2005).

The relationship between IA and emergencies (whether natural or human induced), however, is more complex and subtle that either the blanket assumption that IA is not an appropriate instrument for managing environmental concerns during emergencies or the expectation that the merging of IA requirements and emergency planning/management can be limited to requirements/guidance for a few specific event types and proposal-specific emergency management follow-up procedures. Table 11.6 suggests a range of potential initiatives that would serve to more effectively link and integrate the two fields in a mutually beneficial manner. For example, rather than simply addressing potential risks and accidents on a project-by-project basis, specific requirements and guidelines could be formulated for a list of event types (e.g., terrorist attacks, nuclear material release, major train accident/explosion/chemical release, major flood, tanker sinking, major oil/gas release). Such requirements could be adapted for individual IAs and for emergency planning/management procedures. The event-specific requirements could be modified to suit a list of (GIS cross-referenced) physical/ecological and sociocultural setting types (e.g., arctic, continental shelf, river crossing, environmentally sensitive area, fishery, developing country). Requirements and guidance materials also are likely to be necessary for unique, highly sensitive, and highly valued settings. The matching of requirements and context would be greatly facilitated by regional planning and risk analysis zoning in conjunction with cumulative effects assessment.

IA requirements could be modified to include a rapid IA process stream to be applied in the event of a major incident (Kelly, 2005). IA requirements could be broadened to specify the types of conditions that would have to be met for emergency exemptions to be granted and for the rapid IA process stream to be applied. IA follow-up experiences and insights would inform the evolving set of event-related (generally, by setting types and for specific locations) requirements and guidelines. A general effort could be made to more effectively integrate natural disaster, accident, and intentionally destructive event considerations, together with insights from emergency planning and management, into IA requirements and good practices. Care should be taken to ensure that climate change; social, psychological, and cultural concerns; cumulative environmental effects; uncertainty in all its dimensions and from a precautionary perspective; and links to broader sustainability issues and imperatives are fully reflected in the requirements and guidance materials. Mutual learning among related fields (e.g., emergency planning and management, climate change IA, risk assessment and management, HIA, SIA, SEA, project-level EIA, SA)—all of which are concerned with natural and human-induced disasters—is essential. Resources should be pooled wherever practical, and readily available online. Given the time constraints when emergencies do occur, it is critical that stakeholders have already worked out and agreed upon the responsibilities for avoiding, minimizing, and managing the environmental consequences of major disasters. Research on disaster risk reduction could be coordinated and integrated with the aim of enhancing the capacity to address hazards and to make informed decisions on actions to reduce societal impacts (McBean, 2012). The long-term goal would be, consistent with sustainable development, to shift from response–recovery to prevention–mitigation–resilience building–risk reduction and learning from experience and past mistakes (McBean, 2012). IA could assume a key role in these efforts.

Table 11.6 Examples of Interconnections—IA and Emergencies/Disasters

- An abbreviated, rapid, or "fast-track" IA process track can serve to identify and manage environmental concerns during some types of emergencies and major disasters or during various postdisaster phases.
- The assessment of the effects of proposed actions and their alternatives should include the analysis and management of potential accident-related risks (i.e., emergencies induced by proposed actions).
- In many settings, the environmental baseline capacity is already highly prone/susceptible to emergencies/natural disasters (e.g., severe natural habitat loss, severe limits on ground or surface water supply, limited food supplies, decreased number of wild animal or plant species, near photosynthetic capacity, widespread use of toxic chemicals, spread of alien species, depleted fossil fuel, high level of atmospheric gas emissions). An IA analysis, in such settings, needs to consider the potential that proposed actions might push the environmental capacity beyond sustainability thresholds (i.e., emergency/natural disasters "tipping points"). Arguably, such analyses should be based on highly conservative–precautionary assumptions, and should seek to enhance sustainability rather than to operate within sustainability limits.
- Some environmental settings are highly sensitive and vulnerable to change. Debates regarding the acceptability of risks from accidents or spills in such settings (as part of IA processes regarding major pipelines, for example) often surround questions regarding whether such areas should be "off limits" (i.e., any limits unacceptable) or whether different standards (e.g., design, operations, contingency planning, risk acceptability) should be applied. An SEA approach (undertaken in concert with regional environmental planning), which defines the special "ground rules" concerning, for example, resource development and major projects such as pipelines in such settings, would bring such issues to the fore during the policy/planning stage rather than only addressing them at the project EIA level.
- The potential for terrorist attacks can be an issue for some types of proposed actions subject to IA requirements (e.g., nuclear facilities). Potential terrorist acts can be integrated into IA environmental analyses, wherever pertinent.
- Proposed actions and their alternatives should be designed, managed, and assessed to minimize vulnerability and maximize resistance/resilience in the event of natural disasters, accidents, and terrorist attacks (i.e., effects on proposed actions from emergencies).
- IA represents a decision-making instrument that can help shift the focus from response–recovery to prevention–mitigation, building resilience, reducing risks, learning from experience, and avoiding past mistakes.
- Part of the analysis of IA impacts and options should begin from the assumption that accidents and spills will occur. This suggests, for example, locational decisions that minimize the likelihood and severity of adverse environmental consequences and design-/operational/contingency choices (e.g., double-hulled tankers) premised on "worst-case" scenarios.
- There are close interconnections between acute human health risks analyses (as addressed through emergency planning and management) and both HIA and risk assessment and management.
- IA follow-up should include the monitoring of, for example, accidents and spills and associated contingency and clean-up measures. Part of follow-up planning should include measures to prevent and to minimize the incidence and severity of such incidents.
- The analysis of the likelihood and severity of effects from natural disasters and emergencies is likely to be altered, sometimes profoundly, when climate change considerations are integrated into the analysis.
- An important element of emergency planning and management concerns the division of responsibilities. This can be a problematic issue if it has to be worked out in the midst of the pressures of responding to an emergency or, perhaps years later, in legal proceedings. IA documents, and related follow-up and environmental management planning could clearly define the responsibilities (and associated protocols) of all parties in the event of, for example, an accident or spill.
- The social psychological impacts of natural disasters and deliberate actions can be severe. Emergency management and planning often is poorly equipped to anticipate, analyze, and manage such effects. SIA can make an important contribution to addressing such effects.
- An IA cumulative effects analysis (proposed actions considered in conjunction with other past, present, and likely future actions affecting the same environment) should address the potential for a significant increase in the likelihood of emergencies and/or the likelihood of severe environmental consequences in the event of emergencies or natural disasters.
- IA, because it seeks to predict and manage change well into the future, is fraught with uncertainties. The same could be said for emergency planning and management. The enhanced sharing of knowledge and experience between these two related fields would clearly be mutually beneficial.
- Research on disaster risk reduction could be coordinated and integrated with the aim of enhancing the capacity to address hazards and to make informed decisions on actions to reduce societal impacts. The long-term goal would be, consistent with sustainable development, to shift from response–recovery to prevention–mitigation–resilience building–risk reduction and learning from experience and past mistakes. IA could assume a key role in these efforts.

Sources: Diamond (2005), Eccleston (2003), Kelly (2005), Kumagai et al. (2006), McBean (2012), Utzinger et al. (2005).

Risk Assessment Process Figure 11.3 presents an example of a risk assessment/management process. The process begins by identifying and characterizing the problem and the proposed action. The problem is defined within the context of government requirements, policies and guidelines, and ecological, societal, and political systems (CRAM, 1993). A conceptual model is formulated to provide a framework for generating and evaluating preliminary hypotheses about how and why risk-related effects have or are likely to occur. An analysis plan is prepared describing risk management objectives, options to consider, the scope and focus of analysis, methods, and resource allocation (PCCRARM, 1997b; US EPA, 1998c). The proposed action is scrutinized to identify potential hazards (McCarthy and

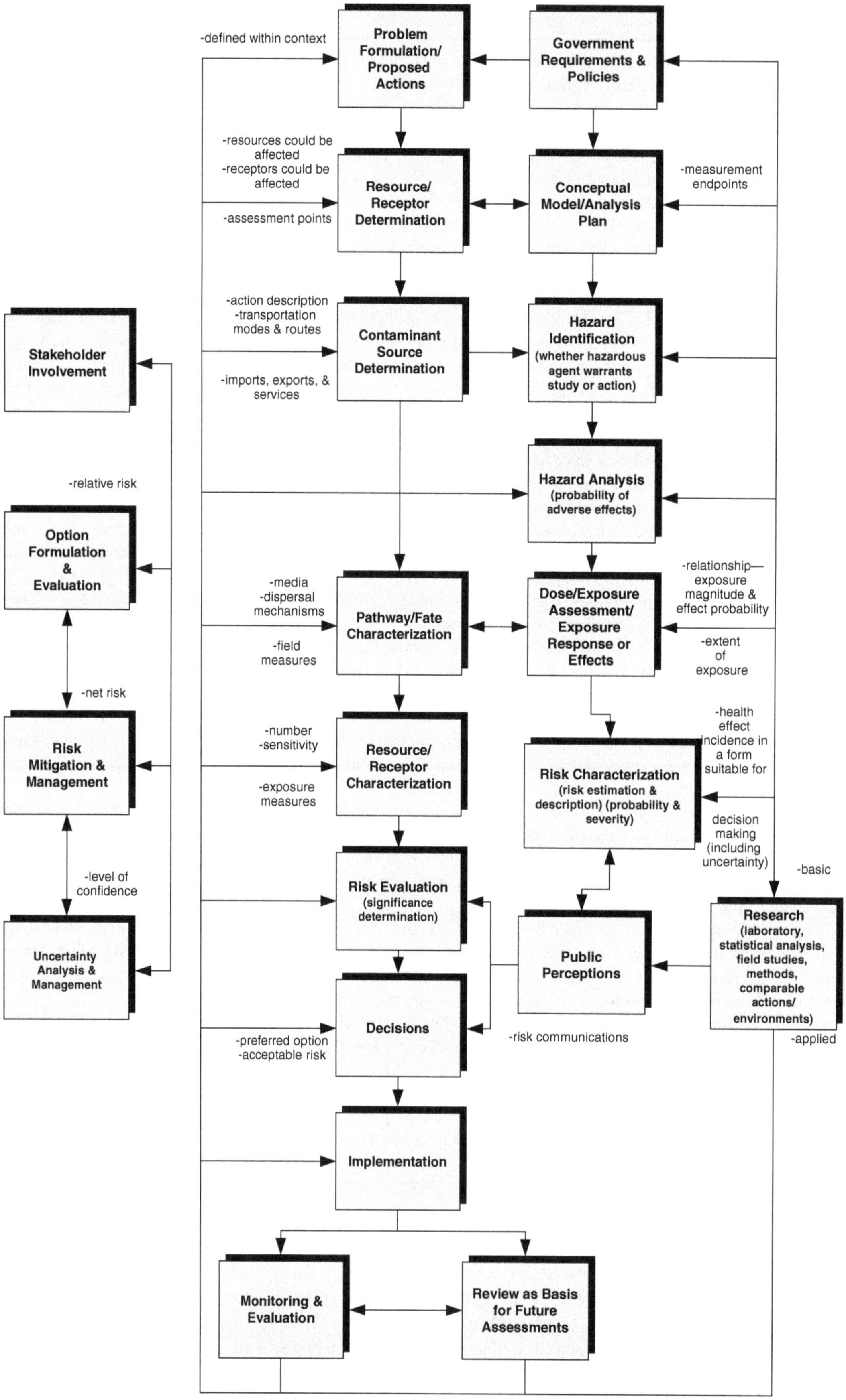

Figure 11.3 Example of a risk assessment management process.

Utley, 2004). Potential health effects associated with the proposed action are identified (Demidova and Cherp, 2005). Hazard identification determines possible sources of harm (usually based on experience with similar technologies, materials, or conditions), explores causal links, identifies the potential adverse effects, and decides whether the effects warrant further study or management action (Canter, 1993b; Carpenter, 1995; CRAM, 1993; Stackelberg and Burmaster, 1994; Yoe, 1996). Contaminant sources; pathways from contaminant sources to resources, receptors, and endpoints; and potentially affected resources, receptors, and endpoints are identified and characterized. A hazards analysis is undertaken to ascertain the probability of adverse events.

The exposure assessment quantifies (e.g., intensity, frequency, duration) the concentrations of contaminants (dose) in the environmental media at the point of human or ecological endpoint contact (Canter, 1993b; Carpenter, 1995; McCarthy and Utley, 2004; Stackelberg and Burmaster, 1994; Yoe, 1996). With ecological risk assessment, it describes the sources of stressors, their distribution in the environment, and their contact or cooccurrence with ecological receptors (US EPA, 1998c). The exposure response assessment determines the relationship of the magnitude of exposure and the probability of effects (Canter, 1993b; CRAM, 1993; Stackelberg and Burmaster, 1994). It involves, in the case of human health effects, evaluating how strongly contaminants elicit health response at various doses (McCarthy and Utley, 2004; Stackelberg and Burmaster, 1994). Ecological risk assessment evaluates stressor–response relationships or evidence that exposure to stressors causes an observed response (US EPA, 1998c). Exposure assessments and exposure response assessments are conducted in parallel and are highly interrelated (CRAM, 1993). Human health and ecological risks, including attendant uncertainties, are presented in a form suitable for public and decision-maker review (CRAM, 1993; Stern and Fineberg, 1996). Risk characterization summarizes the risk analyses, describes the available choices, addresses the implications of uncertainties, and integrates the perspectives and knowledge of interested and affected parties (PCCRARM, 1997b; Stern and Fineberg, 1996; US EPA, 1998c).

Risk evaluation draws upon the risk characterization and public perceptions and is aided by risk communications. It interprets the probability, severity, and significance of estimated and perceived risks (McCarthy and Utley, 2004). Risk acceptability or tolerance and option preference decisions are reached. The decisions are implemented (if approved, often with conditions) and monitored. Monitoring tests the validity of predictions, identifies additional research requirements, contributes to methodological advancements, and identifies the need for management actions (CRAM, 1993). Reviews of the process and of monitoring results are instructive for future assessments (US EPA, 1998c). The process is supported by basic and applied laboratory analyses, statistical analyses, field studies, and comparable actions/environments reviews (CRAM, 1993). Stakeholders are involved in each process activity. Options (e.g., regulatory, nonregulatory) and mitigation measures are formulated and evaluated in an ongoing effort to prevent and reduce risks and uncertainties to acceptable or tolerable levels. Implementation and monitoring are structured and guided by a risk management framework (PCCRARM, 1997b). The analysis and interpretation of qualitative and quantitative uncertainties is consolidated in an uncertainty analysis. The process is open and iterative (PCCRARM, 1997b). Good practice risk assessment and management principles, performance standards, and protocols are formulated, applied, and refined (Canter, 1993b; PCCRARM, 1997b; Steinemann, 2000).

Linking IA and Risk Assessment and Management Risk assessment application in IA practice tends to be confined to large, controversial (high levels of perceived risk) undertakings, usually involving nuclear materials or hazardous chemicals or wastes (Carpenter, 1995). Risk assessment and management has much more to offer. It can supplement regulatory standards and guidelines, which often address risks only partially, indirectly, and qualitatively. It recognizes the limits of deterministic knowledge and the value of probability analysis (Stackelberg and Burmaster, 1994). It provides a systematic, quantitative set of procedures for analyzing, interpreting, and comparing human health and ecological risks and uncertainties (Arquiaga et al., 1992; Suter, 1993). It systematically explores interrelationships that create exposure and effects (Canter, 1993b). It offers an effective bridge to scientific research and to the needs of regulators (Power and Adams, 1997). It appreciates the uncertainties associated with self-organizing and nondeterministic social and ecological systems (Carpenter, 1995). It provides a host of potentially relevant concepts, principles, distinctions, and methods (Canter, 1993b; Erickson, 1994; Grima et al., 1986; Hunsaker and Lee, 1985).

Risk assessment and IA are generally mutually supportive concepts (Grima et al., 1986; Erickson, 1994; Westman, 1985). Risk assessment inputs to IA can aid risk assessment (Barrow, 1997; Ratanachai, 1991). Risk assessment techniques can be incorporated into each IA stage (Demidova and Cherp, 2005). Risk assessment offers a holistic perspective; facilitates the integration of environmental, social, and economic issues; and assists in prioritizing management issues (Hyett, 2010). Principles have been formulated for selectively linking and integrating the two fields (Canter, 1993b). They share a common concern with human health and ecological risks. Both address uncertainty, the role of public perceptions, and the interconnections among science, regulatory requirements, environmental management, and public involvement. Risk assessment and cumulative effects assessment both systematically explore interrelationships. IA and risk assessment processes share many common elements (e.g., problem definition, baseline analysis, impact prediction, mitigation, monitoring) (Dooley, 1985). There are, however, differences also. Risk assessment and management only deal with probabilistic risks. IA addresses risks and impacts and considers certain, uncertain, and probabilistic effects (Dooley, 1985). Risk assessment tends to place less emphasis on alternatives (Barrow, 1997). It is more oriented toward internal management. It is less prone to consider opportunities as well as threats. It is more often applied to regulate industrial and other activities (Barrow, 1997). Enforcement with IA tests for compliance. Risk management determines whether the event probabilities are greater than those agreed to (Dooley, 1985).

Before strengthening the links between IA and risk assessment and management, the many criticisms of the latter must be considered. Some question whether available health and environmental risk data can support the assumptions, models, probability distributions, interpretations, and conclusions (i.e., the pretense of knowledge) (Gee and Stirling, 2003; Heiman, 1997; Montague, 2004; Power and Adams, 1997; SEHN, undated). Concern is raised about the adequacy of risk assessment methods to properly address complex environmental conditions, new technologies, synergistic relationships, latent, indirect and cumulative effects, exogenous events, vulnerable populations, interdisciplinary connections, processes with large geographical and temporal reaches, and carrying and assimilative capacity (Banken, 1998; Davies and Sadler, 1997; Power and Adams, 1997; Tickner and Raffensperger, 1998). The field is criticized for insufficient consideration of public concerns, values, perspectives and perceptions, cultural differences, the social context, nontechnological options, and ecological and biospheric limits (Davies and Sadler, 1997; Fischer, 1996; Kamrin, 1993; Raffensperger and deFur, 1997). It is portrayed as relying too heavily on inadequately supported technical and scientific interpretations and opinions (Hardstaff, 2000). It is described as biased in favor of quantitative scientific methods, rational and centralized decision making, technological "solutions," short-term and local effects, and the analysis of individual environmental components rather than whole systems (Fischer, 1996; Heiman, 1997; Montague, 2004; Raffensperger and deFur, 1997). It is criticized for its failure to warn users about inherent limitations and potential misuses and for not adequately considering cumulative and synergistic risks (Hyett, 2010; Montague, 2004). It is pointed out that it tends to consider only a single option, assesses the risks to others without obtaining their informed consent, ignores benefits or the lack thereof to exposed individuals, and gives people a false sense of safety (Montague, 2004). It is prone to jargon, an unsupportable "aura" of objectivity, and a reliance on the current distribution of power and resources (Heiman, 1997). These shortcomings, where valid, could inhibit democratic debate, heighten public fear and mistrust, exacerbate conflict, reinforce power inequities, undermine political legitimacy, detract from the authority and credibility of associated institutions and science in general, and divert attention and resources away from fundamental social and ethical questions such as acceptable levels of risk, uncertainty, and environmental disruption (Fischer, 1996; Gee and Stirling, 2003; Power and Adams, 1997; Raffensperger and deFur, 1997; SEHN, undated; Tickner and Raffensperger, 1998).

To respond to these concerns, IA practice could tightly circumscribe the application of risk assessment and management to proposals, settings and effects where technological and environmental databases are adequate. Risk assessment and management can be supplemented or combined with other approaches and methods (e.g., consequence analysis, semiquantitative hazards analysis, performance standards, the precautionary principle) (US NRC, 1994). The limitations of risk assessment need to be clearly acknowledged and appropriate steps taken to avoid and minimize its negative tendencies (Montague, 2004). Instances when risk assessment is inappropriate need to be acknowledged (e.g., cannot objectively determine risk acceptability, poorly suited for determining compliance, not well suited for addressing cumulative or synergistic impacts (Hyett, 2010). Governments and regulators could provide more good practice requirements and guidance (Hood and Nicholl, 2002; Hyett, 2010). Greater care could be taken not to oversimplify system interactions or to assume linear risk pathways (Hyett, 2010). More emphasis could be placed on defining and resolving the problem rather than on adapting a predefined set of methods (US NRC, 1994). Unnecessary exposures should be avoided whenever possible, exposures should be periodically reviewed with the aim of eliminating or reducing exposures to the lowest feasible levels, and human rights should be respected (Montague, 2004). More stress could be placed on integrating IA-related risk assessment and management efforts with organizational risk reporting, assessment, and management procedures and practices (Hood and Nicholl, 2002). Additional consideration could be given to risk–sustainability links (Hyett, 2010). Modifications and refinements could be made to minimize potential deficiencies. A greater effort could be made to link the treatment of risk in IA to other planning processes, requirements, and institutional arrangements (Watson, 2010).

11.4.4 The Precautionary Principle

Defining the Precautionary Principle Decision makers face a dilemma. Scientific knowledge of complex environmental and social systems is far from definitive. There will always be scientific uncertainties and varying interpretations of what represents adequate evidence to support a scientific conclusion. Scientists are understandably cautious in coming to firm conclusions. But serious, potentially catastrophic, environmental and health consequences have, and can occur as a result of individual and cumulative human actions. It may be too late to avoid such consequences if no actions are taken until scientific standards of proof are satisfied. In the meantime, decisions must be made on a host of proposed activities that have the potential for environmental and health harm. It is not sufficient simply to approve all activities except those where scientific evidence demonstrates the likelihood of serious or irreversible harm. Nor is it appropriate to automatically reject all proposed actions where there are uncertainties about harm potential and severity. Alternative or supplementary standards of evidence and decision rules are needed to provide a sound and consistent decision-making basis (CEC, 2000). The precautionary principle (PP) is one way to meet this need.

There is no commonly accepted definition of the PP or the precautionary approach. Most definitions begin with the *threat* or risk of harm from a proposed activity to the environment or human health. The threat is based on preliminary scientific evaluations that provide reasonable grounds for concern about the potential for dangerous effects on the environment or on human, animal, or plant health (CEC, 2000). Although possible harm is known, the probability of the harm is not known (WHOROE, 2001c). There may be *shortcomings* (e.g., lack of, inconclusive, or insufficient evidence), *uncertainties* (e.g., lack of certainty, some cause and effect relationships not fully understood) or *divisions* (e.g., lack of consensus) in the scientific knowledge base (CEC, 2000; Hardstaff, 2000; Wingspread Statement on the Precautionary Principle). The conclusion is drawn that scientific knowledge limitations should not preclude or postpone *actions* to prevent the harm when the failure to act would result in serious or irreversible environmental damage (Brown, 2003; Craik, 2008).

The PP shifts the burden of proof to the proponent to demonstrate that proposed action will not lead to serious or irreversible environmental damage, stresses prevention over mitigation, sees uncertainty as a threat (i.e., absence of evidence is not evidence of the absence of a threat), and seeks the elimination of harmful chemicals rather than risk management (Craik, 2008; Quijano, 2003; Tickner, 2004). The moral component of precaution is consistent with the common law obligation of due care and the duty of private actors to behave as society has a right to expect from reasonable people (Jasonoff, 2003). The PP accepts the uncertainty involved in assessing complex systems and challenges government and business to rethink their responsibility to society (Edwards, 2005; Tickner, 2003c). It demands a more dynamic and transparent interaction between science and policy, suggests that there are prudent ways to behave within the limits of available knowledge and experience, and requires an acknowledgement that decision making under uncertainty is often messy (e.g., precise point estimates may not be available) and value based (Jasonoff, 2003; Tickner, 2003c).

The PP is suitable for application when there exists considerable scientific uncertainty, when scientifically reasonable models or scenarios of possible harm are inappropriate, when uncertainties cannot be reduced without increasing ignorance at the same time, when the potential harm is sufficiently serious or even irreversible for present or future generations, and when delays make effective countermeasures more difficult (Kaiser, 2003). Claims of scientific certainty must be critically examined if the PP is to be effectively applied (Barrett and Lee, 2003). The PP is not relevant in cases of ignorance (impacts and probabilities are unknown) or when causal relationships are established (certain and preventable or probabilities can be estimated) (WHOROE, 2001c). Decisions pertaining to uncertainty and regarding when the PP is relevant and appropriate are strongly influenced by institutional, political, and scientific factors (i.e., the context within which knowledge claims are generated, contested, and accepted) (Barrett and Lee, 2003).

Interpreting the Precautionary Principle There are multiple interpretations regarding the harm that should trigger the PP (e.g., harm alone, serious or irreversible harm from proposed action, serious or irreversible harm from cumulative actions, varying interpretations of serious) (Tickner and Raffensperger, 1998). Opinions vary concerning scientific evidence standards. Action has been variously interpreted as (1) deciding that inaction to ameliorate harm is not justified by scientific uncertainty (i.e., action generally proceeds but with mitigation to reduce the threat of harm); (2) deciding that the proposed activity is unacceptable because the scientific evidence is inadequate or because the scientific evidence warrants rejection; (3) only proceeding with the proposed activity if it is proven safe scientifically (i.e., reversing the burden of proof and requiring a level of certainty); (4) only proceeding if a reasonably convincing case can be made that the action is safe (i.e., reversing the burden of proof, acknowledging uncertainties, requiring a weight of evidence argument); and (5) proceeding very carefully (i.e., balancing the burden of proof by adopting prudent decision-making criteria such as safety factors, no or least regrets, best available technology, stringent monitoring) (Gullett, 1997, 1998; Hardstaff, 2000; Wiener and Rogers, 2002). The standards of proof are generally greater if the proposed action is "a priori" hazardous or "new," as in a new technology. Qualifications can be added when applying any interpretation (e.g., proportionality, relative to alternatives, consideration of benefits, additional measures to cope with uncertainties) (Wiener and Rogers, 2002).

These varying interpretations imply thresholds or criteria for threat or risk of harm (which infers a combination of likelihood and severity), thresholds or criteria for deficiencies in scientific knowledge and rules, and principles and procedures for applying the thresholds or criteria. The PP needs to be supported by regulatory authority levels of protection and evidence standards of unacceptable harm (CEC, 2000; Gullett, 1997). Terms such as threat, harm, serious or irreversible, definitive, fully, lack, environment, health, and burden of proof require definition and interpretation, overall, for classes of situations or on a case-by-case basis. A mechanism for determining whether the PP is to be applied is required. Criteria, procedures, decision rules, and institutional arrangements for applying the PP are needed (CEC, 2000). The relationship of PP requirements to risk regulation (e.g., an overarching principle, a risk acceptability criterion applied after risk assessment), and to IA requirements needs to be addressed (Wiener and Rogers, 2002). Some argue that the PP also necessitates a reversed burden of proof from victims to proponents, an open, transparent, and democratic decision-making process, a systematic analysis of all alternatives for reducing (to acceptable levels) or eliminating the harm, a greater weight to "ignorance" in decision making, preventative anticipation or risk avoidance as decision norms, and a proactive effort to safeguard ecological space, to minimize serious or irreversible environmental damage, to avoid social deprivation, to operate within ecological and biosphere limits, to pay past ecological debts, and to protect the interests of future generations (O'Riordan and Cameron, 1994; Porritt, 2000; SEHN, undated; Raffensperger and deFur, 1997; Tickner and Raffensperger, 1998; WHOROE, 2001b,c).

An Example Precautionary IA Process Figure 11.4 is an example of a precautionary IA process. A decision is first made regarding whether the precautionary trigger applies to the proposed action. The PP is commonly triggered when there is a potential for serious or irreversible environmental or human health harm, a scientific evaluation, and a scientific uncertainty (CEC, 2000). The strength of the connection between harm and evidence ranges from significant risk, through likelihood of damage, to reasonable grounds for concern that harm may be caused, to potential for damage and no proof of harmlessness (Gullett, 1997). Very general PP requirements and guidelines maximize the ability to make commonsense adjustments to individual circumstances but increase the potential for arbitrary, biased, and inconsistent interpretations and judgments. A scientific evaluation identifies the potential threat, characterizes the problem, and assesses knowledge and uncertainty levels (CEC, 2000). Qualifications to the principle are added, where appropriate (CEC, 2000; EnHealth Council, 2001a; Government of Canada, 2001). Key terms are defined. Links to IA, to risk management (a framework for or a tool within), and to other environmental management requirements are identified (CEC, 2000; Government of Canada, 2001). Cross-disciplinary approaches and multiple lines of inquiry are favored (Guillette, 2003; Tickner, 2003b). Relevant implications are noted. A clear rationale is provided for each interpretation. The input requirements to apply the PP are specified. The overall precautionary approach is consolidated. Precautionary goals are set (Tickner and Raffensperger, 1998). The precautionary elements of the study design are prepared. A precautionary perspective is applied to the proposal purpose, to the assessment of need, and to the identification of alternatives (Gullett, 2000).

What is known (certainties) and what is not known (types and sources of uncertainties) are determined (Tickner and Raffensperger, 1998). Harm and scientific evidence thresholds and criteria are established. The harm criteria include such considerations as magnitude, temporal and spatial scale, reversibility, degree of complexity and connectivity, vulnerable environments and populations, error friendliness, catastrophic potential, and availability of alternatives to reduce or eliminate harm (Tickner, 1998). The scientific or causal inference criteria pertain to such matters as amount, strength, and consistency of evidence across a wide range of circumstances, knowledge coherence, plausibility of effect, consideration of all evidence and plausible hypotheses, study power to detect effect, statistically significant evidence, public health significance, and causal relatedness based on previous experience (Tickner, 1998). Qualitative and quantitative knowledge is fully and explicitly discussed (Tickner, 2003b). Precautionary decision rules, thresholds and criteria, and application principles and procedures are formulated. Prevention, not mitigation after the damage is done, is emphasized (Quijano, 2003). The decision rules determine what, for example, represents a basis for a moratorium on all actions: phasing out, action rejection, action deferral, additional study (e.g., a risk assessment), incremental approval and implementation, and specific approval conditions (Kaiser, 2003; O'Brian, 2003). Further refinements to the precautionary approach occur through the balance of the IA process.

The customary IA process activities are undertaken. The PP contributes to assessing need (need for reassessment); which alternatives are acceptable, which elements of the proposed action could pose an unacceptable harm; how uncertainties are to shape the application of the principle; which predicted risks and impacts could represent an unacceptable harm; which options are likely to be more harmless; whether mitigation and management measures are likely to reduce the harm to acceptable levels; whether the anticipated risks, impacts, and uncertainties are significant; and whether the proposed action is acceptable (Gee and Stirling, 2003; Gullett, 1997, 1998, 2000; Quijano, 2003). Precautionary measures to manage anticipated impacts, risks and uncertainties are integrated into management strategies (Gullett, 1998). The PP affects decision making (e.g., the taking of precautionary action, the weight of uncertainties in final decisions) (Gullett, 1999, 2000). It influences implementation, monitoring (e.g., early warning, contact protocols, precautionary actions), follow-up, and evaluation (i.e., precautionary measures to be followed unless compelling reason for not doing so) (Gee and Stirling, 2003; Gullett, 1998; ten Hallers-Tjabbes, 2003; Kaiser, 2003; Tickner, 2003a). Care is taken to ensure that the action stays within the precautionary acceptability levels. The PP application is evaluated, both to facilitate postapproval adaptations and to assist in future applications. The IA process is open, transparent, iterative, participatory, and democratic (Gullett, 2000; Quijano, 2003; Tickner, 2003a; Tickner and Raffensperger, 1998). The public and government agencies are involved in scoping, adapting, and applying the PP. The precautionary analyses draw upon multiple disciplines, sources of information, forms of expertise (including local, lay, and traditional knowledge), values, goals, and ways of reasoning (Harremoës et al., 2002). They

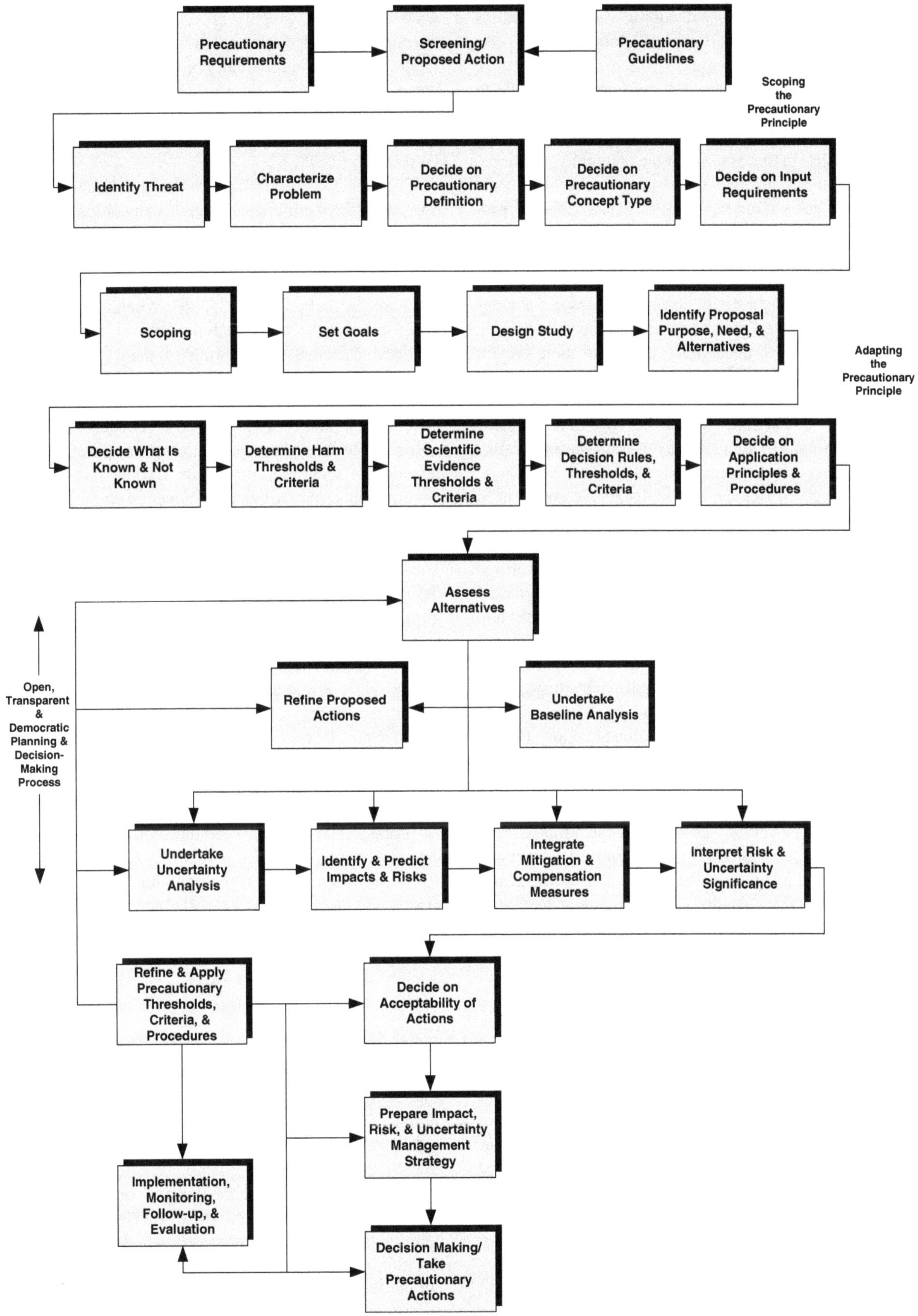

Figure 11.4 Example of a precautionary IA process.

also contribute to reducing and coping with uncertainties and errors (especially Type II errors) (Gullett, 2000).

Potential Contributions of the Precautionary Principle to IA Practice The PP can reduce the incidence and severity of serious, irreversible, and catastrophic environmental harm—a response to the "tragedy of the commons" dilemma, and facilitate the realization of positive public health and environmental goals (Gullett, 1997; O'Brian, 2003). It provides a philosophical framework and an operational guideline for consistently addressing uncertainties in decision making (Gullett, 1997; Seidler and Bawa, 2003). It tempers rational/synoptic IA approaches with corrective measures such as monitoring and feedback, favors the environmentally benign, links potential harm to uncertainty, permits a more careful consideration of available options, requires scientists and other analysts to clearly identify all the plausible effects of a proposed course of action, expands the range of participants in risk decisions, and recognizes the inherent limits in acquiring and applying scientific knowledge (Brown, 2003; Craik, 2008; O'Brian, 2003; Tickner, 2003e). It underscores the need to systematically consider all available options for avoiding harm, helps identify Type 3 errors (good research but the wrong problem), and necessitates the expanded use of science to monitor the integrity of health and the environment (Gullett, 2000; Kaiser, 2003; Thornton, 2003; Tickner and Raffensperger, 1998). It reshapes decision making by shifting the burden of proof from the public to the proponent, by seriously assessing proposal acceptability, by broadening decision making beyond science, and by contributing to more environmentally prudent decision making (Craik, 2008; Tickner and Raffensperger, 1998). It reinforces democratic and substantive environmental and social values and imperatives (Government of Canada, 2001). It is sufficiently flexible to adapt to varying contexts and to proposal and action-specific circumstances. It helps identify research and policy needs (e.g., integrative assessment frameworks, early warning funding, better methods for assessing technologies and activities, means of addressing the fragmented knowledge base, improved language for expressing conclusions and discussing science limits and uncertainties, education for interdisciplinary problem solving) (Tickner, 2003e).

The PP, depending on how it is interpreted and applied, can have serious drawbacks. There will always be risks and uncertainties when seeking to predict and manage environmental change. Few, if any, IA proposed actions would be acceptable if the lack of "proof" of safety or acceptable levels of environmental impact is sufficient grounds for rejecting or deferring a proposed action (Holm and Harris, 1999; Bailey, 1997a). A decision not to act creates risks of its own (Farber, 2003–2004). IA practice and the PP both operate somewhere between the two extremes of scientific certainty of no harm and scientific certainty of harm. Presumably, the goal of the PP is to place more weight on scientific uncertainty about harm potential and less on scientific uncertainty about no harm. Insisting on more decision-making weight for uncertainty, however, could oversimplify and distort an evaluation (Holm and Harris, 1999). Other, perhaps equally or more valid and compelling perspectives, values, and positions could receive no, minimal, or insufficient consideration (Bailey, 1997b). Decision making might be reduced to a hedging mechanism (Wildavsky, 1995).

The breadth of PP interpretations could be used to justify everything from minimal changes in conventional approaches (e.g., identify uncertainties and be careful) to rejecting almost any proposed action, many with significant environmental, social, and economic benefits (Appell, 2001; EnHealth Council, 2001a; Farber, 2003–2004; Whelen, 1996). Potential negative outcomes could include arbitrary, inconsistent, and distorted decisions; the stifling of innovation, application abuses (e.g., trade protectionism); the advancement of agendas of dubious validity and with limited public support; unwarranted public and private costs and delays; the neglect of legitimate risks; the exacerbation of unwarranted fears; the rejection of scientific knowledge; and the misuse of scarce environmental management resources (Appell, 2001; Bailey, 1997a; Foster et al., 2000; Government of Canada, 2001; Hardstaff, 2000; Holm and Harris, 1999; Whelen, 1996). These potential shortcomings point to the need to recognize the valid concerns underlying the principle while avoiding more extreme interpretations and being wary of overly vague and discretionary requirements and guidelines. A prudent, open, and democratic process for both formulating and applying the PP is essential. Vigilance is required to prevent and minimize abuses, distortions, and negative propensities (Wiener and Rogers, 2002). The risks and uncertainties of action and inaction both require consideration (Wiener and Rogers, 2002). Specific requirements and guidelines are needed concerning how to apply the principle within the IA process. The limited experience with the PP underscores the need to compile a good practice knowledge base (Gullett, 1997). More emphasis, when applying the PP, needs to be placed on generating and assessing alternatives, on facilitating greater participation and burden shifting, on reducing multiple risks over the long term, and on stimulating innovation toward sustainability (Tickner, 2004).

11.4.5 Human Health

Definitions and Distinctions A further major uncertainty approach is human health impact assessment. HIA considers human health effects resulting from certain, probabilistic, and uncertain risks and impacts from proposed actions (BMA, 1998; US NRC, 2011a). Health is defined as, "a complete state of physical, mental, and social well-being and not merely the absence of disease" (WHO, 1967). HIA is a combination of methods, procedures, and tools that can help decision makers identify the magnitude and distribution of public health consequences of proposed policies, plans, programs, and projects, together with appropriate actions to measure those effects (IAIA, 2006a; Kemm and Parry,

2004a; US NRC, 2011a). The demand for and practice of HIA has expanded dramatically in recent years, including the development and refinement of good practice standards (Bhatia et al., 2010; Cole and Fielding, 2007; Dannenberg et al., 2006; IAIA, 2006a; US NRC, 2011a).

HIA minimizes the negative and accentuates the positive impacts on the health and well-being of a specified population from a proposed action (Kemm and Parry, 2004a; McIntyre and Petticrew, 1999; NYPHO, 2001). It can be applied to a project, a policy, a program, or a plan. It encompasses the consideration of both the magnitude of health effects on a population and the distribution of effects within the population (Mackenbach et al., 2004). It provides a means by which health hazards, risks, and opportunities can be identified and addressed early in the policy/planning/development process (IAIA, 2006a). It can be a component of SEA or EIA or a "standalone" evaluation for an action subject or not subject to IA requirements. It is generally instigated when there is uncertainty or concern about possible health risks of a proposal or possible opportunities to increase health gain (Scottish Needs Assessment Programme, 2000). HIA overlaps with and is closely connected with SEA, SA, EIA, SIA, risk assessment, and management and health planning, management, and services (EnHealth Council, 2001a). It integrates knowledge and methods from psychology, sociology, economics, toxicology, and epidemiology (Erickson, 1994). It incorporates personal, social, cultural, economic, and environmental factors and considers the opinions, experience, and expectations of potentially affected parties (Davies and Sadler, 1997; EnHealth Council, 2001a; Lehto and Ritsatakis, 1999). It includes the consideration of individual (e.g., genetic, biological, lifestyle, behavioral, circumstantial), social/environmental (e.g., physical, community conditions, economic–financial conditions), and institutional/public policy (e.g., capacities, capabilities, jurisdictions, policies) health determinants (IAIA, 2006a). It recognizes that human health and environmental integrity are interdependent and essential for sustainability (Davies and Sadler, 1997).

Two broad HIA approaches, at the regulatory/legislative level, include (1) requiring, supporting, and promoting the use of HIA (e.g., required as part of IA processes, legislating health authorities to require an HIA at their discretion, legislating that potentially affected communities can request discrete HIAs and be involved in process, regulations or policies that support but do not require the use of HIA) and (2) health within government processes (e.g., requiring a health review or screening for all government policies, discretionary use of non-HIA processes to look at health issues) (Harris-Roxas et al., 2012). At the applied level, HIA can take the form of a quick screening, checklist, or audit (to determine if analysis is warranted), a rapid appraisal, desktop, or mini-impact assessment (available data, minimal quantification, single meeting, structured workshop), an intermediate or standard HIA (standard practice, limited literature review, largely reliant on routine data, impacts quantified, stakeholder participation, nonrigorous, sampling methods, threshold analyses), or a comprehensive or maxi-HIA (extensive literature search, primary and secondary data, rigorous with controlled populations where possible, extensive quantification, sampling and stakeholder participation) (Forsyth et al., 2010; Lehto and Ritsatakis, 1999; Mindell et al., 2004; Parry and Stevens, 2001; Scott-Samuel et al., 1998).

The level of HIA undertaken depends on the proposal timescale, available resources, the importance of the proposal, and the potential health effects (Mindell et al., 2004). HIA can be prospective (potential health impacts), retrospective (impacts after implementation), or concurrent (assessed during implementation) (NYPHO, 2001). It can adopt a broad (holistic view of health, sociological roots, democratic, general quantification, evidence from key informants and popular concerns, low precision) or a tightly defined (defined and observable aspects, epidemiology and toxicology roots, quantification, measurement evidence, high precision) perspective (EnHealth Council, 2001a; Mindell et al., 2004; Palmer, 2004). It can focus on health or disease, be based on a participatory process or on expert opinion, or can adopt a qualitative or quantitative approach (Ison, 2004; McCarthy and Utley, 2004). It can be part of policy preparation, an IA component, or independent from IA (e.g., voluntary, health advocacy) (Cole and Fielding, 2007; Lehto and Ritsatakis, 1999). It can be mandated, serve a decision-support role (voluntary or partnership) or represent a form of advocacy (Harris-Roxas and Harris, 2011). It can be proponent/government or community-driven (Cameron et al., 2011). It can occur at an international, national, subnational, regional, or local level (Birley, 2007; Gunning et al., 2011; Tugwell and Johnson, 2011). It can emphasize health equity or focus on the health impacts of climate change (Gunning et al., 2011; Simpson et al., 2005; Spickett et al., 2011). Often these varying approaches are combined in a complementary manner and are adjusted to suit the context (e.g., simple problems, complex problems, simple and complex problems) (Harris and Spickett, 2011; Harris-Roxas and Harris, 2011; McCarthy and Utley, 2004; Putters, 2005).

HIA Process Figure 11.5 is an example of a comprehensive HIA process. More abbreviated forms (e.g., screening and appraisal only) are also possible (Ison, 2004). The overall process is guided by principles such as democracy, equity, sustainability, the ethical use of evidence, and a comprehensive approach to health (IAIA, 2006a). Screening determines which actions require further review or more detailed health-related analysis, which actions have clearly negligible impacts or produce well understood and easily controllable health effects, and which actions require more information (Bhatia et al., 2010; EnHealth Council, 2001a; IPHI, 2001; Lehto and Ritsatakis, 1999). Particular emphasis is placed on the potential for substantial health effects, especially avoidable, irreversible, catastrophic, and unequally distributed effects (Bhatia et al., 2010). Scoping

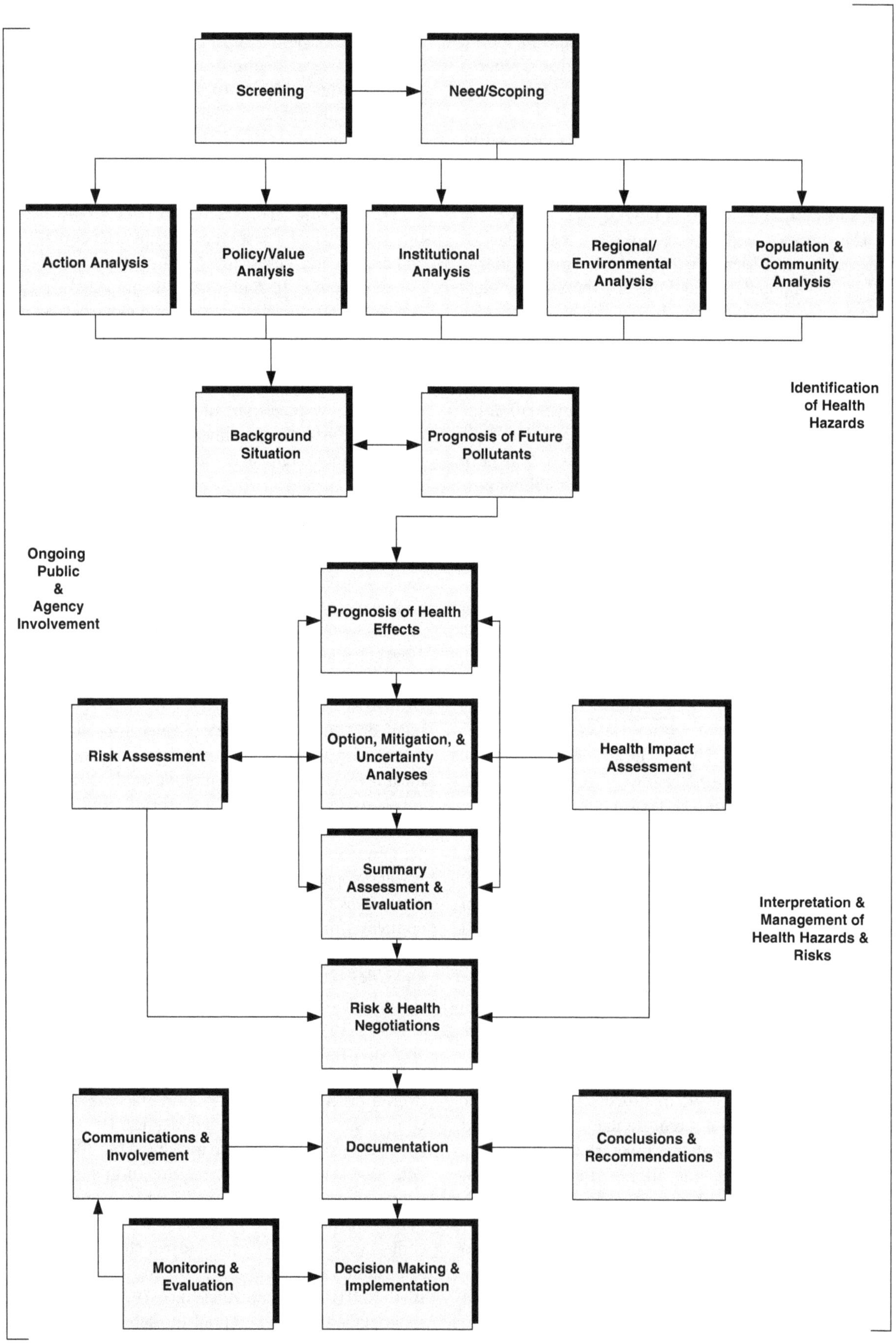

Figure 11.5 Example of a human health impact assessment process.

(a rapid appraisal process) is based on an overview analysis, the application of tools such as checklists and workshops, and extensive stakeholder discussions (Slotterback et al., 2011; Winkler et al., 2011). It confirms and refines need; identifies issues, priorities, alternatives, potential pathways, and standards; specifies potential health concerns and hazards; identifies data requirements and gaps; designs monitoring surveillance systems; and determines the type of HIA (Bhatia et al., 2010; EnHealth Council, 2001a; Winkler et al., 2011). It also bounds the analysis, determines the level of detail, establishes the schedule, selects the study team, designs the approach, decides on consultation procedures, determines documentation requirements, and allocates resources (Bhatia et al., 2010; IPHI, 2001; WHOROE, 2001a). Decision-making links are identified and clarified from the outset (Elliott and Francis, 2005).

Several analyses establish the basis for assessing health effects. Proposal characteristics (e.g., emissions, effluents) that could induce health effects in target populations (e.g., workers, nearby residents) are identified. Relevant policies are determined. Goals and objectives to guide the process are set. The capacity and capability of health protection agencies to prevent and ameliorate acute and chronic health concerns are determined (Arquiaga et al., 1994; Lehto and Ritsatakis, 1999). The physical, natural, resource, built environmental, and land use conditions (community profiling) likely to affect the incidence, dispersion, severity, and management of health effects are identified (Bhatia et al., 2010; Hansell and Aylin, 2003; Lehto and Ritsatakis, 1999). Population (e.g., levels, geographic distribution, food sources and eating habits, age distribution, socioeconomic status, health status, educational levels, genetic endowment) and community (e.g., social support networks, lifestyle and behaviors, community structure, working conditions) characteristics, likely to influence the incidence, severity, and distribution of human health effects, are determined (i.e., health determinants) (Ali et al., 2008; Bhatia et al., 2010; Davies and Sadler, 1997; EnHealth Council, 2001a; Health Canada, 2000; Kwiatkowski, 2004; Lehto and Ritsatakis, 1999). Preexisting health hazard sources (e.g., surface and groundwater water pollution, air pollution, soil and crop contamination, noise, odors, radiation) and health hazards (e.g., communicable diseases, noncommunicable diseases, inappropriate nutrition, injuries, mental disorder) are determined. Methods are justified; data gaps, quality, validity, and statistical stability are assessed; uncertainties are identified; sources of best available evidence are determined; and assumptions are established and substantiated (Bhatia et al., 2010; Hansell and Aylin, 2003).

The prognosis of future conditions establishes how health risks and hazards could change through the duration of the proposed action. It considers projected population levels and characteristics, planned and anticipated land and resource uses, expectations regarding the dispersion and dilution of pollutants, and projections of the exposure of the target populations to health effects from expected future background health hazards. The estimation of potential health effects from the proposed action involves a risk assessment (where probability distributions for health effects can be predicted), a health impact assessment (where potential health hazards and benefits are qualitatively and semiquantitatively described in terms of potential pathways and outcomes) and an uncertainty analysis (Bhatia, 2007; Tickner, 2003d). All three identify and assess options and mitigation measures as means for avoiding and reducing adverse health effects and for enhancing benefits. They also integrate stakeholder perceptions, knowledge, experiences, and perspectives (IPHI, 2001; WHOROE, 2001a). The risk assessment predicts the probability of acute and chronic (cancer-related and noncancer-related) health effects. The health impact assessment identifies and predicts direct and indirect health effects upon exposed segments of the target populations (Davies and Sadler, 1997). Potential chemical, radiological, biological, physical, and psychological health effects are considered (Arquiaga et al., 1994). Causal links between exposure or determinants and physical or mental health are traced (Harris, 2009). Links to physical, natural, social, economic, and service impacts are taken into account (EnHealth Council, 2001a). Hazard agents, exposure conditions, physical health effects, beneficial health effects, effects on health care services, social well-being, social and community health, and psychological well-being are all considered (Davies and Sadler, 1997; Health Canada, 2000). The equity in the distribution of positive and negative health effects is considered (Simpson et al., 2005).

The risk assessment and health impact assessment are combined in a summary assessment of individual and cumulative effects (Kreig and Faber, 2004). Changes caused by proposed actions and other societal changes are clearly differentiated (Kauppinen and Nelimarkka, 2004). Health risk acceptability and impact significance are evaluated, recognizing mitigation and enhancement potential (Davies and Sadler, 1997). Uncertainties, the limitations of the HIA, and methods for coping with uncertainties are specified (Ardern, 2004). A health impact management strategy is devised encompassing such matters as objectives, policies, tactics, priorities, roles and responsibilities, contingency and emergency response procedures, mitigation commitments, compensation criteria and procedures, research and information needs, resources for postproposal management, and monitoring requirements (EnHealth Council, 2001a; Winters and Scott-Samuel, 1997). Documents detail and summarize all aspects of the process, including conclusions and recommendations (WHOROE, 2001b). Decisions are made based on the documentation and on the consultation activities. Decision makers are fully involved in the process from the outset (Elliott and Francis, 2005). Public and agency involvement occurs in all HIA process activities. Agency involvement occurs through an agency steering committee and by means of contacts with individual agencies. Public involvement and participation is critical (Tamburrini et al., 2011). Measures employed include stakeholder and key informant interviews, surveys of potentially affected populations, a public liaison committee, and periodic consultation events (e.g., open houses). The desirability of reducing tensions and conflicts and of obtaining public acceptance by enhancing positive impacts is emphasized (Tamburrini et al., 2011). Opportunities to engage health professionals, policy makers, and affected communities are strengthened through working partnerships (Tugwell and Johnson, 2011). The proposed action is implemented (if approved) with environmental and health control conditions (EnHealth Council, 2001a; IPHI, 2001). Health indicators are identified. Key health conditions, hazards, consequences, and compliance with conditions are monitored for a predetermined period (WHOROE, 2001b). The HIA process, methods, consultation and communications procedures, and databases are evaluated. Adjustments are made to the proposed action and to post-approval procedures based on monitoring results (EnHealth Council, 2001a). The process evaluation is widely distributed to assist other HIA processes.

The typical IA can be broadened and reoriented to systematically address health impacts. This approach treats HIA as a subset of IA. HIA could, in turn, be either a subset of SIA or a field that partially overlaps with SIA (Lehto and Ritsatakis, 1999). Alternatively, the HIA process or process activities can be partially integrated, fully integrated, or simply linked with parallel IA activities (NYPHO, 2001). This approach treats HIA and IA as separate fields that have the potential for linkage, partial integration, or full integration, depending on the circumstances. Or a more targeted approach can be adopted where selective HIA activities or effects are integrated at key points in the IA process. This approach views HIA and IA as selectively and periodically partially overlapping fields. The final approach choice is to address health concerns through risk assessment and management (Arquiaga et al., 1994). The selected approach should suit the circumstances.

Potential Contributions of HIA to IA Practice HIA has many potential benefits for IA. It ensures greater prominence for human health concerns and benefits and responds to identified IA practice deficiencies and major public concerns (Erickson, 1994; IAIA, 2006a; NYPHO, 2001). An enhanced decision-making weight for health is valuable intrinsically, it avoids the transfer of hidden costs and contributes to broader social, equity, and sustainability objectives (Davies and Sadler, 1997; IAIA, 2006a; NYPHO, 2001). At the SEA level, it provides an opportunity to refine screening and causal pathways between policies and health impacts (Wright et al., 2005). HIA addresses more health effects and uncertainties than can be considered in risk assessment and management. The stress on health benefits counterbalances the IA and risk assessment preoccupation with minimizing the negative (Davies and Sadler, 1997; NYPHO, 2001). HIA provides a framework for integrating quantitative and qualitative health concerns. It helps bridge EIA and risk assessment and management, IA and health care planning and services, and EIA and SEA (NYPHO, 2001; WHOROE, 2001b). HIA provides a means of involving health professionals in IA practice (IAIA, 2006a). It helps place public health on the agenda, facilitates the consideration of health inequities, helps legitimize public and private bodies that incorporate health and social concerns, and potentially reduces the burden on the health services sector (IAIA, 2006a). It provides a mechanism for addressing the environmental and social determinants of health from a cross-sectoral perspective (Wernham, 2007). It fosters IA institutional capacity building for addressing health concerns. IA provides an established set of institutional arrangements for implementing HIA. HIA is an additional evaluation tool for the health care community and for public policy development. HIA (in common with IA) also contributes to more open, participatory, transparent, systematic, and substantiated planning and policy making.

There remains considerable room for improvement in HIA practice. More emphasis could be placed on identifying and selecting options rather than just on mitigating the effects of already determined developments (Fischer et al., 2010). Greater consideration could be given to social and behavioral health aspects, the equity of health-related consequences, the role of HIA in public policy making, the management of uncertainties, community-based participation in HIA research and analysis, and HIA capacity building (Bhatia and Seto, 2011; Fischer et al., 2010; Harris and Spickett, 2011; Harris, 2009; Kemm, 2005; Kwiatkowski, 2011; Kwiatkowski et al., 2009; Simpson et al., 2005; Utzinger et al., 2005; Wright et al., 2005). More attention could be paid to the refinement and testing of HIA methods under the umbrella of SEA and EIA (Cole, 2004). A more proactive effort could be made to address the concerns of IA professionals and government officials who have a tendency to resist the integration of HIA into IA practice (Wernham, 2007). Health professionals need to be engaged more effectively (Bond et al., 2011).

Further consideration, at the regulatory level, could be given to institutionalizing HIA either as part of IA (e.g., the nature and rationale for HIA triggers) or as a self-standing set of requirements (Birley, 2007; Bond et al., 2011; Fischer et al., 2010; Harris, 2009; Wright et al., 2005). There continues to be a lively debate concerning whether HIA is more effective when integrated (e.g., could help institutionalize HIA, shared resources, helps build constituency, ensures legitimacy, consistent health trigger, greater standing of EIA/SEA influences policymakers and proponents, helps promote awareness and analysis of social and health effects, action forcing, can build on IA procedural frameworks knowledge and guidelines, mutual learning and skills enhancement, greater transparency, encourages community involvement, facilitates the broader consideration of health determinants, separate HIA runs risk of marginalizing health issues) or independent (e.g., IA—laborious, gives health limited consideration, is project oriented, characterized by low effectiveness and weak implementation, and emphasizes legal defensibility, HIA as part of EIA/SEA will end up conforming to rather than expanding EIA/SEA, fear of losing focus on health and compromising health model, when independent HIA is more open-ended, flexible, and independent, provides opportunity to further develop methods and demonstrate value) from IA (Ahmed, 2004; Atkinson and Cooke, 2005; Bhatia, 2007; Cole and Fielding, 2007; Dannenberg et al., 2006; Harris-Roxas et al., 2012; Morgan, 2011; Slotterback et al., 2011; Wright et al., 2005).

Sometimes integration is clearly inappropriate (e.g., no statutory IA process, severe limitations with existing IA processes) (Morgan, 2011). Generally speaking, the most appropriate choice in terms of full, partial, or no integration is likely to vary depending on the context and could evolve over time. Regardless of the degree of integration, ultimate effectiveness will depend on such factors as government commitment to promoting public health; intersectoral linkages; HIA capacity building initiatives; clear criteria for initiating, conducting, and completing HIA; good practice guidance; and applied research that demonstrates HIA benefits (Cole, 2004; Cole and Fielding, 2007; Dannenberg et al., 2006; Krieger et al., 2003).

There are already numerous demands on hard-pressed health planning and management budgets. A desire to assess health impacts more systematically is of little value if necessary expertise and financial resources are not available. HIA must, therefore, be focused, practical, and realistic. Overlaps, duplication, coordination, and integration with related fields such as EIA, SEA, SIA, and risk assessment and management need to be addressed systematically, without diminishing the genuine need to more effectively address human health concerns (Davies and Sadler, 1997; IPHI, 2001). In some cases HIA may operate more effectively independently from IA (e.g., when health concerns are minor and secondary concerns, when institutional arrangements are too rigid) (McCaig, 2005). A stronger interface between health and other sectors is needed (Bond et al., 2011). Capacity building, networking, methodological development, refinements to theoretical frameworks, critical reviews of HIA approaches, applied research, education and awareness, quality assurance, the clarification of terminology and roles, measures to enhance awareness, measures to facilitate more inclusive participation, the revisiting of HIA governing values and standards, and the creative use of limited available resources are all required (Bhatia, 2007; Birley, 2007; Davies and Sadler, 1997; Harris-Roxas et al., 2012; IPHI, 2001; Krieger et al., 2003; Morgan, 2003b; US NRC, 2011a; WHOROE, 2001a,b). HIA could benefit from more comprehensive and readily accessible HIA information sources (e.g., gateway web sites, HIA good practice repositories) (Dannenberg et al., 2006; Quigley and Taylor, 2003). Greater attention needs to be paid to interactions between HIA and decision making (Becker, Putters, van der Grinton, 2005). A particular effort is needed to apply HIA at the policy (e.g., refining screening and causal pathways between policy and health impacts) and program/plan (e.g., addressing need/opportunity, identifying alternatives) levels (Fischer et al., 2010; McCaig, 2005; Wright et al., 2005). There is a continuing need to institutionalize HIA in many jurisdictions, especially in developing countries (Erlanger et al., 2008). This limitation reflects the tendency of government health agencies to view HIA as a novel activity rather than as a core capacity (Harris-Roxas et al., 2012).

Examples of emerging issues in HIA practice include broad social health determinants (e.g., gender), the role of HIA in facilitating sustainability, links between health threats/risk distribution and environmental justice/injustice, the role of health in cumulative effects assessment, connections between the politics of health and HIA, other forms of alternatives such as knowledge, institutional and goal alternatives, and the health effects of climate change (Harris-Roxas et al., 2012; Kreig and Faber, 2004; Pennock and Uma, 2011; Utzinger et al., 2005; Villani, 2011). The many uncertainties associated with HIA needed to be acknowledged and considered. Insight from related fields in uncertainty management will be essential.

Examples of other challenges facing HIA include defining health and HIA boundaries, balancing the need for valid, timely information with varying data quality realities, balancing the need for credible processes with the need to operate within budgetary requirements and to be responsive to decision making, producing quantitative health effect estimates, synthesizing conclusions on dissimilar health effects, enabling stakeholder participation, ensuring the quality and credibility of HIA, managing expectations, closing institutional gaps and removing institutional barriers, and integrating HIA and EIA/SEA (Becker et al., 2005; Harris-Roxas et al., 2012; US NRC, 2011a).

The continuing resistance to the integration of HIA and EIA/SEA (including the reluctance of IA professionals) underscores the need to better understand the reasons for resistance (e.g., methodological complexities, resource limitations, data controversy or confidentiality, lack of legal framework, professional bias) and to develop and refine strategies for ameliorating resistance, facilitating stakeholder ownership, and illustrating the added value, costs, and negative impacts of including and not including HIA (e.g., demonstration projects that test, refine, and demonstrate different models and methods to maximize utility and acceptance, applied research that shows how decision making is informed and affected, evidence of the predictive validity of HIA, case studies and evidence of health benefits/disbenefits and reduced/increased health inequities, organizational partnership approaches, HIA support systems and capacity building) (Atkinson and Cooke, 2005; Becker et al., 2005; Cole, 2004; Elliott and Francis, 2005; Harris and Spickett, 2011; Harris-Roxas and Harris, 2011; Harris-Roxas et al., 2012; Quigley and Taylor, 2003; Tugwell and Johnson, 2011; Wright et al., 2005; Petticrew et al., 2007).

11.5 INSTITUTING AN ADAPTIVE IA PROCESS

11.5.1 Management at the Regulatory Level

The four jurisdictions address uncertainty. Table 11.7 summarizes positive and negative regulatory level examples in

Table 11.7 Positive and Negative Adaptive Examples at the Regulatory Level

United States	Canada	Europe	Australia
(+) Necessary to identify when information is incomplete or unavailable, together with its relevance and implications (±) Provisions for emergency and major disaster exemptions; more explicit criteria needed (+) NEPA refers to risk to health or safety; regulations identify unique or unknown risks, from both natural hazards and accidents, as significance determination factors (+) Regulations refer to impacts with catastrophic consequences and to a requirement for a reasonably foreseeable analysis (+) NEPA-related requirements and guidelines address such risk-related matters as intentionally destructive acts, the relationship of NEPA and emergency actions, floodplain management, the protection of children from environmental health and safety risks, offshore oil and gas minerals management, the analysis of accidents, pollution prevention, nuclear activities, emergency response actions, and disaster response, recovery, and mitigation (US CEQ, 2010c; US EPA, 1999) (+) Federal agencies have issued a host of guidelines, research reports, case studies, and forum reports on human health and ecological risk assessment (+) Accident and intentionally destructive acts analysis guidance (US DOE, 2002b, 2006) (+) Social and cumulative effects guidelines refer to managing uncertainty by monitoring, mitigation, and adaptive management, and to the importance of communicating uncertainties (US CEQ, 2011) (+) Uncertainty management is addressed in individual agency guidelines (+) Climate change is addressed through regulations and draft guidance on the effects of climate change and reducing	(+) Reference to precautionary principle and to considering in a careful and precautionary manner (+) If responsible authority concludes not sufficient information may require collection (±) Provisions to exclude projects in event of national emergency or would prevent damage to property or environment or in interest of public health or safety; important that explicit criteria; preferable if abbreviated process rather than no process (+) Effects must take into account environmental effects of malfunctions or accidents (+) The probability or predictability of an impact occurring is a significance determination criterion (+) Follow-up a requirement; identified as a means of verifying the accuracy of an EA and determining the effectiveness of mitigation measures (+) Provision for compliance enforcement, including fines ranging from \$100,000 to \$400,000. (+) The EA Agency is responsible for monitoring and facilitating compliance with the Act, promoting and monitoring EA quality, and taking the lead regarding EA quality assurance (+) Adaptive management guidance and research (CEAA, 2009e; ESSA, 1982) (+) Reference is made to follow-up, mitigation effectiveness, and follow-up implementation when determining assessment equivalency (±) SEA guidance refers to considering the need for follow-up but not a requirement	(+) Proposed Project Directive (PPD) addresses environmental issues such as resource efficiency, biodiversity, climate change, and disaster risks; links to disaster (natural and man-made) risk prevention and management concerns (±) PPD includes provisions for exclusions in cases of civil emergency compliance; case by case exemption—civil emergencies and national defense; question of criteria and potential for abbreviated process (+) PPD include health and climate change effects (+) PPD—required to consider exposure, vulnerability, and resilience of factors to natural and man-made disasters (+) PPD—monitoring requirement—part of role of monitoring to identify any unforeseen adverse effects (+) PPD—competent authority to verify whether information up to date, especially regarding mitigation measures (+) PPD—Annex IV refers to describing forecasting methods and accounting for main uncertainties, assessing natural and man-made risks and accident risks (and, where appropriate, measures to prevent), indicating difficulties (technical deficiencies or lack of know-how) encountered by developer in compiling information, sources used and main uncertainties and their influence on effects estimates and selection of preferred alternative (+) SEA directives refer to risks to human health or the environment from accidents and reversibility of effects and requires monitoring and the consideration of human health effects (+) EIA guidance documents refer to projects involving unique or unknown risks, to environmental damage and risks from natural disasters, to actual or perceived human health risks, to risks of accidents and abnormal events, to the occurrence of disease or	(+) Requirement that each IA document describe information timeliness, reliability, and uncertainties (+) Reference is made to indicating whether relevant impacts are likely to be unknown, unpredictable, or irreversible (+) The administrative guidelines on significance identify degree of confidence with which impacts are known or understood as a significance factor (+) General references made to the relationship of risk, especially regarding accidental events, and significance determination; several general health and environmental risk assessment and management guidelines have been prepared (±) Provisions included for exempting actions for defense or security reasons or when dealing with a national emergency (−) The independent review of the Australian IA legislation recommended the inclusion of a greenhouse trigger; not accepted (+) Health effects considered as part of effects on people (+) HIA, including equity-related HIA and health effects from climate change, have been addressed through an Australian and New Zealand collaboration, and through various state, regional, and local level initiatives (+) There have been numerous strategies, partnerships,

Table 11.7 *(Continued)*

United States	Canada	Europe	Australia
greenhouse gas emissions, in policy, program, and planning initiatives for better climate impact preparation, and in requirements for addressing energy requirements and conservation potential (+) Climate change adaptation task force (progress reports, crosscutting strategies, agency climate change adaptation planning) (+) Guidance for GHG emissions and for carbon capture and storage (+) Courts increasingly require that NEPA documents address climate change (Draper, 2010; Eccleston, 2010; Smith, 2010) (±) HIA, as a tool for identifying and managing health effects, has received considerable attention, in terms of both guidance and good practice examples; less evident in project-level EIA documents (Cole, 2004) (±) Precaution is not explicitly mentioned, although reference is made to dealing with unknown information and uncertainties; selective adoption of precautionary approaches by Federal agencies (Graham, 2002) (+) Guidance—EIA and EMS links (US CEQ, 2007c) (+) Adaptation addressed through EIA/EMS and climate change guidance (US ICCATF, 2011)	(+) The relationship of EIA and climate change, follow-up, and adaptive management have been addressed through procedural guidance and through EA Agency sponsored research (±) Changes caused to the health of aboriginal people and health effects directly or indirectly linked to projects, or resulting from the exercise of federal powers must be considered, if they result from project-induced environmental changes (i.e., indirect health effects only) (+) EIA and SEA guidelines refer to health effect type, to cumulative health effects, to the role of health effects, uncertainty and risks as significance factors, and to measures to mitigate significant adverse health effects (+) Mention is made in guidance documents to Type 1 and Type 2 statistical errors; the differences among certain, reasonably foreseeable, and hypothetical future actions; uncertainty sources; and methods for addressing uncertainty (+) Health Canada has prepared an HIA handbook	disease vectors, and to especially vulnerable groups of people (+) Individual European states have sought to integrate health considerations into SEA and into spatial and community planning (+) IA guidelines elaborate on types of potential health effects (+) Ensuring a high level of health protection is prominently featured in European legislation (+) The World Health Organization (WHO) Regional Office for Europe has assumed a lead role in developing and promoting HIA as a policy measure to facilitate health protection (WHOROE, 1999, 2001a,b,c) (+) The European Public Health Strategy also identifies HIA as a means to promote health protection (+) Several European countries (e.g., Netherlands, Ireland, Britain, Sweden) have prepared HIA guidelines (+) The precautionary principle is endorsed in the 1992 Maastricht Treaty of the European Union, the European Court of Justice has defined the conditions for applying precautionary measures in community law and has explained the rationale for precautionary measures, and The Commission of the European Communities has issued a Communications on the Precautionary Principle (+) The Habitats Directive adopts a precautionary approach by ensuring that strategic actions do not exceed limits beyond which irreversible damage can occur (+) Guidelines for European Commission actions and guidance for individual jurisdictions refer to the precautionary principle, HIA, and to risk and uncertainty assessment (EC, 2009a; UK Department of Health (DH), 2007)	committees and research, and guidance documents, at the national/state/territorial level, directed toward progressively integrating (broadly defined) health-related concerns into public policy (Harris and Spickett, 2011) (+) IA legislation includes a definition of the precautionary principle; the Act specifies the decisions for which the Minister must take account of the precautionary principle (+) The government, in response to the independent review of the Australian legislation, committed to produce guidelines regarding the application of the precautionary principle (Australian Government, 2011d) (+) The Minister has the authority to require an environmental audit; provision is made for the preparation of environmental management plans (framework, mitigation, monitoring program, responsibilities) (+) The Hawke report called for the development of foresight reports to help governments manage emerging environmental threats

the United States, Canada, Europe, and Australia. Numerous general and specific references are made to uncertainty and adaptation in the four jurisdictions. More extensive references are made (a different combination in each jurisdiction) to risk assessment and management, to the precautionary principle or approach, to human HIA, to climate change impact assessment, to emergency planning and management, and to AEAM. The only difficult problems specifically referred to are transboundary effects, terrorist attacks, and catastrophes. Minimal references are made to chaos and complexity.

The four jurisdictions collectively address many aspects of uncertainty and uncertainty management. All the jurisdictions provide for follow-up as a means of managing uncertainties. All sponsor research and provide for effectiveness analysis, auditing, and quality assurance to enhance practice and as means of more effectively managing uncertainties. All stress the need to substantiate methods and assumptions and to explain the implications of uncertainties and data gaps. A more systematic effort could be made to identify relevant uncertainty forms and sources, to describe key uncertainty concepts, to identify where and how uncertainties arise in the IA process, and to provide examples of uncertainty management methods. Further direction and advice could be provided regarding documenting uncertainties and concerning the role of uncertainty in IA-related decision making (e.g., screening, scoping, significance determination, option rejection and comparison, the triggering of mitigation, and monitoring requirements). Each jurisdiction could consider the specific sources of uncertainties in IA requirements and guidelines (from the perspective of stakeholders) that have led to unwarranted inconsistencies.

All jurisdictions refer to health risks in IA requirements. More attention is devoted to human health risks and to risks from accidents and natural disasters than to chronic health risks, ecological risks, and perceived risks. The United States provides detailed human health and ecological risk assessment and management requirements and guidance. The other jurisdictions concentrate more on risk management. All jurisdictions could devote more attention to potential risk assessment and management roles in the IA process. Risk assessment and management strengths and limitations, the measures introduced to ameliorate limitations, and similarities and interconnections between IA and risk assessment and management should be addressed. All jurisdictions provide for emergency exemptions from IA requirements. A greater effort could be made to develop and apply abbreviated IA forms for emergency situations and to integrate lessons and insights from emergency planning and management into IA risk management requirements and guidance.

The four jurisdictions have all gradually moved toward the greater integration of climate change considerations and IA requirements. So far, the focus has been on general guidance, applied research, and institutional capacity building. This may lead, over time, to a refinement of general and proposal specific IA requirements to more fully integrate climate change considerations, and to more closely link climate change and sustainability requirements and limits. IA jurisdictions and other stakeholders could benefit, along the way, from sharing experiences and insights (IAIA, 2010). There is likely to be considerable resistance to possible climate change requirements if they could lead to the rejection of major resource development proposals. It is likely that progress in this area will be incremental and tentative at best.

Health effects are mentioned in the IA requirements of all four jurisdictions. But the treatment of health effects is general and fragmentary. The numerous recent HIA initiatives and guidelines are correcting this deficiency. Experiences at the national, subnational, and local level in all four jurisdictions could be instructive. There remains considerable ambivalence regarding whether HIA can operate more effective on its own or under the IA umbrella. More attention needs to be devoted to the interrelationships between HIA and risk assessment and management. The role of the health community in HIA also requires additional consideration. The effectiveness of HIA requirements and guidelines needs to be monitored and evaluated. The many uncertainties associated with identifying, predicting, and managing human health effects should receive particular consideration.

There is considerable variation among the jurisdictions in if and how the precautionary approach or principle is addressed in environmental requirements. When the precautionary principle is inappropriate, IA regulations and guidelines should explain how other mechanisms are to address the relationship of uncertainty and potentially severe consequences. If the principle could be applied, IA requirements should define the principle and specify which harmful effects, uncertainties, and actions trigger its application. Conditions for applying the principle and the decisions to which it applies should be indicated. Guidelines can provide more specific advice regarding roles within the IA process, possible criteria, thresholds, and decision rules, and links to risk assessment and management, human health impact assessment, and uncertainty analyses. The principle's strengths, potential drawbacks, and means of avoiding and reducing potential shortcomings should be assessed. Additional applied research would be helpful.

IA requirements could more explicitly provide for AEAM, appreciating the differences between IA and AEAM. IA guidelines could address potential AEAM and IA interrelationships. The potential role of AEAM in identifying and coping with uncertainties should receive particular attention. More consideration could be given to AEAM strengths, limitations, and measures to reduce limitations. Applied research could explore AEAM adaptations for addressing social and economic concerns, AEAM roles in assessing nonresource management proposals, and steps to ameliorate organizational resistance and inflexibility.

Aside from catastrophic and climate change effects, IA requirements and guidelines in the four jurisdictions largely do not refer to other types of difficult problems or to chaos and complexity. There are scattered references to unusual and complex interactions and effects. Such conditions can affect impact significance interpretations. IA guidelines could devote more attention to other types of difficult problems (e.g., trans-scientific, latent time bombs) in IA practice. Guidelines and applied research could identify insights and implications from chaos and complexity theories for IA practice. At a broader level, risk assessment and management, adaptive management, climate change assessment, HIA, and the precautionary principle collectively represent tools and perspectives for analyzing and coping with uncertainty. They clearly overlap, and they should be complementary. This suggests that there is a need, at the regulatory level, to integrate these individual elements within broader uncertainty management strategies. These strategies, in turn, could guide and bound overall uncertainty-related adaptive public policies, plans, and programs, and provide the foundation for the treatment of uncertainty within IA requirements and guidelines. Mutual learning among IA jurisdictions represents an important knowledge source for enhancing IA uncertainty management capacity.

11.5.2 Management at the Applied Level

Designing an Uncertainty Management Approach This chapter presents a range of concepts and processes dealing with uncertainties—problem types, environment types, uncertainties, general adaptation strategies and tactics, risk assessment and management, health impact assessment, the precautionary principle, adaptive environmental assessment and management, emergency planning and management, and climate change impact assessment. An adaptive IA process could, as illustrated in Figure 11.6, combine elements from some or all of these concepts and processes.

The process first determines whether a risk and uncertainty management framework is warranted. An analysis of the problem is undertaken next. A conventional IA approach is applied if the problem and environmental conditions appear simple and manageable. If the problem and/or environment are determined to be short-term crises, then abbreviated versions of or alternatives to IA are instigated. Uncertainties are considered regardless of the approach adopted. Uncertainty is a central feature of the process if a complex or metaproblem are involved and/or if environmental conditions are complex or chaotic. Pertinent uncertainty forms and sources are identified and analyzed. Uncertainty concepts are applied, where appropriate. An adaptive IA process is designed. Uncertainties are considered for every IA activity and for every interconnection among activities. The process is iterative, flexible, heuristic, open, continuous, cyclical, interactive, and boundary spanning. Institutions are modified and reformed to facilitate and accommodate adaptive IA processes.

Consideration is given to whether one or more of (1) risk assessment and management (RAM); (2) the PP; (3) HIA; (4) AEAM; (5) emergency planning and management; and (6) climate change impact assessment also should be applied. RAM is more suited to situations where human health and/or ecological risks are major concerns and uncertainties are conducive to probabilistic analyses. The PP is usually instigated when there are potentially severe or irreversible adverse environmental impacts, a need for action and significant scientific uncertainties. HIA is undertaken when significant positive and/or negative health effects are likely, there are adequate health-related resources available, and other approaches are unlikely to adequately address health concerns. AEAM is more commonly applied in environmental and resource management situations involving complex but not unique situations, where ecological systems are resilient and conducive to modeling and where stakeholders are willing to engage in workshops. Emergency planning is appropriate when proposed actions could induce environmental hazards and/or are susceptible to adverse environmental effects if natural or human-induced (accidental or deliberate) disasters could occur. Climate change impact assessment is appropriate if proposed actions could have a significant effect on climate change and/or are susceptible to climate change effects.

The characteristics, variations, procedures, methods, IA links, strengths, limitations, and measures to reduce limitations of each approach are considered before it is decided if it should be applied. A clear rationale is prepared for whether the approach is to be applied and, if so, how it is to be linked to or combined with IA. A rationale also is provided for how uncertainty management approaches are combined when more than one approach is used and how the approach(es) are matched to the relevant problem, environment, and uncertainty types. The overall uncertainty management approach integrates all elements into a coherent whole. The approach is applied, monitored, and adapted as needed. Several iterations are required before the approach is finalized. Provision is made for early and ongoing stakeholder involvement in the host of uncertainty management interpretations and judgments.

11.5.3 Adaptive IA Practice by IA Type

Crosscutting Characteristics Table 11.8 provides examples of adaptive IA practice characteristics for various IA types (SA, SEA, EIA, EcIA, SIA, and HIA). All of the IA types recognize the inevitability of risks and uncertainties, the limits of knowledge and the need for an adaptive IA process that can effectively identify, interpret, and manage risks and uncertainties. All acknowledge the need to explain and interpret assumptions and limitations; to justify methods, interpretations, and conclusions; and to seek resilient, robust, and adaptable solutions to complex problems. All emphasize the importance of monitoring and feedback, the need for a plurality of approaches and methods, and the necessity of designing and adapting the process to a complex, uncertain, and fluid context. All stress the importance

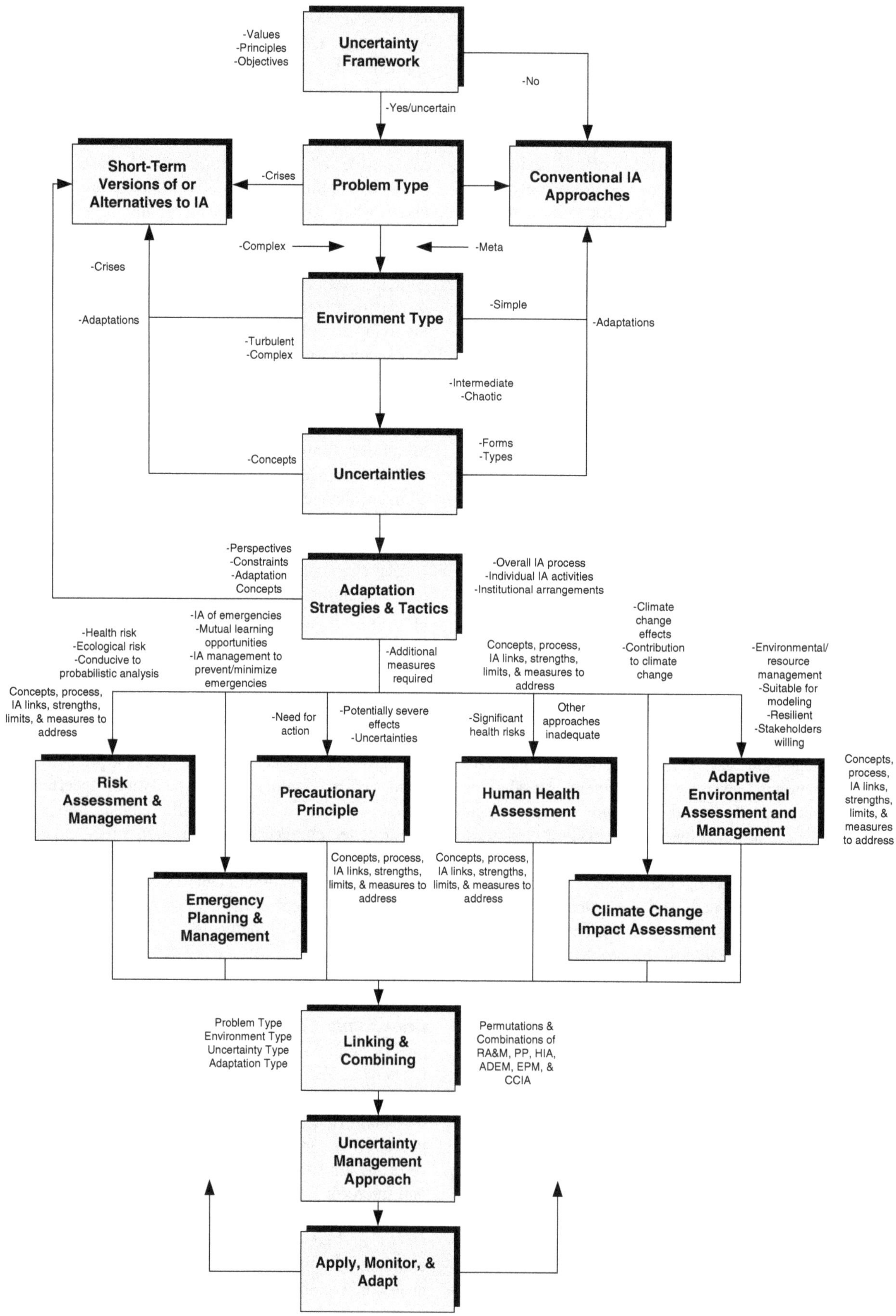

Figure 11.6 Designing an IA uncertainty management approach. Adapted from Lawrence (2005a).

Table 11.8 Adaptive IA Practice by IA Type

Adaptive SA Practice	Adaptive SEA Practice	Adaptive EIA Practice
Emphasizes the importance of continuous learning, feedback, and coordination; process open to critical discussion and local knowledge Sees world as complex, adaptive, self-organizing systems in equilibrium Stresses flexibility, innovation, adaptability, diversity, resilience, and search for creative and pluralistic opportunities and learning Acknowledges and accommodates uncertainties (e.g., sensitivity analyses, scenario analysis) Recognizes many change elements cannot be measured; brings together best of qualitative and quantitative Recognizes interdependencies; seeks mutually reinforcing gains; seeks to preserve and enhance adaptive capacity and human capabilities Recognizes limits and dangers of reductionist science and planning Applies precautionary approach Sustainability definition tailored to context; ensures not so flexible that manipulated or unsustainable outcomes; protects future Integrates climate change concerns Gradually and progressively defines complex and collective vision of what to sustain Seems transformative outcomes (e.g., radically different futures, desirable, resilient, and lasting futures) Employs systems–holistic–dynamic perspective; constant sustainability improvements Health an imperative Characterized by methodological pluralism Integrates postnormal science and complexity theory; transcends individual disciplines Fosters nonlinear knowledge generation, social learning, and system innovation Stresses follow-up, verification, and adaptive management; connects monitoring to adaptive design and management	Seeks an SEA that is as robust as possible; a process that is iterative, flexible, adaptive, dynamic, open, creative, iterative, resilient, communicative, inclusive, continuous, and learning based Sees SEA as a diverse family of approaches that can be adapted to a variety of decision-making settings; no one way or even one best way, at times chaotic; recognizes that many actors with multiple goals in a "messy" reality Combines adaptive management with policy making; is sufficiently flexible that decision makers can take useful elements and apply them Is conscious of and sensitive to the influence of context on aims and outcomes Clearly states assumptions and related predictions, addresses probabilities and confidence levels, carries out sensitivity analyses, uses different scenarios, and predicts in ranges Role for SEA in protecting and improving health, in preventing ill health and in addressing health inequities; consults with health organizations/experts Integrates HIA; treats health as a separate element Acknowledges and exposes uncertainties, risks, and ambiguities; recognizes must deal with an uncontrolled and unpredictable environment Data collection incremental and spread through SEA process; takes care in data and scale choices Recognizes that policies, plans, and programs only partially influenced by environmental analysis; focuses on producing and communicating strategic knowledge Addresses both emergent and deliberate strategies Recognizes that need for flexibility at policy level with limited role for detailed regulation Employs a preventative/precautionary approach Takes care to ensure that environmental protection and sustainability not compromised Integrates comparative risk assessment Assesses effects on and from climate change Emphasizes effective monitoring; specifies organizations responsible for monitoring; monitors outcomes of similar actions; adapts to setting Clearly identifies risks and uncertainties; manages rather than trying to eliminate Integrates climate change concerns, stresses systems resilience and determinants, emphasizes monitoring rather than certainty, utilizes adaptive management and governance, and prioritizes early warning Integrates resilience thinking to address uncertainty and complexity Fosters adaptive SEA systems; designs for resilience and adaptability in the face of uncertainties and risks Prepares uncertainty report, detailing sources and means to reduce, different viewpoints, contingency plans, and postponed decisions	Process adapted to realities, issues, and circumstances of proposal without compromising environmental interest Employs uncertainty assessment methods (e.g., probability analysis, sensitivity analysis, confirmatory analysis) Applies the precautionary principle and risk assessments where needed Locates uncertainty blind spots and vulnerabilities in predictive models; assesses ramifications Assesses alternative future scenarios and reversibility Combines multiple information sources Integrates consideration of natural and human-induced disasters, spills, and accidents Emphasizes prevention over mitigation Tests sensitivity of alternatives ranking and conclusions Estimates the uncertainty factors affecting the impact evaluation and implications; fosters risk aversive decisions Seeks lasting gains and resilient projects and environments; seeks to build desirable and resilient futures Stresses the effective communications of uncertainties Seeks to build to more lasting options with proposed action as catalyst Seeks positive results Designs for surprises and manages for adaptation Assesses effects on and from climate change; addresses the links between climate change mitigation measures and climate change adaptive capacity Recognizes that effectiveness only meaningful in socioeconomic, political, and cultural context of country or countries concerned Integrates mitigation strategies for managing unanticipated outcomes Arranges for monitoring and evaluation; adapts protocols to local needs Recognizes that need for follow-up greatest when inherent uncertainty; employs adaptive management as part of follow-up to address highly unpredictable uncertainties; community fully engaged Tracks cumulative effects Incorporates lessons throughout the project's life cycle

(continued)

Table 11.8 (*Continued*)

Adaptive EcIA Practice	Adaptive SIA Practice	Adaptive HIA Practice
Seeks to foster flexibility, diversity, and resilience of ecological components and systems Recognizes complexity of contemporary ecosystems; more complex than we think or can think Recognizes that ecosystem processes and functions are complex and variable; level of uncertainty increased by social constructs Recognizes that change inevitable Adaptive management a key element of ecosystem approach Estimates uncertainty factors and their effect on interpretations of results and decision-making implications Tests sensitivity of results Acknowledges limits of understanding Integrates precautionary principle; assesses biodiversity threats, and seeks to avoid irreversible biodiversity loss Addresses climate change and disaster impacts on biodiversity Assesses cumulative effects Sees ecosystem management as a learning process that adapts methods and practices to ways in which systems are managed and monitored Recognizes need for flexibility in implementation and policy making Recognizes ecosystem management as a long-term experiment (learning as doing) Includes detailed monitoring and follow-up plans	Recognizes that knowledge of social world and social processes incomplete; social processes constantly changing and vary over space and time Acknowledges that social impacts cannot be precisely defined or quantitatively valued, and that SIA cannot provide definitive answers Acknowledges the gulf between information and knowledge Employs holistic research methods Aware of differential distribution of impacts among groups and burden on vulnerable Incorporates scientific and local and indigenous knowledge, experience, and epistemologies Integrates uncertainty, prevention, and precautionary principles Develops a rich picture of the local community context, including relationship of local community values to planned intervention Addresses impacts from multiple perspectives using multiple methods (e.g., literature, surveys, interviews with community leaders, public consultation events, analyses of comparable actions/environments) Acknowledges complexities and limits Assesses change with and without proposed action against control study Recognizes that impact pathways cut across domains, are iterative, and are direct, indirect, and cumulative Addresses risk perceptions/attitudes and risk communications Recognizes and preserves diversity and seeks to build social resilience and sustainable livelihoods Promotes health and safety; broadly defines health and health determinants Provides for monitoring and managing intended and unintended consequences and social change processes invoked by interventions Addresses uncertainties through contingency planning, adaptive management, and capacity building Provides for social follow-up surveys and research	Recognizes that IA is about decision making in the face of uncertainty and complexity Broadly defines health aspects (e.g., social and behavioral aspects) Flexibility a guiding principle Clearly explains and substantiates models, assumptions, and limitations; data quality, validity, and statistical stability are assessed Recognizes the complexity of multiple levels and causal pathways; acknowledges prediction limitations and implications Differentiates between changes caused by program and other societal changes Precautionary approach: reduces and prevents significant exposures before occur Seeks robust predictions; experimental and adaptive Assesses health impacts of disasters and climate change Identifies controlling and coping strategies Develops a range of adaptation responses Sees health and sustainability as inextricably linked Employs methodological triangulation (statistics and literature, stakeholder inputs, direct observation) Includes an uncertainty analysis; identifies confidence levels Identifies, reduces, and manages uncertainties Stresses the importance of monitoring, evaluation, and follow-up Includes follow-up plan, goals, roles, triggers, reporting, and resources Employs retrospective validation to assess predictions and effectiveness

Sources: Adelle and Weiland (2012), Ahmed (2004), Ali et al. (2008), Ardern (2004), Ayre and Calloway (2005), Bhatia (2007), Bhatia and Seto (2011), Bhatia et al. (2010), Bond (2010), Bond and Morrison-Saunders (2011), Bond et al. (2011, 2012), Burdge (2003b, 2004), Byer et al., (2009), Canter (1993a), Cherp et al. (2007), Clark et al. (2011), Croal et al. (2010), Dannenberg et al. (2006), Devlin (2011), Donnelly et al. (2007), Dovers (2005), Duncan (2008), Duncan and Hay (2007), EC (2009a), Esteves et al. (2012), Faber et al. (2010), Feldman and Khademian (2008), Fischer et al. (2010), Fischer (2003, 2007b), Gardner (2010), Gasparatos et al. (2007), Geneletti (2002, 2003, 2005), Gibson (2006a, 2011), Graham (2002), Grinde and Khare (2008), Hacking and Guthrie (2008), Hanna (2009b), Hermans and Knippenberg (2006), Hindling-Rydevik and Bjarnadóttar (2007), Hodge (2004), Hunsberger et al. (2005), IAIA (1999, 2003, 2005, 2006a), ICPGSIA (2003), João (2007); Khera and Kumar (2010); Kumagai et al. (2006); Kauppinen and Nelimarkka (2004); Kemm (2005); Kørnøv (2009); Kørnøv and Thissen (2000), Krieger et al. (2003), Lane et al. (2003), Mandelik et al. (2005), Morgan (2011, 2012), Morganstern et al. (2008), Marshall (2005), Lobos and Partidário (2010), Meynell (2005), Nilsson and Dalkmann (2010), Noble (2009a), Partidário and Coutinho (2011), Persson and Nilsson (2007), Rajvanshi et al. (2011), Rotmans (2006), Seidler and Bawa (2003), Shepherd (2008), Spickett et al. (2011), Slootweg et al. (2010), Steinemann (2004), Stoeglehner and Wegerer (2006), Storey and Jones (2003), Tetlow and Hanusch (2012), Thérivel et al., (1992); Thérivel (2010), Thompson (2002b), Tickner (2003b,d), Tuinstra et al. (2008), Utzinger et al. (2005), Vanclay (2003, 2010), Wiek and Binder (2005), Winkler et al. (2011), Wilson (2010), Wlodarczyk and Tennyson (2003), Wright et al. (2005).

of enhancing system capacity to adapt to and manage positive and negative change, as interpreted from multiple perspectives. All point to the need for flexibility, creativity, openness, and inclusiveness. All suggest the desirability of an IA process that is iterative, dynamic, innovative, transformative, precautionary, and learning oriented. All emphasize the need for approaches and methods that are holistic, transcend individual disciplines, and systematically address indirect and cumulative effects. There is a general recognition of the overlaps and interdependencies among IA types, the need to systematically explore interconnections and to assess the merits of various integration approaches, and the necessity of integrating and contributing to the IA knowledge base. Many substantive themes (e.g., health, emphasis on the most vulnerable, climate change, contribution to sustainability) are shared among IA types. Differences among the process characteristics for the various IA types are largely a question of degree, emphasis, and orientation.

Adaptive EIA, EcIA. and HIA practice tends to place more emphasis on rational, scientific, quantitative, and technically oriented approaches to uncertainty and risk management. Adaptive SA, SEA, and SIA practice tends to stress nonrational, trans-scientific, qualitative, and collaborative approaches to uncertainty and risk management. Given the many shared themes, overlaps, and interconnections among the adaptive IA process type characterizations, the sharing of knowledge and experiences would seem highly desirable, appreciating the implications of differences among the IA types and in terms of context.

Adaptive SA Practice Adaptive SA practice emphasizes the search for transformative change and the realization of substantive sustainability outcomes. It is context dependent. It fosters methodological pluralism, nonlinear knowledge generation, social learning, and system innovation. It stresses follow-up and adaptive management. It recognizes the limits of reductionist science, applies a precautionary approach, and acknowledges and accommodates uncertainties. It is flexible, innovative, and creative. It employs a holistic systems vision, recognizes interdependencies, and seeks pluralist and resilient solution to complex and uncertain environmental problems.

Adaptive SEA Practice Adaptive SEA practice recognizes the need for a plurality of approaches that can be adapted to and evolve in concert with changing decision-making needs and requirements at multiple levels. It focuses on the search for proposal options and mitigation/enhancement measures that are resilient, contribute to lasting positive change, avoid and prevent significant negative, especially catastrophic change, and can adaptively respond to and manage surprise. It is highly conscious of contextual variation; acknowledges and exposes risks, uncertainties, and ambiguities; and recognizes the limits of knowledge and control. It fosters adaptive SEA systems, designs for resilience, and employs a preventative–precautionary approach.

Adaptive EcIA Practice Adaptive EcIA practice focuses on fostering diverse and resilient ecological components and systems, and on adaptive and effective ecosystem management. It recognizes the complexity of ecosystem processes and functions, acknowledges and estimates uncertainties and associated implications, and tests the sensitivity of results. It fully integrates the precautionary principle, assesses biodiversity threats, and seeks to avoid irreversible biodiversity loss. It appreciates the limits of knowledge, the importance of flexibility and resilience, and the critical role of follow-up.

Adaptive SIA Practice Adaptive SIA practice stresses the limits of knowledge and control, the importance of bottom-up participation, and the value of local and indigenous knowledge and perspectives. It emphasizes the need to differentiate the distribution of impacts among population groups and to ameliorate the burden on and enhance the lives of the most vulnerable. It appreciates the necessity of a sound understanding of community context. It emphasizes the importance of preserving and enhancing social diversity, resilience, and sustainable livelihoods. It employs holistic research methods; integrates the uncertainty, prevention, and precautionary principles; and addresses impacts from multiple perspectives. It seeks to recognize and preserve diversity, build capacity, promote health and safety, and build social resilience and sustainable livelihoods.

Adaptive HIA Practice Adaptive HIA practice emphasizes the importance of broadly defining health and health determinants and the need to explore the complex web of causal pathways that stretch across multiple levels. It fully substantiates methods and assumptions. It acknowledges the inevitability of limits in the ability to predict changes and to differentiate impacts from other health-related societal changes. It appreciates the importance of robust predictions and management approaches that facilitate resilience. It employs multiple and effective coping strategies. It seeks to systematically integrate retrospective analyses.

11.6 CONTEMPORARY CHALLENGE—CLIMATE CHANGE

11.6.1 Definitions and Distinctions

Climate change is a change in the climate that is directly or indirectly attributable to human activity that alters the global atmosphere composition over observed time periods, and is in addition to natural variation (IAIA, 2012). It can be identified by changes in the mean of variability of its properties and persists over an extended period (IAIA, 2012). Climate change is a multidimensional issue (e.g., development, security, equity) (IAIA, 2012; IAIA and World Bank, 2010). IA can help integrate climate change considerations into policies, plans, programs, and projects in a manner that reduces the contributions to, impacts from, and vulnerability of natural and human systems against actual and expected climate change.

Climate change IA (CCIA) addresses the climate change implications of greenhouse gas (GHG) emissions from a proposed action and the effects on a proposed action and the environment from climate change (FPTCCCEA, 2003; US CEQ, 2010a). CCIA integrates climate change considerations into each IA activity (e.g., scoping, alternatives analysis, mitigation, follow-up). Related concepts include (1) adaptation (initiatives and measures to reduce human and natural system vulnerability against actual or expected climate change effects), (2) resilience (capability to anticipate, prepare for, respond to, and to recover from climate-related threats), (3) adaptive capacity (ability of system to adapt to climate change, moderate potential damages, take advantage of opportunities, or cope with consequences), (4) risk (combines magnitude of consequences from climate change impacts with likelihood of occurrence), (5) vulnerability (degree system is susceptible to or unable to cope with climate change adverse effects including climate variability and extremes), (6) mitigation measures (measures such as technological change and substitution that directly or indirectly reduce resource inputs and greenhouse gas emissions and enhance sinks), (7) adaptation measures (measures that increase resistance to withstand, recover from and adapt to climate change), and (8) uncertainty (degree to which a future value such as the future state of the climate system is unknown) (IAIA, 2012; US NRC, 2011a).

11.6.2 Climate Change and the IA Process

Integrating climate change into the IA process has, as summarized in Table 11.9, implications for every IA activity. A basic distinction is drawn, in IA process design and management, between impacts on (avoided and minimized through mitigation) and from (avoided and minimized through adaptation) climate change. Often, proposed actions entail both mitigation and adaptation. The integration of climate change operates at both the SEA and project EIA level. SEA tends to be especially important in integrating climate change considerations into public decision making and in framing the treatment of climate change considerations in public and private EIA project-level decision making (IAIA, 2012).

The effective management of uncertainties and risks, the importance of adaptability and resilience, and the need to fully consider vulnerability and equity are central features of CCIAs. Other key aspects of the IA process, when addressing climate change, include the importance of linking climate change effects and sustainability principles, the need to draw upon applied climate change research, resources and good practice guidance, and the need to place climate change predictions and management measures within the context of cumulative effects on resources and biodiversity, larger scale inventories, targets and thresholds, and a range of climate change scenarios.

11.6.3 Climate Change Good Practices

The integration of climate change and IA practice is arguably still in its infancy (Yi and Hacking, 2012). Although the role of climate change in IA practice has been an issue for more than a decade, good practice examples are still limited in number and scope. As summarized in Table 11.10, considerable work remains to build the capacity to undertake and effectively participate in CCIA. Major gaps and deficiencies remain regarding key data sources, especially at the subnational and regional levels. A considerable effort is needed to enlarge the level of climate-related expertise and to foster communications, mutual learning, and networking among stakeholders. Education and the building of political will and support are essential, given the likelihood that resistance to change will be significant. As the capacity to undertake and participate effectively in climate change impact assessment is enhanced, requirements and guidelines can be refined, harmonized, and more broadly and systematically applied.

Good practice climate change impact assessment guidance still remains quite general. Pilot projects, applied research, and the refinement of frameworks and methods can help extend such guidance, with due allowance for regional and sectoral variations. Standard protocols and guidance should help to reduce the gaps and inconsistencies in the treatment of climate change in IA practice (Yi and Hacking, 2012). Particular care should be taken to fully, explicitly, and systematically address risks, uncertainties, and adaptive capacity, to thoroughly address issues related to the vulnerability of natural and human systems, to favor mitigation, enhancement, and adaptation measures that can flexibly respond to a range of potential future conditions, and to systematically integrate equity-related concerns into all aspects of climate change impact analysis and management. CCIA should not be viewed as an expert-driven procedure. Community and indigenous knowledge, understanding, and support are crucial. The assessment of individual proposals is unlikely to be effective if not framed within broader scale climate change databases, strategies, goals, requirements, initiatives, and policies.

11.7 SUMMING UP

This chapter addresses how IA processes adaptively anticipate and respond to the uncertainties associated with difficult problems in chaotic and complex environments. The three stories address uncertainty in different ways. The first story describes a strategic cumulative environmental assessment that adaptively evolved in response to unanticipated methodological challenges and the changing needs and expectations of the major partners in the process. The second story illustrates the risks associated with a technically driven IA process that does not appreciate or mitigate potential sources of proposal failure, does not recognize changing institutional arrangements, is insensitive to varying stakeholder perspective and

Table 11.9 Climate Change and the IA Process

Activity	General	Impact on Climate Change (GHG)	Impact from Climate Change
Screening and scoping	Establish boundaries and focus assessment Scope uncertainties Identify proposal and climate variables that should include Determine who to conduct assessment	Determine if GHG considerations an issue and should be addressed in greater detail; provide rationale Determine whether qualitative or quantitative assessment appropriate Identify GH-related policies, plans, programs, and targets	Determine if impact from climate change an issue (proposal or environment vulnerable to climate change, beneficially or adversely affected); provide rationale Decide approach for addressing climate change impact
Baseline analysis	Cross-reference research Consider links between climate change and resources Identify a reasonable and credible range of climate change scenarios (e.g., minimum, intermediate, maximum) Consider current and future climate with and without climate change	Link to national, subnational, and regional GHG inventories Determine if GHG emissions have already breached a cumulatively significant level, could manifest in regional or local geographic area, and could be affected by proposed action	Describe how global climate changes could manifest in regional or local geographic area and could affect proposed action Clarify changing climatic parameters (magnitude, distribution, rate of change) Assess impacts of current land use and policies
Proposal characteristics	Identify how climate change and GHG can be integrated into proposed action (e.g., reason for action) Identify proposal characteristics potentially linked to climate change	Identify and quantify (if possible) direct and indirect GHG emissions (composition, magnitude, intensity) Design/operate to avoid and minimize GHG emissions Determine if medium to high emissions and if diverges from industry or government best practices and from reduction targets or objectives	Identify proposal characteristics potentially vulnerable to climate change impacts Assess adaptive capacity Design for resilience
Alternatives analysis	Explore combinations of climate scenarios and options At SEA level, seek to develop and evaluate climate friendlier policy, planning, and program alternatives	Identify alternatives that meet need and reduce GHG emissions Compare alternatives under climate scenarios Select preferred alternatives	Identify avoidance, reduction, and compensations options Identify alternatives that facilitate adaptation to/resilience from climate change Assess options and selected preferred options
Impact identification and prediction	Assess social, economic and, environmental impacts Capture and combine impact ranges Consider inter- and intragenerational equity Identify urgent large risks to determine whether impact management is necessary	Clarify magnitude, intensity, and timing of proposal emissions Analyze climate change effects on environment (individual, cumulative, direct, indirect, large-scale impacts on carbon sinks) Determine possible consequences of accidents or spills Assess degree policies, plans, programs, and projects promote reduction in GHG emissions	Identify climate change risks Analyze cumulative climate change effects on proposed actions (positive or negative, feasibility, viability, sustainability) Analyze effects of climate change on impacts resulting from proposal Analyze extent areas, water resources, land use types, communities, and socioeconomic groups vulnerable or at risk to climate change
Significance determination	Cross-reference government (international, national, subnational, regional) standards and policies Identify impact significance	Set threshold for GHG emissions Assess whether the change in cumulative impact is significant Assess whether net emissions consistent with industry/government best practices and reduction targets or thresholds	Assess significance of climate change impacts on proposed actions and on impacts resulting from proposal
Mitigation, enhancement, and adaptation	Identify mitigation, enhancement, and adaptation objectives	Determine whether additional GHG management necessary Identify additional measure to prevent/reduce GHG emissions (e.	Identify mitigation measures to reduce proposal vulnerability and increase resistant to extreme circumstances and resilience to

(continued)

Table 11.9 (*Continued*)

Activity	General	Impact on Climate Change (GHG)	Impact from Climate Change
	Distinguish between adaptation (anticipated and potential) and adaptive capacity Ensure broad range of mitigation and adaptation strategies (e.g., pricing, technology, behavioral change, ecosystem management)	g., design, operations, energy efficiency measures); directly or indirectly Address how to enhance or impede attainment of GHG targets, policies, goals, and regulations Identify compensatory mitigation (e. g., carbon offsets)	recover quickly after extreme conditions Identify adaptation opportunities; adjustments in human and natural systems to climate stimuli and effects Clarify adaptive management plan to reduce climate change risks Identify measures to enhance resilience
Uncertainty management	Explicitly identify uncertainties (e. g., regarding proposed actions, modeling, future state of climate, and effects) Incorporate ongoing information gathering and risk assessment Apply precautionary principle Identify uncertainty decision rules (e.g., maximin, minimax regret)	Apply methods to address uncertainties (e.g., sensitivity analyses, scenarios, probabilistic analyses, simulation studies) in GHG emissions forecasts	Apply methods to address uncertainties in climate change forecasts and related impacts on proposals and environment (e.g., scenarios, story lines) Identify how adaptive planning can be used to address uncertainties Seek to make actions more adaptable and flexible
Public participation	Identify questions for public review; integrate public concerns, preferences, and knowledge Ensure scenarios credible and acceptable to stakeholders Communicate results and uncertainties to stakeholders	Integrate public concerns and preferences regarding options and proposal acceptability Integrate public concerns and preferences regarding measures to avoid, minimize, and offset GHG emissions	Integrate public concerns and preferences regarding options and proposal acceptability Integrate public concerns and preferences regarding climate proofing measures and climate resilience enhancement measures
Decision making	Explain and justify how results obtained and degree of confidence Confirm consistency with jurisdictional requirements and initiatives Decide based on precautionary principle (do no harm) and sustainability principles Determine if proposed risks and impacts acceptable Identify conditions of approval	Address GHG emissions through established jurisdictional policies or regulations Determine conditions that ensure acceptable GHG emissions Determine if GHG emission levels, after mitigation, acceptable	Determine if proposed action too vulnerable to climate change to proceed Determine if adaptive strategies and methods acceptable Determine conditions that ensure acceptable climate proofing and resilience
Follow-up and management	Monitor, evaluate, manage, and communicate Identify management responsibilities; distinguish between public and private sector risks and responsibilities Incorporate lessons learned Address evolving proposal and climate change knowledge, technology, policy, and legislation (pre- and postapproval) Include adaptive management plan	Clarify how design/operations takes GHG considerations into account Identify performance measures to monitor GHG emissions and links to climate change Verify GHG emissions forecasts and mitigation effectiveness Implement remedial actions as needed	Monitor status of proposal and effectiveness of mitigation measures Monitor climate change and climate proofing Implement remedial actions as needed

Sources: Agrawala (2010), Byer and Yeomans (2007), Byer et al. (2011), CDFAIT (2002), CEAA (2003c), CFTPC (2003), Draper (2010), Eccleston (2008), Farber (2003–2004), FPTCCCEA (2003), IAIA and World Bank (2010), Smith (2010), Spickett et al. (2011), US CEQ (2010a).

Table 11.10 Examples of Climate Change Good Practices

Institutional Arrangements and Capacity Building	Requirements and Guidelines	IA Process Design and Management
Institute or supplement GHG inventories at national, subnational, and regional levels	Establish GHG targets, policies, and goals at national, subnational, and regional levels	Consider vulnerability of human and natural systems and groups; consider potential health effects
Institutionalize climate change integration into policies, plans, programs, and projects at multiple levels; ensure adequate resources	Clarify legal requirements (e.g., required monitoring of carbon emissions)	Use latest credible scientific information and projections
Supplement climate change expertise (e. g., education, training) and climate science capacity	Close decision-making gaps	Link time horizons to national and international climate scenarios
Institute/supplement climate monitoring network	Foster political will	Explore synergies between negative and positive climate and environmental interactions
Institute pilot program to demonstrate how to achieve greater climate protection	Seek government and business commitments to address climate change before making decisions	Place within context of international agreements and national and subnational strategies
Enhance links to related planning processes and legislation	Strengthen legislation and regulations	Look at climate change in context of other pressures on food, water, and biodiversity
Support climate-related information and good science	Provide updated and improved guidance for addressing and managing climate change at SEA and project EIA levels, and for different proposal types, sectors, and regional settings	Link to broader inventories and targets
Foster climate-related education, networking, and communications between scientific community and IA practitioners	Provide climate change guidance on methods, significance determination, mitigation, and adaptation	Distinguish between adaptation and adaptive capacity; importance of flexibility regarding future changes
Foster cross-disciplinary and cross-sectoral coordination	Promote use of guidelines	Explicitly identify uncertainties (e.g., methods, future conditions, impacts, mitigation effectiveness) and implications
Facilitate public access to climate change information	Ensure adequate monitoring of current emissions	Embrace complexity, give more serious attention to worst-case scenarios, always cope with uncertainty, and clearly explain uncertainties
Structure and post climate change information on line in searchable ways; linked to GIS	Emphasize role of monitoring and adaptation	Ensure indirect negative impacts of secondary proposal activities addressed with, for example, renewable energy projects
Document best case examples of addressing climate change in different settings (e.g., developing countries) and at different levels (e.g., SEA)	Require consideration of sustainability at all levels	Integrate uncertainty considerations into each IA activity (e.g., scoping, alternatives, impact management)
Foster private sector capacity for companies to understand and integrate climate change impacts	Harmonize protocols for analysis and for incorporating climate change into IA	Use risk management frameworks to understand implications of climate change impacts and uncertainties for planning, investments, and operational decisions
Seek to develop and refine spatially specific climate change scenarios	Harmonize climate change data gathering through regulations	Consider equity (sociocultural, socioeconomic, inter- and intragenerational), vulnerabilities, and adaptive capacities
Use and promote sustainability principles as a means of framing consideration of climate change		Link climate change IA and sustainability
		Seek no regrets measures that generate net social, ecological, and economic benefits regardless of extent of climate change
		Utilize assistance from experienced communications specialists to help inform decision makers of relevance and implications of climate change to proposal
		Integrate local and indigenous knowledge
		Assess interactions between mitigation and adaptation measures and consequences
		Integrate climate change considerations into life cycle of infrastructure and business investment/performance
		Develop and test different planning, policy, and program options against different climate scenarios
		Explain and justify how results obtained and degree of confidence
		Draw upon IA and climate change information and resources including literature, guidelines, and best practices

Sources: Brown (2003), Byer et al. (2009, 2011), Burdge (2008), Draper (2010), Farber (2003–2004), FPTCCCEA (2003), Furlow (2010), Gardner (2010), IAIA (2012), IAIA and World Bank (2010), Larsen (2012), Larsen et al. (2012), Smith (2010), Spickett et al. (2011), US CEQ (2010a), Watson (2010), Wende et al. (2012), Wilson (2010), Yi and Hacking (2012).

interests, and is unwilling or unable to adapt to an evolving regional context. The third story describes how a proponent learned from past mistakes and built on past successes. The three stories underscore the need for IA processes to be open and adaptive if they are to operate effectively in complex, rapidly changing decision-making environments fraught with uncertainty.

The problem is a combination of confusion regarding the nature of uncertainty, risk and health effects, and ambivalence concerning the most appropriate approach, or combination of approaches, for managing IA process uncertainties. The direction involves an enhanced understanding of uncertainty, difficult problems, chaotic and complex environments, and adaptation, coupled with a selective blending of general adaptation strategies and tactics, risk assessment and management, the precautionary principle, human health impact assessment, emergency planning and management, adaptive environmental assessment and management, and climate change impact assessment.

Problems are triggered by a question or a situation, are negative, and need to be addressed. Problem-solving processes identify, define, bound, and state the problem. The problem is then progressively refined and addressed. There are simple or tame problems, compound or semistructured problems, complex or ill-structured problems, and crises or metaproblems. Simple and compound problems can be addressed by routine and conventional IA procedures, respectively. Complex and metaproblems are more difficult. They are real, complex, messy, transcend boundaries, and disciplines, are prone to dilemmas, impossibilities, and crises, and require ingenuity. An adaptive IA process is needed to properly cope with difficult problems.

IA processes should suit the *environment* or context. There are many environmental components or systems (e.g., ecological, social, economic, institutional, technological). There are simple, moderately complex, and highly chaotic and/or complex environmental systems. Command and control and conventional IA processes, respectively, operate effectively in simple and moderately complex environments. Chaotic or complex environmental systems are more problematic. They exhibit such properties as self-organizing, emergent, turbulent, nonlinear, irreducible, random, incoherent, unpredictable, interdependent, resilient, and unstable. Adaptive IA processes and organizations can operate more effectively in chaotic or complex environments.

Uncertainty, broadly defined, is any situation where we are not absolutely sure. There are many uncertainty forms (e.g., quantitative, qualitative, objective, subjective, methodological, perceived, knowledge, values, past, present, or future conditions). Uncertainties can pertain to any IA process activity. There are many possible uncertainty sources (e.g., data or knowledge deficiencies, theoretical or methodological deficiencies, resource limits, poor communications, natural variations, novel situations). Several uncertainty-related concepts are potentially relevant (e.g., ignorance, errors, indeterminism, vagueness, ambiguity, doubt, confusion, surprise, deep/extreme).

There are many ways of addressing *uncertainty in the IA process*. A perspective change is first required (e.g., uncertainty as a fundamental attribute of the process). Measures can be introduced into each IA activity to anticipate, cope with, learn from, and manage uncertainties. Uncertainty management measures can be integrated into problem definition, scoping, proposed action determination, option identification and evaluation, individual and cumulative impact identification, prediction and interpretation, mitigation and compensation, impact and uncertainty management, public and agency consultation, study team management, decision making, monitoring, and IA institutional arrangement reform. Uncertainty management is facilitated by insights and lessons from design, ingenuity, creativity, strategic choice, consilience, and holistic science.

Adaptive environmental assessment and management treats environmental management as a quasiexperiment (i. e., probing ecosystem responses to human activities). The AEAM process is an iterative cycle of planning, implementation, monitoring, research, and re-examination. AEAM processes are typically built around a series of workshops. The workshops construct and apply a model that characterizes critical environmental conditions and interactions and tests possible management actions and alternative assumptions. The periods between workshops are devoted to consolidation and refinement. The process is guided by a core group and by specialist support staff. Workshops involve policy people, managers, and a diversity of stakeholders. Key indicators are monitoring throughout implementation. Data obtained during monitoring are analyzed, documented, and fed into each process activity. The process is open, continuous, cyclical, evolving, and highly iterative. AEAM has much to offer IA but there also are important differences between the two fields. These differences and AEAM strengths and shortcomings should be carefully considered when connecting or integrating IA and AEAM.

Risk combines frequency or probability with a harmful environmental consequence. There are many risk types (e.g., economic, health, environmental, from natural or human sources, chronic, acute, for overall and for sensitive populations, deterministic, and probabilistic). Potentially relevant risk-related concepts include risk assessment, perceived risks, risk communications, comparative risk assessment, risk acceptability or tolerance, risk management, disasters and hazards, human health risk assessment, and ecological risk assessment. Risk assessment processes include, for example, problem and analysis plan formulation, receptor determination, pathway and receptor characterization, hazard identification and analysis, exposure assessment and response, risk characterization, risk evaluation, decision making, implementation, and monitoring. Risk assessment processes integrate research, public perceptions, stakeholder concerns and preferences, option analyses, mitigation measures, and uncertainty analyses. There is considerable potential to more effectively link and integrate IA and emergency planning and management. There are many similarities but also important differences between IA and risk assessment and management. Similarities, differences, and risk assessment and management strengths, deficiencies, and measures to address deficiencies, all should be considered when linking and integrating IA and risk assessment and management.

The *precautionary principle* responds to the dilemma of what to do when there is a need to take action because of potentially severe environmental consequences but shortcomings in the scientific knowledge base. There are multiple interpretations of what represents severe harm potential, inadequate scientific evidence, and the basis for action (e.g., inaction not justified, rejection of proposal, only proceed if proven safe, proceed if reasonable case can be made, proceed with caution). Applying the precautionary principle requires thresholds, criteria, decision rules, definitions for key terms, and institutional arrangements. Some argue that the principle also requires a reversed burden of proof, open transparent and democratic decision making, systematic alternatives analyses and greater decision-making weight on prevention, risk avoidance, ignorance, and environmental values. A precautionary IA process involves screening (whether the principle is to be applied), scoping, goal setting, study design, an analysis of need and alternatives, adaptations to the principle to suit the situation, refining, and applying precautionary thresholds, criteria, and procedures, precautionary decision making, the taking of precautionary action, implementation, monitoring, follow-up, and evaluation, all within an open, transparent, and democratic IA process. The precautionary principle is highly controversial, but addresses a valid concern. Ascribed strengths and drawbacks need to be carefully considered. A clear rationale should be presented for if and how the principle (or an alternative approach) is applied in IA practice.

Human health impact assessment is concerned with positive and negative, certain and uncertain human health effects. HIA is closely connected with other types of IA, draws upon an interdisciplinary knowledge base, and can contribute to sustainability. It assumes many forms (e.g., quick screening, rapid appraisal, standard HIA, comprehensive HIA). It can be prospective, retrospective, or concurrent. It can be broadly or narrowly defined. HIA processes tend to begin with screening, scoping, a background analysis, and a prognosis of future health-related environmental conditions. Health effects associated with options and before and after mitigation are predicted, summarized, and evaluated. HIA is supported by quantitative (e.g., risk assessment) and qualitative (e.g., health impact assessment) procedures. The health and risk analyses provide the basis for management measures, documentation, conclusions, recommendations, and decision making. Results are monitored and evaluated. Agencies and the public are heavily involved in the process. Health effects are addressed by HIA being undertaken independently, broadening IA, merging IA and HIA, selectively integrating IA and HIA, or broadening risk assessment and management. HIA is a newly emerging and rapidly expanding field of IA practice. It has many attributes, strengths, and limitations that should be carefully considered, especially its uncertainty management procedures.

The *four jurisdictions* (United States, Canada, Europe, Australia) address many aspects of uncertainty. IA guidelines could devote more attention to difficult problems (e.g., trans-scientific, latent time bombs) in IA practice. Guidelines and applied research also could identify insights and implications from chaos and complexity theories for IA practice. A more systematic effort could be make to identify relevant uncertainty forms and sources, to describe key uncertainty concepts, to identify where and how uncertainties arise in the IA process, and to provide examples of uncertainty management methods. Further direction and advice could be provided regarding documenting uncertainties and concerning the role of uncertainty in IA-related decision.

All jurisdictions refer to health risks in IA requirements. More attention could be devoted to chronic health risks, ecological risks, and perceived risks. All jurisdictions could devote more attention to potential risk assessment and management roles in the IA process. A greater effort could be made to develop and apply abbreviated IA forms for emergency situations and to integrate lessons and insights from emergency planning and management into IA risk management requirements and guidance. The four jurisdictions have all gradually moved toward the greater integration of climate change considerations and IA requirements. IA jurisdictions and other stakeholders could benefit from sharing experiences and insights.

Health effects are mentioned in the IA requirements of all four jurisdictions. But the treatment of health effects is general and fragmentary. The numerous recent HIA initiatives and guidelines are correcting this deficiency. The role of the health community in HIA requires additional consideration. The effectiveness of HIA requirements and guidelines needs to be monitored and evaluated. The many uncertainties associated with identifying, predicting, and managing human health effects should receive particular consideration. There is considerable variation among the jurisdictions in if and how the precautionary approach or principle is addressed in environmental requirements. If the principle is to be applied, IA requirements should define the principle and specify which harmful effects, uncertainties, and actions trigger its application. Conditions for applying the principle and the decisions to which it applies should be indicated. IA requirements could more explicitly provide for adaptive management in general and AEAM in particular, appreciating the differences between IA and AEAM. IA guidelines could address potential AEAM and IA interrelationships. The potential role of AEAM in identifying and coping with uncertainties should receive particular attention.

At a broader level, risk assessment and management, adaptive management, climate change assessment, HIA, and the precautionary principle collectively represent tools and perspectives for analyzing and coping with uncertainty. They clearly overlap, and they should be complementary. This suggests that there is a need, at the regulatory level, to integrate these individual elements within broader uncertainty management strategies. Mutual learning among IA jurisdictions represents an important knowledge source for enhancing IA uncertainty management capacity.

Uncertainty management at the applied level involves selectively combining concepts (related to problem types, environment types, and uncertainties) and approaches (general adaptation strategies and tactics, risk assessment and management, health impact assessment, the precautionary principle, emergency planning and management, adaptive environmental assessment and management, climate change impact assessment), within an adaptive IA process. Designing an IA uncertainty management approach entails formulating an uncertainty framework (to identify relevant values, principles, and objectives); identifying the applicable problem and environment type (to determine the appropriate IA approach); characterizing uncertainties; formulating and applying general adaptation strategies and tactics; determining whether and how risk assessment and management, the precautionary principle, health impact assessment, emergency planning and management, climate change impact assessment, and adaptive environmental assessment and management could be used to manage risks, uncertainties, and health effects; linking and combining the concepts and approaches; formulating an overall uncertainty management approach; and applying, monitoring, and adapting the approach.

All the IA process types (SA, SEA, EIA, EcIA, SIA, HIA) recognize the inevitability of risks and uncertainties, the limits of knowledge, and the need for an adaptive IA process that can effectively identify, interpret, and manage risks and uncertainties. All suggest the desirability of an IA process that is iterative, dynamic, innovative, transformative, precautionary, and learning oriented. Many substantive themes (e.g., health, emphasis on the most vulnerable, climate change, contribution to sustainability) are shared among IA types. Differences among the process characteristics for the various IA types are largely a question of degree, emphasis, and orientation. Given the many shared themes, overlaps, and interconnections among the adaptive IA process type characterizations, the sharing of knowledge and experiences would seem highly desirability, again, however, appreciating the implications of differences among the IA types and in terms of context.

Climate change is a change in the climate that is directly or indirectly attributable to human activity that alters the global atmosphere composition over observed time periods, and is in addition to natural variation (IAIA, 2012). It can be identified by changes in the mean of variability of its properties and persists over an extended period (IAIA, 2012). Climate change addresses both the climate change implications of greenhouse gas emissions and the effects on a proposed action and the environment from climate change.

Climate change considerations can be integrated into each IA process activity. It tends to operate more effectively when applied at both the SEA and project EIA levels. Climate change impact assessment incorporates and applies such concepts as adaptation, adaptive capacity, risk, vulnerability, mitigation measures, adaptation measures, equity, resilience, and uncertainty. The uncertainties with climate change predictions and management and links to sustainability need to be fully and systematically considered. Consideration should be given to climate change IA good practices, appreciating the need for contextual adjustments.

Chapter 12

How to Connect and Combine IA Processes

12.1 HIGHLIGHTS

In this chapter, we explore how the individual IA processes, the IA process types, and the approaches for addressing contemporary challenges presented in Chapters 3–11 might be connected and combined at both the regulatory and applied levels. We also address the contemporary challenge of matching process and context, identify the limits of synthesis, and suggest future action priorities.

- The analysis begins in Section 12.2 with three applied anecdotes concerned with the role of synthesis in IA practice. The first story describes an approach for bridging the gap between SEA and project EIA. The second story uses an example to illustrate the role of capacity building in establishing a foundation for integrated good IA practice. The third story describes how IA can serve both its traditional purposes, and provide a means of achieving broader institutional, social, and environmental ends.
- In Section 12.3 we define the problem, which is how to adapt, connect, and combine IA processes, IA types, and approaches for addressing contemporary challenges to suit the situation. The direction is frameworks and procedures for matching IA processes and contexts, for connecting IA types, and for connecting and integrating IA processes at the regulatory and applied levels, and in a manner that effectively addresses both recurrent problems and contemporary challenges.
- In Section 12.4 we explore how IA legislation, regulations, and guidelines could be reformed and refined to better address the regulatory deficiencies and opportunities described in Chapters 2–11.
- In Section 12.5 we address interconnections among the recurrent problems and the IA processes that seek to more effectively manage those problems.
- In Section 12.6 we identify links, overlaps, and middle-ground concepts between and among various IA types (SA, SEA, EIA, EcIA, SIA, and HIA).
- In Section 12.7 we address the contemporary challenge of matching process and context. We identify relevant IA process attributes, and describe ways of facilitating the fit between process and context.
- In Section 12.8 we address interconnections among contemporary challenges and ways of integrating responses to contemporary challenges into IA process design and management.
- In Section 12.9 we present examples of how composite IA processes could be formulated and applied.
- In Section 12.10 we address the limits of synthesis. We also identify priorities for future action.
- In Section 12.11 we provide an overview of the major insights and lessons derived from the analysis.

12.2 INSIGHTS FROM PRACTICE

12.2.1 SEA and EIA—Bridging the Gap

The Austrian transport infrastructure is relevant to the European Union as major European road and rail transit lines, as well as the waterway on the Danube River, cross the country. Therefore, the transport infrastructure network is regularly expanded not only for the national transport demand, but also for European transit. The capacity of existing transport lines is regularly increased and new transport lines are added. Despite the huge strategic importance of the transport network a legally binding intermodal national transport plan does not exist. Only a development strategy for the national transport infrastructure (in German: "Ausbauplan Bundesverkehrsinfrastruktur," BMVIT, 2012) is published by the Ministry of Transport, Innovation and Technology that explains the major road and railway projects for the coming years.

When the SEA directive of the European Union went into force in 2001, there was a chance that only the network of national roads and the waterways might have been affected by SEA (Weber and Stöglehner, 2001): because of the special regulations of Austrian transport planning and the screening conditions of the SEA directive. The scope of SEA in Austrian transport planning might, therefore, have been rather limited. Yet, the Austrian government decided to implement a so-called "strategic assessment- transport"

Impact Assessment: Practical Solutions to Recurrent Problems and Contemporary Challenges, Second Edition. David P. Lawrence.

(SA-T, 2005; in German: "Strategische Prüfung Verkehr"—SP-V) for all changes of the transport infrastructure network including railways, waterways, and national roads. This initiative can be seen as an attempt to introduce not only environmental assessment, but also a means of integrating the first steps toward sustainability assessment into transport planning.

If a change of the transport infrastructure network is planned a process according to the SEA directive has to be carried out. An environmental report has to be drafted, consultations with the public and environmental authorities and in defined cases neighbor states have to be held, the results have to be taken into account in the decision, information about the decision has to be provided, and monitoring has to be implemented. Additionally, the SA-T law introduces objectives that have to be addressed by the change of the national transport network: inter alia, sustainable transport of persons and goods, aiming at a high level of environmental protection, securing a high quality transport network, intermodal transport organization, and high cost–benefit relation. Furthermore, it is stated that in the environmental report the changes in the transport network and their expected benefits have to be reasoned, and that intermodal alternatives have to be included.

The SA-T has some conceptual pitfalls as it is organized similar to EIA: a proponent hands in an environmental report, which is then approved by the Ministry of Transport. As proponents are mainly state-owned companies that are responsible for highway, railroad, or waterway construction the issue of bias come into question. This causes the problem that when one asks a highway company to carry out an assessment of alternatives across transport modes, one will very likely get a road as a result, and not a railway. In this way, the state transfers the responsibility of intermodal, strategic transport planning on the national level to the infrastructure companies that have the purpose to build and maintain specific transport infrastructures. For this reason, one can, inter alia, be skeptical if reducing the transport demand or developing serious intermodal alternatives are options fairly enough discussed in such a planning culture.

So far, seven SA-Ts were completed, six for national roads and one for a railway project (BMVIT, n.y.). All SA-Ts had a wider scope than environmental concerns, for example, a substantiation of the respective transport lines and an intermodal assessment of the transport demands. The assessment of roads regularly comprises a cost–benefit analysis, an environmental assessment, and a sustainability appraisal as stated in a guidance for national road SA-Ts (BMVIT, 2006). All SA-Ts led to a result in favor of the proposed infrastructure expansion. Inter alia, for this reason, critics call SA-T ineffective (Mittendorfer, 2008) and claim a revisal of the SA-T regulation is needed.

It can be concluded that the SA-T idea has some advantages as it closes a gap in environmental decision making addressing the strategic meaning of transport infrastructure projects on the one hand, and introduces elements of sustainability assessment on the other hand. Yet, the conceptual pitfalls and the current methodological practice leave considerable room for quality enhancement. As the SA-T is rather close to EIA and legally binding intermodal strategic transport planning is still underdeveloped in Austria, additional improvements to the transport planning and assessment system should be further instituted.

In terms of potential broader implications, the bridging of SEA and project level EIA can clearly be problematic because of issues associated with the division of responsibility and the independence of those responsible for the preparation of IA documents. Examples of possible approaches for addressing such concerns can be identified on two levels: (1) the strategicness of the planning system and (2) the improvement of the SEA–EIA system. As demonstrated on the SA-T approach, elements of strategic planning like intermodal assessment of alternatives are introduced on a level closer to project implementation than transport infrastructure strategies. In doing so, certain gaps in environmental decision making addressing the demand—which is one of the most important strategic questions in SEA (Stoeglehner et al., 2010)—can be closed. Yet, this still does not lead to full-scale strategic infrastructure planning. In order to establish such a planning and assessment system, a legally binding intermodal national transport strategy/plan would have to be introduced that is also coordinated with urban and regional planning. In my opinion, such a plan would best address strategic questions like demand and most favorable means of transport. An environmental or sustainability assessment on such a scale would have the potential to become "really" strategic (according to the definition of Noble (2000a)).

In general, without such a strategic planning approach the existing system could still be enhanced, for example, by including (1) independent effectiveness reviews of system-level IA documents and processes; (2) the formal and independent peer review of such IA documents; (3) the preparation of the IA document for system-level improvements by an independent third party; (4) the creation of an informal joint body for the purpose of supervising the preparation of system-level EIA–SEA documents; (5) the creation of a permanent body to be responsible for the planning and assessment of system-level infrastructure improvements; (6) a combination of options 1–5. Parties in different jurisdictions struggling with how best to address the difficult middle ground between SEA and project-level EIA also could benefit from the sharing of insights and experiences.

GERNOT STÖGLEHNER

Department of Spatial, Landscape and Infrastructure Sciences, Institute of Spatial Planning and Rural Development, University of Natural Resources and Life Sciences, Vienna, Austria

DAVID LAWRENCE

Lawrence Environmental, Langley, British Columbia, Canada

12.2.2 Capacity Building—Building the Foundation for Integrated Good IA Practice

The World Bank is focusing much of its work on community-driven development (CDD) as a better tool for poverty reduction. There is now the realization that communities do have capacity that can be used, that communities are not homogeneous, and many local institutions and NGOs can be better used to assist in development efforts. Efficient CDD requires an enabling policy and legal environment, downward accountability, capacity building at the local level, community empowerment, sustainability, and social inclusion. Effective CDD means communities have control and can influence the decisions that directly affect them. At the core of CDD is environmental sustainability and capacity building. However, many communities in the Southern African Development Community lack the capacity, and opportunity, to participate in the public participation (PP) element of environmental impact assessment (EIA). The Southern African Institute for Environmental Assessment (SAIEA), through support provided by the World Bank and Canadian CIDA, undertook a 2-year project to develop a process to enhance participation in decision making in Southern Africa. Calabash was not designed to actually do public participation, but rather was structured so that regulators, private sector, practitioners, and civil society had the capacity, knowledge, and tools to better undertake respective public participation programs on individual projects and programs. How a respective government, with respect to decision making, engages civil society is one measure of how a government is reforming its governance and democratic reform processes. Environmental impact assessment is one window through which the public has the opportunity to engage a government in decision making. However, in southern Africa, there is insufficient public access to information and there are inadequate mechanisms for public participation in decision making. EIA has existed for 30 years. During its evolution, public involvement in the EIA process has become a key criterion that distinguishes EIA as a participatory decision support tool. Consequently, EIA is a suitable and appropriate platform from which to build participatory approaches for the southern African region. Governance in its simplest forms describes the relationship among institutions, processes, and ideas. It is about the exercise of power, accountability, and relationships in pursuit of an organization's mission or a nation's goals. In Africa, achievement of a country's goals are severely challenged due to such issues as resource degradation, HIV/AIDS, water scarcity, and conflict. Environmental assessment of projects and strategic environmental assessment (SEA) of policies, plans, and programs, are evolving rapidly to address wider sustainability objectives beyond biophysical concerns. The evolution of these planning tools is recognizing that the public has a significant role to play in the EIA or SEA process to assist a government to achieve its objectives, while at the same time advancing democratic reform and good governance practices. Many African countries have well-written EIA statutes that require the involvement of the public or civil society in the project decisions that affect them. To date, the application and success of public involvement in EA have been most variable due to lack of capacity, information, knowledge, and networks in many stakeholder groups. Regardless, EIA presents a very effective and practical tool for African governments to show to the international investment community and the African democratic review teams that democratic principles at the project/program level are being applied. EIA is one big "window" through which democratic reform can be realized by more participation. This project is but one step of many to assist the SADC region to move forward on democratic reform by using EIA as a catalyst for participatory decision making. Core to the success of the project were four preliminary activities. These were:

- The development of a 600-person Contact Group in the region who represent those involved in EIA and public participation (regulators, private sector, civil society, and practitioners). These people were to be the key "levers" of change for the 14 countries of the SADC region.
- The creation of a Project Advisory Team of 30 people from across the SADC region who advised Calabash on a regular basis on how the program should evolve to meet the needs of the region. These 30 people were drawn from the 600-person Contact List.
- The research and publication of a Situation Assessment on the status of EIA and PP in the SADC region. This document proved to be the foundation upon which the program would be built as it identified the strengths, weaknesses, opportunities, and threats to PP in the region.
- The development of a Communications Strategy that outlined how the outputs of the Calabash program were to be broadcast and made accessible to the SADC region and abroad.

The other key element of Calabash's success was based on the development of partnerships with institutions, agencies, and persons who were also working on public participation issues, but perhaps not from the EIA window, but through, for example, democratic reform, the democratic peer review process, United Nations multilateral agreements on the environment and natural resource management. These partnerships were not "formalized" but were purposely kept informal, light, and reactive so that Calabash and the other respective parties could work together in a mutually beneficial manner to get the most development and public participation "reach."

Calabash had to meet PP needs of the region in a practical and accessible way. All Calabash products have had input from a range of stakeholders who are actively involved in community engagement to ensure the tools are practical and relevant to the social, cultural, environmental, and economic realities of the SADC region.

Below is a brief summary of these products.

- *A Guide to Opportunities for Public Participation in Environmental Assessment Processes in the Southern Africa Development Community*. This handbook contains a clear description of all the rights that communities and the public have to participation in environmental decision making as conferred by international, regional, and SADC region conventions, laws and policies related to environmental impact assessment and decision making. This document has approached the rights issue from "an opportunity to participate" perspective, and is formatted around several key questions for each SADC country.
- Generic Public Participation Terms of Reference for civil society engagement involved in a point EIA (e.g., mine site), linear EIA (e.g., pipeline), or a regional strategic environmental assessment. Also included in the document are guidelines and tips on how to develop Terms of Reference that ensure that all contracting parties achieve maximum satisfaction and results throughout the duration of the relationship for a particular project.
- A Situation Assessment that describes and analyses the status of public participation and EIA for all countries of the SADC region.
- Research into six case studies in the SADC region where environmental assessment and public participation were done with distinction. Time and time again, studies show that when civil society has a chance to contribute to development planning, the end result is a project or program that has more far reaching direct and indirect development benefits than were originally planned. This is the first time that six projects from the SADC region have gone through such detailed analysis.
- *A One-Stop Participation Guide: A Handbook of Public Participation in Environmental Assessment in Southern Africa*. The lessons learned from the six case studies were integrated into the handbook methodologies. The handbook is unique in that it offers tips and the process to follow for public participation from the perspective of the four key stakeholders who are part of any public participation process: regulators, industry, practitioners, and civil society.
- The handbook also contains a PP Best Practice Model, a PP Review Template that can be used during the review or planning of a PP program, and a series of template letters that civil society can use to ensure that their voices are heard and respected in a respective public participation process.
- Calabash has also developed an Electronic Library of Public Participation and Civil Society Engagement tools from around the world. The library has grouped materials from the SADC region, Africa, and International. Over 250 resources and manuals exist on the Calabash site.
- Also available is the 2005 Desktop calendar that devotes text for each month to the benefits of civil society engagement.
- A course on EIA and PP.
- The Calabash web site also has published newsletters, Chat Forum, Calabash Planning workshop proceedings and a Poverty Reduction Strategy Paper (PRSP) analysis tool to determine how well environment is being integrated into the PRSP process.
- Lastly Calabash has a wall poster in cooperation with the Centre for Public Participation, Durban.

Calabash respected the needs of the region and the existing PP capacity that could be built upon. Also, the tools and outputs coming from Calabash were designed to be practical and able to be applied directly to ongoing public participation projects in the region. Calabash was designed to be a capacity building node for PP in the region, and it did not waiver from that core objective. It also sought every opportunity to integrate the outputs and news of Calabash wherever possible through word of mouth, corridor conversations, television and print media, and other development interventions in the region. A project such as Calabash cannot have success unless it gets buy-in from local stakeholders, and the outputs match required needs. This was done, the result of which is a PP capacity building program that well reflects the status, and potential, for PP and EIA in the SADC region.

Calabash had a vision and objective to ensure that the voices of the poor were better reflected in the decision-making process central to EIA and public participation. Therefore, the case studies, the "right to participation" handbook, the procedural handbook, and the template PP Terms of reference handbook emphasized the critical requirement to ensure that the poor and marginalized are brought into the public participation process through such techniques as providing food, transportation, skills, money, use of appropriate language, community radio, and community theatre. These tools also ensured that communities' voices would actually be heard. To do this, Calabash developed practical template letters that communities and Community-Based Organizations could write to ensure their issues were respected by regulators and the proponent. These letters also included templates on how communities would like to be consulted and also letters of appeal if communities felt their rights were not being respected. This very much reflects a direct technique and approach in defining how a community or group of people wishes to participate. With respect to meeting the process needs of all participants, the Calabash program developed an EIA/PP procedural handbook that is unique in Africa. The roles and responsibilities for PP and EIA are not the same for all stakeholder groups in the process. Therefore, the handbook has been designed so that there are four key chapters on how PP and EIAs are to be done if one is either a: regulator, private sector proponent, civil society, or practitioner. This approach has been well received in the region. The handbook also ensures that novel and cultural appropriate techniques are used to ensure those affected participate. In this vein, community radio, use of local language, community theatre, and providing the capacity for communities to participate, are directly emphasized. Many communities in the SADC region are barely surviving. If participation is to be ensured, then one must use techniques that are quite different for participation of wealthy and educated communities.

Calabash also ensures that all products were geared as much as possible to community needs. The "rights to participation handbook" is an example of this. It is quite true that citizens of most countries do not know their "rights" whether they are educated, poor, or wealthy. This is especially true for poorer communities who have lived under the yoke of poverty and possible colonial oppression. The "rights" handbook therefore provides CBOs and communities with a simple yet clear identification of the rights they have to participate in the EIA process that might be affecting them. Once they can articulate their rights to regulators, the regulators develop more respect and patience with the communities. In this way, more of an equal footing is achieved. And once the rights are known, Calabash then provides the information, knowledge, and skills on how to participate in a meaningful way. With respect to how communities' involvement affected a decision, Calabash tools, in particular the training course and the procedural handbook detail how postdecision follow-up must still be part of the PP process. In fact, Calabash emphasizes that community involvement and participation must parallel the project for its entire life. Again, Calabash tools are designed to instruct the regulator, practitioner or private sector on how the communities and local people are to be kept informed and involved during the whole life cycle of the project through culturally, socially, and economically appropriate methods.

PETER CROAL

Environmental Integration, Environmental Sustainability and Responsible Investment (ESRI), Strategic Policy and Performance Branch, Canadian International Development Agency, Gatineau, Quebec, Canada

12.2.3 The Biogas Support Program in Nepal—Impact Assessment as an End and a Means

The Government of Nepal and the Government of the Netherlands worked together to implement the Biogas Support Program (BSP) in Nepal during the period 1992–2010. The Ministry of Science and Technology (MOSTE) authorized the implementation of BSP, and until 2003 the Netherlands development organisation (SNV) was responsible for the implementation of the program.

The objective of BSP is to promote the wide-scale use of biogas through the establishment of small-scale biogas plants at the household level. The majority of those plants are fed by the dung of two to four cows. The latrine can be attached to the plant, as well. Such a plant produces sufficient biogas for cooking for an average household of six persons and one gas lamp. Biogas is used as a substitute for fuel wood (4 tonnes per year per plant), agricultural residues, animal dung, and fossil fuels (kerosene 32 l per year per plant). On average, a plant produces 5 tonnes of organic fertilizer annually (slurry) that can replace chemical fertilizer. In addition, indoor air pollution-related health problems decrease and time saving for fuel wood collection is about 1000 h per household per year.

The BSP was undertaken in four phases:

- Phases I–III: 1992–2003; 91,200 plants
- Phase IV: 2003–2010; 120,000 plants
- Total number 231,200 plants (about 20% of the national potential)

MOSTE and SNV prepared an EIA to achieve the following objectives:

a. BSP wanted an independent evaluation of the impacts of the BSP that they considered very positive. In addition, they wanted to further improve the performance of the program.
b. BSP wanted to secure funding from the involved donors for the fourth phase that was planned to start in the summer of 2003.
c. An EIA could help secure new source of funding through the Clean Development Mechanism (CDM).

The independent Netherlands Commission for Environmental Assessment (NCEA) was requested to draft the terms of reference for the EIA study, and to review the final study.

According to the interviews and desk review the EIA has had a significant positive influence.[1] The positive effects of the EIA result directly from the NCEAs involvement. The following was concluded regarding the extent to which the three EIA objectives were achieved (as interpreted by the author on a four-point scale of influence: irrelevant—little—medium—large).

Objective (a): large influence—BSP program impacts and improvement of performance

The BSP annual progress report 2010, not only considered 2010 but also provided an overview of the results achieved during phase IV (2003–2010). In this report (p. 26), the following is stated about the EIA:

> *"The results of the EIA study then served as a basis for providing recommendations for the fourth phase of BSP.*

[1] Three key resource persons were interviewed in December 2010 in Nepal, and a desk review was made of available documents.

And . . . The EIA carried out with extensive household survey was considered helpful to more clearly understand and define the positive as well as the potentially negative impacts of biogas technology in Nepal. This study is useful with respect to clarifying interrelations between thematic fields, spelling out underlying causes of the impacts, offering an opportunity of balancing and prioritizing of the impacts and contributing an opportunity of a financial – economic evaluation of the impacts."

In the advisory report of the NCEA, the following new or important issues were recommended to be studied in the EIA:

- Potential direct effects: latrine construction, slurry for agricultural purposes, and pathogens in slurry; mosquito problem; biogas/methane leakage; water availability and consumption.
- Potential indirect effects: forest degradation; reduction in GHG; reduction of smoke-related injuries or diseases and translating the effects toward national level.

Latrine construction: the positive effects of combining latrine construction with a biogas plant have been recognized, and promotion activities resulted in a considerable growth of the percentage of latrines connected currently (about 69% as compared with about 50% in 2000).

Pathogens in slurry: pathogens originating from attached latrines may be harmful to public health. A study showed that this, now applied, design change is cost-effective. The combination of the design change and training in the safer handling of the slurry significantly reduces the risks of pathogens and, in turn, diseases.

Use of bioslurry: another benefit has been the increased use and better management of slurry as compost for agricultural production and as a substitute for chemical fertilizer.

Mosquito problem: the observed increase of mosquito problems is the result of a combination of (partly biogas plant-related) factors such as less smoke in house, too much water in the feeder, and thin slurry when latrines are attached. Adaptation of the plant design, together with awareness training, can and has tackled this problem.

Methane—biogas leakage: measurements showed that, on average, about 9.5% of the methane is leaking from the plant. Measures have been taken to reduce this emission.

Water availability and consumption: connection of the latrine helps to reduce the relatively large amount of water that needs to be feed into the plant—a time consuming task in remote and water scarce areas. In addition, rain water harvesting multiple use systems have been developed, which serve other purposes such as drinking water and drip irrigation.

Objective (b): small influence—secure funding for the fourth phase, 2003–2010

The evaluation confirmed that the main reason for not finalizing the EIA report was the lack of secure funding from the main donors. BSP management also was already satisfied with the output of the EIA study (objective a); after approval of the fourth phase, the BSP management shifted their priorities and finalizing the EIA was considered a time consuming exercise. During the process contact was made with the German donor but they stated that they had their own evaluation method.

Objective (c): large influence—funding through CDM

BSP has used the draft EIA as a starting document to solicit CDM funding. According to the people interviewed, BSP was the first project asking for this type of funding. The draft EIA has played a crucial role in obtaining approval for two biogas CDM projects, with about 20 plants being registered in December 2005. Approval of the entire BSP program, under the Certified Emission Reductions, required an improved monitoring system. The program was approved in 2010.

Conclusions. As a result of the EIA, the performance of the biogas plants has improved considerably during the fourth phase. There have been more positive impacts and less negative impacts for each newly constructed plant. These improvements are mainly the result of a learning process by the BSP staff—a process facilitated and supported by EIA through both agenda setting and supplementary research on some issues.

At this time, 1.300 million inhabitants benefit directly from biogas plants in Nepal. This is about 20% of the potential expected to benefit in the coming decades.

The BSP is considered a success story from the perspective of renewable energy in rural areas in developing countries. It has won two global awards in 2005 and 2009 in the field of renewable energy, and it is used as a model for the setup of comparable programs in other countries such as Vietnam, India and, recently, in West-Africa.

Sources: BSP (2002, 2010), Bajgan et al. (2005).

AREND KOLHOFF

Technical Secretary International Cooperation, Netherlands Commission for Environmental Assessment, Utrecht, The Netherlands

12.3 DEFINING THE PROBLEM AND DECIDING ON A DIRECTION

The three stories address the role of synthesis in different ways. In the first story, an example middle-ground approach to bridging the gap between SEA and project-level EIA is described. In the second story, a major IA capacity building initiative is described. The story illustrates that effective IA capacity building provides a mechanism for addressing multiple recurrent problems and contemporary challenges. In the third story, the role of EIA was broadened beyond that of a decision-making aid to that of a vehicle for helping to realize broaden institution building, social learning, and secondary environmental enhancement purposes.

This book identifies (in Chapter 2) numerous choices for managing conventional IA processes. It then presents (in

Chapters 3–11) nine different IA processes, each responding to a different recurrent problems often encountered in IA practice. Variations in responses to the recurrent problems among six different IA types are highlighted. Approaches for addressing 10 different contemporary challenges also are addressed (the contemporary challenge of matching process and context is addressed in this chapter). Chapters 2–11 provide an array of potentially valuable IA process management tools. But, they do not address overall process management regulatory roles, how to deal with overlaps among IA process types, how to deal with interconnections among IA types, when to apply the tools (i.e., matching process to context), and how to respond when there are multiple problems and contemporary challenges (i.e., formulating composite IA processes).

More specifically, it is necessary to consider (1) how far to go, at the regulatory level, in directing and guiding IA process management (given the range of processes and contexts); (2) how to address interconnections among IA processes that seek to manage recurrent problems; (3) how to address interconnections among IA types; (4) how to match process and contextual characteristics; (5) how to address interconnections among approaches for addressing contemporary challenges; (6) how to design and manage composite IA processes; and (5) how to cope with the limits of synthesis and to establish priorities.

12.4 COMPOSITE REGULATORY FRAMEWORKS

IA regulatory practitioners are engaged in a delicate "balancing act." They define, through legislation and regulations, minimum levels of adequate IA practice. IA legislation and regulations also contain broad goals and principles to provide a rationale for requirements and a direction for current and future regulatory and applied practice. Both minimum standards and ideal characteristics change and evolve. IA guidelines and applied research help ensure that requirements are achieved. Because of their greater flexibility, they also contribute to practice levels that often exceed the minimum, thereby narrowing the gap between the adequate and the ideal. IA requirements and guidelines need to be neither too general (which could result in a low and inconsistent level of practice) nor too precise (which could unduly restrict and limit adaptation and innovation). IA requirements operate within the context of a complex set of related environmental requirements, policies, and objectives, a rapidly evolving field of theory and practice, permeable boundaries with a host of related field of theory and practice, and a multiplicity of ecological, social, economic, cultural, and political conditions, events, patterns, trends, and uncertainties.

How then should the IA process be addressed as part of this "balancing act" at the regulatory level? The IA process should not be ignored, by concentrating exclusively on document content and on administrative procedures. The IA process provides the framework for conducting all IA activities and for applying all IA methods. A poorly designed and executed IA process (a circumstance that occurs all too frequently in practice) can readily undermine individual IA activities and methods. IA documents should be outputs from the IA process rather than ends in themselves.

Chapter 2 demonstrates that a single standardized IA process is a dangerous myth. This does not preclude identifying core IA process attributes, if considerable discretion is left to IA process participants in choosing among process management choices and if adequate provision is made for contextual adjustments. Guidelines could provide participants with a sense of the range of potentially appropriate IA process choices. Chapter 2 offers an initial sense of core IA process attributes that could be incorporated into IA requirements and guidelines. It provides an overview of generic regulatory choices for directing, bounding, and guiding SEA and project-level EIA processes, and processes for other IA types (EcIA, SIA, HIA, and SA). It describes "good regulatory EIA practice" for undertaking screening (proponent-driven, action-driven, environment-driven, combinations of proponent, environment- and action-driven, significance determination), for conducting individual EIA activities (general, scoping, proposal characteristics, baseline analysis, proposal characteristics, impact analysis and synthesis, alternatives analysis, mitigation and enhancement, methods, documents, management, auditing, participation, review, and decision making), and for addressing interrelationships (among EIA activities, with international EIA activities, among government levels, with related governmental requirements and action, with the IA knowledge base). Chapter 2 also provides an overview of available EIA process management choices, of possible SEA, SA, EcIA, SIA, and HIA regulatory and process management choices, and of potential SEA good-practice guidance approaches. IA requirements could specify the core IA process elements that must be addressed in IA documents. IA guidelines could elaborate on "good-practice" IA process management and could provide examples of available management choices.

Regulatory measures to address the IA process, as identified in Chapter 2, although necessary, are unlikely to fully respond to the recurrent problems, to the contemporary challenges, to the differences among IA types, and to the need to adapt to and influence context. Chapters 3–11 describe in detail the many measures introduced by the four jurisdictions to avoid and minimize the recurrent problems. Directions for future regulatory reform also are identified. Chapters 3–12 describe potential regulatory level approaches for addressing the various contemporary challenges. Numerous good practices and potential pitfalls are identified. Each of Chapters 3–11 address how the various IA types (SA, SEA, EIA, EcIA, SIA, HIA, SIA) have addressed the recurrent problems at the regulatory and applied levels. Several relevant practice-based anecdotes are also presented.

IA requirements and guidelines should be broad enough to allow for a diversity of approaches for ameliorating the shortcomings and for addressing the challenges, while still ensuring consistency with the purpose and objectives of IA requirements and facilitating an appropriate match between process and context. Approaches applied in other jurisdictions should not be borrowed uncritically. Contextual factors vary greatly among jurisdictions. It is foolhardy to apply a measure simply because it has been applied elsewhere, especially if the effectiveness of the measure has not been assessed. Considerable benefits, however, are likely to accrue from jurisdictions sharing experiences and coordinating applied research efforts.

Regulators should proceed cautiously in controlling and guiding how IA process types, IA types, and approaches for addressing recurrent problems and contemporary challenges are linked and combined. Several examples of composite processes are described briefly later in this chapter. IA requirements and guidelines can help ameliorate the recurrent problems and address the contemporary challenges with individual measures and by recognizing that it often is necessary to "juggle" multiple, sometimes conflicting, values, and approaches. Care must be taken regarding such matters as the appropriate mix of process and substance (arguably greater emphasis needs to be placed on substance and links to related substantive environmental requirements need to be more effective), vertical coordination initiatives (e.g., multijurisdictional IA coordination, harmonization and integration, IA tiering, connections to international environmental and IA treaties and initiatives, connections to local IA and environmental requirements, planning and management), horizontal coordination initiatives (e.g., links to decision making, connections to other measures, requirements seeking to minimize negative and enhance positive environmental effects, administrative and institutional interactions), links to related fields of theory and practice, interconnections among measures for addressing transcending environmental issues (e.g., climate change), and efforts to balance the need for greater convergence among IA requirements and the enhanced application of IA good practices with the importance of allowing for, for example, cultural, political, environmental, institutional, and administrative contextual differences.

Over time it could be possible to identify, from applied research, which approach combinations are best suited to which categories of situations. These patterns can be noted in IA guidelines, while acknowledging the need to make IA type and contextual adjustments. It is unlikely that the evidence from effectiveness reviews (appreciating that effectiveness also is IA type and context-dependent) will be sufficiently definitive to enshrine the circumstances under which particular IA process combinations must and must not be applied to prevent and ameliorate various mixes of recurrent problems and contemporary challenges. It may be possible for individual agencies to establish, as a policy, based on their experiences in dealing with the recurrent problems, the contemporary challenges and various mixes of IA types, the IA process approaches that they will generally take for different classes of situations. Even so, each IA process and context is unique, fluid and uncertain, and will require initial and ongoing, collaboratively generated, adjustments to match the unique and evolving circumstances surrounding each IA process.

12.5 RECURRENT PROBLEMS

The nine IA process types are based on a desire to prevent and minimize specific recurrent problems. Matching process and context, therefore, entails applying an influential IA process when enhanced political influence is a priority, a rigorous IA process when scientific rigor is a priority, a rational IA process when rationality is a priority, and so on. More specifically:

- An *influential* IA process is preferred when the information and analysis generated by more conventional IA requirements and processes are likely to have little, if any, influence over decision makers or prevailing decision-making processes and practices. A more proactive effort is needed to understand, connect to, influence, reform, and ultimately transcend existing administrative, institutional, and political structures and patterns of interaction. Stakeholders, who currently exert political influence within the administrative–political system, need to "buy in" to and actively participate in the process and its outcomes. The range and role of stakeholders participating in and influencing IA-related decisions needs to broadened and redefined in a manner that ensures IA-related decisions are made within rather than external to the IA process. The basis for such decisions needs to be explicit and to flow from the IA process.
- A *rigorous* IA process is more appropriate in situations where the environment is amenable to scientific analysis; where causal webs of direct and indirect effects can be identified, measured, predicted and monitored; where scientific knowledge and methods are likely to make a significant contribution to decision making; and where there are adequate resources and stakeholder support for scientific analyses.
- A *rational* IA process is well suited to stable systems with well-defined proposals, options, and effects. The systematic screening and comparison of multiple options is a priority. It should be possible to aggregate preferences and effects, either quantitatively or qualitatively. This suggests an open and nonoppressive environment where stakeholders are willing to engage, to communicate, and to be "reasonable." Positions, values, and interests are not polarized. A high premium is placed on scientific and technical knowledge and evidence.
- A *substantive* IA process is well adapted for comprehensive medium- to long-term efforts to advance

environmental or sustainability objectives. Such processes may function more effectively at a strategic (policy, plan, program) level, where the scale of analysis is regional or greater. Major project-level EIAs can apply this process but are even more effective if undertaken within the context of an array of larger scale environmental management frameworks and indicator systems.

- A *practical* IA process is well matched to short-term, politicized environments, characterized by limited resources, a limited ability to control the environment, and high levels of uncertainty beyond the immediate future. Stakeholders are generally resistant to change but are prepared to bargain. Changes generally take the form of mutually beneficial, incremental adjustments from the status quo. A heavy emphasis is placed on satisfying decision-making requirements, on efficiently working within available resources, and on ensuring implementation.
- A *democratic* IA process tends to work especially well when clearly identifiable stakeholders wish to "control their own destinies." Proponents must be willing and able to delegate, for selective decisions, their decision-making authority. Local stakeholders must be willing and able to select, maintain contact with, and support representatives. Stakeholder representatives must be prepared for and capable of participating in a time-consuming, demanding, complex, and sometimes controversial planning and decision-making process with other parties. It is essential that there be sufficient time and resources to support the process.
- A *collaborative* IA process seems best suited to situations where major stakeholders can work together collaboratively to achieve mutually agreed upon ends by mutually agreed upon means. It should be possible for stakeholders to select and support representatives. Stakeholder representatives should be willing and able to participate in a time-consuming and often protracted joint planning endeavor. They should be prepared to maintain close contact with their constituents and to accept that other parties retain final decision-making authority. There should be sufficient resources to support the process and sufficient time for the process to proceed "at its own pace."
- An *ethical* IA process is especially appropriate when issues of fairness, equity, and social and/or environmental justice predominate. There should be both procedural and distributional ethical concerns. All major parties should desire that ethical rights and duties be identified and formalized. There should be a willingness to identify and reconcile conflicting ethical procedural and distributional principles, rules, rights, and duties. All parties should be comfortable with ethical concerns taking a lead role in each IA activity.
- An *adaptive* IA process is more effective for complex problems in turbulent, unstable, and complex environments. Concerns with risk, uncertainty and human and ecological health should predominate. There should be a general acceptance of the need to anticipate and to manage uncertainties rapidly and effectively. There should be a willingness to recognize knowledge limits, to ensure that uncertainty-related concerns assume a lead decision-making role and to commit to an iterative, learning, and adaptive IA process and organizational structure.

Several additional refinements are necessary to reflect the complexities of both the processes and the context.

1. The IA processes, described in Chapters 3–11, include numerous subsets and variations. Part of matching process to context necessitates selecting the process variations that best match the context.
2. Part of adapting the process to the context is the determination of the appropriate extent of horizontal integration among IA types and vertical integration among IA levels.
3. Matching process and context also entails integrating and addressing, to the extent practical, the contemporary challenges.
4. The boundaries between core process elements and IA process types are "fuzzy" and permeable. Iterative adjustments to both the process and the core elements are necessary to optimize responses to recurrent problems.
5. Each IA process has strengths and limitations. It is necessary to ameliorate the relevant limitations and to reinforce the relevant strengths. This can entail drawing upon other IA process types, concepts, and methods, as appropriate.
6. More than one recurrent problem and contemporary challenge is likely in any given situation. This means that aspects of more than one IA process will need to be incorporated into most overall IA processes.
7. IA process types and IA types often overlap and sometimes conflict. Interactions among process types will need to be considered. There should be minimal duplication (unless deliberate) and individual process elements should be mutually supportive or counterbalancing.
8. Matching process to context is not a one-step procedure. There should be an initial matching of process to context type. Proposal- and environment-specific adjustments should then be made. Ongoing adjustments will be necessary as the process and context coevolve.

IA process types can be connected in many ways, as highlighted in Table 12.1. One process could provide inputs

Table 12.1 Examples of Recurrent Problem Process Interconnections, Overlaps, and Middle-Ground Concepts

Influential and Rigorous	Influential and Rational	Influential and Substantive	Influential and Practical
Sound science can inform and influence decision making Decision makers can bound scientific analysis and determine scientific priorities Integration of science and politics Environmental advocacy	Rational analysis can inform and influence decision making Decision makers can bound and temper rational analysis Rational analysis taking into account political realities Critical theory	IA can influence decision makers toward substantive outcomes Decision makers can facilitate or inhibit the realization of substantive outcomes Social/ecological action SA encompasses action and substance	Practical IA approaches tend to have greater decision-making influence SA encompasses influential and practical
Influential and Democratic	**Influential and Collaborative**	**Influential and Ethical**	**Influential and Adaptive**
More democratic IA practice can open up and influence decision making Community-controlled IA Shared and delegated decision making in IA practice enhances the decision-making role of interested and affected parties Middle ground: community empowerment, IA as a social struggle, conflict IA	A collaborative IA approach enhances the potential for "buy in" by decision makers Collaborative IA processes do not always lead to greater decision-making influence Middle ground: political–dialogical IA approaches, pluralist political IA model	Appeals to ethical principles and standards tend to have a broad public appeal and the potential for greater decision-making influence SA encompasses political action and procedural and substantive ethical principles	Decision-making processes tend to involve a high degree of uncertainty and change rapidly; adaptive IA processes share such characteristics and enhance the potential for greater decision-making influence SA encompasses adaption/precaution and political action
Rigorous and Rational	**Rigorous and Substantive**	**Rigorous and Practical**	**Rigorous and Democratic**
Complementary methods Reliance on specialist advisors Emphasis on comprehensiveness, explicitness, consistency, and transparency Both stress analysis more than synthesis and seek systematic, traceable process Middle ground: scientific rationality	New or holistic science Emphasis on experimentation and modeling Scientific knowledge informs substantive process Substantive goals guide applied science Middle ground: holistic science, AEAM, ecological theory, positivistic social science, interpretative social science	Both grounded in reality and inductive (i.e., experience-based) Balance of rigor and relevance and of explanation and prescription Applied science informs decision making; practicalities bound scientific analysis Middle ground: phenomenological sociology, interpretative social science, pragmatism, empiricism	Scientific analyses and knowledge inform democratic processes Both explicit and traceable Scientific interpretations informed and facilitated Middle ground: civic, lay, and holistic science, critical social science

Rigorous and Collaborative	Rigorous and Ethical	Rigorous and Adaptive	Rational and Substantive
Scientific analyses inform collaborative processes Scientific interpretations informed and facilitated Scientific analysis and collaborative synthesis complementary Middle ground: AEAM, civic science, peer review, scientific community, SIA (collaborative)	Both seek to avoid bias Ethical standards for scientific research Complementary interpretations Middle ground: research ethics, holistic science	Both explicitly address risks and uncertainties and stress importance of monitoring Adaptation recognizes scientific limits; precautionary principle combines scientific limits with severe harm potential Shared interest in complex and chaotic systems Shared methods (e.g., modeling) Middle ground: AEAM, new science, social learning, uncertainty analysis, uncertainty principle, postnormal science	Rational methods inform substantive processes Substantive framework for rational methods Rational analysis and substantive synthesis complementary Middle ground: ecological and social rationality, rational systems, substantive/purposive rationality, ecological impact assessment, SIA (technical), sustainability assessment
Rational and Practical	**Rational and Democratic**	**Rational and Collaborative**	**Rational and Ethical**
Rationality informed by experience Practical bounds to rationality Middle ground: mixed scanning, strategic planning, purposive incrementalism, bounded rationality, effective planning, practical rationality, legal rationality, knowledge and reflection in action, prescriptive pragmatism	Rational analyses inform democratic processes Democratic offsets autocratic tendencies of rational methods; basis for interpretations and conclusions Middle ground: political rationality, advocacy planning, critical, social, political, structural, and reform rationalism	Rational analyses inform collaborative processes Rational argumentation in collaboration Collaborative inputs to rational Middle ground: communicative rationality, rational argumentation, procedural rationality, rational discourse, communicative planning, joint fact-finding, collaborative planning	Ethical conscience to rationality Rational analysis of distributional consequences Rational testing of ethical assumptions and assertions Ethical goals and principles to guide rational process Middle ground: professional ethics, value or normative rationality, equity or progressive planning
Rational and Adaptive	**Substantive and Practical**	**Substantive and Democratic**	**Substantive and Collaborative**
Integration of risk and uncertainty into rational analysis (e.g., risk assessment) Uncertainties about rational processes Middle ground: strategic choice, strategic planning and management, risk assessment and management	Both real, collective, and experience-based Short-term steps toward long-term needs Practical bounds on substantive aspirations Middle ground: purposive incrementalism, environmental management, integrated environmental, and resource management Shared methods (e.g., rapid rural appraisal)	Substantive analyses inform democratic processes Substance guides and frames procedures Both stress value of local knowledge Middle ground: co-management, bioregionalism, environmental and social movements and advocacy, traditional knowledge, political SIA, deep ecology, ecological politics, eco-feminism	Substantive analyses inform collaborative processes Substance guides and frames procedural Both stress dialogue, local knowledge, and consensus building Both elements of sustainability Shared methods (e.g., scenario writing, storytelling) Middle ground: co-management, shared vision planning

(continued)

Table 12.1 (*Continued*)

Substantive and Ethical	Substantive and Adaptive	Practical and Democratic	Practical and Collaborative
Both stress procedural equity Both concerned with intergenerational equity and needs of disadvantaged Distributional equity conducive to sustainability Middle ground: social and environmental justice, traditional knowledge, ecofeminism, teleological or consequentialist ethics, distributional or outcome equity or fairness, environmental ethics, sustainability ethics, normative ethics, human rights, and gender IA	Both seek to avoid crises and recognize need for adaptive processes and organizations Substantive uncertainties Ecological, social, and health all part of sustainability Middle ground: AEAM, ecosystem approach, social learning, latent time bombs, human health and ecological risk assessment, disaster and hazard analysis Shared methods (e.g., modeling, visioning, life-cycle analysis)	Both are procedural, assume conflicting interests, are overtly political and emphasize dialogue, bargaining, and experience-based knowledge Practical bounds to democratic processes Democratic principles to guide and limit practical processes Middle ground: critical pragmatism, capacity building, participant and intervener funding	Both are procedural and involve bargaining and dialogue Practical bounds to collaborative processes Collaborative principles to guide and limit practical processes Middle ground: dialogical incrementalism, communicative action, practical communications, practical deliberative learning Shared methods (e.g., participatory rural appraisal)
Practical and Ethical	**Practical and Adaptive**	**Democratic and Collaborative**	**Democratic and Ethical**
Ethics a conscience for practical processes Practicality ensures that ethical standards and decision rules can be implemented Middle ground: critical pragmatism, purposive incrementalism, applied or practical ethics	Both recognize major uncertainties and control limits, stress flexibility and hedge away from disasters Effectiveness reviews to anticipate and adapt to uncertainties Practical bounds to adaptation Middle ground: practical, deliberative learning, real problems, ingenuity gaps	Both procedural and participatory Both value local and traditional knowledge and involve consultation, communications, mutual education, negotiations, and collaboration Collaboration as a means for delegation Middle ground: critical dialogue, coalition building and networking, active mediation, co-management, constructive engagement	Procedural equity facilitates democratic processes Both concerned with rectifying power inequities Ethical analyses inform democratic processes Ethical analysis informed by democratic perspectives Middle ground: communitarian principles, traditional knowledge, community development and empowerment, critical, egalitarian, and feminist ethics
Democratic and Adaptive	**Collaborative and Ethical**	**Collaborative and Adaptive**	**Ethical and Adaptive**
Both iterative and interactive Risk and uncertainty analyses inform democratic processes Local and traditional knowledge consistent with a learning process Uncertainties about procedures Middle ground: critical and transformative learning, critical IA education, traditional ecological and indigenous knowledge	Procedural equity facilitates collaboration Ethical analyses inform collaboration Open collaborative process consistent with multiple ethical perspectives Negotiations about procedural and distributional rights and duties Middle ground: communicative or discourse ethics, procedural justice, communitarian ethics, participatory poverty assessment	Both iterative, interactive and stress mutual education Risk and uncertainty analyses inform collaborative processes Collaboration conducive to addressing risk and uncertainty perceptions, evaluation and management Uncertainties about collaborative procedures and outcomes Middle ground: collaborative learning, linguistic imprecision and confusion, risk communications, perceived risk	Uncertainties about values and ethical interpretations Ethical perspectives inform adaptive processes Ethical processes can be informed by risk and uncertainty analyses Middle ground: precautionary principle, risk assessment, professional standards, human and ecological health standards

to or derive outputs from another process. Both processes could provide elements that fit within a larger framework (e.g., sustainability). Combining two processes can offset the negatives or reinforce the positives associated with individual processes. Where two processes are combined, areas of duplication can be eliminated (to facilitate efficiency) or retained (to provide multiple perspectives). Processes can be applied simultaneously (separately, partially integrated, or fully integrated) or sequentially (perhaps applying to different IA activities).

Complementary process subsets, variations, methods, procedures, concepts, frameworks, and institutional arrangements can be partially or completely integrated. Middle-ground concepts provide an opportunity to build outward from the overlaps to take in appropriate IA process-type elements. Conflicts can sometimes be reduced through role definition and by applying potentially conflicting process elements to different IA activities. Or it may be helpful to retain conflicting perspectives in a "dramatic tension," consistent, for example, with societal divisions.

Process integration can involve tracing how each process evolved; identifying the needs met by the processes individually and collectively; determining the advantages and disadvantages of partial or complete integration; focusing on major process characteristics, similarities, differences, and interactions; identifying potential forms of synthesis; strengthening mutually complementary relationships; offsetting individual and mutual weaknesses; and retaining diversity where supportive of individual or joint process needs (Boothroyd, 1995; Nooteboom and Wieringa, 1999; Yiftachel, 1989). The end result could be separated or partially or completely integrated composite IA processes.

A composite IA process for addressing two or more recurrent problems could be formulated in any one of many ways. Elements from other IA process types could then be selectively added if they are mutually supportive and appropriate to the situation. This approach, although attractive in its simplicity, will likely result in only modest improvements to current IA "good practice." It is not as if current good practice does not address the recurrent problems. But the frequency and intensity with which the problems continue to be referenced suggest that more fundamental IA process management reforms are required.

Another, relatively simple approach can take place when the major issues are consistent with the conditions most suited to one of the nine IA process types. If, for example, the issues are largely ethical, then the point of departure would be an ethical IA process. The ethical process would be supplemented by core process choices (as described in Chapter 2) and other IA process elements, as warranted and appropriate. The IA process would focus on identifying and managing ethical issues and trade-offs. The same core—supplemental pattern would be followed if the issues primarily revolved around risk and uncertainty management, the rational evaluation of available choices, contributions to substantive environmental objectives, the practical resolution of approval and implementation-related issues, and so on. The situations where issues are so highly focused are likely to be rare. A variation of this approach would be to undertake two or more processes simultaneously or successively to provide varying perspectives at each decision.

A more realistic combination would be to build the process around clearly complementary (e.g., rational and scientific or collaborative and democratic) or counterbalancing (e.g., substantive and practical or rational and adaptive) IA process types. Other IA process types would assume a support role. Figure 12.1 portrays a composite IA process with a collaborative–democratic-influential core. Some IA decisions would be delegated. Some would be shared. Proponents and regulators would retain final decision-making authority for the remaining decisions, although there would be extensive consultation. This type of composite process is best suited to situations where issues are highly clustered. There should be broad agreement concerning issue clusters and the rationale for distinguishing between core and support roles.

A more complex and perhaps more realistic composite IA process is illustrated in Figure 12.2. The process begins from core IA process elements. The procedural elements are structured around a combination of collaborative, influential, and democratic decision making. Practical perspectives, strategies, and constraints are counterbalanced against substantive visions, goals, and indicators. The process is supported and informed by rational analyses; scientific analyses; adaptive, risk, and uncertainty perspectives and analyses; and ethical perspectives, principles, and analyses. It is structured within public participation, ethical, risk and uncertainty, and environment and sustainability frameworks. It is adapted to an array of contexts. It draws upon a variety of tools, fully involves all interested and affected participants, and produces both direct (e.g., documents, conclusions) and indirect (e.g., environmental quality changes, institutional changes, mutual education, contributions to the state of IA practice) products.

Another approach, illustrated in Figure 12.3, structures the process around temporal and spatial distinctions. Practical considerations predominate in the short term. The focus during this period is at the micro level (i.e., incremental adjustments from current practices). Rigorous, rational, adaptive, ethical, and substantive perspectives, methods, and analyses all contribute to the analysis, but are filtered through a practical perspective. Both micro and macro analyses occur in the medium term. Rigorous, rational, adaptive, and ethical IA processes all directly contribute to medium-term analyses. Practical and substantive IA processes make indirect contributions (i.e., filtered through the directly contributing approaches) during the medium term. Substantive methods, insights and ideals take the lead in the long term—macro analysis. Ethical and adaptive perspectives, methods and analyses assume a support role. The practical, rigorous, and rational IA processes make a

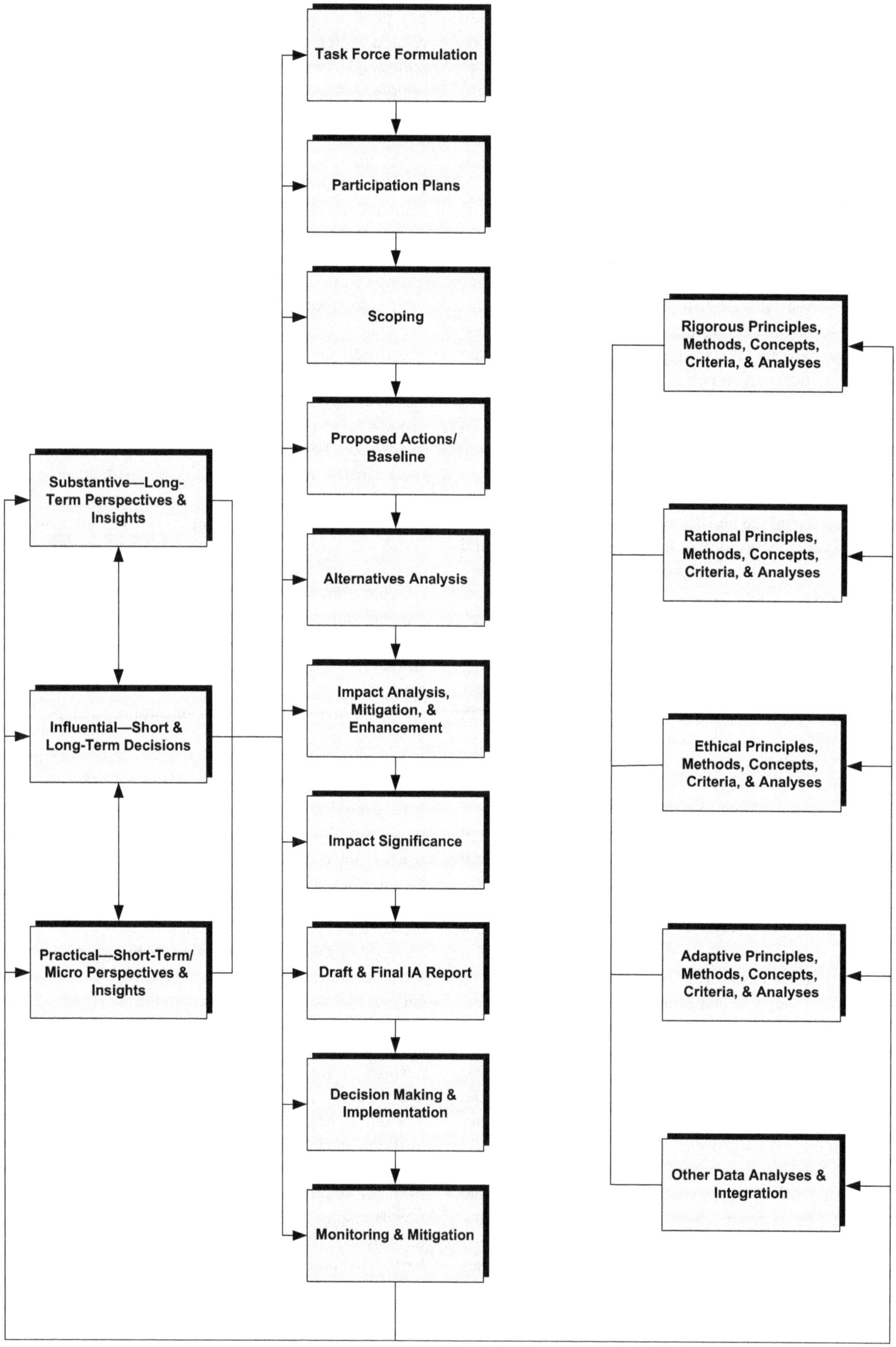

Figure 12.1 Recurrent problems—composite IA process: collaborative, democratic/influential core.

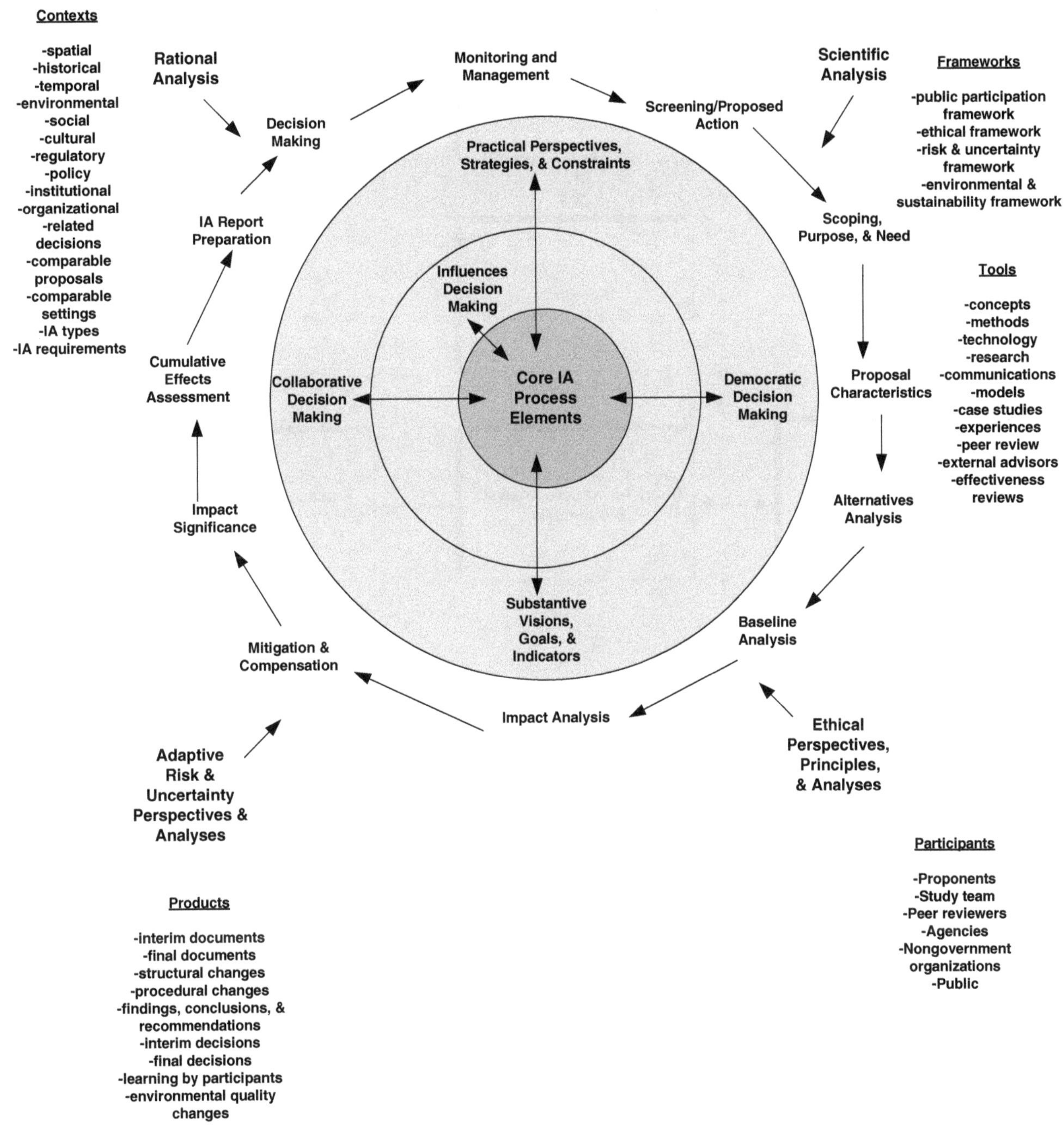

Figure 12.2 Recurrent problems an example composite IA process—IA core. Adapted from Lawrence (2005b).

minor contribution. A blending of democratic, influential, and collaborative decision making occurs for all time horizons and for all scales of analysis. Review and approval steps occur in association with each time horizon.

Suggested IA process characterizations increasingly combine elements that address two or more recurrent problems. Some examples include:

- Combining and counterbalancing collaborative/democratic/influential IA approaches with scientific/rational analysis approaches (Bina et al., 2011; Devlin, 2011; Fischer, 2005; McCluskey and João, 2011; Walker, 2003).
- Integrating collaborative/democratic/participatory approaches, ethical performance standards (e.g., inter- and intragenerational equity) and the adaptive management of risks and uncertainties under the umbrella of sustainability assessment (Bond and Morrison-Saunders, 2011; Craik, 2008; Sinclair et al., 2008; Kates, 2000; O'Brian, 2003).

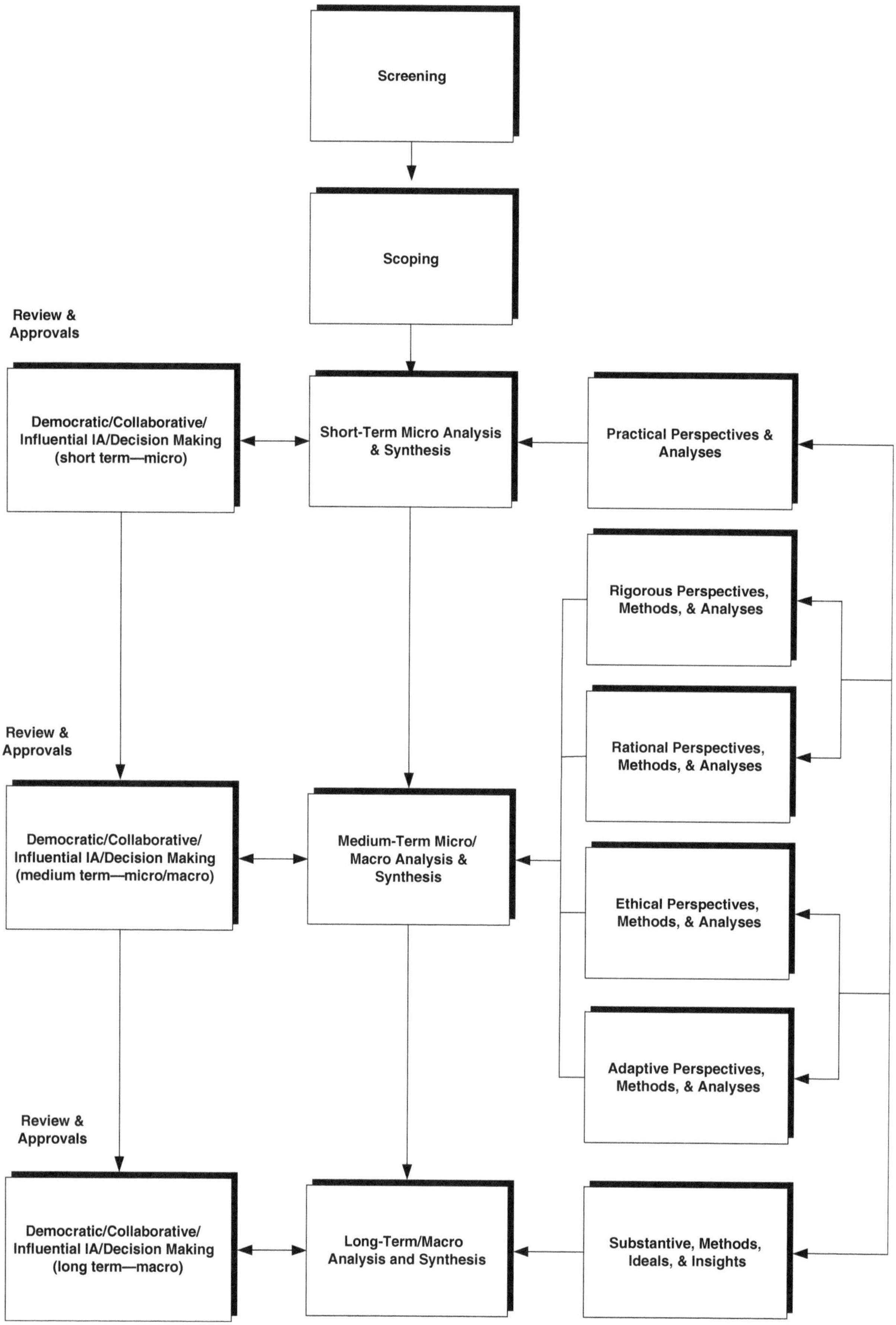

Figure 12.3 Recurrent problems—example of a composite IA process: timing and scale.

- The tempering of scientific/rational analyses with practical limits, political and economic realities, contextual variations, disputed values, and a precautionary approach to risks, uncertainties, and the unknown (Duncan, 2008; Gasparatus et al., 2007; Jasonoff, 2003; Kaiser, 2003; Levins, 2003; Morrison-Saunders and Sadler, 2010).
- The counterbalancing and constructive interaction of science and politics, science and adaptive management, and science within and outside IA (Devlin, 2011; Greig and Duinker, 2011; Tuinstra et al., 2008).
- The integration of the rational, the moral/ethical, and the practical (e.g., as in critical theory) (Elling, 2007).
- The recognition that the integration of environmental substance (e.g., sustainability) into IA is unlikely to be effective without a sound understanding of decision-making processes, the identification of critical decision windows, the integration of sustainability considerations into each decision, and a proactive effort to influence the sustainability of outcomes (i.e., combining environmental substance and decision-making influence) (Pischke and Cashmore, 2006).
- The acknowledgment that a plurality of IA frameworks, processes, and methods are needed when there are multiple contextual variations, diverse perspectives, and a variety of available approaches to knowledge acquisition and application (Gasparatus et al., 2007).
- The recognition of the need to more systematic address procedural and substantive integration, vertical integration (among IA levels), horizontal integration (among IA types), and integration with decision making, using approaches and tools such as integrated IA and CEA (Eales et al., 2005; Fischer, 2006).

12.6 IA TYPES

The appropriate environmental scope of IA is a recurrent theme in IA literature and practice (Bisset and Sadler, 2004; Eales et al., 2005; Milner et al., 2005; Orenstein et al., 2005; Ortolano, 2008). Each of Chapters 3–11 addresses how six different IA types (SA, SEA, EIA, EcIA, SIA and HIA) address the recurrent problems. The contemporary challenge addressed in Chapter 6 is horizontal integration (i.e., how widely should the environment be defined). The Chapter 6 analysis explores the case for and against integration, as well as considering partial integration options.

Table 12.2 identifies a range of potential interactions between various pairs of IA types. As is evident from Table 12.2, there are a great many potential interconnections among IA types. A failure to at least partially integrate IA types runs the risk of failing to address or adequately address these interactions. Duplication, IA requirements and processes operating at cross-purposes and a failure to address potentially significant environmental effects would appear to be the inevitable result of the absence of even partial integration. At the very least overlaps and interactions between and among IA types must be addressed if the overall IA regulatory and applied system, however informal, is to have at least some semblance of coherence, efficiency and effectiveness. At the same time, as detailed in Chapter 6, there needs to be considerable sensitivity to the potential drawbacks associated with partial or complete horizontal integration approaches.

The real choice would, therefore, seem be among various forms of weak and strong horizontal integration rather than between complete and no integration (Jiliberto, 2004). The often suggested reorientation of IA processes and procedures towards substantive purposes and outcomes suggests that the partial or complete integration of substantive IA types (e.g., EcIA, SIA, and HIA) into SEA (at the policy, plan, and programs) and project level EIA would appear to be a priority (Bond and Morrison-Saunders, 2011; Cashmore et al., 2004; Doelle and Sinclair, 2006). This is not to suggest that there will not be occasions when substantive IA types, such as HIA and SIA, cannot function effectively outside formal SEA/EIA requirements or that there will not be circumstances when it is better, given the limitations of those requirements, that separate procedures be retained. Nevertheless, it is still arguably necessary to systematically consider the interactions of the various elements of the IA system, however loosely and informally the system is defined.

Increasingly, it is being suggested that SA is either the next stage in the evolution of IA and/or represents an umbrella concept/framework encompassing and integrating all IA types (Bisset and Sadler, 2004; Gibson, 2010; Gibson and Hanna, 2009; Pope and Dalal-Clayton, 2011). Care, however, must be taken to ensure that the SA is designed and managed such that the integration of social, economic, and ecological considerations is systematically addressed from the outset (not as a bottom-line trade-off) and sustainable outcomes are clearly demonstrated (e.g., net positive sustainability benefits, avoidance of undesirable trade-offs, explicit trade-off rules) (Gibson, 2010). The illusion and pretense of sustainability remains all too common in contemporary IA requirements and practices (Bond and Morrison-Saunders, 2009).

Partial or complete horizontal integration among aspects of the environment can be facilitated by drawing upon and effectively utilizing a range of integrative forms of IA (e.g., integrative IA, CEA, territorial IA), specialized IA types (e.g., gender IA, poverty IA, disaster IA), related tools (e.g., risk assessment, LCA, multicriteria analysis, cost–benefit analysis), integrative frameworks, and related fields of theory and practice (e.g., EMS, ecological landscape planning) (Bina et al., 2011; Eales et al., 2005; Fischer, 2011; Hacking and Guthrie, 2008; Milner et al., 2005; Orenstein et al., 2010; Rotmans, 2006). It also can be furthered by addressing interactions with other forms of integration) (e.g., methodological, procedural, institutional, vertical among IA tiers,

Table 12.2 Examples of Interconnections Among IA Types

SA and SEA	SA and EIA	SA and EcIA
SA evolution of SEA SA frames SEA; SA as upper tier SA provides environmental substance for SEA (aspirations, environmental values, shared visions and limits); blending of substance and process SA principles and decision rules (e.g., contribution to sustainability test) structure SEA SEA as a tool for achieving sustainability SEA contributes to sustainability governance SA improves legitimacy of strategic decisions and broadens the range of participants SA outcome test of SEA	SA evolution of EIA SA frames SEA (sustainability strategy) Sustainability integrated into each EIA activity SA provides environmental substance for EIA (aspirations, environmental values, shared visions and limits) SA outcome test of EIA SA principles and decision rules (e.g., contribution to sustainability test) structure EIA SA improves legitimacy of project-level decisions and broadens range of participants	Ecological sustainability SA frames EcIA; EcIA a subset of SA Shared focus on environmental substance Sustainability integrated into each EcIA activity SA a means of integrating EcIA and habitats assessment into SEA and EIA SA as a means of addressing interconnections among ecological, social, economic, and health impacts (CEA) and integrating into decision making

SA and SIA	SA and HIA	SEA and EIA
Social/collaborative sustainability Sustainability learning Sustainability facilitates greater transparency and increased participation by directly affected individuals SA frames SIA Shared focus on environmental substance Sustainability integrated into each SIA activity SA as a means of integrating social concerns into SEA and EIA SA as a means of addressing interconnections among social, economic, ecological, and health effects (CEA) and integrating into decision making	Health sustainability Environmental sustainability as a prerequisite for long-lasting health Shared focus on environmental substance SA frames HIA Sustainability integrated into each HIA activity SA as means of integrating health concerns into SEA and EIA SA as a means of addressing interconnections among social, economic, ecological, and health effects (CEA) and integrating into decision making	SEA and EIA tiered SEA frames and directs EIA SEA addresses the why; EIA addresses the how, and both address the where EIA informs and refines SEA SEA and EIA connected, framed, and integrated by CEA, Integrated IA and system-level planning Individual project-level EIAs have influenced policy and institutional contexts—a form of social learning

SEA and EcIA	SEA and SIA	SEA and HIA
Integration of ecological effects into SEA EcIA of policies, plans, and programs SEA as a biodiversity instrument SEA as a means of removing barriers to integration of ecological concerns into decision making Blends process and substance CEA complementary (resource/receiving environment focus) to SEA (plan focus) Integration of habitats assessment into SEA Integration of SEA and ecological planning and management EcIA as an instrument for integrating natural sciences and landscape planning into SEA	Integration of social effects into SEA SIA of policies, plans, and programs SEA as a social justice/equity instrument (e.g., gender inequities, poverty reduction, peace, and conflict) SEA a means of removing barriers to integration of social concerns into decision making Blends process and substance SEA as a catalyst for social learning SIA an instrument for integrating social sciences into SEA and integrating SEA and social planning SEA fosters learning and institutional reform Expands SIA focus from community and project level	Integration of health effects into SEA Integration of SEA and health planning and management SEA as a means of removing barriers to integration of health concerns into decision making Blends process and substance HIA of policies, plans, and programs Role of SEA in protecting and promoting public health SEA as a health enhancement instrument HIA as an instrument for integrating health sciences into SEA

Table 12.2 (*Continued*)

EIA and EcIA	EIA and SIA	EIA and HIA
Integration of ecological effects into EIA EcIA of projects and activities Role of EcIA in integrating ecological and biodiversity concepts, principles, and methods into EIA EIA a means of removing barriers to integrating ecological concerns into project planning Blends process and substance EcIA as a means of integrating natural sciences into EIA Role of EIA in addressing ecological risks (integration of ecological risk assessment) EIA and EcIA parallel and connected or partially overlap	Integration of social effects into EIA SIA of projects and activities Integration facilitates social learning Role of SIA in integrating social principles (e.g., free, informed, and prior consent), human rights, and social performance standards into EIA A means of removing barriers to integrating social concerns into project planning A means of integrating social sciences into EIA Role of SIA enhancement in converting project risks to development opportunities (e.g., local content requirements) Blending of process and substance; SIA and EIA parallel and connected or partially overlap	Integration of health effects into EIA HIA of projects and activities Role of HIA in integrating medical science methods and health planning, assessment and management methods, and practices into EIA HIA and EIA parallel and connected or partially overlap EIA a means of removing barriers to integrating health concerns into project planning Blending of process and substance Role of EIA in protecting and promoting public health Role of EIA in addressing calculated (human health risk assessment) and perceived health risks

EcIA and SIA	EcIA and HIA	SIA and HIA
SIA provides understanding of social/cultural context; EcIA provides understanding of ecological context; overlap—mutually dependent Social consequences of ecological effects Ecological consequences of social effects Use of methods, such as landscape management, integrated IA and CEA, to trace ecological/social change, and management interactions Subsumed within SA Shared emphasis on environmental substance (purposes and outcomes) Links between biodiversity and poverty Treating ecological and socioeconomic system as a unit Social and natural science links	Ecological determinants of health Overlap ecological and human health Ecological consequences of health effects Health consequences of ecological effects Connected by CEA, integrated IA and SA Interconnections ecological and human health risks; links to human health and ecological risk assessment, and management Shared emphasis on environmental substance (purposes and outcomes) Coordinated approach to monitoring integrity of environment and health Natural and medical science links	Social determinants of health Broad definition of health (includes social) Health and safety promotion and protection an SIA principle Broad definition of social (includes health) Social consequences of health effects Health consequences from social change Shared emphasis on environmental substance (purposes and outcomes) Connected by CEA, integrated IA and SA Interconnections—calculated and perceived risks Social (e.g., gross national happiness) frames HIA Integration of SIA and HIA helps counterbalance qualitative and quantitative Social and medical science links

Sources: Alshuwaikhat (2005), Bina (2007), Bina et al. (2011), Bond (2004), Burdett (2008b), Burdge et al. (2004), Burdge, (2004), Craik (2008), Demidova and Cherp (2005), Devuyst (2000), Dixon and Thérivel (2011), Doelle and Sinclair (2006), Dora (2004), Esteves et al. (2012), Fischer (2006, 2011), Gibson (2006b, 2011), Guillette (2003), Gunn and Noble (2011), Hanna (2009a,b), Hanusch and Fischer (2011), Harris-Roxas et al. (2012), IAIA (2003), Jiliberto (2011), Kemm and Parry (2004a), Milner et al. (2005), Mishra (2009), Noble and Bronson (2006), OECD (2006), Orenstein et al. (2010), Pennock and Uma (2011), Pope and Dalal-Clayton (2011), Rajvanshi et al. (2011), Rowan and Streather (2011), Sadler (2011a), Seidler and Bawa (2003), Täbara and Pahl-Wostl (2007), Thérivel (1992, 2010), Thornton (2003), Treweek et al. (2011).

integration between the global/transboundary and the local, the integration of IA and project planning, the integration of SEA and policy, program, and plan making, the integration of IA and decision making, integrating the qualitative and the quantitative, the conceptual with the particular, the future with the present) (Pope, 2006). Critical considerations in any integrative initiative include who undertakes and how the integration is undertaken, the adjustments made because of the nature of proposed actions, and contextual adaptations (Morrison-Saunders and Thérivel, 2006).

12.7 CONTEMPORARY CHALLENGE—MATCHING PROCESS AND CONTEXT

The concept of a plurality of approaches or paradigms, where application choices are contingent on contextual characteristics, is "well-trodden ground" in planning, in the social sciences and, to some degree, in the natural sciences (Patterson and Williams, 1998; Ritzer, 1996; Rothman and Sudarshan, 1998; Schön and Rein, 1994). The general thrust of the pluralistic/contingent approach is (1) to differentiate key approach characteristics, strengths, and limitations; (2) to identify contingent factors or classes of situational characteristics; and (3) to match compatible approach and situational characteristics. A variation of the pluralistic/contingent approach is to consider both the similarities and the differences among alternative approaches or processes. The similarities or shared characteristics provide the basis for core elements. The differences provide the basis for matching approach and contextual characteristics.

In the early years of IA practice, there was a tendency to assume that IA requirements and practices were largely context-free. Context was, at best, a backdrop to be influenced but not accommodated. IA requirements could be applied in any setting as could good general IA practices. IA effectiveness could be determined through the application of generic principles and criteria. Process was expected to influence context in the sense that IA-related information and analysis would inform decision making, adverse environmental consequences would be ameliorated, and the positive purposes associated with proposed actions would be realized. Also implicit in this characterization of the relationship between process and context was the expectation of rational, apolitical decision making where IA-related choices would be largely determined by value-free scientific and rational analyses generated by a unitary IA process, and a simple, understandable, certain, manageable, and largely static environmental setting.

Not surprisingly, in retrospect, such assumptions have turned out to be overwhelmingly erroneous and dangerously naïve. The reality is that IA requirements and practices have turned out to highly context-dependent (or more appropriately interdependent), as illustrated in Figure 12.4. This reality has, over the past decade in particular, been widely acknowledged and accepted in IA literature and practice (Bina et al., 2011; Chaker et al., 2006; Fischer, 2006; Noble, 2009a,b; Partidário, 2011). The conversation, as detailed in Table 12.3, has moved on to how IA requirements and processes should more effectively influence context (both in the sense of positively affecting the environment consistent with the purposes of IA and removing the contextual impediments to decision making consistent with intended IA substantive outcomes), and should adapt to context (such that IA can effectively realize its full potential consistent with contextually based characteristics, perspectives, and priorities).

The concept of context, as applied in IA, in general terms, refers to those conditions external to IA that IA requirements and practices can and should consider, integrate, and positively influence. In the process of adapting to and influencing context, contextual elements and IA requirements and practices overlap and converge, preferably in a mutually supportive way. Care needs to be taken to avoid false dichotomies such as context-free or context-dependent. It should be possible to identify broad IA principles and aspirations (and even some general good practices) that transcend most contextual variations, while at the same time being highly conscious of and sensitive to those contextual variations that should necessarily be adapted to and integrated into IA requirements, IA process types, and individual processes. Contextual characteristics are rarely static, simple, or certain. Instead, they more frequently tend to be complex, uncertain, emergent, fluid, heterogeneous, conflicting, value full and difficult to understand much less manage. As listed in Table 12.3, the range of potentially relevant context types and characteristics is immense. It is essential, in the design and application of IA requirements and processes, to obtain a sound understanding of the contextual characteristics relevant to or potentially relevant to IA requirements, guidelines, activities, and potential decision making and environmental outcomes. It also is necessary to have a clear undertaking of the purposes and preferred outcomes of IA requirements and processes, taking contextual characteristics in account, if there is to be any hope of a sound match between process and context.

Blending process and context is far from a simple endeavor. IA requirements/processes and context tend to be highly interdependent. They generally coevolve with considerable iterative feedback. In some cases, as in unique and highly sensitive environments, context should assume a controlling role in IA process design and management. Perspectives will vary among interested and affected parties regarding what is the appropriate blend of process and context. Definitions of the appropriate match of process and context also will vary among IA levels, among IA types, among jurisdictions, across temporal and spatial scales, and as a consequence of a host of geographic, environmental, social, economic, political, and organizational factors. Juxtaposed between IA process and context are a range of middle-ground IA types (e.g., CEA, territorial IA), concepts and frameworks, which provide the opportunity to facilitate the more effective blending of process and context. Even when there appears to be a good match between IA and contextual categories, there will always be a unique mix of circumstances associated with an individual IA process that must be taken into consideration. The fluid and dynamic nature of the relationship between IA requirements/process and context is especially evident at the (especially policy) SEA level, where boundaries between action and IA are highly permeable and subject to change, and where influencing decision making (i.e., enhancing the capacity to alter context) is just or even more important than adapting to context. It is at the SEA level, in particular, where the concept of a plurality of IA processes, approaches, and

Table 12.3 IA and Context

Definitions and Distinctions	IA Characteristics Sensitive to Context	Context Characteristics Pertinent to IA
Context—the set of facts and conditions that have an impact on the chosen IA approaches (ends, means) and outcomes (effects on content, decision making, participants, organizational) in the short and long term	IA legal framework, appropriateness to context and ability to adapt to changing context	General characteristics—complex, uncertain, difficult to control, emergent, value-full, political, heterogeneous, conflicting, and partial
Immediate context (individual actions and related policies, plans, programs, and projects) vs. wider context (political, institutional, social/cultural, organizational)	IA definitions and criteria for evaluating necessarily broad because of contextual differences	Institutional and administrative capacity; available resources
Policy (legislative, regulatory) vs. societal (political, cultural) contexts	Tiering approaches vary by context	Historical, social/cultural, economic, natural environmental, political, organizational and administrative characteristics, conditions, and trends
Structured (identified steps, formalized, centralized), unstructured (fluid, issue-driven, reactive, bottom-up) and blended (elements of each) approaches to matching process and context	Purposes of IA requirements and guidelines	Geographic characteristics (e.g., urban vs. rural)
Regulated vs. unregulated contexts	Purposes of IA process; process fit for purpose	Planning system, planning level, and type of policy, plan, program, or project and IA type
Local/regional vs. international/national/subnational contexts	Problem and opportunity definitions	Formal (rules, division of responsibility, competence) and informal (attitudes, norms, traditions) organizational structures and planning, and IA culture
Context consciousness and sensitivity (awareness and sensitivity to)	IA objectives—context-specific (concrete, meaningful, practical, comprehensive, and credible)	Political/policy goals, priorities, definitions, pressures, and commitments, power structure and constraints, dominant policy styles and discourses; role of IA (existing, potential)
Universal and context-dependent	Purpose of and need/opportunity for proposed actions vary with context	Policy-making context (technical or political rationality); environmental decision making and behavior
Country type contexts (e.g., developed, in transition, developing)	Baseline and impact characterization and forecasts influenced by context (e.g., role of climate change in alternative futures)	Variations in the "structuredness" of problem, openness of decision making to other environmental values and in denial/resistance
Direct and indirect effects of contextual factors	Scoping activities highly interdependent with context	The norms and values that frame, define, and fund public decisions; nature and degree of conflict or consensus
Complex, dynamic, and emergent notions of context	Proposal characteristics; context defines nature and character of policies, plans, programs, and projects	Decision makers understanding of environmental and sustainability issues; political will and commitment to IA
Contextual determinants	Choice of sustainability and environmental indicators dependent on context	Variations in levels of certainty about knowledge base
Current and future (e.g., certain, reasonably foreseeable, hypothetical) contexts	Types, intensities, and interpretations of impact vary by context	Variations in extent and rates of change in context
CEA integrates contextual elements into IA process	Interpretations of significance highly dependent on context	Variations in environmental governance capacity
Fit between process and context	Cumulative effects vary by context	Variations in capacity for organizational and social learning
The contextual setup; important role in IA system performance	Contextually appropriate mitigation and enhancement measures	International, transboundary, national, regional, and local contexts
Diversity of interconnected, overlapping and evolving contexts; can never fully understand or manage (contextual uncertainty)	Options and option evaluation criteria appropriate to context	Diverse forms of knowledge
Influencing vs. adapting to context	Capacity of government and nongovernment sectors to undertake, participate effectively in IA, implement measures, and support public participation	Transcending issues (e.g., climate change)
		Population characteristics
		Public concerns and preferences
		Sources of unsustainability embedded in policy and institutional contexts (e.g., laws, processes, customs, perspectives, values)
		Stakeholders—interests, perspectives, issues, concerns, power, positions, and capacities; stakeholder attitudes toward environment and sustainability

(continued)

Table 12.3 (*Continued*)

Definitions and Distinctions	IA Characteristics Sensitive to Context	Context Characteristics Pertinent to IA
	Meaningful public participation; if poorly adapted public participation weak and ineffectual Uncertainty factors dependent on institutional, political, and scientific factors; assessment of contextual uncertainty Follow-up approach and impact management strategies and measures designed to match context	Proponent type (e.g., private proponents reluctant to consider options outside business interests, corporate environmental policy) Role of professions, participating sectors, and informal networks Values and methodological assumptions; prevailing planning and thinking processes Temporal (generational, decisional) and spatial (global, macroregional, country, regional, municipality, site) scales—multiscale dimensions of context
Linkages—IA and Context	**Regulatory Level—Adapting to and Influencing Context**	**Applied Level—Adapting to and Influencing Context**
Purpose of IA—to positively influence context (environment) IA context-dependent; IA adapts to IA requirements/processes and context are interdependent, coevolve with iterative feedback; should converge over time Legal IA requirements separated from context less likely to be considered legitimate Context controls requirements/process as when unique and highly sensitive environment Capacity of IA system to influence and/or adapt to context Context determines purpose of IA requirements and IA processes Cumulative IA experience as driver of contextual change; responsibility of practitioners Multiple IA processes (centrality of pluralism) to match multiple contextual variations Policy/regional planning sets development context for CEA and EIA Good practices and IA effectiveness are context-dependent Openness of IA requirements and processes to contextual influence Role of prevailing planning and thinking processes (discourses) on IA Different perspectives and perceptions of IA process—context interconnections	Identify—purposes for influencing context through IA requirements and guidelines Decide if guidelines should be a model or a point of reference subject to contextual adjustments Sometimes will need binding requirements for highly sensitive/significant environmental conditions Seek to harmonize IA requirements among levels while allowing for contextual differences Ensure guidance materials specific to planning system, planning level, and type of policy, plan, program, or project Assess gaps in legislation and practice for adapting to and influencing context Monitor changing context and IA regulatory implications Monitor influence of IA requirements on context (intended and unintended) Adapt IA requirements in response to contextual changes; seek to ensure locally appropriate Sponsor case studies and workshops to address relationship of context and effectiveness of IA requirements and guidelines Identify context elements that need to be in place for effective IA (e.g., transparent value frames, time, support) Identify and seek to ameliorate (e.g., capacity building) policy/planning weaknesses	Identify objectives for influencing context Identify IA objectives based on context Facilitate contextual influence from bottom-up Seek to understand, make explicit, and influence values and political goals, pressures and commitments that influence decision making Place IA approval process within policy/regional planning context Seek to identify and map contextual uncertainties Monitor changing context and IA process implications Assess what works, why, under what circumstances, and from whose perspective; emphasize continuous learning Identify pertinent contextual characteristics; tailor process to context Adapt each IA process activity to context; ensure holistic approach and flexible process When addressing process–context connections communicate benefits, focus on key issues and linkages, capitalize on opportunities and learn by doing Combine generic criteria and trade-off rules with case and context-specific concerns Identify informal networks and emergent strategies that pose contextual constraints; seek to influence, alter and redefine; link to related objectives and initiatives; focus on producing and communicating strategically relevant knowledge Specify who is to be involved and how context-related adaptations are to be made; aim to change mind sets and motivations of major stakeholders and decision makers Specify focus, tasks, and roles; focus on critical factors that could make a decision-making difference (decision-oriented)

General principles that transcend context but requirements and practices vary by context Middle-ground IA forms to allow blending of general and context-specific (e.g., territorial sustainability) Extending IA to encompass context (e.g., closer to everyday life); extending context to encompass process (e.g., bottom-up participation) Role of EcIA in characterizing ecological context, SIA in characterizing social/cultural context, HIA in characterizing health context, and SA in establishing sustainability status; role of each form of IA (individually and collectively) to maintain and enhance Adapting to context essential to successful IA implementation Acknowledgment that no one way or one best way to conduct IA	Foster role of SEA in establishing development context for project-level EIA Tailor effectiveness criteria to IA system Assess contextual margin of discretion in IA requirements; too much (poor practice, inconsistent, IA purposes frustrated); too little (rigid, poorly adapted to context); adjust accordingly; ensure ample room for creativity and adaptability without compromising substantive objectives and outcomes Foster methodological pluralism Foster continuous improvement of process–context match by supporting national and international networks Address process–context throughout process Shift focus of IA capacity building from procedural shortcomings to contextual limitations (e.g., political system, capacities of key stakeholders, organizational and administrative system, human and financial resources, interorganizational linkages, leadership, access to information) Seek to accommodate contextual diversity while still enhancing IA quality of practice	Matching process and context highly dependent on effective collaboration, coordination, negotiation, and persuasion; emphasize organizational and social learning and self-reflection Draw upon good practice IA frameworks and practices but be sensitive to IA type and contextual variations Make substantive ends (e.g., sustainability) operational within organizational practice Develop, refine, and use frameworks that explore interconnections between process and context Support with capacity building, including self-help networks Favor IA process and methodological pluralism; multiple approaches and quantitative/qualitative methods to facilitate process–context match Draw upon lessons and insights from related fields of theory and practice regarding process/context relationship (e.g., planning, environmental management); contribute to theory building Seek shared understanding of appropriate substantive outcomes Seek shared understanding (practitioner, researchers, stakeholders) of appropriate match—process and context Seek complementary relationship between IA outcomes and outcomes from related policies, plans, programs, and projects Pilot test and evaluate any proposed approaches; experiment with alternative approaches Compare methods across a broad range of jurisdictions and system types to identify common questions and solutions Assess added value of IA in context

Sources: Adelle and Weiland (2012), Alshuwaikhat (2005), Azcarate and Balfors (2009), Barrett and Lee (2003), Bina (2007), Bina et al. (2011), Bond (2010), Bond and Morrison-Saunders (2009), Bond and Pope (2012), Bond et al. (2012), Briffet et al. (2003), Cashmore et al. (2010), Cherp and Antypas (2003), Cherp et al. (2007), Craik (2008), Dovers (2005), Greig and Duinker (2007), Dusik and Sadler (2004), Égré and Senécal (2003), Faber et al. (2010), Fahy and Cinnéide (2007), Finnveden (2003), Fischer (2002, 2005, 2006, 2007b), Fischer and Gazzola (2006), Gasparatus et al. (2007), Gazzola (2008), Genter et al. (2008), Gertler (2009), Gibson (2006b, 2010), Hacking and Guthrie (2006, 2011), Hansen and Kørnøv (2010), Harris-Roxas et al. (2012), Hartley and Wood (2005), Hemmings and Roura (2003), Hildén et al. (2004), Hindling and Bjarnadóttar (2007), IAIA (2003), Jackson and Dixon (2006), Jay (2005, 2007), Jiliberto (2011), Kiewiet and Vos (2007), Koivurova (2008), Kolhoff et al. (2009), Kørnøv and Dalkmann (2011), Lee (2006), Lemon et al. (2004), Lyhne (2009), Marshall et al. (2005), McCluskey and João (2011), Morgan (2012), Morrison-Saunders et al. (2003), Ng and Hui (2007), Nilsson and Dalkmann (2010), Noble (2009a,b), Noble and Gunn (2009), Partidário (2007, 2011), Partidário and Coutinho (2011), Persson and Nilsson (2007), Péti (2012), Pinho et al. (2010), Pischke and Cashmore (2006), Pope and Grace (2006), Posas (2011), Putters (2005), Risse et al. (2003), Ross et al. (2006), Runhaar (2009), Runhaar and Driessen (2007), Sadler (2005a), Sadler and Jurkeviciute (2011), Schijf (2011), Sheate (2011), Sinclair and Diduck (2009), Slootweg (2005), Smith and Schin (2004), Storey and Jones (2003), Tetlow and Hanusch (2012), Treweek et al. (2011), Vanclay (2006), Weaver et al. (2008), Weiland (2010), Weston (2011), Whitelaw et al. (2009), Wirutskulshai et al. (2011), Zhu et al. (2010).

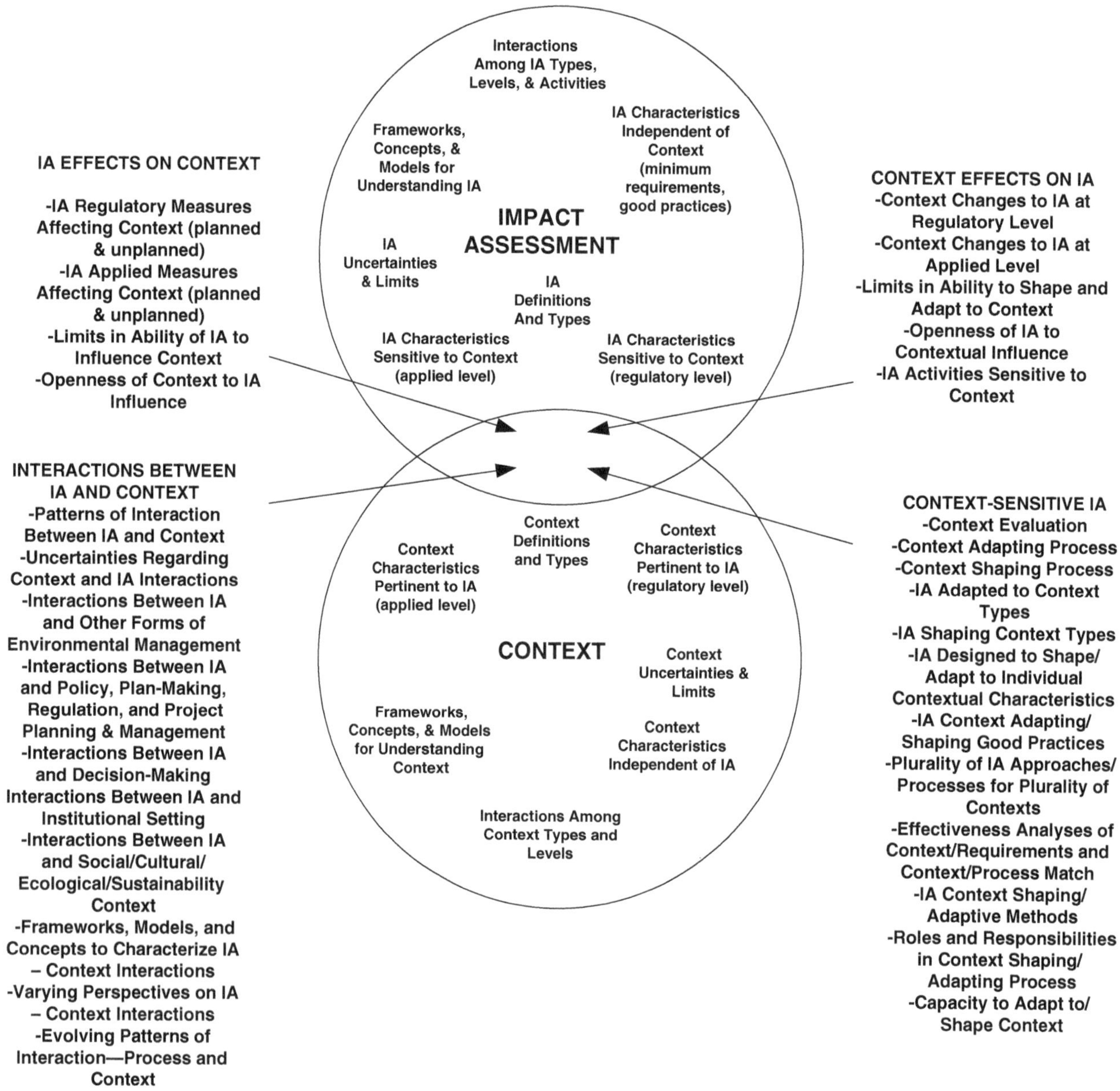

Figure 12.4 Matching process and context.

methods to match a complex array of uncertain and rapidly changing contextual conditions and constraints appears to be most appropriate.

As highlighted in Figure 12.4 and as detailed in Table 12.3, IA affects (sometimes deliberately and sometimes not) and is affected by context. The patterns of interaction between IA and contextual characteristics are many and complex. The process of matching IA and context operates at both the regulatory and applied levels. It encompasses such considerations as characterizing context, IA requirements, and processes designed to influence context, IA requirements, and processes adapted to context, a host of contextual shaping and adapting procedures and methods, an openness to a plurality of approaches and frameworks to match a plurality of contexts, an appreciation that contextual interpretations will vary among perspectives and over time, an acknowledgment of the need to ensure sufficient capacity for contextual adaptation/influence, and a realization of the need to assess the effectiveness of the ongoing fit (again from multiple perspectives) between process and context.

Table 12.3 identifies a range of potential measures, available at the regulatory and applied levels, to facilitate the dynamic and evolving fit between process and context. The appropriate mix of measures will, of course, vary from situation to situation. IA literature has broadly recognized that definitions of IA system and process effectiveness are context-dependent. This suggests that relying on "good-practice" IA guidance should be approached with great caution. Although the need to tailor the process to the context is widely accepted, good practice guidance for undertaking such endeavors is, at best, in its infancy. Arguably, generic IA good-practice guidance and context tailoring could be seen as a contradiction in terms. Possible middle-ground opportunities worth exploring include identifying generic context-matching principles, seeking to narrow the definition of effectiveness to specific IA types, levels and setting types, sponsoring comparative studies across contextual categories to explore the potential for identifying transcending and contexts-specific good practices, sponsoring case and pilot studies for testing various approaches for blending process and context, sponsoring forums and workshops to encompass a range of perspectives on the issue of blending process and context, and devoting greater attention to such middle-ground approaches as CEA and territorial IA, which have already sought to more effectively integrate contextual considerations into IA practice.

12.8 CONTEMPORARY CHALLENGES

Chapters 2–12 each address a contemporary challenge encountered in IA practice. Examples of interconnections among IA contemporary challenges are highlighted in Figure 12.5. As illustrated in Figure 12.5, CEA represents a bridge between SEA and project-level EIA. SA subsumes SEA, CEA, and project-level EIA. Structuring challenges address the integration of environmental disciplines and efforts to match process and context. IA type and activity challenges address issues associated with formulating and applying good practices for various IA types (e.g., SEA, CEA) and for various key IA activities (e.g., significance determination, follow-up). Institutional decision-making challenges are concerned with the relationship of IA and decision making and with IA and related institutional arrangements (e.g., capacity building, multijurisdictional IA). Crosscutting issue challenges are concerned with critical issues (e.g., climate change, the siting of locally unwanted land uses) that affect all types of IA process types, IA types, and IA levels.

IA practitioners must frequently address a multiplicity of contemporary challenges. As detailed in Table 12.4, these contemporary challenges are highly interconnected. To give a hypothetical example, in the course of designing and managing an SEA for an analysis of transportation infrastructure choices:

a. The process should be designed and managed, as appropriate, in accordance with SEA good practices.
b. Any IA process is of little value if it does not effectively inform and influence decision making. This necessitates drawing upon good practice guidance for influencing decision making.
c. It should seek to both influence context (in the sense of affecting decision making and realizing desired environmental context) and adapt to context (in the sense of being appropriate for the institutional, social, economic, political, and environmental setting).
d. Careful consideration should be given to how best to address interactions among environmental components (horizontal integration), and to the analysis and management of cumulative effects (drawing upon CEA good practices).
e. Inasmuch as transportation infrastructure projects (e.g., highways) can be viewed as "locally unwanted land uses" by directly affected individual and communities, the SEA should draw upon good practice LULU siting approaches, and should effectively frame the consideration of major transportation projects at the project EIA level.
f. Transportation infrastructure choices inevitably encompass a diverse array of jurisdictions and stakeholders, some of which may lack the capacity to fully and effectively participate in the SEA process. Consideration will, therefore, need to be given to multijurisdictional IA arrangements (including IA tiering), and to the need for IA capacity building;
g. Significance determination and follow-up are two critical activities in any IA process. Approaches for undertaking these activities should be systematically and collaboratively formulated and applied, with ample thought given to good practice guidance and contextual variation; and
h. Climate change is likely to a major issue with any major transportation infrastructure SEA. Systematic and comprehensive approaches for integrating climate change concerns will need to be integrated into the SEA process.

The same or a similar array of contemporary challenges would apply with a major sectoral SEA, a major pipeline, and so on. IA literature has much to offer in addressing these types of challenges. However, care should be taken to ensure that the approaches taken to address multiple challenges are complementary and appropriate to the setting.

12.9 COMPOSITE IA PROCESSES

Composite IA processes have been a recurrent theme in IA literature in recent years. Largely gone is the view of the IA process as a simplistic, unitary, linear, standardized, and context-free procedure for complying with regulatory requirements. Suggested composite IA approaches have assumed many forms. Examples include:

- The use of *structuring frameworks and models* (e.g., to systematically identify and explore various interconnections among IA process types, IA types, IA levels, decision making, planning policy, and institutional arrangements, qualitative and quantitative tools,

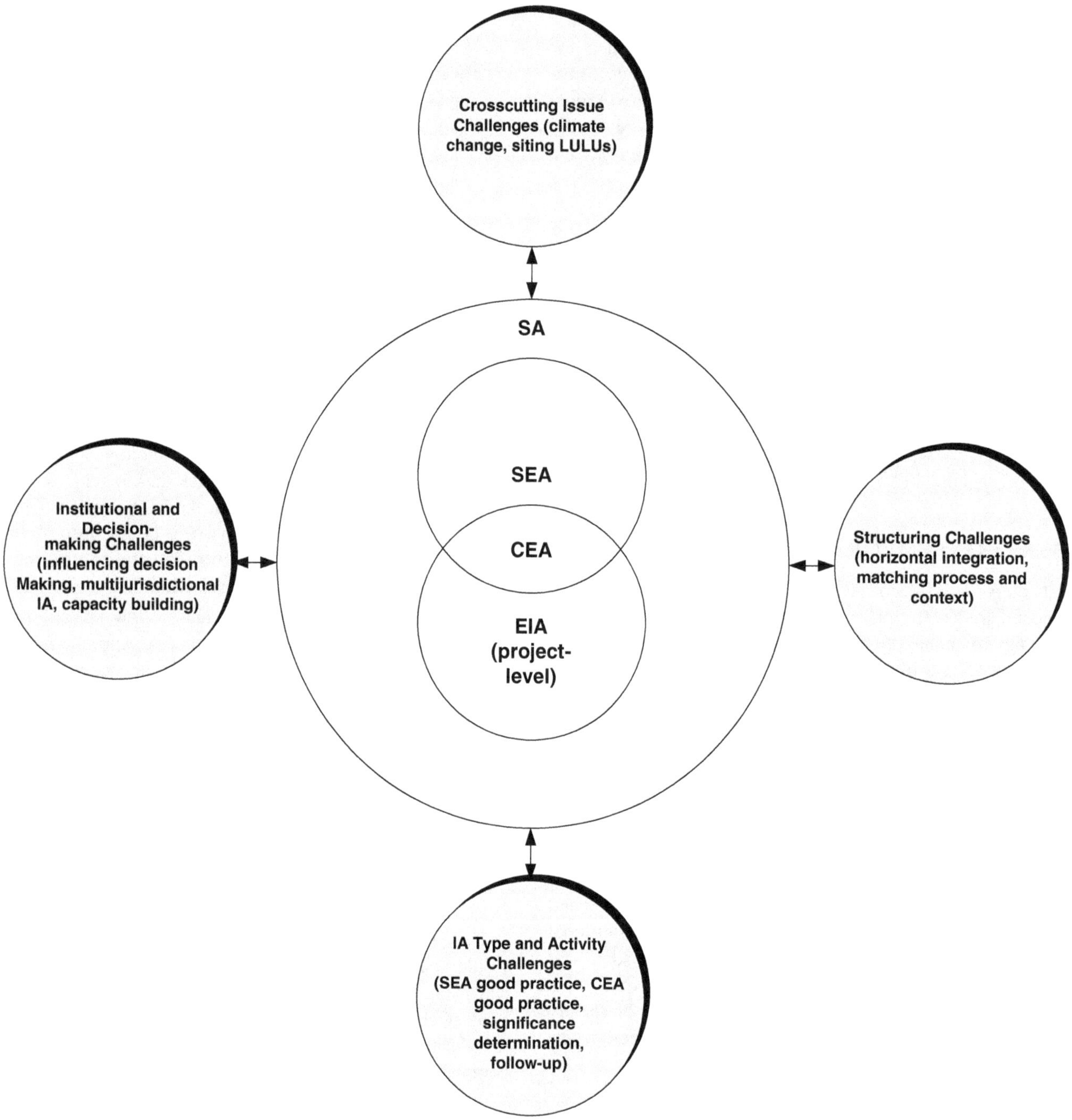

Figure 12.5 IA and contemporary challenges.

stakeholders, spatial and temporal scales, and contextual characteristics, to link and transcend the procedural, the substantive, the normative, and the systemic) (Binder et al., 2010; Fischer, 2005, 2006; Hacking and Guthrie, 2006; Jiliberto, 2004; Kessler and Abaza, 2006; Lee, 2006; Pope, 2006; Wiek and Binder, 2005).

- Approaches that *combine IA process types, IA types, and contemporary challenge approaches*, as described in Chapters 3–12 (e.g., combinations of rational/technical/scientific approaches with participative/communicative/learning oriented/adaptive approaches, a combination of collaboration, adaptive management,

Table 12.4 Examples of Interactions Among Contemporary Challenges

SEA Good Practices and Good Practices—Making IA Influential	SEA Good Practices and Good Practice Follow-up	SEA Good Practices and Siting LULUs	SEA Good Practices and Horizontal Integration	SEA Good Practices and CEA Good Practice
SEA good practices are of little value if they have no influence on decision making Marrying the two types of good practices makes it possible for the level of IA to be raised and for it to make a difference	Good practice SEA and good practice follow-up are complementary Follow-up is a vital component of SEA Effective follow-up makes it possible to progressively improve SEA good practice	Tiered and effective SEA frames the siting of LULUs Effective SEA reduces the likelihood and severity of adverse effects and enhances the potential for community benefits from LULUs SEA provides goals and collaboration momentum	SEA good practice tends to broadly define the environment The scope of effects and the treatment of trade-offs are crucial; need to carefully consider horizontal integration arguments and choices	SEA is the most appropriate level for considering cumulative effects; in practice tends to be poorly done The integration of SEA and CEA good practices has the potential to be complementary and mutually beneficial
SEA Good Practices and Multijurisdictional IA	**SEA Good Practices and Capacity Building**	**SEA Good Practices and Significance Determination**	**SEA Good Practices and Climate Change**	**SEA Good Practices and Context**
SEA is tiered (policies, plans, and programs PPP); also frames projects Many PPPs involve multiple jurisdictions; effective SEA systematically addresses the role of each jurisdiction SEA good practices are adapted to jurisdictional differences	Prevailing institutional arrangements often limit effective SEA practice IA capacity building facilitates effective SEA practice Capacity building is a component of SEA good practice	Significance determinations occur throughout the SEA process Good practice SEA encompasses significance determination; often significance determination is one of the weaker elements of prevailing SEA practice	Climate change is more effectively addressed at the SEA level Climate change IA good practices can be effectively integrated with SEA good practices Climate change should be addressed in every SEA activity	SEA and context are highly interdependent SEA goals seek to influence environmental context A plurality of SEA processes and methods can be adapted to context types An iterative relationship should exist between the SEA process and context
Good Practices—Making IA Influential and Good Practice Follow-up	**Good Practices—Making IA Influential and Siting LULUs**	**Good Practices—Making IA More Influential and Horizontal Integration**	**Good Practices—Making IA More Influential and CEA Good Practices**	**Good Practices—Making IA More Influential and Multijurisdictional IA**
Follow-up is of little value if it does not influence decision making; good practices for making IA more influential address follow-up roles Follow-up can inform and enhance good practices for making IA influential	Good practices for the siting of LULUs are only relevant if they influence decision making Effective LULU siting experiences can contribute to organizational learning Good practices for making IA more influential should encompass LULU siting	The siting of LULUs is unlikely to be effective unless the full range of social, ecological, and economic concerns are addressed The manner in which horizontal integration occurs is crucial to the realization of net benefits at every level	CEA good practices are of little value if decision-making influence is limited CEA good practices should include measures to enhance IA influence Good practices for making IA more influential could further the role of CEA in decision making	Efforts to make IA more influential should apply to all pertinent jurisdictions and should be tailored to each jurisdiction A systematic approach to multijurisdictional IA can enhance the potential for decision-making influence

(continued)

Table 12.4 (*Continued*)

Good Practices—Making IA More Influential and IA Capacity Building	Good Practices—Making IA More Influential and Significance Determination	Good Practices—Making IA More Influential and Climate Change	Good Practices—Making IA More Influential and Context	Good Practice Follow-up and Siting LULUs
IA capacity building efforts are of limited relevance if they do not address the issue of decision-making influence Effective IA capacity building can enhance the potential for greater decision-making influence Efforts to make IA more influential should include IA capacity building	Significance determination good practices are of limited relevance if they do not influence decision making Effective significance determination procedures can better inform and facilitate enhanced IA influence Measures to enhance IA influence should include who decides and how significance determinations are made	Climate change IA good practices are of limited relevance if they do not influence decision making Measures to make IA more influential should include who decides and how climate change is to be addressed Climate change IA good practices can better inform and influence IA-related decision making	Measures to make IA more influential need to be adapted to context Measures to make IA influential include how IA is to reshape context (i.e., aspirations and outcomes) Measures to match process and context should include how process and context can be more effectively blended to better inform and influence decision making	The follow-up stage is an important component of any strategy for siting LULUs Follow-up of LULU siting experiences can contribute to more effective siting approaches Good practice follow-up should include measures for addressing anticipated and unanticipated land use conflicts and conflicts among stakeholders
Good Practice Follow-up and Horizontal Integration	**Good Practice Follow-up and CEA Good Practices**	**Good Practice Follow-up and Multijurisdictional IA**	**Good Practice Follow-up and Capacity Building**	**Good Practice Follow-up and Significance Determination**
Good practice follow-up should address the full range of direct and indirect environmental effects The manner in which environmental disciplines are connected and integrated is crucial to effective follow-up Good practice follow-up can inform and guide IA horizontal integration efforts	Follow-up is a critical component of CEA good practices Follow-up can inform and guide effective CEA Good practice CEA and good practice IA follow-up can be complementary and mutually beneficial	Effective multijurisdictional IA includes a coordinated and consistent follow-up approach Good practice follow-up specifies the roles and responsibilities of each jurisdiction Good practice follow-up can inform and guide multijurisdictional IA	Good practice follow-up is likely to be of little value if there are follow-up capacity deficiencies IA capacity building can ameliorate and enhance the ability to undertake effective follow-up Good practice follow-up can inform and guide IA capacity-building initiatives	Sound procedures for determining the significance of effects are critical in follow-up Good practice follow-up should draw upon and include good practice significance determination Good practice follow-up can inform and guide significance determination methods
Good Practice Follow-up and Climate Change	**Good Practice Follow-up and Context**	**Siting LULUs and Horizontal Integration**	**Siting LULUs and CEA Good Practices**	**Siting LULUs and Multijurisdictional IA**
Follow-up is a critical stage of good practice climate change IA Good practice climate change IA can draw upon and include good practice follow-up Good practice follow-up can inform and guide climate change IA	Good practice IA follow-up and context are interdependent Good practice IA follow-up should both influence and be adapted to context Insights from follow-up can enhance efforts to more effectively blend process and context	The manner in which interconnections among environmental components are addressed can be crucial to success or failure in siting LULUs Insights from LULU siting experiences can inform choices regarding horizontal integration approaches	Any successful approach to siting LULUs must explicitly and systematically analyze and manage cumulative effects CEA good practices can be integrated into LULU siting approaches LULU siting experiences can inform CEA good practice	A tiered IA system is more conducive to the successful siting of LULUs A successful LULU approach clearly defines the roles and responsibilities of each jurisdiction LULU siting experiences can inform multijurisdictional IA good practices

Siting LULUs and Capacity Building	Siting LULUs and Significance Determination	Siting LULUs and Climate Change	Siting LULUs and Context	Horizontal Integration and CEA Good Practice
Good practice LULU siting approaches are likely to be of little value if there are follow-up capacity deficiencies IA capacity building can ameliorate and enhance the ability to successfully site LULUs LULU siting experiences can inform IA capacity building initiatives	Crucial to any LULUs siting approach are the approaches used to interpret significance Sound and appropriate significance determination approaches can facilitate the successful siting of LULUs LULU siting experiences can inform the formulation and refinement of significance determination approaches	Climate change is often a major issue among government bodies and public stakeholders Mutually beneficial solutions include how climate change requirements and objectives are to be satisfied Climate change impacts a potentially transcending factor in the determination of acceptable solutions	The siting of LULUs and context are interdependent; what works in one situation can be completely inappropriate in another A successful LULU siting approach seeks to both positively influence context and to adapt to an evolving context LULU siting experiences can inform efforts to match process and context	CEA good practice includes the systematic consideration of environmental interconnections CEA good practice can draw upon insights derived from the horizontal integration debates Horizontal integration approaches can learn from CEA experiences and good practices
Horizontal Integration and Multijurisdictional IA	**Horizontal Integration and IA Capacity Building**	**Horizontal Integration and Significance Determination**	**Horizontal Integration and Climate Change**	**Horizontal Integration and Context**
Multijurisdictional IA is more likely to be effective if complementary approaches to horizontal integration are adopted by all jurisdictions Horizontal integration good practice includes specifying the roles and responsibilities of each jurisdiction Multijurisdictional IA good practice includes systematic horizontal integration procedures	Good practice horizontal integration is dependent upon the adequate capacity to systematically analyze all environmental components and their interactions Horizontal integration good practice includes capacity building IA capacity building initiatives can inform horizontal integration efforts	Procedures for addressing significance will vary among substantive IA types Good practice significance determination include the significance of interactions among environmental components and cumulative effects Effective horizontal integration is crucial to effective significance determination	Climate change effects cut across environmental components and are highly interdependent An effective horizontal integration approach is crucial to good practice climate change IA Horizontal integration approaches and methods can inform and facilitate good practice climate change IA; the reverse is also true	The effectiveness of horizontal integration IA approaches is highly interdependent with context Effective approaches to matching process and context should draw upon and subsume systematic horizontal integration approaches Effective horizontal integration is crucial when matching process and context
CEA Good Practice and Multijurisdictional IA	**CEA Good Practice and IA Capacity Building**	**CEA Good Practice and Significance Determination**	**CEA Good Practice and Climate Change**	**CEA Good Practice and Context**
CEA invariably involves multiple jurisdictions An effective multijurisdictional IA system is critical to good practice CEA Good practice CEA specifies the roles and responsibilities of each jurisdiction CEA good practices and multi-jurisdictional good practices should draw upon one another and be complementary	CEA good practices are not likely to be effective if there are major capacity deficiencies CEA good practices should include measures to address capacity deficiencies IA capacity building should include measures to enhance the ability to undertake CEA	Significance determination is a critical component of and integral to every CEA activity Significance determinations should address significance of cumulative effects Good practice CEA should draw upon good practice significance determination	CEA good practice should include the consideration of climate change The systematic analysis of cumulative effects provides a sound foundation for the consideration of climate change effects The analysis of climate change effects includes interactions with other effects	CEA good practice is interdependent with context CEA good practice seeks to positively influence and adapt to context CEA good practice experience can facilitate efforts to match process and context Efforts to match process and context can inform CEA good practice

(continued)

Table 12.4 (*Continued*)

Multijurisdictional IA and Capacity Building	Multijurisdictional IA and Significance Determination	Multijurisdictional IA and Climate Change	Multijurisdictional IA and Context	IA Capacity Building and Significance Determination
Often there are major capacity deficiencies in the ability to undertake multijurisdictional IA Multijurisdictional IA initiatives should include a capacity building component IA capacity building initiatives should include the roles and responsibilities of each jurisdiction	Perspectives on significance will vary among jurisdictions Significance determination procedures should be integrated into any multijurisdictional IA system Collaborative approaches across jurisdictions are needed to determine significance	Climate change IA, given the global perspective, is an inherently multijurisdictional form of IA Multijurisdictional IA approaches should include procedures for addressing climate change Climate change IA should specify the roles and responsibilities of each jurisdiction	Approaches for undertaking multijurisdictional approaches and context are interdependent Multijurisdictional IA good practices seek to influence and adapt to context Collaborative approaches across jurisdictions needed to match process and context	IA capacity building may be needed to ensure sound approach for determining significance Significance determinations needed to determine capacity building priorities IA capacity building initiatives should include approaches for addressing significance
IA Capacity Building and Climate Change	**IA Capacity Building and Context**	**Significance Determination and Climate Change**	**Significance Determination and Context**	**Climate Change and Context**
Good practice climate change IA often hampered by capacity limitations IA capacity building initiatives should address ability to undertake good practice climate change IA Good practice climate change IA should include measures for facilitating ability to adapt to and minimize contribution to climate change	IA capacity building highly interdependent with context Efforts to more effectively match process and context can be hampered by capacity limitations; should seek to remedy deficiencies IA capacity building efforts should be designed to positively influence and adapt to context	Significance determination is a critical component of climate change IA Significance determinations occur in every climate change IA activity Climate change impacts can be impacts of transcending significance Significance determination approaches can facilitate the interpretation of climate change impacts	Significance and context are highly interdependent Significance determination approaches should be designed to influence and adapt to context Efforts to match process and context should encompass significance determinations Experience with significance determinations can facilitate efforts to match process and context	Climate change impacts and the effectiveness of measures to ameliorate and adapt to climate change vary with context Climate change IA approaches should be designed to influence and adapt to context Approaches for matching process and context can inform good practice climate change IA

rigor and follow-up, the integration of climate change IA with adaptive management and sustainability assessment, a combination of the political, the democratic and the technical, a combination of citizen-centered follow-up, sustainability and local knowledge, a combination of IA types such as HIA, SIA, EcIA, and gender IA with decision-support tools such as LCA, CBA, RA, and MCA) (Ahmed and Sánchez-Triana, 2008c; Binder et al., 2010; Devlin, 2011; Hacking and Guthrie, 2008; Hunsberger et al., 2005; Lane et al., 2003; Morrison-Saunders and Sadler, 2010; Wilson, 2010).

- The development and refinement of *middle-ground concepts, variations, and subsets* for bridging IA process types and IA types, for integrating IA with related methods, and for matching process and context types (e.g., the use of follow-up to link EIA and LCA, the adaptation of SEA for the private sector) (Noble, 2009b; Sheate, 2011).
- Approaches that seek to delineate the *pattern of current and potential interconnections* between IA and decision making/governance, IA and the policy, planning and project development process, and IA and institutional structures and procedures (Ahmed and Sánchez-Triana, 2008c; Cashmore et al., 2010; Ortolano, 2008).
- Efforts to link and integrate *the IA process and practice with related decision-support methods, instruments, and tools* (e.g., life-cycle assessment, environmental auditing, environmental management systems, substance flow analysis, risk assessment, environmental permitting, initial environmental review, environmental site assessment, environmental indicators, ecosystem management, product and technology assessment, economic valuation, multiattribute evaluation, modeling tools, physical analysis tools, participatory methods) (Benson, 2003; De Ridder et al., 2010; Emilsson et al., 2004; Finnveden, 2003; Noble, 2009b; Porter, 2006; Ridgway, 2005, 2010; Sheate, 2011; Thompson, 2002a; Vanclay, 2010; Van der Vorst et al., 2010).
- Efforts to bridge *IA theory and practice and IA and related theory–practice fields* (e.g., interdisciplinary, transdisciplinary, and local/indigenous knowledge, the reorientation of research to substantive outcomes, connections to related fields such as land-use planning, organizational theory, policy and project appraisal, policy integration, contentious politics, and natural and social science theory and methods) (Benson, 2003; Binder et al., 2010; Cashmore et al., 2010; Cherp et al., 2007; Doelle and Sinclair, 2006; Lima and Marques, 2005; Porter, 2006; Rossou and Makan, 2007; Tang, 2008).
- Approaches that use *sustainability assessment (SA) to frame, direct, and integrate a range of IA elements and types* (e.g., connecting sustainability, participation, learning/education, social action and IA, an SA approach that integrates rationality/rigor, equity, precaution, democratic governance, socioecological integrity, livelihood sufficiency, and contextual influence/adaptation) (Gibson, 2006a; Pope and Dalal-Clayton, 2011; Sinclair et al., 2008).

Collectively, these examples provide a range of integrative possibilities to IA practitioners and other stakeholders, although considerable room remains for refining, combining, and testing them in practice. The examples suggest a field of theory–practice characterized by fluid and permeable internal and external boundaries, a plurality of processes and contexts, and a host of choices for connecting and integrating the procedural and the substantive, the regulatory and the applied, process and context, IA types and levels, and approaches for addressing both recurrent problems and contemporary challenges.

IA process design and management, as illustrated in Figure 12.6, while beginning, albeit tentatively, from core elements and good practices, is or should be framed by sustainability principles and priorities and operate within a tiered structure of SEA/project EIA levels. It should progressively and systematically encompass IA types, approaches for addressing recurrent problems, approaches for addressing contemporary challenges, and means of influencing and adapting to context. All IA generic core elements and good practices should be tested for relevance and capacity to influence decision making, and refined and adapted to more effectively match process to context.

Figure 12.7 illustrates conceptually the need to blend the procedural with the substantive (with a greater emphasis on the substantive), the need to encompass both the regulatory and the applied, the central role of efforts to match and blend IA and context types (including adjustments at the individual process/setting level), the importance of drawing upon and contributing to IA (i.e., theory building) and related fields of practice, the need for IA process types and IA types to be designed and blended to more effectively address recurrent problems and contemporary challenges, and the critical importance of bridging, melding, and transcending individual IA process design elements.

Figure 12.8 combines the substantive (broadly defined but focused on critical needs and aspirations) with the ethical (e.g., intergenerational equity, resource equity, equity in the distribution of power). It fosters planning and decision making that is influential, rational, adaptive, practical, collaborative, and democratic. Both the substantive and the procedural operate within a sustainability framework. Sustainability assessment recognizes the need for institutional reform and for mutually supportive links to other forms of sustainability management, to sustainability goals, objectives, indicators, and decision rules, and to sustainability limits and carrying capacity. A sustainability-oriented composite IA process would selectively draw upon insights, methods, and perspectives from the IA process types, the IA types, and the approaches for addressing

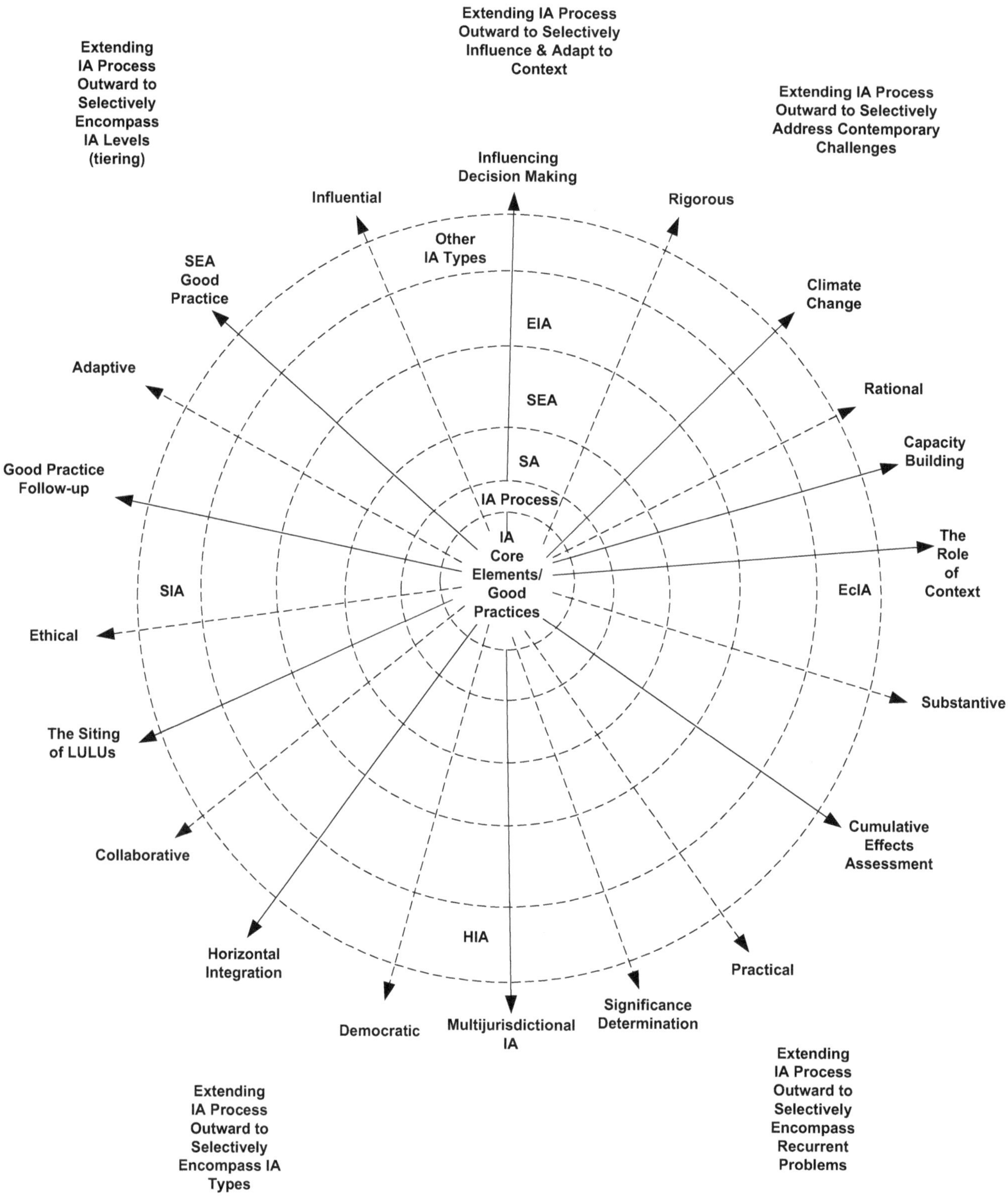

Figure 12.6 IA process design.

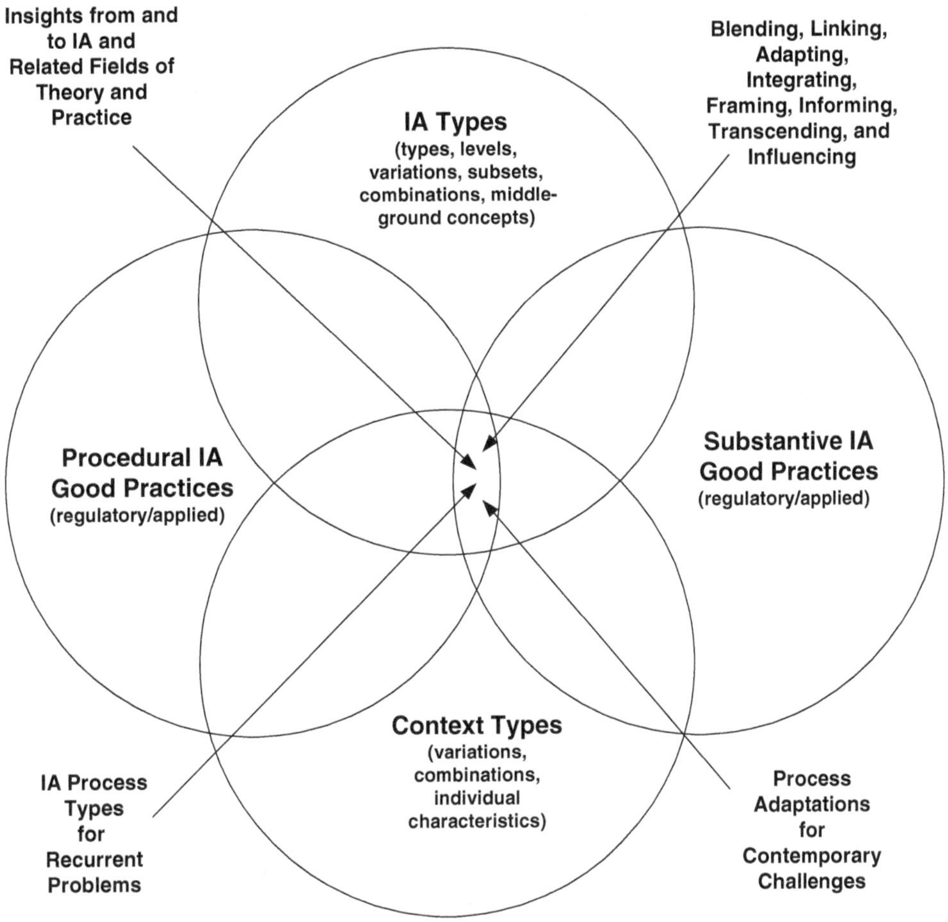

Figure 12.7 Elements of composite IA processes.

recurrent problems and contemporary challenges. The resulting modifications and refinements would reinforce the strengths and help offset the limitations (e.g., politically naïve, limited consideration of uncertainties, weak on consensus building and conflict resolution methods) of sustainability assessment. It also would facilitate a more effective fit between process and context.

An additional candidate would be the ecosystem approach. The ecosystem approach combines substantive environmental visions, principles, and objectives with collaborative decision making, practical political realities, adaptive planning, and a sound foundation of scientific and technical analysis. More consideration could be given to social, cultural, and economic concerns. A greater effort could be made to identify and resolve ethical concerns and trade-offs, to offset power inequities, and to ensure that choices are screened and compared systematically. Particular attention could be devoted to the differences between the ecosystem approach and IA. Two other existing framework possibilities are traditional knowledge (a blending of the ecological, scientific, spiritual, social, cultural, ethical, historic, collaborative, and local control) and integrated environmental assessment (starts from a combination of environment and IA types). Building from existing frameworks is helpful because the integration process has already begun. However, it is still necessary to add missing elements, supplement elements that have received insufficient attention, ameliorate shortcomings, and explore the implications of differences from IA.

12.10 LIMITS AND PRIORITIES

Enhancing IA process management, especially with reference to the recurrent shortcomings, the contemporary challenges, and the IA types, is a daunting task. Although there is considerable potential for improvement, there also are major knowledge gaps regarding the processes and process elements most and least effective in various classes of situations. The process-related knowledge base is incomplete and scattered, both within and outside IA literature. Most IAs involve multiple issues, a diversity of stakeholders, and a complex and evolving environment. IA institutional arrangements are highly interrelated with other regulatory instruments. It is difficult to separate the influence of process characteristics from other variables that affect outcomes. Definitions of process "success" and "failure" vary greatly.

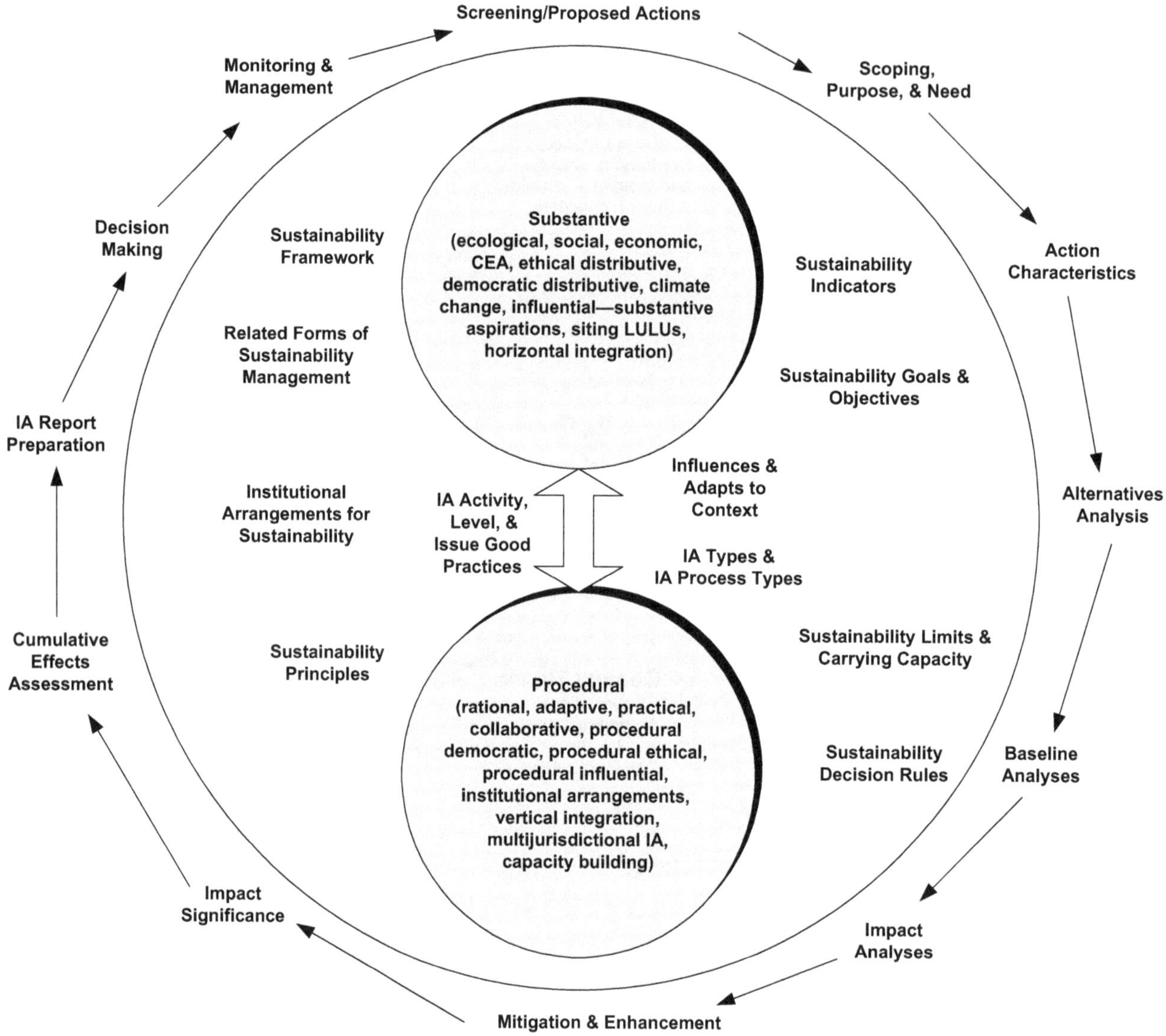

Figure 12.8 Example of a composite IA process: sustainability shaped.

There are generally multiple perceptions and perspectives on what occurred and should have occurred. IA usually occupies a peripheral position relative to project planning, policy making, and decision making. IA is, at best, a secondary mission for most proponents and regulators. IA process management problems and challenges exacerbate the organizational tendency to bypass, inhibit, or "starve for funds" agency IA functions.

Not surprisingly process participants are inclined to and frequently have a vested interest in describing and interpreting the process to their own best advantage. Memories are selective. Documentation is rarely complete. It is never easy to separate the unique characteristics of an IA process from those process attributes with potentially broader implications, either as pitfalls to be avoided or as models to be followed. Effectiveness reviews of IA process experience at both the regulatory and applied levels are limited, often anecdotal, generally are presented from only a single perspective, and tend not to be well adapted to contextual variations. Poorly designed and executed IA processes contribute to a legacy of mistrust and conflict, which, in turn, widens the gulf between IA practitioners and other stakeholders. Breaking out of this cycle requires a major effort.

These constraints are compounded by a persistent but rapidly dimensioning, belief in the myth of a "one-size-fits-all" IA process. The desire for or assumption of a "blueprint" IA process (based on a misplaced wish for certainty in an uncertain world) contributes to a common expectation that IA process management is no more than the scheduling of a standard set of IA activities, in combination with agency consultation, public consultation, study team management, and IA document preparation. The assumed simplicity of the IA process reinforces the belief that IA process management requires no specialized knowledge beyond normal project management skills and a working knowledge of regulatory requirements. Without a perception of either the IA process knowledge base or of recurrent IA process management shortcomings and contemporary challenges, there is minimal incentive for improvement. This reluctance to alter prevailing practice dovetails with the organizational propensity to resist change. When the "recurrent problems" and "contemporary challenges" arise, the tendency is to revert to explanations that focus on unique circumstances, beyond the ability of the IA process manager to either anticipate or control. Flawed practice is repeated. When outcomes continue to disappoint, there are always new reasons for the "new" problems and challenges.

Notwithstanding these impediments, there is considerable potential for improvement if initiatives are focused, efficient, and demonstrably effective. The first priority should be demonstrating the negative consequences associated with failing to adequately address the recurrent shortcomings and contemporary challenges, and to make appropriate adjustments to suit IA level, IA type, and contextual variations. If an effective case can be made for reforming current practice, based on the experiences and perceptions of critical stakeholders, an incentive for action is created. It is next necessary to demonstrate that practical improvements are possible, are not likely to be costly, can be readily implemented, and can produce short-term benefits to the major parties.

The knowledge base for enhanced IA process management, especially concerning the recurrent problems, contemporary challenges, and IA types, is considerable, as demonstrated in the preceding chapters. Possibly this book and related references can spark a dialogue about how to facilitate better IA process management and how best to move forward. Relevant models, concepts, and frameworks can be refined and tested. Process-related regulatory requirements and guidelines can be compared and evaluated. More documentation of IA process management experiences could occur. Greater emphasis can be placed on making IA theory and practice more substantive. Further consideration could be given to the relationship between IA processes and contextual characteristics.

There is a particular need for comparative evaluations of process management experiences, IA process case studies from multiple perspectives, and more effective links to process management efforts in related fields. The net result of such efforts could be an enhanced repertoire of process management tools and skills and a better definition of good and bad process management practice. However, before significant steps forward can be taken, a serious effort must be made to prevent and ameliorate recurrent IA process management problems and to more effectively address the contemporary challenges. Hopefully this book will facilitate such efforts.

12.11 SUMMING UP

This chapter begins with three IA process stories. The three stories address the role of synthesis in different ways. In the first story, an example middle-ground approach to bridging the gap between SEA and project-level EIA is described. In the second story, a major IA capacity building initiative is described. The story illustrates that effective IA capacity building provides a mechanism for addressing multiple recurrent problems and contemporary challenges. The third story describes how IA can serve both its traditional purposes, and provide a means of achieving broader institutional, social, and environmental ends.

This chapter describes how the individual IA processes, the IA types, and the approaches for addressing the contemporary challenges, presented in Chapters 2–12, might be connected and combined at both the regulatory and applied levels. It also identifies residual limitations and priorities for future action.

The problem is when and how to apply the IA process types, the IA types, and the approaches for addressing the contemporary challenges and recurrent problems, both individually and collectively. The direction is (1) how far to go, at the regulatory level, in directing and guiding IA process management (given the range of processes and contexts); (2) how to address interconnections among IA processes that seek to manage recurrent problems; (3) how to address interconnections among IA types; (4) how to match process and contextual characteristics; (5) how to address interconnections among approaches for addressing contemporary challenges; (6) how to design and manage composite IA processes; and (7) how to cope with the limits of synthesis and to establish priorities.

IA requirements and guidelines should define minimum levels of IA process-related practice and establish process-related goals and principles. IA guidelines should narrow the gap between the minimum and the ideal. Core IA process attributes should be identified. A sense should be provided of the available IA process management choices. IA requirements and guidelines should acknowledge recurrent process-related problems and contemporary challenges, and facilitate efforts to avoid and reduce the problems and confront the challenges. The need to adapt the IA process to contextual factors and to link and integrate IA process types and IA types should be recognized. IA requirements and guidelines should draw upon process-related experiences, applied research, and effectiveness reviews.

The IA process should be designed and adapted to suit relevant recurrent problems. Consideration should be given to process subsets, variations, and middle-ground concepts. Interconnections among recurrent problems and alternative approaches for addressing those problems should be identified and explored. Due consideration should be given to whether approaches conflict or are complementary, and to measures for ameliorating conflicts. Care should be taken to ensure that approaches for addressing multiple recurrent problems are appropriate to the context.

The IA process should be designed and adapted to suit the IA types encompassed by the process. Careful consideration should be given to interconnections among IA types and to whether the procedures for integrating IA types are appropriate and suited to the context.

The IA process should be appropriate to the situation. The contextual characteristics best suited to each IA process type and IA type should be identified. Matching process and context should include considering process subsets and variations, links between core process elements and good practices, process-type strengths and limitations, the possibility of multiple problems and challenges, process overlaps and conflicts, possible synthesis forms, measures to strengthen complementary relationships, measures to offset individual and mutual weaknesses and to retain diversity, and process adjustments to match an evolving context. Elements of each IA process type and each IA type need to be integrated into any IA process.

Effective IA process design and management should systematically address structuring challenges such as horizontal integration and matching process and context. Effective use should be made of SEA good practices, CEA good practices, and good practices for determining significance and for undertaking follow-up. IA process design and management should extend to the consideration of measures for influencing decision making, for IA capacity building, and for operating effectively in a sustainability-driven multijurisdictional and multitier IA system. Wherever pertinent, IA process design and management should systematically address climate change impacts and draw upon good practice approaches to the siting of "locally unwanted land uses." The IA should take into account interconnections among contemporary challenges and should systematically combine and integrate approaches for addressing multiple contemporary challenges and recurrent problems.

IA process and management, in the course of formulating and applying composite approaches, should make effective use of structuring frameworks and models, effectively combine elements of IA process types, IA types, and contemporary challenge approaches, effectively apply middle-ground concepts, variations, and subsets, ensure effective links to decision making, to policy, planning, and project development processes, and to institutional structures and arrangements, effectively integrate and apply decision-support methods, instruments and tools, effectively bridge IA theory and practice and related theory–practice fields, and effectively draw upon core IA elements and good practices while allowing for contextual variation. When designing and managing composite IA processes consideration should be given to the role of sustainability and SA in framing, directing, and structuring the process, to procedures for linking and integrating multiple IA types, IA process types, and approaches for addressing contemporary challenges and recurrent problems, to procedures for influencing and matching process and context, and to insights from related integrative concepts and approaches (e.g., the ecosystem approach).

There are major gaps in the IA process management knowledge base. It is very difficult to separate out and interpret the effects of IA process management choices. IA in general and IA process management in particular are rarely an agency priority. Process management limitations and negative repercussions are seldom acknowledged or documented. There continues to be a widespread, but rapidly diminishing, belief in a single, infinitely adaptable, IA process. There is little recognition of the need for specialized IA process management skills and knowledge. The IA process knowledge base is poorly understood. Recurrent IA process management shortcomings and contemporary challenges, although widely acknowledged in IA literature, are recognized to a much lesser degree in IA practice. Notwithstanding these challenges, there remains considerable potential for improvement. The first priority should be demonstrating the negative consequences of failing to adequately address recurrent shortcomings, IA type variations, and contemporary challenges. The IA process knowledge base should then be supplemented with applied examples, methodological refinements, effectiveness assessments, information and perspective exchanges, and better links to related fields.

The types of problems and challenges addressed in Chapters 3–12 can often be avoided, ameliorated, and effectively addressed. A host of procedures and concepts conducive to enhanced IA process management are described. Definitive prescriptions cannot be provided because of the complexity of the field and situation-specific circumstances. But a sufficient knowledge base has been established to suggest numerous improvement possibilities. Enhancing IA process management necessarily begins with an open mind, a willingness to consider the possibility of recurrent, avoidable problems, a recognition of the need to systematically address an array of contemporary challenges, an acknowledgment of differences among IA types and the care that must be taken when integrating IA types, and a commitment to explore how best to prevent, ameliorate, and confront the problems and challenges.

References

Adelle, C., and S. Weiland, "Policy Assessment: The State of the Art," *Impact Assessment and Project Appraisal* **30**, 25–33 (2012).

Agrawala, S., "Incorporating Climate Change Impacts and Adaptation in Environmental Impact Assessments (EIA)—Opportunities and Challenges," *Presentation to IAIA Special Symposium on Climate Change and Impact Assessment,* Washington, DC, 2010.

Ahmadvand, M., and E. Karami, "A Social Impact Assessment of the Floodwater Spreading Project on the Gareh-Bygone Plain in Iran: A Causal Comparative Approach," *Environmental Impact Assessment Review* **29**, 126–136 (2009).

Ahmed, B. S., "Integrating Health into Impact Assessment: Challenges and Opportunities," *Impact Assessment and Project Appraisal* **22**, 2–4 (2004).

Ahmed, K., and E. Sánchez-Triana, eds., *Strategic Environmental Assessment for Policies*, The World Bank, Washington, DC, 2008a.

Ahmed, K., and E. Sánchez-Triana, in K. Ahmed and E. Sánchez-Triana, eds., "SEA and Policy Formulation," *Strategic Environmental Assessment for Policies*, The World Bank, Washington, DC, 2008b.

Ahmed, K., and E. Sánchez-Triana, in K. Ahmed and E. Sánchez-Triana, eds., "Using Strategic Environmental Assessments to Design and Implement Public Policy," *Strategic Environmental Assessment for Policies*, The World Bank, Washington, DC, 2008c.

Albergaria, R., and T. Fidelas, "Transboundary EIA: Iberian Experiences," *Environmental Impact Assessment Review* **26**, 614–632 (2006).

Alberti, M., and J. D. Parker, "Indices of Environmental Quality—The Search for Credible Measures," *Environmental Impact Assessment Review* **11**, 95–101 (1991).

Albrecht, E., "Transboundary Consultations in Strategic Environmental Assessment," *Impact Assessment and Project Appraisal* **26**, 289–298 (2008).

Albrecht, S. L., "Equity and Justice in Environmental Decision-making: A Proposed Research Agenda," *Society and Natural Resources* **8**, 67–72 (1995).

Alexander, C., H. Neis, A. Anninou, and I. King, *A New Theory of Urban Design*, Oxford University Press, New York, 1987.

Alexander, E., *Approaches to Planning*, Gordon and Breach Science Publishers, New York, 1986.

Alexander, E., "Rational Revisited: Planning Paradigms in a Post-modern Perspective," *Journal of Planning Education and Research* **19**, 242–256 (2000).

Ali, S., V. O'Callaghan, and J. D. Middleton, "A Case Study of the 'Towards 2010' Programme Health Impact Assessment," *Journal of Environmental Assessment Policy and Management* **10**, 403–430 (2008).

Allen, G. M., and E. M. Gould, "Complexity, Wickedness and Public Forests," *Journal of Forestry* **84**, 20–23 (1986).

Alshuwaikhat, H. M., "Strategic Environmental Assessment Can Help Solve Environmental Impact Assessment Failures in Developing Countries," *Environmental Impact Assessment Review* **25**, 307–317 (2005).

Alton, C. C., and P. B. Underwood, "Let Us Make Impact Assessments More Accessible," *Environmental Impact Assessment Review* **23**, 141–155 (2003).

Amy, D. J., *The Politics of Environmental Mediation*, Columbia University Press, New York, 1987.

Anderson, D., *Strengthening Environmental Assessment for Canadians*, Report of the Minister of the Environment to the Parliament of Canada on the Review of the Canadian Environmental Assessment Act, Canadian Environmental Assessment Agency, Hull, Québec, 2001.

Andrews, R. N. L., in R. Clark and L. Canter, eds., "The Unfinished Business of National Environmental Policy," *Environmental Policy and NEPA: Past, Present and Future*, St. Lucie Press, Boca Raton, FL, 1997.

Androulidakis, I., and I. Kanakassis, "Evaluation of the EIA Performance in Greece, Using Quality Indicators," *Environmental Impact Assessment Review* **26**, 242–256 (2006).

Ansell, C., and A. Gash, "Collaborative Governance in Theory and Practice," *Journal of Public Administration Research and Theory* **18**, 543–571 (2007).

Antunes, P., P. Santos, and L. Jordão, "The Application of Geographical Information Systems to Determine Impact Significance," *Environmental Impact Assessment Review* **21**, 511–535 (2001).

Appell, D., "The Uncertainty Principle," *Scientific American*, January 2001.

Ardern, K., in J. Kemm, J. Parry, and S. Palmer, eds., "HIA: A Practitioner's View," *Health Impact Assessment*, Oxford University Press, Oxford, UK, 2004.

Armitage, D., "An Integrative Methodological Framework for Sustainability Environmental Planning and Management," *Environmental Management* **19**, 469–479 (1995).

Armitage, D. R., "Collaborative Environmental Assessment in Northwest Territories, Canada," *Environmental Impact Assessment Review* **25**, 239–258 (2005).

Armour, A., "Integrating Impact Assessment in the Planning Process: From Rhetoric to Reality," *Impact Assessment Bulletin* **8**, 3–13 (1990a).

Armour, A., *Socially Responsible Facility Siting*, Doctoral Dissertation, University of Waterloo, Waterloo, ON, 1990b.

Armour, A., and B. Sadler, "Synopsis of Workshop Discussions," *The Place of Negotiations in Environmental Assessment*, Canadian Environmental Assessment Research Council, Hull, Québec, 1990.

Arnstein, S. R., "A Ladder of Citizen Participation," *Journal of the American Institute of Planners* **35**, 216–224 (1969).

Arquiaga, M. C., L. W. Canter, and D. I. Nelson, "Risk Assessment Principles in Environmental Impact Studies," *The Environmental Professional* **14**, 204–219 (1992).

Impact Assessment: Practical Solutions to Recurrent Problems and Contemporary Challenges, Second Edition. David P. Lawrence.

Arquiaga, M. C., L. W. Canter, and D. I. Nelson, "Integration of Health Impact Considerations in Environmental Impact Studies," *Impact Assessment* **12**, 175–198 (1994).

Arts, J., P. Tomlinson, and H. Voogd, in B. Sadler et al., eds., "Planning in Tiers? Tiering as a Way of Linking SEA and EIA," *Handbook of Strategic Environmental Assessment*, Earthscan, London, 2011.

Athanas, A., "The Role of Business in Biodiversity and Impact Assessment," *Impact Assessment and Project Appraisal* **23** (2005).

Atkinson, S. F., and L. W. Canter, "Assessing the Cumulative Effects of Projects Using Geographic Information Systems," *Environmental Impact Assessment Review* **31**, 457–464 (2011).

Atkinson, S. F., L. W. Canter, and M. D. Rowen, "The Influence of Incomplete or Unavailable Information on Environmental Impact Assessment in the USA," *Environmental Impact Assessment Review* **26**, 448–467 (2006).

Atkinson, P., and A. Cooke, "Developing a Framework to Assess Costs and Benefits of Health Impact Assessment," *Environmental Impact Assessment Review* **25**, 791–798 (2005).

Au, E., "Common Sense Means Different Things to Different People," *Impact Assessment and Project Appraisal* **24**, 14–15 (2006).

Auditor General, *Referrals, Assessments and Approvals Under The Environmental Protection and Biodiversity Conservation Act*, Canberra, Commonwealth of Australia, Audit Report No. 2002–2003, Performance Audit, Canberra, Australia, 2002–2003.

Australian Government, *Environment Protection and Biodiversity Conservation Act, 1999 and Regulations 2000*, Office of Legislative Drafting and Publishing, Attorney-General's Department, Canberra, Australia, 1999 and 2000.

Australian Government, *Environment Protection and Biodiversity Conservation Act 1999—Guide to the EPBC Act*, Australian Government Department of the Environment and Water Resources, Canberra, Australia, 2007a.

Australian Government, *Draft Policy Statement: Use of Environmental Offsets Under the Environmental Protection and Biodiversity Conservation Act, 1999*, Department of the Environment and Water Resources, Canberra, Australia, 2007b.

Australian Government, *Use of Environmental Offsets Under the Environment Protection and Biodiversity Conservation Act 1999—Discussion Paper*, Department of the Environment and Water Resources, Canberra, Australia, 2007c.

Australian Government, *Matters of National Environmental Significance*, Department of the Environment, Water, Heritage and the Arts, Canberra, Australia, 2009a.

Australian Government, *Australian Government Biodiversity Policy—Consultation Draft, Department of Sustainability, Environment, Water, Population and Communities*, 2011a.

Australian Government, *Consultation Paper on Cost Recovery Under the Environment Protection and Biodiversity Conservation Act*, Commonwealth of Australia, Canberra, Australia, 2011b.

Australian Government, *Reforming National Environmental Law—An Overview*, Department of Sustainability, Environment, Water, Population and Communities, Canberra, Australia, 2011c.

Australian Government, *Australian Government Response to the Report of the Independent Review of the Environment Protection and Biodiversity Conservation Act 1999*, Department of Sustainability, Environment, Water, Population and Communities, Canberra, Australia, 2011d.

Australian Government, *Australian Government Response to the Senate Standing Committee on Environment, Communications and the Arts Committee Report*, Operations of the Environment Protection and Biodiversity Conservation Act 1999, First, Second and Final Report, Canberra, Australia, 2011e.

Australian Government, *Consultation Draft—Environmental Offsets Policy*, Department of Sustainability, Environment, Water, Population and Communities, Canberra, Australia, 2011f.

Axelrod, R., and M. D. Cohen, *Harnessing Complexity*, The Free Press, New York, 1999.

Ayre, G., and R. Calloway, *Governance for Sustainable Development—A Foundation for the Future*, Earthscan, London, 2005.

Azcarate, J., and B. Balfors, "Participative SEA Approach for Data Collection and Objective Formulation," *Journal of Environmental Assessment Policy and Management* **13**, 189–211 (2009).

Azcona, E. A., and F. Palmada, "Significance Analysis/Vulnerability Maps: A Useful Tool for ESIA," *Presentation to IAIA11 Conference Proceedings—Impact Assessment and Responsible Development for Infrastructure, Business and Industry*, International Association for Impact Assessment (IAIA), Puebla, Mexico, 2011.

Bagri, A., J. McNeely, and F. Vorhies, *Biodiversity and Impact Assessment*, Paper Presented at a Workshop on Biodiversity and Impact Assessment, Christchurch, New Zealand, 1998.

Bailey, J., "Environmental Impact Assessment and Management: An Unexplored Relationship," *Environmental Management* **21**, 317–327 (1997a).

Bailey, P. D., "EIA: A New Methodology for Environmental Policy," *Environmental Impact Assessment Review* **10**, 221–227 (1997b).

Bajgan, S., I. Shakya, and M. S. Mendis, eds., *The Nepal Biogas Support Program: A Successful Model for Rural Household Energy Supply*, ReproVision Press, Kathmandu, Nepal, 2005.

Baker, D. C., and J. N. McClelland, "Evaluating the Effectiveness of British Columbia's Environmental Assessment Process for First Nations' Participation in Mining Development," *Environmental Impact Assessment Review* **23**, 581–603 (2003).

Baker, D., and E. Rappaport, in K. S. Hanna, ed., "The Science of Assessment: Identifying and Predicting Environmental Impacts," *Environmental Impact Assessment in Canada*, Oxford University Press, Oxford, 2009.

Banfield, E. C., in M. Meyerson and E. C. Banfield, eds., "Notes on a Conceptual Scheme," *Politics, Planning and the Public Interest*, Free Press, New York, 1955.

Banken, R., in A. L. Porter and J. J. Fittipaldi, eds., "Public Health in Environmental Assessments," *Environmental Methods Review: Retooling Impact Assessment for the New Century*, Army Environmental Policy Institute and International Association for Impact Assessment, Fargo, ND, 1998.

Banken, R., in J. Kemm, J. Parry, and S. Palmer, eds., "HIA of Policy in Canada," *Health Impact Assessment*, Oxford University Press, Oxford, UK, 2004.

Bardwell, L. V., "Problem-Framing: A Perspective on Environmental Problem-Solving," *Environmental Management* **5**, 603–612 (1991).

Barker, A., and C. Wood, "An Evaluation of EIA Performance in Eight EU Countries," *Environmental Impact Assessment Review* **19**, 387–404 (1999).

Barnes, J. L., L. Hardwick, and C. Chan, "A Review of Project Scope and Environmental Assessment Scope for Energy and Mining Projects Across Canada," *Presentation to IAIA10 Conference Proceedings—The Role of Impact Assessment in Transitioning to the Green Economy*, International Association for Impact Assessment (IAIA), Geneva, Switzerland, 2010.

Barrett, K., and S. Lee, in J. A. Tickner, ed., "Certainty Claimed: Science and Politics in Canadian GMO Regulations," *Environmental Science and Preventive Public Policy*, Island Press, Washington, DC, 2003.

Barrow, C. J., *Environmental and Social Impact Assessment—An Introduction*, Arnold, London, 1997.

Barrow, C. J., "How is Environmental Conflict addressed by SIA?" *Environmental Impact Assessment Review* **30**, 293–301 (2010).

Barrow, J. D., *Impossibility—The Limits of Science and the Science of Limits*, Oxford University Press, New York, 1998.

Barth, R., and A. Fuder, *Impel Project: Implementing Article 10 of the SEA Directive, 2001/42/EC—Final Report*, Dormstadt, Öko-Institut e.V, Brussels, Belgium, 2002.

Bartlett, R. V., *Policy Through Impact Assessment*, Greenwood Press, New York, 1989.

Bartlett, R. V., in R. Clark and L. Canter, eds., "The Rationality and Logic of NEPA Revisited," *Environmental Policy and NEPA—Past, Present and Future*, St. Lucie Press, Boca Raton, FL, 1997.

Bass, R. E., "Evaluating Environmental Justice Under the National Environmental Policy Act," *Environmental Impact Assessment Review* **18**, 83–92 (1998).

Bauer, M. L., and J. Randolph, "Characteristics of Collaborative Planning and Decision-making Processes," *Environmental Practice* **2**, 156–165 (2000).

Baxter, W., W. A. Ross, and H. Spaling, "Improving the Practice of Cumulative Effects Assessment in Canada," *Impact Assessment and Project Appraisal* **19**, 253–262 (2001).

Beanlands, G. E., and P. N. Duinker, *An Ecological Framework for Environmental Impact Assessment in Canada*, Institute for Resource and Environmental Studies and Federal Environmental Assessment Office, Hull, Québec, 1983.

Beatley, T., "Environmental Ethics and Planning Theory," *Journal of Planning Literature* **4**, 1–32 (1989).

Beatley, T., "Planning and Sustainability: The Elements of a New Paradigm," *Journal of Planning Literature* **9**, 383–394 (1995).

Beatley, T., "Preserving Biodiversity," *American Planning Association Journal* **66**, 5–20 (2000).

Beauregard, R. A., "The Object of Planning," *Urban Geography* **8**, 367–373 (1987).

Becker, D. R., et al., "A Participatory Approach to Social Impact Assessment: The Interactive Community Forum," *Environmental Impact Assessment Review* **23**, 367–382 (2003).

Becker, D. R. et al., "A Comparison of a Technical and a Participatory Application of Social Impact Assessment," *Impact Assessment and Project Appraisal* **22**, 177–189 (2004).

Becker, M. P. M., K. Putters, and T. E. D. van der Grinton, "Evaluating the Impact of HIA on Urban Reconstruction Decision Making—Who Manages the Risk?" *Environmental Impact Assessment Review* **25**, 758–771 (2005).

Beder, S., "Bias and Credibility in Environmental Impact Assessment," *Chain Reaction* **6** (1993).

Behn, R. D., "Managing by Groping Along," *Journal of Policy Analysis and Management* **7**, 643–663 (1988).

Bendix, S., in L. L. Hart, G. A. Enk, and W. F. Hornick, eds., "How to Write a Socially Useful EIS," *Improving Impact Assessment*, Westview Press, Boulder, CO, 1984.

Benson, J. F., "What is the Alternative? Impact Assessment Tools and Sustainability Planning," *Impact Assessment and Project Appraisal* **21**, 261–281 (2003).

Benveniste, G., *Mastering the Politics of Planning*, Jossey-Bass, San Francisco, CA, 1989.

Berke, P. R., "Does Sustainability Development Offer a New Direction for Planning? Challenges for the Twenty-First Century," *Journal of Planning Literature* **17**, 21–36 (2002).

Berkes, F., in J. T. Inglis, ed., "Traditional Ecological Knowledge in Perspective," *Traditional Ecological Knowledge: Concepts and Cases*, International Program on Traditional Ecological Knowledge and International Development Research Centre, Ottawa, Canada, 1993.

Bérubé, M., "Cumulative Effects Assessment at Hydro-Québec: What Have We Learned?" *Impact Assessment and Project Appraisal* **25**, 101–109 (2007).

Bhatia, R., "Protecting Health Using an Environmental Impact Assessment: A Case Study of San Francisco Land Use Decision making," *American Journal of Public Health* **97**, 406–413 (2007).

Bhatia, R., and E. Seto, "Quantitative Estimation in Health Impact Assessment: Opportunities and Challenges," *Environmental Impact Assessment Review* **31**, 301–309 (2011).

Bhatia, R., et al., *Minimum Elements and Practice Standards for Health Impact Assessment, Version 2*, North American HIA Practice Standards Working Group, Montreal, Québec, 2010.

Biermann, F., et al., "Transforming Governance and Institutions for Global Sustainability: Key Insights From the Earth Governance Project," *Current Issues in Environmental Sustainability* **4**, 51–60 (2012).

Bina, O., "A Critical Review of the Dominant Lines of Argumentation on the Need for Strategic Environmental Assessment," *Environmental Impact Assessment Review* **27**, 585–606 (2007).

Bina, O., T. Wallington, and W. Thissen, in B. Sadler, et al., eds., "SEA Theory and Research: An Analysis of the Early Discourse," *Handbook of Strategic Environmental Assessment*, Earthscan, London, 2011, pp. 445–471.

Binder, C. R., G. Feola, and J. K. Steinberger, "Considering the Normative, Systematic and Procedural Dimensions in Indicator-Based Sustainability Assessments in Agriculture," *Environmental Impact Assessment Review* **30**, 71–81 (2010).

Biogas Support Programme (BSP), *Annual Progress Report*, Phase IV (July 2003–December 2010), 2010.

Birley, M., "Health Impact Assessment, Integration and Critical Appraisal," *Impact Assessment and Project Appraisal* **21**, 313–321 (2003).

Birley, M., in J. Kemm, J. Parry, and S. Palmer, eds., "HIA in Developing Countries," *Health Impact Assessment*, Oxford University Press, Oxford, UK, 2004.

Birley, M., "A Fault Analysis for Health Impact Assessment: Procurement, Competence, Expectations, and Jurisdictions," *Impact Assessment and Project Appraisal* **25**, 281–289 (2007).

Bingham, G., *Resolving Environmental Disputes: Decades of Experience*, Conservation Foundation, Washington, DC, 1986.

Bingham, G., *What is Consensus-building and Why is it Important for Resource Management?* RESOLVE—Center for Environmental and Policy Dispute Resolution, Washington, DC, 2001.

Bingham, G., and L. M. Langstaff, in R. Clark and L. Canter, eds., "Alternative Dispute Resolution in the NEPA Process," *Environmental Policy and NEPA—Past, Present and Future*, St. Lucie Press, Boca Raton, FL, 1997.

Bird, A., *Philosophy of Science*, McGill–Queen's University Press, Montreal, 1998.

Birkeland, J., "The Relevance of Ecofeminism to the Environmental Professions," *The Environmental Professional* **17**, 55–71 (1995).

Bishop, A. B., *Structuring Communications Programs for Public Participation in Water Resource Planning*, IWR Contract Rep. 75-2, U.S. Army Engineer Institute for Water Resources, Fort Belvoir, Virginia, 1975.

Bishop, A. B., in J. Creighton, J. D. Priscoli, and C. M. Dunning, eds., "Communications in the Planning Process," *Public Involvement Techniques: A Reader of Ten Years Experience at the Institute for Water Resources*, US Army Corps of Engineers, IWR Research Report, 82-R, Washington, DC, 1983.

Bisset, R., *Environmental Impact Assessment: Issues, Trends and Practice*, United Nations Environment Program, Scott Wilson Resource Consultants, Nairobi, Kenya, 1996.

Bisset, R., in N. Lee and C. George, eds., "Methods of Consultation and Public Participation," *Environmental Assessment in Developing and Transitional Countries*, John Wiley & Sons, Chichester, 2000.

Bisset, R., and B. Sadler, *Environmental Impact Assessment and Strategic Environmental Assessment: Towards an Integrated Approach*, United Nations Environment Program (UNEP), Geneva, 2004.

Blake, T., "Informal Partnerships to Facilitate Impact Assessment for Major Projects in Victoria, Australia," *Presentation to IAIA10 Conference Proceedings IAIA10 Transitioning to the Green Economy,* 30th Annual Conference of the International Association for Impact Assessment, 6–11 April 2010, Geneva, Switzerland, 2010.

Blanco, H., *How to Think About Social Problems—American Pragmatism and Idea of Planning*, Greenwood Press, Westport, CT, 1994.

BMVIT, *Bundesministerium für Vekehr, Innovation und Technologie: Leitfaden zur Erstellung des Umweltberichts im Rahmen der strategischen Prüfung—Verkehr für Netzveränderungen im hochrangigen Bundesverkehrswegenetz (SP-V Leitfaden)*, 2006, Available online: http://www.asfinag.net/c/document_library/get_file?uuid=e8282b61-01fb-4af6-b671-ea4400bc488d&groupId=10141 [last accessed October 15, 2012].

BMVIT, *Bundesministerium für Vekehr, Innovation und Technologie: Ausbauplan Bundesverkehrsinfrastruktur*, 2012, Available online: http://www.bmvit.gv.at/verkehr/gesamtverkehr/ausbauplan/downloads/ausbauplaene_2012_8.pdf [last accessed October 15, 2012].

BMVIT, *Bundesministerium für Vekehr, Innovation und Technologie: Finished and ongoing SA-Ts in Austria*, n.y. Available online: http://www.bmvit.gv.at/verkehr/gesamtverkehr/strategische_pruefung/pruefungen/index.html [last accessed October 15, 2012].

Bolan, R. S., "The Practitioner as Theorist," *Journal of the American Planning Association* **46**, 261–274 (1980).

Bond, A., in J. Kemm, J. Parry, and S. Palmer, eds., "Lessons from EIA," *Health Impact Assessment*, Oxford University Press, Oxford, UK, 2004.

Bond, A., "Transition Challenge for Sustainability Assessment," *IAIA10 Transitioning to the Green Economy,* 30th Annual Conference of the International Association for Impact Assessment, 6–11 April 2010, Geneva, Switzerland, 2010.

Bond, A., and G. Stewart, "Environmental Agency Scoping Guidance on the Environmental Impact Assessment of Projects," *Impact Assessment and Project Appraisal* **20**, 135–142 (2002).

Bond, A. J., and A. Morrison-Saunders, "Sustainability Appraisal: Jack of All Trades, Master of None?" *Impact Assessment and Project Appraisal* **27**, 321–329 (2009).

Bond, A. J., and A. Morrison-Saunders, "Re-evaluating Sustainability Assessment: Aligning the Vision and the Practice," *Environmental Impact Assessment Review* **31**, 1–7 (2011).

Bond, A., A. Morrison-Saunders, and J. Pope, "Sustainability Assessment: The State of the Art," *Impact Assessment and Project Appraisal* **30**, 53–62 (2012).

Bond, A., J. Palerm, and P. Haigh, "Public Participation in EIA of Nuclear Power Plant Decommissioning Projects: A Case Study Analysis," *Environmental Impact Assessment Review* **24**, 617–641 (2004).

Bond, A., et al., in B. Sadler, et al., eds., "Addressing Health Impacts in SEA," *Handbook of Strategic Environmental Assessment*, Earthscan, London, 2011a.

Bond, A., et al., "Environmental Impact Assessment and the Decommissioning of Nuclear Power Plants—A Review and Suggestion for a Best Practical Approach," *Environmental Impact Assessment Review* **23**, 197–217 (2003).

Bond, A., et al., "Learning How to Deal with Values, Frames and Governance in Sustainability Appraisal," *Regional Studies* **45**, 1157–1170 (2011b).

Bonifazi, A., C. Rega, and P. Gazzola, "Strategic Environmental Assessment and the Democratization of Spatial Planning," *Journal of Environmental Assessment Policy and Management* **13**, 9–37 (2011).

Bonvoisin, N., in B. Sadler, et al., eds., "Transboundary Issues in SEA," *Handbook of Strategic Environmental Assessment*, Earthscan, London, 2011.

Booth, A. L., and N. W. Skelton, "Involving First Nations' Participation in Environmental Assessment Processes: Recommendations From the Field," *Impact Assessment and Project Appraisal* **29**, 49–58 (2011a).

Booth, A. L., and N. W. Skelton, "Industry and Government Perspectives on First Nations' Participation in the British Columbia Environmental Assessment Process," *Environmental Impact Assessment Review* **31**, 216–225 (2011b).

Boothroyd, P., in F. Vanclay and D. A. Bronstein, eds., "Policy Assessment," *Environmental and Social Impact Assessment*, John Wiley & Sons, New York, 1995.

Boothroyd, P., and W. Rees, "Impact Assessment From Pseudo Science to Planning Process: An Educational Response," *Impact Assessment Bulletin* **2**, 9–17 (1985).

Bowler, P. J., *The Fontana History of the Environmental Sciences*, Fontana Press, UK, 1992.

Boyce, D. E., in A. G. Wilson, ed., "The Metropolitan Plan Making Process; Its Theory and Practical Complexities," *Urban and Regional Planning*, Pion, London, 1971.

Boyce, J. J., "Equity and the Environment," *Alternatives* **21**, 12–17 (1995).

Boyer, M. C., *Dreaming the Rational City*, The MIT Press, Cambridge, MA, 1983.

Brascoupé, S., and J. Mann, *A Community Guide to Protecting Indigenous Knowledge*, Research and Analysis Directorate, Department of Indian Affairs and Northern Development, Canada, June 2001.

Braybrooke, D., and C. E. Lindblom, *A Strategy of Decision*, Free Press, New York, 1963.

Bredariol, C. S., and A. Marini, "Conflicts in Developing Countries: A Case Study from Rio de Janeiro," *Environmental Impact Assessment Review* **23**, 489–513 (2003).

Briassoulis, H., "Theoretical Orientations in Environmental Planning: An Inquiry Into Alternative Approaches," *Environmental Management* **13**, 381–392 (1989).

Briassoulis, H., "Who Plans Whose Sustainability? Alternative Roles for Planners," *Journal of Environmental Planning and Management* **42**, 889–902 (1999).

Briffett, C., J. P. Obbard, and J. Mackee, "Toward SEA for the Developing Nations of Asia," *Environmental Impact Assessment Review* **23**, 171–196 (2003).

British Columbia Environmental Assessment Office (BC EAO), "Part D: Environmental Assessment Aboriginal Issues and Consultation," *Guide to the BC EA Process*, BC EAO, Victoria, BC, Canada, January 2001.

British Medical Association (BMA), *Health and Environmental Impact Assessment: An Integrated Approach*, Earthscan, London, 1998.

Bronfman, L. M., "Setting the Social Agenda: An Organizational Perspective," *Environmental Impact Assessment Review* **11**, 69–79 (1991).

Brooke, C., *Biodiversity and Impact Assessment*, Paper Prepared for the Conference on Impact Assessment in a Developing World, Manchester, England, 1998.

Bro-Rasmussen, F., in J. A. Tickner, ed., "Risks, Uncertainties, and Precautions in Chemical Legislation," *Environmental Science and Preventive Public Policy*, Island Press, Washington, DC, 2003.

Brown, A. L., in A. L. Porter and J. Fittipaldi, eds., "Decision-Scoping," *Environmental Methods Review: Retooling Impact Assessment for the New Century*, Army Environmental Policy Institute and International Association for Impact Assessment, Fargo, ND, 1998.

Brown, D. A., in J. A. Tickner, ed., "The Precautionary Principle as a Guide to Environmental Impact Analysis: Lessons Learned from Global Warming," *Environmental Science and Preventive Public Policy*, Island Press, Washington, DC, 2003.

Brown, J. H., "Commentary 1," *Proceedings of the Workshop on Cumulative Environmental Effects: A Bi-national Perspective,* Canadian Environmental Assessment Research Council and the U.S. National Research Council Board of Basic Biology, Ottawa, Ontario, 1986.

Brown, R. B., et al., "Local Flexibility in Spending Mitigation Monies: A Case Study of Successful Social Impact Mitigation of the Intermountain Power Project in Delta, Utah," *Impact Assessment and Project Appraisal* **21**, 205–213 (2003).

Bruner, N., and M. Starkl, "Decision Aid Systems for Evaluating Sustainability: A Critical Survey," *Environmental Impact Assessment Review* **24**, 441–469 (2004).

BSP/Biogas Support Program, *Integrated Environment Impact Assessment*, Biogas Support Program, 2002.

Buchan, D., "Buy-in and Social Capital: By-products of Social Impact Assessment," *Impact Assessment and Project Appraisal* **21**, 168–172 (2003).

Buckley, R., in A. L. Porter and J. Fittipaldi, eds., "Improving the Quality of Environmental Impact Statements (EISs)," *Environmental Methods Review: Retooling Impact Assessment for the New Century*, Army Environmental Policy Institute and International Association for Impact Assessment, Fargo, ND, 1998.

Burdett, T., "Closing the Loop—Embedding SEA Monitoring in the Plan Making and Review Process," *Presentation to IAIA08 Conference Proceedings—The Art and Science of Impact Assessment*, International Association for Impact Assessment, IAIA, Perth, Australia, 2008a.

Burdett, T., "Integrating Strategic Assessments—What Works and Is It the Right Approach," *Presentation to IAIA08 Conference Proceedings—The Art and Science of Impact Assessment*, International Association for Impact Assessment, IAIA, Perth, Australia, 2008b.

Burdge, R. J., "Why is Social Impact Assessment the Orphan of the Assessment Process?" *Impact Assessment and Project Appraisal* **20**, 3–9 (2002).

Burdge, R. J., "The Practice of Social Impact Assessment—Background," *Impact Assessment and Project Appraisal* **21**, 84–88 (2003a).

Burdge, R. J., "Benefiting From the Practice of Social Impact Assessment," *Impact Assessment and Project Appraisal* **21**, 225–229 (2003b).

Burdge, R. J., "The Focus of Impact Assessment (and IAIA) Must Now Shift to Global Climate Change," *Environmental Impact Assessment Review* **28**, 618–622 (2008).

Burdge, R. J., et al., *The Concepts, Process and Methods of Social Impact Assessment*, Social Ecology Press, Madison, WI, 1994 and 2004.

Burdge, R. J., and R. A. Robertson, "Social Impact Assessment and Public Involvement," *A Conceptual Approach to Social Impact Assessment*, Social Ecology Press, Middleton, WI, 1994.

Burdge, R. J., in F. Vanclay and D. A. Bronstein, eds., "Social Impact Assessment," *Environmental and Social Impact Assessment*, John Wiley & Sons, New York, 1995.

Buuren, A. V., and S. Nooteboom, "Evaluating Strategic Environmental Assessment in the Netherlands: Content, Process and Procedure as Indissoluble Criteria for Effectiveness," *Impact Assessment and Project Appraisal* **27**, 144–154 (2009).

Buuren, A. V., and S. Nooteboom, "The Success of SEA in Dutch Planning Practice—How Formal Assessments Can Contribute to Collaborative Governance," *Environmental Impact Assessment Review* **30**, 127–135 (2010).

Byer, P. H., and J. S. Yeomans, "Methods for Addressing Climate Change Uncertainties in Project Environmental Impact Assessments," *Impact Assessment and Project Appraisal* **25**, 85–99 (2007).

Byer, P. H., M. J. Lalani, and J. S. Yeomans, "Addressing and Communicating Climate Change and its Uncertainties in Project Environmental Impact Assessments," *Journal of Environmental Assessment Policy and Management* **11**, 29–50 (2009).

Byer, P. H., et al., *Decision-making Under Uncertainties for Adapting to Climate Change in Project Environmental Assessments*, Research supported by the Canadian Environmental Assessment Agency's Research and Development Program, Hull, Québec, 2011.

Byron, H., and J. Treweek, "Guest Editorial: Strategic Environmental Assessment—Great Potential for Biodiversity?" *Journal of Environmental Assessment Policy and Management* **7**, v–xiii (2005).

Caldwell, L. K., *The National Environmental Policy Act: An Agenda for the Future*, Indiana University Press, Bloomington, IN, 1988.

Caldwell, L. K., "Analysis-Assessment-Decision: The Anatomy of Rational Policymaking," *Impact Assessment Bulletin* **9**, 81–92 (1991).

Caldwell, L. K., in R. Clark and L. Canter, eds., "Implementing NEPA: A Non-Technical Political Task," *Environmental Policy and NEPA—Past, Present and Future*, St. Lucie Press, Boca Raton, FL, 1997.

Calow, P., and V. E. Forbes, "Science and Subjectivity in the Practice of Ecological Risk Assessment," *Environmental Management* **21**, 803–830 (1997).

Cameron, C., S. Ghosh, and S. L. Eaton, "Facilitating Communities in Designing and Using Their Own Health Impact Assessment Tool," *Environmental Impact Assessment Review* **31**, 433–437 (2011).

Campbell, M. C., "Intractability in Environmental Disputes," *Journal of Planning Literature* **17**, 360–371 (2003).

Campbell, M. C., and D. W. Floyd, "Thinking Critically About Environmental Mediation," *Journal of Planning Literature* **10**, 235–247 (1996).

Canada and the Northwest Territories, *Canada and the Northwest Territories, Governments of Canada and of the Northwest Territories Final Response to the Joint Review Panel Report for the Proposed Mackenzie Gas Project*, November 2010, Available at: http://www.acee-cceaa.gc.ca/default.asp?lang=En&n=71B5E4CF-1.

Canadian Department of Foreign Affairs and International Trade (CDFAIT), *Guidelines for Conducting Environmental Assessment of Projects Outside Canada*, Government of Canada, Ottawa, Canada, 2002.

Canadian Environmental Assessment Agency (CEAA), *Guide on Biodiversity and Environmental Assessment*, CEAA, Hull, Québec, 1996a.

Canadian Environmental Assessment Agency (CEAA), *Reference Guide on Physical and Cultural Heritage Resources*, CEAA, Hull, Québec, 1996b.

Canadian Environmental Assessment Agency (CEAA), *A Reference Guide for the Canadian Environmental Assessment Act: Addressing Cumulative Effects Assessment*, CEAA, Hull, Québec, 1998a.

Canadian Environmental Assessment Agency (CEAA), *Operational Policy Statement—Need, Purpose, Alternatives to and Alternative Means*, 1998b.

Canadian Environmental Assessment Agency (CEAA), *Canadian Environmental Assessment Act—Federal Coordination: An Overview*, CEAA, Hull, Québec, 2003a.

Canadian Environmental Assessment Agency (CEAA), *The Process for Defining the Spatial Boundary of a Study Area During an Environmental Assessment of Offshore Exploratory Drilling Projects*, CEAA, Hull, Québec, 2003b.

Canadian Environmental Assessment Agency (CEAA), *Incorporating Climate Change Considerations in Environmental Assessment: General Guide for Practitioners*, CEAA, Hull, Québec, 2003c.

Canadian Environmental Assessment Agency (CEAA), *Glossary: Terms Commonly Used in Environmental Assessment*, CEAA, Hull, Québec, 2006a.

Canadian Environmental Assessment Agency (CEAA), *Ministerial Guideline on Assessing the Need for and Level of Public Participation in Screenings Under the Canadian Environmental Assessment Act*, CEAA, Hull, Québec, 2006b.

Canadian Environmental Assessment Agency (CEAA), *How to Determine If the Act Applies*, CEAA, Hull, Québec, 2007a.

Canadian Environmental Assessment Agency (CEAA), *Project Description Guide: British Columbia*, CEAA, Hull, Québec, 2007b.

Canadian Environmental Assessment Agency (CEAA), *Federal Screenings—An Analysis Based on Information from The Canadian Environmental Assessment Registry Site*, Report # 1—Government of Canada Quality Assurance Program for Canadian Environmental Assessments Under the Canadian Environmental Assessment Act, CEAA, Hull, Québec, 2007c.

Canadian Environmental Assessment Agency (CEAA), *Operational Policy Statement—Addressing Cumulative Effects Under the Canadian Environmental Assessment Act Updated*, CEAA, Hull, Québec, 2007d.

Canadian Environmental Assessment Agency (CEAA), *Operational Policy Statement—Follow-up Programs Under the Canadian Environmental Assessment Act*, CEAA, Hull, Québec, 2007e.

Canadian Environmental Assessment Agency (CEAA), *Operational Policy Statement—Preparing Project Descriptions Under the Canadian Environmental Assessment Act*, CEAA, Hull, Québec, 2007f.

Canadian Environmental Assessment Agency (CEAA), *Operational Policy Statement—Addressing "Need for", "Purpose of" "Alternatives to" and "Alternative means" Under the Canadian Environmental Assessment Act*, CEAA, Hull, Québec, 2007g.

Canadian Environmental Assessment Agency (CEAA), *An Introduction to Class Screenings Under the Canadian Environmental Assessment Act*, CEAA, Training and Guidance, Hull, Québec, 2008a.

Canadian Environmental Assessment Agency (CEAA), *Guide to the Participant Funding Program Under the Canadian Environmental Assessment Act*, CEAA Training and Guidance, Hull, Québec, 2008b.

Canadian Environmental Assessment Agency (CEAA), *A Guide to the Class Screening Process Under the Canadian Environmental Assessment Act*, CEAA, Hull, Québec, 2009a.

Canadian Environmental Assessment Agency (CEAA), *Public Participation in Screenings—An Analysis of Efforts to Obtain Information and Views of the Public during the Conduct of Screenings*, Government of Canada, Quality Assurance Program for Environmental Assessments Under the Canadian Environmental Assessment Act, CEAA, Hull, Québec, 2009b.

Canadian Environmental Assessment Agency (CEAA), *Memorandum of Understanding for the Cabinet Directive on Implementing the Canadian Environmental Assessment Act*, CEAA, Hull, Québec, 2009c.

Canadian Environmental Assessment Agency (CEAA), *Comprehensive Study Process Guide—Updated*, CEAA, Hull, Québec, 2009d.

Canadian Environmental Assessment Agency (CEAA), *Operational Policy Statement—Addressing Adaptive Management Under the Canadian Environmental Assessment Act*, CEAA, Hull, Québec, 2009e.

Canadian Environmental Assessment Agency (CEAA), *Considering Aboriginal Traditional Knowledge in Conducting Environmental Assessments Conducted Under the Canadian Environmental Assessment Act—Interim Principles*, CEAA, Hull, Québec, 2010a.

Canadian Environmental Assessment Agency (CEAA), *Public Participation Guide*, CEAA, Hull, Québec, 2010b.

Canadian Environmental Assessment Agency (CEAA), *Operational Policy Statement—Establishing the Project Scope and Assessment Type Under the Canadian Environmental Assessment Act*, CEAA, Hull, Québec, 2011a.

Canadian Environmental Assessment Agency (CEAA), *A Framework for the Analysis of the Quality of Screening Reports*, Government of Canada, Quality Assurance Program for Environmental Assessments Under the Canadian Environmental Assessment Act, 2011b.

Canadian Federal-Provincial-Territorial Committee on Climate Change and Environmental Assessment (CFTPC), *Incorporating Climate Change Considerations into Environmental Assessment: General Guidelines for Practitioners*, Catalog No. En106-50/2003, ISBN 0-662-67812-5, Ottawa, Ontario, 2003.

Canadian Standards Association (CSA), *A Guide to Public Involvement*, Canadian Standards Association, Etobicoke, ON, 1996.

Canelas, L. et al. "Quality of Environmental Impact Statements in Portugal and Spain," *Environmental Impact Assessment Review* **25**, 217–225 (2005).

Canter, L. W., *Health Risk Impact in Environmental Impact Statements*, Environmental and Ground Water Institute, University of Oklahoma, Norman, OK, 1990.

Canter, L. W., "The Role of Environmental Monitoring in Responsible Project Management," *The Environmental Professional* **15**, 76–87 (1993a).

Canter, L. W., "Pragmatic Suggestions for Incorporating Risk Assessment Principles in EIA Studies," *The Environmental Professional* **15**, 125–138 (1993b).

Canter, L. W., *Environmental Impact Assessment*, 2nd ed., McGraw-Hill Inc., New York, 1996.

Canter, L. W., "Conceptual Models, Matrices, Networks and Adaptive Management—Emerging Methods for CEA," *Presented at Assessing and Managing Cumulative Effects, Special Topic Meeting,* International Association for Impact Assessment, November 6–9, 2008, Calgary, Alberta, Canada, 2008.

Canter, L., and S. F. Atkinson, "Adaptive Management with Integrated Decision Making: An Emerging Tool for Cumulative Effects Management," *Impact Assessment and Project Appraisal* **28**, 287–297 (2010).

Canter, L. W., and S. F. Atkinson, "Multiple Uses of Indicators and Indices in Cumulative Effects Assessment and Management," *Environmental Impact Assessment Review* **31**, 491–501 (2011).

Canter, L., and B. Ross, "State of Practice of Cumulative Effects Assessment and Management: The Good, the Bad and the Ugly," *Impact Assessment and Project Appraisal* **30**, 261–268 (2010).

Cardinall, D., and J. C. Day, "Embracing Value and Uncertainty in Environmental Management and Planning: A Heuristic Model," *Environments* **25**, 110–125 (1998).

Carmichael, L. et al., "Integration of Health into Urban Spatial Planning Through Impact Assessment: Identifying Governance and Policy Barriers and Facilitators," *Environmental Impact Assessment Review* **32**, 187–194 (2012).

Carpenter, R. A., "Risk Assessment," *Impact Assessment* **13**, 153–187 (1995).

Carpenter, S. L., *Managing Public Disputes*, Jossey-Bass, San Francisco, CA, 1991.

Carpenter, S. L., in R. Clark and L. Canter, eds., "The Case for Continuous Monitoring and Adaptive Management under NEPA," *Environmental Policy and NEPA—Past Present and Future*, St. Lucie Press, Boca Raton, FL, 1997.

Carter, J., and J. Howe, "The Water Framework Directive and the Strategic Environmental Assessment Directive: Exploring the Linkages," *Environmental Impact Assessment Review* **26**, 287–300 (2006).

Cartwright, T. J., "Problems, Solutions and Strategies: A Contribution to the Theory and Practice of Planning," *Journal of the American Institute of Planners* **39**, 179–187 (1973).

Cartwright, T. J., "Planning and Chaos Science," *American Planning Association Journal* **57**, 44–56 (1991).

Cashmore, M., "The Role of Science in Environmental Impact Assessment: Process and Procedure versus Purpose in the Development of Theory," *Environmental Impact Assessment Review* **24**, 403–426 (2004).

Cashmore, M., et al., "The Interminable Issue of Effectiveness: Substantive Purposes, Outcomes and Research Challenges in the Advancement of Environmental Impact Assessment Theory," *Impact Assessment and Project Appraisal* **22**, 295–310 (2004).

Cashmore, M., et al., "Evaluating the Effectiveness of Impact Assessment Instruments: Theorizing the Nature and Implications of Their Political

Constitution," *Environmental Impact Assessment Review* **30**, 371–379 (2010).

Casti, J., *Complexification*, HarperCollins, New York, 1994.

Catchpole, S., and R. Moreno, "Environmental Risks and Natural Hazards in EIA's," *Presentation to the IAIA12 Conference Proceedings, Energy Future—The Role of Impact Assessment,* 32nd Annual Meeting of the International Association for Impact Assessment, Porto, Portugal, 2012.

Cavatassi, R., and G. Atkinson, "Social and Private Determinants of Opposition to Landfill Siting in Italy," *Journal of Environmental Assessment Policy and Management* **5**, 27–43 (2003).

Centre for Indigenous Environmental Resources (CIER), *Meaningful Involvement of Aboriginal Peoples in Environmental Assessment—Final Report*, Catalogue No. En106-84/2009E-PDF, Winnipeg, Manitoba, 2009.

Chaker, A., et al., "A Review of Strategic Environmental Assessment in 12 Selected Countries," *Environmental Impact Assessment Review* **26**, 15–56 (2006).

Chase, A., "Anthropology and Impact Assessment: Development Pressures and Indigenous Interests in Australia," *Environmental Impact Assessment Review* **10**, 11–23 (1990).

Checkoway, B., "Paul Davidoff and Advocacy Planning in Retrospect," *American Planning Association Journal* **60**, 139–143 (1994).

Cherp, A., and A. Antypas, "Dealing with Continuous Reform: Towards Adaptive EA Policy Systems in Countries in Transition," *Journal of Environmental Assessment Policy and Management* **5**, 455–476 (2003).

Cherp, A., M. Partidário, and J. Arts in B. Sadler, et al., eds., "From Formulation to Implementation: Strengthening SEA Through Follow-up," *Handbook of Strategic Environmental Assessment*, Earthscan, London, 2011.

Cherp, A., A. Watt, and V. Vinichenko, "SEA and Strategy Formulation Theories: From Three Ps to Five Ps," *Environmental Impact Assessment Review* **27**, 624–644 (2007)

Canadian International Development Agency (CIDA), *Environment Handbook for Community Development Initiatives: Second Edition of the Handbook on Environmental Assessment of Non-governmental Organizations and Institutions, Programs and Projects*, CIDA, Gatineau, Québec, 2005, Available at: http://www.acdi-cida.gc.ca/CIDAWEB/acdicida.nsf/En/JUD-47134825-NVT

Clark, B., "Improving Public Participation in Environmental Impact Assessment," *Built Environment* **20**, 294–308 (1994).

Clark, R., in R. Clark and L. Canter, eds., "The Rational Approach to Change," *Environmental Policy and NEPA: Past, Present and Future*, St. Lucie Press, Boca Raton, FL, 1997.

Clark, R., L. Mahoney, and K. Pierce, in B. Sadler, et al., eds., "SEA in the US," *Handbook of Strategic Environmental Assessment*, Earthscan, London, 2011.

Cline, K., and B. L. Lamb, "Membership in Voluntary Organizations on the Colorado Plateau: A Re-examination of the Technical Information Quandary," *Environmental Practice* **7**, 143–154 (2005).

Cloquell-Ballester, V. A., R. Monterde-Diaz, and M. C. Santamarina-Siurana, "Systematic Comparative and Sensitivity Analyses of Additive and Outranking Techniques for Supporting Impact Significance Assessments," *Environmental Impact Assessment Review* **27**, 62–83 (2007).

Cockerill, K., et al., "Engaging the Public and Decision Makers in Cooperative Modeling for Regional Water Management," *Environmental Practice* **12**, 316–327 (2010).

Coe, R., "EIA Coordination: A Tool for Decision Making," *Presentation to the IAIA12 Conference Proceedings, Energy Future—The Role of Impact Assessment*, 32nd Annual Meeting of the International Association for Impact Assessment, Porto, Portugal, 2012.

Cole, B. L., "Prospects for Health Impact Assessment in the United States: New and Improved Environmental Impact Assessment or Something Different?," *Journal of Health Politics, Policy and Law* **29**, 1153–1186 (2004).

Cole, B. L., and J. E. Fielding, "Health Impact Assessment: A Tool to Help Policy Makers Understand Health Beyond Health Care," *Annual Review of Public Health* **28**, 393–412 (2007).

Coleman, W. G., "Biodiversity and Industry Ecosystem Management," *Environmental Management* **20**, 815–825 (1996).

Commission for Environmental Cooperation of North America (CECNA), *North American Environmental Law and Policy*, 2nd ed., Éditions Yvon Blais, Montréal, Québec, 2003.

Commission of the European Communities (CEC), *Communications from the Commission on the Precautionary Principle*, COM (2000) 1 Final, Brussels, 2000.

Committee on the Challenges of Modern Society (CCMS), *Evaluation of Public Participation in EIA*, Report 207, North Atlantic Treaty Organization, Brussels, 1995.

Committee on Risk Assessment Methodology (CRAM), *Issues on Risk Assessment*, National Research Council, National Academic Press, Washington, DC, 1993.

Conçalves, M. E., "Implementation of EIA Directives in Portugal—How Changes in Civic Culture Are Challenging Political and Administrative Practice," *Environmental Impact Assessment Review* **22**, 249–269 (2002).

Connelly, R., "Canada and International EIA Frameworks as They Apply to Cumulative Effects," *Environmental Impact Assessment Review* **31**, 453–456 (2011).

Connelly, S., and T. Richardson, "Value-driven SEA: Time for an Environmental Justice Perspective," *Environmental Impact Assessment Review* **25**, 391–409 (2005).

Connor, R., and S. Dovers, *Institutional Change for Sustainable Development*, Edward Elgar Publishing, Cheltenham, UK, 2004.

Constanza, R., "Visions of Alternative (Unpredictable) Futures and Their Use in Policy Analysis," *Conservation Ecology* **4**, 5–24 (2000).

Cooper, L. M., "CEA in Policies and Plans: UK Case Studies," *Environmental Impact Assessment Review* **31**, 465–480 (2011).

Cooper, L. M., and W. R. Sheate, "Integrating Cumulative Effects Assessment into UK Strategic Planning: Implications of the European Union SEA Directive," *Impact Assessment and Project Appraisal* **22**, 5–16 (2004).

Covello, V. T., P. Sandman, and P. Slovic, *Communications, Risk Statistics and Risk Comparisons: A Manual for Plant Managers*, Chemical Manufacturers' Association, Washington, DC, 1988.

Coveney, P., and R. Highfield, *Frontiers of Complexity—The Search for Order in a Chaotic World*, Faber & Faber, London, 1995.

Craib, I., *Modern Social Theory—From Parsons to Habermas*, Wheatsheaf Books Ltd., Sussex, UK, 1984.

Craig, D., "Social Impact Assessment: Politically Oriented Approaches and Applications," *Environmental Impact Assessment Review* **10**, 37–54 (1990).

Craik, N., *The International Law of Environmental Impact Assessment*, Cambridge University Press, Cambridge, UK, 2008.

Cranor, C. F., in J. A. Tickner, ed., "What Could Precautionary Science Be? Research for Early Warnings and a Better Future," *Environmental Science and Preventive Public Policy*, Island Press, Washington, DC, 2003.

Crawford, S., et al., *Translation of Indigenous/Western Science Perspectives on Adaptive Management for Environmental Assessment*, Prepared for the Canadian Environmental Assessment Agency's (CEAA) Research and Development Monograph Series, 2009, Hull, Québec, 2010.

Creasey, R., and W. A. Ross, in K. S. Hanna, ed., "The Cheviot Mine Project: Cumulative Effects Assessment Lessons from Professional

Practice," *Environmental Impact Assessment in Canada*, Oxford University Press, Oxford, 2009.

Creighton, J. L., J. D. Priscoli, and C. M. Dunning, *Public Involvement Techniques: A Reader of Ten Years Experience at the Institute for Water Resources*, The Institute for Water Resources, US Army Corps of Engineers, Ft. Belvoir, VA, May 1983.

Creighton, J. L., in J. L. Creighton, C. M. Dunning, J. D. Priscoli, and D. B. Ayres, eds., "Comparing Public Involvement and Environmental Mediation," *Public Involvement and Dispute Resolution—A Reader on the Second Decade of Experience at the Institute for Water Resources*, Institute for Water Resources, US Army Corps of Engineers, Alexandria, VA, 1999.

Creighton, J. L., C. M. Dunning, J. D. Priscoli, and D. B. Ayres, eds., *Public Involvement and Dispute Resolution—A Reader on the Second Decade of Experience at the Institute for Water Resources*, Institute for Water Resources, US Army Corps of Engineers, Alexandria, VA, October 1999.

Croal, P., et al., "A Decision-maker's Tool for Sustainability-centred Strategic Environmental Assessment," *Journal of Environmental Assessment Policy and Management* **12**, 1–27 (2010).

Cronin, C., in J. Habermas, ed., "Translator's Introduction," *Justification and Application—Remarks on Discourse Ethics*, MIT Press, Cambridge, MA, 1993.

Crookes, D. J., and M. P.de Witt, "An Evaluation of Tools for an Assessment of Cumulative Effects in Socioeconomic Impact Studies," *Journal of Environmental Assessment Policy and Management* **11**, 311–329 (2009).

Crossley, J. W., "Managing Ecosystems for Integrity: Theoretical Considerations for Resource and Environmental Managers," *Society and Natural Resources* **9**, 465–482 (1996).

Culhane, P. J., "Post-EIS Auditing: A First Step to Make a Rational Environmental Assessment a Reality," *The Environmental Professional* **15**, 66–75 (1993).

Culhane, P. J., H. P. Friesema, and J. A. Beecher, *Forecasts and Environmental Decision Making*, Westview Press, Boulder, CO, 1987.

Cumulative Effects Assessment Working Group and AXYS Environmental Consulting Group (CEAWG and AXYS), *Cumulative Effects Assessment Practitioners Guide*, Canadian Environmental Assessment Agency, Hull, Québec, Canada, February 1999.

Curtis, F., and H. Epp, "Role of Deductive Science in Evaluation of Alternatives in Environmental Assessment," *Impact Assessment and Project Appraisal* **17**, 3–8 (1999).

Cuppen, M., B. Broekhans, and B. Enserink, "Public Participation in EIA and Attitude Formation," *Impact Assessment and Project Appraisal* **30**, 63–74 (2012).

Dalal-Clayton, B., and B. Sadler, in A. Donnelly, B. Dalal-Clayton, and R. Hughes, eds., "Strategic Environmental Assessment: A Rapidly Evolving Approach," *A Directory of Impact Assessment Guidelines*, Russell Press, Nottingham, 1998.

Dalal-Clayton, B., and B. Sadler, *Sustainability Appraisal*, (First Draft), International Institute for Environment and Development, London, UK, 2004.

Dalkmann, H., R. J. Herrara, and D. Bondardt, "Analytical Strategic Environmental Assessment (ANSEA) Developing a New Approach to SEA," *Environmental Impact Assessment Review* **24**, 385–402 (2004).

Damasio, A. B., *Descartes' Error*, Avon Books, New York, 1994.

Daniels, S. E., and G. B. Walker, "Collaborative Learning: Improving Public Deliberation in Ecosystem-based Management," *Environmental Impact Assessment Review* **16**, 71–102 (1996).

Dannenberg, A. L., et al., "Growing the Field of Health Impact Assessment in the United States: An Agenda for Research and Practice," *American Journal of Public Health* **96**, 262–270 (2006).

Davidoff, P., "Advocacy and Pluralism in Planning," *Journal of the American Institute of Planners* **3**, 331–338 (1965).

Davidoff, P., R. W. Burchell and G. Sternlieb, eds., "The Redistributive Function in Planning: Creating Greater Equity Among Citizens of Communities," *Planning Theory in the 1980's*, Center for Urban Policy Research, New Brunswick, NJ, 1978.

Davidoff, P., and T. A. Reiner, "A Choice Theory of Planning," *Journal of the American Institute of Planners* **28**, 102–109 (1962).

Davies, K., and B. Sadler, *Environmental Assessment and Human Health: Perspectives, Approaches and Future Directions*, A Background Report for the International Study of the Effectiveness of Environmental Assessment, Health Canada and International Association for Impact Assessment, Ministry of Supply and Services Canada, Ottawa, May 1997.

Dawkins, R., *Unweaving the Rainbow—Science, Delusion and the Appetite for Wonder*, Houghton Mifflin Company, Boston, MA, 1998.

Day, D., "Citizen Participation in the Planning Process: An Essentially Contested Concept? *Journal of Planning Literature* **11**, 421–434 (1997).

Deakin, M., S. Curwell, and P. Lombardi, "Sustainable Urban Development: The Framework and Directory of Assessment Methods," *Journal of Environmental Assessment Policy and Management* **4**, 171–197 (2002).

Dearden, P., and B. Mitchell, *Environmental Change and Challenge*, Oxford University Press, Toronto, 1998.

De Bono, E., *Serious Creativity*, HarperCollins, Toronto, 1992.

Deelstra, Y., et al., "Using Knowledge for Decision-making Purposes in the Context of Large Projects in the Netherlands," *Environmental Impact Assessment Review* **23**, 517–543 (2003).

Demidova, O., and A. Cherp, "Risk Assessment for Improved Treatment of Health Considerations in EIA," *Environmental Impact Assessment Review* **25**, 411–429 (2005).

de Mulder, J., "The Institutional Context of Transboundary Environmental Impact Assessment in Belgium: Multi-level setting—A Matter of Smooth Governance," *Impact Assessment and Project Appraisal* **26**, 277–288 (2008).

Dennis, N. B., R. Clark and L. Canter, eds., "Can NEPA Prevent 'Ecological Train Wrecks'?" *Environmental Policy and NEPA—Past, Present and Future*, St. Lucie Press, Boca Raton, FL, 1997.

Denq, F., and J. Altenhofel, "Social Impact Assessments Conducted by Federal Agencies: An Evaluation," *Impact Assessment* **15**, 209–231 (1997).

Department of Communities and Local Government, *Environmental Impact Assessment—A Guide to Good Practice and Procedures*, DCLG Publishers, London, 2006.

Department of the Environment, Transport and the Regions (DETR), *Public Participation in Making Local Environmental Decisions, The Aarhus Convention, Newcastle Workshop, Good Practice Handbook*, DETR, London, UK, 2000.

De Ridder, W., et al., in W. R. Sheate, ed., "A Framework for Tool Selection and Use in Integrated Assessment for Sustainable Development," *Tools, Techniques and Approaches for Sustainability*, World Scientific, New Jersey, 2010, pp. 125–143.

Dery, D., "Coping With 'Latent Time Bombs' in Public Policy," *Environmental Impact Assessment Review* **17**, 413–425 (1997).

DeSario, J., and S. Langton, eds., *Citizen Participation in Public Decision-making*. Greenwood, Westport, CT, 1987.

Desmond, M., "Decision Criteria for the Identification of Alternatives in Strategic Environmental Assessment," *Impact Assessment and Project Appraisal* **25**, 259–269 (2007).

Devall, B., *Deep Ecology*, Gibbs Smith, Salt Lake City, UT, 1985.

Devlin, J. F., *Community Engagement for Adaptive Management in Environmental Assessment Follow Up—Final Report*, Canadian Environmental Assessment Agency, Hull, Québec, 2011.

Devlin, J. F., and N. T. Yap, "Contentious Politics in Environmental Assessment: Blocked Projects and Winning Coalitions," *Impact Assessment and Project Appraisal* **26**, 17–27 (2008).

Devuyst, D., "Sustainability Assessment: The Application of a Methodological Framework," *Journal of Environmental Assessment Policy and Management* **1**, 459–488 (1999).

Devuyst, D., "Linking Impact Assessment and Sustainable Development at the Local Level: The Introduction of Sustainability Assessment Systems," *Sustainable Development* **8**, 67–78 (2000).

Diamond, J., *Collapse: How Societies Choose to Fail or Succeed*, Penguin Books, New York, 2005.

Dickman, M., "Failure of Environmental Impact Assessment to Predict the Impact of Mine Tailings of Canada's Most Northerly Hypersaline Lake," *Environmental Impact Assessment Review* **11**, 171–180 (1991).

Diduck, A., and B. Mitchell, "Learning, Public Involvement and Environmental Assessment: A Canadian Case Study," *Journal of Environmental Assessment Policy and Management* **5**, 339–364 (2003).

Diduck, A. P., and A. J. Sinclair, in A. J. Sinclair, ed., "Testing the Concept of Critical Environmental Assessment (EA) Education," *Canadian Environmental Assessment in Transition*, Department of Geography, University of Waterloo, Waterloo, ON, 1997.

Diffenderfer, M., and D. Birch, "Bioregionalism: A Comparative Study of the Adirondacks and the Sierra Nevada," *Society and Natural Resources* **10**, 3–16 (1997).

Dimento, J. F. C., and H. Ingram, "Science and Environmental Decision-making: The Potential Role of Environmental Impact Assessment in Pursuit of Appropriate Information," *Natural Resources Journal* **45**, 283–309 (2005).

Dixon, J., and R. Thérivel, in B. Sadler et al., eds., "Managing Cumulative Impacts: Making It Happen," *Handbook of Strategic Environmental Assessment*, Earthscan, London, 2011.

Doberstein, B., "Environmental Capacity-building in a Transitional Economy: The Emergence of EIA Capacity in Vietnam," *Impact Assessment and Project Appraisal* **21**, 25–42 (2003).

Doberstein, B., "EIA Models and Capacity Building in Vietnam: An Analysis of Development Aid Programs," *Environmental Impact Assessment Review* **24**, 283–318 (2004).

Doelle, M., "The Role of Strategic Environmental Assessments (SEAs) in Energy Governance: A Case Study of Tidal Energy in Nova Scotia's Bay of Fundy," *Journal of Energy and Natural Resources Law* **27**, 112 (2009).

Doelle, M., and J. Sinclair, "Time for a New Approach to Public Participation in EA: Promoting Cooperation and Consensus for Sustainability," *Environmental Impact Assessment Review* **26**, 185–205 (2006).

Donahue, J., *On Collaborative Governance*, A Working Paper of the Corporate Social Responsibility Initiative, Working Paper # 2, Harvard University, John F. Kennedy School of Government, 2004.

Donnelly, A., et al., "Decision Support Framework for Establishing Objectives, Targets and Indicators for Use in Strategic Environmental Assessment," *Impact Assessment and Project Appraisal* **24**, 151–157 (2006).

Donnelly, A., et al., "Selecting Environmental Indicators for Use in Strategic Environmental Assessment," *Environmental Impact Assessment Review* **27**, 161–175 (2007).

Donnelly, A., T. Prendergast, and M. Hanusch, "Examining Quality of Environmental Objectives, Targets and Indicators in Environmental Reports Prepared for Strategic Environmental Assessments," *Journal of Environmental Assessment Policy and Management* **10**, 381–401 (2008).

Donnelly, A., and T. O'Mahoney, in B. Sadler et al., eds., "Development and Application of Environmental Indicators in SEA," *Handbook of Strategic Environmental Assessment*, Earthscan, London, 2011.

Dooley, J. E., in V. W. Maclaren and J. B. Whitney, eds., "Risk and Environmental Management," *New Directions in Environmental Impact Assessment in Canada*, Methuen, Toronto, 1985.

Dora, C., in B. Sadler et al., eds., "HIA in SEA and its Application to Policy in Europe," *Handbook of Strategic Environmental Assessment*, Earthscan, London, 2004.

Dovers, S., *Environment and Sustainability Policy: Creation, Implementation, Evaluation*, Federation Press, Sydney, 2005.

Dowlatabadi, H., et al., *Bridging the Gap Between Project-level Assessments and Regional Development Dynamics: A Methodology for Estimating Cumulative Effects*, Prepared for the Canadian Environmental Assessment Agency's Research and Development Program, Hull, Québec, 2003.

Draper, H. M., "Keeping Below the Tipping Point: A Literature Review of Climate Change with Attention to NEPA," *Environmental Practice* **12**, 144–157 (2010).

Dubé, M. G., "Cumulative Effects Assessment in Canada: A Regional Framework for Aquatic Ecosystems," *Environmental Impact Assessment Review* **23**, 723–746 (2003).

Duffy, P., "Agriculture, Forestry and Fisheries: The Orphans of Environmental Impact Assessment," *Impact Assessment and Project Appraisal* **22**, 175–176 (2006).

Duffy, P., "Why is the EIA Model So Widely Used?," *IAIA Newsletter* **20** (2008).

Duinker, P. N., and L. A. Grieg, "The Impotence of Cumulative Effects Assessment in Canada: Ailments and Ideas for Redeployment," *Environmental Management* **37**, 153–161 (2006).

Duinker, P. N., and L. A. Grieg, "Scenario Analysis in Environmental Impact Assessment: Improving Explorations of the Future," *Environmental Impact Assessment Review* **27**, 206–219 (2007).

Duncan, R., "Problematic Practice in Integrated Impact Assessment: The Role of Consultants and Predictive Computer Models in Burying Uncertainty," *Impact Assessment and Project Appraisal* **26**, 53–66 (2008).

Duncan, R., and P. Hay, "A Question of Balance in Integrated Impact Assessment: Negotiating Away the Environmental Interest in Australia's Basslink Project," *Journal of Environmental Assessment Policy and Management* **9**, 273–297 (2007).

Dunning, C. M., in J. L. Creighton, C. M. Dunning, J. D. Priscoli, and D. B. Ayres, eds., "Collaborative Problem Solving for Installation Planning and Decision Making," *Public Involvement and Dispute Resolution, Institute of Water Resources*, U.S. Army Corps of Engineers, Alexandria, VA, 1998.

Dusik, J., and B. Sadler, "Reforming Strategic Environmental Assessment Systems: Lessons from Central and Eastern Europe," *Impact Assessment and Project Appraisal* **22**, 89–97 (2004).

Dutta, P., S. Mahanta, and P. De, "A Methodology for Cumulative Impact Assessment of Open Pit Mining Projects with Special Reference to Air Quality Assessment," *Impact Assessment and Project Appraisal* **22**, 235–250 (2004).

Eales, R. P., and W. R. Sheate, "Effectiveness of Policy Level Environmental and Sustainability Assessment: Challenges and Lessons from Recent Practice," *Journal of Environmental Assessment Policy and Management* **13**, 39–65 (2011).

Eales, R., et al., "Emerging Approaches to Integrated Appraisal in the UK," *Impact Assessment and Project Appraisal* **23**, 113–123 (2005).

Ebrahim, A., in K. Ahmed and E. Sánchez-Triana, eds., "Learning in Environmental Policy Making and Implementation," *Strategic Environmental Assessment for Policies*, The World Bank, Washington, DC, 2008.

Eccleston, C. H., *The NEPA Planning Process: A Comprehensive Guide with Emphasis on Efficiency*, John Wiley & Sons, New York, 1999a.

Eccleston, C. H., "Determining If and When Impacts of Potential Accidents Need to be Evaluated in a NEPA Analysis," *Environmental Practice* **1**, 89–96 (1999b).

Eccleston, C. H., "How NEPA Can be Used to Counter Terrorism," *Environmental Practice* **5**, 302–312 (2003).

Eccleston, C. H., *NEPA and Environmental Planning*, CRC Press, Boca Raton, FL, 2008.

Eccleston, C. H., "Assessing Cumulative Significance of Greenhouse Gas Emissions: Resolving the Paradox—The Sphinx Solution," *Environmental Practice* **12**, 105–115 (2010).

Ecologic et al., *Improving Assessment of the Environment in Impact Assessment*, Final Report, Project No., ENV.G1/FRA/2004/0081, 2007.

ECORYS Research and Consulting in Collaboration with IZA, *Review of Methodologies Applied for the Assessment of Employment and Social Impacts*, for DG Employment, Social Affairs and Equal Opportunities of the European Commission, 2010.

Edelstein, M. R., "Weight and Weightlessness: Administrative Court Efforts to Weigh Psycho-social Impacts of Proposed Environmentally Hazardous Facilities," *Impact Assessment and Project Appraisal* **21**, 195–203 (2003).

Edwards, A. R., *The Sustainability Revolution*, New Society, Gabriola Island, BC, Canada, 2005.

Égré, D., and P. Senécal, "Social Impact Assessments of Large Dams Throughout the World: Lessons Learned Over Two Decades," *Impact Assessment and Project Appraisal* **21**, 215–224 (2003).

Elkin, P., and V. Voiturez, eds., *Trade, Globalization and Sustainability Impact Assessment—A Critical Look at Methods and Outcomes*, Earthscan, London, 2009.

Elling, B., "Integration of Strategic Environmental Assessment into Spatial Planning," *Impact Assessment and Project Appraisal* **18**, 233–244 (2000).

Elling, B., "Rationality and Effectiveness: Does EIA/SEA Treat Them as Synonyms?," *Impact Assessment and Project Appraisal* **27**, 121–131 (2007).

Elling, B., in B. Sadler et al., eds., "Some Wider Reflections on the Challenge of Public Participation in SEA," *Handbook of Strategic Environmental Assessment*, Earthscan, London, 2011.

Elliott, E., and S. Francis, "Making Effective Links to Decision-making: Key Challenges for Health Impact Assessment," *Environmental Impact Assessment Review* **25**, 747–757 (2005).

Elliott, E., G. Williams, and B. Rolfe, in J. Kemm, J. Parry, and S. Palmer, eds., "The Role of Lay Knowledge in HIA," *Health Impact Assessment*, Oxford University Press, Oxford, UK, 2004.

Emilsson, S., S. Tyskeng, and A. Carsson, "Potential Benefits of Combining Environmental Management Tools in a Local Authority Context," *Journal of Environmental Assessment Policy and Management* **6**, 131–151 (2004).

Emond, D. P., "Accommodating Negotiation/Mediation Within Existing Assessment and Approval Processes," *The Place of Negotiation in Environmental Assessment*, Canadian Environmental Assessment Research Council, Hull, Québec, 1990.

EnHealth Council, *Health Impact Assessment Guidelines*, Commonwealth of Australia, Canberra, September 2001a.

EnHealth Council, *Environmental Health Risk Assessment: Guidelines for Assessing Human Health Risks from Environmental Hazards*, Commonwealth of Australia, Canberra, 2001b.

Enk, G. A., and W. F. Hornick in F. A. Rossini and A. L. Porter, eds., "Human Values and Impact Assessment," *Integrated Impact Assessment*, Westview Press, Boulder, CO, 1983.

Ensminger, J. T., and R. B. McLean, "Reasons and Strategies For More Effective NEPA Implementation," *The Environmental Professional* **15**, 46–56 (1993).

Environment Australia, *Report on the Operation of the Environmental Protection and Biodiversity Conservation Act*, Department of the Environment and Heritage, Commonwealth of Australia, Canberra, Australia, December 2001.

Environmental and Social Systems Analysts (ESSA), *Review and Evaluation of Adaptive Environmental Assessment and Management*, Environment Canada, Vancouver, 1982.

Environmental Impact Screening Committee (EISC), "Benefit Guidelines Associated With Mineral Exploration in the Inuvialuit Settlement Region," *Operating Guidelines and Procedures*, EISAC, Inuvik, Northwest Territories, Canada, February 1999.

Environmental Resources Management (ERM), *Revision of EU Guidance Documents on EIA—First Interim Report*, European Commission DGXI, Edinburgh, Scotland, July 2000.

Environmental Resources Management (ERM), *Guidance on EIA Screening*, European Communities, Edinburgh, June 2001a.

Environmental Resources Management (ERM), *Guidance on EIA Scoping*, European Communities, Edinburgh, June 2001b.

Environmental Resources Management (ERM), *Guidance on EIA EIS Review*, European Communities, Edinburgh, June 2001c.

Erickson, P. A., *A Practical Guide to Environmental Impact Assessment*, Academic Press, San Diego, CA, 1994.

Erikstad, L., et al., "Environmental Value Assessment in a Multidisciplinary EIA Setting," *Environmental Impact Assessment Review* **28**, 131–143 (2008).

Erkmann, J., "Commentary II," *Cumulative Environmental Effects: A Binational Perspective,* Canadian Environmental Assessment Research Council and U.S. National Research Council Board of Basic Biology, Ottawa, Canada, 1986.

Erlanger, T. E., et al., "The 6/94 Gap in Health Impact Assessment," *Environmental Impact Assessment Review* **28**, 349–358 (2008).

Esteves, A. M., and M. A. Barclay, "Enhancing the Benefits of Local Content: Integrating Social and Economic Impact Assessment into Procurement Strategies," *Impact Assessment and Project Appraisal* **29**, 205–211 (2011).

Esteves, A. M., and F. Vanclay, "Social Development Needs Analysis as a Tool for SIA to Guide Corporate-Community Investment: Applications in the Mineral Industry," *Environmental Impact Assessment Review* **29**, 137–145 (2009).

Esteves, A. M., D. Franks, and F. Vanclay, "Social Impact Assessment: The State of the Art," *Impact Assessment and Project Appraisal* **30**, 34–42 (2012).

Etzioni, A., "Mixed Scanning: A 'Third' Approach to Decision-making," *Public Administration Review* **27**, 385–392 (1967).

Etzioni, A., "Mixed Scanning Revisited," *Public Administration Review* **46**, 8–14 (1986).

Etzioni, A., *The Spirit of Community*, Simon and Schuster, New York, 1993.

Etzioni, A., *Rights and the Common Good*, St. Martin's Press, New York, 1995.

European Commission (EC), *The Aarhus Convention: An Implementation Guide*, United Nations, New York and Geneva, 2000.

European Commission (EC), *Implementation of Directive 2001/42/EC*, Luxembourg, Official Publication of the European Communities, Luxembourg, 2002.

European Commission (EC), *The SEA Manual—A Sourcebook on Strategic Environmental Assessment of Transport Infrastructure Plans and Programmes*, Official Publication of the European Communities, Luxembourg, 2005.

European Commission (EC), *Clarification of the Application of Article 2(3) of the EIA Directive*, Official Publication of the European Communities, Luxembourg, 2006a.

European Commission (EC), *Guidance on Public Participation in Environmental Impact Assessment in a Transboundary Context*, New York and Geneva, United Nations, 2006b.

European Commission (EC), *Interpretation of Definition of Certain Project Categories of Annex I and II of the EIA Directive*, Official Publication of the European Communities, Luxembourg, 2008a.

European Commission (EC), *Review of Implementation of the Espoo Convention*, New York and Geneva, United Nations, 2008b.

European Commission (EC), *Impact Assessment Guidelines and Compendium of Further Material on Impact Assessment*, SEC (2009) 02., Official Publication of the European Communities, Luxembourg, 2009a.

European Commission (EC), *Guidelines for Assessing Social Impacts Within the Commission Impact Assessment System*, EC, Ref. Ares (2009) 326974. Official Publication of the European Communities, Luxembourg, 2009b.

European Commission (EC), *Report of the Commission to the Council, the European Parliament, the European Economic and Social Committee and the Committee of the Regions on the Application and Effectiveness of the EIA Directive (Directive 85/337/EEC as amended by Directives 97/11/EC and 2003/35/EC)*, Official Publication of the European Communities, Luxembourg, 2009c.

European Commission (EC), *Study Concerning the Report of the Application and Effectiveness of the SEA Directive (2001/42/EC), Final Report*, COWI, Denmark, 2009d.

European Commission (EC), *Guidance on Notification According to the Espoo Convention*, New York and Geneva, United Nations, 2009e.

European Commission (EC), *Environmental Impact Assessment of Projects—Rulings of the Court of Justice*, Official Publication of the European Communities, Luxembourg, 2010.

European Commission (EC), *Operational Guidance on Taking Account of Fundamental Rights in Commission Impact Assessment*, Commission Staff Working Paper, Luxembourg, Official Publication of the European Communities, 2011a.

European Commission (EC), *Commission Staff Working Paper—Impact Assessment Board Report for 2010*, Official Publication of the European Communities, Luxembourg, 2011b.

European Commission (EC), *Commission Staff Working Document—Executive Summary of the Impact Assessment Accompanying the Document Proposal for a Directive of the European Parliament and of the Council*, Amending Directive 2011/92/EU on the Assessment of Effects of Certain Public and Private Projects on the Environment, European Commission, Brussels, Belgium, 2012a.

European Commission (EC), *Commission Staff Working Paper, Impact Assessment Accompanying the Doc Document Proposal for a Directive of the European Parliament and of the Council*, Amending Directive 2011/92/EU on the Assessment of Effects of Certain Public and Private Projects on the Environment, European Commission, Brussels, Belgium, 2012b.

European Commission (EC), *Questions and Answers on the Commission's Proposal for the Revision of European Law on Environmental Impact Assessments*, European Commission, Brussels, Belgium, 2012c.

European Commission (EC), *Proposal for a Directive of the European Parliament and of the Council, Amending Directive 2011/92/EU on the Assessment of Effects of Certain Public and Private Projects on the Environment*, European Commission, Brussels, Belgium, 2012d.

European Court of Auditors, *Impact Assessment in the EU Institutions: Do They Support Decision-making?* Special Report No. 3, Luxembourg, Official Publication of the European Communities, Luxembourg, 2010.

European Union, *"Directive 2003/4/EC of the European Parliament and of the Council of January 2003 on Public Access to Environmental Information and Repealing Directive 90/3/3/EEC,"* Official Journal of the European Union, Brussels, Belgium, 2003.

European Union Committee of the Regions (EUCOTR), *Opinion of the Committee of the Regions—Improving the EIA and SEA Directives*, EUCOTR, 84th Plenary Session, Brussels, Belgium, 2010.

Euston, A. F., in R. Clark and L. Canter, eds., "NEPA's Interdisciplinary Mandate: Redirection for Sustainability," *Environmental Policy and NEPA: Past, Present and Future*, St. Lucie Press, Boca Raton, FL, 1997.

Evaluation Partnership, *Evaluation of the Commission's Impact Assessment System*, Submitted to Secretariat General of the European Commission, Contract # SG-02/2006, Brussels, Belgium, 2007.

Faber, N., R. Jorna, and J.Van Engelen, in W. R. Sheate, ed., "The Sustainability of 'Sustainability,'—A Study into the Conceptual Foundations of the Notion of 'Sustainability'," *Tools, Techniques and Approaches for Sustainability*, World Scientific, New Jersey, 2010.

Fahy, F., and M. Ó. Cinnéide, "Developing and Testing An Operational Framework for Assessing Quality of Life," *Environmental Impact Assessment Review* **28**, 366–379 (2007).

Fainstein, S. S., and N. I. Fainstein, "Economic Restructuring and the Rise of Urban Social Movements," *Urban Affairs Quarterly* **21**, 187–206 (1985).

Faludi, A., *Critical Rationalism and Planning Methodology*, Pion, London, 1986.

Farber, D. A., "Probabilities Behaving Badly: Complexity Theory and Environmental Uncertainty," *U.C. Davis Law Review* **37**, 145–173 (2003–2004).

Farley, H. M., and Z. A. Smith, "Sustainability: Guiding Principle or Broken Compass?," *Environmental Practice* **14**, 85–86 (2012).

Federal-Provincial-Territorial Committee on Climate Change and Environmental Assessment (FPTCCCEA), *Incorporating Climate Change Considerations into Environmental Assessment: General Guidance for Practitioners*, CEAA, Ottawa, ON, 2003.

Feldman, M. S., and A. M. Khademian, in K. Ahmed and E. Sánchez-Triana, eds., "The Continuous Process of Policy Formation," *Strategic Environmental Assessment for Policies*, The World Bank, Washington, DC, 2008.

Fell, A., and B. Sadler, "Public Involvement in Environmental Assessment and Management—Preview of IER Guidelines on Good Practice," *EA* (June 1999).

Filyk, G., and R. Côté, in R. Boardman, ed., "Pressure From Inside: Advisory Groups and the Environmental Policy Community," *Canadian Environmental Policy: Ecosystems, Politics and Process*, Oxford University Press, Toronto, 1992.

Finnveden, G., "Strategic Environmental Assessment Methodologies—Applications Within the Energy Sector," *Environmental Impact Assessment Review* **23**, 91–123 (2003).

Finsterbusch, K., "In Praise of SIA—A Personal Review of the Field of Social Impact Assessment: Feasibility, Justification, History, Methods, Issues," *Impact Assessment* **13**, 229–252 (1995).

Firth, L. J., "Role of Values in Public Decision-making: Where is the Fit?," *Impact Assessment and Project Appraisal* **16**, 325–329 (1998).

Fischer, F., in S. Campbell and S. Fainstein, eds., "Risk Assessment and Environmental Crisis: Toward an Integration of Science and Participation," *Readings in Planning Theory*, Blackwell Publishers Inc., Cambridge, MA, 1996.

Fischer, T. B., "Strategic Environmental Assessment Performance Criteria—The Same Requirements for Every Assessment?," *Journal of Environmental Assessment Policy and Management* **4**, 83–99 (2002).

Fischer, T. B., "Strategic Environmental Assessment in Post-modern Times," *Environmental Impact Assessment Review* **23**, 155–170 (2003).

Fischer, T. B., "Having an Impact? Context Elements for Effective SEA Application in Transport Policy, Plan and Programme Making," *Journal of Environmental Assessment Policy and Management* **7**, 407–432 (2005).

Fischer, T. B., "Conference Report—IAIA Conference on 'International Experience and Perspectives in SEA' 26–30 September 2005, Prague, Czech Republic—Conclusions Stream C: Linkages Between SEA and Other Assessment or Planning Tools," *Journal of Assessment Policy and Management* **8**, 495–504 (2006).

Fischer, T. B., "Strategic Environmental Assessment and Transport Planning: Towards a Generic Framework for Evaluating Practice and Developing Guidance," *Impact Assessment and Project Appraisal* **24**, 183–197 (2007a).

Fischer, T. B., *Theory and Practice of Strategic Environmental Assessment*, Earthscan, London, 2007b.

Fischer, T. B., "Reviewing the Quality of Strategic Environmental Assessment Reports for English Spatial Plan Core Strategies," *Environmental Impact Assessment Review* **30**, 62–69 (2010).

Fischer, T. B., in B. Sadler et al., eds., "Thematic Overview of Linkages Between SEA and Other Instruments," *Handbook of Strategic Environmental Assessment*, Earthscan, London, UK, 2011.

Fischer, T. B., and P. Gazzola, "SEA Effectiveness Criteria—Equally Valid in All Countries? The Case of Italy," *Environmental Impact Assessment Review* **26**, 396–409 (2006).

Fischer, T. B., M. Matuzzi, and J. Nowacki, "The Consideration of Health in Strategic Environmental Assessment (SEA)," *Environmental Impact Assessment Review* **30**, 200–210 (2010).

Fischer, T. B., et al., "Learning Through the EC Directive-based SEA in Spatial Planning? Evidence from the Brunswick Region in Germany," *Environmental Impact Assessment Review* **29**, 421–428 (2009).

Fischler, R., "Communicative Planning Theory: A Foucauldian Assessment," *Journal of Planning Education and Research* **19**, 358–368 (2000).

Fischhoff, B., P. Lichtenstein, P. Slovic, S. L. Derby, and R. L. Keeney, *Acceptable Risk*, Cambridge University Press, Cambridge, 1981.

Fischhoff, B., P. Slovic, and P. Lichtenstein, "Lay Foibles and Expert Fables in Judgments About Risk," *The American Statistician* **30**, 240–255 (1982).

Fitzpatrick, P., "In It Together: Organizational Learning through Participation in Environmental Assessment," *Journal of Environmental Assessment Policy and Management* **8**, 157–182 (2006).

Fitzpatrick, P., and A. J. Sinclair, in K. S. Hanna, ed., "Multi-jurisdictional Environmental Assessment," *Environmental Impact Assessment—Practice and Participation*, Oxford University Press, Oxford, UK, 2009a.

Fitzpatrick, P., and A. J. Sinclair, "Multi-jurisdictional Impact Assessment: Canadian Experiences," *Environmental Impact Assessment Review* **29**, 252–260 (2009b).

Forester, J., "Bounded Rationality and the Politics of Muddling Through," *Public Administration Review* **44**, 23–31 (1984).

Forester, J., *Planning in the Face of Power*, University of California Press, Berkeley, 1989.

Forester, J., *The Deliberative Practitioner*, MIT Press, Cambridge, MA, 1999.

Forsyth, A., C. S. Slotterback, and K. K. Krizek, "Health Impact Assessment in Planning: Development of the Design for HIA Tools," *Environmental Impact Assessment Review* **30**, 42–51 (2010).

Foster, K. R., P. Vecchia, and M. H. Rapacholi, "Science and the Precautionary Principle," *Science* **12**, 979–981 (May 2000).

Franz, J., and C. Kirkpatrick, "Integrating Sustainable Development into European Policymaking: The Role of Impact Assessments," *Journal of Environmental Assessment Policy and Management* **9**, 141–160 (2007).

Freudenburg, W. R., in G. A. Daneke, M. W. Garcia, and J. D. Priscoli, eds., "The Promise and Peril of Public Participation in Social Impact Assessment," *Public Involvement and Social Impact Assessment*, Westview Press, Boulder, CO, 1983.

Freudenburg, W. R., "Social Impact Assessment," *Annual Review of Sociology* **12**, 451–478 (1986).

Friedmann, J., *Planning in the Public Domain: From Knowledge to Action*, Princeton University Press, Princeton, New Jersey, 1987.

Friend, J., and A. Hickling, *Planning Under Pressure—The Strategic Choice Approach*, Butterworth Heinemann, Oxford, 1997.

Fuggle, R., "Have Impact Assessments Passed Their 'Sell By' Date?" *IAIA Newsletter* **16** (2005a).

Fuggle, R. "Are Impact Assessment Practitioners Craftsmen, Tradesmen or Professionals?" *IAIA Newsletter* **16** (2005b).

Fuggle, R., "What is Common Sense in the First World may not be Common Sense in the Third World," *Impact Assessment and Project Appraisal* **24**, 17–18 (2006).

Furlow, J., "Participatory Climate Vulnerability Assessment," *Presentation to IAIA Special Symposium on Climate Change and Impact Assessment*, Washington, DC, 2010.

Gachechiladze, M., B. F. Noble, and B. W. Bitter, "Follow-up in Strategic Environmental Assessment: A Case Study of 20-year Forest Management Planning in Saskatchewan, Canada," *Impact Assessment and Project Appraisal* **27**, 45–56 (2009).

Gadgil, M., F. Berkes, and C. Folke, "Indigenous Knowledge for Biodiversity Conservation," *Ambio* **22**, 151–156 (1993).

Gagnon, C., P. Hirsch, and R. Howitt, "Can SIA Empower Communities?," *Environmental Impact Assessment Review* **13**, 229–254 (1993).

Galbraith, L., and B. Bradshaw, M. Rutherford, "Towards a New Supraregulatory Approach to Environmental Assessment in Northern Canada," *Impact Assessment and Project Appraisal* **25**, 27–41 (2007).

García-Montero, L. G., et al., "Environmental Screening Tools for Assessment of Infrastructure Plans Based on Biodiversity Preservation and Global Warming (PEIT, Spain)," *Environmental Impact Assessment Review* **30**, 158–168 (2010).

Gardner, D., *Future Babble*, McClelland & Stewart, Toronto, 2010.

Gardner, J., and M. Roseland, "Thinking Globally—The Role of Social Equity in Sustainable Development," *Alternatives* **16**, 26–34 (1989).

Garver, G., and A. Podhora, "Transboundary Environmental Impact Assessment as Part of the North American Agreement on Environmental Cooperation," *Impact Assessment and Project Appraisal* **26**, 253–263 (2008).

Gasparatus, A., M. El-Haram, and M. Horner, "The Argument Against a Reductionist Approach for Assessing Sustainability," *International Conference on Whole Life Urban Sustainability and its Assessment*, Glasgow, 2007.

Gauthier, M., L. Simard, and J. P. Waaub, "Public Participation in Strategic Environmental Assessment (SEA): Critical Review and the Québec (Canada) Approach,," *Environmental Impact Assessment Review* **31**, 48–60 (2011).

Gazzola, P., "What Appears to Make SEA Effective in Different Planning Systems," *Journal of Environmental Assessment Policy and Management* **10**, 1–24 (2008).

Gee, D., and A. Stirling, in J. A. Tickner, ed., "Late Lessons from Early Warnings: Improving Science and Governance Under Uncertainty and Ignorance," *Environmental Science and Preventive Public Policy*, Island Press, Washington, DC, 2003.

Geneletti, D., *Ecological Evaluation for Environmental Impact Assessment*, Netherlands Geographical Society, Utrecht, 2002.

Geneletti, D., "Biodiversity Impact Assessment of Roads: An Approach Based on Ecosystem Rarity," *Environmental Impact Assessment Review* **23**, 343–365 (2003).

Geneletti, D., "Multicriteria Analysis to Compare the Impact of Alternative Road Corridors: A Case Study in Northern Italy," *Impact Assessment and Project Appraisal* **23**, 135–146 (2005).

Geneletti, D., et al., "Accounting for Uncertainty Factors in Biodiversity Impact Assessment: Lessons from a Case Study," *Environmental Impact Assessment Review* **23**, 471–487 (2003).

Genter, S., S. A. Moore, and J. Bailey, "Developing and Testing a Policy Environmental Assessment Checklist for Biodiversity

Conservation," *Impact Assessment and Project Appraisal* **26**, 189–200 (2008).

George, C., "Assessing Global Impacts at Sector and Project Levels," *Environmental Impact Assessment Review* **17**, 227–247 (1997).

George, C., and C. Kirkpatrick, "Sustainability Impact Assessment for Trade Agreements: From Public Dialogue to International Governance," *Journal of Environmental Assessment Policy and Management* **10**, 67–89 (2008).

George, C., R. Nafti, and J. Curran, "Capacity Building for Trade Impact Assessment: Lessons from the Development of Environmental Impact Assessment," *Impact Assessment and Project Appraisal* **19**, 311–319 (2001).

Gerrard, M. B., *Whose Background, Whose Risk*, The MIT Press, Cambridge, MA, 1995.

Gershberg, S., "Sustainable Development, Sustainable Governance and Environmental Impact Assessments in Canada," *Intergovernmental Policy Forum June 17, 2002*, Held in Conjunction with 22nd Annual Conference of the International Association for Impact Assessment, The Hague, Netherlands, 2002.

Gertler, L., in K. S. Hanna, ed., "The Hearing Process in Environmental Impact Assessment," *Environmental Impact Assessment—Practice and Participation*, Oxford University Press, Oxford, UK, 2009.

GHK, *Collection of the Information and Data to Support the Impact Assessment Study of the Review of the EIA Directive*, A Study for DG Environment, Clerkenwell House, London, 2010.

Ghanimé, L., et al., in B. Sadler et al., eds., "Using SEA to Enhance Poverty Reduction Strategies," in *Handbook of Strategic Environmental Assessment*, Earthscan, London, UK, 2011.

Gibson, R. B., in E. Lykke, ed., "Respecting Ignorance and Uncertainty," *Achieving Environmental Goals: The Concept and Practice of Environmental Performance Review*, Belhaven, London, 1992.

Gibson, R. B., "Environmental Assessment Design: Lessons From the Canadian Experience," *The Environmental Professional* **15**, 12–24 (1993).

Gibson, R. B., "Favouring the Higher Test: Contribution to Sustainability as the Central Criterion for Reviews and Decisions Under the Canadian Environmental Assessment Act," *Journal of Environmental Law and Practice* **10**, 39–55 (2000).

Gibson, R. B., *Specification of Sustainability-based Environmental Assessment Decision Criteria and Implications for Determining "Significance" in Environmental Assessment*, Canadian Environmental Assessment Agency, Hull, Québec, Canada, 2001.

Gibson, R. B., et al., *Sustainability Assessment—Criteria and Processes*, Earthscan, London, UK, 2005.

Gibson, R. B., "Sustainability Assessment: Basic Components of a Practical Approach," *Impact Assessment and Project Appraisal* **24**, 170–182 (2006a).

Gibson, R. B., in K. S. Hanna, ed., "The Canadian Federal EIA System," *Environmental Impact Assessment in Canada*, Oxford University Press, Oxford, 2006b.

Gibson, R. B., "Sustainability Assessment," *Impact Assessment and Project Appraisal* **24**, 170–182 (2009).

Gibson, R. B., in W. R. Sheate, ed., "Beyond the Pillars: Sustainability Assessment as a Framework for Effective Integration of Social, Economic and Ecological Considerations in Significant Decision-making," *Tools, Techniques and Approaches for Sustainability*, World Scientific, New Jersey, 2010.

Gibson, R. B., "Application of a Contribution to Sustainability Test by the Joint Review Panel for the Canadian Mackenzie Gas Project," *Impact Assessment and Project Appraisal* **29**, 231–244 (2011).

Gibson, R. B., and K. S. Hanna, in K. S. Hanna, ed., "Progress and Uncertainty: The Evolution of Federal Environmental Assessment in Canada," *Environmental Impact Assessment in Canada*, Oxford University Press, Oxford, UK, 2009.

Gibson, R. B., "In Full Retreat: The Canadian Government's New Environmental Assessment Law Undoes Decades of Progress," *Impact Assessment and Project Appraisal* **30**, 179–188 (2012).

Giere, R. N., *Science Without Limits*, University of Chicago Press, Chicago, 1999.

Gilpin, A., *Environmental Impact Assessment—Cutting Edge for the Twenty-First Century*, Cambridge University Press, Cambridge, UK, 1995.

Glasson, J., and C. Bellanger, "Divergent Practice in a Converging System? The Case of EIA in France and the UK," *Environmental Impact Assessment Review* **23**, 605–624 (2003).

Glasson, T., R. Thérivel, and A. Chadwick, *Introduction to Environmental Impact Assessment*, 3rd ed., Taylor & Francis, New York, 1999 and 2005.

Gleick, J., *Chaos—Making a New Science*, Viking Penguin Inc., New York, 1988.

Goldstein, H. A., "Planning as Argumentation," *Environment and Planning B: Planning and Design* **11**, 297–312 (1984).

Gontier, M., B. Balfors, and U. Mörtberg, "Biodiversity in Environmental Assessment—Current Practice and Tools for Prediction," *Environmental Impact Assessment Review* **26**, 268–286 (2006).

Goodland, R., "Definition of Environmental Sustainability," *IAIA Newsletter* **5**, 1–2 (1993).

Gordon, W. J. J., *Synectics: The Development of Creative Capacity*, Collier Books, New York, 1961.

Gorz, A., *Ecology as Politics*, Black Rose Books, Montréal, Québec, 1980.

Govender, K., R. Hounsome, and A. Weaver, "Sustainability Assessment: Dressing up SEA?—Experiences from South Africa," *Journal of Environment Policy and Management* **8**, 321–340 (2006).

Government of British Columbia (GBC), *An Introduction Guide to Adaptive Management*, Forest Practice Branch, BC Forest Services, Ministry of Forests, Victoria, 1999.

Government of Canada, *A Canadian Perspective on the Precautionary Approach/Principle—Discussion Document*, Ottawa, Canada, September 2001.

Government of Canada, *Strategic Environmental Assessment—The Cabinet Directive on the Environmental Assessment of Policy, Plans and Program Proposals—Guidelines for Implementing the Cabinet Directive*, Privy Council and Canadian Environmental Assessment Agency, Minister of Public Works and Government Services Canada, Ottawa, Ontario, 2010a.

Government of Canada, *Effectiveness of the Environmental Assessment Track Process Under the Canadian Environmental Assessment Act*, Quality Assurance Program for Environmental Assessments Under the Canadian Environmental Assessment Act, Ottawa, Ontario, 2010b.

Gower, B., *Scientific Method—A Historical and Philosophical Introduction*, Routledge, London, 1997.

Graham, J. D., "The Role of Precaution in Risk Assessment and Management: An American's View," *Presentation to the Conference on the US, Europe, Precaution and Risk Management: A Comparative Case Study Analysis of the Management of Risk in a Complex World*, Administrator, Office of Information and Regulatory Affairs, OMB, 2002.

Gray, B., *Collaborating: Finding Common Ground for Multiparty Problems*, Jossey-Bass, San Francisco, 1989.

Greenall, W., "Organizing and Managing the Planning Team," *Presentation to the Seminar on Integrated Approaches to Resource Planning*, Banff School of Management, Banff, Alberta, 1985.

Greene, B., *The Elegant Universe*, Vantage Press, New York, 1999.

Greer-Wooten, B., in A. J. Sinclair, ed., "Contexualizing the Environmental Impact Assessment Process," *Canadian Environmental Assessment in Transition*, Canadian Association of Geographers, Waterloo, 1997.

Greig, L., and P. Duinker, *Scenarios of Future Developments in Cumulative Effects Assessment: Approaches for the Mackenzie Gas Project*, Prepared for the Mackenzie Gas Project—Joint Review Panel, Inuvik, Canada, 2007.

Greig, L. A., and P. N. Duinker, "A Proposal for Further Strengthening Science in Environmental Impact Assessment in Canada," *Impact Assessment and Project Appraisal* **29**, 159–165 (2011).

Greig, L., K. Pawley, and P. Duinker, *Alternative Scenarios of Future Development: An Aid to Cumulative Effects Assessment*, Research Supported by the Canadian Environmental Assessment Agency's Research and Development Program, Hull, Québec, 2004.

Grima, A. P., P. Timmerman, C. D. Fowle, and P. Byer, *Risk Management and EIA: Research Needs and Opportunities*, Canadian Environmental Assessment Research Council, Ministry of Supply and Services, Ottawa, 1986.

Grinde, J., and A. Khare, "The Ant, the Grasshopper or Schrödinger's Cat: An Exploration of Concepts of Sustainability," *Journal of Environmental Assessment Policy and Management* **10**, 115–141 (2008).

Guillette, E. A., in J. A. Tickner, ed., "The Children of the Yaqui Valley: Precautionary Science and Communities," *Environmental Science and Preventive Public Policy*, Island Press, Washington, 2003.

Gullett, W., "Environmental Protection and the 'Precautionary Principle': A Response to Scientific Uncertainty in Environmental Management," *Environmental and Planning Law Journal* **14**, 52–69 (1997).

Gullett, W., "Environmental Impact Assessment and the Precautionary Principle: Legislating Caution in Environmental Protection," *Australian Journal of Environmental Management* **5**, 146–158 (1998).

Gullett, W., "The Precautionary Principle in Australian Policy, Law and Potential for Precautionary EIAs," *Paper Presented at the Risk Assessment and Policy Association Meeting*, Washington, DC, 1999.

Gullett, W., "Environmental Decision Making in a Transboundary Context: Principles and Challenges for the Denmark-Sweden Øresund Fixed Link," *Journal of Environmental Assessment Policy and Management* **2**, 529–559 (2000).

Gunderson, L., "Resilience, Flexibility and Adaptive Management—Antidotes for Spurious Certitude?" *Conservation Ecology* **3**, 7 (1999).

Gunn, J. H., and B. F. Noble, "A Conceptual Basis and Methodological Framework for Regional Strategic Environmental Assessment (R-SEA)," *Impact Assessment and Project Appraisal* **27**, 258–270 (2009a).

Gunn, J. H., and B. F. Noble, "Integrating Cumulative Effects in Regional Strategic Environmental Assessment Frameworks," *Journal of Environmental Assessment Policy and Management* **11**, 267–290 (2009b).

Gunn, J., and B. F. Noble, "Conceptual and Methodological Challenges to Integrating SEA and Cumulative Effects Assessment," *Environmental Impact Assessment Review* **31**, 154–160 (2011).

Gunning, C., P. Harris, and J. Mallett, "Assessing the Health Equity Impacts of Regional Land-Use Planning: An Equity Focused Health Impact Assessment of Alternative Patterns of Development of the Whitsunday Hinterland and Mackay Regional Plan, Australia," *Environmental Impact Assessment Review* **31**, 415–419 (2011).

Habermas, J., *Justification and Application—Remarks on Discourse Ethics*, The MIT Press, Cambridge, MA, 1993.

Hacking, T., and P. Guthrie, "Sustainable Development Objectives in Impact Assessment: Why are They Needed and Where Do They Come From?," *Journal of Environmental Assessment Policy and Management* **8**, 341–371 (2006).

Hacking, T., and P. Guthrie, "A Framework for Clarifying the Meaning of Triple Bottom-Line, Integrated and Sustainability Assessment," *Environmental Impact Assessment Review* **28**, 73–89 (2008).

Hacking, T., and P. Guthrie, in B. Sadler et al., eds., "Assessment for Sustainable Development: Theoretical Framework and Application to the Mining Sector," *Handbook of Strategic Environmental Assessment*, Earthscan, London, 2011.

Hainer, R. M., in E. Glatt and M. V. Shelly, eds., "Rationalism, Pragmatism and Existentialism: Perceived But Undiscovered Multicultural Problems," *The Research Society*, Gordon and Breach, New York, 1968.

Halstead, J. H., R. A. Chase, S. H. Murdock, and F. L. Leistritz, *Socio-economic Impact Management*, Westview Press, Boulder, CO, 1984.

Hance, B. J., C. Chess, and P. M. Sandman, *Industry Risk Communications Manual: Improving Dialogue with Communities*, Lewis Publishers, Chelsea, MI, 1990.

Hanna, K. S., "Planning for Sustainability: Experiences in Two Contrasting Communities," *Journal of American Planning Association* **71**, 27–40 (2005).

Hanna, K. S., ed., *Environmental Impact Assessment—Practice and Participation*, Oxford University Press, Oxford, UK, 2009a.

Hanna, K. S., in K. S. Hanna, ed., "Environmental Impact Assessment: Process, Setting and Efficacy," *Environmental Impact Assessment—Practice and Participation*, Oxford University Press, Oxford, UK, 2009b.

Hansell, A., and P. Aylin, "Use of Health Data in Health Impact Assessment," *Impact Assessment and Project Appraisal* **21**, 57–64 (2003).

Hansen, A. M., and L. Kørnøv, "A Value-Rational View of Impact Assessment of Mega Industry in a Greenland Planning and Policy Context," *Impact Assessment and Project Appraisal* **28**, 135–145 (2010).

Hansen, R. P., and T. A. Wolff, "Reviewing NEPA's Past: Improving NEPA's Future," *Environmental Practice* **13**, 235–249 (2011).

Hanusch, M., and T. B. Fischer, in B. Sadler et al., eds., "SEA and Landscape Planning," *Handbook of Strategic Environmental Assessment*, Earthscan, London, 2011.

Hanusch, M., and J. Glasson, "Much Ado about SEA/SA Monitoring: The Performance of English Regional Spatial Strategies, and Some German Comparisons," *Environmental Impact Assessment Review* **28**, 601–617 (2008).

Harashina, S., "Environmental Dispute Resolution Process and Information Exchange," *Environmental Impact Assessment Review* **15**, 69–80 (1995).

Hardstaff, P., "The Precautionary Principle, Trade and the WTO," *Discussion Paper for the European Commission Consultation on Trade and Sustainable Development, Royal Society for the Protection of Birds*, Bedfordshire, UK, 2000.

Harper, T. L., and S. M. Stein, "The Centrality of Normative Ethical Theory to Contemporary Planning Theory," *Journal of Planning Education and Research* **11**, 105–116 (1992).

Harremoës, P., D. Gee, M. MacGarvins, J. Stirling, J. Keys, B. Wynee, and S. Guedes Vaz, eds., *Late Lessons from Early Warnings: The Precautionary Principle 1896–2000*, Environmental Issues Report # 22, European Environmental Agency, Copenhagen, EEA, 2002.

Harriman, J. A. E., and B. F. Noble, "Characterizing Project and Strategic Approaches to Regional Cumulative Effects Assessment in Canada," *Journal of Environmental Assessment Policy and Management* **10**, 25–50 (2008).

Harris, C. H., et al., "Community-based Social Impact Assessment: The Case of Salmon-recovery on the Lower Snake River," *Impact Assessment and Project Appraisal* **21**, 109–118 (2003).

Harris, P. J., "Human Health and Wellbeing in Environmental Impact Assessment in New South Wales, Australia: Auditing Health Impacts within Environmental Assessments of Major Projects," *Environmental Impact Assessment Review* **29**, 310–318 (2009).

Harris, P. J., and J. Spickett, "Health Impact Assessment in Australia: A Review and New Directions," *Environmental Impact Assessment Review* **31**, 425–432 (2011).

Harris-Roxas, B., and E. Harris, "Differing Forms, Differing Purposes: A Typology of Health Impact Assessment," *Environmental Impact Assessment Review* **31**, 396–403 (2011).

Harris-Roxas, B., et al., "Health Impact Assessment: The State of the Art," *Impact Assessment and Project Appraisal* **30**, 43–52 (2012).

Harrop, D. O., and J. A. Nixon, *Environmental Assessment in Practice*, Routledge, London, 1999.

Hartley, N., and C. Wood, "Public Participation in Environmental Impact Assessment—Implementing the Aarhus Convention," *Environmental Impact Assessment Review* **25**, 319–340 (2005).

Harvey, S., and A. Usher, "Communities and Natural Resource Management: Bridging the Gap," *Plan Canada* **36**, 34–38 (1996).

Hassan, O. A., "Assessing the Sustainability of a Region in the Light of Composite Indicators," *Journal of Environmental Assessment Policy and Management* **10**, 51–65 (2008).

Hawke, A., EPBC Act Reviewer, 2009, *Report of the Independent Review of the Environmental Protection and Biodiversity Act 1999—The Australian Environment Act—Final Report*, Commonwealth of Australia, Canberra Australia, 2009.

Hayes, N., and A. Morrison-Saunders, "Effectiveness of Environmental Offsets in Environmental Impact Assessment: Practitioner Perspectives from Western Australia," *Impact Assessment and Project Appraisal* **25**, 209–218 (2007).

Healey, P., *Collaborative Planning—Shaping Places in Fragmented Societies*, UBC Press, Vancouver, 1997.

Health Canada, *Canadian Handbook on Health Impact Assessment, Volume 1—The Basics; Volume 2—Decision making in Environmental Health Impact Assessment; Volume 3—Roles for the Health Practitioner*, Health Canada, 2004.

Health Canada, *Public Involvement Framework and Guidelines*, Minister of Public Works and Government Services Canada, Ottawa, Canada, 2000.

Hegmann, G. H., and G. A. Yarranton, *Cumulative Effects and the Energy Resources Conservation Board's Review Process*, Macleod Institute for Environmental Analysis, Working Paper #1, Calgary, Alberta, Canada, July 1995.

Hegmann, G. H., and G. A. Yarranton, "Alchemy to Reason: Effective Use of Cumulative Effects Assessment in Resource Management," *Environmental Impact Assessment Review* **31**, 484–490 (2011).

Heinma, K., and T. Põder, "Effectiveness of Environmental Impact Assessment System in Estonia," *Environmental Impact Assessment Review* **30**, 272–277 (2010).

Heiman, M. K., "Science by the People: Grassroots Environmental Monitoring and the Debate over Scientific Expertise," *Journal of Planning Education and Research* **16**, 291–299 (1997).

Hemmings, A. D., and R. Roura, "A Square Peg in a Round Hole: Fitting Impact Assessment Under the Antarctic Environmental Protocol to Antarctic Tourism," *Impact Assessment and Project Appraisal* **21**, 13–24 (2003).

Hendler, S., "Feminist Planning Ethics," *Journal of Planning Literature* **9**, 115–127 (1994).

Henriques, A., and J. Richardson, eds., *The Triple Bottom Line—Does it Add Up?* Earthscan, London and Nevada, 2004.

Hermans, F., and L. Knippenberg, "A Principles-based Approach for the Evaluation of Sustainable Development," *Journal of Environmental Assessment Policy and Management* **8**, 299–319 (2006).

Herring, R. J., in K. S. Hanna, ed., "The Canadian Federal EIA System," *Environmental Impact Assessment in Canada*, Oxford University Press, Oxford, 2009.

Hettwer, R., "Ethical Factors in Environmental Negotiation," *Environmental Impact Assessment Review* **11**, 197–202 (1991).

Hildén, M., E. Furman, and M. Kaljonen, "Views on Planning and Expectations of SEA: The Case of Transport Planning," *Environmental Impact Assessment Review* **24**, 519–536 (2004).

Hindling-Rydevik, T., and H. Bjarnadóttar, "Context Awareness and Sensitivity in SEA Implementation," *Environmental Impact Assessment Review* **27**, 666–684 (2007).

Hinte, T. V., T. I. Gunton, and J. C. Day, "Evaluation of the Assessment Process for Major Projects: A Case Study of Oil and Gas Pipelines in Canada," *Impact Assessment and Project Appraisal* **25**, 123–137 (2007).

Hirschmann, D. J., "Peer Review and the Contract Between Science and Society," *The Environmental Professional* **16**, 79–89 (1994).

Hoch, C., "Doing Good and Being Right," *Journal of the American Planning Association* **50**, 335–345 (1984).

Hoch, C., "Racism and Planning," *Journal of the American Planning Association* **59**, 451–460 (1993).

Hodge, R. A., "Tracking Progress Toward Sustainability: Linking the Power of Measurement and Story," *Paper Presented at the Symposium—Sustainable Development from the Ground Up: Measuring Progress Toward Sustainability*, Denver, CO, 2004.

Hodgson, P., and R. P. White, *Relax, Its Only Uncertainty*, Pearson Education Limited, London, 2001.

Hofstadter, D. R., *Metamagical Themes: Questing for the Essence of Mind and Pattern*, Basic Books, New York, 1985.

Hollick, M., "Self-organizing Systems and Environmental Management," *Environmental Management* **17**, 621–628 (1993).

Holling, C. S., ed., *Adaptive Environmental Assessment and Management*, John Wiley & Sons, Chichester, 1978.

Holm, S., and J. Harris, "Precautionary Principle Stifles Discovery," *Nature* **400**, (July 29, 1999).

Homer-Dixon, T., *The Ingenuity Gap*, Vintage Canada, Toronto, 2000.

Hood, J., and S. Nicholl, "The Role of Environmental Risk Management and Reporting: An Empirical Analysis," *Journal of Environmental Assessment Policy and Management* **4**, 1–29 (2002).

Hooper, B. P., G. T. McDonald, and B. Mitchell, "Facilitating Integrated Resource and Environmental Management: Australian and Canadian Perspectives," *Journal of Environmental Planning and Management* **42**, 747–766 (1999).

Howe, E., "Normative Ethics in Planning," *Journal of Planning Literature* **5**, 123–149 (1990).

Howell, R. E., M. E. Olsen, and D. Olsen, *Designing a Citizen Involvement Program: A Guidebook for Involving Citizens in the Resolution of Environmental Issues*, Oregon State University, Convallis, Oregon, 1987.

Hughes, R., in A. Donnelly, B. Dalal-Clayton, and R. Hughes, eds., "Environmental Impact Assessment and Stakeholder Involvement," *A Directory of Impact Assessment Guidelines*, 2nd ed., International Institute for Environment and Development, Russell Press, Nottingham, 1998.

Hummel, R. P., *The Bureaucratic Experience*, St. Martin's Press, New York, 1977.

Hunsaker, Jr. D. B., and D. W. Lee, in B. Sadler, ed., "Environmental Impact Assessment of Abnormal Events: A Follow-up Study," *Auditing and Evaluation in Environmental Assessment and Management: Canadian and International Experiences*, Environmental Protection Services of Environment Canada and the Banff School of Management, Banff, Alberta, Canada, 1985.

Hunsberger, C. A., R. B. Gibson, and S. K. Wismer, "Citizen Involvement in Sustainability-centred Environmental Assessment Follow-up," *Environmental Impact Assessment Review* **25**, 609–627 (2005).

Huxley, M., and O. Yiftachel, "New Paradigm or Old Myopia? Unsettling the Communicative Turn in Planning Theory," *Journal of Planning Education and Research* **19**, 333–342 (2000).

Hyett, D., "Environmental Risk Assessment in Environmental Impact Assessment—Optional or Mandatory?," *Presentation to IAIA'10 Conference Proceedings, The Role of Impact Assessment in Transitioning to a Green Economy*, International Association for Impact Assessment (IAIA), Geneva, Switzerland, 2010.

Hyman, E. L., B. Stiftel, D. H. Moreau, and R. C. Nichols, *Combining Facts and Values in Environmental Impact Assessment*, Westview Press, Boulder, CO, 1988.

IGC et al., *Environmental Impact Statement Terms of Reference for the Mackenzie Gas Project*, Inuvialuit Game Council, Mackenzie Valley Environmental Impact Review Board and the Minister of the Environment (Canada), August 2004, available at http://www.acee-ceaa.gc.ca/default.asp?lang=En&n=155701CE-1.

Ignatieff, M., *The Rights Revolution*, Canadian Broadcasting Corporation, Etobicoke, Ontario, 2000.

Ijäs, A., M. T. Kuitunen, and K. Jalava, "Developing the RIAM Method (Rapid Impact Assessment Matrix) in the Context of Impact Significance Assessment," *Environmental Impact Assessment Review* **30**, 82–89 (2010).

Imperial College Consultants Ltd., (ICCL), Babtie Allott and Lomax, Austrian Institute for the Development of Environmental Assessment, ECA, CESAM, University of Aarhus, *SEA and Integration of the Environment into Strategic Decision-making*, CEC Contract No. B4-3044/99/136634/MAR/B4, London, UK, May 2001.

Imperial College London Consultants (ICLC), *The Relationship Between the EIA and SEA Directives—Final Report to the European Commission*, Contract # ENV.G.4./ETU/2004/0020r, Imperial College, London, UK, 2005.

Innes, J. E., "Planning Theory's Emerging Paradigm: Communicative Action and Interactive Practice," *Journal of Planning Education and Research* **14**, 183–190 (1995).

Innes, J. E., "Planning Through Consensus Building," *American Planning Association Journal* **62**, 460–472 (1996).

Innes, J. E., "Information in Communicative Planning," *American Planning Association Journal* **64**, 52–63 (1998).

Innes, J. E., and D. E. Booher, "Consensus Building and Complex Adaptive Systems," *Journal of the American Planning Association* **65**, 412–423 (1999).

Institute for European Environmental Policy (IEEP), *Sustainable Development in the European Commission's Integrated Impact Assessment for 2004*, Final Report, IEEP, London, UK, 2004.

Institute of Public Health in Ireland (IPHI), *Health Impact Assessment—An Introductory Paper*, Dublin, Ireland, September 2001.

Interagency Ecosystem Management Task Force (IEMTF), *The Ecosystem Approach: Health Ecosystems and Sustainable Economics: Volume I—Overview; Volume II—Implementation Issues*, IEMTF, Washington, DC, November 1995.

International Associates, *Conflict Resolution in Organizations*, Handout from the Third National Conference in Peace-making and Conflict Resolution, Denver, CO, 1986.

International Association for Impact Assessment (IAIA), in cooperation with Institute of Environmental Assessment, UK, *Principles of Environmental Impact Assessment Best Practice*, IAIA, Fargo, ND, USA, 1999.

International Association for Impact Assessment (IAIA), *Strategic Environmental Assessment*, IAIA, Special Publications Series No. 1, IAIA, Fargo, ND, USA, 2002a.

International Association for Impact Assessment (IAIA), *Impact Assessment in the Corporate Context*, Business and Industry Series No.1, IAIA, Fargo, ND, USA, 2002b.

International Association for Impact Assessment (IAIA), *Social Impact Assessment International Principles*, IAIA, Special Publications Series No. 2, IAIA, Fargo, ND, USA, 2003.

International Association for Impact Assessment (IAIA), *Biodiversity in Impact Assessment*, IAIA, Special Publications Series No. 3, IAIA, Fargo, ND, USA, 2005.

International Association for Impact Assessment (IAIA), *Health Impact Assessment—International Best Practice Principles*, Special Publications Series No. 5, IAIA, Fargo, ND, USA, 2006a.

International Association for Impact Assessment (IAIA), *Public Participation International Best Practice Principles*, IAIA, Special Publications Series No. 4, IAIA. Fargo, ND, USA, 2006b.

International Association for Impact Assessment (IAIA), *Guidance Standard for Environmental Professionals*, IAIA, 2010.

International Association for Impact Assessment (IAIA), *Climate Change in Impact Assessment*, Special Publication No. 5, IAIA, Fargo, ND, USA, 2012.

International Association for Impact Assessment (IAIA), *Respecting Indigenous Peoples and Traditional Knowledge—International Best Practices Principles*, IAIA, undated a.

International Association for Impact Assessment (IAIA), *What is Impact Assessment?* IAIA, undated b.

International Association for Impact Association, in cooperation with the Institute of Environmental Assessment (IAIA and IEA), *Principles of Environmental Impact Assessment Best Practice*, Fargo, North Dakota and Lincoln, UK, undated.

International Association for Impact Assessment (IAIA) and World Bank, *Climate Change and Impact Assessment Special Symposium Report*, IAIA and World Bank, Washington, DC, 2010.

Interorganizational Committee on Guidelines and Principles, "Guidelines and Principles for Social Impact Assessment," *Impact Assessment* **12**, 107–152 (1994).

Institute for European Policy (IEPP), *Review of Methodologies Applied for the Assessment of Employment and Social Impacts*, for DG Employment, Social Affairs and Equal Opportunities of the European Commission, 2004.

Institute of Ecology and Environmental Management (IEEM), *Guidelines for Ecological Impact Assessment in the United Kingdom*, IEEM, UK, 2006.

International Finance Corporation (IFC), *Stakeholder Engagement—A Good Practice Handbook for Companies Doing Business in Emerging Markets*, IFC, Washington, DC, 2007.

International Finance Corporation (IFC), *Labor Toolkit*, IFC, Washington, DC, 2008.

International Finance Corporation (IFC), *Good Practice Note—Addressing Grievances from Project-Affected Communities*, Guidance for Projects and Companies on Designing Grievance Mechanisms, #7, IFC, World Bank Group, 2009.

International Institute for Sustainable Development (IISD), *IEA Training Manual—A Training Manual on Integrated Environmental Assessment and Reporting*, United Nations Environment Program (UNEP) and IISD, Winnipeg, Canada, 2008.

Interorganizational Committee on Principles and Guidelines for Social Impact Assessment (ICPGSIA), "Principles and Guidelines for Social Impact Assessment in the USA," *Impact Assessment and Project Appraisal* **21**, 231–250 (2003).

Ison, E., in J. Kemm, J. Parry, and S. Palmer, eds., "Rapid Appraisal Techniques," *Health Impact Assessment*, Oxford University Press, Oxford, UK, 2004.

Jackson, T., and J. Dixon, "Applying Strategic Environmental Assessment to Land Use and Resource Management Plans in Scotland and New Zealand: A Comparison," *Impact Assessment and Project Appraisal* **24**, 89–101 (2006).

Jackson, T., and B. Illsley, "An Analysis of the Theoretical Rationale for Using Strategic Environmental Assessment to Deliver Environmental Justice in the Light of the Scottish Environmental Assessment Act," *Environmental Impact Assessment Review* **27**, 607–623 (2007).

Jalava, K., et al., "Quality of Environmental Impact Assessment: Finish EISs and the Opinions of EIA Professionals," *Impact Assessment and Project Appraisal* **28**, 15–27 (2010).

Jantsch, E., "Inter and Transdisciplinary University: A Systems Approach to Education and Innovation," *Ekistics* **32**, 433 (1971).

Jasonoff, S., in J. A. Tickner, ed., "A Living Legacy: The Precautionary Ideal in American Law: Ethics, Science and Precaution: A View from Norway," *Environmental Science and Preventive Public Policy*, Island Press, Washington, 2003.

Jasonoff, S., *Risk, Precaution and Environmental Values*, Carnegie Council on Ethics and International Affairs, New York, undated.

Jay, S., "The Place of Strategic Environmental Assessment in the Privatised Electricity Industry," *Impact Assessment and Project Appraisal* **23**, 315–324 (2005).

Jay, S., "Customers as Decision-makers: Strategic Environmental Assessment in the Private Sector," *Impact Assessment and Project Appraisal* **25**, 75–84 (2007).

Jay, S., et al., "Environmental Impact Assessment: Retrospect and Prospect," *Environmental Impact Assessment Review* **27**, 287–300 (2007).

Jepson, E. J., "Sustainability and Planning: Diverse Concepts and Close Associations," *Journal of Planning Literature* **14**, 499–510 (2001).

Jha-Thakur, U., et al., "Effectiveness of Strategic Environmental Assessment—The Significance of Learning," *Impact Assessment and Project Appraisal* **27**, 133–144 (2009).

Jiliberto, H. R., "Decisional Environmental Values as the Object of Analysis by Strategic Environmental Assessment," *Impact Assessment and Project Appraisal* **20**, 61–70 (2002).

Jiliberto, H. R., "A Holarchical Model for Regional Sustainability Assessment," *Journal of Environmental Assessment Policy and Management* **6**, 511–538 (2004).

Jiliberto, H. R., "Recognizing the Institutional Dimension of Strategic Environmental Assessment," *Impact Assessment and Project Appraisal* **29**, 133–140 (2011).

Jiricka, A., and U. Pröbstl, "SEA in Local Land Use Planning—First Experience in Alpine States," *Environmental Impact Assessment Review* **28**, 328–337 (2008).

Jiricka, A., and U. Pröbstl, "One Common Way—The Strategic and Methodological Influence on Environmental Planning Across Europe," *Environmental Impact Assessment Review* **29**, 379–389 (2009).

João, E. M., in A. L. Porter and J. Fittipaldi, eds., "Use of Geographical Information Systems in Impact Assessment," *Environmental Methods Review: Retooling Impact Assessment for the New Century*, Army Environmental Policy Institute and International Institute for Impact Assessment, Fargo, ND, 1998.

João, E. M., "A Research Agenda for Data and Scale Issues in Strategic Environmental Assessment," *Environmental Impact Assessment Review* **27**, 479–491 (2007).

João, E. M., and A. Mclauchlan, "Strategic Environmental Assessment as a Tool to Contribute to High Level Policy Objectives," *Journal of Environmental Assessment Policy and Management* **13**, 1–7 (2011).

João, E. M., F. Vanclay, and L. den Broede, "Emphazing Enhancement in All Forms of Impact Assessment: Introduction to a Special Issue," *Impact Assessment and Project Appraisal* **29**, 170–181 (2011).

Johannes, R. E., in J. T. Inglis, ed., "Integrating Traditional Ecological Knowledge and Management with Environmental Impact Assessment," *Traditional Ecological Knowledge: Concepts and Cases*, International Program on Traditional Ecological Knowledge and International Development Research Centre, Ottawa, 1999.

Johnson, B. L., "The Role of Adaptive Management as an Operational Approach for Resource Management Agencies," *Conservation Ecology* **3**, 8 (1999).

Johnson, D., et al., "Improving Cumulative Effects Assessment in Alberta: Regional Strategic Assessment," *Environmental Impact Assessment Review* **31**, 481–483 (2011).

Johnson, S., *Emergence*, Scribner, New York, 2001.

Jones, C. E., and P. Slinn, "Cultural Heritage in EIA—Reflections on Practice in North West Europe," *Journal of Environmental Assessment Policy and Management* **10**, 215–238 (2008).

Jones, M. L., and L. A. Greig, in V. W. Maclaren and J. B. Whitney, eds., "Adaptive Environmental Assessment and Management: A New Approach to Impact Assessment," *New Directions in Environmental Impact Assessment*, Methuen, Toronto, 1988.

Julien, B., "Current and Future Directions for Structured Impact Assessments," *Impact Assessment* **13**, 403–432 (1995).

Kain, J. H., and H. Söderberg, "Management of Complex Knowledge in Planning for Sustainable Development: The Use of Multi-criteria Decision Aids," *Environmental Impact Assessment Review* **28**, 7–21 (2008).

Kaiser, M., in J. A. Tickner, ed., "Ethics, Science and Precaution: A View from Norway," *Environmental Science and Preventive Public Policy*, Island Press, Washington, 2003.

Kakonge, J. O., "Environmental Impact Assessment in Sub-Saharan Africa: The Gambian Experience," *Impact Assessment and Project Appraisal* **24**, 57–64 (2006).

Kamrin, M. A., "The Need for Risk by Consensus," *Impact Assessment* **11**, 167–174 (1993).

Karjalainen, T. J., and T. Järvikoski, "Negotiating River Ecosystems: Impact Assessment and Conflict Mediation in the Cases of Hydro Power Construction," *Environmental Impact Assessment Review* **30**, 319–327 (2010).

Kasperson, R. E., P. G. Derr, and R. W. Kates, in M. J. Pasqualetti and K. D. Pijawka, eds., "Confronting Equity in Radioactive Waste Management: Modest Proposals for a Socially Just and Acceptable Program," *Nuclear Power—Assessing and Managing Hazardous Technology*, Westview Press, Boulder, Colorado, 1984.

Kasperson, R. E., S. Ratick, and O. Renn, *Assessing the State/National Distributional Equity Issues Associated with the Proposed Yucca Mountain Repository: A Conceptual Approach*, State of Nevada Agency for Nuclear Projects/Nuclear Waste Project Office, Carson City, NV, 1988.

Kates, R., "Sustainability Science," *World Academies Conference—Transition to Sustainability in 21st Century*, Tokyo, Japan, 2000.

Kaufman, H., in R. Clark and L. Canter, eds., "The Role of NEPA in Sustainable Development," *Environmental Policy and NEPA*, St. Lucie Press, Boca Raton, FL, 1997.

Kauppinen, T., and K. Nelimarkka, "A Review of Finnish Social and Health Impact Assessment," *Journal of Environmental Assessment Policy and Management* **6**, 1–18 (2004).

Keen, M., and Mahanty, S., "Learning in Sustainable Natural Resources Management: Challenges and Opportunities in the Pacific," *Society and Natural Resources* **19**, 497–513 (2006).

Kelly, C., *Guidelines for Rapid Environmental Assessments in Disasters*, Benfield Hazard Research Centre, University College London and CARE International, 2005.

Kemm, J., "The Future Challenges for HIA," *Environmental Impact Assessment Review* **25**, 799–807 (2005).

Kemm, J., and J. Parry, in J. Kemm, J. Parry, and S. Palmer, eds., "What is HIA? Introduction and Overview," *Health Impact Assessment*, Oxford University Press, Oxford, UK, 2004a.

Kemm, J., and J. Parry, in J. Kemm, J. Parry, and S. Palmer, eds., "The Development of HIA," *Health Impact Assessment*, Oxford University Press, Oxford, UK, 2004b.

Kende-Robb, C., and W. A. Van WicklinIII, in K. Ahmed and E. Sánchez-Triana, eds., "Giving the Most Vulnerable A Voice," *Strategic Environmental Assessment for Policies*, The World Bank, Washington, DC, 2008.

Kennedy, A. J., and W. A. Ross, "An Approach to Integrate Impact Scoping with Environmental Impact Assessment," *Environmental Management* **16**, 475–484 (1992).

Kent, R. B., and R. E. Klosterman, "GIS and Mapping: Pitfalls for Planning," *Journal of the American Planning Association* **66**, 189–198 (2000).

Kessler, J. J., and H. Abaza, "United Nations Environment Programme's Approach to Integrated Assessment of Trade-related Policies: Evolution and Recent Progress," *Impact Assessment and Project Appraisal* **24**, 273–283 (2006).

Ketelsen, L., in J. A. Tickner, ed., "Scientists and the Real World," *Environmental Science and Preventive Public Policy*, Island Press, Washington, 2003.

Keys, D. L., L. Canter, and R. Senner, "Strengthening the National Environmental Policy Act of 1969," *Environmental Practice* **13**, 216–226 (2011).

Keysar, E., and A. Steinemann, "Integrating Environmental Impact Assessment with Master Planning: Lessons from the US Army," *Impact Assessment Review* **22**, 583–611 (2002).

Khera, N., and A. Kumar, "Inclusion of Biodiversity in Environmental Impact Assessments (EIA): A Case Study of Selected EIA Reports in India," *Impact Assessment and Project Appraisal* **28**, 189–200 (2010).

Kiewiet, D. J., and J. F. J. Vos, "Organizational Sustainability: A Case of Formulating a Tailor-made Definition," *Journal of Environmental Assessment Policy and Management* **9**, 1–18 (2007).

Kim, J. O., "A Study on Facilitating Factors and Barriers in Addressing Cumulative Effects," *Presentation to the IAIA12 Conference Proceedings, Energy Future—The Role of Impact Assessment*, 32nd Annual Meeting of the International Association for Impact Assessment, Porto, Portugal, 2012.

King, T. F., "Cultural Resources in an Environmental Assessment Under NEPA," *Environmental Practice* **4**, 137–144 (2002).

King, T. F., "EIA: A Corrupted System," *NAEP National E-News* 8–12 (March–April 2012).

Kirchhoff, D., et al., "A Policy Window Opens: Strategic Environmental Assessment in York Region, Ontario, Canada," *Journal of Environmental Assessment Policy and Management* **12**, 333–354 (2009).

Kirk, D., *A Decision Support Tool to Aid in Evaluating Significance of Adverse Effects on Birds for Environmental Assessment*, Canadian Environmental Assessment Agency Research Program, Hull, Québec, 2000.

Kirkpatrick, C., and C. George, "Methodological Issues in the Impact Assessment of Trade Policy: Experience from the European Commission's Sustainability Impact Assessment (SIA) Programme," *Impact Assessment and Project Appraisal* **24**, 325–334 (2006).

Knaus, M., D. Löhr, and B. O'Reagan, "Valuation of Ecological Impacts—A Regional Approach Using the Ecological Footprint Concept," *Environmental Impact Assessment Review* **26**, 156–169 (2006).

Kobus, D., "Development and Testing of a Conceptual Framework for Assessment of Progress Towards Achieving Sustainable Development in Countries in Transition," *Journal of Environmental Assessment Policy and Management* **7**, 457–491 (2005).

Koestler. A., *The Act of Creation*, Pan Piper, London, 1964.

Koivurova, T., "Transboundary Environmental Assessment in the Arctic," *Impact Assessment and Project Appraisal* **26**, 265–275 (2008).

Kolhoff, A., and R. Slootweg, "Biodiversity in SEA for Spatial Plans—Experiences from the Netherlands," *Journal of Environmental Assessment Policy and Management* **7**, 267–286 (2005).

Kolhoff, A. J., H. A. C. Runhaar, and P. J. Driessen, "The Contribution of Capacities and Context to EIA System Performance and Effectiveness in Developing Countries: Towards a Better Understanding," *Impact Assessment and Project Appraisal* **27**, 271–282 (2009).

Kolkman, M. J., A. van der Veen, and P. A. T. M. Geurts, "Controversies in Water Management," *Environmental Impact Assessment Review* **27**, 685–706 (2007).

Kontić, B., and D. Kontić, "A Viewpoint on the Approval Context of Strategic Environmental Assessments," *Environmental Impact Assessment Review* **32**, 151–155 (2012).

Kørnøv, L., "Strategic Environmental Assessment and the Limits of Rationality in Decision Making Processes," *Presentation to the International Association for Impact Assessment,* Christchurch, New Zealand, 1998.

Kørnøv, L., "Strategic Environmental Assessment as Catalyst of Healthier Spatial Planning: The Danish Guidance and Practice," *Environmental Impact Assessment Review* **29**, 60–65 (2009).

Kørnøv, L., P. Christensen, and E. H. Nielson, "Mission Impossible: Does Environmental Impact Assessment in Denmark Secure a Holistic Approach to the Environment?" *Impact Assessment and Project Appaisal* **23**, 303–314 (2005).

Kørnøv, L., and H. Dalkmann, in B. Sadler, ed., "Institutional Challenges for SEA Implementation and Decision-making: Search for Appropriate Organizations," *Handbook of Strategic Environmental Assessment*, Earthscan, London, 2011.

Kørnøv, L., and W. A. H. Thissen, "Rationality in Decision and Policy-making: Implications for Strategic Environmental Assessment," *Impact Assessment and Project Appraisal* **18**, 191–200 (2000).

Kozlowski, J., "Toward Ecological Orientation of the Planning Process: A Planner's Perspective," *Impact Assessment Bulletin* **8**, 47–68 (1990).

Kreig, E. J., and D. R. Faber, "Not So Black and White: Environmental Justice and Cumulative Impact Assessments," *Environmental Impact Assessment Review* **24**, 667–694 (2004).

Kreske, D. L., *Environmental Impact Statements: A Practical Guide for Agencies, Citizens and Consultants*, John Wiley & Sons, New York, 1996.

Krieger, M. H., *Advice and Planning*, Temple University Press, Philadelphia, 1981.

Krieger, N., et al., "Assessing Health Impact Assessment: Multidisciplinary and International Perspectives," *Journal of Epidemiological Community Health* **57**, 659–662 (2003).

Kumagai, Y., J. Edwards, and M. S. Carroll, "Why are Natural Disasters Not 'Natural' for Victims?" *Environmental Impact Assessment Review* **26**, 106–119 (2006).

Kvaener, J., D. Swenson, and L. Erikson, "Assessing Environmental Vulnerability in EIA: The Content and Context of the Vulnerability Concept in an Alternative Approach to Standard EIA Procedure," *Environmental Impact Assessment Review* **26**, 511–527 (2006).

Kwasniak, A. L., "Use and Abuse of Adaptive Management in Environmental Assessment Law and Practice: A Canadian Example and General Lessons," *Journal of Environmental Assessment Policy and Management* **12**, 425–468 (2010).

Kwiatkowski, R. E., in J. Kemm, J. Parry, and S. Palmer, eds., "Impact Assessment in Canada: An Evolutionary Process," *Health Impact Assessment*, Oxford University Press, Oxford, UK, 2004.

Kwiatkowski, R. E., "Indigenous Community-based Participatory Research and Health Impact Assessment: A Canadian Example," *Environmental Impact Assessment Review* **31**, 445–450 (2011).

Kwiatkowski, R. E., et al., "Canadian Indigenous Engagement and Capacity Building in Health Impact Assessment," *Impact Assessment and Project Appraisal* **27**, 57–67 (2009).

Lajoie, G., and M. A. Bouchard, "Native Involvement in Strategic Assessment of Natural Resource Development: The Example of the Crees

Living in the Canada Taiga," *Impact Assessment and Project Appraisal* **24**, 211–220 (2006).

Lal, P., H. Lim-Applegate, and M. Scoccimarro, "The Adaptive Decision-making Process as a Tool for Integrated Natural Resource Management: Focus, Attitudes and Approach," *Conservation Ecology* **5** (2001).

Landry, V., et al., *Aboriginal Land Planning in Canada: The Role of Strategic Environmental Assessment in Adaptive Co-management*, Research Supported by the Canadian Environmental Assessment Agency's Research and Development Program, CEAA, Hull, Québec, 2009.

Landsberg, F., et al., *Ecosystem Services Review for Impact Assessment—Introduction and Guide to Scoping*, World Resources Institute, Washington, DC, 2011.

Lane, M. B., et al., "Sacred Land, Mineral Wealth, and Biodiversity at Coronation Hill, Northern Australia: Indigenous Knowledge and SIA," *Impact Assessment and Project Appraisal* **21**, 89–98 (2003).

LaPierre, L., "Development of a Consensus Model to Address Traditional Environmental Knowledge Process," *Presentation to the IAIA12 Conference Proceedings, Energy Future—The Role of Impact Assessment,* 32nd Annual Meeting of the International Association for Impact Assessment, Porto, Portugal, 2012.

Larsen, S. V., "Climate Change in Environmental Impact Assessments of Renewable Energy Projects," *Presentation to the IAIA12 Conference Proceedings, Energy Future—The Role of Impact Assessment,* 32nd Annual Meeting of the International Association for Impact Assessment, Porto, Portugal, 2012.

Larsen, S. V., L. Kørnøv, and A. Wejs, "Mind the Gap in SEA: An Institutional Perspective on Why Assessment of Synergies Among Climate Change Mitigation, Adaptation and Other Policy Areas are Missing," *Environmental Impact Assessment Review* **33**, 32–40 (2012).

Lauria, M., and M. J. Soll, "Communicative Action, Power and Misinformation in a Site Selection Process," *Journal of Planning Education and Research* **15**, 199–211 (1996).

Lavallée, L., and P. André, "Social Impact Follow-up in Québec, Canada: 25 Years of Environmental Impact Assessment Practice," *Impact Assessment and Project Appraisal* **23**, 241–245 (2005).

Law, L. B., et al., "Cumulative Effects Assessment and EIA-based Follow-up: A Proposed Community-based Monitoring Program in the Oil Sands Region, Northeastern Alberta," *Impact Assessment and Project Appraisal* **23**, 205–209 (2005).

Lawrence, D. P., "Quantitative Versus Qualitative Evaluation: A False Dichotomy?" *Environmental Impact Assessment Review* **13**, 3–11 (1993).

Lawrence, D. P., "Designing and Adapting the EIA Planning Process," *The Environmental Professional* **16**, 2–21 (1994).

Lawrence, D. P., "Approaches and Methods of Siting Locally Unwanted Waste Facilities," *Journal of Environmental Planning and Management* **39**, 165–187 (1996).

Lawrence, D. P., "Quality and Effectiveness of Environmental Impact Assessment: Lessons and Insights From Ten Assessments in Canada," *Project Appraisal* **12**, 219–232 (1997a).

Lawrence, D. P., "Integrating Sustainability and Environmental Impact Assessment," *Environmental Management* **21**, 23–42 (1997b).

Lawrence, D. P., "EIA—Do We Know Where We Are Going?" *Impact Assessment* **15**, 3–13 (1997c).

Lawrence, D. P., "The Need for EIA Theory-building," *Environmental Impact Assessment Review* **17**, 79–107 (1997d).

Lawrence, D. P., *Multi-jurisdictional Environmental Assessments*, Canadian Environmental Assessment Agency (CEAA), Hull, Québec, 1999.

Lawrence, D. P., "Choices for EIA Process Design and Management," *Journal of Environmental Assessment and Policy Management* **3**, 437–464 (2001).

Lawrence, D. P., *The Significance of Social and Economic Impacts in Environmental Assessment*, Canadian Environmental Assessment Agency (CEAA), Hull, Québec, 2004.

Lawrence, D. P., "Environmental Impact Assessment: Practical Solutions to Recurrent Problems, Part 1," *Environmental Quality Management* Summer, 39–62 (2005a).

Lawrence, D. P., "Environmental Impact Assessment: Practical Solutions to Recurrent Problems, Part 2," *Environmental Quality Management* Autumn, 27–51 (2005b).

Lawrence, D. P., "Impact Significance Determination—Back to Basics," *Environmental Impact Assessment Review* **27**, 755–769 (2007a).

Lawrence, D. P., "Impact Significance Determination—Designing an Approach," *Environmental Impact Assessment Review* **27**, 730–754 (2007b).

Lawrence, D. P., "Impact Significance Determination—Pushing the Boundaries," *Environmental Impact Assessment Review* **27**, 770–788 (2007c).

Lawrence, P., "Equator Principles: Or How I Learned to Stop Worrying and Love Sustainability," *Impact Assessment and Project Appraisal* **27**, 3–6 (2009).

Lawrence, R. L., S. E. Daniels, and G. H. Stankey, "Procedural Justice and Public Involvement in Natural Resources Decision Making," *Society and Natural Resources* **10**, 577–589 (1997).

Laws, D., "The Practice of Fairness," *Environmental Impact Assessment Review* **16**, 65–70 (1996).

Lee, B., L. Hawath, and C. Brunk, "Value and Science in Impact Assessment," *Environments* **23**, 93–100 (1992).

Lee, K. N., "Appraising Adaptive Management," *Conservation Ecology* **3**, 2 (1999).

Lee, N., in N. Lee and C. George, eds., "Reviewing the Quality of Environmental Assessments," *Environmental Impact Assessment in Developing and Transitional Countries—Principles, Methods and Practice*, John Wiley & Sons, Chichester, 2000.

Lee, N., "Bridging the Gap between Theory and Practice in Integrated Assessment," *Environmental Impact Assessment Review* **26**, 57–78 (2006).

Lee, N., and C. Kirkpatrick, "Evidence-based Policy-making in Europe: An Evaluation of European Commission Integrated Impact Assessment," *Impact Assessment and Project Appraisal* **24**, 23–33 (2006).

Lehto, J., and A. Ritsatakis, "Health Impact Assessment as a Tool for Intersectoral Health Policy," *Discussion Paper for the Conference on Health Impact Assessment: From Theory to Practice*, European Centre for Health Policy, WHO Regional Office for Europe, Gothenburg, 1999.

Lein, J. K., "Expressing Environmental Risk Using Fuzzy Variables: A Preliminary Examination," *The Environmental Professional* **14**, 257–267 (1992).

Leith, B., "The Social Cost of Sustainability," *Alternatives* **21**, 18–24 (1995).

Lemon, M., et al., "Understanding Perceptions of Change: A Pathways Contribution to Community Consultation and Environmental Decision-making," *Journal of Environmental Assessment Policy and Management* **6**, 189–211 (2004).

Lemonick, M. D., "Climate Heretic—Can We Have a Civil Conversation About Climate?" *Scientific American* **303**, 78–83 (2010).

Lemons, J., and D. A. Brown, "The Role of Science in the Decision to Site a High-level Nuclear Waste Repository at Yucca Mountain, Nevada, USA," *The Environmentalist* **10**, 3–24 (1990).

Letsela, L., A. Pelser, and M. Seaman, "A Participatory Sustainability Assessment Process for Biodiversity Conservation in Lesotho," *Journal of Environmental Assessment Policy and Management* **12**, 107–129 (2010).

Levins, R., in J. A. Tickner, ed., "Whose Scientific Method? Scientific Methods in a Complex World," *Environmental Science and Preventive Public Policy*, Island Press, Washington, 2003.

Lien, K., et al., "Cumulative Effects at the National Energy Board of Canada," *Presentation to IAIA11 Conference—Impact Assessment and Responsible Development for Infrastructure, Business and Industry*, Puebla, Mexico, 2011.

Lima, M. L., and S. Marques, "Toward a Successful Social Impact Assessment Follow-up: A Case Study of Psychosocial Monitoring of a Solid Waste Incinerator in the North of Portugal," *Impact Assessment and Project Appraisal* **23**, 223–233 (2005).

Lindblom, C., *The Intelligence of Democracy*, Free Press, New York, 1965.

Liu, F., "Dynamics and Causation of Environmental Equity, Locally Unwanted Land Uses, and Neighborhood Changes," *Environmental Management* **21**, 643–656 (1997).

Lobos, V., and M. Partidário, "Rationale Behind the Current Practice in Strategic Environmental Assessment," *Presentation to IAIA10 Conference Proceedings—The Role of Impact Assessment in Transitioning to the Green Economy*, 30th Annual Meeting of the International Association for Impact Assessment, IAIA, Geneva, Switzerland, 2010.

Lockie, S., "SIA in Review: Setting the Agenda for Impact Assessment in the 21st Century," *Impact Assessment and Project Appraisal* **19**, 277–287 (2001).

Lockie, S., et al., "Democratization Versus Engagement? Social and Economic Impact Assessment and Community Participation in the Coal Mining Industry of the Bowen Basin, Australia," *Impact Assessment and Project Appraisal* **26**, 177–187 (2008).

Looney, P. B., "What Kind of Environmental Professional Are You?" *Environmental Practice* **13**, 167–171 (2011).

Lord, F., "Understanding Social Impacts Using New Variables and a Causal Diagram in New England Fisheries," *Impact Assessment and Project Appraisal* **29**, 59–68 (2011).

Lou, D. K., and E. J. Rykiel, Jr., "Integrated Resource Management Systems: Coupling Expert Systems with Database Management and Geographic Information Systems," *Environmental Management* **16**, 167–177 (1992).

Lowry, K., P. Adler, and N. Milner, "Participating the Public: Group Process, Politics and Planning," *Journal of Planning Education and Research* **16**, 177–187 (1997).

Lundberg, K., et al., "SEA Monitoring in Swedish Regional Transport Infrastructure Plans—Improvement Opportunities Identified in Practical Experience," *Environmental Impact Assessment Review* **30**, 400–406 (2010).

Luther, L., *The National Environmental Policy Act: Streamlining NEPA*, CRS Report for Congress, Order Code RL33267, 2007.

Lyhne, I., "SEA and Transboundary Energy Infrastructure," *Presentation to IAIA09 Conference Proceedings—Impact Assessment and Human Well-being*, International Association for Impact Assessment (IAIA), Accra, Ghana, 2009.

Macintosh, A., "The Australian Government's Environmental Impact Assessment (EIA) Regime: Using Surveys to Identify Proponent Views on Cost-effectiveness," *Impact Assessment and Project Appraisal* **28**, 175–188 (2010).

Mackenbach, J. P., et al., in J. Kemm, J. Parry, and S. Palmer, eds., "Health Inequities and HIA," *Health Impact Assessment*, Oxford University Press, Oxford, UK, 2004, pp. 25–37.

Mackenzie Gas Project Joint Review Panel, *Determination on Sufficiency*, July 18, 2005.

Mackenzie Gas Project Joint Review Panel, *Foundation for a Sustainable Northern Future*, December 2009. Available at http://www.ceaa-acee.gc.ca/default.asp?lang=En&n=71B5E4CF-1

MacNiven, C. D., *The Moral Question: Ethical Theory*, Ontario Educational Communications Authority, Toronto, 1982.

Malik, M., and R. Bartlett, "Formal Guidance For the Use of Science in EIA: Analysis of Agency Procedures for Implementing NEPA," *The Environmental Professional* **15**, 34–45 (1993).

Mandelbaum, S. J., "Telling Stories," *Journal of Planning Education and Research* **10**, 209–214 (1991).

Mandelik, Y., T. Dyan, and E. Feitelson, "Issues and Dilemmas in Ecological Scoping: Scientific, Procedural and Economic Perspectives," *Impact Assessment and Project Appraisal* **23**, 55–63 (2005).

Manou, D., and J. Papathanasiou, "Exploring the Potential Failure of the Regulatory Framework and Management Tools Which Govern the Conservation of Biodiversity: The Case of Artificial Lake Kerkiki in Greece," *Journal of Environmental Assessment and Management* **11**, 213–243 (2009).

Manring, N., P. C. West, and P. Bidol, "Social Impact Assessment and Environmental Conflict Management: Potential for Integration and Application," *Environmental Impact Assessment Review* **10**, 247–252 (1990).

March, F., *NEPA Effectiveness—Measuring the Process*, Government Institutes, Inc., Rockville, MD, 1998.

Margerum, R. D., "Integrated Approaches to Environmental Planning and Management," *Journal of Planning Literature* **11**, 459–475 (1997).

Marris, P., *The Politics of Uncertainty: Attachment in Public and Private Life*, Routledge, London, 1996.

Marshall, R., "Environmental Impact Assessment Follow-up and Its Benefits to Industry," *Impact Assessment and Project Appraisal* **23**, 191–196 (2005).

Marshall, R., J. Arts, and A. Morrison-Saunders, "International Principles for Best Practice EIA Follow-up," *Impact Assessment and Project Appraisal* **23**, 175–181 (2005).

Mas, D. M. L., "A Comparison of Three Key Elements of National and State Environmental Policy Acts," *Environmental Practice* **5**, 349–362 (2003).

Maser, C., *Resolving Environmental Conflict*, St. Lucie Press, Delray Beach, FL, 1996.

Mayda, J., "Reform Impact Assessment: Issues, Premises and Elements," *Impact Assessment* **14**, 87–96 (1996).

Maynes, C., *Public Consultation: A Citizens Handbook*, Ontario Environmental Network, Toronto, ON, 1989.

Mazmanion, D. A., and D. Morell, in N. J. Vig and M. E. Kraft, eds., "The 'NIMBY' Syndrome: Facility Siting and the Failure of Democratic Discourse," *Environmental Policy in the 1990s*, 2nd ed., Congressional Quarterly Inc., Washington, DC, 1994.

McArdie, J., "EPA: Agency Creates Mapping Tool Aimed at Identifying 'Environmental Justice' Hot Spots," *The NAEP National Desk*, August (2012).

McBean, G. A., "Integrating Disaster Risk Reduction Towards Sustainable Development," *Current Opinion in Environmental Sustainability* **4**, 122–127 (2012).

McCabe, M., and B. Sadler, *Studies of EIA Practice in Developing Countries*, United Nations Environment Program (UNEP), Geneva, Switzerland, undated.

McCaig, K., "Canadian Insights: The Challenge of an Integrated Environmental Assessment Framework," *Environmental Impact Assessment Review* **25**, 737–746 (2005).

McCarthy, M., and M. Utley, in J. Kemm, J. Parry, and S. Palmer, eds., "Quantitative Approaches to HIA," *Health Impact Assessment*, Oxford University Press, Oxford, UK, 2004.

McClendon, B. W., "The Paradigm of Empowerment," *Journal of the American Planning Association* **59**, 145–147 (1993).

McCluskey, D., and E. João, "The Promotion of Environmental Enhancement in Strategic Environmental Assessment," *Environmental Impact Assessment Review* **31**, 344–351 (2011).

McGlennon, J. A. S., and L. Susskind, "Responsibility, Accountability, and Liability in the Conduct of Environmental Negotiations," *The Place of Environmental Negotiations in Environmental Assessment*, Canadian Environmental Assessment Research Council, Hull, Québec, Canada, undated.

McIntyre, L., and M. Petticrew, *Methods of Health Impact Assessment: A Literature Review*, Occasional Paper, Medical Research Council Social and Public Health Services Unit, University of Glasgow, Glasgow, Scotland, 1999.

McLain, R. J., and R. G. Lee, "Adaptive Management: Promises and Pitfalls," *Environmental Management* **20**, 437–448 (1996).

McLauchlan, A., and E. João, "The Inherent Tension Arising from Attempting to Carry Out Strategic Environmental Assessments of All Policies, Plans and Programmes," *Environmental Impact Assessment Review* **36**, 23–33 (2012).

McLoughlin, J. B., *Urban and Regional Planning: A Systems Approach*, Faber and Faber, London, 1969.

McNeil, D., and P. Freiberger, *Fuzzy Logic*, Simon and Schuster, New York, 1994.

Mebratu, D., "Sustainability and Sustainable Development: Historical and Conceptual Review," *Environmental Impact Assessment Review* **18**, 493–520 (1998).

Melo-Escrihuela, C., "Promoting Ecological Citizenship: Rights, Duties and Political Agency," *ACME Editorial Collective*, Keele University, Staffordshire, UK, 2008.

Menard, L., *Pragmatism—A Reader*, Vintage Books, New York, 1997.

Meredith, T., "Environmental Impact Assessment, Cultural Diversity, and Sustainable Rural Development," *Environmental Impact Assessment Review* **12**, 125–138 (1992).

Metzger, J. T., "The Theory and Practice of Equity Planning: An Annotated Bibliography," *Journal of Planning Literature* **11**, 112–126 (1996).

Meynell, P. J., "Use of IUCN Red Listing Process as a Basis for Assessing Biodiversity Threats and Impacts in Environmental Impact Assessment," *Impact Assessment and Project Appraisal* **23**, 65–72 (2005).

Mezirow, J., "Understanding Transformation Theory," *Adult Education Quarterly* **44**, 222–232 (1994).

Michael, D. N., "Forecasting and Planning in an Incoherent Context," *Technological Forecasting and Social Change* **36**, 79–87 (1989).

Miller, A., "The Role of Analytical Science in Natural Resource Decision Making," *Environmental Management* **17**, 563–574 (1993).

Milner, S. J., et al., "Integrated Impact Assessment in the UK—Use, Efficacy and Future Development," *Environmental Impact Assessment Review* **25**, 47–61 (2005).

Mindell, J., M. Joffe, and E. Ison, in J. Kemm, J. Parry, and S. Palmer, eds., "Planning an HIA," *Health Impact Assessment*, Oxford University Press, Oxford, UK, 2004.

Mintzberg, H., *The Rise and Fall of Strategic Planning*, Free Press, New York, 1994.

Mishra, P. K., "Integrated Impact Assessment Model for Explaining Differential Impacts of Watershed Development Projects," *Impact Assessment and Project Appraisal* **27**, 175–184 (2009).

Mittelmark, M. B., D. E. Gillis, and B. Hsu-Hage, in J. Kemm, J. Parry, and S. Palmer, eds., "Community Development: The Role of HIA," *Health Impact Assessment*, Oxford University Press, Oxford, UK, 2004.

Mittelstaedt, G., T. Williams, and K. Ordon, in R. Clark and L. Canter, eds., "NEPA and Tribal Matters," *Environmental Policy and NEPA—Past, Present and Future*, St. Lucie Press, Boca Raton, FL, 1997.

Mittendorfer, C., ed., Die Strategische Umweltprüfung im Verkehrsbereich. Tagungsband. Informationen zur Umweltpolitk Nr. 176 der Bundeskammer für Arbeiter und Angestellte. 2008. Available online: http://www.arbeiterkammer.at/bilder/d73/InformationenzurUmweltpolitik_176.pdf [last accessed October 15, 2012].

Moles, R. et al., "Practical Appraisal of Sustainable Development—Methodologies for Sustainability Measurement at Settlement Level," *Environmental Impact Assessment Review* **28**, 144–165 (2008).

Momtaz, S., "The Practice of Social Impact Assessment in a Developing Country: The Case of Environmental and Social Impact Assessment of Khulna–Jessore Drainage Rehabilitation Project in Bangladesh," *Impact Assessment and Project Appraisal* **21**, 125–132 (2003).

Montague, P., "Reducing the Harms Associated with Risk Assessments," *Environmental Impact Assessment Review* **24**, 733–748 (2004).

Moore, C. W., *Negotiating Bargaining and Conflict Management*, Prepared by CDR Associates for US Army Corps of Engineers, Huntsville, Alabama, 1986.

Moore, C. W., in J. L.Creighton, C. M. Dunning, J. D. Priscoli, and J. D. Ayres, eds., "Negotiations," *Public Involvement and Dispute Resolution—A Reader on the Second Decade of Experience at the Institute for Water Resources*, Institute for Water Resources, US Army Corps of Engineers, Alexandria, VA, 1999.

Moore, C. W., and P. Woodrow, in L. Susskind, S. McKearnen, and J. Thomas-Lamar, eds., *The Consensus-Building Handbook—A Comprehensive Guide to Reaching Agreements*, Sage, Thousand Oaks, California, 1997.

Moore, S. N., "A View of the History of the National Environmental Policy Act," *NAEP Newsletter* **17**, 8–9 (1992).

Morell, D., "Siting and the Politics of Equity," *Hazardous Waste* **1**, 555–571 (1984).

Moreno, R., and S. Catchpole, "EIA Review Methodology," *Presentation to the IAIA12 Conference Proceedings, Energy Future—The Role of Impact Assessment*, 32nd Annual Meeting of the International Association for Impact Assessment, Porto, Portugal, 2012.

Morgan, R. K., *Environmental Impact Assessment—A Methodological Perspective*, Kluwer Academic Press, Boston, 1998.

Morgan, R. K., "From the President," *IAIA Newsletter* **15**, 2 (2003a).

Morgan, R. K., "Health Impact Assessment: The Wider Context," *Bulletin of the World Health Organization* **81**, 390 (2003b).

Morgan, R. K., "Solutions to Environmental Impact Assessment Failure Require More than Common Sense," *Impact Assessment and Project Appraisal* **24**, 15–16 (2006).

Morgan, R. K., "Health and Impact Assessment: Are We Seeing Closer Integration?" *Environmental Impact Assessment Review* **31**, 404–411 (2011).

Morgan, R. K., "Environmental Impact Assessment: The State of the Art," *Impact Assessment and Project Appraisal* **30**, 5–14 (2012).

Morgan, R. K., et al., "Practitioners, Professional Cultures, and Perceptions of Impact Assessment," *Environmental Impact Assessment Review* **32**, 11–24 (2012).

Morganstern, R. D., K. Ahmed, and E. Sánchez-Triana, eds., "Toward Environmental Priority Setting in Development," *Strategic Environmental Assessment for Policies*, The World Bank, Washington, DC, 2008.

Morris, P., and R. Thérivel, *Methods of Environmental Impact Assessment*, Routledge, New York, 1995 and 2009.

Morrison-Saunders, A., and J. Arts, "Learning from Experience: Emerging Trends in Environmental Assessment Follow-up," *Impact Assessment and Project Appraisal* **23**, 170–174 (2005).

Morrison-Saunders, A., and M. Bailey, "Practitioner Perspectives on the Role of Science in Environmental Impact Assessment," *Environmental Management* **31**, 683–695 (2003).

Morrison-Saunders, A., and M. Bailey, "Appraising the Role of Relationships Between Regulators and Consultants for Effective EIA," *Environmental Impact Assessment Review* **29**, 284–294 (2009).

Morrison-Saunders, A., J. Baker, and J. Arts, "Lessons from Practice: Towards Successful Follow up," *Impact Assessment and Project Appraisal* **21**, 43–56 (2003).

Morrison-Saunders, A., and G. Early, "What is Necessary to Ensure Natural Justice in Environmental Impact Assessment Decision Making?" *Impact Assessment and Project Appraisal* **26**, 29–42 (2008).

Morrison-Saunders, A., and T. B. Fischer, in W. R. Sheate, ed., "What is Wrong with EIA and SEA Anyway? A Sceptic's Perspective on Sustainability Assessment," *Tools, Techniques and Approaches for Sustainability*, World Scientific, New Jersey, 2010.

Morrison-Saunders, A., and N. Hodgson, "Applying Sustainability Principles in Practice: Guidance for Assessing Individual Proposals," *Presentation to IAIA09 Conference Proceedings—Impact Assessment and Human Well-being,* 29th Annual Conference of the International Association for Impact Assessment (IAIA), Accra, Ghana, 2009.

Morrison-Saunders, A., and B. Sadler, "The Art and Science of Impact Assessment: Results of a Survey of IAIA Members," *Impact Assessment and Project Appraisal* **28**, 77–82 (2010).

Morrison-Saunders, A., and R. Thérivel, "Sustainability Integration and Assessment," *Journal of Environmental Assessment Policy and Management* **8**, 281–298 (2006).

Morrone, M., "Walking A Tight Rope: Planning and the Environmental Movement," *Journal of Planning Literature* **7**, 100–107 (1992).

Mosley, D., C. Moore, M. Stagle, and D. Burns, in J. L. Creighton, C. M. Dunning, D. Priscoli, and D. B. Ayres, eds., "Partnering Win-Win Strategic Management in Action," *Public Involvement and Dispute Resolution—A Reader on the Second Decade of Experience at the Institute of Water Resources*, Institute of Water Resources, U.S. Army Corps of Engineers, Alexandria, VA, 1999.

Mostert, E., "Subjective Environmental Impact Assessment: Causes, Problems and Solutions," *Impact Assessment* **14**, 191–214 (1996).

Motion, J., "Participative Public Relations: Power to the People or Legitimacy for Government Discourse?" *Public Relations Review* **31**, 505–512 (2005).

Mulvihill, P. R., "Expanding the Scoping Community," *Environmental Impact Assessment Review* **23**, 39–49 (2003).

Mulvihill, P. R., and R. F. Keith, "Institutional Requirements for Adaptive EIS: The Kativik Environmental Quality Commission," *Environmental Impact Assessment Review* **9**, 399–412 (1989).

Nagel, J. H., *Participation*, Prentice-Hall Inc., Englewood Cliffs, 1987.

Neefjes, K., in B. Vira and R. Jeffery, eds., "Learning from Participatory Environmental Impact Assessment of Community-centred Development: The Oxfam Experience," *Analytical Issues in Natural Resources Management*, Palgrave, New York, 2001.

Negev, M., "Knowledge, Data and Interests: Challenges in Participation of Diverse Stakeholders in HIA," *Environmental Impact Assessment Review* **33**, 48–54 (2012).

Neuman, M., "Communicate This! Does Consensus Lead to Advocacy and Pluralism?" *Journal of Planning Education and Research* **19**, 343–350 (2000).

Newman, W. L., *Social Research Methods*, 3rd ed., Allyn and Bacon, Boston, 1997.

Ng, M. K., and D. L. L. Hui, "Practice Report: Sustainability Impact Assessment in Hong Kong and the Pearl River Delta: 'Both Necessary and Impossible'," *Impact Assessment and Project Appraisal* **25**, 233–238 (2007).

Nicolis, G., and I. Prigogine, *Exploring Complexity—An Introduction*, Freeman and Company, New York, 1989.

Nilsson, M., "On (Re)focusing SEA on the Natural Environment," *Environmental Impact Assessment Review* **29**, 76–77 (2009).

Nilsson, M., and H. Dalkmann, "Decision Making and Strategic Environmental Assessment," *Journal of Environmental Assessment Policy and Management* **3**, 305–328 (2001).

Nilsson, M., and H. Dalkmann, in W. R. Sheate, ed., "Decision Making and Strategic Environmental Assessment?" *Tools, Techniques and Approaches for Sustainability*, World Scientific, New Jersey, 2010.

Nilsson, M., et al., "Analytical Framework and Tool Kit for SEA Follow up," *Environmental Impact Assessment Review* **29**, 186–199 (2009).

Nitz, T., and A. L. Brown, "SEA Must Learn How Policy Making Works," *Journal of Environmental Assessment Policy and Management* **3**, 329–342 (2001).

Noble, B. F., "Strategic Environmental Assessment: What Is It and What Makes It Strategic?" *Journal of Environmental Assessment Policy and Management* **2**, 203–224 (2000a).

Noble, B. F., "Strengthening EIA Through Adaptive Management: A Systems Perspective," *Environmental Impact Assessment Review* **20**, 97–112 (2000b).

Noble, B., "Auditing Strategic Environmental Assessment Practice in Canada," *Journal of Environmental Assessment Policy and Management* **5**, 127–148 (2003).

Noble, B., "Strategic Environmental Assessment Quality Assurance: Evaluating and Improving the Consistency of Judgments in Assessment Panels," *Environmental Impact Assessment Review* **24**, 3–25 (2004a).

Noble, B., "A State-of-Practice Survey of Policy, Plan and Program Assessment in Canadian Provinces," *Environmental Impact Assessment Review* **24**, 351–361 (2004b).

Noble, B., "Strategic Environmental Assessment Quality Assurance: Evaluating and Improving the Consistency of Judgments in Assessment Panels," *Environmental Impact Assessment Review* **24**, 3–25 (2004c).

Noble, B., "Strategic Approaches to Regional Cumulative Effects Assessment: A Case Study of the Great Sand Hills, Canada," *Impact Assessment and Project Appraisal* **26**, 78–90 (2008).

Noble, B., "Promise and Dismay: The State of Strategic Environmental Assessment Systems and Practices in Canada," *Environmental Impact Assessment Review* **29**, 66–75 (2009a).

Noble, B., *Introduction to Environmental Impact Assessment*, 2nd ed., Oxford University Press, Oxford, UK, 2009b.

Noble, B., and J. Birk, "Comfort Monitoring? Environmental Assessment Follow-up Under Community-negotiated Environmental Agreements," *Environmental Impact Assessment Review* **31**, 17–24 (2011).

Noble, B., and J. Bronson, "Practitioner Survey of the State of the Art of Health Integration in Environmental Assessment: The Case of Northern Canada," *Environmental Impact Assessment Review* **26**, 410–424 (2006).

Noble, B., and J. H. Gunn, in K. S. Hanna, ed., "Strategic Environmental Assessment," *Environmental Impact Assessment—Practice and Participation*, Oxford University Press, Oxford, UK, 2009.

Noble, B. F., J. Gunn, and J. Martin, "Survey of Current Methods and Guidance for Strategic Environmental Assessment," *Impact Assessment and Project Appraisal* **30**, 139–147 (2012).

Noble, B., and K. Storey, "Toward Increasing the Utility of Follow-up in Canadian EIA," *Environmental Impact Assessment Review* **25**, 163–180 (2005).

Noorbakhsh, F., and S. Ranjan, "A Model of Sustainable Development: Integrating Environmental Assessment and Project Planning," *Impact Assessment and Project Appraisal* **17**, 283–294 (1999).

Nooteboom, S., "Impact Assessment Procedures for Sustainable Development: A Complexity Theory Perspective," *Environmental Impact Assessment Review* **27**, 645–665 (2007).

Nooteboom, S., and K. Wieringa, "Comparing Strategic Environmental Assessment and Integrated Environmental Assessment," *Journal of Environmental Assessment Policy and Management* **1**, 441–457 (1999).

Northern and Yorkshire Public Health Observatory (NYPHO), *An Overview of Health Impact Assessment*, Occasional Paper # 1, NYPHO, Stockton on Tees, UK, May 2001.

Novek, J., "Environmental Impact Assessment and Sustainable Development: Case Studies of Environmental Conflict," *Society and Natural Resources* **8**, 145–159 (1995).

Nykvist, B., and M. Nilsson, "Are Impact Assessment Procedures Actually Promoting Sustainable Development? Institutional Perspectives on Barriers and Opportunities Found in the Swedish Committee System," *Environmental Impact Assessment Review* **29**, 15–24 (2009).

O'Brian, M., in J. A. Tickner, ed., "Science in the Service of Good: The Precautionary Principle and Positive Goals," *Environmental Science and Preventive Public Policy*, Island Press, Washington, 2003.

O'Faircheallaigh, C., "Environmental Agreements, EIA Follow-up and Aboriginal Participation in Environmental Management: The Canadian Experience," *Environmental Impact Assessment Review* **27**, 319–342 (2006).

O'Faircheallaigh, C., "Effectiveness in Social Impact Assessment: Aboriginal Peoples and Resource Development in Australia," *Impact Assessment and Project Appraisal* **27**, 95–110 (2009).

O'Faircheallaigh, C., "Public Participation and Environmental Impact Assessment: Purpose, Implications and Lessons for Public Policy Making," *Environmental Impact Assessment Review* **30**, 19–27 (2010).

Office of Management and Budget and the Secretariat General of the European Commission, *Review of the Application of EU and US Regulatory Impact Assessment Guidelines on the Analysis of Impacts of International Trade and Investment—Final Report and Conclusions*, Brussels/WA, 2008.

Office of the Auditor General of Canada (AGC), *Report of the Commissioner of the Environment and Sustainable Development to the House of Commons—Chapter 4—Assessing the Environmental Effects of Policies, Plans and Programs*, Minister of Public Works and Government Services Canada, Ottawa, 2004.

Office of the Auditor General of Canada (AGC), *Status Report of the Commissioner of the Environment and Sustainable Development to the House of Commons—Chapter 9—Strategic Environmental Assessment*, Minister of Public Works and Government Services Canada, Ottawa, 2008.

Office of the Auditor General of Canada (AGC), *Report of the Commissioner of the Environment and Sustainable Development—Chapter 1—Applying the Canadian Environmental Assessment Act*, Minister of Public Works and Government Services Canada, Ottawa, 2009.

Offringa, P., in R. Clark and L. Canter, eds., "Creating a User-friendly NEPA," *Environmental Policy and NEPA—Past, Present and Future*, St. Lucie Press, Boca Raton, FL, 1997.

Okello, N., et al., "The Doing and Un-doing of Public Participation During Environmental Impact Assessments in Kenya," *Impact Assessments and Project Appraisal* **27**, 217–226 (2009).

Orenstein, M., et al., "Case Study of an Integrated Assessment: Shell's North Field Test in Alberta, Canada," *Impact Assessment and Project Appraisal* **28**, 147–157 (2010).

Organization for Economic Cooperation and Development (OECD), *Applying Strategic Environmental Assessment—Good Practice Guidance for Development Cooperation, OECD—DAC Guidelines and Reference Series*, OECD, Paris, France, 2006.

O'Riordan, T., ed., *Environmental Science for Environmental Management*, John Wiley & Sons, New York, 1995.

O'Riordan, T., and J. Cameron, *Interpreting the Precautionary Principle*, Earthscan, London, 1994.

Ortolano, L., and A. Shepherd, in F. Vanclay and D. A. Bronstein, eds., "Environmental Impact Assessment," *Environmental Impact Assessments*, John Wiley & Sons, New York, 1995.

Ortolano, L., "Controls on Project Proponents and Environmental Impact Assessment Effectiveness," *The Environmental Professional* **15**, 352–363 (1993).

Ortolano, L., in K. Ahmed and E. Sánchez-Triana, eds., "Policy-level Strategic Environmental Assessments: Process Integration and Incentives of Policy Proponents," *Strategic Environmental Assessment for Policies*, The World Bank, Washington, DC, 2008.

Orwell, D., *Apollo's Arrow—The Science of Prediction and the Future of Everything*, Harper Perennial, Toronto, 2007.

Ostrom, E., "A Behavioral Approach to the Rational Choice Theory of Collective Action: Presidential Address, American Political Science Association," *American Political Science Review* **92**, 1–22 (1998).

Ozawa, C. P., *Recasting Science: Consensual Procedures in Public Policy Making*, Westview Policy Making, Boulder, CO, 1991.

Paci, C., A. Tobin, and P. Robb, "Reconsidering the Canadian Environmental Impact Assessment Act: A Place for Traditional Environmental Knowledge," *Environmental Impact Assessment Review* **22**, 118–128 (2002).

Page, J., "Make It Easy on Your Readers: Ideas on Environmental Impact Document Focus, Organization and Style," *Impact Assessment and Project Appraisal* **24**, 235–245 (2006).

Pallen, D., *Environmental Assessment Manual for Community Development Projects*, Canadian International Development Agency (CIDA), Asia Branch, Ottawa, 1996.

Palmer, S., in J. Kemm, J. Parry, and S. Palmer, eds., "Causal Mechanisms for HIA: Learning from Epistemology," *Health Impact Assessment*, Oxford University Press, Oxford, UK, 2004.

Parashar, A., R. Paliwal, and P. Rambabu, "Utility of Fuzzy Cross-impact Simulation in Environmental Assessment," *Environmental Impact Assessment Review* **17**, 427–447 (1997).

Parenteau, R., *Public Participation in Environmental Decision-making*, Federal Environmental Assessment Review Office, Hull, Québec, 1988.

Parkin, J. B., "When Science Is Not Enough: A Case Study in Social Impact Mitigation," *Impact Assessment* **14**, 321–327 (1996).

Parry, J., and J. Kemm, in B. Sadler et al., eds., "Future Directions for HIA," *Handbook of Strategic Environmental Assessment*, Earthscan, London, 2004.

Parry, J., and A. Stevens, "Prospective Health Impact Assessment: Pitfalls, Problems and Possible Ways Forward," *BMJ* **323**, 1177–1182 (2001).

Partidário, M. R., "Strategic Environmental Assessment: Key Issues Emerging from Recent Practice," *Environmental Impact Assessment Review* **16**, 31–55 (1996).

Partidário, M. R., "Improving EIA to Enable SEA," *Paper Presented at the Annual Meeting of the International Association for Impact Assessment,* Stavanger, Norway, May 23–26, 2006.

Partidário, M. R., "Scales and Associated Data—What is Enough for SEA Needs?" *Environmental Impact Assessment Review* **27**, 460–478 (2007).

Partidário, M. R., in B. Sadler, et al., eds., "SEA Process Development and Capacity-building," *Handbook of Strategic Environmental Assessment*, Earthscan, London, 2011.

Partidário, M. R., *Strategic Environmental Assessment (SEA)—Current Practices, Future Demands and Capacity-building Needs—Course Manual*, International Association for Impact Assessment—IAIA Training Courses, Lisbon, Portugal, undated.

Partidário, M. R., and J. Arts, "Exploring the Concept of Strategic Environmental Assessment Follow-up," *Impact Assessment and Project Appraisal* **23**, 246–257 (2005).

Partidário, M. R., and M. Coutinho, "The Lisbon New International Airport: The Story of a Decision-making Process and the Role of Strategic Environmental Assessment," *Environmental Impact Assessment Review* **31**, 360–367 (2011).

Partidário, M. R., and L. Wilson, in B. Sadler, et al., eds., "Professional and Institutional Capacity-building for SEA," *Handbook of Strategic Environmental Assessment*, Earthscan, London, 2011.

Pateman, C., *Participation and Democratic Theory*, Cambridge University Press, London, 1970.

Patterson, M. E., and D. R. Williams, "Paradigms and Problems: The Practice of Social Science in Natural Resource Management," *Society and Natural Resources* **11**, 279–295 (1998).

Patton, B. R., K. Giffin, and E. N. Patton, *Decision-making Group Interaction*, Harper Collins, New York, 1989.

Patton, C. V., and D. S. Sawicki, *Basic Methods of Policy Analysis and Planning*, Prentice Hall, New Jersey, 1993.

Peche, R., and E. Rodríguez, "Environmental Impact Assessment by Means of a Procedure Based on Fuzzy Logic: A Practical Approach," *Environmental Impact Assessment Review* **31**, 87–96 (2011).

Peirson-Smith, T. J., "Embracing Social Media to Enhance Public Participation," *Presentation to the IAIA12 Conference Proceedings, Energy Future—The Role of Impact Assessment,* 32nd Annual Meeting of the International Association for Impact Assessment, Porto, Portugal, 2012.

Peltonen, L., and R. Sairinen, "Integrated Impact Assessment and Conflict Management in Urban Planning: Experiences from Finland," *Environmental Impact Assessment Review* **30**, 328–337 (2010).

Pennock, M., and K. Uma, "Gross National Happiness as a Framework for Health Impact Assessment," *Environmental Impact Assessment Review* **31**, 61–65 (2011).

Perdicoúlis, A., and J. Glasson, "Causal Networks in EIA," *Environmental Impact Assessment Review* **26**, 553–569 (2006).

Perdicoúlis, A., and J. Glasson, "The Causality Premise of EIA in Practice," *Impact Assessment and Project Appraisal* **27**, 247–250 (2009).

Perdicoúlis, A., et al., "The Handling of Causality in SEA Guidance," *Environmental Impact Assessment Review* **27**, 176–187 (2007).

Persson, A., and M. Nilsson, "Towards a Framework for SEA Follow-up: Theoretical Issues and Lessons from Policy Evaluation," *Journal of Environmental Assessment Policy and Management* **9**, 473–496 (2007).

Persson, J., "Theoretical Reflections on the Connection Between Environmental Assessment Methods and Conflict," *Environmental Impact Assessment Review* **26**, 605–617 (2006).

Petäjäjärvi, R., "Follow-up of Socio-economic Aspects of a Road Project in Finland," *Impact Assessment and Project Appraisal* **23**, 234–240 (2005).

Peter, F., in E. N. Zalta, ed., "Political Legitimacy," *Stanford Encyclopedia of Philosophy*, Stanford University, Stanford, CN, 2010.

Peterlin, M., B. C. Kross, and B. Kontic, "Information in an EIA Process and the Influence Thereof on Public Opinion," *Journal of Environmental Assessment Policy and Management* **8**, 183–204 (2006).

Peterlin, M., B. C. Kross, and B. Kontic, "A Method for the Assessment of Changes in Environmental Perception During an EIA Process," *Environmental Impact Assessment Review* **28**, 533–545 (2008).

Peterson, K., "The Role and Value of Strategic Environmental Assessment in Estonia," *Impact Assessment and Project Appraisal* **22**, 159–165 (2004).

Péti, M., "A Territorial Understanding of Sustainability in Public Development," *Environmental Impact Assessment Review* **32**, 61–73 (2012).

Petticrew, M., et al., "Validating Health Impact Assessment: Prediction Is Difficult (Especially About the Future)," *Environmental Impact Assessment Review* **27**, 101–107 (2007).

Petticrew, M., S. Macintyre, and H. Thomson, in J. Kemm, J. Parry, and S. Palmer, eds., "Evidence and HIA," *Health Impact Assessment*, Oxford University Press, Oxford, UK, 2004.

Petts, J., ed., *Handbook of Environmental Impact Assessment—Process, Methods and Potential*, Blackwell Science, Oxford, UK, 1999.

Petts, J., "Barriers to Deliberative Participation in EIA: Learning from Waste Policies, Plans and Projects," *Journal of Environmental Assessment Policy and Management* **5**, 269–293 (2003).

Pickering, A., *The Mangle of Practice—Time, Agency and Science*, University of Chicago Press, Chicago, 1995.

Pickvance, C., "The Rise and Fall of Urban Movements and the Role of Comparative Analysis," *Environment and Planning D: Society and Space* **3**, 31–53 (1985).

Pinho, P., S. McCallum, and A. S. Cruz, "A Critical Appraisal of EIA Screening Practice in EU Member States," *Impact Assessment and Project Appraisal* **28**, 91–107 (2010).

Pisani, J. A. D., and L. A. Sandham, "Assessing the Performance of SIA in the EIA Context: A Case Study of South Africa," *Environmental Impact Assessment Review* **26**, 707–724 (2006).

Pischke, F., and M. Cashmore, "Decision-oriented Environmental Assessment: An Empirical Study of Its Theory and Methods," *Environmental Impact Assessment Review* **26**, 643–662 (2006).

Põder, T., and T. Lukki, "A Critical Review of Checklist-based Evaluation of Environmental Impact Assessment," *Impact Assessment and Project Appraisal* **29**, 27–36 (2011).

Polido, A., and T. B. Ramos, "Evaluation of Strategic Environmental Scoping Effectiveness," *Presentation to IAIA11 Conference Proceedings, Impact Assessment and Responsible Development for Infrastructure, Business and Industry,* International Association for Impact Assessment (IAIA), Puebla, Mexico, 2011.

Pölönen, I., "Quality Control and the Substantive Influence of Environmental Impact Assessment in Finland," *Environmental Impact Assessment Review* **26**, 481–491 (2006).

Pölönen, I., P. Hokkanen, and K. Jalava, "The Effectiveness of the Finnish EIA System—What Works, What Doesn't, and What Could Be Improved," *Environmental Impact Assessment Review* **31**, 120–128 (2011).

Pope, J., "What's So Special About Sustainability Assessment?" *Journal of Environmental Assessment Policy and Management* **8**, v–x (2006).

Pope, J., and B. Dalal-Clayton, in B. Sadler, et al., eds., "From SEA to Sustainability Assessment?" *Handbook of Strategic Environmental Assessment*, Earthscan, London, 2011.

Pope, J., and W. Grace, "Sustainability Assessment in Context: Issues of Process, Policy and Governance," *Journal of Environmental Assessment Policy and Management* **8**, 373–398 (2006).

Pope, J., and D. Klass, "Decision Quality for Sustainability Assessment," *Presentation to IAIA10 Conference Proceedings—The Role of Impact Assessment in Transitioning to the Green Economy,* 30th Annual Meeting of the International Association for Impact Assessment, Geneva, Switzerland, 2010.

Pope, J., A. Morrison-Saunders, and D. Annandale, "Applying Sustainability Assessment Models," *Impact Assessment and Project Appraisal* **23**, 293–302 (2005).

Pope, J., et al., "Conceptualizing Sustainability Assessment," *Environmental Impact Assessment Review* **24**, 595–616 (2004).

Porritt, J., *Playing Safe: Science and the Environment*, Thames and Hudson Inc., New York, 2000.

Porter, A. L., "Impact Assessment Methodology is too Soft," *Impact Assessment and Project Appraisal* **24**, 86–88 (2006).

Posas, P. J., "The UKs Draft Guidance for Health in SEA in Light of HIA Community Priorities and the UNECE SEA Protocol," *Environmental Impact Assessment Review* **31**, 320–327 (2011).

Poulton, M. C., "Rationalism Versus Radicalism: The Debate Revisited," *Plan Canada* **30**, 37–40 (1990).

Powell, D., *The Role of Risk Perception and the Public in Decision-making on Risk, Solid and Hazardous Waste Management Series*, Publication WM-84-12, University of Toronto, Toronto, Canada, 1984.

Power, M., and S. M. Adams, "Perspectives on the Scientific Community on the Status of Ecological Risk Assessment," *Environmental Management* **21**, 803–830 (1997).

Power, M., G. Power, and D. G. Dixon, "Detection and Decision-making in Environmental Effects Monitoring," *Environmental Monitoring* **19**, 629–639 (1995).

Praxis, *Public Involvement—Planning and Implementing Public Involvement Programs*, Prepared for the Canadian Federal Environmental Assessment Review Office, Calgary, Alberta, 1988.

Presidential/Congressional Commission on Risk Assessment and Risk Management (PCCRARM), *Framework for Environmental Health Risk Management*, Final Report, Volume 1, Washington, DC, 1997a.

Presidential/Congressional Commission on Risk Assessment and Risk Management (PCCRARM), *Risk Assessment and Risk Management in Regulatory Decision-making*, Final Report, Volume 2, Washington, DC, 1997b.

Presidential Council on Sustainable Development (PCSD), *Lessons Learned from Collaborative Approaches*, Final Draft, New National Opportunities Task Force, Washington, DC, April 1997.

Priscoli, J. D., "The Enduring Myths of Public Involvement," *Citizen Participation* **3**, 5–7 (1982).

Priscoli, J. D., and P. Homenuck, in R. Lang, ed., "Consulting the Publics," *Integrated Approaches to Resource Planning and Management*, Banff Centre School of Management, 1986.

Priscoli, J. D., in J. L. Creighton, C. M. Dunning, J. D. Priscoli, and D. B. Ayres, eds., "Public Involvement; Conflict Management; and Dispute Resolution in Water Resources and Environmental Decision-making," *Public Involvement and Dispute Resolution—A Reader on the Second Decade of Experience at the Institute for Water Resources*, Institute for Water Resources, U.S. Army Corps of Engineers, Alexandria, VA, 1999.

Pritchard, D., "International Biodiversity-related Treaties and Impact Assessment—How Can They Help Each Other?" *Impact Assessment and Project Appraisal* **23**, 7–16 (2005).

Pullin, A. S., and T. M. Knight, "Support for Decision Making in Conservation Practice," *Journal of Nature Conservation* **11**, 83–90 (2003).

Purnama, D., "Review of a Transboundary Environmental Impact Assessment: A Case Study from the Timor Gap," *Impact Assessment and Project Appraisal* **22**, 17–35 (2004).

Puschchak, R., and A. M. Farrugia-Uhalde, in K. S. Hanna, ed., "Social Impact Assessment and High-level Radioactive Waste Disposal: The Canadian Concept and Aboriginal Responses," *Environmental Impact Assessment in Canada*, Oxford University Press, Oxford, 2009.

Putters, K., "HIA, the Next Step: Defining Models and Roles," *Environmental Impact Assessment Review* **25**, 693–701 (2005).

Quigley, R. J., and L. C. Taylor, "Evaluation as a Key Part of Health Impact Assessment," *Bulletin of the World Health Organization* **81**, 415–419 (2003).

Quijano, R. F., in J. A. Tickner, ed., "Elements of the Precautionary Principle," *Environmental Science and Preventive Public Policy*, Island Press, Washington, 2003.

Rabe, B. G., *Beyond NIMBY—Hazardous Waste Siting in Canada and the United States*, Brookings Institution, Washington, DC, 1994.

Radford, K. J., in B. H. Massam, ed., "Complex Decision Situations," *Complex Locational Problems: Interdisciplinary Approaches*, Institute for Social Research, York University, North York, Canada, 1988.

Raffensperger, C., and P. deFur, "A Paradigm Shift: Rethinking Environmental Decision Making and Risk Assessment," *A Paper Presented to Risk Assessment and Policy Association Meeting, Virginia, March 6, 1997.*

Rahm-Crites, L., in A. L. Porter and J. J. Fittipaldi, eds., "Risk Communication in Environmental Assessment," *Environmental Methods Review: Retooling Impact Assessment for the New Century*, Army Environmental Policy Institute and International Association for Impact Assessment, Fargo, ND, 1998.

Rajvanshi, A., et al., "Maximizing Benefits for Biodiversity: The Potential of Enhancing Strategies in Impact Assessment," *Impact Assessment and Project Appraisal* **29**, 181–193 (2011).

Ramos, T. B., S. Caeiro, and J. Joanaz de Melo, "Environmental Indicator Frameworks to Design and Assess Environmental Monitoring Programs," *Impact Assessment and Project Appraisal* **22**, 47–62 (2004).

Raphael, C., "Collaborative Sustainability Assessment," *Presentation to the IAIA12 Conference Proceedings, Energy Future, The Role of Impact Assessment*, 32nd Annual Meeting of the International Association for Impact Assessment, Porto, Portugal, 2012.

Ratanachai, C., "Environmental Impact Assessment as a Tool for Risk Management," *Toxicology and Industrial Health* **7**, 379–391 (1991).

Rauschmayer, F., and N. Risse, "A Framework for the Selection of Participatory Approaches for SEA," *Environmental Impact Assessment Review* **25**, 650–666 (2005).

Ravetz, J., "Integrated Assessment Models—from Global to Local," *Impact Assessment and Project Appraisal* **16**, 147–154 (1998).

Rawls, J., *A Theory of Justice*, Harvard University Press, Cambridge, 1971.

Rawls, J., *Justice as Fairness: A Restatement*, Harvard University Press, Cambridge, MA, 2001.

Reckhow, K. E., "Importance of Scientific Uncertainty in Decision Making," *Environmental Management* **18**, 161–166 (1994).

Reinke, D. C., and L. L. Swartz, *The NEPA Reference Guide*, Battelle Press, Columbia, Ohio, 1999.

Reinstein, M., *Research Report and Discussion Paper on the Regulation of Planners*, Prepared for the PIBC Professional Legislation and Certification Task Force, Reinstein and Associates, Victoria, BC, 2010.

Renda, A., *Impact Assessment in the EU—The State of the Art and the Art of the State*, Center for European Policy Studies, Brussels, 2006.

Retief, F., "Effectiveness of Strategic Environmental Assessment (SEA) in South Africa," *Journal of Environmental Assessment Policy and Management* **9**, 83–101 (2007a).

Retief, F., "A Quality and Effectiveness Review Protocol for Strategic Environmental Assessment (SEA) in Developing Countries," *Journal of Environmental Assessment Policy and Management* **9**, 443–471 (2007b).

Retief, F., C. Jones, and S. Jay, "The Emperor's New Clothes—Reflections on Strategic Environmental Assessment (SEA) Practice in South Africa," *Environmental Impact Assessment Review* **28**, 504–514 (2008).

Richardson, T., "Environmental Assessment and Planning Theory: Four Short Stories About Power, Multiple Rationality and Ethics," *Environmental Impact Assessment Review* **25**, 341–365 (2005).

Rickson, R. E., R. B. Burdge, T. Hundloe, and G. T. McDonald, "Institutional Constraints to Adoption of Social Impact Assessment as a Decision-making and Planning Tool," *Environmental Impact Assessment Review* **10**, 233–243 (1990a).

Rickson, R. E., J. S. Western, and R. J. Burdge, "Social Impact Assessment: Knowledge and Development," *Environmental Impact Assessment Review* **10**, 1–10 (1990b).

Ridgway, B., "The Project Cycle and the Role of EIA and EMS," *Journal of Environmental Assessment and Management* **3**, 93–106 (1999).

Ridgway, B., "Environmental Management System Provides Tools for Delivering on Environmental Impact Assessment Commitments," *Impact Assessment and Project Appraisal* **23**, 325–331 (2005).

Ridgway, B., in W. R. Sheate, ed., "The Project Cycle and the Role of EIA and EMS," *Tools, Techniques and Approaches for Sustainability*, World Scientific, New Jersey, 2010.

Risse, N., et al., "Implementing the SEA Directive: The Member States' Margin of Discretion," *Environmental Impact Assessment Review* **23**, 453–470 (2003).

Ritsatakis, A., in J. Kemm, J. Parry, and S. Palmer, eds., "HIA at the International Policy-making Level," *Health Impact Assessment*, Oxford University Press, Oxford, UK, 2004.

Rittel, H. W. J., and M. W. Webber, "Dilemmas in a General Theory of Planning," *Policy Sciences* **4**, 155–169 (1973).

Ritzdorf, M., in S. Campbell, and S. Fainstein, eds., "Feminist Thoughts in Theory and Practice of Planning," *Readings in Planning Theory*, Blackwell Publishers Inc., Cambridge, MA, 1996.

Ritzer, G., *Modern Social Theory*, 4th ed., McGraw-Hill Companies, Inc., New York, 1996.

Robinson, J., G. Francis, R. Legge, and S. Lerner, "Defining a Sustainable Society," *Alternatives* **17**, 36–46 (1990).

Robinson, M., and A. Bond, "Investigation of Different Stakeholder Views of Local Resident Involvement During Environmental Impact Assessments in the UK," *Journal of Environmental Assessment Policy and Management* **5**, 45–82 (2003).

Rocha, E. M., "A Ladder of Empowerment," *Journal of Planning Education and Research* **17**, 31–44 (1997).

Rodwin, M. A., "Can Bargaining and Negotiation Change the Administrative Process?" *Environmental Impact Assessment Review* **3**, 373–386 (1982).

Rorty, R., *Objectivity, Relativism and Truth*, Cambridge University Press, Cambridge, UK, 1991.

Rose, L., et al., "Environmental Justice Analysis: How Has It Been Implemented in Draft Environmental Impact Statements," *Environmental Practice* **7**, 235–245 (2005).

Ross, H., and T. McGee, "Conceptual Frameworks for SIA Revisited: A Cumulative Effects Study on Lead Contamination and Economic Change," *Impact Assessment and Project Appraisal* **24**, 139–149 (2006).

Ross, W. A., A. Morrison-Saunders, and R. Marshall, "Common Sense in Environmental Impact Assessment: It Is Not as Common as it Should Be," *Impact Assessment and Project Appraisal* **24**, 3–10, (2006a).

Ross, W. A., A. Morrison-Saunders, and R. Marshall, "Riposte: A Common Sense Approach to the Use of Common Sense in EIA," *Impact Assessment and Project Appraisal* **24**, 20–22, (2006b).

Ross, W. A., and D. Thompson, in D. Thompson, ed., "Environmental Impact Assessment," *Tools for Environmental Management*, New Society Publishers, Gabriola Island, BC, Canada, 2002.

Rossou, R., and S. Makan, "The Importance of Theory in Shaping Social Impact Monitoring: Lessons from the Berg River Dam, South Africa," *Impact Assessment and Project Appraisal* **25**, 291–299 (2007).

Rothblatt, D. N., "Multiple Advocacy," *Journal of the American Institute of Planners* **44**, 193–199 (1978).

Rothman, T., and G. Sudarshan, *Doubt and Certainty*, Perseus Books, Reading, MA, 1998.

Rotmans, J., "Tools for Integrated Sustainability Assessment: A Two-track Approach," *Bridging Sciences and Policy* **6**, 35–57 (2006).

Rowan, M., "Refining the Attribution of Significance in Social Impact Assessment," *Impact Assessment and Project Appraisal* **27**, 185–191 (2009).

Rowan, M., and T. Streather, "Converting Project Risks to Development Opportunities Through SIA Enhancement Measures," *Impact Assessment and Project Appraisal* **29**, 217–230 (2011).

Rowe, P. G., *Design Thinking*, MIT Press, Cambridge, MA, 1991.

Rowe, W. D., "Understanding Uncertainty," *Risk Analysis* **14**, 41–50 (1994).

Rowson, J., in J. A. Sinclair, ed., "The Appraisal of Environmental Impact Assessment Processes: Towards a 'Situated' Approach," *Canadian Environmental Assessment in Transition*, University of Waterloo, Waterloo, ON, 1997.

Rozema, J., et al., "An Investigation of Environmental and Sustainability Discourses Associated with the Substantive Purposes of Environmental Assessment," *Environmental Impact Assessment Review* **33**, 80–90 (2012).

Ruddy, T. F., and L. M. Hilty, "Impact Assessment and Policy Learning in the European Commission," *Environmental Impact Assessment Review* **28**, 90–105 (2008).

Runhaar, H., "Putting SEA in Context: A Discourse Perspective on How SEA Contributes to Decision Making," *Environmental Impact Assessment Review* **29**, 200–209 (2009).

Runhaar, H., and P. P. J. Driessen, "What Makes Strategic Environmental Assessment Successful Environmental Assessment? The Role of Context in the Contribution of SEA to Decision Making," *Impact Assessment and Project Appraisal* **25**, 2–14 (2007).

Runhaar, H., P. R. Runhaar, and T. Oegema, "Food for Thought: Conditions for Discourse Reflection in the Light of Environmental Assessment," *Environmental Impact Assessment Review* **30**, 339–346 (2010).

Ryu, J., et al., "The Role of Environmental Non-governmental Organizations in the Environmental Impact Assessment Process in Japan," *Impact Assessment and Project Appraisal* **22**, 283–293 (2004).

Saarikowski, H., "Environmental Impact Assessment as a Collaborative Learning Process," *Environmental Impact Assessment Review* **20**, 681–700 (2000).

Sadler, B., "Environmental Assessment in a Changing World; Evaluating Practice to Improve Performance (final report)," *International Study of the Effectiveness of Environmental Assessment*, Canadian Environmental Assessment Agency and International Association for Impact Assessment, Hull, Québec, 1996.

Sadler, B., in B. Sadler, ed., "Initial Perspectives on SEA at the Policy Level," *Strategic Environmental Assessment at the Policy Level: Recent Progress, Current Status and Future Prospects*, Ministry of the Environment, Czech Republic, Prague, 2005a.

Sadler, B., in B. Sadler, ed., "The Status of SEA Systems with Application to Policy and Legislation," *Strategic Environmental Assessment at the Policy Level: Recent Progress, Current Status and Future Prospects*, Ministry of the Environment, Czech Republic, Prague, 2005b.

Sadler, B., in B. Sadler, et al., eds., "Taking Stock of SEA," *Handbook of Strategic Environmental Assessment*, Earthscan, London, 2011a.

Sadler, B., in B. Sadler, et al., eds., "SEA in Canada," *Handbook of Strategic Environmental Assessment*, Earthscan, London, 2011b.

Sadler, B., "Building Mediation into the Federal Environmental Assessment Process," *The Place of Negotiations in Environmental Assessment*, Canadian Environmental Assessment Agency, Hull, Québec, Canada, undated.

Sadler, B., and A. Jurkeviciute, in B. Sadler, et al., eds., "SEA in the European Union," *Handbook of Strategic Environmental Assessment*, Earthscan, London, 2011.

Sager, T., *Communicative Planning Theory*, Avebury, Brookfield, VA, 1994.

Sallenave, J., "Giving Traditional Ecological Knowledge Its Rightful Place in Environmental Impact Assessment," *CARC—Northern Perspectives* **22**, 1–7 (1994).

Sánchez, L. E., and A. L. C. F. Gallardo, "On the Successful Implementation of Mitigation Measures," *Impact Assessment and Project Appraisal* **23**, 180–192 (2005).

Sánchez, L. E., and S. S. Silva-Sánchez, "Tiering Strategic Environmental Assessment and Project Environmental Impact Assessment in Highway Planning in São Paulo, Brazil," *Environmental Impact Assessment Review* **28**, 515–522 (2008).

Sandham, L. A., and H. M. Pretorious, "A Review of EIA Quality in the North West Province of South Africa," *Environmental Impact Assessment* **28**, 229–240 (2008).

Sandman, P. M., *Responding to Community Outrage: Strategies for Effective Risk Communications*, American Industrial Hygiene Association, Fairfax, VA, 1992.

Saskatchewan Environment and Resource Management (SERM), *Public Involvement Policy Framework and Guidelines*, Saskatchewan Environment and Resource Management, Public Involvement Working Group, undated.

Saskatchewan Indian Federated College (SIFC), *Co-managing Natural Resources with First Nations*, Guidelines to Reaching Agreements and Making Them Work, Department of Indian Affairs and Northern Development, April 1996.

SA-T Regulation, *Bundesgesetzüber die strategische Prüfung im Verkehrsbereich. BGBl.* I Nr. 96/2005, 2005.

Saul, J. R., *Voltaire's Bastards*, Penguin Books, London, 1992.

Scanlon, J., and A. Davis, "The Role of Sustainability Advisors in Developing Sustainability Outcomes for an Infrastructure Project: Lessons from the Australian Urban Rail Sector," *Impact Assessment and Project Appraisal* **29**, 121–133 (2011).

Schijf, B., in B. Sadler, et al., eds., "Developing SEA Guidance," *Handbook of Strategic Environmental Assessment*, Earthscan, London, 2011.

Schirmer, J., "Scaling Up: Assessing Social Impacts at the Macro-Scale," *Environmental Impact Assessment Review* **31**, 382–391 (2011).

Schmidt, O., "NEPA Myths: Things We Tell Ourselves About NEPA That Aren't Necessarily True," *Environmental Practice* **13**, 157–160 (2011a).

Schmidt, O., "Saying What We Mean," *NAEP National E-News*, The National Association of Environmental Professionals, September–October 2011b.

Schön, D. A., *The Reflective Practitioner*, Basic Books, New York, 1983.

Schön, D. A., and M. Rein, *Frame Reflection: Toward the Resolution of Intractable Policy Disputes*, Basic Books, New York, 1994.

Science and Environmental Health Network (SEHN), *Risk Assessment and Management*, Ames, IN, undated.

Scott, C., "Governmentality and Strategic Environmental Assessment: Challenging the SEA/Good Governance Nexus," *Journal of Environmental Assessment Policy and Management* **13**, 67–100 (2011).

Scott-Samuel, A., M. Birley, and K. Ardern, *The Merseyside Guidelines for Health Impact Assessment*, Merseyside Health Impact Assessment Steering Group, Liverpool, 1998.

Scottish Needs Assessment Programme, *Health Impact Assessment: Piloting the Process in Scotland*, Scottish Needs Assessment Programme, Glasgow, 2000.

Scrase, R., and G. Sheate, "Integration and Integrated Approaches to Assessment: What Do They Mean for the Environment?" *Journal of Environmental Policy and Planning* **4**, 275–294 (2002).

Seidler, R., and R. Bawa, in J. A. Tickner, ed., "Uncertainty and Biological Diversity," *Environmental Science and Preventive Public Policy*, Island Press, Washington, 2003.

Seitz, N. E., C. J. Westbrook, and B. F. Noble, "Bringing Science into River Systems Cumulative Effects Assessment Practice," *Environmental Impact Assessment Review* **31**, 172–179 (2011).

Seley, J. E., *The Politics of Public-facility Planning*, Lexington Books, Lexington, MA, 1983.

Senner, R., "Assessing the Sustainability of Project Alternatives: An Increasing Role for Cumulative Effects Assessment," *Presented at Assessing and Managing Cumulative Effects, Special Topic Meeting*, International Association for Impact Assessment, November 6–9, 2008, Calgary, Alberta, Canada, 2008.

Serafin, R., G. Nelson, and R. Butler, "Post Hoc Assessment in Resource Management and Environmental Planning—A Typology and Three Case Studies," *Environmental Impact Assessment Review* **12**, 271–294 (1992).

Sharma, S., "The Impact of Mining on Women: Lessons from the Coal Mining Bowen Basin of Queensland, Australia," *Impact Assessment and Project Appraisal* **28**, 201–210 (2010).

Shearman, R., "The Meaning and Ethics of Sustainability," *Environmental Management* **14**, 1–8 (1990).

Sheate, W. R., in B. Sadler, et al., eds., "SEA and Environmental Planning and Management Tools," *Handbook of Strategic Environmental Assessment*, Earthscan, London, 2011.

Sheate, W. R., "Purposes, Paradigms and Pressure Groups: Accountability and Sustainability in EU Environmental Assessment, 1985–2010," *Environmental Impact Assessment Review* **33**, 91–102 (2012).

Sheate, W. R., and M. R. Partidário, "Strategic Approaches and Assessment Techniques—Potential for Knowledge Brokerage Towards Sustainability," *Environmental Impact Assessment Review* **30**, 278–288 (2010).

Shepherd, A., and C. Bowler, "Beyond the Requirements: Improving Public Participation in EIA," *Journal of Environmental Planning and Management* **40**, 739–750 (1997).

Shepherd, G., ed., *The Ecosystem Approach—Learning from Experience*, International Union for Conservation of Nature (IUCN), *Ecosystem Management Series #5* World Conservation Union, Gland, Switzerland, 2008.

Sherrington, M., "Biodiversity Assessment in the Oil Sands Region, Northeastern Alberta, Canada," *Impact Assessment and Project Appraisal* **23**, 73–81 (2005).

Shippey, K., "Ex Africa Semper Aliquid Novi," *IAIA Newsletter* **15** (April) (2004).

Shoemaker, D. J., *Cumulative Environmental Assessment, Waterloo, Ontario, Department of Geography Publications Series # 42*, University of Waterloo, Waterloo, ON, 1994.

Siebenhűner, B., and V. Barth, "The Role of Computer Modeling in Participatory Integrated Assessment," *Environmental Impact Assessment Review* **25**, 367–389 (2005).

Simon, H., "A Behavioral Model of Rational Choice," *Models of Man, Social and Rational: Mathematical Essays on Rational Human Behavior in a Social Setting*, John Wiley & Sons, New York, 1957.

Simon, H., *Administrative Behaviour: A Study of Decision-making Procedures in Administrative Organizations*, Free Press, New York, 1976.

Simpson, S., et al., "Equity-focused Health Impact Assessment: A Tool to Assist Policy Makers in Addressing Health Inequalities," *Environmental Impact Assessment Review* **25**, 772–782 (2005).

Sinclair, A. J., and A. P. Diduck, "Public Involvement in EA in Canada: A Transformative Learning Perspective," *Environmental Impact Assessment Review* **21**, 113–136 (2001).

Sinclair, A. J., A. Diduck, and P. Fitzpatrick, "Conceptualizing Learning for Sustainability Through Environmental Assessment: Critical Reflections on 15 years of Research," *Environmental Impact Assessment Review* **28**, 415–428 (2008).

Sinclair, A. J., and A. Diduck, in K. Hanna, ed., "Public Participation in Canadian Environmental Assessment: Enduring Challenges and Future Directions," *Environmental Impact Assessment—Practice and Participation*, Oxford University Press, Oxford, UK, 2009.

Sinclair, A. J., G. Schneider, and L. Mitchell, "Environmental Impact Assessment Process Substitution: Experiences with Public Participants," *Impact Assessment and Project Appraisal* **30**, 85–93 (2012).

Sinclair, A. J., L. Sims, and H. Spaling, "Community-based Approaches to Strategic Environmental Assessment: Lessons from Costa Rica," *Environmental Impact Assessment Review* **29**, 147–156 (2009).

Sinclair, P., et al., "A Case Study in Participatory Environmental Systems Assessment with the Use of Multimedia Materials and Quantitative LCA," *Journal of Environmental Assessment Policy and Management* **9**, 399–421 (2007).

Singh, R. K., et al., "An Overview of Sustainability Assessment Methodologies," *Ecological Indicators* **9**, 189–212 (2009).

Sipe, N. G., "An Empirical Analysis of Environmental Mediation," *Journal of the American Planning Association* **64**, 275–285 (1998).

Sipe, N. G., and B. Stiftel, "Mediating Environmental Enforcement Disputes: How Well Does It Work?" *Environmental Impact Assessment Review* **15**, 139–156 (1995).

Siting Process Task Force (SPTF), *Options for Co-operation—Report of the Siting Process Task Force on Low-level Radioactive Waste Disposal*, Energy, Mines and Resources Canada, Ottawa, Canada, undated.

Sköllerhorn, E., "Habermas and Nature: The Nature of Communicative Action for Studying Environmental Policy," *Journal of Environmental Planning and Management* **41**, 555–573 (1998).

Slocombe, D. S., "Environmental Planning, Ecosystem Science and Ecosystem Approaches for Integrating Environment and Development," *Environmental Management* **17**, 289–303 (1993).

Slootweg, R., "Biodiversity Assessment Framework: Making Biodiversity Part of Corporate Responsibility," *Impact Assessment and Project Appraisal* **23**, 37–46 (2005).

Slootweg, R., and A. Kolhoff, "A Generic Approach to Integrate Biodiversity Considerations in Screening and Scoping for EIA," *Environmental Impact Assessment Review* **23**, 657–682 (2003).

Slootweg, R., et al., *Biodiversity in EIA and SEA—Background Document to CBN Decision VIII/28: Voluntary Guidelines on Biodiversity-inclusive Impact Assessment*, Commission for Environmental Assessment, The Netherlands, 2006.

Slootweg, R., et al., "Resilience Thinking Improves SEA: A Discussion Paper," *IAIA10 Conference Proceedings—The Role of Impact Assessment in Transitioning to the Green Economy*, 30th Annual Meeting of the International Association for Impact Assessment, Geneva, Switzerland, 2010.

Slotterback, C. S., "Evaluating the Implementation of Environmental Review Mitigation in Local Planning and Development Processes," *Environmental Impact Assessment Review* **28**, 546–561 (2008).

Slotterback, C. S., et al., "Testing Three Health Impact Assessment Tools in Planning: A Process Evaluation," *Environmental Impact Assessment Review* **31**, 144–153 (2011).

Slovic, P., "Perception of Risk," *Science* **236**, 280–285 (1987).

Smit, B., and H. Spaling, "Methods for Cumulative Effects Assessment," *Environmental Impact Assessment Review* **15**, 81–106 (1995).

Smith, E. T., and A. Schin, "An Evolution of Public Consultation Requirements Within the Environmental Impact Assessment Process at the European Bank for Reconstruction and Development," *Impact Assessment and Project Appraisal* **22**, 153–157 (2004).

Smith, H., *Administrative Behavior—A Study of Decision-making Processes in Administrative Organizations*, 3rd ed., Free Press, New York, 1976.

Smith, L. G., *Impact Assessment and Sustainable Resource Management*, John Wiley & Sons, New York, 1993.

Smith, L. G., C. Y. Nell, and M. V. Prystupa, "The Converging Dynamics of Interest Representation in Resources Management," *Environmental Management* **21**, 139–146 (1997).

Smith, M. D., "A Review of Recent NEPA Alternatives Analysis Case Law," *Environmental Impact Assessment Review* **27**, 136–144 (2007).

Smith, M. D., "NEPA and Climate Change, Part 2: Ten Steps to Taking a Hard Look," *Environmental Practice* **12**, 183–186 (2010).

Smythe, R., and C. Isber, "NEPA in the Agencies: A Critique of Practice," *Environmental Practice* **5**, 290–297 (2003).

Snell, T., and R. Cowell, "Scoping in Environmental Impact Assessment," *Environmental Impact Assessment Review* **26**, 339–358 (2006).

Söderman, T., and T. Kallio, "Strategic Environmental Assessment in Finland: An Evaluation of the SEA Act Application," *Journal of Environmental Assessment Policy and Management* **11**, 1–28 (2009).

Söderman, T., and S. R. Saarela, "Biodiversity in Strategic Environmental Assessment (SEA) of Municipal Spatial Plans in Finland," *Impact Assessment and Project Appraisal* **28**, 117–133 (2010).

Solomon, R. M., S. Yonts-Shepard, and W. T. SupulskilI, in R. Clark and L. Canter, eds., "Public Involvement Under NEPA—Trends and Opportunities," *Environmental Policy and NEPA: Past, Present and Future*, St. Lucie Press, Boca Raton, FL, 1997.

Soneryd, L., and S. Weldon, "Noise and Newts: Public Engagement in the UK and Sweden," *Environmental Impact Assessment Review* **23**, 17–37 (2003).

Sorenson, A. D., and M. L. Auster, "Fatal Remedies—The Sources of Ineffectiveness," *Town Planning Review* **60**, 29–43 (1989).

Spaling, H., "Innovations in Environmental Assessment of Community-based Projects in Africa," *The Canadian Geographer* **47**, 151–168 (2003).

Spickett, J. T., H. L. Brown, and D. Katscherian, "Adaptation Strategies for Health Impacts of Climate Change in Western Australia: Application of a Health Impact Assessment Framework," *Environmental Impact Assessment Review* **31**, 297–300 (2011).

Spooner, B., in A. Donnelly, B. Dalal-Clayton, and R. Hughes, eds., "Review of the Quality of EIA Guidelines, Their Use and Circumnavigation," *A Directory of Impact Assessment Guidelines*, 2nd ed., International Institute for Environment and Development, Russell Press, Nottingham, 1998.

Stackelberg, K. V., and D. E. Burmaster, "A Discussion of the Use of Probabilistic Risk Assessment in Human Health Risk Assessment," *Environmental Impact Assessment Review* **14**, 385–401 (1994).

Stec, S., ed. *Handbook on Access to Justice Under the Aarhus Convention*, Regional Environmental Center for Central and Eastern Europe, Szentendre, Hungary, 2003.

Steinemann, A., "Rethinking Human Health Impact Assessment," *Environmental Impact Assessment Review* **20**, 627–645 (2000).

Steinemann, A., "Improving Alternatives for Environmental Impact Assessment," *Environmental Impact Assessment Review* **21**, 13–22 (2001).

Steinemann, A., "Human Exposure, Health Hazards and Environmental Regulations," *Environmental Impact Assessment Review* **24**, 695–710 (2004).

Stern, M. J., et al., "Visions of Success and Achievement in Recreation-related USDA Forest Service NEPA processes," *Environmental Impact Assessment Review* **29**, 220–228 (2009).

Stern, M. J., and S. A. Predmore, "Decision Making, Procedural Compliance, and Outcome Definition in U.S. Forest Service Planning Processes," *Environmental Impact Assessment Review* **31**, 271–278 (2011).

Stern, P. C., and H. V. Fineberg, *Understanding Risk: Informing Decisions in a Democratic Society*, National Research Council, National Academic Press, Washington, DC, 1996.

Stewart, J. M. P., and A. J. Sinclair, "Meaningful Public Participation in Environmental Assessment: Perspectives from Canadian Participants, Proponents and Government," *Journal of Environmental Assessment Policy and Management* **9**, 161–183 (2007).

Stoeglehner, G., "Enhancing SEA Effectiveness: Lessons Learnt from Austrian Experiences in Spatial Planning," *Impact Assessment and Project Appraisal* **28**, 217–221 (2010).

Stoeglehner, G., A. L. Brown, and L. B. Kørnøv, "SEA and Planning: 'Ownership' of Strategic Environmental Assessment by the Planners is the Key to Its Effectiveness," *Impact Assessment and Project Appraisal* **27**, 111–120 (2009).

Stoeglehner, G., A. Morrison-Saunders, and G. Early, "Comparing Legislative Mechanisms for SEA Screening and Decision-making: Austrian and Australian Experiences," *Journal of Environmental Assessment Policy and Management* **12**, 399–423 (2010).

Stoeglehner, G., and G. Wegerer, "The SEA Directive and SEA-Protocol Adapted to Spatial Planning—Similarities and Differences," *Environmental Impact Assessment Review* **26**, 586–600 (2006).

Storey, K., and P. Jones, "Social Impact Assessment, Impact Management and Follow-up: A Case Study of the Construction of the Hibernia Offshore Platform," *Impact Assessment and Project Appraisal* **21**, 99–107 (2003).

Storey, K., and B. Noble, "Socio-economic Effects Monitoring: Toward Improvements Informed by Bio-physical Effects Monitoring," *Impact Assessment and Project Appraisal* **23**, 210–214 (2005).

Susskind, L., in L. Susskind, S. McKearnen, and J. Thomas-Lamar, eds., "Part One: A Short Guide to Consensus Building," *The Consensus Building Handbook—A Comprehensive Guide to Reaching Agreement*, Sage, Thousand Oaks, CA, 1999.

Susskind, L., and J. Cruikshank, *Breaking the Impasse: Consensual Approaches to Resolving Public Disputes*, Basic Books, New York, 1987.

Susskind, L., and D. Madigan, "New Approaches to Resolving Disputes in the Public Sector," *Justice Systems Journal* **9**, 179–203 (1984).

Susskind, L., S. McKearnen, and J. Thomas-Larmar, *The Consensus Building Handbook—A Comprehensive Guide to Reaching Agreement*, Sage, Thousand Oaks, CA, 1999.

Suter, G. W. II., *Ecological Risk Assessment*, Lewis Publishers, Chelsea, UK, 1993.

Swangjang, K. B., P. Wathern, and T. Rochanaburanon, "Ecological Issues in Thai Environmental Assessment Scoping Guidance," *Impact Assessment and Project Appraisal* **22**, 271–282 (2004).

Swor, T., and L. Canter, "Promoting Environmental Sustainability via an Expert Elicitation Process," *Environmental Impact Assessment Review* **31**, 506–514 (2011).

Täbara, J. D., and C. Pahl-Wostl, "Sustainability Learning in Natural Resource Use and Management," *Ecology and Society* **12**, 3 (2007).

Tamas, A., *Capacity Development Analysis Framework*, Tamas Consultants, 2008, http:www.tamas.com/samples/sourcesdocs/capdev.pdf

Tamburrini, A-L., K. Gilhuly, and B. Harris-Roxas, "Enhancing Benefits in Health Impact Assessment Through Stakeholder Consultation," *Impact Assessment and Project Appraisal* **29**, 195–204 (2011).

Tang, Z., "Integrating the Principles of Strategic Environmental Assessment into Local Comprehensive Land Use Planning," *Journal of Environmental Assessment Policy and Management* **10**, 143–171 (2008).

Tang, Z., "Assessing Sustainable Development Gaps Between the State and Local Jurisdictions," *Journal of Environmental Assessment Policy and Management* **12**, 263–289 (2010).

Taylor, C. N., C. H. Bryan, and C. Goodrich, *Social Assessment: Theory, Process and Techniques*, Social Ecology Press, Middleton, WI, 2004.

Taylor, N., "Mistaken Interests and the Discourse Model of Planning," *Journal of the American Planning Association* **64**, 64–76 (1998).

Taylor, N., W. McClintock, and B. Buckenham, "Social Impacts of Out-of-centre Shopping Centres on Town Centres: A New Zealand Case Study," *Impact Assessment and Project Appraisal* **21**, 147–153 (2003).

Taylor, P. W., *Respect for Nature—A Theory of Environmental Ethics*, Princeton University Press, Princeton University, 1986.

Teller, J., and A. Bond, "Review of Present Environmental Policies and Legislation Involving Cultural Heritage," *Environmental Impact Assessment Review* **22**, 611–632 (2002).

ten Hallers-Tjabbes, C. C., in J. A. Tickner, ed., "Science Communication and Precautionary Policy: A Marine Case Study Community," *Environmental Science and Preventive Public Policy*, Island Press, Washington, DC, 2003.

Tennøy, A., J. Kvaerner, and K. I. Gjerstad, "Uncertainty in Environmental Impact Assessment Predictions: The Need for Better Communications and More Transparency," *Impact Assessment and Project Appraisal* **24**, 45–56 (2006).

Tetlow, M., and M. Hanusch, "Strategic Environmental Assessment: The State of the Art," *Impact Assessment and Project Appraisal* **30**, 15–24 (2012).

Theophilou, V., A. Bond, and M. Cashmore, "Application of the SEA Directive to EU Structural Funds: Perspectives on Effectiveness," *Environmental Impact Assessment Review* **30**, 136–144 (2010).

Thérivel, R., and A.. L. Brown, in J. Petts, ed., "Methods of Strategic Environmental Assessment," *Handbook of Environmental Impact Assessment*, Blackwell Science Ltd., UK, 1999.

Thérivel, R., and D. Minas, "Ensuring Effective Sustainability Appraisal," *Impact Assessment and Project Appraisal* **29**, 81–91 (2002).

Thérivel, R., et al., *Strategic Environmental Assessment*, Earthscan, London, 1992 and 2010.

Thérivel, R., and M. R. Partidario, eds., *The Practice of Strategic Environmental Assessment*, Earthscan, London, 1996.

Thérivel, R., and B. Ross, "Cumulative Impact Assessment: Does Scale Matter?" *Environmental Impact Assessment Review* **27**, 365–385 (2007).

Thérivel, R., et al., "Sustainability-focused Impact Assessment: English Experience," *Impact Assessment and Project Appraisal* **27**, 155–168 (2009).

Thissen, W. A. H., and D. B. Agusdinata, "Handling Deep Uncertainties in Impact Assessment," *Presentation to IAIA08 Conference Proceedings, The Art and Science of Impact Assessment,* International Association for Impact Assessment, IAIA, Perth Australia, 2008.

Thomas, I., and M. Elliott, *Environmental Impact Assessment—Theory and Practice in Australia*, 4th ed., Federation Press, Annandale, NSW, 2005.

Thompson, D., in D. Thompson, ed., "Introduction: The Environmental Manager's Toolbox," *Tools for Environmental Management*, New Society Publishers, Gabriola Island, BC, Canada, 2002a.

Thompson, D., in D. Thompson, ed., "Where Do We Go From Here?" *Tools for Environmental Management*, New Society Publishers, Gabriola Island, BC, Canada, 2002b.

Thompson, J. G., and G. Williams, "Vertical Linkage and Competition for Local Political Power: A Case of Natural Resource Development and Federal Land Policy," *Impact Assessment Bulletin* **10**, 33–58 (1992).

Thornton, J., in J. A. Tickner, ed., "Chemicals Policy and the Precautionary Principle: The Case of Endocrine Disruption," *Environmental Science and Preventive Public Policy*, Island Press, Washington, 2003.

Throgmorton, J. A., "On the Virtues of Skillful Meandering," *American Planning Association Journal* **66**, 367–383 (2000).

Tickner, J., "A Commonsense Framework for Operationalizing the Precautionary Principle," *Paper Presented at the Wingspread Conference, Strategies for Implementing the Precautionary Principle*, Racine, Washington, 23–25 January, 1998.

Tickner, J., and C. Raffensperger, *The Precautionary Principle in Action—A Handbook*, 1st ed., Science and Environmental Health Network, Ames, IN, 1998.

Tickner, J. A., in J. A. Tickner, ed., "The Role of Environmental Science in Precautionary Decision making," *Environmental Science and Preventive Public Policy*, Island Press, Washington, 2003a.

Tickner, J. A., in J. A. Tickner, ed., "The Implications of Uncertainty on Science and Public policy: Case Studies from the Field," *Environmental Science and Preventive Public Policy*, Island Press, Washington, 2003b.

Tickner, J. A., in J. A. Tickner, ed., "Science in Governance and Governance of Science," *Environmental Science and Preventive Public Policy*, Island Press, Washington, 2003c.

Tickner, J. A., in J. A. Tickner, ed., "Precautionary Assessment: A Framework for Integrating Science, Uncertainty and Preventative Public

Policy," *Environmental Science and Preventive Public Policy*, Island Press, Washington, 2003d.

Tickner, J. A., in J. A. Tickner, ed., "Precaution, Environmental Science, and Preventative Public Policy," *Environmental Science and Preventive Public Policy*, Island Press, Washington, 2003e.

Tickner, J. A., "The Precautionary Principle Stimulus for Solutions and Alternatives-based Environmental Policy," *Environmental Impact Assessment Review* **24**, 801–824 (2004).

Tinker, L., et al., "Impact Mitigation in Environmental Impact Assessment: Paper Promises on the Basis of Consent Conditions?" *Impact Assessment and Project Appraisal* **23**, 265–280 (2005).

Todd, S., "Building Consensus on Divisive Issues: A Case Study of the Yukon Wolf Management Team," *Environmental Impact Assessment Review* **22**, 655–684 (2002).

Tomlinson, P., "The Role of Significance Criteria in SEA," *Presentation to the IAIA04 Practitioners Workshop*, Vancouver, Canada, 2004.

Tomlinson, P., and S. F. Atkinson, "Environmental Audits: Proposed Terminology," *Environmental Monitoring and Assessment* **8**, 187–198 (1987).

Tonn, B. E., "Environmental Decision Making in the Face of Uncertainty," *Environmental Practice* **2**, 188–202 (2000).

Torgerson, D., *Industrialization and Assessment*, York University Publications in Northern Studies, Toronto, 1980.

Torgerson, D., in F. J. Tester and W. Mykes, eds., "SIA as a Social Phenomenon: The Problem of Contextuality," *Social Impact Assessment*, Detselig Enterprises, Calgary, AB, 1981.

Treweek, J., "Ecological Impact Assessment," *Impact Assessment* **13**, 289–315 (1995).

Treweek, J., *Ecological Impact Assessment*, Blackwell Science, Oxford, 1999.

Treweek, J., and P. Hankard, in A. L. Porter and J. Fittipaldi, eds., "Ecological Impact Assessment," *Environmental Methods Review: Retooling Impact Assessment for the New Century*, Press Club, Fargo, ND, 1998.

Treweek, J., et al., "Principles for the Use of Strategic Environmental Assessment as a Tool for Promoting the Conservation and Sustainable Use of Biodiversity," *Journal of Environmental Assessment Policy and Management* **7**, 173–199 (2005).

Treweek, J., et al., in B. Sadler, ed., "SEA as a Tool for the Conservation and Sustainable Use of Biodiversity in Developing Countries," *Handbook of Strategic Environmental Assessment*, Earthscan, London, 2011.

Trist, E., "The Environment and System-response Capability," *Futures* **12**, 113–127 (1980).

Tsuji, L. J. S., et al., "Getting Back to Basics: The Victor Diamond Mine Environmental Assessment Scoping Process and the Issue of Family-based Traditional Lands Versus Registered Traplines," *Impact Assessment and Project Appraisal* **29**, 37–47 (2011).

Tugwell, A., and P. Johnson, "The Coffs Harbour 'Our Living City Settlement Strategy' Health Impact Assessment," *Environmental Impact Assessment Review* **31**, 441–444 (2011).

Tuinstra, W., J. Jäger, and P. M. Weaver, "Learning and Evaluation in Integrated Sustainability Assessment," *International Journal of Innovation and Sustainable Development* **3**, 128–152 (2008).

Tzoumis, K., "Comparing the Quality of Draft Environmental Impact Statements by Agencies in the United States Since 1998 to 2004," *Environmental Impact Assessment Review* **27**, 26–40 (2007).

Ugoretz, S. M., in S. G. Hildebrand and S. G. Cannon, eds., "Risk Assessment and Environmental Impact Assessment," *Environmental Analysis: The NEPA Experience*, Lewis Publishers, Boca Raton, FL, 1993.

UK Department of Health (DH), in C. Williams and P. Fischer, eds., *Draft Guidance on Health in Strategic Environmental Assessment*, UK Department of Health, London, UK, 2007.

United Nations Development Program and Regional Environmental Center for Central and Eastern Europe (UNDP and REC), *Capacity Development Needs for the Implementation of the UNECE SEA Protocol: Sub-regional Overview of Armenia, Belarus, Georgia, Republic of Moldova and Ukraine*, 2nd ed., Regional Environmental Center, 2006.

United Nations Economic Commission for Europe (UNECE), "Convention on Access to Information, Public Participation in Decision-making and Access to Justice in Environmental Matters," *Fourth Ministerial Conference—Environment for Europe*, Aarhus, Denmark, 1998.

United Nations Economic Commission for Europe (UNECE), *Guidance on Public Participation in an Environmental Impact Assessment in a Transboundary Context*, New York and Geneva, United Nations, 2006.

United Nations Environment for Europe (UNECE), "Initiative on Strategic Environmental Assessment," *Sixth Ministerial Conference—Environment for Europe*, Belgrade, Serbia, 2007.

United Nations Environment Program (UNEP), *Environmental Impact Assessment Training Manual*, Prepared for the UNEP by Environment Australia, Canberra, Australia, 1997.

United Nations Environment Program (UNEP), *Guidelines for Incorporating Biodiversity-related Issues into Environmental Impact Assessment Legislation and/or Process and in Strategic Environmental Assessment*, Secretariat of the Convention on Biological Diversity, The Hague, Netherlands, 2002.

United Nations Environment Program (UNEP), *Training Module—Introduction to Capacity Building for Environment, Trade and Sustainable Development*, Economics and Trade Branch, UNEP, Geneva, 2004.

United Nations Environment Program (UNEP), "Ways to Increase the Effectiveness of Capacity Building for Sustainable Development," *Discussion Paper Presented in 2006 IAIA Conference*, Stavenger, Norway, Geneva, Switzerland, 2006.

US Administration of Barack H. Obama, *Memorandum on Scientific Integrity*, Memorandum for Heads of Executive Departments and Agencies,Washington DC, 2009.

US Administration of Barack H. Obama, "The President's Proclamation Reaffirming NEPA's Role in Protecting Public Health, Safety and Environmental Quality and in Ensuring Transparency, Accountability and Public Involvement in Our Government," *Federal Register* **75**, 4 (2010).

U.S. Army Corps of Engineers (US ACE), *Guidelines for Risk and Uncertainty Analysis in Water Resources Planning*, Volume 1, Principles with Technical Appendices, Water Resources Support Center, Institute for Water Resources, IWR Report 92-R-1, March, 1992.

U.S. Army Corps of Engineers (US ACE), *Ohio River Mainstem Systems Study (ORMSS) System Investment Plan/Programmatic Environmental Impact Statement—Illinois, Indiana, Kentucky, Ohio, Pennsylvania, and West Virginia*, Includes the Integrated Main Report, Engineering Appendix, Economics Appendix, and Volumes 1 and 2 of an Environmental Appendix, Pittsburgh District, Pittsburgh, Pennsylvania, 2011, http://www.lrp.usace.army.mil/pm/ormss.htm.

US Committee on Resources (US COR), U.S. House of Representatives, *Task Force on Improving the National Environmental Policy Act and Task Force on Updating the National Environmental Policy Act: Initial Findings and Draft Recommendations*, Washington, DC, Committee on Resources, U.S. House of Representatives, 2005.

US Council on Environmental Quality (US CEQ), *Incorporating Biodiversity Considerations into Environmental Impact Analysis Under the National Environmental Policy Act*, CEQ, Executive Office of the President, Washington, DC, 1993.

US Council on Environmental Quality (US CEQ), *The National Environmental Policy Act—A Study of its Effectiveness After Twenty-five Years*, CEQ, Executive Office of the President, Washington, DC, January, 1997a.

US Council on Environmental Quality (US CEQ), *Considering Cumulative Effects Under the National Environmental Policy Act*, CEQ, Executive Office of the President, Washington, DC, January, 1997b.

US Council on Environmental Quality (US CEQ), *Memorandum to Federal Agency NEPA Contacts and Tribal Coordinators—Improving the NEPA Process Through Enhanced Tribal Capacity*, Washington DC, 2004.

US Council on Environmental Quality (US CEQ), *Emergency Actions and NEPA*, From H. G. Grezmiel, Associate Director for NEPA Oversight, Memorandum for Federal NEPA Contacts, 2005a.

US Council on Environmental Quality (US CEQ), *Guidance on the Consideration of Past Actions in Cumulative Effects Assessment*, Connaughton, J. L., to Heads of Federal Agencies, 2005b.

US Council on Environmental Quality (US CEQ), *Collaboration in NEPA—A Handbook for NEPA Practitioners*, 2007a.

US Council on Environmental Quality (US CEQ), *A Citizen's Guide to the NEPA—Having Your Voice Heard*, 2007b.

US Council on Environmental Quality (US CEQ), *Aligning National Environmental Policy Act Processes with Environmental Management Systems—A Guide for NEPA and EMS Practitioners*, 2007c.

US Council on Environmental Quality (US CEQ), *Draft NEPA Guidance on Consideration of the Effects of Climate Change and Greenhouse Gas Emissions*, N. H. Sutter, Chair, Council on Environmental Quality, Memorandum for Heads of Federal Departments and Agencies, 2010a.

US Council on Environmental Quality (US CEQ), *Establishing, Applying and Revising Categorical Exclusions Under the National Environmental Policy Act*, N. H. Sutter, Chair, Council on Environmental Quality, Memorandum for Heads of Federal Departments and Agencies, 2010b.

US Council on Environmental Quality (US CEQ), *Report Regarding the Mineral Management Service's National Policy Act Policies, Practices and Procedures as They Relate to Outer Continental Shelf Oil and Gas Exploration and Development*, 2010c.

US Council on Environmental Quality (US CEQ), *Appropriate Use of Mitigation and Monitoring and Clarifying the Appropriate Use of Mitigated Findings of No Significant Impact*, N. H. Sutley, Memorandum for Heads of Federal Departments and Agencies, 2011.

US Council on Environmental Quality (US CEQ), "Improving the Process for Preparing Efficient and Timely Environmental Reviews Under the National Environmental Policy Act," *Memorandum for Heads of Departments and Agencies*, N. H. Sutley, Executive Office of the President, Council on Environmental Quality, Washington, DC, 2012.

US Department of Energy (US DOE), *NEPA Environmental Impact Statement Checklist*, US DOE, Office of Environment, Safety and Health, Office of NEPA Policy and Assistance, Washington, DC, November 1997.

US Department of Energy (US DOE), *Effective Public Participation Under the National Environmental Policy Act*, Environment, Safety and Health Office of NEPA Policy and Assistance, Washington, DC, August 1998.

US Department of Energy (US DOE), *Directory of Potential Stakeholders for Department of Energy Actions Under the National Environmental Policy Act*, 17th ed., US Department of Energy, Washington, DC, January 2002a.

US Department of Energy (US DOE), *Recommendations for Analyzing Accidents Under the National Environmental Policy Act*, 2002b.

US Department of Energy (US DOE), *Need to Consider Intentional Destructive Acts in NEPA Documents*, Office of NEPA Policy and Compliance to DOE NEPA Community, 2006.

US Department of Energy (US DOE), *Environmental Justice Strategy*, 2007.

US Department of Energy (US DOE), *Public Involvement in the Environmental Assessment Process*, D. B. Poneman, Memorandum for Heads of Department Elements, 2010.

US Environmental Protection Agency (US EPA), *Final Guidance for Incorporating Environmental Justice Concerns in EPA's NEPA Compliance Analysis*, Washington, DC, April 1998a.

US Environmental Protection Agency (US EPA), *Principles of Environmental Impact Assessment—An International Training Course*, Washington, DC, April 1998b.

US Environmental Protection Agency (US EPA), Guidelines for Ecological Risk Assessment, EPA/630/R-95/002F, Risk Assessment Forum, *Federal Register*, **63** (93), 26845–26924 Washington, DC, April 1998c.

US Environmental Protection Agency (US EPA), *Guidelines for Carcinogen Risk Assessment*, Risk Assessment Forum, NCEA-F-0644, Review Draft, Washington, DC, July 1999.

US Environmental Protection Agency (US EPA), *The Model Plan for Public Participation*, Developed by the Public Participation and Accountability Subcommittee of the National Environmental Justice Advisory Council, Washington, DC, February 2000a.

US Environmental Protection Agency (US EPA), *Resource Guide: Resolving Environmental Conflicts in Communities*, EPA 360-F-00–001, Washington, DC, May 2000b.

US Environmental Protection Agency (US EPA), *Stakeholder Involvement and Public Participation at the US EPA: Lessons Learned, Barriers and Innovative Approaches*, Washington, DC, January 2001a.

US Environmental Protection Agency (US EPA), *Public Involvement in EPA Decisions*, A National Dialogue Convened by the US Environmental Protection Agency, Washington, DC, July 2001b.

US Environmental Protection Agency (US EPA), *Scientific Integrity Policy—Draft*, August 5, 2011.

US Interagency Climate Change Adaptation Task Force (US ICCATF), *Federal Actions for a Climate Resilient Nation*, Report of the Interagency Climate Change Adaptation Task Force, 2011.

US National Research Council (US NRC), *Risk Assessment in the Federal Government: Managing the Process*, National Academy Press, Washington, DC, 1983.

US National Research Council (US NRC), *Science and Judgment in Risk Assessment*, National Academy Press, Washington, DC, 1994.

US National Research Council (US NRC), *Understanding Risk: Informing Decisions in a Democratic Society*, National Academy Press, Washington, DC, 1997.

US National Research Council (US NRC), *Improving Health in the United States—The Role of Health Impact Assessment*, National Academic Press, Washington, DC, 2011a.

US National Research Council (US NRC), *Sustainability and the U.S. EPA*, National Academic Press, Washington, DC, 2011b.

US NEPA Task Force (US NTF), *Modernizing NEPA Implementation*, Report to the Council on Environmental Quality, 2003.

US Office of Management and Budget (US OMB), "Guidelines for Ensuring and Maximizing the Quality, Objectivity, Utility, and Integrity of Information Disseminated by Federal Agencies; Notice; Republication," *Federal Register* **67** (36), 8452–8460 (2002).

US Office of Science and Technology Policies (US OST), *Memorandum for the Heads of Executive Departments and Agencies—Scientific Integrity*, J. P. Holden, Assistant to the President for Science and Technology and Director of the Office of Science and Technology Policies, 2010.

Utzinger, J., et al., "Assessing Health Impacts of the Chad-Cameroon Petroleum Development and Pipeline Project: Challenges and a Way Forward," *Environmental Impact Assessment Review* **25**, 63–94 (2005).

Vanclay, F., "Conceptualizing Social Impacts," *Environmental Impact Assessment Review* **22**, 183–212 (2002).

Vanclay, F., "International Principles for Social Impact Assessment," *Impact Assessment and Project Appraisal* **21**, 5–11 (2003).

Vanclay, F., "Principles of Social Impact Assessment: A Critical Comparison Between the International and US Documents," *Environmental Impact Assessment Review* **26**, 3–14 (2006).

Vanclay, F., and S. A. Bronstein, *Environmental and Social Impact Assessment*, John Wiley & Sons, New York, 1995.

Vanclay, F., in W. R. Sheate, ed., "The Triple Bottom Line and Impact Assessment: How the TBL, EIA, SEA, and EMS Relate to Each Other?" *Tools, Techniques and Approaches for Sustainability*, World Scientific, New Jersey, 2010.

Vanderhaegen, M., and E. Muro, "Contribution of a European Spatial Data Infrastructure to the Effectiveness of EIA and SEA Studies," *Environmental Impact Assessment Review* **25**, 123–142 (2005).

Van Der Vorst, R., A. Grafé-Buckens, and W. R. Sheate, in W. R. Sheate, ed., "A Systematic Framework for Environmental Decision-making," *Tools, Techniques and Approaches for Sustainability*, World Scientific, New Jersey, 2010.

Van Gent, P., in B. Sadler et al., eds., "SEA Knowledge and Its Use in Information Sharing, Training and Learning," *Handbook of Strategic Environmental Assessment*, Earthscan, London, 2011.

VanGundy, A. B., *Techniques of Structured Problem Solving*, 2nd ed., Van Nostrand Reinhold Company Inc., New York, 1988.

Varghese, J., and S. Nadeau, "Critical Exposé of the Relationship Between Local Ownership and Community Sustainability," *Presentation to IAIA04 SIA Session*, 2004.

Verloo, M., and C. Roggeband, "Gender Impact Assessment: The Development of a New Instrument in the Netherlands," *Impact Assessment* **14**, 3–20 (1996).

Verma, N., *Similarities, Connections and Systems: The Search for a New Rationality for Planning and Management*, Lexington Books, Lanham, MD, 1998.

Verma, V. K., *Organizing Projects for Success, The Human Aspects of Project Management*, Project Management Institute, Upper Darby, PA, 1995.

Vicente, G., and M. Partidário, "SEA—Enhancing Communications for Better Environmental Decisions," *Environmental Impact Assessment Review* **26**, 696–706 (2006).

Villani, F., "Gender as a Determinant of Health in HIA," *Presentation to the Impact Assessment and Responsible Development for Infrastructure, Business and Industry IAIA11 Conference Proceedings,* 31st Annual Meeting of the International Association for Impact Assessment, Centro de Convenciones, Puebla, Mexico, 2011.

Virji, H., J. Padgham, and C. Seipt, "Capacity Building to Support Knowledge Systems for Resilient Development," *Current Opinion in Environmental Sustainability* **4**, 115–121 (2012).

Voituriez, T., et al., "Making Trade Sustainability Impact Assessment More Relevant to Trade Negotiations," *Impact Assessment and Project Appraisal* **24**, 335–342 (2006).

Wackernagel, M., and W. Rees, *Our Ecological Footprint*, New Society Publishers, Gabriola Island, BC, 1996.

Waldeck, S., A. Morrison-Saunders, and D. Annandale, "Effectiveness of Non-legal EIA Guidance from the Perspective of Consultants in Western Australia," *Impact Assessment and Project Appraisal* **21**, 251–256 (2003).

Wale, E., and A. Yalew, "On Biodiversity Impact Assessment: The Rationale, Conceptual Challenges, and Implications for Future EIA," *Impact Assessment and Project Appraisal* **28**, 3–13 (2010).

Walker, A., "Restoring Flows on Australia's Snowy River: Assessing the Impact of Local Amenity," *Impact Assessment and Project Appraisal* **21**, 119–124 (2003).

Walker, G., "Environmental Justice, Impact Assessment and the Politics of Knowledge: The Implications of Assessing the Social Distribution of Environmental Outcomes," *Environmental Impact Assessment Review* **30**, 312–318 (2010).

Walker, G., H. Fay, and G. Mitchell, *Environmental Justice Impact Assessment—An Evaluation of Requirements and Tools for Distributional Analysis*, Institute for Environment and Sustainability Research, Report for Friends of the Earth, Stoke-on-Trent, Staffordshire University, 2005.

Walker, L. J., and J. Johnston, *Guidelines for the Assessment of Indirect and Cumulative Impacts as well as Impact Interactions*, EC DG XI, Environment, Nuclear Safety and Civil Protection, Hyder, European Commission, Brussels, Belgium, May 1999.

Walters, C. J., "Challenges in Adaptive Management of Riparian and Coastal Ecosystems," *Conservation Ecology* **1** (1997).

Wandesforde-Smith, G., in R. V. Bartlett, ed., "Environmental Impact Assessment, Entrepreneurship and Policy Change," *Policy Through Impact Assessment*, Greenwood Press, New York, 1989.

Warawa, M., Chair, *Report of the Standing Committee on Environment and Sustainable Development*, Statutory Review of the Canadian Environmental Assessment Act: Protecting the Environment, Managing Our Resources, 41st Parliament, 1st Session, House of Commons, Public Works and Government Services Canada, Ottawa, Ontario, 2012.

Watson, R. T., "Risk in the Context of Climate Change," *Presentation to Climate Change and Environmental Assessment Special Symposium, International Association for Impact Assessment (IAIA) and World Bank, Washington, DC*, 2010.

Weaver, C., and A. M. Cunningham, "Social Theory, Impact Assessment and Northern Native Communities," *Berkeley Planning Journal* **2**, 6–29 (1985).

Weaver, A., et al., "Contributing to Sustainability as an Environmental Impact Assessment Practitioner," *Impact Assessment and Project Appraisal* **26**, 91–98 (2008).

Webber, M. M., "The Myth of Rationality: Development Planning Reconsidered," *Environment and Planning B: Planning and Design* **10**, 89–99 (1983).

Weber, E. P., "Successful Collaboration—Negotiating Effective Regulations," *Environment* **40**, 10–37 (1998).

Weber, G., and G. Stöglehner, *Integrationsmöglichkeiten der strategischen Umweltprüfung in die nominelle und funktionelle Raumordnung—dargestellt an ausgewählten Beispielen*, 2001, Available online: http://www.lebensministerium.at/umwelt/betriebl_umweltschutz_uvp/uvp/sup/supoesterreich.html [last accessed October 15, 2012].

Webler, T., H. Kastenholz, and O. Renn, "Public Participation in Impact Assessment: A Social Learning Perspective," *Environmental Impact Assessment* **15**, 443–464 (1995).

Webster, R., in R. Clark and L. Canter, eds., "Increasing the Efficiency and Effectiveness of NEPA Through the Use of Technology," *Environmental Policy and NEPA—Past, Present and Future*, St. Lucie Press, Boca Raton, FL, 1997.

Wegner, A., S. A. Moore, and J. Bailey, "Consideration of Biodiversity in Environmental Impact Assessment in Western Australia," *Environmental Impact Assessment Review* **25**, 143–162 (2005).

Weiland, U., "Strategic Environmental Assessment in Germany," *Environmental Impact Assessment Review* **30**, 211–217 (2010).

Weinberg, A. S., "The Environmental Justice Debate: New Agendas for a Third Generation of Research," *Society and Natural Resources* **11**, 605–614 (1998).

Weiner, K. S., in R. Clark and L. Canter, eds., "Basic Purposes and Policies of the NEPA Regulations," *Environmental Policy and NEPA: Past Present and Future*, St. Lucie Press, Boca Raton, FL, 1997.

Weiss, E. H., "An Unreadable EIS is an Environmental Hazard," *Environmental Professional* **11**, 236–240 (1989).

Welles, H., in R. Clark and L. Canter, eds., "The CEQ NEPA Effectiveness Study: Learning From Our Past and Shaping Our Future," *Environmental Policy and NEPA—Past, Present and Future*, St. Lucie Press, Boca Raton, FL, 1997.

Wende, W., "Evaluation of the Effectiveness and Quality of Environmental Impact Assessment in the Federal Republic of Germany," *Impact Assessment and Project Appraisal* **29**, 93–100 (2002).

Wende, W., A. Herberg, and A. Hertzberg, "Mitigation Banking and Compensation Pools: Improving the Effectiveness of Impact Mitigation Regulation in Project Planning Procedures," *Impact Assessment and Project Appraisal* **23**, 101–111 (2005).

Wende, W., et al., "Climate Change Mitigation and Adaptation in Strategic Environmental Assessment," *Environmental Impact Assessment Review* **32**, 88–93 (2012).

Werling, B., and D. Turner, "A Method for Preparing and Executing Conflict of Interest Assessments," *Environmental Practice* **12**, 168–174 (2010).

Wernham, A., "Inupiat Health and Proposed Alaskan Oil Development: Results of the First Integrated Health Impact Assessment/Environmental Impact Statement for Proposed Oil Development on Alaska's North Slope," *Ecohealth* **4**, 500–513 (2007).

Westman, W. E., *Ecology, Impact Assessment and Environmental Planning*, John Wiley & Sons, New York, 1985.

Weston, J., "Common Sense in Environmental Impact Assessment—It Cannot Be Avoided," *Impact Assessment and Project Appraisal* **24**, 12–13 (2006).

Weston, J., "EIA Theories—All Chinese Whispers and No Critical Theory," *Journal of Environmental Assessment Policy and Management* **12**, 357–374 (2010).

Weston, J., "Screening for Environmental Impact Assessment Projects in England: What Screening?" *Impact Assessment and Project Appraisal* **29**, 90–98 (2011).

Whelen, E. H., "Our 'Stolen Future' and the Precautionary Principle," *Priorities* **8**, 3 (1996).

Whitelaw, G. S., D. D. McCarthy, and L. J. S. Tsuji, "The Victor Diamond Mine Environmental Assessment Process: A Critical First Nations Perspective," *Impact Assessment and Project Appraisal* **27**, 205–215 (2009).

Whitney, J. B. R., in V. W. Maclaren and J. B. Whitney, eds., "Integrated Economic Models in Environmental Impact Assessment," *New Directions in Environmental Impact Assessment in Canada*, Methuen, Toronto, 1986.

Whyte, A. V., and I. Burton, eds., *Environmental Risk Assessment*, SCOPE 15, John Wiley & Sons, New York, 1980.

Wiek, A., and C. Binder, "Solution Spaces for Decision-making—A Sustainability Assessment Tool for City-regions," *Environmental Impact Assessment Review* **25**, 589–608 (2005).

Wiener, J. B., and M. D. Rogers, *Comparing Precaution in the United States and Europe*, WP 2002-01, Duke Center for Environmental Solutions, Duke University, Durham, NC, 2002.

Wieringa, M. J., and A. G. Morton, "Hydropower, Adaptive Management and Biodiversity," *Environmental Management* **20**, 831–840 (1996).

Wikland, H., "In Search of Arenas for Democratic Deliberation: A Habermasian Review of Environmental Assessment," *Impact Assessment and Project Appraisal* **23**, 281–292 (2005).

Wildavsky, A., *But Is It True? A Citizen's Guide to Environmental Health and Safety Issues*, Harvard University Press, Cambridge, MA, 1995.

Wiles, A., J. McEwen, and M. H. Sadar, "Use of Traditional Ecological Knowledge in Environmental Assessment of Uranium Mining in Saskatchewan," *Impact Assessment and Project Appraisal* **17**, 107–114 (1999).

Wilkins, H., "The Need for Subjectivity in EIA: Discourse as a Tool for Sustainable Development," *Environmental Impact Assessment Review* **23**, 401–414 (2003).

Wilson, J., in R. Boardman, ed., "Green Lobbies: Pressure Groups and Environmental Policy," *Canadian Environmental Policy: Ecosystems, Politics and Process*, Oxford University Press, Toronto, 1992.

Wilson, E. O., *Consilience—The Unity of Knowledge*, Alfred A. Knopf, New York, 1998.

Wilson, E., "Regional Sustainability Appraisal and Climate Change," *Presentation to the IAIA10 Conference Proceedings—The Role of Impact Assessment in Transitioning to the Green Economy*, 30th Annual Meeting of the International Association for Impact Assessment (IAIA), Geneva, Switzerland, 2010.

Winds and Voices Environmental Services, Inc., *Determining Significance of Environmental Effects: An Aboriginal Perspective*, Canadian Environmental Assessment Agency's Research and Development Program, CEAA, Hull, Québec, 2000.

Winkler, M. S., et al., "Assessing Health Impacts in Complex Eco-epidemiological Settings in the Humid Tropics: The Centrality of Scoping," *Environmental Impact Assessment Review* **31**, 310–319 (2011).

Winters, L., and A. Scott-Samuel, *Health Impact Assessment of Community Safety Projects*, Huyton SRB Area, Merseyside, UK, Observatory Report Series # 38, 1997.

Wirutskulshai, U., E. Sajor, and N. Coowanitwong, "Importance of Context in Adoption and Progress in Application of Strategic Environmental Assessment: Experience of Thailand," *Environmental Impact Assessment Review* **31**, 352–359 (2011).

Witty, D., "The Practice Behind the Theory—Co-management as a Community Development Tool," *Plan Canada* **36**, 22–27 (1994).

Wlodarczyk, T., and J. Tennyson, "Social and Economic Effects from Attitudes Towards Risk," *Impact Assessment and Project Appraisal* **21**, 179–185 (2003).

Wolfe, A. K., N. Kerchner, and T. Wilbanks, "Public Involvement on a Regional Scale," *Environmental Impact Assessment Review* **21**, 431–448 (2001).

Wolfe, L. D. S., *Methods of Scoping Environmental Impact Assessments—A Review of Literature and Experience*, Federal Environmental Assessment Review Office, Vancouver, BC, January 1987.

Wolsink, M., "Contested Environmental Policy Infrastructure: Socio-political Acceptance of Renewable Energy, Water and Waste Facilities," *Environmental Impact Assessment Review* **30**, 302–311 (2010).

Wood, C., *Environmental Impact Assessment—A Comparative Review*, Longman Scientific and Technical, Essex, 1995.

Wood, C., in N. Lee and C. George, eds., "Screening and Scoping," *Environmental Impact Assessment in Developing and Transitional Countries—Principles, Methods and Practice*, John Wiley & Sons, Chichester, 2000.

Wood, C., A. Barker, C. Jones, and J. Hughes, *Evaluation of the Performance of the EIA Process—Final Report, Volume 1: Main Report*, EIA Centre, University of Manchester, Manchester, October 1996.

Wood, G., "Thresholds and Criteria for Evaluating and Communicating Impact Significance in Environmental Statements: 'See No Evil, Hear No Evil, Speak No Evil'?" *Environmental Impact Assessment Review* **28**, 22–38 (2008).

Wood, G., J. Glasson, and J. Becker, "EIA Scoping in England and Wales: Practitioner Approaches, Perspectives and Constraints," *Environmental Impact Assessment Review* **26**, 221–241 (2006).

Woodward, A., in J. A. Tickner, ed., "Uncertainty and Global Climate Change: The Case of Mosquitoes and Mosquito-borne Disease," *Environmental Science and Preventive Public Policy*, Island Press, Washington, 2003.

World Commission on Environment and Development, *Our Common Future*, Oxford University Press, Oxford, 1987.

World Health Organization (WHO), "The Constitution of the World Health Organization," *World Health Organization Chronicles* **1**, 29 (1967).

World Health Organization (WHO), *Health and Safety Component of Environmental Impact Assessment*, *Environmental Health Series No. 15*, WHO Regional Office for Europe, Copenhagen, 1987.

World Health Organization Regional Office for Europe (WHOROE), *Gothenburg Consensus Paper, Health Impact Assessment: Main Concepts and Suggested Approach*, European Centre for Health Policy, Brussels, 1999.

World Health Organization Regional Office for Europe (WHOROE), *HIA in Strategic Environmental Assessment Brochure*, Copenhagen, Denmark, 2001a.

World Health Organization Regional Office for Europe (WHOROE), *HIA in Strategic Environmental Assessment—Current Policy Issues for Debate and Action by the Health Sector*, Copenhagen, Denmark, 2001b.

World Health Organization Regional Office for Europe(WHOROE), *Precautionary Policies and Health Protection: Principles and Applications*, Report on a WHO Workshop, Rome, Italy, 2001c.

Wright, J., J. Parry, and E. Scully, "Institutionalizing Policy-level Health Impact Assessment in Europe: Is Coupling Health Impact Assessment with Strategic Environmental Assessment the Next Step Forward?" *Bulletin of World Health Organization* **83**, 472–477 (2005).

Yi, J., and T. Hacking, "Gaps in EIA in Incorporating Climate Change," *Presentation to the IAIA12 Conference Proceedings, Energy Future—The Role of Impact Assessment,* 32nd Annual Meeting of the International Association for Impact Assessment, Porto, Portugal, 2012.

Yiftachel, O., "Towards a New Typology of Urban Planning Theories," *Environment and Planning B: Planning and Design* **16**, 23–39 (1989).

Yoe, C. H., *An Introduction to Risk and Uncertainty in the Evaluation of Environmental Investments*, Evaluation of Environmental Investments Research Program, US Army Corps of Engineers, IWR Report 96-R-8, Alexandria, VA and Vicksburg, MS, March 1996.

Yoe, C. H., and L. Skaggs, *Risk and Uncertainty Analysis Procedures for the Evaluation of Environmental Outputs*, Evaluation of Environmental Investments Research Program, US Army Corps of Engineers, IWR Report 97-R-7, Alexandria, VA and Vicksburg, MS, August 1997.

Youngkin, D., et al., "The Place of Social Impacts in the Iterative Assessment Process: A Case Study of a Highway Project in the US State of Georgia," *Impact Assessment and Project Appraisal* **21**, 173–177 (2003).

Zhou, K. Y., and W. R. Sheate, "Comparative Analysis of SEA Legal Requirements and Institutional Structure in China (Mainland), Canada, and the UK (England)," *Journal of Environmental Assessment Policy and Management* **11**, 387–426 (2009).

Zhu, Z., et al., "An Alternative Approach to Institutional Analysis in Strategic Environmental Assessment in China," *Journal of Environmental Assessment Policy and Management* **12**, 155–183 (2010).

Ziller, A., and P. Phibbs, "Integrating Social Impacts into a Cost–benefit Analysis: A Participative Method: Case Study: The NSW Area Assistance Scheme," *Impact Assessment and Project Appraisal* **21**, 141–146 (2003).

Index

Impact Assessment: Practical Solutions to Recurrent Problems and Contemporary Challenges, Second Edition. David P. Lawrence.

www.ingramcontent.com/pod-product-compliance
Lightning Source LLC
LaVergne TN
LVHW062137200926
848484LV00007B/1826
* 9 7 8 1 1 1 8 0 9 7 3 7 3 *